Progress in Atomic Spectroscopy

Part B

PHYSICS OF ATOMS AND MOLECULES

Series Editors:

P.G. Burke, *Queen's University of Belfast, Northern Ireland*
and
H. Kleinpoppen, *Institute of Atomic Physics, University of Stirling, Scotland*

Editorial Advisory Board:

R.B. Bernstein *(New York, U.S.A.)*
J.C. Cohen-Tannoudji *(Paris, France)*
R.W. Crompton *(Canberra, Australia)*
J.N. Dodd *(Dunedin, New Zealand)*
G.F. Drukarev *(Leningrad, U.S.S.R.)*

W. Hanle *(Giessen, Germany)*
W.E. Lamb, Jr. *(Tucson, U.S.A.)*
P.-O. Löwdin *(Gainesville, U.S.A.)*
M.R.C. McDowell *(London, U.K.)*
K. Takayanagi *(Tokyo, Japan)*

1976:
ELECTRON AND PHOTON INTERACTIONS WITH ATOMS
Edited by H. Kleinpoppen and M.R.C. McDowell

1978:
PROGRESS IN ATOMIC SPECTROSCOPY, Parts A and B
Edited by W. Hanle and H. Kleinpoppen

In preparation:
ATOM–MOLECULE COLLISION THEORY: A Guide for the Experimentalist
Edited by Richard B. Bernstein

THEORY OF ELECTRON–ATOM COLLISIONS
By P.G. Burke and C.J. Joachain

A Continuation Order Plan is available for this series. A continuation order will bring delivery of each new volume immediately upon publication. Volumes are billed only upon actual shipment. For further information please contact the publisher.

Progress in Atomic Spectroscopy

Part B

Edited by

W. Hanle
Ist Institute of Physics
Justus Liebig University
Giessen, Germany

and

H. Kleinpoppen
Institute of Atomic Physics
University of Stirling
Stirling, Scotland

Plenum Press · New York and London

Library of Congress Cataloging in Publication Data

Main entry under title:

Progress in atomic spectroscopy.

(Physics of atoms and molecules)
Includes bibliographical references and index.
1. Atomic spectra. I. Hanle, Wilhelm, 1901-   II. Kleinpoppen, Hans.
QC454.A8P76   539.7   78-18230
ISBN 0-306-31116-X (Part B)

© 1979 Plenum Press, New York
A Division of Plenum Publishing Corporation
227 West 17th Street, New York, N.Y. 10011

Printed in the United States of America

Contents of Part B

Chapter 19

Multiphoton Spectroscopy
Peter Bräunlich

Chapter 20

Fast-Beam (Beam-Foil) Spectroscopy
H. J. Andrä

Chapter 21

Stark Effect
K. J. Kollath and M. C. Standage

Chapter 22

Stored Ion Spectroscopy
Hans A. Schuessler

Chapter 23
The Spectroscopy of Atomic Compound States
J. F. Williams

Chapter 24
Optical Oscillator Strengths by Electron Impact Spectroscopy
W. R. Newell

Chapter 25
Atomic Transition Probabilities and Lifetimes
W. L. Wiese

Chapter 26

Lifetime Measurement by Temporal Transients
Richard G. Fowler

Chapter 27

Line Shapes
W. Behmenburg

Chapter 30
X-Ray Spectroscopy
K.-H. Schartner

Chapter 31
Exotic Atoms
G. Backenstoss

Contents of Part A

I. BASIC PROPERTIES OF ATOMS AND PERTURBATIONS

Chapter 1
Atomic Structure Theory
A. Hibbert

Chapter 2
Density Matrix Formalism and Applications in Spectroscopy
K. Blum

Chapter 3

Perturbation of Atoms
Stig Stenholm

Chapter 4

Quantum Electrodynamical Effects in Atomic Spectra
A. M. Ermolaev

Chapter 5

Inner Shells

B. Fricke

Chapter 6

Interatomic Potentials for Collisions of Excited Atoms

W. E. Baylis

II. METHODS AND APPLICATIONS OF ATOMIC SPECTROSCOPY

Chapter 7

New Developments of Classical Optical Spectroscopy
Klaus Heilig and Andreas Steudel

Chapter 8

Excitation of Atoms by Impact Processes
H. Kleinpoppen and A. Scharmann

Chapter 9
Perturbed Fluorescence Spectroscopy
W. Happer and R. Gupta

Chapter 10
Recent Developments and Results of the Atomic-Beam Magnetic-Resonance Method
Siegfried Penselin

Chapter 11

The Microwave–Optical Resonance Method
William H. Wing and Keith B. MacAdam

Chapter 12

Lamb-Shift and Fine-Structure Measurements on One-Electron Systems
H.-J. Beyer

Chapter 13

Anticrossing Spectroscopy

H.-J. Beyer and H. Kleinpoppen

Chapter 14

Time-Resolved Fluorescence Spectroscopy

J. N. Dodd and G. W. Series

Chapter 15
Laser High-Resolution Spectroscopy
W. Demtröder

Introduction

W. HANLE and H. KLEINPOPPEN

In 1919, in the first edition of *Atombau and Spektrallinien,* Sommerfeld referred to the immense amount of information which had been accumulated during the first period of 60 years of spectroscopic practice. Sommerfeld emphasized that the names of Planck and Bohr would be connected forever with the efforts that had been made to understand the physics and the theory of spectral lines. Another period of almost 60 years has elapsed since the first edition of Sommerfeld's famous monograph. As the editors of this monograph, *Progress in Atomic Spectroscopy,* we feel that the present period is best characterized by the large variety of new spectroscopic methods that have been invented in the last decades.

Spectroscopy has always been involved in the field of research on atomic structure and the interaction of light and atoms. The development of new spectroscopic methods (i.e., new as compared to the traditional optical methods) has led to many outstanding achievements, which, together with the increase of activity over the last decades, appear as a kind of renaissance of atomic spectroscopy.

The substantial improvement of the spectroscopic resolving power by many orders of magnitude has prepared the testing ground for the foundations of quantum mechanics and quantum electrodynamics. Traditional optical spectroscopy provided sets of measurements of wavelengths that were ordered by their representation in spectral series; modern atomic spectroscopy and its applications have links with many other branches of fundamental physics. High-precision spectroscopy, as a test for quantum electrodynamics, links atomic spectroscopy with the physics of leptonic interactions. The remarkable and continued success of quantum electrodynamics in predicting precisely the properties of atoms is a most impressive testament to the applicability of physical theory and mathematics to experimental physics. The fascination which results from precision measurements in atomic spectroscopy, and which matches the highly accurate quantum-electrodynamical predictions of atomic structure, still remains an important attraction to physicists working in that field. While links between atomic spectroscopy and the physics of leptonic interactions have been well established over the last two decades, present precision experiments with dye lasers on parity-nonconserving effects in atoms are about to become most sensitive test experiments of theoretical models of weak neutral currents in particle physics.

A wide range of fundamental processes of physics is applied in modern atomic spectroscopy. Quantum-mechanical coherence and perturbation effects in atomic states are involved in various methods such as level crossing and anticrossing spectroscopy, fast-beam spectroscopy, and other time-resolving methods. Laser applications have had, and certainly will have, of course, a most profound impact not only on the range of spectroscopic applications but also on the precision that can be obtained. Higher-order interactions between light and atoms have been applied not only as tools for precision measurements of atomic structures but also to studies of higher-order effects in intense electromagnetic fields. There are, however, other less spectacular but nevertheless important areas that have a significant impact on the overall field of applications in atomic spectroscopy; for instance, the field of interatomic interactions has successfully been studied with methods of atomic spectroscopy. The amount of data collected by atomic spectroscopists has vastly increased; new methods and improved old ones have provided a huge amount of information on parameters characterizing atomic states, e.g., lifetimes (or oscillator strengths), level splittings, multipole moments, Stark shifts, anisotropic magnetic susceptibilities, coupling coefficients, etc. Atomic compound state spectroscopy, which originated from atomic collision studies, requires both representation in spectral series and a detailed analysis of the configuration interactions, which links atomic collision physics with atomic spectroscopy. Perturbations have not only been used as spectroscopic tools but also play a decisive role in the interpretation of collisional depolarization and energy transfer from one excited state to another. Most impressive progress has been made in neighboring areas such as x-ray spectroscopy and the spectroscopy of exotic atoms and positronium. The traditional links between astronomy and atomic spectroscopy are as strong as ever, leading to the advancement of knowledge in both of these branches of science. Because of extreme conditions that are possible in space but not obtainable in laboratory investigations, data from astronomical observations often find their interpretation through knowledge based upon atomic structure.

Of course, this book can only attempt to summarize the most important methods and applications of atomic spectroscopy that are prevalent at present. We will be happy if the scientific community appreciates the efforts made by the distinguished authors in this book; these authors have enthusiastically taken up the laborious but important task of describing their fields of interest within the severe limitation of available space. In this connection we are most grateful to the publishing company for their willingness to extend the book to two volumes. We would also like to express our indebtedness to our secretaries Mrs. A. Dunlop (Stirling), Mrs. A. Füchtemann (Bielefeld), Mrs. H. Schaefferling (Bielefeld), and Ms. H. Wallbott (Giessen) for their assistance in communicating with the authors and for many very careful considerations. We acknowledge the benefit of discussions and advice from colleagues and students including in particular Professor G. W. Series, Dr. H.-J. Beyer, Dr. K. Blum, Dr. K. J. Kollath, Dr. M. Standage, Mr. W. Johnstone, Mr. H. Jakubowicz, Mr. N. Malik, and Mr. A. Zaidi. One of us (H.K.) is very grateful for the hospitality of the University of Bielefeld, where parts of the book were prepared.

16

The Spectroscopy of Highly Excited Atoms*

DANIEL KLEPPNER

1. Introduction

Rydberg atoms—atoms in highly excited states—have become the focus of attention of a number of research groups, though approaches to the field have been along a variety of routes. Some studies, for example the precision measurements of fine structure in helium described by Wing and MacAdam[1] elsewhere in this volume, grew naturally out of systematic studies of low-lying states. Much of the work, however, approaches highly excited atoms as a more-or-less separate species with its own characteristic phenomena. Ionization by a static electric field, for instance, plays little if any role in conventional spectroscopy but can be of central importance to the study of highly excited atoms. Because field ionization encompasses both discrete state and continuum behavior, photoabsorption in a strong field can simultaneously display characteristics of bound-state resonant absorption, photoionization, and photoabsorption. As another example, the phenomenology of thermal collisions between molecules and Rydberg atoms can be dominated by inelastic processes between quasidegenerate levels, or by electron attachment, processes which are entirely absent in ground-state collisions.

* A portion of this material was presented at the Fifth International Conference on Atomic Physics, Berkeley, 1976 and appears in the paper, "Highly Excited Atoms," by D. Kleppner in *Atomic Physics*, Vol. 5, Proceedings of the Fifth International Conference on Atomic Physics, Eds. R. Marrus, H. A. Shugart, and M. H. Prior, Plenum Press, New York (1977).

DANIEL KLEPPNER • Research Laboratory of Electronics and Department of Physics, Massachusetts Institute of Technology, Cambridge, Massachusetts 02139.

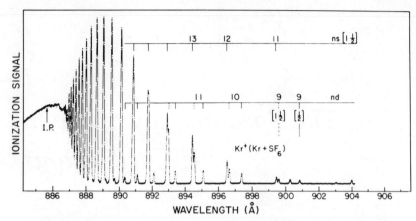

Figure 1. Kr^+ signal vs. wavelength for Rydberg state of Kr (Kr^{**}) detected by $Kr^{**} + SF_6 \rightarrow Kr^+ + SF_6^-$ (from Chupka[2]).

This brief review describes some recent progress on the spectroscopy of highly excited atoms. The review is by no means comprehensive, and the author apologizes in advance to his colleagues whose work has been omitted.

2. Experimental Methods

The tunable laser has played a key role in most of the recent studies of Rydberg atoms. Chupka,[2] however, in a largely unpublished series of experiments, has successfully populated Rydberg states of atoms and molecules by optical excitation using a continuous ultraviolet source and a vacuum monochromator. Figure 1 shows excitation spectra for Kr detected by collisional ionization with SF_6, according to the reaction $Kr^{**} + SF_6 \rightarrow Kr^+ + SF_6^-$. ($Kr^{**}$ is a Rydberg state of Kr.) The decreasing signal for low n states is due to the competition between the electron affinity of the SF_6 and the binding energy of Kr^{**}; the decrease for high n is due to loss of oscillator strength.

The alkalis, which offer the double attraction of experimental and theoretical simplicity, have played a principal role in many studies of highly excited atoms. Although alkali Rydberg states can be easily observed by absorption spectroscopy, modern work requires techniques for efficiently populating selected states which had to await the advent of the tunable laser. Stepwise excitation—absorption on a principal line followed by excitation to a Rydberg level—is most commonly employed. Svanberg *et al.*[3] combined an rf discharge lamp with a cw dye laser to stepwise populate s and d states of Rb and Cs. By applying optical double resonance and level-crossing methods, they were able to measure hyperfine dipole and quadrupole constants in terms up to $7d$ for Rb and $10d$ for Cs. Haroche et al.[4] populated $9d$ and $10d$ levels of sodium using pulsed dye lasers and measured the fine-structure intervals by observing the quantum beats. These techniques have since been applied to a considerable body of work on fine structure, hyperfine structure, and Stark shifts of alkali s, p, and d states.[5]

Two-photon Doppler-free spectroscopy has also been applied to Rydberg state spectroscopy. Harvey and Stoicheff[6] applied this technique to study the fine structure of Rb for levels up to $32d$, while Harper and Levenson[7] applied it to a study of K fine structure for levels up to $19d$.

Lifetimes for many alkali Rydberg states have recently been measured. Gallagher et al.[8] have determined lifetimes for the s, p and d Rydberg levels of Na, while Landberg and Svanberg[9] and Gounand et al.[10] have measured lifetimes for Rb and Cs. For the light alkalis, oscillator strengths and lifetimes can be reliably calculated from the Coulomb approximation. For Rb and Cs, however, spin–orbit effects are sufficiently large to require more refined methods. High-angular-momentum states of atoms have remained largely unstudied in the past owing to the difficulty of populating or observing them by conventional spectroscopic methods. A number of techniques have now been developed for working with high-angular-momentum Rydberg states. Gallagher, Hill, and Edelstein[11] have measured the fine structures of $l = 2–5$ states of sodium by radio-frequency spectroscopy. A Rydberg d state is populated by stepwise excitation with pulsed lasers and the $d–f$ dipole transition is induced by an applied rf field. Resonance is detected by observing cascade fluorescence on the $3d–3p$ transition, 8197 Å. The $d–g$ and $d–h$ transitions are induced by multiphoton resonance. (The large dipole matrix elements between Rydberg states permits multiphoton processes with only modest power.) The f, g, and h fine-structure splittings were all found to be hydrogenic within experimental error. The microwave spectra also yield values for the quantum defects.

Freeman and Kleppner[12] have pointed out that the quantum defects for high-angular-momentum states of alkalis can be related to the polarization interaction between the valence electron and the ionic core. The perturbation is of the form

$$V_{\text{pol}} = -\tfrac{1}{2}\alpha'_d \langle 1/r^4 \rangle - \tfrac{1}{2}\alpha'_Q \langle 1/r^6 \rangle$$

where α'_d and α'_Q are, respectively, the dipole and quadrupolar polarizabilities of the alkali ion. They have shown that this interaction leads to a quantum defect

$$\delta_l \simeq \tfrac{3}{4}\alpha'_d l^{-5} + \tfrac{35}{16}\alpha'_Q l^{-9}$$

Using the data of Gallagher et al.,[11] they obtained $\alpha'_d = 1.0015(15)$ a.u., and $\alpha'_Q = 0.48(15)$ a.u. These values differ from the static polarizabilities because of nonadiabatic effects, which are at present not well understood. The opportunity for precision measurement of the core polarizabilities offers stimulus for further theoretical work in this area.

A second technique that has been successfully applied is Stark spectroscopy, which utilizes the large Stark effects possible in Rydberg states. (The extreme shift, $\delta W_s \simeq \pm \frac{3}{2} n^2 E$, can be much larger than the separation between Rydberg terms $\Delta W_0 \simeq 1/n^3$.) Atoms are excited from some convenient initial state to an energy lying near a Rydberg state using a fixed, precisely known, frequency. A gradually increasing electric field is applied. An excitation signal appears whenever a Stark subcomponent is shifted into resonance. Ducas and Zimmerman[13] have used this method to observe levels near the $n = 16$ state of sodium using lines from a CO_2 laser. The energy levels are shown in Figure 2. Atoms were prepared in the $10s$ state and excited by the 9.6μ $P(20)$ CO_2 line to an energy slightly above the $n = 16$ manifold (and the nearby $17p$ state). A Stark excitation spectrum for $|m| = 1$ is shown in Figure 3a. (Some $m = 0$ features are visible due to imperfect polarization of the CO_2 radiation. The separation between the

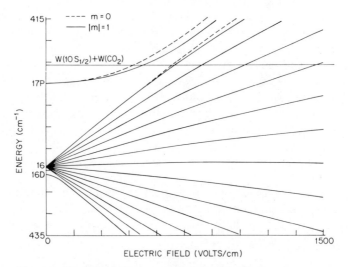

Figure 2. Theoretical Stark structure showing binding energy vs. electric field for the $16d$, $n = 16$ manifold, and $17p$ levels of sodium. All of the $|m| = 1$ features and two of the $m = 0$ features are shown. The horizontal line indicates the energy satisfying the resonance condition from the $10s$ level for the CO_2 laser operating on the $P(20)$ line of the $9.6\ \mu$ band. (From Ducas and Zimmerman.[13])

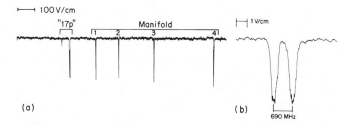

Figure 3. Experimental resonance spectrum of signal vs. electric field for situation depicted in Figure 2. (a) Excitation with polarization perpendicular to dc electric field ($\Delta m_l = \pm 1$). (b) High-resolution scan of $|m| = 1$ component of the "17p" feature showing fine-structure splitting of this Stark state. (From Ducas and Zimmerman.[13])

$m = 0$ and $|m| = 1$ components of the "17p" state is visible in Figure 3a.) Each $|m| = 1$ state is actually a fine-structure doublet; the splitting of the "17p" state (i.e., the level that adiabatically connects to $17p$), barely discernible in Figure 3a, is clearly displayed in the high resolution sweep of the "17p" line in Figure 3b. The linewidth is due to electric field inhomogeneity, estimated at about 1%. The Stark spectrum can be unfolded to yield values for the fine-structure splittings and quantum defects, as well as to yield a more precise value for the term value of the initial $10s$ state.

3. Field Ionization

Field ionization has proved to be a versatile and efficient tool for the detection of Rydberg state atoms.[14–16] Early work using fast beams has already been mentioned; what has given the technique new importance is the ability to discriminate individual Rydberg states, providing at once both a detector and a spectrometer. Field ionization can occur by two somewhat related processes. In the presence of an electric field, E, the potential, $V = -1/r - Ez$, has a maximum value $V_{max} = -(2E)^{1/2}$. Any state with energy $W > V_{max}$ ionizes essentially instantaneously; this process is sometimes called classical ionization. If W is less than, but close to, V_{max}, the electron can escape by tunneling through the potential barrier. The tunneling rate increases with field so rapidly, however, that for many purposes it is sufficient to take the rate as zero for $W < V_{max}$, and infinite for $W > V_{max}$. This leads to the idea of a "threshold field" for ionization given by $E_0 = W^2/4$. If we take W to be the unperturbed energy of a Rydberg state with principal quantum number n, we have

$$E_0 = 1/16n^4 \text{ a.u.} = 3.2 \times 10^8/n^4 \text{ V/cm}$$

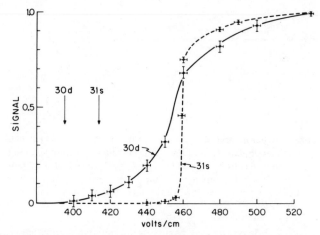

Figure 4. Ionization signal as a function of electric field for the 30d and 31s levels of Na. The broadening of the d-state threshold is believed to be due to unresolved fine structure. (From Ducas et al.[14])

The value $1/16n^4$ must be regarded as a characteristic unit rather than as an accurate threshold value. Herrick[17] has developed the idea of threshold ionization in hydrogen along more rigorous lines.

In practice, a threshold for ionization is clearly exhibited for many states. The $1/n^4$ scaling law has so far been found to hold well,[13,14] though

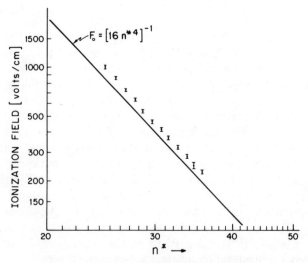

Figure 5. Log–log plot of threshold ionization field for Na s states. (From Ducas et al.[14])

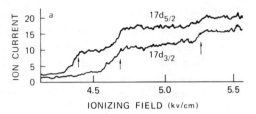

Figure 6. Ionization signal vs. electric field for $17d_{3/2}$ and $17d_{5/2}$ states of Na. Approximate threshold locations for the $|m_l| = 0, 1$, and 2 levels are indicated by arrows. (From Gallagher et al.[18])

the coefficient varies slightly from system to system. Figure 4 shows a typical threshold ionization case for the $31s$ and $30d$ levels of Na. Figure 5 shows the $(n^*)^{-4}$ variation of the treshold for the sodium S states. Gallagher et al.[18] have observed differential ionization rates for the different $|m_l|$ and $|m_j|$ levels of Na d states. Figure 6 presents data for the $17d_{3/2}$ and $17d_{5/2}$ levels.

In order to understand field ionization it is essential to understand precisely how the ionization rate varies with electric field. Figure 7, taken from the calculations of Bailey, Hiskes, and Riviere[19] (BHR), shows representative curves for hydrogen, $n = 13$–15. The edges of the shaded area are the ionization curves for the extreme Stark components of the $n = 14$ manifold. Curves for the intermediate Stark levels lie between in order of increasing Stark energy.

Littman et al.[20] have developed techniques for populating and identifying individual Stark sublevels of Rydberg states of sodium, and have measured the ionization rates of selected levels. Using single-ion timing

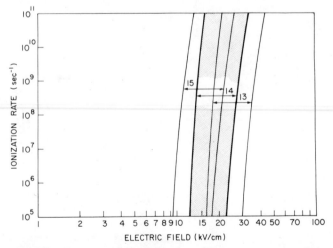

Figure 7. Ionization rate curves for hydrogen, $n = 13$–15. The shaded area represents the family of curves for the $n = 14$ manifold of Stark states. (Based on BHR.[19])

methods, Littman *et al.*[21] have measured the ionization rate of a number
of Stark levels in the vicinity of $n = 14$. The lowest sublevels of each term
were found to be in excellent agreement with BHR. Figure 8 shows the
ionization rate of the level (14, 0, 11, 2) [the indices are the parabolic
quantum numbers $(n, n_1, n_2, |m|)$]. The data agree with the calculations of
BHR to within the uncertainty in calibration of the electric field, namely,
2%.

The results for other Stark sublevels, however, can be in serious
disagreement with BHR. Figure 9 shows the ionization rate of the level
(12, 6, 3, 2). The ionization curve is not even monotonic and reaches a rate
of $10^7 \, \text{sec}^{-1}$ at less than half the field predicted by BHR. As Littman *et
al.*[21] discuss, the anomalous behavior arises from effects of level mixing.
Higher-lying levels, which ionize at lower fields than the level of interest,
are mixed with the latter by the dipole interaction. The effect of the state
mixing is to decrease the radiation rate of the upper level, and to increase
the rate of the lower level. The solid curve of Figure 9 is calculated from
this model with no adjustable parameters. [Two crossings were included in
the calculation with level (14, 0, 11, 2) at 15.7 kV/cm and with level (15, 0,
12, 2) at 17.3 kV/cm.] These results show that the theory developed for
field ionization of hydrogen cannot be applied simply to other atoms. Also,

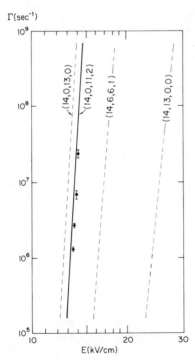

Figure 8. Measured ionization rates vs. field for the
state $(n = 14, \ n_1 = 0, \ n_2 = 11, \ |m| = 2)$ in Na.
The solid line is extrapolated from the calculations of BHR.[19] (From Littman *et al.*[20])

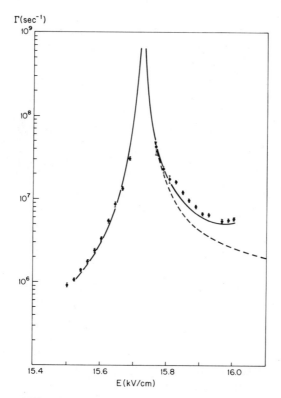

Figure 9. Ionization rate for the level ($n = 12$, $n_1 = 6$, $n_2 = 3$, $|m| = 2$) in the region of crossing with the level (14, 0, 11, 2). (From Littman et al.[20])

there is some question about the effects of level mixing in hydrogen; in the nonrelativistic theory Stark levels of different terms can cross, but relativistic effects, or an external perturbation such as a magnetic field, can produce level-mixing effects which dominate the ionization rates.

4. Two-Electron Systems

Rydberg states of two-electron atoms are of considerable physical interest, and are attracting increasing attention. The most fundamental atom, and the atom most systematically studied, is helium. Because the work is described elsewhere in this volume by Wing and MacAdam,[1] and also in a review by Miller and Freund,[22] we shall forego describing it here.

Aside from helium, our understanding of two-electron atoms is relatively primitive. Although principal lines have been studied in some detail,

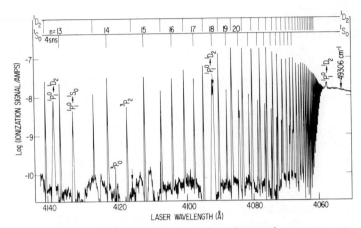

Figure 10. Multiphoton ionization spectrum of $^1S_0$ and $^1D_2$ states of Ca converging on the ionization limit 49300 cm^{-1}. Two-photon transitions $3d^2\ ^3p_0$ and $^3p_2$ are also indicated. The lines indicated by $^1P_1^0 \to {}^1S_0$ or $^1P_1^0 \to {}^1D_2$ arise from transitions from the $4s4p\ ^1P_1^0$ level. (From Armstrong et al.[24])

the spectroscopy of many series is still relatively incomplete. Level identification is often difficult because of fine-structure and configuration interactions. There are compensations for these difficulties, however, for these same effects lead to interesting features such as selection rule anomalies, autoionization, and series interactions.

Esherick et al.[23] recently made an important advance in the spectroscopy of the alkaline earths using a novel method of two-photon spectroscopy. Radiation from a pulsed tunable dye laser was focused on calcium vapor in a heat-pipe oven. Two-photon transitions from the $4s^2,\ ^1S_0$ ground state to high-lying $^1S_0$ and $^1D_2$ state were detected by an ionization probe. (The ionization mechanism is as yet not understood.) Well-resolved multiphoton spectra for levels up to $4s62d$ were obtained; Figure 10 shows a portion of the data. The laser was calibrated by doubling a portion of the output and exciting the single-photon transition $4s^2\ ^1S_0 \to 4snp\ ^1P_1$. The $^1P_1$ levels, which are accurately known and converge to the same limit as the $^1S_0$ and $^1D_2$ lines, provide convenient reference frequencies.

Armstrong et al.[24] have successfully analyzed the Ca $J = 0$ and $J = 2$ spectra using the multichannel quantum defect theory (MCQDT) of Seaton[25] in a particularly convenient elaboration due to Fano and Lu.[26] Quantum defect theory starts from the observation that the potential for a Rydberg electron is essentially Coulombic at large distances where the electron spends most of its time. The electron can be pictured as continually scattering from the core. For a tightly bound core, as in the alkalis, the effect of the non-Coulombic part of the scattering process is to introduce a phase shift into the quantum phase integral. This leads to an energy expression as the form $E(n, l) = -1/[2(n - \delta_l)^2]$, where the quantum defect δ_l is a slowly varying function of energy. The quantum defect is related to the scattering phase shift Δ_l by $\delta_l = \Delta_l/\pi$. The phase shift and quantum defect are so intimately related that the distinction between free and bound states loses significance.

In the case of two-electron atoms such as the alkaline earths, the Rydberg electron again experiences a Coulombic potential at large distances. Scattering from the open shell core, however, may introduce important dynamical effects. The inner electron is easily excited. The inner and outer electron can exchange angular momentum and create a virtual state of double excitation before the Rydberg electron emerges. The resulting phase shift leads to a quantum defect which may vary appreciably with energy, displaying resonance behavior as the energy approaches that of a doubly excited state. Often a number of doubly excited states can contribute to the phase shift, and in extreme cases the Rydberg series may be so heavily perturbed that it is difficult to identify its fundamental character.

Multichannel quantum defect theory offers a systematic procedure for analyzing and interpreting such series. Essentially it provides a method for sorting contribution to the phase shifts from the various core states, allowing one to describe the eigenstates of the scattering process in terms of angular momentum eigenstates of the two-electron system.

The methods used to analyze $J = 0, 2$ states of Ca have recently been applied to Sr by Esherick.[27] Using three-photon excitation, Esherick et al.[28] have also studied the $^3P_1$ series of Ca, Ba, and Sr.

Rydberg progressions in uranium have been studied by Solarz et al.[29] by combining three-step excitation and photoionization. Two fixed-frequency dye lasers and one variable-frequency laser were pulsed simultaneously, exciting odd-parity Rydberg progressions and also numerous valence states (states having multiply excited electrons). Rydberg state lifetimes are generally long compared to those of valence states and the two species were distinguished by applying a delayed ionizing pulse from a CO_2 laser. Series belonging to $5f^37s^2np$ and ^2nf progressions were identified, and the ionization limit was determined to 5×10^{-4} eV.

S and D Rydberg series in Ba have recently been measured with high precision by Rubbmark et al.[30] by absorption spectroscopy. The $6s6p$ level was excited by a pulsed dye laser. A second, broadband, dye laser provided the background continuum. $^1S_0$, $^1D_2$, and $^3D_2$ were observed with a resolution of 0.001 cm^{-1}. Interactions of doubly excited states with Rydberg states is very pronounced in barium, and our laboratory has undertaken a study of the effect of the region from 41,140 cm^{-1} to 41,320 cm^{-1}. Some initial results are presented in Figure 11. The energy assignments are from the data of Rubbmark et al.[30] A particularly inter-

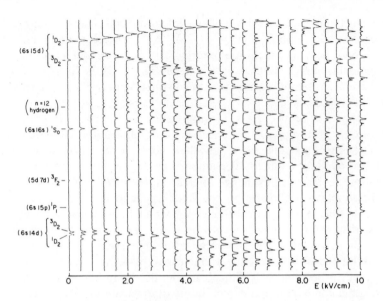

Figure 11. Stark structure of $|m| = 0$ states of Ba in the range 41150–41320 cm^{-1}. A tunable laser was scanned across the energy range displayed vertically. The horizontal peaks are generated by ionization of the excited atoms with a pulsed electric field. The level assignments are based on the data of Rubbmark *et al.*[30]

esting feature is the $(5d7d)^3F_2$ level at 41,205 cm^{-1}, which strongly perturbs the $^1D_2$ series. A preliminary analysis which neglects Stark coupling between the singlet and triplet series is in reasonable agreement with the main features, but detailed understanding requires the evaluation of Stark matrix element between the Rydberg level and the doubly excited states. [For more on the Stark effect of Rydberg levels, see Chapter 21 (Kollath and Standage) of this work.]

Acknowledgments

The author's work was supported by the U.S. Air Force Office of Scientific Research, the National Science Foundation, and the Joint Services Electronics Program.

References

1. W. H. Wing and K. B. MacAdam, Chapter 11 of this volume.
2. W. A. Chupka, private communication.

3. S. Svanberg, P. Tsekeris, and W. Happer, *Phys. Rev. Lett.* **30**, 817 (1973).
4. S. Haroche, M. Gross, and M. P. Silverman, *Phys. Rev. Lett.* **33**, 1063 (1974).
5. G. Fabre, M. Gross, and S. Haroche, *Opt. Commun.* **13**, 393 (1975); K. Fredrickson and S. Svanberg, *J. Phys. B* **9**, 1237 (1976).
6. K. C. Harvey and B. P. Stoicheff, *Phys. Rev. Lett.* **38**, 537 (1977).
7. C. D. Harper and M. D. Levenson, *Phys. Lett.* **56A**, 361 (1976).
8. T. F. Gallagher, S. A. Edelstein, and R. M. Hill, *Phys. Rev. A* **11**, 1504 (1975).
9. H. Landberg and S. Svanberg, *Phys. Lett.* **56A**, 31 (1976).
10. F. Gounand, D. F. Fournier, J. Cuvellier, and J. Berlande, *Phys. Lett.* **59A**, 23 (1976).
11. T. F. Gallagher, R. M. Hill, and S. A. Edelstein, *Phys. Rev. A* **14**, 744 (1976).
12. R. R. Freeman and D. Kleppner, *Phys. Rev. A* **14**, 1614 (1976).
13. T. W. Ducas and M. L. Zimmerman, *Phys. Rev. A* **15**, 1523 (1977).
14. T. W. Ducas, M. G. Littman, R. R. Freeman, and D. Kleppner, *Phys. Rev. Lett.* **35**, 366 (1975).
15. R. F. Stebbings, C. J. Latimer, W. P. West, F. B. Dunning, and T. B. Cook, *Phys. Rev. A* **12**, 1453 (1975).
16. A. F. J. van Raan, G. Braum, and W. Raith, *J. Phys. B* **9**, L173 (1976).
17. D. Herrick, *J. Chem. Phys.* **65**, 3529 (1976).
18. T. F. Gallagher, L. M. Humphrey, R. M. Hill, and S. A. Edelstein, *Phys. Rev. Lett.* **37**, 1465 (1976).
19. D. S. Bailey, J. R. Hiskes, and A. C. Riviere, *Nucl. Fusion* **5**, 41 (1965).
20. M. G. Littman, M. L. Zimmerman, T. W. Ducas, R. R. Freeman, and D. Kleppner, *Phys. Rev. Lett.* **36**, 788 (1976).
21. M. G. Littman, M. L. Zimmerman, and D. Kleppner, *Phys. Rev. Lett.* **37**, 486 (1976).
22. T. A. Miller and R. S. Freund, in *Advances in Magnetic Resonance*, Vol. 9, Ed. J. S. Waugh, Academic Press, New York (1977), p. 49.
23. D. Esherick, J. A. Armstrong, R. W. Dreyfus, and J. J. Wynne, *Phys. Rev. Lett.* **36**, 1296 (1976).
24. J. A. Armstrong, D. Esherick, and J. J. Wynne, *Phys. Rev. A* **15**, 180 (1977).
25. M. J. Seaton, *Comments At. Mol. Phys.* **D2**, 37 (1970).
26. U. Fano, *J. Opt. Soc. Am.* **65**, 979 (1975); K. T. Lu and U. Fano, *Phys. Rev. A* **2**, 81 (1970).
27. P. Esherick, *Phys. Rev. A* **15**, 1920 (1977).
28. P. Esherick, J. J. Wynne, and J. A. Armstrong (to be published).
29. R. W. Solarz, C. A. May, L. R. Carlson, E. F. Worden, S. A. Johnson, J. A. Paisner, and L. J. Radziemski, Jr., *Phys. Rev. A* **14**, 1129 (1976).
30. J. R. Rubbmark, S. A. Borgström, and K. Bockasten, *J. Phys. B* **10**, 421 (1977).

17

Optical Spectroscopy of
Short-Lived Isotopes

H.-Jürgen Kluge

1. Introduction

This chapter will concentrate on a special branch of the large field of optical spectroscopy, namely, the study of hyperfine structure (hfs) and isotope shift (IS) of short-lived isotopes. Such investigations give information on some basic properties of these nuclei: the spin (I), the magnetic moment (μ_I), the spectroscopic quadrupole moment (Q_s), and the change of the mean square charge radius between different isotopes ($\delta\langle r^2\rangle$).

Optical spectroscopy is among the oldest experimental techniques in nuclear research and has been widely applied in the past to stable and long-lived isotopes. It is still the only available experimental method for the determination of charge radii of short-lived isotopes. The theory of hfs and IS has been understood in principle since the early days of quantum mechanics. The richness of information obtainable by optical spectroscopy as well as the safe theoretical basis have therefore made it desirable to extend this method to short-lived isotopes which lie far outside the valley of nuclear stability. Attempts to do this have, however, failed in the past because of the use of insufficient radioactive material and/or the isotopic impurity in the samples. The situation partly changed when powerful accelerators became available and mass separators were set up with them. In addition, however, more sensitive spectroscopic methods were

H.-Jürgen Kluge • Institut für Physik, Universität Mainz, Postfach 3980, D-6500 Mainz, Germany.

developed. Apart from atomic beam resonance (ABR),* which will not be considered here because it is restricted to atomic ground or metastable states, these are nuclear-radiation-detected optical pumping (RADOP) and laser spectroscopy. It is easy to forecast that optical spectroscopy will be revived as an important tool for nuclear research. The first steps have already been made. Today it is possible to produce almost any isotope in an isotopically clean form. Almost any element is accessible if tunable cw lasers that cover the whole spectral region from the ultraviolet to the infrared are available. Thus, for the first time, systematic studies of hfs and IS are possible for long chains of isotopes; such cuts through the chart of nuclei along constant proton number Z will considerably enlarge the knowledge and, it is to be hoped, the understanding of the various phenomena exhibited by nuclear matter.

2. Information on Nuclear Properties of Short-Lived Isotopes from hfs and IS

In Chapter 7 of this work, "New Developments of Classical Optical Spectroscopy," Heilig and Steudel treat the general features of hfs and IS and give references to recent reviews. This section can thus be restricted to the question how information on spins, moments, and changes of charge radii can be extracted from the hfs and IS of short-lived isotopes, assuming that an analysis of hfs and IS has been performed for the stable isotopes.

2.1. Hyperfine Structure

The Hamiltonian of the magnetic dipole and the electric quadrupole hfs interaction under the influence of an external magnetic field H_0 can be written in the $|I, J, m_I, m_J\rangle$ representation as[1,2]

$$\hat{H} = A(\hat{I}\hat{J}) + B\{3(\hat{I}\hat{J})^2 + \tfrac{3}{2}(\hat{I}\hat{J}) - \hat{I}^2\hat{J}^2\}$$
$$\times \{2I(2I-1)J(2J-1)\}^{-1} - g_J\mu_B(\hat{J}\mathbf{H}_0) - g_I\mu_N(\hat{I}\mathbf{H}_0) \tag{1}$$

with

$$A = \langle I, J, m_I = I, m_J = J|\hat{\mu}\hat{H}_e|I, J, m_I = I, m_J = J\rangle/(IJ) \tag{2}$$

$$B = e\langle I, J, m_I = I, m_J = J|\hat{Q}\,\partial^2 V/\partial z^2|I, J, m_I = I, m_J = J\rangle \tag{3}$$

$\hat{\mu}$ is the operator corresponding to the nuclear magnetization, $\hat{H}_e$ is the operator of the magnetic hfs field produced by the electrons at the site of

---

* ABR has been applied very successfully to isotopes far from stability. See Chapter 10 of this work, by Penselin, and Ref. 53.

the nucleus, $\hat{Q}$ is the operator of the nuclear quadrupole moment Q_s, and $\partial^2 V/\partial z^2$ is the operator of the electric field gradient at the origin produced by the electrons; I is the nuclear, J the atomic spin; g_J and g_I are, respectively, the electronic and nuclear g factors. At zero magnetic field, Eq. (1) gives the hfs energy of a state of total angular momentum $\mathbf{F} = \mathbf{I} + \mathbf{J}$ as

$$W(F, I, J, A, B) = \frac{AK}{2} + B \frac{\frac{3}{4}K(K+1) - I(I+1)J(J+1)}{2I(2I-1)J(2J-1)} \tag{4}$$

with

$$K = F(F+1) - I(I+1) - J(J+1)$$

Thus I, A, and B can be determined from a measurement of the hfs splitting of an optical line. To derive the values of μ_I and Q_s, the matrix elements for A and B are factorized into $\langle \mu \rangle_{II} \langle H_e(0) \rangle_{JJ}$ and $\langle Q \rangle_{II} \langle \partial^2 V/\partial z^2 \rangle_{JJ}$ by neglecting the spread of the nucleus. In doing this, the error made in the quadrupole interaction is small because of the vanishing electron density at the nucleus for electrons with orbital angular momentum $l > 1$. The error introduced to the magnetic hfs interaction is of the order of the hfs anomaly. Values for $\langle H_e(0) \rangle_{JJ}$ and $\langle \partial^2 V/\partial z^2 \rangle_{JJ}$ can be taken as those of stable isotopes provided that the value of μ_I is known by, e.g., NMR, and that of Q_s by, e.g., mesic atoms. A compilation of nuclear spins and moments has been performed recently by Fuller.[3] For the determination of the nuclear spin and magnetic moment by magnetic resonance see Section 3.1.4.

2.2. Isotope Shift

The centers of gravity of the optical transitions with frequencies ν_0 of two isotopes with mass numbers A and A' are shifted by the sum of a field shift and a mass shift

$$IS^{AA'} = \nu_0^{A'} - \nu_0^A = \delta\nu_{\text{field}}^{AA'} + \delta\nu_{\text{mass}}^{AA'} \tag{5}$$

The field shift accounts for the influence of the extended nuclear charge distribution on the electronic binding energy and therefore depends on the change of the nuclear size and angular shape when neutrons are added or subtracted. This is the information that is of interest when dealing with very unstable isotopes. In order to extract this information from experimental IS data, the mass shift has to be isolated. The latter originates from the difference in nuclear mass between two isotopes, whereby the center of mass of the electron–nucleus system is changed.

As discussed in Chapter 7 of this work, the IS can be written as

$$IS^{AA'} = M(A' - A)/AA' + F^{AA'}\lambda^{AA'} \tag{6}$$

where M is a constant depending only on the energies and the wave functions of the states of the optical transition. F is essentially an electronic factor; it depends on Z and, in addition, weakly on A and, if realistic nuclear shapes are considered, weakly on the model of charge distribution. λ^{AA} is the sum of radial moments:

$$\lambda^{AA'} = \delta\langle r^2\rangle^{AA'} + \sum_{k=2}^{\infty} \frac{c_k}{c_1} \delta\langle r^{2k}\rangle^{AA'} \tag{7}$$

The parameters c_k/c_1 have been computed by Seltzer.[4] Since the variation of the electron wave function over the nuclear volume is small, the contribution of higher moments to the leading term $\delta\langle r^2\rangle$ is also small (e.g., -7% of $\delta\langle r^2\rangle$ in the case of Hg). Hence, $\lambda^{AA'} \simeq \delta\langle r^2\rangle^{AA'}$ is still a fair approximation even for heavier elements. Most IS data are treated in terms of $\delta\langle r^2\rangle$ values.

The various methods that are used to extract $\delta\langle r^2\rangle$ from experimental IS data gave been discussed in Chapter 7 of this work and also in a number of reviews.[5-8] The most straightforward method is to calibrate the optical data with respect to those obtained by measurements of muonic[9] or electronic x-ray shifts,[10] or by electron scattering.[11] Once the change of $\langle r^2\rangle$ between a pair of isotopes A, A' is known from an analysis of the data of stable atoms, the radial change of any other pair A'', A''' can be calculated from the relation

$$\delta<r^2>^{A''A'''} = \delta<r^2>^{AA'} \frac{\delta\nu_{\text{field}}^{A''A'''}}{\delta\nu_{\text{field}}^{AA'}} \left(\frac{A''+A'''}{A+A'}\right)^{(2\sigma-2)/3} \tag{8}$$

where σ equals $(1-\alpha^2 Z^2)^{1/2}$, and α is the fine-structure constant.

Since the calculation of $\delta\langle r^2\rangle$ from electronic IS data is (to a good approximation) model independent, nuclear theory should compute the observable $\delta\langle r^2\rangle$ rather than model parameters. Microscopic calculations are, however, often missing, and phenomenological models[6,12-14] have to be used which involve parameters such as deformation parameters,[15] isotope shift discrepancy,[6] compressibility under deformation,[16] neutron skin thickness,[14] etc.

3. Nuclear-Radiation-Detected Optical Pumping of Short-Lived Isotopes (RADOP)

In the past, most spectroscopic methods, including the purely optical and the radio-frequency techniques, required samples of about 10^{12}–10^{14} atoms. This number can be regarded as the minimum quantity that is necessary to form a vapor of reasonable optical thickness. Such amounts of

Figure 1. OP cycle for a Hg isotope with $I = 1/2$. The thick arrows represent $\Delta m_F = +1$ transitions. In the presence of a buffer gas the sublevels of the nonspherical $^3P_1$ state are mixed (wavy line), and the spontaneous emission takes place via all indicated transitions. The numbers between are the relative absorption and emission rates starting from an unpolarized atom.

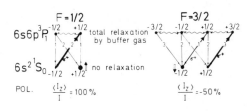

atoms can, however, be produced only for isotopes with half-lives longer than about 1 hr. For example, in the case of Hg, where the first optical pumping experiment on a radioactive isotope was performed by Stavn and Walter,[17] the isotopic string investigated by optical spectroscopy (OS), level crossing (LC), optical double resonance (ODR), or conventional optical pumping (OP), spans from ^{204}Hg (stable) down to ^{192}Hg ($T_{1/2} = 5$ h).* For isotopes with half-lives of the order of minutes or seconds or less, the number of atoms obtainable drops by many orders of magnitude (e.g., to 10^4 atoms for ^{181}Hg, see below). In these cases, the lack of sufficient material can be compensated for by combining the sensitivity of radioactive tracing techniques with the detection of nuclear orientation as obtained by OP.

3.1. The Method

Optical pumping has proved to be a very successful spectroscopic method for the determination of spins and magnetic moments of atoms with spherical symmetric (atomic) ground state. The process produces an orientation of the atom by absorption of polarized light, and via the hfs interaction an orientation of the nuclear spin. In standard optical spectroscopy the orientation is monitored by its influence on the absorption of the pumping light, but for lower vapor densities this absorption becomes nondetectable. In the RADOP method, the β-decay asymmetry or γ-decay anisotropy replace the photons of the optical transition as monitor for the nuclear orientation obtained by OP.

3.1.1. Principle of Optical Pumping

Optical pumping was proposed by Kastler[19] in 1950 as a method for producing orientation in the total angular momentum $F = I + J$ (for reviews of OP see Refs. 20–23). For a qualitative explanation of OP, the

---

* A compilation of hfs and IS data of Hg and references are given in Ref. 18.

case of a Hg atom, which has a diamagnetic $^1S_0$ ground state, will be chosen.* Figure 1 shows the relative transition probabilities A_i between the Zeeman sublevels of the $6s^2\,^1S_0$ and the excited $6s6p\,^3P_1$ state (normally used for OP) of Hg for an $I = 1/2$ isotope. The transitions involved in the OP process are (i) optical absorption and emission, (ii) transitions between Zeeman levels in the $^1S_0$ ground state induced by relaxation or radio frequency, (iii) transitions between Zeeman levels in the excited state induced by relaxation during the lifetime of the $^3P_1$ state.

Without a buffer gas the relaxation in the excited state is negligible. Using right-handed circularly polarized light incident parallel to the quantization axis ($\Delta m_F = +1$), the atoms are pumped into the sublevels with highest projection $m = I$, and a polarization of 100% is achieved (neglecting relaxation in the $^1S_0$ state). The situation changes for OP in the presence of buffer gas. Because of the large difference in the cross sections for the disorientation of the $^3P_1$ and $^1S_0$ atomic state (about $10^3\,\text{Å}^2$ compared to $10^{-6}\,\text{Å}^2$) a buffer-gas pressure can be chosen that leads to a complete mixing of the sublevels of the $^3P_1$ state (≈ 10 Torr He) but hardly increases the ground-state relaxation. In fact it reduces relaxation at the walls by increasing the diffusion time. Assuming complete relaxation in the excited state, the atom will return to either of the two substates with equal probability. Thus, the steady state of the population numbers n_i is reached when the absorption probabilities from each sublevel are equal, i.e.,

$$n_i A_i = n_j A_j \qquad \text{for all } i, j \qquad (9)$$

OP via the hfs level with the highest F number has a special feature: Pumping in vacuum and in the presence of a buffer gas (complete mixing in the excited state) leads to an inversion of the polarization. This can be used to identify the order of the hfs levels.

A quantitative description of the OP cycle for RADOP can be obtained by setting up rate equations with the appropriate matrix elements for the electric dipole transitions (optical absorption and reemission) and those for the magnetic dipole and electric quadrupole transitions which govern the relaxation phenomena. The rate equations form a system of linear differential equations, the solution of which is a sum of exponential factors. A simple analytic expression can be obtained only for isotopes with a single $F = 1/2$ ground state. This solution, however, is also a satisfying approximation for systems with higher angular momenta.

After switching on the OP light, the polarization will increase exponentially with a time constant

$$\tau_1^F = (1/T_1 + p_c^F/T_c)^{-1} \qquad (10)$$

---

* The case of Hg was extensively treated by Cagnac[24] and Cohen-Tannoudji.[25]

where T_1 is the longitudinal relaxation time, T_c is the average time between subsequent absorptions of a light quantum per atom, and p_c^F is the average polarization achieved by the first pumping cycle and depends on the relative absorption probabilities A_i. A polarization of

$$P^F = P_0^F T_1 / (T_1 + T_c/p_c^F) \tag{11}$$

is reached at equilibrium, where P_0^F is the polarization achieved without relaxation. Equation (11) shows the critical parameters of an OP experiment. Maximum polarization is obtained when the intensity of the OP light is so high that the pumping process is much faster than the relaxation. In addition, pumping is more efficient for isotopes with a low spin because in this case fewer OP cycles are needed to concentrate the population in the Zeeman state with highest $|m|$ number.

The nuclear orientation obtained by OP can be described by moments of orientation, which are defined by[26]

$$f_k(I) = \binom{2k}{k}^{-1} I^{-k} \sum_m \sum_{\nu=0}^k (-1)^\nu \frac{(I-m)!(I+m)!}{(I-m-\nu)!(I+m-k+\nu)!} \binom{k}{\nu}^2 n_m \tag{12}$$

where n_m is the population number of the substrate with projection $m_I = m$. The maximum order is restricted by the condition $0 \leq k \leq 2I$. f_0 accounts for the isotropic population, and f_1 describes the polarization:

$$f_1 = P_I = \frac{\langle I_z \rangle}{I} = I^{-1} \sum_m m n_m \tag{13}$$

For $I > 1/2$, alignment can also occur with

$$f_2(I) = I^{-2} \left[\sum_m m^2 n_m - \tfrac{1}{3} I(I+1) \right] \tag{14}$$

This is the highest order of orientation that is detectable in the angular distribution of optical radiation.

3.1.2. Polarization through Spin Exchange

In OP, a nuclear polarization can only be obtained if the emission profile of the light source matches the absorption profile of the isotope under investigation to within the width of the spectral line. If isotopes far from stability are optically pumped by use of a normal light source as described below, the difference in IS and hfs between the stable and radioactive isotope will normally hinder such a coincidence. In this case the Zeeman effect can be used to tune the frequency of the light source (see, e.g., the RADOP experiments on Hg, Section 3.4). Another possibility is to use the spin exchange process.[27-29] This plays an important role in OP experiments, especially in the case of alkali atoms. Colliding paramagnetic atoms exchange their valance electrons and hence their spin polarization with cross sections, σ_{ex}, of the order of 100 $Å^2$. The reaction may be symbolized as

$$A(\uparrow) + B(\downarrow) \rightarrow A(\downarrow) + B(\uparrow) \tag{15}$$

If the species A is polarized by a dynamical process, such as OP, the polarization is transferred to the spin system B. The exchange rate is given by

$$1/T_{ex} = N_A v_{rel} \sigma_{ex} \tag{16}$$

where v_{rel} is the mean relative velocity between an atom A and an atom B, and N_A is the density of atoms of species A. For the D_1 or D_2 line of the alkalis, a vapor density of $N_A = 10^{11}$ cm^{-3} is sufficient to obtain a sizable absorption of the OP light in a usual resonance cell which has a diameter of some centimeters. The exchange rate is then of the order of 100 s^{-1} and hence comparable to the pumping time. The electronic polarizations of both paramagnetic species can consequently become equal and the relative population numbers n_{m_F} can be characterized by

$$n_{m_F} \sim e^{-\beta m_F} \tag{17}$$

where β^{-1} is a fictitious spin temperature that depends on the pumping and relaxation times.

3.1.3. Angular Distribution of Nuclear Radiation from Oriented Nuclei

The angular distribution of nuclear radiation emitted by an oriented ensemble of nuclei is given by[26]

$$W(\theta) = \sum_k A_k f_k P_k(\cos \theta) \tag{18}$$

where f_k is the orientation parameter [defined by Eq. (12)], A_k a correlation factor of the nuclear decay, and P_k the Legendre polynomial of order k.

(a) β Asymmetry. As a consequence of parity violation in the weak interaction the electrons or positrons emitted in β decay have an asymmetric angular distribution given in lowest order by[30]

$$W_\beta(\theta) = 1 + (v/c) A_\beta f_1 \cos \theta \tag{19}$$

where v is the velocity of the emitted β particle and A_β is the asymmetry parameter, which depends on the type of the β-decay and ranges from $-1 \le A \le +1$. f_1 is the nuclear polarization obtained by OP [Eq. (13)]. Higher-order terms with $k > 1$ are not important.

The observation of the angular distribution is a very elegant and efficient method of observing nuclear polarization and has been used in many experiments. In order to monitor the polarization achieved by OP, two detectors are placed at 0° and 180° with respect to the polarization axis. Their unequal efficiencies can be canceled out since the polarization, and hence the asymmetry, can be inverted by simply turning the polarization of the light from right-handed (σ^+) to left-handed (σ^-) and vice versa. When N_0^+ is the count rate of the 0° detector with σ^+ polarization, etc., the experimental asymmetry, $2a$, is approximately given by

$$2a = \frac{N_0^+ N_{180}^- - N_0^- N_{180}^+}{N_0^+ N_{180}^- + N_0^- N_{180}^+} \approx \overline{2f_1 \frac{v}{c} A_\beta} \tag{20}$$

The average accounts (i) for different β decay channels with individual asymmetry parameters, (ii) for different continuous velocity spectra v/c, and (iii) for nuclear polarization, which is the average determined by the ratio of polarized nuclei and unpolarized background.

(b) γ *Anisotropy.* The angular distribution of γ radiation of multipolarity L emitted by an oriented sample is given by Eq. (18). Only the correlation coefficients A_k with even k contribute here since parity is conserved in the electromagnetic interaction. As a consequence the angular distribution is symmetric with respect to a plane perpendicular to the quantization axis, and the appropriate placing of the two γ detectors that are needed to monitor the nuclear orientation is generally under 0° and 90° (another choice might be better for a special γ transition). Usually, the lowest-order alignment, f_2 [Eq. (14)], gives the dominant contribution to the anisotropy that is observable by the optical radiation (pure $E1$). In principle, however, higher orders of orientation might be observed since k ranges up to $2I$ or $2L$, depending on which is smaller. This offers the possibility of studying the influence of spin-dependent interactions such as OP, relaxation, radio frequency (rf) transitions, spin exchange, etc. on f_k with $k \geq 2$.

In γ RADOP, the anisotropy is used solely to monitor the alignment built up by OP. By analogy with Eq. (20) an anisotropy signal can be defined by

$$S_\gamma = (N_0^a N_{90}^b - N_0^b N_{90}^a)/(N_0^a N_{90}^b + N_0^b N_{90}^a) \tag{21}$$

where N_0^a is the count rate in the 0° detector with pumping mode a, etc. Considering only the leading f_2 term, Eq. (14) shows that a concentration of population into states with high or low $|m|$ leads to $f_2 > 0$ or $f_2 < 0$, respectively. This can be achieved by different OP modes (see Figure 2):

(i) The geometry used in β RADOP, i.e., circularly polarized light (either σ^+ or σ^-) with the direction of the magnetic field H_0 parallel to the

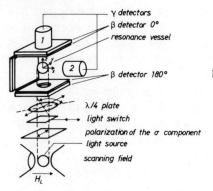

Figure 2. Geometry for β or γ RADOP. Different pumping modes are achieved by switching the direction of the static magnetic field over the resonance vessel into the x, y, or z axis, or by turning the quarter wave plate. The direction of the scanning field (see Section 3.4) and hence the polarization of the σ component of the light source are fixed in space.

light propagation (z axis). Hence the $\Delta m = +1$ (or -1) transitions lead to an increase of population in states with high $|m|$ irrespective of the sense of circular polarization (σ^+ or σ^- pumping).

(ii) With unpolarized light propagating along H_0, $\Delta m = \pm 1$ transitions are induced (σ pumping).

(iii) With linearly polarized light propagating perpendicular to H_0 and with the direction of the polarization parallel to H_0 (y axis), $\Delta m = 0$ transitions are induced (π pumping).

(iv) With the direction of polarization perpendicular to H_0 (x axis) $\Delta m = \pm 1$ transitions are induced (σ pumping).

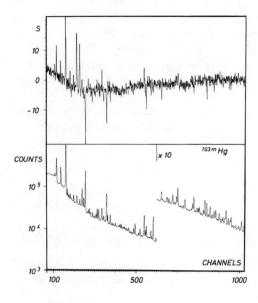

Figure 3. *Bottom*: Spectrum of the γ decay of ^{193}Hg and ^{193m}Hg taken by a 35-cm^3 Ge(Li) detector. This spectrum is obtained for σ pumping. A spectrum obtained for π pumping is stored in another subgroup of the memory of the analyzer. *Top*: The normalized difference spectrum given by Eq. (22). As the ^{193m}Hg isomeric state is selectively oriented, the γ lines showing statistically significant anisotropy signals are fed by the decay of ^{193m}Hg.[31]

It is not obvious which pumping mode will yield the maximum aniso-tropy. It depends on the nuclear states involved (spins of the initial and final state and multipolarity of the γ transition) and the conditions of optical pumping (OP in buffer gas or vacuum, OP via different hfs state). A detailed calculation has been made for the $I = 13/2$ isomeric state of ^{199}Hg.[31] When investigating nuclei that are far from stability, the decay schemes are usually not sufficiently known to enable a calculation to be done before applying γ RADOP. In this case, the best pumping modes and the best-suited γ transition can be found experimentally, since switching between the different pumping geometries can be done easily and quickly. Figure 3 shows as an example the γ spectrum of ^{193}Hg $(I = 3/2)$ and its isomer $(I = 13/2)$ under conditions where ^{193m}Hg is oriented by OP via the $F = 13/2$ hfs state. The calculation of the anisotropy signal (21) for each γ line of the multichannel spectrum (shown for one detector at the bottom of Figure 3) would require a careful peak analysis (the 0° and 90° detectors differ in resolution, spectral response, etc.). For a fast survey it is preferable to calculate for each channel i an anisotropy signal, which is given by

$$S'_\gamma(i) = [N(i)^a - N(i)^b]/[N(i)^a + N(i)^b]^{1/2} \qquad (22)$$

from the count rates N of one detector only. The integral number of counts taken in both OP modes a and b is chosen to be equal. The square root in the denominator of (22) is the statistical error of the difference in count rate of each channel. Thus, $S'_\gamma(i)$ is thus the anisotropy in units of one statistical standard deviation. A plot of S'_γ (top of Figure 3) immediately shows the γ lines that are best suited for γ RADOP. In addition, an anisotropy spectrum, such as Figure 3, may give valuable information for nuclear spectroscopists and enable spins, multipolarities, and branching ratios in complex decay schemes to be determined.

3.1.4. Magnetic Resonance in β and γ RADOP

The results of resonance experiments in the $^2S_{1/2}$ ground states of alkali atoms are described by the Breit–Rabi formula,[2] which describes the dependence of the hfs levels $|F, m_F\rangle$ on the magnetic field:

$$W(F = I \pm 1/2, m_F) = -\delta W_0/[2(2I+1)] - m_F \mu_N g_I H_0$$
$$\pm \tfrac{1}{2}\delta W_0[1 + 4m_F x/(2I+1) + x^2]^{1/2} \qquad (23)$$

where

$$\delta W_0 = (2I+1)A/2 \qquad (24)$$

is the zero-field hfs separation and

$$x = (\mu_B g_J - \mu_N g_I)H_0/\delta W_0 \qquad (25)$$

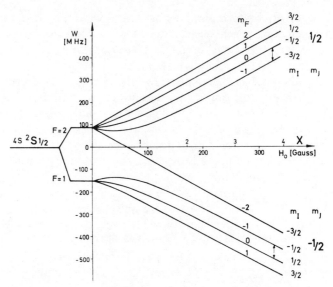

Figure 4. Hyperfine-structure level scheme of the ground state of ^{37}K in a magnetic field. The transitions used for the determination of μ_I by the doublet method are indicated.

The other symbols are defined in Section 2. Up to second order in the external field, the frequencies for ($\Delta F = 0$, $\Delta m_F = \pm 1$) transitions with $F = I \pm 1/2$ are given approximately by

$$\nu_i(m_F, m_F - 1) = g_J \mu_B H_0/[h(2I + 1)]$$

$$\mp [(2m_F - 1)(\mu_B g_J H_0)^2]/[h(2I + 1)\mu_N g_I \langle H_e(0)\rangle] \quad (26)$$

At low magnetic field the quadratic term of (26) vanishes and the nuclear spin can be determined directly by inducing rf transitions.

If the nuclear spin is known, the transition frequencies for higher H_0 depend essentially on the single unknown parameter g_I since the other parameters can be taken from measurements of the stable isotopes. As the change of the distribution of nuclear magnetism between different isotopes is neglected in this treatment, the nuclear Landé factor can be determined only to within an uncertainty of the order of the hfs anomaly (10^{-2}–10^{-4}, Ref. 32). An exact value of the magnetic moment can be obtained by the doublet method.[2] Figure 4 shows as an example the hfs scheme of ^{37}K, and Figure 5 the corresponding ($\Delta F = 0$) transition frequencies.[33] The thick lines in Figure 5 represent doublets separated by

$$\Delta \nu = 2\mu_I H_0/(Ih) \quad (27)$$

which is exactly twice the NMR frequency.

Since ν_i ($\Delta F = 0$, $\Delta m_F = \pm 1$) has a flat maximum [in the case of ^{37}K at 320 G for the doublet ($F = 1$, $m_F = 0$) $\leftrightarrow$ (1, −1); (2, 0) $\leftrightarrow$ (2, −1)], a precision measurement of A and a direct measurement of μ_I can be performed.

In diamagnetic atomic ground states as in the case of Hg, the nuclear g_I factor can be determined directly by NMR. The theory of NMR has been treated extensively by Matthias *et al.*,[34] who derived formulas for the line shape under various experimental conditions. For the usual geometry (H_0 along the z axis; rotating rf field in the x–y plane; photo, β, or γ detector at 0° or 90° with respect to the z axis) the resonance curve splits for high rf power into a k-fold curve, where k is the order of orientation observed. For high rf power the orientation will be completely destroyed at frequencies

$$\omega = \omega_0\{1 - [\cos \beta_k/(1 - \cos^2 \beta_k)^{1/2}]H_1/H_0\} \qquad (28)$$

ω_0 is the resonance frequency, ω and H_1 are, respectively, the frequency and field strength of the rf field. β_k are the angles for which the Legendre polynomial P_k ($\cos \beta_k$) is zero, e.g., $\beta_{k=1} = 90°$ and $\beta_{k=2} = 54.7°$. The interesting feature of Eq. (28) is that the signal does not vanish at resonance for k even, but instead has a hard core. Away from resonance, the signal can actually reach zero for sufficiently large amplitudes of H_1. This effect was first observed by Brossel and Bitter in optical double-resonance experiments on Hg[35] and in the special case of γ RADOP by Cappeller and Mazurkewitz.[36]

The line shapes of resonance signals in β RADOP are mainly determined by the term in $k = 1$. This is strictly true for an allowed β transition and still a very good approximation for the general case of β RADOP. The

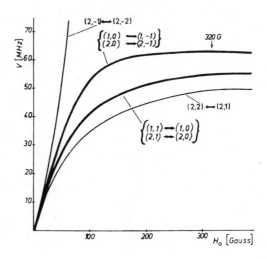

Figure 5. $\Delta F = 0$ transition frequencies (26) in the ground state of ^{37}K as function of magnetic field. The thick lines represent doublets. The magnetic field for maximum transition frequency is marked.

Table 1. Short-Lived Alkali Isotopes Investigated by In-Beam RADOP

Isotope	$T_{1/2}$	Reaction	Particle energy (MeV)	I	$\mu_i(\mu_N)$[a]	δW_0 (MHz)	Ref.
^{8}Li	850 ms	^{7}Li$(d,p)^8$Li	10	2	1.6532[c]	382.543(7)	42
^{20}Na	445 ms	^{20}Ne$(p,n)^{20}$Na	20	2	0.3694(2)[d]	276.855(3)[b]	44
^{21}Na	23 s	^{20}Ne$(d,n)^{21}$Na	7.5	3/2	2.46(8)[e]		40, 41
^{25}Na	60 s	^{22}Ne$(\alpha,p)^{25}$Na	17	5/2	3.683(4)[d]	2648.5(3.0)	43, 45
^{36}K	340 ms	^{36}Ar$(p,n)^{36}$K	17	2	0.548(2)[d]	604.5(1.1)	44
^{37}K	1.2 s	^{36}Ar$(d,n)^{37}$K	10.5	3/2	0.20320(6)[f]	240.2672(7)[b]	41, 33
^{80}Rb	30 s	^{80}Kr$(p,n)^{80}$Rb	12	1	−0.0836(6)[f]	233.936(2)[b]	43, 46
^{82}Rb	75 s	^{82}Kr$(p,n)^{82}$Rb	10	1	0.554(3)[d]	1549(8)	46

[a] Corrected for diamagnetism.   [b] From δW_0; μ_i is corrected by an estimated correction for hyperfine anomaly. This correction is added to the experimental error.   [c] Measured by NMR.[3]

[d] From $\Delta F = 1$ rf transitions.

[e] The RADOP technique was first tested on ^{21}Na. Hence a precision measurement was not attempted and a more precise value of μ_I can be found in Ref. 3.

[f] Measured directly by the doublet method (see Section 3.1.4).

line shape of the asymmetry signal (20) as function of ω is thus a Lorentzian as derived from Bloch's equations for the longitudinal magnetization.

3.2. In-Beam Experiments on Alkali Isotopes

The first group of β RADOP experiments that will be reported are in-beam measurements on short-lived alkali isotopes with half-lives ranging from 340 ms to 75 s (Table 1). With the exception of ^{8}Li, they are produced from noble-gas targets by a proton, deuteron, or alpha reaction. Earlier reviews on the subject were given by Otten.[37-39]

Figure 6 shows the principle of the experimental setup used in these experiments. The particle beam from a cyclotron passes through a resonance cell, which is made of glass, via thin entrance and exit windows. The bulb is filled with a stable alkali isotope for spin exchange and a noble gas of a pressure of 100–760 Torr. This gas fulfills several purposes: It serves (i) as target material for the production of the investigated isotope, (ii) as stopping material for the radioactive isotope, which is produced with a considerable momentum, (iii) as buffer gas for OP (see Section 3.1.1), and (iv) to broaden the optical absorption line D_1 of stable Rb, which is generally used for spin exchange. This broadening reduces the optical density of the resonance cell so that the density of polarized Rb vapor can

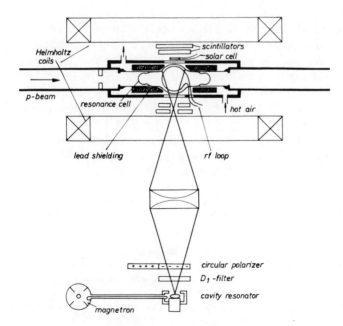

Figure 6. Experimental setup for in-beam RADOP experiments.

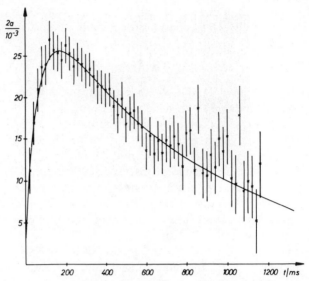

Figure 7. Time dependence of the β-asymmetry (20) of ^{36}K after a short activation pulse at $t = 0$. The solid line represents a fit taking into account the recombination time for the recoil ions of ^{36}K and the diffusion time to the glass walls of the resonance vessel.[44]

be enlarged without affecting the optical thickness. The D_1 resonance line from an electrodeless microwave-discharge lamp is selected by an interference filter, circularly polarized and focused on the cell. A fast change of the sense of circular polarization is made possible by means of a pneumatically driven carriage on which two polarizers are mounted. β telescopes are positioned at 0° and 180° with respect to the magnetic field produced by Helmholtz coils.

After optimizing the energy of the particle beam for maximum selective production of the isotope under investigation, the experiment is per-

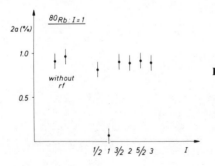

Figure 8. Spin determination of ^{80}Rb in low magnetic field.[46] The two values of the average β asymmetry (20) to the left were taken without rf. The other points were obtained by inducing rf transitions according to the first term on the right side of Eq. (26).

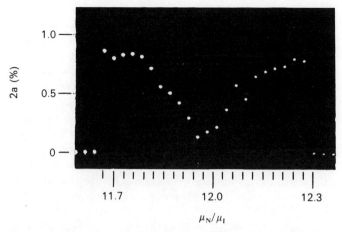

Figure 9. On-line display of the computer used for the determination of μ_I of ^{80}Rb. Three programmable frequency synthesizers were set simultaneously to four different transitions (26) as function of μ_N/μ_I. The steps were 0.03 μ_N/μ_I, and the static magnetic field was fixed to 20 G.

formed as follows: after an activation period of about one half-life a counting period follows in which the cyclotron is switched off since otherwise the background of the β detectors would be too high and OP would be disturbed by the plasma created by the charged beam in the resonance vessel. Even after switching off the beam it takes up to 1 s until the plasma is completely neutralized and maximum asymmetry is reached. Figure 7 shows as an example the asymmetry signal (20) for the case of ^{36}K. The rise time is determined by the recombination rate of the produced ions and the rate of spin exchange collisions, especially with free electrons. The decay time is determined by the time required for diffusion timc to the walls of the resonance vessel where the atoms get trapped.

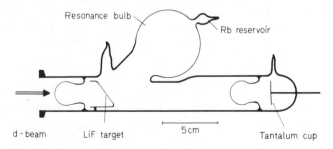

Figure 10. Target and resonance vessel for β-RADOP experiments on ^{8}Li.[42]

On averaging the asymmetry over the whole counting window for several activation and counting cycles, an asymmetry signal of about 2% is obtained. This is the order of magnitude that is observed in all other cases, too. If now rf transitions corresponding to the different possible spin values are induced in low magnetic field, the polarization is destroyed, e.g., for $I = 1$ in the case of ^{80}Rb (Figure 8). As described above, the magnetic moment can be determined in a higher magnetic field where the Zeeman effect becomes nonlinear; this is shown in Figure 9. In a magnetic field of 20 G a set of rf frequencies is applied corresponding to different values of the magnetic moment. for $|\mu_I| = 0.084\mu_N$ maximum resonance is achieved.

The results of the in-beam experiments are shown together in Table 1. A detailed discussion of the data with respect to nuclear structure and to the question of moments of mirror pairs can be found in the original literature.

The application of in-beam RADOP as described above is rather limited. Figure 10 shows the first attempt to use a solid target instead of a noble-gas target. ^{8}Li is produced from LiF evaporated on the entrance window of the OP vessel. By the recoil of the (d, p) reaction ^{8}Li is ejected out of the target and stopped in the He-filled vessel, where it is polarized by spin exchange with stable Rb.[42]

3.3. RADOP Experiments on Mass-Separated Alkali Isotopes

In order to investigate highly unstable isotopes production schemes other than the low-energy reactions of Table 1 have to be used. These are high-energy proton (or deuteron, or alpha particle) reaction, fission, and high-energy heavy ion reaction. Until now, only the first two of these methods have been used to produce short-lived isotopes for investigation by optical spectroscopy.

The most powerful means for a large-scale production of neutron-deficient nuclei is spallation as induced by high-energy protons. However, the large energy deposited in the nucleus opens many reaction channels and a mixture of elements and isotopes is obtained. Figure 11 shows as a typical example the mass yield for the reaction La + p (600 MeV) calculated according to the semiempirical formula of Rudstam,[47,48] which agrees well with experiment. On-line mass separators with special target and ion-source systems have been successfully applied to the isolation of single isotopes.[48,49]

The on-line separator with the broadest spectrum of produced isotopes is the ISOLDE facility at CERN.[48] Since its construction in 1969, it has provided 20 different elements in a chemically clean form and an additional 20 as chemical mixtures. Since 20–30 isotopes are available for each element (see Figure 12) several hundred nuclear species are accessible

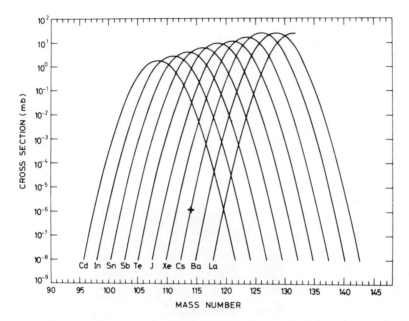

Figure 11. Isotopic yield distribution for the spallation of La by 600-MeV protons, calculated according to the semiempirical formula of Rudstam.[47,48] ^{114}Cs (marked by a cross) is produced with a cross section of about 1 n.b. After mass separation, one ^{114}Cs atom/s is obtained (Figure 12).

to investigation. In favorable cases such as the alkali or Hg isotopes, the yield reaches up to 10^{10} atoms per sec and per mass number (Figure 12).

Fission can be used to obtain short-lived, neutron-rich isotopes and several on-line mass separators have been built up at reactors.[48,49] Such a facility has been set up at the TRIGA reactor in Mainz for use in RADOP experiments and investigations by laser spectroscopy (Figure 13).[50] The target-ion-source system of the Bernas type[51] is placed close to the core of the reactor in a flux of 2×10^{11} neutrons/s cm^2. Alkali isotopes produced by fission of ^{235}U are selectively ionized by surface ionization. The ion beam is mass-separated and focused on a foil which is mounted in the center of a glass bulb. After a collection time of about one half-life the resonance vessel is filled with some hundred Torr of He and small amounts of stable Rb. The helium serves as buffer gas for OP and is also used to prolong the diffusion time to the walls. The Rb atoms are polarized by OP and transfer their polarization by spin-exchange collisions to the radioactive ones. Compared to the in-beam RADOP experiments (Figure 7), the rise time of asymmetry signal (20) is much faster in this setup because it is essentially determined by the evaporation time of the heated foil. The decay time is again given by the diffusion time because the alkali atoms

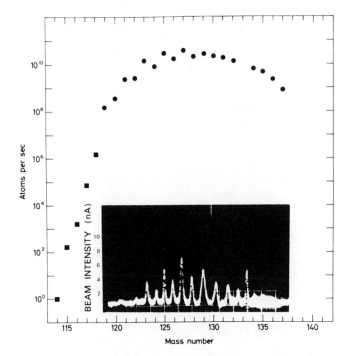

Figure 12. Experimental yield curve for Cs given as atoms per
second in the mass separated ion beams of the ISOLDE separator. The target was $140 \, g/cm^2$ lanthanum placed in a proton
beam of $0.7 \, \mu A$. The inset shows a scan of the current of the ion
beams after mass separation.[48]

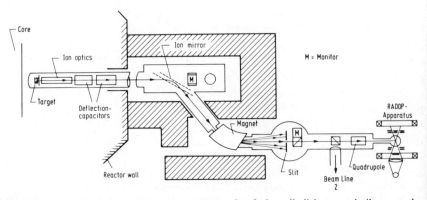

Figure 13. Layout of the on-line mass separator for fission alkali isotopes built up at the
TRIGA reactor in Mainz.[50]

Table 2. Spin Determination of Short-Lived, Mass-Separated Alkali Isotopes by β RADOP

Isotope	$T_{1/2}$ (s)	Reaction	Facility	I	Ref.
^{76}Rb	37	spallation of Nb by p (600 MeV)	ISOLDE/CERN	1	52
^{91}Rb	58	fission of ^{235}U	TRIGA/Mainz	3/2	50
^{94}Rb	2.7	fission of ^{235}U	TRIGA/Mainz	3	50
^{119}Cs	38	spallation of La by p (600 MeV)	ISOLDE/CERN	9/2	52

become trapped at the walls of the resonance vessel (although they are coated with paraffin or silan). At ISOLDE, where a similar OP apparatus was used for the investigation of neutron-deficient Rb and Cs isotopes, asymmetry signals of up to 30% were obtained in the cases of ^{80}Rb and ^{124}Cs.[52] This value has to be compared to the asymmetry signal of the in-beam experiment on ^{80}Rb (Section 3.2, Figures 8 and 9, and Ref. 46), where the high background inherent to these experiments diminishes the asymmetry by more than an order of magnitude.

The first results from these experiments are listed in Table 2. Together with the data on longer-lived and stable atoms obtained by conventional optical methods,[3] by in-beam RADOP (Table 1), and by ABR, which has been applied very successfully to on-line investigation of the hfs of atomic ground states at ISOLDE,[53] a rather complete set of data set on spins and magnetic moments of Rb and Cs isotopes has been accumulated.

3.4. RADOP Experiments on Mass-Separated Hg Isotopes

Mercury is ideally suited for the application of RADOP for the following reasons: (i) Hg is volatile and chemically inert, thus the problem of long wall-adsorption times as in the case of alkalies does not exist; (ii) owing to the diamagnetic ground state the relaxation time of nuclear orientation is long (of the order of seconds) compared to the pumping time (about 100 ms), this enables a very high degree of nuclear polarization to be obtained; (iii) the $^1S_0$–$^3P_1$, $\lambda = 2537$ Å line, which is used for OP, has a hfs and IS much larger than the Doppler width, hence, in addition to a determination of g_I by NMR in the ground state (as in the case of RADOP of alkalis), the hfs of the excited state and the IS of the optical line can be determined; (iv) numerous investigations have been performed for stable and long-lived Hg isotopes and even before the RADOP experiments these measurements yielded the widest and most complete set of hfs and IS data for any single element.[54–57]

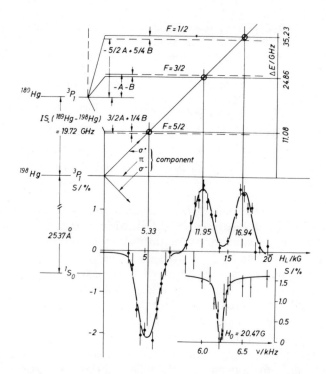

Figure 14. Determination of IS, hfs, and g_I of ^{189}Hg by β-RADOP. *Top*: Excited $^3P_1$ state of ^{189}Hg in zero magnetic field. The light source filled with ^{198}Hg is placed in a magnetic field H_L. When the hfs of ^{189}Hg is scanned by varying the magnetic field H_L, OP takes place at the three matching points within the Doppler width. *Bottom*: Asymmetry of the β-decay (20) of ^{189}Hg as function of H_L. The solid curve is the theoretical asymmetry calculated from the steady-state solution of the OP rate equation. The free parameters of the least-squares-fit procedure are A, B, IS, Doppler width, and the ratio of pumping to relaxation rates. *Inset*: NMR resonance of ^{189}Hg. The (uncorrected) parameters deduced by fitting the scanning and the NMR signals are $I = 3/2$, $A = -5.86(6)$ GHz, $B = 0.71(9)$ GHz, IS($\text{Hg}^{198,189}$) = 19.72(10) GHz, $g_I = -0.3994(5)$.

β RADOP has been applied to the neutron-rich isotope ^{205}Hg ($T_{1/2} = 5.2$ min)[58] and to the odd, neutron-deficient isotopes ^{191}Hg ($T_{1/2} = 51$ min) $- ^{181}$Hg ($T_{1/2} = 3.6$ s).[59,60,18,31] The IS and quadrupole moment of ^{199m}Hg were determined by γ RADOP observing the anisotropy of the internal $M4$–$E2$ γ cascade.[61] The results of these investigations have been summarized and discussed in several reviews.[62–65]

The principle of RADOP in Hg can be seen from Figure 14, which shows the measurement on ^{189}Hg as an example. Its hyperfine structure is scanned by the σ^+ Zeeman component of the $\lambda = 2537$ Å line of the light source which is filled with stable ^{198}Hg and placed in the pole gap of a magnet. When the Zeeman component matches a hfs level of the unstable isotope within the Doppler width then OP leads to a nuclear polarization and hence to a β asymmetry (or γ anisotropy). The variation of the obtained signal as function of the magnetic field H_L over the light source immediately yields hfs and IS of the $^1S_0-^3P_1$ line. g_I is obtained independently with high precision by NMR in the oriented $^1S_0$ ground state (inset of Figure 14).

The experimental apparatus is shown in Figure 15. It differs from the in-beam setup (Figure 6) in the magnet used for Zeeman scanning and in the mechanism by which the resonance vessel is filled with the unstable isotope. The transfer system consists of two collector foils mounted on either side of a rotatable and movable flange which separates the resonance

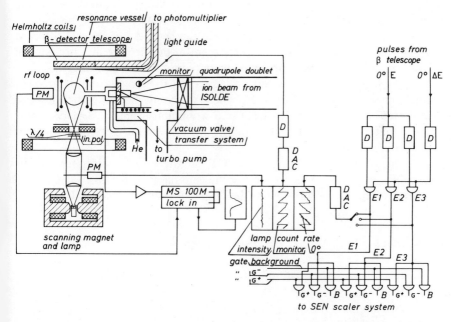

Figure 15. Setup for β RADOP of mass-separated Hg isotopes. The Ge(Li) or Na(J) detectors used for γ-RADOP are not shown here. The signals from the photomultiplier (PM) of the thick scintillator are fed into discriminators (D) set to different thresholds (E_1–E_3). Selectivity to β events is achieved by requiring coincidence with the signals of the thin scintillator. The scalers are gated synchronously with the rotation of the quarter-wave plate which determines the sense of circular polarization of the OP light. A frequency synthesizer (MS 100) is used for NMR.

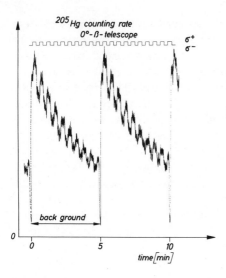

Figure 16. Count rate of β particles emitted from polarized ^{205}Hg $(T_{1/2} = 5.2m)$ under $0°$. Every five minutes the resonance vessel is filled with a fresh probe. The chart recording shows the exponential decay and the sharp drop to the background level, when the remaining activity is pumped off. The count rate is modulated by the periodical inversion of the circular polarization of the OP light and hence of the nuclear polarization.[58]

volume from the vacuum tank. One of the foils is exposed to the ion beam and collects the ions for about one half-life; the two foils are then interchanged by a mechanical system. While the second foil collects the next probe, the first one is heated and a stream of purified He at a pressure of 100 Torr transports the evaporated activity into the resonance cell.

The principle of the measuring sequence can be seen from the trace of a strip chart recorder in Figure 16, where the count rate of the $0°$ telescope is plotted. In this favorable case, the β-asymmetry signal is a macroscopic effect that can be recognized instantly without major electronic tricks or computer work.

The results obtained by RADOP on the odd Hg isotopes will be discussed in Section 4.2, where the measurement of the IS of even–even neutron-deficient Hg isotopes by laser spectroscopy is described.

4. Laser Spectroscopy on Short-Lived Isotopes

The development of tunable dye lasers has allowed the spectroscopist's dream of an ideal light source to come almost true. Today, commercial systems have started to cover the whole visible spectrum and, by frequency doubling, also the ultraviolet spectral range down to $\lambda \sim$ 2100 Å. High output power, small bandwidth, and little beam divergence enable a very sensitive trace analysis to be obtained. In a favorable case, vapor densities as low as 100 atoms/cm^3 can be optically detected.[66] These properties make tunable lasers especially suitable for the investigation of hfs and IS of short-lived isotopes, where the number of atoms obtainable by nuclear reactions is the main restriction to experimentation. Moreover, statistics favor laser spectroscopy in comparison to RADOP: whereas γ RADOP deals with few detectable quanta and β RADOP with even only one detectable particle per nucleus and lifetime, a cw laser can, in principle, excite an atom up to about 10^8 times per second.

Until recently, physicists working with dye lasers have mainly concentrated on the development of the instrument itself and on the new spectroscopic methods the excellent properties of this light source have made possible. In the meantime, however, tunable lasers have gained ground in almost every field of natural science, and their use is rapidly increasing.

In the following sections the first two applications of lasers to the study of the nuclear properties of short-lived isotopes will be reported. For a general review of tunable lasers and their applications see Chapter 15 of this work, "Laser High-Resolution Spectroscopy," by W. Demtröder, and see also Refs. 67–76.

4.1. On-Line Laser Spectroscopy of Na Isotopes

Laser spectroscopy was first applied to short-lived isotopes at Orsay by Huber *et al.* They measured the hfs of the D_1 (Ref. 77) and D_2 (Ref. 78) line of ^{21}Na–^{25}Na. This complex experiment involves OP by a cw dye laser, state selection by a six-pole magnet, and on-line mass separation.

Figure 17 shows the experimental setup. Radioactive Na isotopes are produced from a molten aluminum target by spallation with 150-MeV protons. For ^{25}Na the obtained intensities were as high as 10^8 atoms/s. The outlet of the target is used as the oven of an atomic beam which passes through the field of a six-pole magnet. The atoms with high field projection $m_s = +1/2$ are focused and enter a hot rhenium tube, which acts as the ion source of a magnetic mass separator. The mass-separated ions are counted

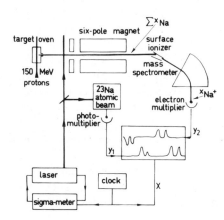

Figure 17. Experimental setup for production and on-line laser spectroscopy of ^{21}Na–^{25}Na.[77]

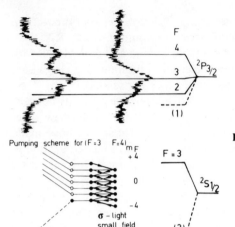

Figure 18. Measurement of the hfs of ^{25}Na in the D_2 line by laser spectroscopy. Using σ^+ light for the $(^2S_{1/2}, F = 3)\leftrightarrow (^2P_{3/2}, F = 4)$ transition, OP leads to an increase of the population of the $(^2S_{1/2}, F = 3, m_F = +4)$ level. With σ^- light, the $m_F = -4$ state is over-populated and defocused by the six-pole magnet.[78]

by using an electron multiplier. Count rates of up to 3000 ions of ^{25}Na per second were obtained. Before entering the six-pole magnet, the atomic beam is illuminated at right angles by a single-mode dye laser. With the laser frequency tuned to the D_1 or D_2 line and to one of the two hfs components of the ground state, about 10 excitations take place in the interaction region, and the atoms are pumped into the other F state (hyperfine pumping). As a consequence of the selectivity of the six-pole magnet with respect to m_s, there is an increase or decrease of the count rate of the electron multiplier according to whether the laser is tuned to the $(^2S_{1/2}, F = I + 1/2)\leftrightarrow^2P_{1/2,3/2}$ transition or to that starting from the $F = I - 1/2$ state. The scan of the D_1 line (see Figure 9 of Chapter 15 of this work, by Demtröder) immediately gives the magnetic hfs in the ground state and in the $^2P_{1/2}$ excited state. The IS is determined from the center of gravity of the hfs components that are well resolved, the simultaneously re-corded fluorescence intensity of a ^{23}Na atomic beam serving as a reference.

In order to determine the spectroscopic quadrupole moment from the hfs of the D_2 line, circularly polarized light and a small magnetic field are used in the interaction region. With this geometry, OP leads to an elec-tronic polarization that can be probed by the six-pole magnet. Figure 18 shows the count rates of ^{25}Na for OP with σ^+ light (left) and σ^- light (right). From this measurement the quadrupole moment has been deter-mined as $Q(^{25}$Na$) = 0.23(8)$ b. The value of the magnetic moment $[\mu_I = 3.685(22)\mu_N]$ is in excellent agreement with the result of the β-RADOP experiment (Table 1). The IS data of the isotopes ^{21}Na–^{25}Na have been determined with an average accuracy of 6 MHz; they are completely dominated by the mass effect with $M = 3.87 \times 10^5$ MHz, whereas the volume effect contributes only with $F = -26.4$ MHz/fm^2 [see Eq. (6)].

The same group plans to extend the measurements, in the middle of 1977, to very-neutron-rich Na isotopes produced by high-energy fission at the 22-GeV proton synchrocyclotron at CERN. Major improvements of the setup should enable optical information on spins and moments up to ^{31}Na. ^{31}Na is especially interesting because mass measurements[79] and theoretical calculations[80] give indications of a shape transition at this mass number. If this explanation is correct, the IS between ^{30}Na and ^{31}Na is expected to deviate by 10 MHz from the pure mass shift, which is of the order of 390 MHz.

4.2. On-Line Laser Spectroscopy of Hg Isotopes

In Section 3.4 RADOP experiments on neutron-deficient Hg isotopes have been reviewed. In these experiments an unexpected break in the IS

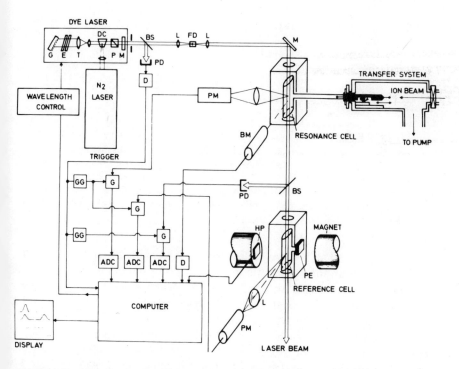

Figure 19. Experimental setup for on-line laser spectroscopy of short-lived Hg isotopes. E: etalon, T: beam-expanding telescope; P: polarizer; M: mirror; BS: beam splitter; L: lens; FD: frequency-doubling crystal; PD: photodiode; PM: photomultiplier; BM: β (or γ) detector; PE: Peltier element; HP: Hall probe; GG: gate generator; G: gate; D: discriminator; ADC: analog-to-digital converter.

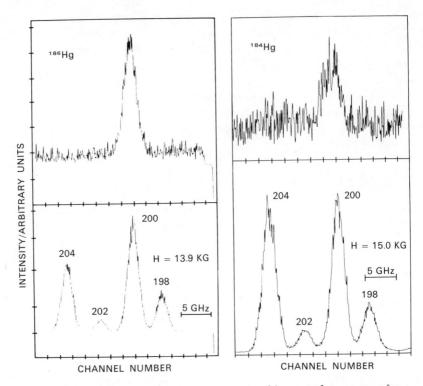

Figure 20. Intensity of the fluorescence light in the $6s^2\,{}^1S_0\text{--}6s6p\,{}^3P_1$, $\lambda = 2537$ Å line of ^{186}Hg (left, upper part) and of ^{184}Hg (right, upper part) versus the frequency of exciting laser light. For a calibration of the channel numbers in frequency units, the resonance intensity of the σ^+ Zeeman components of the even–even stable Hg isotopes in a magnetic field are simultaneously recorded (lower parts). Each measurement took roughly 10 min, and about 10 sweeps of the frequency of the laser were summed up.[84]

between ^{187}Hg and ^{185}Hg was found (see below). This discovery made it desirable to extend the systematic investigations in the Hg chain to even–even neutron-deficient Hg isotopes, which, however, are only accessible by purely optical methods. For this reason, a laser experiment[81] has been set up at the up-graded ISOLDE facility at CERN (Figure 19). The dye laser used is based on the Hänsch design.[82] It is pumped by a 400-kW pulsed nitrogen laser. Laser light for the excitation of the $6s^2\,{}^1S_0\text{--}6s6p\,{}^3P_1$, $\lambda = 2537$ Å transition is generated by frequency doubling in an ADP crystal.* The laser beam passes a resonance cell which is periodically filled with the isotope under investigation by means of a system similar to that of Figure

---

* A first laser experiment on the $\lambda = 2537$ Å line of stable isotopes was reported by Wallenstein and Hänsch.[83]

Table 3. Data of Neutron-Deficient Even–Even Hg Isotopes Obtained by Laser Spectroscopy and $B(E2)$ Measurements, and Comparison with Theory (from Ref. 84)

1	2	3	4	5	6	7	8
xHg	$T_{1/2}$	Yield[a] (ions/sec)	$IS^{204/x}$ (GHz)	$\lambda^{204/x}$ (fm^2)[b]	$\langle\beta\rangle_{x\mathrm{Hg}} - \langle\beta^2\rangle_{x-1\mathrm{Hg}}$ from IS[c]	$\langle\beta^2\rangle_{B(E2,2^+\to0^+)} - \langle\beta^2\rangle_{B(E2,6^+\to4^+)}$	$\langle\beta^2\rangle_{x\mathrm{Hg}} - \langle\beta\rangle_{x-1\mathrm{Hg}}$ Theory[d]
^{184}Hg	31 s	4×10^7	43.1(2)	−0.962(4){106}	−0.060(6)	−0.055(30)[e]	−0.051
^{186}Hg	1.4 min	2×10^8	39.4(2)	−0.881(4){97}	−0.053(5)	−0.056(31)[f]	−0.040
^{188}Hg	3.3 min	2×10^9	35.8(2)	−0.800(4){88}	−0.002(1)		−0.001
^{190}Hg	20 min	1×10^9	31.8(2)	−0.711(4){79}	0.000(1)		

[a] As obtained in the actual experiment by ISOLDE.
[b] Two different uncertainties are given. The first one includes only the error of the IS measurement (preceding column) and the uncertainty in the mass shift. The second one (quoted in curly brackets) includes in addition the uncertainty of F [Eq. (6)].
[c] Obtained by use of Eq. (30).
[d] Frauendorf and Pashkevich, Ref. 96.
[e] Rud et al., Ref. 94.
[f] Proetel et al., Ref. 92.

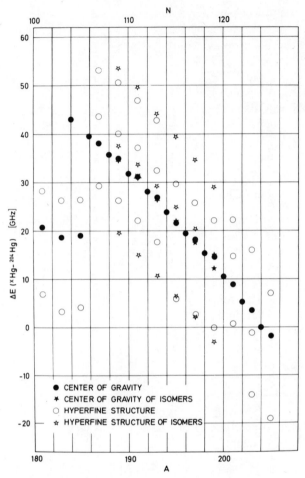

Figure 21. Hyperfine-structure splitting and IS of Hg isotopes relative to ^{204}Hg in the spectral line $6s^2\,^1S_0$–$6s6p\,^3P_1$, $\lambda =$ 2537 Å. The data on ^{191m}Hg and ^{189m}Hg are preliminary.

15. The IS of the spectral line $\lambda = 2537$ Å of the radioactive atoms in the resonance cell is measured by tuning the frequency of the laser over the Doppler-broadened absorption profile and by observing the intensity of the fluorescence light by means of a photomultiplier. Typical scanning patterns for ^{186}Hg and ^{184}Hg are shown in the upper parts of Figure 20. To obtain a calibration of the channel numbers in frequency units, the laser beam is sent through a second resonance cell containing stable even–even Hg isotopes and placed in the pole gap of a magnet. It is thus possible to shift the Zeeman components into the region of the signal corresponding to

unstable isotopes (lower parts of Figure 20), and the IS can be determined from the magnetic field values with the help of the known Landé factor.

Results have been obtained by laser spectroscopy for ^{190}Hg, ^{188}Hg, ^{186}Hg, ^{184}Hg (Table 3), and preliminary measurements have been performed on ^{191m}Hg and ^{189m}Hg (see Figure 21). Together with the results for g_I, I, hfs, and IS, which have been obtained by classical optical spectroscopy and RADOP, a rather complete set of data is obtained for the Hg chain spanning from ^{181}Hg to ^{205}Hg (for a list of data and references see Refs. 18 and 84). Hfs and IS data are shown in Figure 21 as a function of the neutron number.

4.2.1. Isotope Shift and Charge Radii of Hg Isotopes

The most interesting data obtained are those of IS's; they will be discussed in the following. With $M = 6(3) \times 10^{11}$ Hz and $F = 4.5(5) \times 10^{10}$ Hz/fm^2 (Ref. 18) they can be converted, by means of Eqs. (6) and (8), to changes of radial moments $\lambda^{AA'}$ [Eq. (7)] as shown in Figure 22. Apart from some minor irregularities, the even–even Hg isotopes and the odd ones with $A \geqslant 187$ follow an almost straight line from the nearly double magic (and therefore spherical) nucleus ^{205}Hg down to ^{184}Hg. For the light isotopes a strong discontinuity is observed. Although this staggering attracted most interest, the almost straight line provides valuable information, too: (i) It is surprising that the overall trend of the shrinking of a Hg nucleus with decreasing neutron number is so smooth, although 21 neutrons belonging to different nuclear subshells are removed between ^{205}Hg and ^{184}Hg; (ii) the slope in Figure 22 is by a factor of 2 too flat as compared to the prediction of the spherical uniformly charged, and incompressible liquid drop:

$$\delta \langle r^{2n} \rangle_{\text{unif}} = [2n/(2n+3)] R_0^{2n} \, \delta A/A \qquad (29)$$

with $R_0 = 1.2 A^{1/3}$ fm. Part of this discrepancy can be explained by a small but steadily growing deformation $\langle \beta^2 \rangle$ with decreasing neutron number which leads to an increase of the radial moments. This increase is given in first approximation by

$$\delta \langle r^{2n} \rangle_{\text{def}} = 3n/(4\pi) R_0^{2n} \, \delta \langle \beta^2 \rangle \qquad (30)$$

Nevertheless, the greatest part of this so-called IS discrepancy still remains. It is observed in other mass regions too,[85] but in Hg the advantage of an extremely long isotopic chain eliminates misinterpretation which is otherwise possible due to local irregularities from shell or deformation effects. The IS discrepancy is explained phenomenologically by nuclear compressibility[6,85] and/or by a neutron nuclear skin.[13] The first microscopic calculation has recently been performed by Beiner and Lombard,[86] who

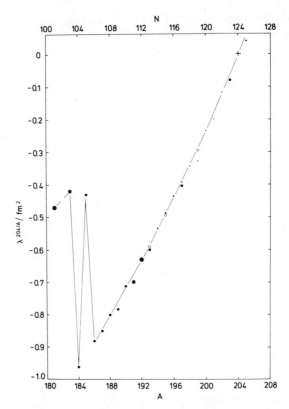

Figure 22. Changes of charge radii of Hg isotopes relative to ^{204}Hg. $\lambda = \delta\langle r^2\rangle - 1.1 \times 10^{-3}\ \delta\langle r^4\rangle + \cdots$. Filled circles represent charge radii of ground states, and open circles indicate those of isomers. For easy recognition they are connected by a line. The statistical errors of λ caused by the experimental uncertainty is given by the diameter of the symbols. An additional (scaling) error of 10% arises from the uncertainty of the electron density at the nucleus [F in Eq. (6)].

calculated the average slope of $\delta\langle r^2\rangle$ between ^{204}Hg and ^{187}Hg using the Hartree–Fock method with density-dependent forces (and thereby introducing nuclear compressibility). Their result is in perfect agreement with the experimental values.

Finally, the sudden change of the IS between ^{186}Hg and ^{185}Hg and the huge odd–even staggering in the light Hg isotopes will be discussed. This break, which was observed by RADOP,[60] was completely unexpected and its explanation as deformation effect[60] was initially questioned because

Hg has an almost magic proton shell. In the meantime, this interpretation has been accepted and completed in the sense that the increase of the nuclear charge radius of ^{185}Hg is most probably caused by a transition from a small oblate ($\beta \simeq -0.15$) to a strong prolate deformation ($\beta \simeq 0.3$). By means of Eq. (30) this change in deformation can be calculated from the IS data to be $\delta\langle\beta^2\rangle^{185/186} = 0.053(5)$ (see column 6 of Table 3). It is similar in size to that found by Brix and Kopfermann[87] in 1949 in the rare-earth region between $N = 88$ and $N = 90$ and which first led to the concept of nuclear deformation. In the case of Hg, however, the nuclear shape swings back to a nearly spherical shape at $A = 184$ and becomes deformed again at $A = 183$. This is a very peculiar behavior, which has not been observed in other mass regions. The calculation of the staggering parameter yields

$$\gamma^{185} = (2 \cdot \text{IS}^{185/184})/\text{IS}^{186/184} = 13(1) \tag{31}$$

which is an order of magnitude larger than that normally observed as, for example, in the case of the heavier Hg nuclei. The effect there is explained by the pairing of high spin states, which is energetically favored and gives rise to a larger charge radius of the even isotopes[54]; otherwise it is interpreted as being associated with the zero-point motion of quadrupole vibrations which produce a mean-square deformation, $\langle\beta^2\rangle$, which is larger for even isotopes than for odd ones.[88] In the case of the Hg nuclei around ^{185}Hg, the large odd–even staggering is caused by the coexistence of two different nuclear shapes which are energetically almost degenerate. This has been proved by γ spectroscopy of ^{188}Hg,[89,90] ^{186}Hg,[91–93] and ^{184}Hg,[94,95] which gave evidence of a coexistence and crossing of two bands in these nuclei, one built on an almost spherical shape and the other on a strongly deformed shape. The difference of the $B(E2)$ values measured in both bands agrees very well with the optical data between neighboring isotopes (columns 6 and 7 of Table 3). Unfortunately, the uncertainties of the $B(E2)$ measurements are by far too large to use this agreement as a proof of the correctness of Eq. (30).

Out of the large number of theoretical papers* that are available only that of Frauendorf and Pashkevich[96] gives numerical results for the odd–even effect in the light Hg isotopes (column 8 of Table 3). The calculated changes in deformation between the even and the neighboring odd isotope are in good agreement with the experiment. The other theoretical calculations were performed only for the even Hg isotopes and placed the shape transition between ^{188}Hg and ^{186}Hg.† Although this turned out to be

---

* For complete references on the theoretical work related to the IS of light Hg isotopes see Ref. 18.

† Kolb and Wong,[97] using a Hartree–Fock calculation, found an oblate shape for the even–even isotopes with $A \geqslant 184$ and a prolate shape for ^{182}Hg. By using a phenomenological two-center model they obtained qualitative agreement with the γ spectra of ^{188}Hg–^{184}Hg.

wrong, the qualitative agreement should be regarded as a great success: The calculation of the odd–even prolate–oblate staggering involves (i) complex heavy nuclei, (ii) interaction parameters deduced from the valley of nuclear stability, and (iii) critical energy differences of less than 1 MeV in contrast to the total binding energy of about 1400 MeV.

4.2.2. Monopole Transition Probability and Isotope Shift

The band head of the deformed band of ^{184}Hg has recently been found.[95] It decays by an electric monopole transition to the 0^+ ground state with a decay energy $E_\gamma = 375$ keV and a lifetime of $\tau = 0.9(3)$ nsec. Since the $E0$ matrix element is (to a good approximation) an off-diagonal element of the r^2 operator, it represents the dynamic analog of the field shift in atomic spectra.

The transition probability is given by[98]

$$W(E0)_{i \to f} = \Omega\rho^2 \simeq \Omega(Z, E_\gamma)(Z/R_0^2)|\langle\phi_f|r^2|\phi_i\rangle|^2 \tag{32}$$

where $\Omega(Z, E_\gamma)$ is essentially an electronic factor tabulated in Ref. 99. ϕ_f is the wave function of the 0^+ ground state of ^{184}Hg and ϕ_i that of the 0^+ shape-isomeric state. The IS measures

$$\delta\langle r^2\rangle^{204/184} = \langle\phi_f|r^2|\phi_f\rangle^{184} - \langle r^2\rangle^{204} \tag{33}$$

The two physical 0^+ wave functions ϕ_i and ϕ_f can be written as mixtures of the pure oblate and prolate solutions

$$\phi_i = a\psi_{\text{prol}} + b\psi_{\text{obl}} \quad \text{and} \quad \phi_f = -b\psi_{\text{prol}} + a\psi_{\text{obl}} \tag{34}$$

Thus, by combination of the experimental data from both sources, IS and lifetime measurements, information can be obtained on the mixing parameters. Using the usual argument for shape isomerism, the nondiagonal matrix element $\langle\psi_{\text{prol}}|r^2|\psi_{\text{obl}}\rangle$ is expected to be small because of the large difference in shape; by neglecting it completely the mixing parameter can be determined as $b = 0.085(15)$ (Ref. 84). This is a quite accurate determination of such a small mixing parameter, provided that the neglection is justified. This will be proved by calculations.[100] Indeed, monopole strength and IS provide a sensitive text for the theories.

5. Discussion

The RADOP measurements described in Section 3 and the laser spectroscopy experiments discussed in Section 4 have demonstrated for the examples of alkali and Hg isotopes that optical spectroscopy can provide valuable information on nuclear properties far from the valley of β stability

Table 4. Comparison of Different Techniques for Optical Spectroscopy of Short-Lived Isotopes

	Optical spectroscopy with			β RADOP	γ RADOP
	Conventional light source	Pulsed laser	cw laser		
Restrictions	not Doppler free unless high-resolution techniques are used	$\Delta\nu_{laser} \geq$ 100–1000 MHz low-duty cycle	only restricted region of wavelength available today	not Doppler-free S ground states $I \geq 1/2$ $A_\beta \neq 0$ cell technique	not Doppler-free S ground states $I \geq 1$ limited count rate cell technique
Number of emitted quanta per atom	up to $\sim 10^4$/s typically 10–100/s	up to 1000/s typically 30/s	up to 10^8/s typically 10^5/s	1 per nuclear lifetime	1–10 per nuclear lifetime
Typical detector efficiency times solid angle	10^{-2}–10^{-3}	10^{-2}–10^{-3}	10^{-2}–10^{-3}	up to 0.5	10^{-3}
Source of background (possibility of discrimination)	scattered light from walls (no; yes for observation of cascades)	scattered light from entrance and exit windows for the laser beam (yes if laser pulse lengths $\ll$ atomic lifetime)	scattered light from entrance and exit windows for the laser beam (no; yes for observation of cascades)	daughters, and atoms trapped at the surface of the resonance vessel (no)	daughters (yes, by γ spectroscopy); atoms trapped at the surface of the vessel (no)
Minimum number of atoms in vapor phase in experiments on Hg[a]	10^{11}	10^9	10^4	10^4	10^8
Investigated isotope with shortest half-life	^{63}Zn (38 m)[b]	^{184}Hg (31 s)[c]	^{21}Na (23 s)[d]	^{36}K (340 ms)[e]	^{199m}Hg (43 m)[f]

[a] References 8, 84, 101.　[b] Reference 102.　[c] Reference 18.　[d] Reference 77.　[e] Reference 44.　[f] Reference 61.

and, with some luck, can yield surprising results on new phenomena of nuclear matter such as the shape transition and the shape staggering in the light Hg isotopes.

This chapter is intended as a discussion of the advantages and limitations of the different methods as well as a brief outline of the further developments of these techniques. A survey is given in Table 4. The main disadvantage of RADOP in comparison to pure optical methods is that it is confined to the investigation of atoms with spherical atomic ground states and to nuclei with $I > 0$. In addition, the element has to be volatile and as chemically inert as possible. On drawing up a list of elements according to their suitability for RADOP, one obtains the following order: members of the IIb group of the periodic table; alkalis (wall adsorption problems); noble gases (ground-state transition frequencies in the far ultraviolet); alkali earth, Yb, and Eu (chemically aggressive); Ib elements (low vapor pressure). Fortunately, the physicochemical requirements of target–ion source systems for on-line mass separation are very similar to those of RADOP. Hence, the list of elements that can be produced at on-line mass separators agrees quite well with the order given above.

Compared to γ RADOP, β RADOP has the advantage of having a high detection efficiency, in particular when the γ decay branches to many nuclear levels. In addition, the γ count rate is limited to about 30 kHz or 1 MHz, since Ge(Li) or Na(J) detectors have to be used. On the other hand, β RADOP requires a sizable β asymmetry coefficient and is not selective against neighboring isotopes. The latter may be important for in-beam RADOP experiments.

In comparison to optical detection RADOP has the disadvantage of having a lower number of detectable quanta per atom. This represents a drawback, especially for isotopes with long nuclear half-lives. The limiting factor for optical detection is the background of the light scattered by the walls of the resonance vessel or by the windows for the laser beam. Using atomic beams, this background can be reduced by careful construction of the apparatus without a significant loss in the fluorescence signal. For closed cells, special shapes of the resonance cell are required to keep down the stray light intensity (Brewster window, Wood horn, diaphragms inside the cell, etc.). This, in turn, increases the dead volume of the resonance vessel.

In comparison to a cw laser, a pulsed laser has the advantage of covering the greatest part of the interesting spectral range, but it is limited with respect to resolution and duty cycle. The consequence of the poor duty cycle is that pulsed lasers cannot be used for the investigation of short-lived isotopes in atomic beams.

In Table 4 the sensitivity of the different methods is compared for the special case of Hg measurements in closed resonance cells. A value for cw lasers is missing, but it can be deduced from an experiment on Na performed by Fairbank et al.[66] They achieved a reduction of the background stray light reaching the photomultiplier by a factor of 3×10^{-10} of the incident light. This corresponds to the fluorescence of about 3000 Na atoms/cm^3. By using a lock-in technique 100 atoms/cm^3 could be detected. This gives a minimum number of 10^4 atoms if a typical volume of the resonance vessel of 100 cm^3 is assumed and wall adsorption is neglected. The same order of magnitude has been reached by β RADOP on ^{181}Hg. In this case, however, several hours of measuring time were needed to obtain sufficient statistics. γ RADOP has not yet been tested with respect to sensitivity, but in favorable cases it may be as sensitive as β RADOP. With optical detection and a conventional light source 10^{11} Hg atoms is the minimum quantity necessary, as has been tested in the case of ^{192}Hg.[101] In the experiment on ^{184}Hg with a pulsed laser, about 10^9 atoms were contained in the resonance cell. As shown in Figure 20, this represents roughly the minimum number needed in the actual setup. By using some modifications (smaller volume of the resonance vessel, larger angle of detection, gating of the photomultiplier) and by taking a longer measuring time, an improvement by a factor of 100 seems to be feasible; this method should then finally work with a minimum number of about 10^7 atoms.

A very promising alternative to the closed-cell technique is the use of an atomic beam as shown by the experiment reported in Section 4.1. Good signal-to-noise ratios were obtained within 10 min of measuring time and with a yield of 3×10^6 atoms/sec of ^{21}Na measured at the outlet of the oven. In a purely optical fluorescence experiment, Nowicki et al.[103] succeeded in performing hfs and IS measurements on ^{128}Ba ($T_{1/2} = 2.4d$) and

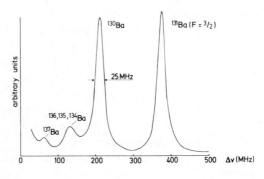

Figure 23. Intensity of the fluorescence light in the $6s^2\ ^1S_0$–$6s6p\ ^1P_1$, $\lambda = 5535$ Å line of Ba versus the frequency of exciting laser light. Two cw dye lasers are used, one is locked to the transition frequency of ^{138}Ba in a reference atomic beam and the other is frequency programmable by detecting the beat frequency. The frequency steps are 10 MHz. The intensity pattern shows the signal of the $F = 3/2$ component of ^{131}Ba, the other hfs component is masked by the signal of ^{138}Ba at $\Delta\nu = 0$ MHz.

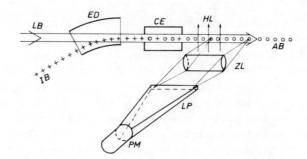

Figure 24. Setup of fast-beam laser spectroscopy with optical detection. AB: fast atomic beam; CE: charge exchange cell; ED: electrostatic deflector; HL: magnetic field for level crossing; IB: mass-separated ion beam; LB: laser beam; LP: light pipe; PM: photomultiplier; ZL: cylindric lens.[65]

^{131}Ba $(T_{1/2} = 11.5d)$ by using a collimated atomic beam of an intensity of 10^5–10^6 atoms/s (Figure 23). With a collimation ratio of $1:200$ this corresponds to 10^7–10^8 atoms/sec evaporated out of the oven. The Ba isotopes are produced by deuteron bombardment (or in a reactor) from enriched Ba isotopes; they are then off-line mass-separated and implanted into a Si backing, which is put into the oven of the atomic beam apparatus.

Kaufman, Neugart, and Otten have proposed to perform optical experiments directly in the fast, mass-separated beam. Such a technique would be an ideal adaption of optical spectroscopy to the situation met at on-line mass separators. Figure 24 shows the setup for fast-beam laser spectroscopy. The beam of a cw laser is sent parallel to the ion beam of a mass separator, or, if the ionic spectrum is inconvenient for laser spectroscopy, to the fast atomic beam which is obtained by charge exchange. Although the absorption lines are considerably shifted in this geometry, the Doppler broadening is reduced by a factor of $2\ (eU/kT)^{1/2}$, which is about 500 for an ion-source temperature of 2000 K and beam energies of 10 keV. A laser power of a few milliwatts is sufficient to excite an atom once along the path of about 20 cm so that it can be observed by a photomultiplier via a light pipe. With the usual efficiency for optical detection (Table 4), the resulting signal is $\geqslant 10^{-3}$ photoelectrons per incident ion. Kaufman has made a calculation for the 4555-Å line of Cs taking into account OP effects and efficiency of charge exchange, and arrived at 5×10^{-4} photoelectrons per incident ion.[104] A test experiment on Na has recently been performed yielding a linewidth of 30 MHz and a good signal-to-noise ratio at intensities of 10^7 atoms/s within a measuring time of some minutes.[105] A similar experiment was reported on metastable Xe ions produced by a plasma ion source.[106]

References

1. H. Kopfermann, *Nuclear Moments*, Academic Press, New York (1956).
2. N. F. Ramsey, *Molecular Beams*, Oxford Clarendon Press, Oxford (1972).
3. G. H. Fuller, *J. Phys. Chem. Ref. Data* **5**, 835 (1976).
4. E. C. Seltzer, *Phys. Rev.* **188**, 1916 (1969).
5. P. Brix and H. Kopfermann, *Rev. Mod. Phys.* **30**, 517 (1958).
6. D. N. Stacey, *Rep. Prog. Phys.* **29**, 171 (1966).
7. K. Heilig and A. Steudel, *At. Data Nucl. Data Tables* **14**, 613 (1974).
8. J. Bauche and R. J. Champeau, *Adv. At. Mol. Phys.* **12**, 39 (1976).
9. R. Engfer, H. Schneuwly, J. L. Vuillenmier, H. K. Walter, and A. Zehnder, *At. Data Nucl. Data Tables* **14**, 509 (1974).
10. F. Boehm and P. L. Lee, *At. Data Nucl. Data Tables* **14**, 605 (1974).
11. C. W. de Jager, H. de Vries, and C. de Vries, *At. Data Nucl. Data Tables* **14**, 479 (1974).
12. M. Brack, J. Damgaard, H. C. Pauli, A. S. Jensen, V. M. Strutinsky, and C. Y. Wong, *Rev. Mod. Phys.* **44**, 320 (1972).
13. W. D. Myers, *Phys. Lett.* **30B**, 451 (1969).
14. W. D. Myers, *At. Data Nucl. Data Tables* **17**, 411 (1976).
15. K. E. G. Löbner, M. Vetter, and V. Hönig, *Nucl. Data Tables* **A7**, 495 (1970).
16. E. E. Fradkin, *Sov. Phys.–JETP* **15**, 550 (1962) [*Zh. Eksper. Teor. Fiz.* **42**, 787 (1962)].
17. M. Stavn and W. T. Walter, *Bull. Am. Phys. Soc.* **9**, 10 (1964).
18. J. Bonn, G. Huber, H.-J. Kluge, and E.-W. Otten, *Z. Phys.* **A276**, 203 (1976).
19. A. Kastler, *J. Phys. Radium* **11**, 225 (1950).
20. J. Brossel, *Quantum Optics and Electronics*, Gordon and Breach, New York (1964).
21. R. A. Bernheim, *Optical Pumping*, Benjamin, New York (1965).
22. C. Cohen-Tannoudji and A. Kastler, in *Progress in Optics*, Ed. E. Wolf, Vol. V, p. 1, North-Holland, Amsterdam (1966).
23. W. Happer, *Rev. Mod. Phys.* **44**, 170 (1972).
24. B. Cagnac, *Ann. Phys. Paris* **6**, 467 (1961).
25. C. Cohen-Tannoudji, *Ann. Phys. Paris* **7**, 423, 469 (1962).
26. S. R. de Groot, H. A. Tolhoek, and W. J. Huiskamp, in *Alpha-, Beta- and Gamma-Ray Spectroscopy*, Ed. K. Siegbahn, Chapter XIXB, North-Holland, Amsterdam (1966).
27. E. M. Purcell and G. B. Field, *Astrophys. J.* **124**, 542 (1956).
28. J. P. Wittke and R. H. Dicke, *Phys. Rev.* **103**, 620 (1956).
29. H. G. Dehmelt, *Phys. Rev.* **109**, 381 (1958).
30. C. S. Wu and S. A. Moszowski, *Beta Decay*, Interscience Publishers, Wiley, New York (1966).
31. G. Huber, J. Bonn, H.-J. Kluge, and E.-W. Otten, *Z. Phys. A* **276**, 187 (1976).
32. A. Bohr and V. F. Weisskopf, *Phys. Rev.* **77**, 94 (1950).
33. Ch. v. Platen, J. Bonn, U. Köpf, R. Neugart, and E.-W. Otten, *Z. Phys.* **244**, 44 (1971).
34. E. Matthias, B. Olsen, D. A. Shirtley, J. E. Templeton, and R. M. Steffen, *Phys. Rev. A* **4**, 1626 (1971).
35. J. Brossel and F. Bitter, *Phys. Rev.* **86**, 308 (1952).
36. U. Cappeller and W. Mazurkewitz, *J. Mag. Res.* **10**, 15 (1973).
37. E.-W. Otten, in *Proceedings of the International Conference on Properties of Nuclei Far from the Region of β-Stability*, Leysin 1970, Vol. 1, p. 361, CERN Yellow Report No. 70-30, Geneva (1970).
38. E.-W. Otten, in *Atomic Physics 2*, Proceedings of the International Conference on Atomic Physics, Oxford 1970, Ed. P. G. H. Sandars, p. 113, Plenum Press, New York (1971).

39. E.-W. Otten, in *Hyperfine Interaction in Excited Nuclei*, Proceedings of the International Conference on Hyperfine Interaction, Rehovot and Jerusalem 1970, Eds. G. Goldring and R. Kalish, p. 363, Gordon and Breach, New York (1971).

40. H. J. Besch, U. Köpf, and E.-W. Otten, *Phys. Lett.* **25B**, 120 (1967).

41. H. J. Besch, U. Köpf, E.-W. Otten, and Ch. v. Platen, *Phys. Lett.* **26B**, 721 (1968); *Z. Phys.* **226**, 297 (1969).

42. R. Neugart, *Z. Phys.* **261**, 237 (1973).

43. M. Deimling, J. Dietrich, R. Neugart, and H. Schweickert in "Abstracts of Contributed Papers of the Fourth International Conference on Atomic Physics," Heidelberg, 1974, Eds. J. Kowalski and H. G. Weber, p. 207 (1974).

44. H. Schweickert, J. Dietrich, R. Neugart, and E.-W. Otten, *Nucl. Phys.* **A246**, 187 (1975).

45. M. Deimling, R. Neugart, H. Schweickert, *Z. Phys.* **A273**, 15 (1975).

46. R. Neugart, private communication.

47. G. Rudstam, *Z. Naturforsch.* **21A**, 1027 (1966).

48. H. L. Ravn, in *Proceedings of the Third International Conference on Nuclei Far from Stability*, Cargèse 1976, p. 22, CERN Yellow Report (1976); *Nucl. Instr. Methods* **139**, 282, 267 (1976).

49. W. L. Talbert, in *Proceedings of the International Conference on the Properties of Nuclei Far from the Region of Beta-Stability*, Leysin 1970, Vol. 1, p. 109, CERN Yellow Report No. 70-30, Geneva (1970).

50. J. Bonn, F. Buchinger, P. Dabkiewicz, H. Fischer, S. L. Kaufman, H.-J. Kluge, H. Kremmling, L. Kugler, R. Neugart, E.-W. Otten, L. von Reisky, J. M. Rodriguez-Giles, H.-J. Steinacher, and K. P. C. Spath, to be published in *Hyperfine Interactions* (1978).

51. R. Klapisch, C. Thibault-Phillipe, C. Detraz, J. Chaumont, R. Bernas, and E. Beck, *Phys. Rev. Lett.* **23**, 652 (1969).

52. H. Fischer, P. Dabkiewicz, P. Freilinger, H.-J. Kluge, H. Kremmling, R. Neugart, and E.-W. Otten, *Z. Phys.* **A284**, 3 (1978).

53. C. Ekström, S. Ingelman, G. Wannberg, and M. Skarestad, in *Proceedings of the Third International Conference on Nuclei Far from Stability*, Cargèse 1976, p. 193, CERN Yellow Report No. 76-13, Geneva (1976).

54. W. J. Tomlinson III and H. H. Stroke, *Nucl. Phys.* **60**, 614 (1964).

55. S. P. Davis, T. Aung, and H. Kleiman, *Phys. Rev.* **147**, 861 (1966).

56. P. A. Moskowitz, C. H. Liu, L. Fulop, and H. H. Stroke, *Phys. Rev. C* **4**, 620 (1971).

57. R. J. Reiman and M. N. McDermott, *Phys. Rev. C* **7**, 2065 (1973).

58. J. M. Rodriguez, J. Bonn, G. Huber, H.-J. Kluge, and E.-W. Otten, *Z. Phys.* **A272**, 369 (1975).

59. J. Bonn, G. Huber, H.-J. Kluge, K. Köpf, L. Kugler, and E.-W. Otten, *Phys. Lett.* **36B**, 41 (1971).

60. J. Bonn, G. Huber, H.-J. Kluge, L. Kugler, and E.-W. Otten, *Phys. Lett.* **38B**, 308 (1972).

61. J. Bonn, G. Huber, H.-J. Kluge, E.-W. Otten, and D. Lode, *Z. Phys.* **A272**, 375 (1975).

62. E.-W. Otten, *J. Phys. (Paris)*, **34**, 63 (1973) (Colloque **C4**, suppl. to No. 11-12).

63. E.-W. Otten, *Hyperfine Interactions* **2**, 127 (1976).

64. H.-J. Kluge, in *Proceedings of the Third International Conference on Nuclei Far from Stability*, Cargèse 1976, p. 177, CERN Yellow Report No. 76-13, Geneva (1976).

65. E.-W. Otten, in *Atomic Physics*, Vol. 5, Eds. R. Marrus, M. Prior, and H. Shugart, Plenum Publishing Corp., New York (1977).

66. W. M. Fairbanks, Jr., T. W. Hänsch, and A. L. Shalow, *J. Opt. Soc. Am.* **65**, 199 (1975).

67. F. P. Schäfer, Ed., *Dye Lasers*, Topics in Applied Physics, Vol. 1, Springer-Verlag, Berlin (1973).

68. W. Demtröder, *Phys. Rep.* **7**, 223 (1973).
69. W. Lange, J. Luther, and A. Steudel, *Adv. At. Mol. Phys.* **10**, 173 (1974).
70. H. Walther, *Phys. Scr.* **9**, 297 (1974).
71. H. Walter, Ed., *Laser Spectroscopy of Atoms and Molecules*, Topics in Applied Physics, Vol. 2, Springer Verlag, Berlin (1975).
72. S. Jacobs, M. Sargent, J. Scott, and M. Scully, *Laser Applications to Optics and Spectroscopy*, Physics of Quantum Electronics, Vol. 2, Addison-Wesley, Reading, Massachusetts (1975).
73. M. J. Colles and C. R. Pidgeon, *Rep. Progr. Phys.* **38**, 329 (1975).
74. R. A. Smith, Ed., *Very High Resolution Spectroscopy*, Academic Press, London (1976).
75. K. Shimoda, Ed., *High Resolution Laser Spectroscopy*, Topics in Applied Physics, Vol. 13, Springer-Verlag, Berlin (1976).
76. R. Baliau, S. Haroche, and S. Liberman, Eds., *Frontiers in Laser Spectroscopy*, Les Houches Summer School Proceedings, North-Holland, Amsterdam (1977).
77. G. Huber, C. Thibault, R. Klapisch, H. T. Duong, J. L. Vialle, J. Pinard, P. Juncar, and P. Jacquinot, *Phys. Rev. Lett.* **34**, 1209 (1975).
78. G. Huber, R. Klapisch, C. Thibault, H. T. Duong, P. Juncar, S. Liberman, J. Pinard, J.-L. Vialle, and P. Jacquinot, in *Proceedings of the Third International Conference on Nuclei Far from Stability*, Cargèse 1976, p. 188, CERN Yellow Report No. 76-13, Geneva (1976).
79. C. Thibault, R. Klapisch, C. Rigaud, A. M. Poskanzer, R. Prieels, L. Lessard, and W. Reisdorf, *Phys. Rev. C* **12**, 644 (1975).
80. X. Campi, H. Flocard, A. K. Kernan, and S. Koonin, *Nucl. Phys.* **A251**, 193 (1975).
81. C. Duke, H. Fischer, H.-J. Kluge, H. Kremmling, T. Kühl, and E.-W. Otten, *Phys. Lett.* **60A**, 303 (1977).
82. T. W. Hänsch, *Appl. Opt.* **11**, 895 (1972).
83. R. Wallenstein and T. W. Hänsch, *Opt. Commun.* **4**, 353 (1975).
84. T. Kühl, P. Dabkiewicz, C. Duke, H. Fischer, H.-J. Kluge, H. Kremmling, and E.-W. Otten, *Phys. Rev. Lett.* **39**, 180 (1977).
85. H. R. Collard, L. R. V. Elton, and R. Hofstadter, in *Kernradien*, Landolt-Börnstein, Neue Serie, Vol. 2, Eds. K. H. Hellwege and H. Schopper, Springer-Verlag, Berlin (1967).
86. M. Beiner and R. J. Lombard, *Phys. Lett.* **47B**, 399 (1973).
87. P. Brix and H. Kopfermann, *Z. Phys.* **126**, 344 (1949).
88. B. S. Reehal and R. A. Sorensen, *Nucl. Phys.* **A161**, 385 (1971).
89. J. H. Hamilton *et al.*, *Phys. Rev. Lett.* **35**, 562 (1975).
90. C. Bourgeois *et al.*, *J. Phys.* **37**, 49 (1976).
91. D. Proetel, R. M. Diamond, P. Kienle, J. R. Leigh, K. H. Maier, and F. S. Stephens, *Phys. Rev. Lett.* **31**, 896 (1973).
92. D. Proetel, R. M. Diamond, and F. S. Stephens, *Phys. Lett.* **48B**, 102 (1974).
93. R. Béraud, C. Bourgeois, M. G. Desthuilliers, P. Kilcher, and J. Letessier, *J. Phys. Colloque* **C5**, 101 (1975).
94. N. Rud, D. Ward, H. R. Andrews, R. L. Graham, and J. S. Geiger, *Phys. Rev. Lett.* **31**, 1421 (1975).
95. J. D. Cole *et al.*, *Phys. Rev. Lett.* **37**, 1185 (1976).
96. S. Frauendorf and V. V. Pashkevich, *Phys. Lett.* **55B**, 365 (1975).
97. D. Kolb and C. Y. Wong, *Nucl. Phys.* **A245**, 205 (1975).
98. E. L. Church and J. Wesener, *Phys. Rev.* **103**, 1035 (1956).
99. R. S. Hager and E. C. Seltzer, *Nucl. Data Tables* **A6**, 1 (1969).
100. D. Kolb, private communication.
101. G. Huber, H.-J. Kluge, L. Kugler, and E.-W. Otten, *Z. Phys.* **A272**, 381 (1975).

102. N. S. Laulainen and M. N. McDermott, *Phys. Rev.* **177**, 1606 (1969).
103. G. Nowicki, K. Bekk, S. Göring, A. Hanser, H. Revel, and G. Schatz, private communication and *Phys. Rev. Lett.* **39**, 332 (1977).
104. S. L. Kaufman, *Opt. Commun.* **17**, 309 (1976).
105. K.-R. Anton, S. L. Kaufman, W. Klempt, G. Moruzzi, R. Neugart, E.-W. Otten, and B. Schinzler, submitted for publication in *Phys. Rev. Lett.*; Proceedings of the Third International Conference of Laser Spectroscopy, Jackson Lake Lodge, Wyoming 1977, *Laser Spectroscopy III*, p. 446, Eds. J. L. Hall and J. L. Carlsten, Springer-Verlag, Berlin (1977).
106. Th. Meier, H. Hühnermann, and H. Wagner, *Opt. Commun.* **20**, 397 (1977).

18

Spin and Coherence Transfer in Penning Ionization

L. D. SCHEARER and W. F. PARKS

1. Introduction

The Penning ionization process has been effectively used as a mechanism for energy transfer from one species to another. Penning processes are special cases of collisions of the second kind in which a portion of the internal energy of the first species is utilized to ionize and excite the second species. This mechanism has been successfully applied in the He–Cd and He–Zn lasers, in which the laser levels are excited states of the Cd or Zn ions which have been populated by the Penning collision between helium metastable atoms and the Cd or Zn ground-state neutral atom. The Penning process has also been used as a convenient and efficient source of ground-state ions. In a recent series of measurements[1] ensembles of ground-state ions of Mg, Ca, Zn, Sr, Cd, and Ba were formed with sufficient densities to perform conventional Hanle experiments. In this way the radiative lifetimes and depolarization cross sections of the first excited $^2P_{3/2}$ states of the ions were determined.

In this article we discuss the phenomenon of spin exchange and coherence transfer in the population of the Zeeman levels of ions by the Penning ionization process.

2. Spin Transfer

Consider the Penning collision between a triplet metastable helium atom possessing an excitation energy of 19.8 eV and a strontium atom:

$$He(2^3S_1) + Sr(^1S_0) \rightarrow Sr^+(^2P_{3/2,1/2}) + He(^1S_0) + e \tag{1}$$

L. D. SCHEARER AND W. F. PARKS • University of Missouri, Rolla Rolla, Missouri.

In this particular reaction the ejected electron has approximately 11 eV kinetic energy and the ion is left in an excited state. It has been established that the Wigner spin rule is satisfied for Penning collisions. It is through the conservation of spin that an orientation of the metastable helium atoms can induce an orientation of the resulting ions. (See also Section 3.3.6 of Chapter 9 of this work.)

To illustrate this point we assume that an ensemble of $He(2^3S_1)$ atoms has been prepared in which all the atoms have a total angular momentum of $\hbar$ projected onto some fixed-space z axis, commonly the direction of an applied static magnetic field. This means, of course, that both electrons on a $He(2^3S_1)$ atom have a z component of spin angular momentum $+\hbar/2$. In the Penning collision the $He(2^3S_1)$ atom is de-excited to its ground state $He(1^1S_0)$, an electron is ejected, and the strontium ion is formed in a spin doublet. In order to satisfy the Wigner spin rule the ejected electron will have z component of spin angular momentum of $+\hbar/2$. The electrons in the ion, however, will have orbital angular momentum to which the spin is coupled. In the cited reaction we are dealing with a ^2P ion. We will assume here that the $\pm\hbar$ and 0 states of orbital angular momentum are randomly populated in this process. The states of the ion are then obtained by combining $+\hbar/2$ z component of electron spin with each possible orbital angular momentum. In general any coherence between $^2P_{3/2}$ and $^2P_{1/2}$ multiplets is quickly destroyed and the resulting fractional populations of the Zeeman levels for $^2P_{3/2}$ are

$$n_{3/2} = 1/3, \qquad n_{1/2} = 2/9, \qquad n_{-1/2} = 1/9, \qquad n_{-3/2} = 0$$

and for $^2P_{1/2}$ they are

$$n_{1/2} = 1/9, \qquad n_{-1/2} = 2/9$$

and the ions are therefore oriented. It should be noted that since an electron can be oriented but not aligned, an alignment of the initial metastable helium atom will have no effect on the states of the ions produced in Penning collisions.

3. Experimental Conditions for Observing Spin Transfer

The fast-flowing helium afterglow is a convenient environment in which to observe the effects of Penning collisions between $He(2^3S_1)$ atoms and an impurity species such as strontium since the reaction takes place in an electric-field-free region. Thus, the optical emission from levels of the ion represent the effects of the Penning process uncomplicated by radiation

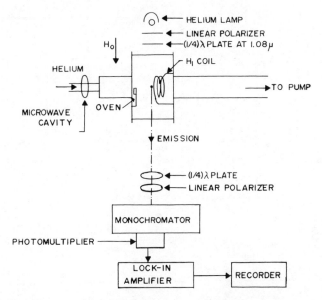

Figure 1. Schematic representation of apparatus. The expanded region shown is used to inhibit condensation of metal atoms on glass surfaces. H_0 is a static magnetic field and the H_1 coil provides an oscillating transverse field as in magnetic resonance experiments.

which might be produced by excitation with hot electrons. A schematic representation of the experimental apparatus is shown in Figure 1.

It is also possible to orient the $He(2\,^3S_1)$ atoms in the afterglow by optical pumping techniques. In this case a spin polarization of this system on the order of 30%–40% is routinely obtainable. In the flowing afterglow there is no preferred direction for the Penning collisions between the $He(2\,^3S_1)$ atom and the impurity atom. In the absence of any preferred collision axis it is possible to assume that the various orbital angular momentum states are randomly populated in the formation of the ensemble of Penning ions. It follows then from our earlier discussion that this experimentally induced orientation of the $He(2\,^3S_1)$ atoms will result in the spin polarization of the ejected electrons and the orientation of the Penning ions. Since the excited ion is oriented its emission is polarized and the polarization is readily detected.

The orientation of the Penning ions also makes it possible to perform magnetic resonance experiments.[2] Figure 2 is the magnetic resonance spectrum obtained by viewing the polarized emission from the $Cd^+(5\,^2D_{5/2}\text{–}5\,^2P_{3/2})$ transition at 4416 Å. The magnetic resonance signal at $g = 2$ is due to the $He(2\,^3S_1)$ atoms; at $H = 12.0$ Oe, the $g = 1.2$ resonance is due to the even isotope of $Cd^+(5\,^2D_{5/2})$ level; the resonance at $H = 10.7$ Oe is due to the $F = 2$ component of the odd isotope of the

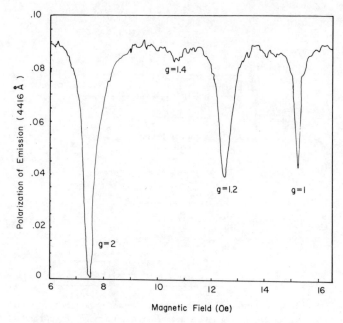

Figure 2. Magnetic resonance signals from Cd^+; $V_0 \approx 21$ MHz.

$Cd^+(5\,^2D_{5/2})$ level for which $I = \frac{1}{2}$ and occurs with a natural abundance of 25%. The $F = 3$ resonance at $g = 1$ is partially masked by the double quantum resonance signal of the helium metastable atoms. The width of the magnetic resonance signal yields the radiative lifetime of the excited state. From measurements of this type we have obtained the results shown in Table 1.

Hamel and Vienne[3] have also used this technique to observe the hyperfine resonance of the $5\,^2S_{1/2}$ ground state of the odd Cd^+ isotopes. From this they obtained a hyperfine-structure constant for this level of $|A| = 14{,}700 \pm 700$ MHz.

The polarization of the Penning electrons in an optically pumped flowing afterglow has been demonstrated by Walters *et al.*[4] They have

Table 1. Lifetime Measurements

Ion	Level	Lifetime (μs)
Cd^+	$5\,^2D_{5/2}$	0.775 ± 0.027
Zn^+	$4\,^2D_{5/2}$	1.61 ± 0.11
Zn^+	$4\,^2D_{3/2}$	2.22 ± 0.11

extracted the Penning electrons from an optically pumped flowing helium afterglow and obtained electron beam currents of $1 \mu A$ with a spin polarization of 30%. Their technique presents a new and practical source of polarized electrons.

4. Coherence Transfer in Penning Collisions: Hanle Signals in a Rotating Reference Frame

The discussion up to this point has only dealt with a static orientation of the metastable helium atoms. As shown by Dehmelt[5] and Bell and Bloom,[6] it is also possible by optical pumping techniques to prepare and observe the helium metastable atoms in a coherent superposition of its Zeeman levels. Since, by this method, all the atoms of the ensemble are "in phase," this corresponds to a macroscopic spin polarization which is precessing about the external applied field, H_0, with an angular frequency given by $\omega_0 = g\mu_B H_0/\hbar$, where μ_B is the Bohr magneton and g is the Landé factor.

For Penning collisions occurring at a given time t the orientation of the metastable helium atoms at that time induces a corresponding orientation in the ions formed. The orientation of these ions then precesses at the ion's natural frequency ($\omega' = g'\mu_B B_0/\hbar$ with the appropriate g factor) about the static field. At each instant in time, of course, a new ensemble of Penning ions will be formed with their initial orientation determined by that of the helium metastable atoms. The mathematical equations describing the complete ensemble of ions can be developed, as shown by the authors,[7] using a suitable modification of the model presented by Elbel in Chapter 29, Section 11.2, of this work. We shall present here however, a more intuitive approach.

The coherence induced in the $He(2\,^3S_1)$ system plays the same role in Penning excitation as polarized light does in the optical excitation, i.e., the excited ion levels are coherently excited. In the case of optical excitation in an applied static magnetic field the Hanle signals would be observed. Consequently if we consider the systems from the frame of reference rotating with the orientation of the metastable helium atoms, the Hanle signals should be observed from the excited ions. This frame of reference will be rotating with angular frequency ω_0. The Hanle signal depends on the precession rate of the orientation of the excited state. In this rotating coordinate system the magnetic moment of the excited ion will be observed to precess about the axis of the external magnetic field with an angular frequency

$$\omega = (g'\mu_B H_0/\hbar) - (g\mu_B H_0/\hbar) \tag{2}$$

i.e., the difference of the natural precession frequencies of the orientations of the ions and the metastable helium atoms. The (circularly polarized) Hanle signals, as observed in the rotating reference frame, are given by

$$P = A/(1 + \omega^2 \tau^2) \tag{3}$$

$$P' = A\omega\tau/(1 + \omega^2 \tau^2) \tag{4}$$

where P is the signal in phase with the orientation of the helium metastable atom and P' that leading this orientation. τ is the relaxation time for the orientation of the ions from which the radiative lifetime can be extracted and A is a constant determined by the particular system and apparatus.

On transforming to the laboratory frame and utilizing Eq. (2) we obtain

$$P_L = A \cos (\omega_0 t)/[1 + \omega_0^2 (1 - g'/g)^2 \tau^2] \tag{5}$$

where we have arbitrarily chosen this signal to be maximum at $t = 0$. The other component of the signal is

$$P'_L = A\omega_0(1 - g'/g)\tau \sin \omega_0 t/[1 + \omega_0^2(1 - g'/g)^2\tau^2] \tag{6}$$

P_L and P'_L are the signals obtained from the components in phase and in quadrature with the orientation of the metastable helium atoms, respectively.

5. Experimental Observations of Coherence Transfer

In the laboratory frame of reference the Hanle signal is modulated due to the precession of the orientation of the metastable helium atoms and damped due to the loss of phase coherence in the ion ensemble. Experimental observations of the coherence transfer proceeds in a straightforward manner with apparatus essentially identical to that shown in Figure 1. The single important difference is that ion emission is observed through a circular analyzer perpendicular to the applied magnetic field. The observation of the transfer of the phase coherence (or equivalently, the transfer of the transverse component of spin angular momentum) to the excited ion was first reported by Schearer.[8]

Hamel et al.,[9] in an elegant experiment, have recently reported the observation of the magnetic field dependence of the amplitude of these Penning ion Hanle signals. The results they obtained are shown in Figure 3 for the amplitudes of P_L and P'_L for the $He^m + Cd \to Cd^+(^2D_{5/2}) + He + e$ Penning process. The ion emission is at 4416 Å ($5\,^2D_{5/2}-5\,^2P_{3/2}$). The experimental points are labeled $+$ and $\bigcirc$ and the solid curve is a plot of the amplitudes appearing in Eqs. (5) and (6) with $g' = 1.1998 \pm 0.0001$ and

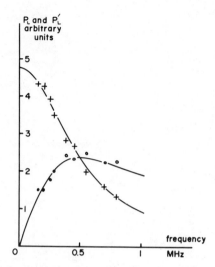

Figure 3. Hanle signals from the $(4d^9 5s^2)$ $^2D_{5/2}$ level of Cd^+. $+$: experimental amplitude of P_L, $\bigcirc$: experimental amplitude of $P_L{}'$.

$\tau = 0.800 \pm 0.040\ \mu s$. Using this technique Hamel *et al.*[10] have also obtained radiative lifetimes for the $(4f)^2F$ levels of Mg^+.

This technique holds considerable promise for the measurement of radiative levels of ions that are energetically accessible in a Penning collision with He^m. The method effectively extends the use of the Hanle effect to levels of ions that are not necessarily optically connected to a ground or metastable level of the ion.

Acknowledgment

The authors wish to acknowledge the support of the Office of Naval Research, which has led to the development of much of the work reported here.

References

1. F. H. K. Rambow and L. D. Schearer, *Phys. Rev. A* **14**, 1735 (1976).
2. L. D. Schearer, *Phys. Rev. A* **10**, 1380 (1974).
3. J. Hamel and J.-F. Vienne, *Opt. Commun.* **7**, 83085 (1973).
4. P. J. Keliher, R. E. Gleason, and G. K. Walters, *Phys. Rev.* **11**, 1279 (1975).
5. H. G. Dehmelt, *Phys. Rev.* **105**, 1924 (1957).
6. W. E. Bell and A. L. Bloom, *Phys. Rev.* **107**, 1559 (1957).
7. W. F. Parks and L. D. Schearer, *Phys. Rev. Lett.* **29**, 531 (1972).
8. L. D. Schearer and L. A. Riseberg, *Phys. Rev. Lett.* **26**, 599 (1971).
9. J. Hamel, J. Margerie, and J.-P. Barrat, *Opt. Commun.* **12**, 409 (1974).
10. J. Hamel and J.-P. Barrat, *Opt. Commun.* **18**, 357 (1976).

19

Multiphoton Spectroscopy

PETER BRÄUNLICH

1. Introduction

In this chapter, a class of electronic transitions in atoms involving the simultaneous absorption or emission of more than one photon is discussed. In describing Raman scattering in atoms as a two-photon process, in which one photon of frequency ω_1 is absorbed and one of frequency ω_2 ($\neq \omega_1$) is emitted, Kramers and Heisenberg[1] pointed out that an entirely new process may take place, namely, singly stimulated two-photon emission. A bound electron in an excited state $|i\rangle$ can reach a lower state $|f\rangle$ via the emission of two photons. The presence of a photon field of frequency ω_1 stimulates the simultaneous emission of one photon having the same frequency and of a second photon of frequency $\omega_2 = \omega_{fi} - \omega_1$. This process is greatly enhanced in the presence of real intermediate states and was therefore observed first under these conditions by Yatsiv et al.[2] In the absence of intermediate states, singly stimulated two-photon emission and anti-Stokes Raman scattering in atoms were reported by Bräunlich et al.[3-5]

Göppert-Mayer[6] presented a detailed theoretical treatment of two-photon excitation and emission processes based on time-dependent perturbation theory. In the former process, the electron undergoes a transition from its ground state to an excited state by simultaneously absorbing two photons. The first observation of this transition was reported by Kaiser and Garrett[7] in a solid and by Abella[8] in atoms. Spontaneous two-photon emission was found by Breit and Teller[9] to be the dominant

PETER BRÄUNLICH • Bendix Research Laboratories, Southfield, Michigan 48076. Present address: Department of Physics, Washington State University, Pullman, Washington 99163.

mechanism in the decay of metastable $2S$ levels in hydrogen as well as most hydrogenlike atoms and was first observed by Lipeles et al.[10]

A variety of other nonlinear electronic transitions may be observed in atoms at sufficiently large photon fluxes. They are all, strictly speaking, multiphoton processes which can be calculated with the aid of higher-order perturbation theory. The following are examples: the simultaneous absorption of more than two photons, leading to highly excited states or even to the emission of an electron (multiphoton ionization, Bunkin and Prokhorov,[11] Voronov and Delone,[12] and Bakos[13]); the absorption of three photons of equal frequency ω and the emission of one photon of frequency 3ω (third harmonic generation, Ward and New,[14] Young et al.,[15] Leung et al.[16]); the absorption of three photons of different frequencies and the emission of one photon of the sum frequency (infrared up-conversion, Harris and Bloom[17]); or the absorption of one photon of frequency ω and the emission of another photon of frequency ω' [electronic Raman scattering, inelastic photon scattering, resonance scattering (if $\omega = \omega'$), or elastic photon scattering, Zernik,[18] Bräunlich and Lambropoulos[3]].

Most of these multiphoton processes were amenable to experimental observation only after the development of the laser. This work was initially performed with pulsed, high-power solid-state lasers with linewidths of typically $1\ \text{cm}^{-1}$ or more. Precision atomic multiphoton spectroscopy was therefore impossible and the experiments concentrated instead on the demonstration of the existence or technical feasibility of certain multiphoton processes (by measuring their efficiency or merely the dependence of the transition rate on the photon flux) and on the confirmation of theoretically predicted effects. For example, Lambropoulos et al.[19] found that multiphoton transition rates depend on the statistical properties of the photon field; this was verified by a number of subsequent experiments (e.g., Shiga and Iamura,[20] Lecompte et al.[21]). Of course, experiments have also led to more detailed theoretical investigations. A classic example was the discovery by Fox et al.[22] that the cross sections for the two and three-photon ionization of atomic cesium depend on the polarization of the absorbed light. This was later explained by Lambropoulos[23] and by Klarsfeld and Maquet.[24]

Until narrow-linewidth tunable lasers became available, the exploitation of multiphoton transitions in high-resolution atomic spectroscopy was limited to the radio-frequency region of the electromagnetic spectrum (Brossel et al.,[25] Cohen-Tannoudji and Kastler,[26] Hughes and Grabner[27]). The first application of a two-photon transition in precision spectroscopy was not reported before 1974 (Pritchard et al.[28]). In the same year, a suggestion by Vasilenko and co-workers[29] to utilize two oppositely traveling light waves for the elimination of Doppler broadening was implemented by Biraben *et al.*[30] Since then, high-resolution multiphoton spectroscopy has developed into an exciting and rapidly progressing field. Most certainly, by the time this article is in press, many more important contributions will have been added to the already impressive list of achievements which were only possible with the advent of novel multiphoton techniques.

Two-photon absorption offers several advantages for precision spectroscopy: (1) access to levels via transitions that are parity-forbidden by single-photon dipole radiation, e.g., to S and D levels from an S ground state, (2) extension of the accessible frequency range by combining tunable lasers or by using a combination of fixed-frequency and tunable lasers, and (3) high-resolution spectroscopy of highly excited states which were previously only accessible by stepwise-excitation experiments in which the linewidth of the observed transition depends on the width of the intermediate states.

This article on multiphoton processes concentrates on two aspects of multiphoton spectroscopy, namely, the investigations of transition probabilities and of the frequency spectra associated with the transition. We will first deal with the former. The necessary theoretical background is provided in Section 2. Additional information may be found in the original literature as referenced in recent reviews of multiphoton ionization processes by Bakos[13] and Lambropoulos.[31] Subsequently, a number of newly developed experimental techniques and their applications to precision frequency measurements are described, and finally a review of important results is given.

The material is presented with the intention of providing an introduction to multiphoton spectroscopy. A great deal of detail is given in some parts (e.g., the derivation of the transition probability for two-photon absorption), while other parts are superficial or sketchy, such as the brief descriptions of multiphoton ionization and the newly discovered quadrupole and polarization effects.

2. Transition Probabilities for Multiphoton Processes

2.1. Time-Dependent Perturbation Theory

The interaction of the radiation field with an atom is usually described by using the formalism of time-dependent perturbation theory.[32] The Hamiltonian of the system may be written as

$$H = H^A + H^R + V = H^0 + V \tag{1}$$

where H^A is the Hamiltonian of the free atom, H^R is the Hamiltonian of the radiation field, and V is the interaction between the two.

The formal solution of the Schrödinger equation

$$\dot{\psi}(t) = iH\psi(t), \qquad \hbar \equiv 1 \tag{2}$$

may be expressed with the aid of the time evolution operators

$$U(t, t_0) = \exp[-i(t - t_0)H]$$

and

$$\tag{3}$$

$$U^0(t, t_0) = \exp[-i(t - t_0)H^0]$$

as

$$\frac{\partial U}{\partial t} = iHU = iH^0U - iVU \tag{4}$$

Equation (4) can be rewritten in the form

$$\frac{\partial U^I(t)}{\partial t} = -iV^I(t)U^I(t) \tag{5}$$

where

$$U^I(t) = U^{0\dagger}U(t) = e^{iH^0(t - t_0)}U(t)$$

and

$$V^I(t) = U^{0\dagger}VU^0 \tag{6}$$

With the initial condition $U^I(t_0) = 1$, the solution of the Schrödinger equation (5) is

$$U^I(t) = 1 - i\int_{t_0}^{t} V^I(\tau)U^I(\tau)\,d\tau \tag{7}$$

This integral equation can be solved by iteration:

$$U^I(t, 0) = 1 - \int_0^t V^I(\tau)\,d\tau + i^2\int_0^\tau V^I(\tau_1)\,d\tau_1\int_0^{\tau_1} V^I(\tau_2)\,d\tau_2 + \cdots$$

which leads to the following expression for the time evolution operator as a power series in the coupling parameter V:

$$U(t, t_0) = U^0(t, t_0)\left[1 + \sum_{N=1}^{\infty} U^{(N)}(t, t_0)\right] \tag{8}$$

with

$$U^{(N)}(t, t_0) = i^N\int_{t > \tau_N > \tau_{N-1} > \cdots > t_0} d\tau_N\, d\tau_{N-1} \cdots d\tau_1 U^{0\dagger}(\tau_N)VU^0(\tau_N) \cdots$$

$$\tag{9}$$

This expression for the time evolution operator is the basis of time-dependent perturbation theory. It has proved to be a powerful tool in the calculation of transition probabilities of an atomic system during the interaction with a perturbation V. For example, the interaction of matter with radiation fields is satisfactorily described with this perturbation expansion. N-photon transitions are obtained, in the lowest order, from the Nth-order terms. High-order contributions are often required to be included to account for resonance effects, thereby necessitating the consideration of linewidths and shifts. Perturbation theory is valid as long as V does not exceed the binding of the electron to the nucleus. This condition is fulfilled in most experiments,[33] and only at laser powers above 10^{16} W/cm^2 (or 10^{35} optical photons/cm^2/s) does it become invalid. These high-photon fluxes are within the realm of present experiments. They are used routinely in investigations of laser-triggered thermonuclear fusion. However, the vast majority of spectroscopic data involving multiphoton transitions has been obtained under conditions that are within the region of perturbation theory (an important exception being the experiments by Bayfield and Koch on multiphoton ionization of order $N \geq 76$ of highly excited Rydberg states[34]); on the other hand, two-photon spectroscopy is possible at laser powers as low as 1 mW.

The transition probability, P_{ba}, for the system in a state $|a\rangle$ to reach a state $|b\rangle$ in time t, is given by Fermi's golden rule,

$$P_{ba} = |\langle b|U(t)|a\rangle^2 \equiv U_{ba}^2 \tag{10}$$

which is obtained as follows:

Assume $|a\rangle, |b\rangle, \ldots$ is a complete set of eigenstates of H^0. At $t = 0$, the system is in the state $\psi(0) = |a\rangle$. Since the time evolution of ψ is given by $\psi(t) = U(t)\psi(0)$, we have $\psi(t) = U(t)|a\rangle$. Expressing $\psi(t)$ in a series of the form

$$\psi(t) = \sum_b \langle b|\psi(t)|b\rangle = \sum_b \langle b|U(t)|a\rangle|b\rangle$$

$$= \sum_b U_{ba}|b\rangle$$

the time-dependent expansion coefficients are expressed in terms of the time evolution operator. The square of these coefficients is the transition probability P_{ba}.

One can arrive at the perturbation expansion of P_{ba} in many other ways. Goldberger and Watson,[35] for example, take the Laplace transform of $U(t)$ and introduce the resolvent operator $G(z) \equiv 1/(z - H)$, where z is a complex variable. The matrix elements G_{ba} of G are then calculated and transformed back to U_{ba} via an inverse Laplace integral. The resolvent operator formalism enables one to incorporate level broadening and shifts in a more systematic way and thus to directly treat resonance effects. It is therefore preferred by some authors (e.g., Lambropoulos[36,37]).

Combining Eqs. (9) and (10), we find in first and second order for $N = 1$ (single-photon transition)

$$U_{ba}^{(1)} = -i \int_0^t d\tau\, e^{-i\omega_b(t-\tau)} V_{ba}\, e^{-\omega_a\tau} \tag{11a}$$

and for $N = 2$ (two-photon transition):

$$U_{ba}^{(2)} = i^2 \sum_\mu \int_0^t d\tau \int_0^\tau d\tau_1\, e^{-i\omega_b(t-\tau)} V_{b\mu}\, e^{-\omega_\mu(\tau-\tau_1)} V_{\mu a}\, e^{-i\omega_a\tau_1} \tag{11b}$$

The summation over the intermediate states $|\mu\rangle$ originates from the use of the following operator relations:

$$\langle b|VW|a\rangle = \sum_\mu \langle b|V|\mu\rangle\langle\mu|W|e\rangle$$

where the states $|\mu\rangle$ are a complete set, and

$$\langle b|e^{-iH^0(t-\tau)}|\mu\rangle = e^{-\omega_b(t-\tau)}\langle b|\mu\rangle$$

which is obtained because $|b\rangle$ and $|\mu\rangle$ are the eigenstates of H^0.

For calculations of the transition probabilities P_{ba}^2, the integrals in Eqs. (11a) and (11b) are evaluated explicitly and the following expression for $U_{ba}^{(2)}$ is obtained:

$$U_{ba}^{(2)} = \frac{e^{-i\omega_b t}}{\omega_b - \omega_a}(e^{-i(\omega_b-\omega_a)t} - 1)\sum_\mu \frac{V_{b\mu}V_{\mu a}}{\omega_\mu - \omega_a} \tag{12}$$

Integrals of exponential of the form $e^{-i(\omega_\mu-\omega_a)t}$ have been neglected in Eq. (12). This can be done if we postulate $\omega_\mu \neq \omega_a, \omega_b$ (absence of resonances). Since $e^{-i\omega_b t}$ becomes averaged over time, we finally have for the transition probability

$$W_{ba}^{(2)} \equiv \lim_{t\to\infty} \frac{1}{t}|U_{ba}^{(2)}|^2$$

$$= \left|\sum_\mu \frac{V_{b\mu}V_{\mu a}}{\omega_\mu - \omega_a}\right|^2 2\pi\delta(\omega_b - \omega_a) \tag{13}$$

where the relation

$$\lim_{t\to\infty} \frac{1}{x}(e^{-ix} - 1) = 2\pi\delta(x)$$

has been used. The δ function expresses the fact that the total energy must be conserved during the transition of the total system ("atom plus photon field") from the initial state $|a\rangle$ to the final state $|b\rangle$.

After algebraic manipulations similar to those that lead to Eq. (13), one obtains for *nonresonant N-photon* transitions the transition probability

$$W_{ba}^{(N)} = \left| \sum_{\mu_{N-1} \cdots \mu_1} \frac{V_{b\mu_{N-1}} V_{\mu_{N-1}\mu_{N-2}} \cdots V_{\mu_1 a}}{(\omega_{\mu_{N-1}} - \omega_a)(\omega_{\mu_{N-1}} - \omega_a) \cdots (\omega_{\mu_1} - \omega_a)} \right|^2 2\pi\delta(\omega_b - \omega_a)$$

(14)

where the $(N-1)$fold summation is to be taken over all intermediate states of the "atom plus photon field" system and includes the continuum states of the atom.

Let us now calculate the transition probabilities more explicitly. The usual expansion of the Hamiltonian of the free radiation field H^R in terms of the modes of the radiation field is[38]

$$H^R = \sum_{k\lambda} \omega_k (a_{k\lambda}^{\dagger} a_{k\lambda} + \tfrac{1}{2})$$

(15)

where $\mathbf{k}$ denotes the photon wave vector, λ is the polarization index (taking the values 1 and 2), $a_{k\lambda}^{\dagger}$ and $a_{k\lambda}$ are the creation and annihilation operators, and $\omega_k = ck$ is the frequency of the kth mode of the photon field.

The Hamiltonian of the free atom plus the interaction Hamiltonian V is given by

$$H^A + V = \frac{1}{2m} \left[\mathbf{p} - \frac{e}{c} \mathbf{A}(\mathbf{r}) \right]^2$$

(16)

where $\mathbf{p}$ is the electron momentum operator and $\mathbf{A}(\mathbf{r})$ is the vector potential of the radiation field.

In the *frequency range of interest* (infrared to vacuum uv), atomic dimensions are much smaller than the wavelengths of the photons. We can therefore neglect the volume dependence of A and set $e^{\pm ikr} \approx 1$ (dipole approximation). Neglecting terms in A^2 (which would contribute to Rayleigh scattering but not to multiphoton transitions in lowest-order perturbation theory), one obtains for the interaction between the atomic system and the radiation field

$$V = -\xi L^{3/2} \sum_{k\lambda} \mathbf{p} \cdot \boldsymbol{\varepsilon}_{k\lambda} (a_{k\lambda}^{\dagger} + a_{k\lambda}) \omega_k^{-1/2}$$

(17)

where $\xi = (2\pi e^2/m^2\hbar)^{1/2}$ and $\boldsymbol{\varepsilon}_{k\lambda}$ is the unit polarization vector of the $(k\lambda)$th mode.

The states of the total system may be expressed as products of the eigenstates of H^A and the single-mode states $|\cdots n_{k_1\lambda} \cdots n_{k_2\lambda} \cdots\rangle$ of the radiation field inside a large box of volume L^3 with periodic boundary conditions,[39] where $n_{k_1\lambda}$, etc., are the number of photons occupying the $(k_1\lambda)$th, etc., mode. We now apply the developed formalism to two-photon absorption. This will provide some of the theoretical basis for the discussion of Doppler-free two-photon spectroscopy in Section 7.

By *simultaneously absorbing two photons* of frequency ω_{k_1} and ω_{k_2}, a bound electron in the ground state $|i\rangle$ of the free atoms reaches the excited state $|f\rangle$ with $\omega_f - \omega_i \equiv \omega_{f_i} = \omega_{k_1} + \omega_{k_2}$. The product states of the system in the ground and excited states are, respectively,

$$|a\rangle = |i\rangle| \cdots n_1 \cdots n_2 \cdots \rangle$$
$$|b\rangle = |f\rangle \cdots n_1 - 1 \cdots n_2 - 1 \cdots \rangle \tag{18}$$

where for brevity, the notation n_1 and n_2 has been used in the place of $n_{k_1\lambda}$ and $n_{k_2\lambda}$. The final state of the total system can be reached via two different types of intermediate states:

$$|\mu_1\rangle = |c_1\rangle| \cdots n_1 - 1 \cdots n_2 \cdots \rangle$$
$$|\mu_2\rangle = |c_2\rangle| \cdots n_1 \cdots n_2 - 1 \cdots \rangle \tag{19}$$

where $|c_1\rangle$ and $|c_2\rangle$ are intermediate states of the free atom.

The interaction matrix elements assume the forms

$$V_{\mu_1 a} = -L^{-3/2}\xi\left(\frac{n_1}{\omega_{k_1}}\right)^{1/2} \langle\mu_1|\mathbf{p}\cdot\mathbf{\varepsilon}_1|a\rangle$$

$$V_{b\mu_1} = -L^{-3/2}\xi\left(\frac{n_2}{\omega_{k_2}}\right)^{1/2} \langle b|\mathbf{p}\cdot\mathbf{\varepsilon}_2|\mu_1\rangle \tag{20}$$

Similar expressions are obtained for $V_{\mu_2 a}$ and $V_{b\mu_2}$. To arrive at Eq. (20), we used the well-known relations

$$a| \cdots n_1 \cdots \rangle = n_1^{1/2}| \cdots n_1 - 1 \cdots \rangle$$
$$a^\dagger| \cdots n_1 \cdots \rangle = (n_1 + 1)^{1/2}| \cdots n_1 + 1 \cdots \rangle \tag{21}$$

with $n_1 + 1 \approx n_1$. Note that terms in $a^\dagger$ are neglected in two-photon absorption since they account for the generation of photons. In the case of *spontaneous or stimulated two-photon emission*, they alone are retained and expressions similar to Eqs. (20) are obtained.

The frequencies appearing in the denominators of Eq. (13) become

$$\omega_a = \omega_i + \omega_{k_1}n_1 + \omega_{k_2}n_2$$

$$\omega_b = \omega_t + \omega_{k_1}(n_1 - 1) + \omega_{k_2}(n_2 - 1)$$

$$\omega_{\mu_1} = \omega_{c_1} + \omega_{k_1}(n_1 - 1) + \omega_{k_2}n_2$$

$$\omega_{\mu_2} = \omega_{c_2} + \omega_{k_1}n_1 + \omega_{k_2}(n_2 - 1) \tag{22}$$

The transition probability can then be written as

$$W_{ba}^{(2)} = 2\pi\delta(\omega_{fi} - \omega_{k_1} - \omega_{k_2})\xi^4 L^{-6} \sum_{\mathbf{k}\lambda} \frac{n_1 n_2}{\omega_{k_1}\omega_{k_2}} |M^{(2)}|^2 \qquad (23)$$

with

$$|M^{(2)}|^2 = \left| \sum_c \left\{ \frac{(\mathbf{p}_{fc} \cdot \boldsymbol{\varepsilon}_1)(\mathbf{p}_{ci} \cdot \boldsymbol{\varepsilon}_2)}{\omega_{ci} - \omega_{k_1}} + \frac{(\mathbf{p}_{fc} \cdot \boldsymbol{\varepsilon}_2)(\mathbf{p}_{ci} \cdot \boldsymbol{\varepsilon}_1)}{\omega_{ci} - \omega_{k_2}} \right\} \right|^2 \qquad (24)$$

where, again, abbreviated notations are used, such as ω_{fc} for $\omega_f - \omega_c$, $\boldsymbol{\varepsilon}_1$ instead of $\boldsymbol{\varepsilon}_{k_1}$ etc., and $\mathbf{p}_{fc}$ for $\langle f|\mathbf{p}|c\rangle$ etc. The two terms in Eq. (24) arise from the fact that the excited state of the atoms can be reached via the two different intermediate states of the photon field, i.e., absorption of ω_1 first and subsequent absorption of ω_2 and vice versa [Eq. (19)]. In an N-photon transition $N!$ different combinations of this type are possible. The dipole matrix elements in Eq. (24) are, of course, nonzero only for states of odd parity and, as a consequence, the two photon transition can occur only when the states $|i\rangle$ and $|f\rangle$ have the same parity. This important property is one of the attractive features of two-photon spectroscopy.

Since photon beams used in experiments have a narrow but nevertheless continuous spectrum, the summation over the discrete modes in Eq. (23) is replaced by an integration:

$$\sum_{\mathbf{k}\lambda} \to \frac{L^3}{8\pi^3 c^3} \int_0^\infty \omega_k^2 \, d\omega_k \int_{\Omega_{\mathbf{k}}} d\Omega_{\mathbf{k}} \qquad (25)$$

where $\Omega_{\mathbf{k}}$ is the solid angle of incidence of the photon beam. The number of photons n_k per mode is related to the photon flux $I(\omega_k)$ (in units of photons per square centimeter per second and unit frequency) through the equation

$$I(\omega_k) = \frac{\omega_k^2}{8\pi^3 c^3} \int_{\Omega_{\mathbf{k}}} n_k \, d\Omega_{\mathbf{k}} \qquad (26)$$

With these relations the transition probability per atom becomes

$$W_{fi}^{(2)} = \frac{2\pi\xi^4}{c^2} \int d\omega_1 \, d\omega_2 \, \frac{I(\omega_1)I(\omega_2)}{\omega_1\omega_2} |M^{(2)}|^2 \delta(\omega_{fi} - \omega_1 - \omega_2)$$

$$= 2\pi\xi^4 c^{-2} |M^{(2)}|^2 \frac{I_1 I(\omega_{fi} - \omega_1)}{\omega_1\omega_2} \qquad (27)$$

where I_1 is now the total flux of one of the two photon beams in photons per square centimeter per second. Note that we have changed the notation for the transition probability from $W_{ga}^{(2)}$ to $W_{fi}^{(2)}$; this is possible after the integration indicated by Eq. (25). W_f refers to the transition of the electron from state $|i\rangle$ to state $|f\rangle$.

It may now be helpful to point out an important assumption that is made in arriving at Eq. (27): In using relation (26), it is assumed that each of the two beams (e.g., laser beams), I_1 and I_2, is strictly monochromatic since it contains only photons of one single mode of the radiation field. We will see in Section 3 that the transition probability for multimode beams requires special attention. In general, it is sufficient to assume that the spectral distributions of $I(\omega_1)$ and $I(\omega_2)$ are narrow and that all quantities depending on ω_1 and ω_2 vary slowly over the frequency intervals. This precludes, of course, resonances in the frequency denominators of Eq. (24). The frequency interval of $I(\omega_{fi} - \omega_1)$ is determined by the requirement of energy conservation: $\omega_2 = \omega_{fi} - \omega_1$.

In many experiments, the spectral width of the laser beam is much narrower than the widths of the atomic levels involved. It is then convenient to introduce the normalized spectral width $g(\omega_{fi})$ [with $\int_0^\infty g(\omega_{fi})\, d\omega = 1$] so that $I(\omega_{fi} - \omega_c) \rightarrow I_2 g(\omega_{fi})$, where I_2 now has the dimensions of photons per square centimeter per second.

We will now generalize the problem to the case of an N-photon transition. However, instead of using N different monochromatic photon sources, we will assume that all photons are provided by just one single mode of the radiation field (e.g., one narrow-frequency laser beam) of frequency ω. By using the relation $\langle \alpha | \mathbf{p} \cdot \boldsymbol{\varepsilon} | \beta \rangle = \mathrm{Im}\, \omega_{\alpha\beta} \langle \alpha | \mathbf{r} \cdot \boldsymbol{\varepsilon} | \beta \rangle$, we obtain

$$W_{fi}^{(N)} = 2\pi \frac{(m\xi)^{2N}}{(c\omega)^N} I^N |K^{(N)}|^2 g(\omega_{fi}) \tag{28}$$

and

$$|K^{(N)}|^2 = \sum_{c_1 \cdots c_{N-1}} \omega_{fc_1} \cdots \omega_{c_{N-1}i} \left| \frac{\langle f | \mathbf{r} \cdot \boldsymbol{\varepsilon} | c_1 \rangle \langle c_1 | \mathbf{r} \cdot \boldsymbol{\varepsilon} | c_2 \rangle \cdots \langle c_{N-1} | \mathbf{r} \cdot \boldsymbol{\varepsilon} | i \rangle}{[\omega_{c_1 i} - (N-1)\omega] \cdots (\omega_{c_{N-1} i} - \omega)} \right|^2 \tag{29}$$

Peticolas *et al.*[40] have shown that Eq. (29) may be rewritten in the following simpler form[41]:

$$|K^{(N)}|^2 = \omega^{2N} \left| \sum_{c_1 \cdots c_{N-1}} \frac{\langle f | \mathbf{r} \cdot \boldsymbol{\varepsilon} | c_1 \rangle \cdots \langle c_{N-1} | \mathbf{r} \cdot \boldsymbol{\varepsilon} | i \rangle}{[\omega_{c_1 i} - (N-1)\omega] \cdots (\omega_{c_{N-1} i} - \omega)} \right|^2$$

$$= \omega^{2N} |T^{(N)}|^2 \tag{30}$$

so that finally we have

$$W_{fi}^{(N)} = (2\pi)^{N+1} \left(\frac{r_0 c m}{\hbar} \right)^N \omega^N I^N |T^{(N)}|^2 g(\omega_{fi}) \tag{31}$$

where $r_0 = e^2 / mc^2$ is the classical electron radius.

Experimentalists find it convenient to introduce the cross section σ or the so-called "generalized cross section" $\sigma^{(N)}$, which are defined by

$$W_{fi}^{(N)} = \sigma I = \sigma^{(N)} I^N \tag{32}$$

where $\sigma^{(N)}$ has the dimensions $cm^{2N} s^{N-1}$.

The dependence of the multiphoton process on the atomic structure and on the polarization of the incident light is contained in $T^{(N)}$. In principle, calculations of transition probabilities or generalized cross sections are evaluations of this term. The final state $|f\rangle$ may be a bound state of the atom or, as in multiphoton ionization, the continuum state. In this case, the line-shape function $g(\omega_{fi})$ is replaced by the density of states in the continuum. Most of the investigations of multiphoton processes of order $N > 2$ have been concerned with the ionization of the atoms. A detailed review of the results is given in a recent paper by Bakos.[13] In this article we are mainly interested in the bound–bound transitions. Up to this time, the experimental work reported in the literature has dealt almost exclusively with two-photon transitions.

2.2. Calculation of Multiphoton Cross Sections

The calculation of cross sections or transition probabilities involves the evaluation of infinite sums over intermediate states and the continuum. Slow convergence may lead to highly inaccurate results; nevertheless, truncation of the sum has been used quite successfully to obtain estimations of low-order multiphoton cross sections. It is advantageous to evaluate the sum in Eq. (31) rather than the one in Eq. (29) because the former should, in general, converge faster.[42] Truncation is possible since, starting with the lowest intermediate state, the matrix elements decrease as one goes to higher states and the energy denominators increase at the same time. The contributions of higher states thus decrease. Retention of only a finite number of terms and neglect of the entire continuum contribution will often give quite reasonable approximations. The problem of calculating the dipole matrix elements remains, however, except for hydrogen or hydrogenlike ions (Honig and Jortner[42]). They may be obtained from experimental determinations of some matrix elements (Kovner and Parshkov[43]). The relative amplitudes of additional matrix elements may be found using angular momentum relations (Bjorkholm and Liao,[44] Vriens[45]) or from any direct calculations, e.g., from quantum defect theory[46,47] or direct summation with Hartree–Fock states.[48-50]

The infinite sum may be circumvented by the use of the "mean-energy" approximation of Bebb and Gold[51-55] or the "mean-matrix-element" approximation of Morton.[56] Bebb and Gold define an average

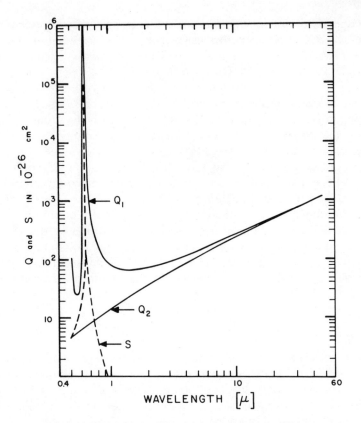

Figure 1. Graphs of the differential cross-section coefficients

$$Q_1 = \frac{1}{1+\cos^2\theta}\frac{d\sigma_r}{d\Omega}, \quad Q_2 = \frac{1}{1+\cos^2\theta_2}\frac{d\sigma^{(2)}}{d\Omega}, \quad S = \frac{1}{1+\cos^2\theta_3}\frac{d\sigma_S}{d\Omega}$$

for metastable hydrogen atoms as a function of the wavelengths of the incident light. The angles θ are between the wave vectors of the incident and the scattered photons. Q_1: Anti-Stokes Raman scattering; Q_2: singly stimulated two-photon emission; S: elastic photon scattering; Ω: solid angle of detection (Zernik, 1963[18]).

frequency $\overline{(\omega_c - \omega_i)}$ via

$$\sum_c \frac{\langle f|\mathbf{r}\cdot\boldsymbol{\varepsilon}_1|c\rangle\langle c|\mathbf{r}\cdot\boldsymbol{\varepsilon}_2|i\rangle}{\omega_c - \omega_i - \omega} \equiv \frac{\langle f|\mathbf{r}_1\cdot\mathbf{r}_2|i\rangle}{\overline{(\omega_c - \omega_i)} - \omega} \tag{33}$$

where $\mathbf{r}_i$ denotes the $\boldsymbol{\varepsilon}_i$ component of $\mathbf{r}$. Since $|c\rangle$ is a complete set of eigenstates of H^0, $\Sigma_c|c\rangle\langle c| = 1$. Equation (33) may be generalized to the N-photon problem.

The exact calculation of $\overline{\omega_{ci}}$ involves also infinite summations. However, good approximations are frequently obtained if one particular state dominates the series expansion by having either a large matrix element or a small-frequency denominator, e.g., one close to but not exactly at a resonance of the frequency as a transition between the initial state and an intermediate state. The average frequency should be chosen to correspond to this dominant term. Consequently, and also because the continuum contributions are always neglected in this approximation, the results are quite poor in the valleys between resonances, but they are rather reasonable close to a resonance. The accuracy decreases with increasing order N of the multiphoton process.

The most accurate results are obtained by using closed-form solutions of the perturbation equation. Sum-rule techniques, which were first developed for time-independent problems by Dalgarno and Lewis,[57] Allison and Dalgarno,[58] and Dalgarno,[59] and which were applied using Laplace-transform techniques to the calculation of the Lamb shift in helium (Schwartz and Tiemann[60,61]) have been extended by Zernik[18,62,63] to time-dependent perturbation problems in hydrogen atoms. He calculated the optical quenching[18] (via singly stimulated two-

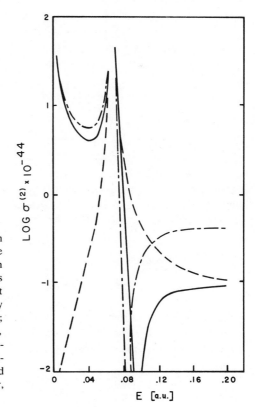

Figure 2. Theoretical two-photon absorption cross sections in the hydrogen atom for the transition $1S \to 3S$. The photon energy is given in atomic units. ———— exact "sum-rule" calculation (method by Schwartz and Tiemann[60,61]); • – • – truncated sum technique, continuum contributions neglected; – – – average energy calculation according to Bebb and Gold[52] (Honig and Jortner, 1967[42]).

photon decay as well as inelastic and elastic photon scattering) and multiphoton ionization[62] of metastable 2S hydrogen.

For two-photon processes in more complex systems it is necessary to use variational techniques whereby the true wave functions, which are required for the calculation of the dipole matrix elements, are replaced by a set of variational functions.[64,65] The summation is then carried out implicitly by solving a differential equation (Karplus and Kolker,[66] Dalgarno and Victor,[67] Honig and Jortner,[42] Jacobs and Mizuno,[68] Mizuno,[69] Chang and Poe,[70] and Choudhury and Gupta[71]). The extension of these implicit summation techniques to N-photon processes has been described by Gontier and Trahin.[72]

Transition probabilities for multiphoton processes of any order N in hydrogen or hydrogenic atoms can be expressed in closed form in terms of hypergeometric functions with the aid of the Coulomb Green's-function technique. Vetchinkin and Khristenko,[73] Klarsfeld,[74] and Au[75] calculated second-order transition probabilities, and Karule[76] extended this method to the evaluation of multiphoton ionization in hydrogen. In atoms more complex than hydrogen the Green's function cannot be expressed analytically and must therefore be approximated. A phenomenological Green's function may be constructed on the basis of a quantum defect theory (Zon et al.[77,78], Davydkin,[79] Manakov et al.[80]). Results of several different calculations are presented in Figures 1 and 2.

3. Experimental Investigation of Multiphoton Transition Probabilities

Before we present experimental results of measurements of transition probabilities or generalized cross sections and their comparison with theory, we will describe a number of important experimental peculiarities that are characteristic of multiphoton spectroscopy. These have not always been recognized, and, as a consequence, a large number of experiments have yielded only qualitative rather than quantitative results.

3.1. Photon Statistics

In contrast to linear absorption and emission, which are proportional to the time- and space-averaged number of photons (or average intensity of the light), the number of multiphoton events occurring in a volume V during time t depends on the spatiotemporal distribution $I(r, t)$ of the photon flux in the interaction volume:

$$\int_0^V \int_0^t I^N(\mathbf{r}, t) \, dr \, dt$$

The exact instantaneous flux distribution $I(r, t)$ is not known for any light source. It has become apparent in the last few years, that only low-order transverse mode laser beams or thermal light beams with known and reproducible spatial distributions are suitable for quantitative multiphoton experiments. For example, the flux distributions across a TEM_{00} mode laser beam is Gaussian; thus, the spatial part $D(r)$ of $I(r, t)$ can be assumed to be separable from the temporal part $g(t)$: $I(r, t) = D(r)g(t)$. In high-order multiphoton processes, large photon fluxes are required in order to induce a conveniently detectable number of transitions. For this purpose, the beam is usually focused into the vessel containing the atoms or onto the atomic beam, thereby resulting in a double-cone-shaped interaction volume along the laser beam axis (z direction). If the minimum beam cross section is located at $z = 0$ and its radius is w_0, the spatial flux distribution is given by[81]

$$D = \frac{I_0}{1 + \xi^2} \exp\left[-\frac{2(x^2 + y^2)}{w_0^2(1 + \xi^2)} \right] \tag{34}$$

where $\xi = \lambda z / \pi w_0^2$. The radius of the beam waist depends on the initial beam diameter and on the focal length of the lens and may be calculated in the case where the spherical focusing lens is free of aberrations. In most situations, however, aberration-free optics is not used and w_0 should therefore be determined experimentally.

The Nth order multiphoton transition probability per atom (at a given spatial coordinate r_0) is proportional to the Nth power of the instantaneous flux $D(r_0)g(t)$. The evaluation of the temporal part of the integral would therefore require an exact knowledge of $g(t)$. Because of the statistical nature of the photon flux only average intensities $D(r_0)\langle g(t)\rangle = \langle I(r_0, t)\rangle$ are accessible to experimental determination and the transition probability must therefore be written in the form

$$W_{fi}^{(N)} = G^N \sigma^{(N)} \langle I^N \rangle \tag{35}$$

where G^N is the Nth-order normalized correlation function of the laser intensity $I(t)$ (see, e.g., Debethune[82]). This dependence of the multiphoton transition probability on the temporal properties of the photon field is known as the "field correlation" or "photon statistics" effect and was first described by Lambropoulos[19] as well as Teich and Wolga[83] and later extended by Lambropoulos,[84,85] Chen,[86] Mollow,[87] Agrawal,[88] Sanchez,[89] and Armstrong *et al.*[90]

The temporal fluctuations of a multimode laser beam are quite obvious from oscilloscope displays of the pulse shapes. Even a single-mode cw laser beam exhibits statistical properties, however, and cannot be approximated by the pure number state $|\cdots n \cdots\rangle$, the initial state of the photon field used for the derivation of Eq. (23). All real light sources have a finite bandwidth and the number of photons in the initial state is not well-defined. The correlation function, G^N, has been calculated only for a limited number of model photon distributions. When the initial state is a pure coherent state in the Glauber sense (Glauber[91,92]), the photon statistics is given by a Poisson distribution and $G^N = 1$. Thermal or "chaotic" light sources have a Bose–Einstein photon distribution[93] for which $G^N = N!$

Experiments on multiphoton transitions are generally conducted with laser beams and their statistical properties are not well known. An important contribution to the understanding of photon statistics of pulsed multimode beams from solid-state lasers is the investigation by Lecompte et al.,[21] who compared G^N for a variety of multimode TEM_{00} pulses, both with random distributions of the mode phases and with correlated phase relations (mode-locked pulses). Their work shows how multiphoton ionization of order $N > 2$ can be utilized to investigate high-order correlation functions of the photon flux of laser beams. They find a $10^{6.9 \pm 3}$ times larger eleven-photon ionization yield in Ne with a multimode (≈ 100 modes) laser pulse as compared to a single-mode pulse of the same average photon flux, which is in agreement with the theoretical expectation of $11! = 4 \times 10^7$. As a result, a multimode pulse from a solid-state laser having random pulse relations between the modes must be considered for all practical purposes a "chaotic" light source. Surprisingly, mode locking, which leads to a train of well-separated light pulses and correlated phase relations between the modes, was found to even increase G^N over the value obtained for the same number of modes with random phase relations. Indeed, the "chaotic" light source in the Glauber sense does not have the maximum conceivable Nth-order correlation function. G^N may be larger than $N!$ when the light intensity fluctuates on a time scale comparable to or smaller than the lifetime of the nonstationary states formed during the multiphoton process. This can be the case when mode-locked picosec pulses are used (Simaan and Loudon[93]). These considerations, manifested also in a number of additional experimental investigations (Shiga and Iamura,[20] Carusotto et al.,[94] Krasinski et al.[95,96]), underline the fact that either the statistical properties of the light beam must be known or else methods that are independent of the photon statistics must be used for the absolute measurement of the multiphoton cross sections. One such method was described by Herman and Ducuing[97]; it is, however, applicable only to molecular systems.

3.2. Measurements of Multiphoton Cross Sections

In addition to the difficulties arising from the photon correlation effect, absolute measurements of $\sigma^{(N)}$ usually require well-calibrated equipment and the detailed knowledge of $D(r)$. Because of the nonlinearity of the multiphoton process, accurate experimental values for $\sigma^{(N)}$ are extremely difficult to obtain (especially for $N > 2$). This is probably one of the reasons for the very large discrepancies between measured and calculated generalized cross sections reported on numerous occasions in the literature (Chin et al.,[98] Held et al.,[99] Bakos et al.,[100] Fox et al.,[22] Delone et al.,[101] Granemann and Van der Wiel,[102] and Granemann et al.[103]).

We will present here only two types of experimental results: One concerns the work of Cervanen et al.[104] on three-photon ionization of Cs atoms, and the other involves spontaneous two-photon decay of simple metastable atoms or ions. Both examples have been selected in order to demonstrate that, on the one hand, the experimental and theoretical complexities of the problem prevented, up to this point, satisfactory agreement between measured and calculated generalized cross sections for multiphoton excitation process of order $N > 2$ and that, on the other hand, it is relatively easy to obtain exact two-photon decay rates of metastable hydrogenic atoms or metastable helium produced data that are not only in excellent agreement with theoretical calculations but are accurate enough to permit conclusions concerning basic quantum mechanical principles.

We will refrain from listing experimental and theoretical two-photon absorption cross sections here, despite the fact that their knowledge seems rather important when one considers the role of such transitions in high-resolution spectroscopy (see Section 7). In general, remarkable agreement between calculated and measured values is obtained. Further, the experimentalist finds that the truncated summation method can be used in most situations (Section 2.2) to arrive at an estimate that is sufficiently accurate to design a spectroscopy experiment.

3.2.1. Multiphoton Ionization Cross Sections

The experiment of Cervanen and co-workers[104] utilizes a method developed by Cervanen and Isenor[81] which permits the determination of $\sigma^{(N)}$ for multiphoton ionization processes by simply measuring normalized yield curves as a function of the peak flux, I_0, and the effective temporal width, τ, defined via $\langle I(t) \rangle \, dt = D(r)\tau I_0$, of single mode TEM$_{00}$ pulses for a Q-switched ruby laser. Their method is distinguished from others inasmuch as no absolute calibration of the ion detection efficiency or the density of the atomic beam is required, thus eliminating significant contributions to the overall experimental error. The results are compared with a number of different calculations in Table 1. They agree to within a factor of 2 with the most recently generated theoretical values by Flank and Rachman[105] and Teague and Lambropoulos.[106] Since the ratio $R = \sigma_c^{(3)}/\sigma_l^{(3)}$

Table 1. The Cross Section for Three-Photon Ionization of Cs Atoms for Linearly $(\sigma_l^{(3)})$ and Circularly $(\sigma_c^{(3)})$ Polarized Single-Mode Ruby Laser Light[a]

	$\sigma_l^{(3)}$	$\sigma_c^{(3)}$	R
Experiment			
Fox et al. (1971)[22]	—	—	2.15 ± 0.4
Cervanen et al. (1975)[104]	1.8 ± 1.1	4.1 ± 2.46	2.24 ± 0.11
Theory			
Bebb (1967)[55]	~1	—	—
Morton (1967)[56]	10.8	—	—
Manakov et al. (1973)[80]	95.7	136	1.42
Rapoport (1973)[112]	14.6	19.2	1.32
Teague and Lambropoulos (1975)[46,106]	1.17	2.59	2.22
Flank and Rachmann (1975)[105]	0.957	2.345	2.45

[a] $\lambda = 6943$ Å; $R = \sigma_c^{(3)}/\sigma_l^{(3)}$ (units 10^{-77} cm^6 s^2).

of the cross section for circularly and linearly polarized light can be measured with better accuracy than absolute cross sections, theoretical and experimental data are often in good agreement.

In general, experimental values of $\sigma^{(N)}$ for multiphoton excitation processes are larger than theoretical calculations. Owing to the difficulties in obtaining exact experimental data, it has been impossible up to now to test the accuracy of the various methods employed to calculate generalized cross sections.

3.2.2. Measurement of Two-Photon Decay Times

The situation is radically different when one examines available data on spontaneous two-photon decay. The enormous experimental problems encountered in Nth-order excitation measurements (e.g., characterization of the light source to determine G^N, absolute detector calibrations, etc.) are absent. Furthermore, experiments are possible with simple atomic systems for which low-order multiphoton transitions may either be calculated or approximated to a high degree of accuracy within nonrelativistic perturbation theory.

The $2^2S_{1/2}$ states of hydrogenic atoms and the 2^1S_0 states of helium and its isoelectronic sequence decay to the $1S$ ground state primarily by spontaneous emission of two simultaneous photons.[9] Spitzer and Greenstein[107] calculated both the two-photon decay time as well as the continuous emission spectrum of H (2S). More exact nonrelativistic calculations by Shapiro and Breit,[108] and later by Klarsfeld,[74] who obtained a solution in closed form, resulted in

$$\Gamma_{2S_{1/2}} = 8.2283 Z^6 \, \text{s}^{-1}$$

for the two-photon decay rate of hydrogenic atoms. Other closed-form solutions have been presented by Vetchinkin and Khristenko,[73] Zon and Rapoport,[109,110] and Johnson.[111]

The first laboratory observation of a two-photon emission process from a metastable hydrogenlike state was reported by Novick and co-workers[10,113,114] who measured the emission spectrum was well as the angular distribution of photons emitted in the spontaneous decay of the $2S$ state of He II. Later, Elton et al.[115] obtained the continuous two-photon spectrum of metastable Ne IX. The spontaneous two-photon decay of metastable hydrogen atoms was observed only recently by O'Connell et al.[116] First experimental values for $\Gamma_{2S_{1/2}}$ were reported by Schmieder and Marrus,[117] who produced metastable Ar XVIII by stripping electrons from argon ions. Their experiment is of interest because it presents evidence for contributions of relativistic effects to the total decay rate via magnetic dipole radiation which increases with Z^{10} (Drake et al.,[65] Metz[118]). At $Z = 18$ (argon), 4% of the measured decay rate stems from this mode of decay of the $2^2S_{1/2}$ state (Table 2).

Two-photon decay processes in atomic systems other than hydrogenic atoms are of fundamental importance to theoreticians, since the decay time and the spectral distribution of spontaneous decay or the transition probability for stimulated decay are very sensitive to the wave functions used in calculations. The dipole matrix elements in Eq. (24) are, of course, exactly known only for hydrogenlike atoms, and therefore the comparison of precise experimental results with calculated values for nonhydrogenlike systems provide important clues as to the accuracy of the employed

Table 2. Two-Photon Decay Rates Γ_{2s} (s^{-1})

Metastable system	Experiment		Theory	
	Value	Reference	Value	Reference
He$^+$ ($2^2S_{1/2}$)	525.4 ± 1.4	Hinds et al. (1970)[120]	526.01	Klarsfeld (1969)[74]
			526.66	Johnson (1972)[111]
Ar XVIII ($2^2S_{1/2}$)	(3.54 ± 0.25) $\times 10^{-9}$	Marrus and Schmieder (1970)[117]	3.57 $\times 10^{-9}$	Klarsfeld (1969)[74]
				Johnson (1972)[111]
He (2^1S_0)	50.8 ± 2.6	Van Dyck et al. (1971)[118]	45.5	Dalgarno (1966)[59]
			83.3	Dalgarno and Victor (1966)[67]
			50	Victor (1967)[64]
			51.3	Drake et al. (1969)[65]
			51.0	Jacobs (1971)[48]

theoretical method. The most thoroughly investigated system is the iso-electronic sequence of metastable 2^1S_0 helium. In Table 2, we also compare theoretical lifetimes of He (2^1S_0), which were obtained using a variety of different approximations, with an accurate experimental result measured by Van Dyck et al.[119] The first estimate of $\Gamma_{2^1S_0}$ of helium was provided by Dalgarno,[59] who used sum rules of oscillator strengths that had been obtained both experimentally and theoretically. Using sum rules and wave functions obtained from the time-dependent uncoupled Hartree–Fock approximation, Dalgarno and Victor[67] obtained a less accurate value, and later Victor[64] repeated this calculation with a coupled Hartree–Fock approximation. The most accurate calculations were performed by Drake et al.[65] and Jacobs.[48] As can be seen from Table 2, the agreement between theory and experiment is remarkable.

Very recent measurements by Hinds et al.[120] of the lifetime of the metastable $2^2S_{1/2}$ state of the He$^+$ are so accurate ($\Gamma_{2^2S_{1/2}} = 525.4 \pm 1.4 \text{ sec}^{-1}$) that they permit an upper limit of $|\delta| \leq 1.2 \times 10^{-5}$ to be assigned to the amplitude of any parity-forbidden $2P$ admixture to the two-photon decay rate. The first quantitative measurement of a spontaneous $H(2S)$ decay rate has now been reported by Krüger and Oed,[121] who find for the spectral decay probability $A(\lambda)\, d\lambda = 1.5 \text{ s}^{-1} \pm 43\%$ for $d\lambda = 2554$–2320 Å, which is in agreement with calculations by Spitzer and Greenstein.[107]

4. Survey of Multiphoton Emission Phenomena

The experimental investigation of stimulated two-photon emission processes was originally spurred not only by the desire to explore their experimental accessibility (Yatsiv et al,[2] Ducuing et al.,[122] Bräunlich and Lambropoulos,[3,5] Platz,[123] and Fornaca et al.[124]), but also because such processes held promise for the construction of laser oscillators and quantum amplifiers with new desirable qualities (Sorokin and Braslau,[125] Prokhorov,[126] Garwin,[127] Selivanenko,[128] Letokhov[129]). Since the gain in an inverted two-photon medium depends on both the inversion as well as the photon flux, the proposed lasers are highly nonlinear systems. A strong pulse propagating in such a material will increase in amplitude, decrease in width, and acquire a sharp leading edge.[128,130,131] It also appears possible to utilize stimulated two-photon emission for the construction of vacuum uv (Bräunlich et al.[5]) as well as tunable lasers (Kirsanov and Selivanenko[132]).

With the development of the H_2 laser and efficient new techniques for generating vacuum uv coherent light by thrid and fifth harmonic generation in atomic gases (Harris,[133] Harris et al.,[134] Kung et al.[135,136]), the interest in two-photon lasers had temporarily diminished. Only recently has the possibility of light amplification via two-photon energy-extraction schemes again been discussed. This time, the impetus was given by the newly discovered advantages two-photon amplifiers have as light sources for laser-triggered thermonuclear fusion (Carman[137]). The attractive features of two-photon lasers include the possibility of tailoring the output pulse in time and space and, because amplification can take place at a frequency different from that of the input driving laser, achieving effective optical isolation of the fusion target from the laser. In addition, so-called "frequency tailoring" (Kirsanov and Selivanenko[132]), that is, frequency change with time evolution of the laser pulse, appears feasible with these proposed lasers. In principle, this will permit the initiation of target heating in the infrared and will aid, later in the pulse, hydrodynamic shock and adiabatic decompression, thus eliminating several implosion symmetry problems.[137]

4.1. Stimulated Two-Photon Emission and Raman Scattering

Two somewhat different modes of stimulated decay are possible from an upper (initial) level to the ground state: (a) two-photon emission (TPE) and (b) anti-Stokes Raman Scattering (ASRS). In the first process, the presence of photons ω_1, or ω_1 and ω_2, triggers the emission of photons of the same frequency and phase (Figures 3a and 3b). In the second, the presence of ω_1 results in the emission of photons ω_2 such that $\omega_{fi} + \omega_1 = \omega_2$ (Figures 3d and 3c). Both processes are of the same order of nonlinearity within the time-dependent, nonrelativistic perturbation expansion. To date, singly stimulated or enhanced two-photon emission (only ω_1 originally present, e.g.,in the form of a triggering laser beam, Figure 3a) has been observed[2,3,5,124] as well as spontaneous[4,5,127] and stimulated[138,139] anti-Stokes Raman scattering. The probability $W_{fi}^{(2)}$ for TPE from an upper (usually metastable) level $|i\rangle$ to the ground level $|f\rangle$ is proportional to $[n_1(\omega_1; \mathbf{k}_1) + 1][n_2(\omega_2; \mathbf{k}_2) + 1]$, where n_1 and n_2 are the number densities of photons in the photon field. Similarly, ASRS is proportional to $n_1(\omega_1; \mathbf{k}_1)[n_2(\omega_2; \mathbf{k}_2) + 1]$. By using a derivation analogous to that used in obtaining Eqs. (23), (27), and (30), one obtains

$$W_{fi}^{(2)} = \frac{2\pi\xi^4 m^4 \omega_1 \omega_2}{c^2} \frac{I_1 I_2(\omega_2)}{\Delta\omega} |T^{(2)}|^2 \qquad (36)$$

with

$$|T^{(2)}|^2 = \left| \sum_c \frac{\langle f|\mathbf{r}\cdot\boldsymbol{\varepsilon}_1|c\rangle\langle c|\mathbf{r}\cdot\boldsymbol{\varepsilon}_2|i\rangle}{\omega_{ci} \pm \omega_1} + \frac{\langle f|\mathbf{r}\cdot\boldsymbol{\varepsilon}_2|c\rangle\langle c|\mathbf{r}\cdot\boldsymbol{\varepsilon}_1|i\rangle}{\omega_{ci} - \omega_2} \right|^2$$

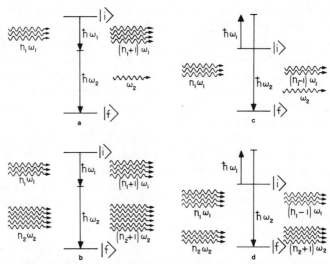

Figure 3. Schematic representation of (a) spontaneous ($n_1 \equiv 0$) and singly stimulated two-photon emission ($n_1 \neq 0$); (b) doubly stimulated two-photon emission; (c) anti-Stokes Raman scattering; and (d) stimulated anti-Stokes Raman scattering.

The minus and plus signs in the denominator of the first term designates ASRS and TPE, respectively.

The change of the population densities N_i and N_f of the two levels may be expressed as

$$-\frac{dN_i}{dt} = E(\omega_1, \omega_2)[(n_1 + \gamma)(n_2 + 1)N_i - (n_1 + 1)(n_2 + \gamma)N_f] \quad (37)$$

where $\gamma = 1$ for TPE and $\gamma = 0$ for ASRS. $E(\omega_1, \omega_2)$ is readily obtained from Eq. (36). The first term in Eq. (37) represents decay of the density N_i due to emission processes and the second arises from two-photon absorption or Stokes Raman scattering ($\omega_1 > \omega_2$) leading to an increase of N_i. Other competing processes (here neglected) are elastic photon scattering, the usual cavity losses, and linear absorption at the two frequencies. This latter loss mechanism is very small in atomic systems unless one of the involved photon energies happens to be in resonance with a dipole-permitted transition of the atom. Equation (37) is instructive since it permits a simple explanation of the various terms used in this and the previous sections.

We can distinguish between the following second-order emission processes:

1. $n_1 = n_2 = 0$:   spontaneous two-photon decay (has no equivalent in Raman scattering)

2. $n_1 \gg 1$; $n_2 = 0$:   singly stimulated or enhanced two-photon
(or $n_1 = 0$; $n_2 \gg 1$)   decay; spontaneous anti-Stokes Raman scattering

3. $n_1 \gg 1$; $n_2 \gg 1$:   doubly stimulated (sometimes simply referred to as "stimulated") two-photon emission; stimulated anti-Stokes Raman scattering

In many atomic systems (e.g., H, He$^+$, etc.), ASRS is the dominant process[18,137] However, Carman[137] has indicated that, under certain conditions, two-photon amplification may dominate, and experiments are presently in progress to verify this.

The gain for these processes is proportional to $n_1 n_2 E(\omega_1, \omega_2)(N_i - N_f)$. This term is usually quite small and this appears to be the reason why doubly stimulated two-photon emission has not been observed as yet. The first observation of stimulated ASRS in an atomic system was reported by Sorokin et al.[138] and later by Carman and Lowdermilk.[139] The small gain makes it mandatory to use high-intensity laser beams to initiate the stimulation process, a requirement which is sometimes called "hard triggering." Various schemes for "soft triggering" of stimulated two-photon emission were recently discussed by Krochik and Khronopulo.[140] In view of the above-mentioned special properties of two-photon amplifiers and their anticipated usefulness as "tailored" light sources for laser-triggered thermonuclear fusion, we expect considerable experimental progress toward their development in the not too distant future.

We will not discuss here the extensive literature on Stokes Raman scattering but rather refer to the review articles by Bloembergen[141] and Shen.[142]

4.2. Optical Mixing

Optical mixing in atomic vapors does not require population-inverted levels. It has been developed to a powerful technique for efficient generation of coherent uv and vacuum uv radiation by third and fifth harmonic frequency conversion (Figure 4). Presently, atomic systems appear to be the only media to generate far-uv or even soft x-ray radiation by multiphoton processes.[15,133,134,136,143] Tunable uv generators,[144] infrared

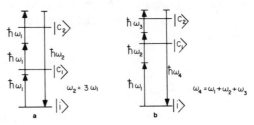

Figure 4. Schematic representation of (a) third harmonic generation and (b) sum frequency generation.

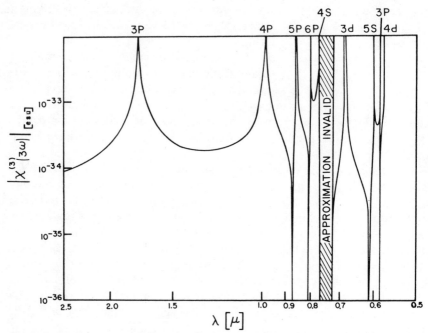

Figure 5. Nonlinear susceptibility of sodium versus incident wavelength, calculated by Miles and Harris, 1973.[143]

sources,[145,146] and infrared up-converters have been constructed as well.[17,147]

The efficiency of $(N-1)$th harmonic generation is proportional to $\chi^{(N-1)}$, the optical susceptibility, which is related to the transition amplitude $|T^{(N)}|$, defined in Eq. (30), via

$$\chi^{(N-1)} = \frac{e^N}{\hbar^{N-1}} T^{(N)} \tag{38}$$

The concept of $\chi^{(N-1)}$ stems from the classical theory of nonlinear processes which treats them in terms of a radiation-induced polarization.[148]

In the language of multiphoton transitions, $(N-1)$ harmonic generation corresponds to an N-photon process, e.g., in third harmonic generation three photons of frequency ω are absorbed and one of frequency 3ω is emitted:

$$\chi^{(3)} = \frac{e^4}{\hbar^3} \sum_{i,a,b,c} \langle i|\mathbf{r}|a\rangle\langle a|\mathbf{r}|b\rangle\langle b|\mathbf{r}|c\rangle\langle c|\mathbf{r}|i\rangle A_{abc}$$

$$A_{abc} = [(\omega_{ai}-3\omega)(\omega_{bi}-2\omega)(\omega_{ci}-\omega)]^{-1} + [(\omega_{ai}+\omega)(\omega_{bi}+2\omega)(\omega_{ci}+3\omega)]^{-1}$$

$$+ [(\omega_{ai}+\omega)(\omega_{bi}+2\omega)(\omega_{ci}-\omega)]^{-1} + [(\omega_{ai}+\omega)(\omega_{bi}-2\omega)(\omega_{ci}-\omega)]^{-1} \tag{39}$$

where $|i\rangle$ is the ground (initial as well as final) state, ω is the frequency of the laser photons, and the subscripts a, b, c denote the intermediate states involved. Again, the sum is over all intermediate states including the continuum. In Figure (5), we show results of the calculation by Miles and Harris[143] of $\chi^{(3)}$ for sodium as a function of ω. Strong resonances exist when ω, 2ω, or 3ω coincide with resonance transitions of the atom. Naturally, for practical exploitation of this effect as a frequency-tripled coherent light source, ω is selected for a near-resonant condition to occur. In this way, Young et al.[15] and Bloom et al.[149] have achieved 10%–20% conversion efficiency in a mixture of Rb vapor and Xe. High conversion efficiencies have also been obtained in third-harmonic generation of vacuum uv radiation in Cd : Ar and Xe : Ar mixtures.[135,136] The addition of inert buffer gases is required to adjust the refractive index of the mixture to $n_r(\omega) = n_r(3\omega)$ so that both the first and third harmonics travel at the same velocity through the entire length of the gas cell (phase matching). The efficiency of conversion is limited by single- and multiphoton absorption as well as ionization, the Kerr effect, optical breakdown (multiphoton assisted avalanche breakdown due to inverse Bremsstahlung[150]), thermal defocusing and breakdown of the phase-matching condition as a result of atomic saturation.[143] Gases have several advantages over solid harmonic generating media. Breakdown thresholds are usually higher, the break-

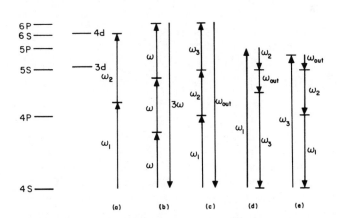

Figure 6. Energy level diagram of potassium and schematic of some nonlinear optical near-resonance processes. (a) two-photon absorption; (b) third-harmonic generation; (c) sum-frequency generation with ω_1, $\omega_1 + \omega_2$, and $\omega_1 + \omega_2 + \omega_3$ close to resonance levels; (d) infrared generation with ω_1 and $\omega_1 - \omega_2$ close to resonance levels; (e) infrared generation with ω_1 and $\omega_1 + \omega_2$ close to resonance levels (adapted from Shen, 1976[142]).

down is nondestructive, and large cells may be constructed to handle large incident optical energies. The maximum laser powers in gaseous alkali systems are limited to $10^{10} - 10^{12}$ W/cm^2. $\chi^{(3)}$ may be well above the values obtained with a single frequency of the incident laser beam by utilizing two or even three different incident light sources (e.g., dye laser) so that multiple resonances occur.[144] A number of other variations of optical mixing processes in atomic vapors have been reviewed by Shen[142] (Figure 6).

5. Resonance Effects

So far we have only occasionally touched upon the possibility that one or more of the involved photons in a multiphoton transition are in resonance with dipole-permitted transitions or that one or more of the resonance denominators in Eq. (30) approach zero because the energy of one or several of the intermediate levels is equal to one or the sum of several photons involved in the process. Some resonance or near-resonance situations are shown in Figure 6, e.g., the two-photon absorption process reduces to a resonance or two-step cascade transition when $\omega_{4P,4S} = \omega_1$ and $\omega_{5S,4P} = \omega_2$.

In the Nth-order multiphoton transition it may even be possible to have an Mth-order resonance $(M < N)$. An example, third harmonic generation with three simultaneous resonances, was discussed by Shen.[142] Choosing the frequencies of up to three different dye lasers such that ω_1, $\omega_1 + \omega_2$, and $\omega_1 + \omega_2 + \omega_3$ are simultaneously close to resonances with intermediate levels of an atomic system (Figure 6c) the third-order susceptibility is dominated by the resonance terms in the infinite sum and

$$\chi^{(3)} = \frac{e^4}{\hbar^3} \sum_{S,P} A_{S,P} \langle 4S|\mathbf{r}|4P\rangle\langle 4P|\mathbf{r}|5S\rangle\langle 5S|\mathbf{r}|5P\rangle\langle 5P|\mathbf{r}|4S\rangle$$

$$A_{S,P} = [(\omega_{4P,4S} - \omega_1 - S_{4P} + i\Gamma_{4P})(\omega_{5S,4P} - \omega_1 - \omega_2 - S_{5S} + i\Gamma_{5S}) \tag{40}$$

$$\times (\omega_{5P,4P} - \omega_1 - \omega_2 - \omega_3 - S_{5P} + i\Gamma_{5P})]^{-1}$$

The summation extends over S and P states only; because of the dominant resonance terms contributions of other states and the continuum may be neglected. A phenomenological shift-width function, $-S_c(\omega_c) + i\Gamma_c(\omega_c)$, has been introduced to remove the singularities at resonance. In weak field processes, the shift can be neglected and the width of the level is due to its spontaneous decay:

$$\Gamma_c(\omega_c) = \frac{r_0}{3_c} \sum_{b<c} \omega_{bc}^2 f_{bc} \tag{41}$$

where f_{bc} is the oscillator strength of the dipole transition from the excited state $|c\rangle$ to the lower state $|b\rangle$ and r_0 is the classic electron radius. The summation is taken over all states to which $|c\rangle$ decays spontaneously. In strong fields, the lifetimes of atomic states are influenced by induced processes and one has intensity-dependent widths are well as shifts (Goldberger and Watson,[34] Cohen-Tannoudji,[151] Cohen-Tannoudji and Haroche,[152] Cohen-Tannoudji and Dupont-Roc,[153] Lambropoulos[36]). We will not review the numerous theoretical treatments of photon-induced width-shift functions, but rather refer to the articles by Stenholm[154] and by Behmenburg[155] in this work and to the review articles by Lambropoulos[31] and Bakos[13] in which their role in high photon flux multiphoton ionization is described in detail.

Resonances effects in multiphoton processes manifest themselves in a variety of experimentally observable phenomena. There is, of course, the often dramatic variation of the transition probability as the photon frequency is scanned through a resonance. Such resonance effects are of considerable practical importance not only in increasing the efficiency of optical mixing but also in other selective transition processes, e.g., selective

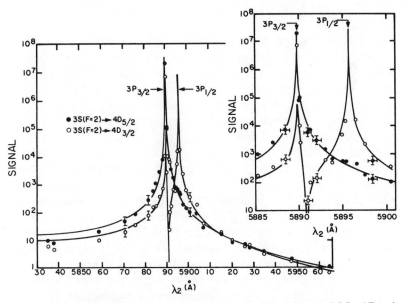

Figure 7. Normalized two-photon transition rates for $3S \to 4D_{5/2}$ and $3S \to 4D_{3/2}$ in Na versus the wavelengths of the fixed-frequency laser, λ_2. (Note that $\nu_1 = \nu_2$ for $\lambda_2 = 5787$ Å.) The points are experimental and the curves are calculated with the truncated sum technique. The insert gives the detailed behavior around the resonances (Bjorkholm and Liao, 1974[44]).

two-photon excitation for isotope separation. As an example, we mention a recent investigation of two-photon absorption in sodium vapor by Bjorkholm and Liao.[44] They observed a resonance enhancement of over seven orders of magnitude in the two-photon transition probability $3S \rightarrow 4D_{5/2}$ and $3S \rightarrow 4D_{3/2}$ by scanning the frequencies ω_1 and ω_2 so that ω_2 moves through resonance with the $3P_{3/2}$, $3P_{1/2}$ states. Also of interest is the observed destructive interference when ω_c is between the $3P$ levels. The measured line shapes agree well with calculations based on the truncated summation method. No photon-induced shift or broadening is observed at the low laser powers used for this experiment (Figure 7).

Historically, the first line shape calculations were performed by Zernik[18] and Bebb and Gold.[50-55] A large number of such calculations is now available (e.g., Gontier and Trahin,[72] Zon et al.,[78] Miles and Harris,[143] Chang and Stehle,[156] Teague and Lambropoulos.[46,47]

The presence of resonances may also lead to an intensity- and frequency-dependent deviation from the I^N dependence of the transition probability[13,31,100,101,156-162] as well as influencing the statistical effect discussed in Section 3.[90]

6. Polarization Effects

Single- and multiphoton transitions are affected by the polarization of the electromagnetic radiation. However, while only differential cross sections (e.g., the angular dependence of emitted photoelectrons) and not the total transition probability are polarization dependent in single-photon processes, there is a distinct polarization effect in multiphoton transitions which is conveniently measured as the ratio of the cross section for circularly and linearly polarized light $R = \sigma_c^{(N)}/\sigma_l^{(N)}$. This can be understood by considering the succession of dipole matrix elements in Eq. (30) and by applying successively known selection rules for the involved transitions to intermediate states.

Let $\langle c_N | \mathbf{r} \cdot \boldsymbol{\varepsilon}_{k\lambda} | c_{N-1} \rangle$ be one of the dipole matrix elements in the products under the sum in Eq. (30). Choosing the coordinate system along the z direction of the unit vector $\boldsymbol{\varepsilon}_{k\lambda} = (0, 0, 1)$ one obtains for linearly polarized light $\mathbf{r} \cdot \boldsymbol{\varepsilon}_{k,\lambda} = r(\frac{4}{3}\pi)^{1/2} Y_{10}(\theta, \phi)$, where $Y_{10}(\theta, \phi)$ is the angular part (spherical harmonics) of the wave function.[163] For bound–bound transitions, the angular part of the dipole matrix elements becomes proportional to $\langle Y_{l_N m_N} | Y_{10} | Y_{l_{N-1} m_{N-1}} \rangle$ which does not vanish for $\Delta l \pm 1$ and 0 (with $0 \rightarrow 0$ forbidden) and $\Delta m = 0$. As a result, in each step of the multiphoton transition, as described in the dipole approximation by a product of matrix elements, a unit of angular momentum is transferred between the photon field and the electron.

Figure 8. Angular momentum channels available for four-photon ionization of an S state with linearly ($\rightarrow$) and circularly ($--\rightarrow$) polarized light (Lambropoulos, 1977[31]).

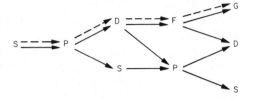

The selection rules are different for circularly polarized light. Choosing the wave vector of the photons in the z direction we have $\varepsilon_{k,\lambda} = \frac{1}{2}(e_x \pm ie_y)$ and $\mathbf{r} \cdot \varepsilon_{\bar{k}\lambda} = \frac{1}{2}(x \pm iy)$. [The positive sign refers to right-hand circularly polarized light (RCPL) and the negative to left-hand circularly polarized light (LHCPL).] Expressing it again in terms of spherical harmonics, we have $\mathbf{r} \cdot \varepsilon_{\bar{k}\lambda} = -r(\frac{4}{3}\pi)^{1/2} Y_{1 \pm 1}(\theta, \phi)$ and the angular part of the matrix elements becomes proportional to $\langle Y_{l_N m_N} | Y_{1 \pm 1} | Y_{l_{N-1} m_{N-1}} \rangle$, which disappears unless $\Delta l \pm 1$ and $\Delta m = \pm 1$. (Again, the signs apply to right- and left-hand polarized light, respectively.) It is instructive to display the transitions permitted by these selection rules as possible "channels" through which a particular multiphoton transition can take place.[164] For example, transitions from the S state of hydrogenic atoms are shown in Figure 8. Clearly, two-photon or four-photon absorption to the higher S levels is not possible with either left- or right-hand circularly polarized light. However, absorption of one left-hand polarized and one right-hand polarized photon would result in $\Delta m = 0$, and, therefore, the transition to the first excited S level is permitted in this case.[165]

Bound–free transitions are possible with any polarization. There are no selection rules, and any difference in the multiphoton ionization probability must therefore originate in differences of available channels for the intermediate bound–bound transitions and the magnitude of the involved matrix elements.

Light polarization effects in multiphoton ionization manifest themselves also in the angular distribution as well as in the spin polarization of the emitted electrons; they are one of the reasons for the enormous upsurge in theoretical and experimental work in this field. The new experimental possibilities are of interest not only as a test for the theory but also for the practical implications of these effects for the production of spin-polarized electrons and the understanding of multiphoton assisted breakdown of gases as well as solids.

The influence of light polarization on the total transition rate was first discovered by Fox et al.,[22] who measured $\sigma_c^{(N)}/\sigma_l^{(N)}$ for two- and three-photon ionization of Cs.[166] Thereafter, the effect has been studied in a number of other multiphoton ionization experiments (Cervanen and Isenor,[81] Cervanen et al.,[104] Agostini and Bensoussan,[167] Agostini et *al.*,[168] and Delone,[169] see Table 1). The first theoretical interpretations were given by Lambropoulos[23] and Klarsfeld and Maquet.[24] The latter authors give an upper bound for $\sigma_c^{(N)}/\sigma_l^{(N)}$, however, Lambropoulos[31] points out that for each case of N-photon ionization $\sigma_c^{(N)}/\sigma_l^{(N)}$ has to be calculated individually as it depends on the frequency as well as on resonances with intermediate states; no least upper bound can therefore be found for this ratio since, as shown above, it can even be zero for certain bound–bound conditions and can certainly be smaller than unity for bound–free transitions. A number of additional calculations are now available (Gontier and Trahin,[170] Rapoport,[112] Mizumo,[69] Jacobs,[171] Lambropoulos and Teague,[46] and Teague and Lambropoulos[47]).

Of considerable theoretical interest is the investigation of the angular distribution of the photoelectrons generated by N-photon ionization because they offer potentially attractive possibilities for the study of excited states of atoms and molecules.[172] The first calculations of angular distributions in two-photon ionization were carried out by Zernik[62] and later by Cooper and Zare[173] as well as Tully *et al.*[174] Recently, initial

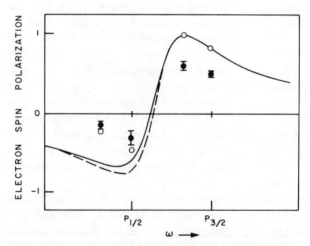

Figure 9. Experimentally determined spin polarization (solid circles) in three-photon ionization of sodium. The frequency of the first laser was tuned around the $3S \rightarrow$ $3P_{3/2,1/2}$ transition while the second was adjusted so that $\omega_1 + \omega_2 = \omega_{4D} - \omega_S$ (M. Lambropoulos *et al.*, 1973[180]). The solid and dashed lines are calculations by P. Lambropoulos,[172] obtained with $\langle n^1 P_{3/2}|r|nS_{1/2}\rangle/$ $\langle n^1 P_{1/2}|r|nS_{1/2}\rangle = 1$. (————) two-photon ionization; (– – –) three-photon ionization. The open circles are a plot of the experimental data normalized to the maximum of the theoretical curves.

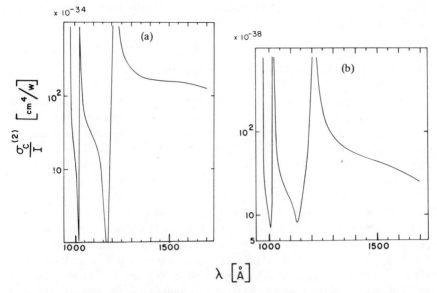

$$\lambda \ \left[\overset{\circ}{A}\right]$$

Figure 10. (a) Dipole–dipole contributions to the two-photon ionization cross section of hydrogen atoms in the $1S$ state versus the wavelength of the incident photons. (b) Dipole–quadrupole plus quadrupole–dipole plus quadrupole–quadrupole contributions to the two-photon process in Fig. 10a (Laplanche et al., 1975[162]).

experiments were performed by Lambropoulos and Berry[175] and Edelstein et al.[176]; a number of additional calculations of angular distributions were also published.[23,171,69,177,178] At this point, however, it is difficult to judge whether the considerable experimental difficulties associated with the measurement of angular distributions can be overcome so that the full potential of this new experimental technique can be utilized for the exploration of the properties of excited states.

As mentioned before, another reason for the recent interest in multiphoton ionization is its potential application to the generation of spin-polarized electrons (Figure 9). Calculations of spin polarization of multiphoton-generated photoelectrons have been performed by a number of authors.[37,46,106,164,171,179] The first experiments were conducted by Lambropoulos et al.[180] (Figure 9) and are in fair agreement with theory.

Under certain conditions, the dipole approximation is insufficient for the calculation of the probabilities for multiphoton transitions. As has been shown by Lambropoulos et al.,[181] at photon frequencies for which the values of the dipole matrix elements and the frequency denominators result in extremely small transition probabilities (e.g., the valleys between resonance peaks generated by destructive interference, see for example Figure 10), contributions from quadrupole transitions can dominate.[162]

This has recently been demonstrated experimentally by Lambropoulos *et al.*,[182] who studied the three-photon ionization of Na.

7. High-Resolution Multiphoton Spectroscopy

In the previous section we have emphasized methods of calculation and measurement of multiphoton transition probabilities as a function of light intensity, polarization, statistical properties of the light source, and, to some extent, frequency. We followed, more or less, the historical development of the experimental aspects of the field, which were dictated by the initial availability of only fixed frequency lasers or lasers with very limited tunability. It is, therefore, not surprising that until very recently the majority of investigations dealt with either bound–free transitions (multiphoton ionization) or with those bound–bound transitions, which do not require one or several laser photons to be in resonance with narrow excited levels (e.g., third- and fifth-harmonic generation, singly stimulated two-photon emission, and anti-Stokes Raman scattering). High-precision multiphoton spectroscopy in the optical region of the electromagnetic spectrum became feasible only with the development of narrow-band tunable dye lasers. An early application of two-photon absorption techniques was reported by Burrell and Kunze.[183] These authors investigated oscillating electric fields in a helium plasma by measuring two-photon absorption and stimulated Raman scattering using simultaneously a beam from a tunable dye laser and microwave radiation.

7.1. Multiphoton Ionization Spectroscopy

The potential of multiphoton techniques for the study of highly excited levels was convincingly demonstrated by Popescu *et al.*,[184] Esherick *et al.*,[185] and by Bensoussan.[186] Popescu and co-workers excited the $n^2D_{5/2,3/2}$ and $n^2S_{1/2}$ levels ($n = 9$–14) of cesium vapor in a one-laser experiment by two-photon absorption and detected their energies by measuring the yield of ions obtained by subsequent one-photon ionization of the excited atom. The process may be looked upon as resonance three-photon ionization. The frequency of the dye laser is swept through the two-photon resonances of the n^2D and n^2S levels.

Esherick and co-workers[185] employed a similar technique for their observation of 72 previously unknown even-parity $J = 0$ and $J = 2$ states of calcium. The beam of a nitrogen-laser-pumped dye laser, tunable over a range from 4025–4325 Å, was focused into a heated pipe containing Ca vapor and Ne or Kr as buffer gas to prevent diffusion of Ca to the cold ends of the container. Laser power densities of up to 10^8 W/cm^2 were used

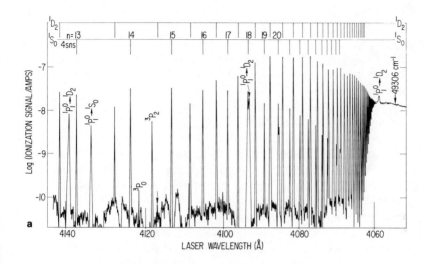

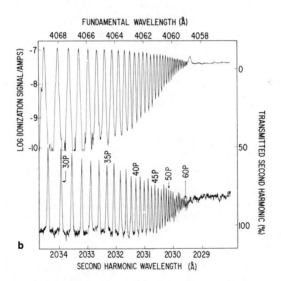

Figure 11. (a) Three-photon ionization spectra of calcium. The structure arises from two-photon absorption resonances at the $n\,^1S_0$ and $n\,^1D_2$ states, which converge to the ionization that limit at 49306 cm^{-1}. A tunable dye laser (linearly polarized, 10^7 W/cm^2) was focused into a cell containing 0.1 Torr Ca and 10 Torr Kr as a buffer gas. Some two-photon resonances at $3P_0$ and $3P_2$ and single-photon resonances originating at $4s4p\,^1P_1^0$ states are indicated. (b) Simultaneous recording of the high-resolution three-photon ionization signal (top) and the two-photon ionization signal obtained with the frequency-doubled laser beam (bottom) (Esherick *et al.*, 1976[185]).

without observable ac Stark shifts. The newly identified $4sns\,^1S_0$ states
($n = 13$–30) and $4sns\,^1D_2$ states were measured to within $\pm 0.1\ \mathrm{cm}^{-1}$.
Single-photon absorption to the known $4snp\,^1P_1^0$ levels of the frequency-
doubled fundamental laser photons provided a series of calibration points.
The obtained ionization yields are shown in Figure 11a. In Figure 11b, the
absorption edge of the $^1P_1^0$, $^1S_0$, and $^1D_2$ states is given in greater detail. All
three states converge to the same limit. Some two-photon transitions to $^3P_0$
and $^3P_2$ states are also observed.

These recent experiments are expected to be only the start of what
may be described as multiphoton ionization spectroscopy of high-lying
states in atomic vapor. The potential of this technique has now been
recognized and this author anticipates its wide application in the not too
distant future.

The experimental scheme employed by Bensoussan[186] is somewhat
different. In a two-photon absorption process, he populated the $^2D_{5/2,3/2}$
levels of sodium vapor and, using the beam of a second dye laser, probed
the single-photon absorption resonances $3^2D_{5/2,3/2} \to n^2F_{5/2,7/2}$ for $n = 10$–
14. In this way, the first fairly accurate determination of the energy of these
highly excited levels was achieved with a resolution limited only by the
Doppler effect.

7.2. Doppler-Free Spectroscopy

The linewidths of tunable dye lasers can be reduced to a few MHz,
which is much smaller than the homogeneous linewidth of many atomic
levels. To fully utilize this remarkable resolution, inhomogeneous line
broadening, mainly due to the Doppler effect, has to be circumvented. In
single-photon spectroscopy this is successfully done by using saturation
spectroscopy, also known as Lamb-dip spectroscopy,[187] or by using the
method of Schlossberg and Javan,[188] which is called "double-resonance
saturation spectroscopy." These and several other methods are described
in Chapter 15 of this volume.[189] In the present chapter, we will discuss in
detail only those methods that involve multiphoton transitions in atoms.

Pritchard et al.[28] succeeded in reducing the observed linewidth of a
two-photon transition to 120 MHz by tightly focusing a laser beam onto a
supersonic beam of sodium atoms. The dominant contribution to the
linewidth was due to the small focal volume of the laser beam, which
limited the interaction time of the atoms with the field to about $2 \times$
10^{-9} sec. With this linewidth the hyperfine structure of the $3^2S_{1/2}$ ground
level could be resolved and the fine structure of the $4d\,^2D$ state was
measured to be $\Delta_{\mathrm{fs}} = 1025 \pm 6$ MHz. Salomaa and Stenholm[190] have
recently suggested an extension of the fast-beam method that permits the

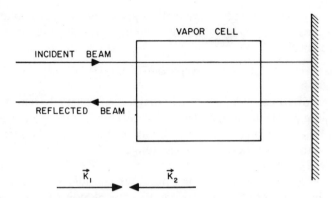

Figure 12. Schematic of counterpropagating beam arrangement for Doppler-free two-photon spectroscopy.

elimination of Doppler brodening for even larger interaction volumes (e.g., nearly collinear atomic and laser beams).

Precision multiphoton spectroscopy received renewed attention with the discovery of Doppler-free two-photon absorption spectroscopy and several related techniques, such as N-photon absorption and Raman scattering, in which Doppler broadening can be at least partially eliminated.

Since their conception by Vasilenko et al.[29] and the discussions of possible experimental demonstration by Cagnac et al.[191] and by Roberts and Fortson,[192] multiphoton Doppler-free techniques have been developed to an extremely useful spectroscopic tool that is particularly suited for the investigation of the structure of highly excited atomic and molecular states as well as metastable levels. The wealth of experimental results obtained in the last three years is impressive in view of the fact that the field must still be considered to be in its infancy.

The principle of Doppler-free two-photon spectroscopy is based on the simultaneous absorption of two counter-propagating photons by the atom. Consider two oppositely traveling waves of frequency ω such that $2\omega = \omega_{fi}$ for a two-photon permitted transition from the initial state $|i\rangle$ to the final or excited state $|f\rangle$ (Figure 12). In a gas the atom is not at rest and it possesses a velocity component v in the propagation direction of the photons. Therefore, in the rest frame of the atom, the frequencies of the two photons will be Doppler-shifted and the condition for the absorption can be expressed as

$$|\omega_{fi} - \omega(1 - v/c) - \omega(1 + v/c)| \leq 2\Gamma_f \tag{42}$$

where $2\Gamma_f$ is the homogeneous width (FWHM) of the resonance transition. Since Eq. (42) is independent of v, the width of the transition is given by the homogeneous linewidth centered around $\omega = \omega_{fi}/2$.

In a two-laser experiment with oppositely traveling photons of different frequencies, ω_1 and ω_2, the condition for the two-photon transition becomes

$$|\omega_{fi} - (\omega_1 + \omega_2) + |\omega_2 - \omega_1| v/c| \le 2\Gamma_f \qquad (43)$$

This case, for which Doppler broadening is only partially removed, is of considerable practical interest because it permits (via the proper selection of ω_1 and ω_2) the choice of optimal conditions for a particular experiment, e.g., a compromise between the strength of the transition and the frequency resolution.

For the calculation of the linewidth, we consider the transition probability given by Eq. (23). We assume that the photon frequency is not in resonance with one of the intermediate states [the line-shift function in the resonance denominator of Eq. (24) can be neglected] and that at the same time $\omega_{ci} - \omega(v)$ is large compared to the homogeneous width and Doppler width of the two-photon transition. Under these conditions, the velocity dependence of $W_{fi}^{(2)}$ occurs only in the linewidth function,

$$g(\omega_{fi}) = \frac{\Gamma_f}{\pi} \frac{1}{(\delta\omega)^2 + \Gamma_f^2}$$

where $\delta\omega$ is the difference between ω_{fi} and the sum of the two-photon frequencies (see below). One has to distinguish between two cases:

I.   Two-photon absorption of counter-propagating photons and of photons propagating in the same direction is possible. The photon frequencies must then be the same. All photons must have the same polarization.

II.   Two-photon absorption occurs only from counter-propagating beams. The photons must then have either different frequencies or, e.g., in the case the ground level as well as the excited level have the same magnetic quantum number, different polarization (σ^+ and σ^-) and the same frequency.

The above requirements on polarization stem from the selection rules.

We first discuss case I, for which $\delta\omega$ assumes the values $\omega_{fi} - 2\omega$ (counterpropagating photons) and $\omega_{fi} - 2\omega(1 \pm v/c)$, (photons propagating in the same direction) and recall the derivation of Eq. (24). The initial state of the photon field contains two modes, one for each counterpropagating beam: $|\cdots n_1 \cdots n_2 \cdots\rangle$. The intermediate states of the photon field for two-photon absorption of two counterpropagating photons are $|\cdots n_1 - 1 \cdots n_2 \cdots\rangle$ and $|\cdots n_1 \cdots n_2 - 1 \cdots\rangle$. For absorption of two photons having identical propagation directions only one of these intermediate states is possible. The total transition probability then becomes[(29,191)]

$$W_{fi}^{(2)} = F(\omega)I^2 \left| \sum_c \langle f | \mathbf{r} \cdot \mathbf{\varepsilon} | c \rangle \langle c | \mathbf{r} \cdot \mathbf{\varepsilon} | i \rangle \right|^2 \left\{ \frac{4\Gamma_f}{(\omega_{fi} - 2\omega)^2 + \Gamma_f^2} \right.$$

$$\left. + \frac{\Gamma_f}{[(\omega_{fi} - 2\omega(1 + v/c)]^2 + \Gamma_f^2} + \frac{\Gamma_f}{[(\omega_{fi} - 2\omega(1 - v/c)]^2 + \Gamma_f^2} \right\} \quad (44)$$

$F(\omega)$ is proportional to ω^2 [see Eq. (30)]. Since it is a relatively slowly varying function of ω, it may be considered a constant over the limited tuning range around two-photon resonances. The first term in the curly brackets of Eq. (44) is a Lorentzian and represents the Doppler-free contribution to the total transition rate. The second and third terms are the Doppler-broadened contributions. In order to calculate the line shape, one has to average over the velocities using the thermal velocity distribution function. In thermal equilibrium, the density of atoms having a velocity v is given by

$$n(v) = \left(\frac{m}{2\pi kT} \right)^{1/2} \exp\left(-\frac{mv^2}{2kT} \right) \quad (45)$$

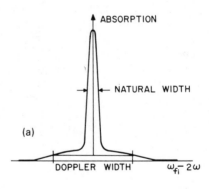

(a)

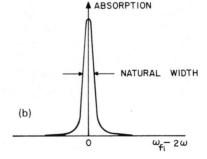

(b)

Figure 13. Absorption line profiles obtained with (a) counterpropagating beams of equal frequency and polarization; (b) counterpropagating beams of different helicity. The line shape is a pure Lorentzian in case (b) and Lorentzian superimposed on a residual Doppler profile in case (a).

and the line-shape function becomes

$$g(\omega) = \frac{2c}{\pi \omega \bar{v}} \exp \left[-\left(\frac{\omega_{fi} - 2\omega}{2\omega \bar{v}/c}\right)^2 \right] + \frac{4\Gamma_f}{\pi[(\omega_{fi} - 2\omega)^2 + \Gamma_f^2]} \qquad (46)$$

where $\bar{v}$ is the average thermal velocity. It is shown in Figure 13a that the area under the Lorentzian is twice that under the Gaussian and the ratio of their peaks is $2\omega \bar{v}/c\Gamma_f$, which can be much larger than unity.

Case II was discussed in detail by Bjorkholm and Liao.[44,193] If $\omega_1 = \omega_2$ the line-shape function reduces to the Lorentzian contribution (Figure 13b) in Eq. (46), thus permitting the highest frequency resolution. This case has been implemented in the arrangement described by Levenson and Bloembergen.[165] A single, circularly polarized laser beam traverses the absorption cell and is reflected upon itself by a 100% mirror. A $\lambda/4$ retardation plate changes the polarization of the reflected photons from σ^+ to σ^-.

If $\omega_1 \neq \omega_2$, the line shape of the transition is more complex. It has the form of the well-known Voigt profile.[194] Bjorkholm and Liao show that the efficiency of the two-photon interaction decreases and the linewidth increases as the two frequencies involved are chosen to be more and more unequal. In a given experimental situation the best combination of wavelengths is dictated by the required frequency resolution versus the signal strength (which can be greatly enhanced by choosing one of the photon frequencies close to a resonance with an intermediate state).

Doppler-free spectroscopic techniques are not restricted to two-photon absorption. They may be applied, in a general sense, to a number of nonlinear transitions to atoms and molecules. Examples are multiphoton absorption of the order $N > 2$[191] and stimulated Raman scattering.[195] The principles of these techniques are derived directly from the requirements for conservation of energy and momentum:

$$\sum_i^N c|\mathbf{k}_1| = \sum_i^N \omega_i = \omega_{fi} \qquad (47)$$

$$\sum_i^N \mathbf{k}_i = 0 \qquad (48)$$

In the rest frame of the atom, this leads, for two-photon absorption, directly to Eq. (42) (Γ_f is neglected). Complete elimination of Doppler broadening requires Eq. (42) to be independent of v, which is the case because of conditions (47) and (48) or, in other words, for $\omega_1 = \omega_2$ and counterpropagating beams. The extension to Doppler-free three-photon absorption is straightforward, e.g., for the case shown in Figure 14, we have

$$\omega_1\left(1 + \frac{v}{c}\right) + \omega_2\left(1 - \frac{v}{c}\cos\beta\right) + \omega_3\left(1 - \frac{v}{c}\cos\alpha\right) = \omega_{fi} \qquad (49)$$

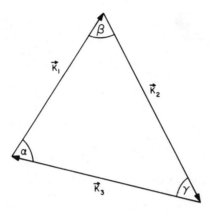

Figure 14. Wave-vector conditions for a Doppler-free three-photon transition.

The frequencies of the three beams can be selected with a wide freedom of choice. In order to maximize the transition probability, which will permit the use of low-power tunable lasers, the energy of the photons can be selected to be close to resonances with intermediate states. The angles α and β must then be chosen so as to satisfy

$$|k_1| = |k_2| \cos \beta + |k_3| \cos \alpha \tag{50}$$

The potential advantages of Doppler-free three-photon spectroscopy are obvious. With the aid of three conventional dye lasers, tunable to the visible region of the spectrum, high-resolution spectroscopic investigations are possible of levels which otherwise would be accessible only with tunable vacuum uv sources.

The first experimental implementation of two-photon Doppler-free spectroscopy was reported by Biraben et al[30] and almost simultaneously and independently by Levenson and Bloembergen[165] and Hänsch et al.[196]

Calculations by Cagnac et al.[191] had shown in 1973 that the technique proposed by Vasilenko and co-workers[29] is feasible with cw laser power densities on the order of 1 W/mm². Readily detectable signals were indeed observed in sodium vapor with a cw dye laser pumped by an argon laser, providing 10 mW output power at about 6000 Å.[196,197] A typical experimental arrangement is shown in Figure 15. A sodium vapor cell is placed in a standing wave field outside the cavity of the continuously tunable single-mode dye laser. Transitions from the $3S$ ground level to the $4D$ level and $5S$ level are observed by monitoring the uv radiation produced by the decay of these levels via the $4P–3S$ transition in the first case, or the red emission from the $5S–3P$ transition in the latter. Since the width

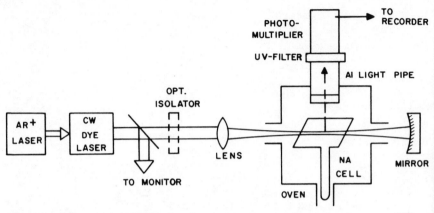

Figure 15. Schematic diagram of the experimental arrangement used by Hänsch *et al.*[196] to measure the Doppler-free two-photon spectrum of the Na $3s$–$4d$ transition.

of the laser line can be reduced to 20 MHz or less and frequency jitter is minimized by reducing mechanical vibration in the laser system, resolutions and absolute accuracies of a few MHz have been achieved. This is sufficient for an exact experimental determination of the fine structure and hyperfine structure of certain atomic levels, of the Zeeman and Paschen–Back effects, of the Stark effect, of light, Lamb, and isotope shifts, and of collision broadening. Improved experimental techniques, such as laser frequency locking and, in particular, optical heterodyning, should increase the resolution even further.[198]

7.3. Applications of Two-Photon Doppler-Free Spectroscopy

The bulk of the experimental data collected with the aid of two-photon Doppler-free absorption spectroscopy stems from investigations with dilute alkali vapors (10^7–10^{12} atoms/cm^3) where collision broadening is absent. With low-power cw laser sources, light shifts can be kept below several MHz and are therefore of no consequence.[198]

By measuring the relative frequencies between the Na transitions

$$3S\ (F=2) \rightarrow 4D_{5/2} \tag{51a}$$

$$3S\ (F=2) \rightarrow 4D_{3/2} \tag{51b}$$

$$3S\ (F=1) \rightarrow 4D_{5/2} \tag{51c}$$

$$3S\ (F=1) \rightarrow 4D_{3/2} \tag{51d}$$

the fine-structure splitting of the $4D$ level has been measured to be 1028.5 ± 3 MHz (Biraben *et al.*[199]), which is in good agreement with theoretical predictions and independent measurements by Hänsch *et al.*,[196] Bjorkholm and Liao,[44] and Pritchard *et al.*[28] A typical scan is shown in Figure 16. The width of these spectral lines is still larger than the natural linewidth of about 3 MHz, corresponding to the 5-nsec lifetime of the 4D level.

An experimental determination of the sodium fine-structure intervals of n^2D ($n = 3, 6, 7$, and 8) using Doppler-free two-photon spectroscopy was recently reported by Salour.[200] He shows that the structures are inverted and measures the following absolute values (in MHz):

$$\Delta_{fs}(n^2D) = \begin{cases} -1528 \pm 8\ (n = 3) \\ -385 \pm 5\ (n = 6) \\ -253 \pm 5\ (n = 7) \\ -173 \pm 10\ (n = 8) \end{cases}$$

In the experiment, a high-power N_2-laser-pumped dye laser system is used consisting of one oscillator and two dye amplifiers.

Further measurements of fine-structure splittings [$\Delta_{fs}(4^2F) = 229 \pm 4$ MHz for sodium] were reported by Liao and Bjorkholm,[201] who employed a two-laser arrangement with counterpropagating beams of

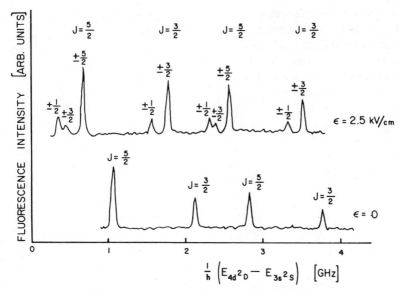

Figure 16. Typical fluorescence emission intensity versus energy of the laser beam which is scanned through the two-photon resonances from Eq. (51). The lower trace is obtained without external field and the upper one with a field of 2.5 kV/cm, showing the Stark splitting of the $4d^2D_{5/2}$ and $4d^2D_{3/2}$ levels. (Harvey *et al.*, 1975[198].)

different frequencies. As mentioned before, Doppler broadening is not expected to be eliminated completely because $\omega_1 \neq \omega_2$; however, the authors demonstrate that the absorption linewidth is Doppler-free when ω_2 is chosen to be in resonance with an intermediate state. The photons ω_2 will then strongly interact with a velocity group of the atoms in the vapor that is exactly resonant, i.e., $|\omega_{ic} - (1 - v/c)\omega_2| \leq \Delta\omega_c$, where ω_{ic} is the frequency separation of the ground state and the intermediate state $|c\rangle$ and $\Delta\omega_c$ is the natural width of the intermediate state. The two-photon absorption rate for this velocity group is, owing to the resonance, greatly enhanced over all that for other velocities, thus effectively narrowing the velocity distribution and resulting in a Doppler-free excitation line on top of a broad but weak background. This method is technically of great importance as it permits high absorption rates as well as Doppler-free spectra to be simultaneously obtained.

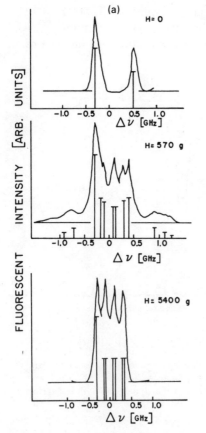

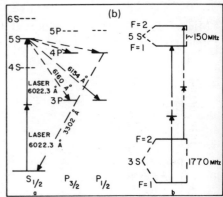

Figure 17. (a) Hyperfine splitting of the two-photon $3S \to 5S$ transition in ^{23}Na in magnetic fields of various strengths (Zeeman effect). The vertical lines indicate the theoretical positions and their relative intensities. The experimental traces record the observed fluorescence intensities of the $4P \to 3S$ transition following the two-photon excitation process (Bloembergen et al., 1974[202]). (b) Energy levels of sodium at $H = 0$ g.

The hyperfine splitting $\Delta(5S)_{\text{hfs}} = 150 \pm 10$ MHz of the $5S$ level of ^{23}Na was measured by Biraben et al.,[30,197] again with a cw dye laser and similar techniques as employed in the previous work (Figure 15). An interesting variation of the experimental setup was used by Levenson and Bloembergen,[165] who measured the hyperfine splitting of this level with a pulsed dye laser having a laser linewidth of about 80 MHz. As mentioned earlier, the residual Doppler background was eliminated in this experiment by using two counterpropagating waves of opposite helicity (circular polarization) with respect to the laboratory frame. The selection rule $\Delta m = 0$ prevents the absorption of photons of the same direction of propagation and of the same circular polarization.[191] Their results of $\Delta_{\text{hfs}}(5S) = 156 \pm 5$ MHz are in good agreement with the value of 159 ± 6 MHz obtained by Duong et al.[202] in a two-step excitation experiment and with the theoretical prediction by Rosen and Lindgren.[203]

Precise measurements of the dc Stark effect,[198] Zeeman and Paschen–Back effects[199,204] of the ac Stark effect (light shift),[205,206] and of isotope shifts[207] are further evidence of Doppler-free two-photon spectroscopy being a relatively simple but yet very elegant and powerful new technique for high-precision atomic spectroscopy[208-210] (see Figures 16 and 17). High-resolution multiphoton spectroscopy of highly excited atomic states, pioneered by Popescu et al.,[184] Bensoussan,[186] and Esherick et al.,[185] has now been extended to Doppler-free two-photon spectroscopy of even-parity states with a principal quantum number $n > 20$ of alkali vapors. Stoicheff et al.[211] studied the $5s \rightarrow nd$ transition in rubidium for $n = 1$–30. The resolution was sufficient to resolve the $30d$ state which is only 134 cm^{-3} below the ionization limit. Recently they have observed even states up to $n = 56$ or only 16 cm^{-1} below the ionization limit.[211] They measured the Rb^{85}–Rb^{87} isotope shift of the ground state to be 160 ± 12 MHz. Levenson et al.[212] used a similar technique to study quantum defects of even-parity Rydberg states of potassium which begin to systematically decrease for $n > 20$.

Doppler-free two-photon absorption has been possible in the vacuum uv region of the spectrum as well. Lee et al.[213] and Hänsch et al.[214] performed elegant experiments in which they not only very accurately measured the Lyman-α isotope shift of atomic hydrogen (670.933 ± 0.096 GHz) for the first time but also the Lamb shift on the hydrogen $1S$ state (8.20 ± 0.10 GHz for H and 8.25 ± 0.11 GHz for D), using Doppler-free two-photon spectroscopy of the $1S$–$2S$ transition with a frequency-doubled pulsed dye laser near 2430 Å and observing simultaneously the Balmer-β line with the fundamental laser output at 4860 Å in two separate hydrogen discharge vessels. The resolution of these experiments was primarily limited by the relatively broad linewidth which is characteristic of pulsed lasers. Refinements in the techniques (e.g., the use of a cw laser and improvements in the resolution of the Balmer-β line by two-photon spectroscopy instead of saturation spectroscopy) appear possible so that measurements of the Lamb shift of the $1S$ state may be performed with a precision exceeding present calculations, which are limited by our knowledge of the fine-structure constant α and of nuclear structure effects. It may even be possible to discriminate experimentally between several different theoretical approaches[215] to the calculation of Lamb shifts. An accurate determination of the ratio of the electron mass m_e and the proton mass m_p from precise measurements of the hydrogen isotope shift is expected to provide new criteria for the validity of quantum electrodynamics.

In this connection it is important to discuss briefly a very recent new development in the field: Doppler-free polarization spectroscopy. Wiemann and Hänsch[216] replace the conventional fluorescent decay technique of detecting the final state by a technique that is sensitive to the change in polarization of the two counterpropagating beams. A linearized probe beam traverses the sample and only a small fraction of it reaches a photodetector after passing through a crossed polarizer. Any change in the probe polarization will alter the light flux through the blocking polarizer and is detected with high sensitivity. Such a change is induced by sending a circularly polarized beam of the same frequency in nearly opposite directions (to avoid reflection into the detector) through the sample. Two-photon absorption of oppositely circularly polarized counter-propagating photons removes photons from the probe beam (which may be considered to be decomposed into two oppositely circularly polarized beams) and thus produces a slightly elliptically polarized beam and a signal on the detector. The polarization spectra of the Balmer-β line revealed the Stark splitting of the fine-structure components caused by the electric field in a Wood deutrium or hydrogen discharge. The authors expect a substantial improvement in the accuracy of the $1S$ lamb shift if a Balmer-β polarization spectrum is used as a reference in the experiments described in Refs. 213 and 214. In addition, new precision measurements of the ratio m_e/m_p as well as of the Rydberg constant should be possible with this technique.

Finally, we briefly discuss the first experimental demonstration of a Doppler-free three-photon transition by Grynberg et al.,[217] who observed the hyperfine structure of the transition $3S_{1/2}-3P_{1/2}$ in sodium vapor. Two different lasers are used in an angular arrangement shown in the **k**-vector diagram of Figure 18a. In this case, Eq. (47) yields $2\mathbf{k}_1 = \mathbf{k}_2$ and $\omega_{fi} = 2c|\mathbf{k}_1| - c|\mathbf{k}_2|$. By scanning the frequency of one laser and keeping that of the second fixed, the Doppler-free three-photon spectrum of Figure 18b is obtained. This experiment is of considerable importance as it demonstrates that Doppler-free multiphoton spectroscopy is not restricted to two-photon transition. By absorption of an odd number of photons, it becomes

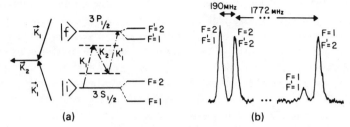

Figure 18. (a) Spatial orientation of the wave vectors used to eliminate Doppler broadening in three-photon absorption and schematic energy level diagram of sodium. (b) Doppler-free three-photon spectrum of sodium obtained by fixing ω_2 30 GHz below the sodium D_1 resonance and scanning ω_1 (Grynberg et al., 1976[216]).

possible to investigate highly excited states whose parity is opposite of that of the initial state. However, as the order N of the multiphoton transition increases, the transition probability decreases and increasingly higher laser power densities are required to obtain an adequate signal. AC Stark broadening and light shifts must then be carefully considered.

8. Concluding Remarks

In summary, we have discussed a number of multiphoton phenomena in atoms that can be described with perturbation theory, Specifically, the photon fluxes required in high-resolution multiphoton spectroscopy are so small ($< 10^8$ W/cm^2) that in most cases the experimental conditions can even be chosen to prevent detectable light shifts and broadening. Multiphoton ionization and emission processes have been observed at power densities up to almost 10^{16} W/cm^2 without producing evidence to show that perturbation theory is invalid.[33] It is clear, however, that with increasing light intensity, perturbation theory will eventually become inadequate.

Several attempts have been made to treat the ultrastrong field case. Apparently, all of these are of somewhat controversial nature and were therefore not discussed here.

These notes were not intended to be an exhaustive and complete account of all aspects of multiphoton phenomena. They should serve as an introduction to the basic theory and to some of the experimental topics that presently seem to be of major interest. None of the implications for molecular and solid state spectroscopy have been included. Applications to laser isotope separation have been omitted for the same reason.

A considerable amount of work has been carried out in this field in the last 15 years but, nevertheless, it is apparently still in its early stages. Many interesting subjects remain to be investigated. These include not only the numerous unsolved theoretical problems pertaining to high-order and strong-flux multiphoton ionization, but also the experimental investigations of such phenomena as the angular distribution of multiphoton-generated photoelectrons, the production of spin-polarized electrons on a routine basis as an electron source for the study of various collision phenomena, etc., and the construction of soft x-ray photon sources via multiphoton emission phenomena.

The development of new tunable uv, vacuum uv, and x-ray lasers will undoubtedly contribute greatly toward wide applications of multiphoton spectroscopy in the near future.

Acknowledgments

The author gratefully acknowledges stimulating discussions with Professor P. Lambropoulos and the support of Ms. K. Rankin and Drs. B. Rosenblum and J. P. Carrico in preparing the manuscript. He is indebted to Dr. J. A. Armstrong, who kindly provided Figure 11.

References

1. H. A. Kramers and W. Heisenberg, *Z. Phys.* **31**, 681 (1925).
2. S. Yatsiv, M. Rokni, and S. Barak, *Phys. Rev. Lett.* **20**, 1282 (1968).
3. P. Bräunlich and P. Lambropoulos, *Phys. Rev. Lett.* **25**, 135 (1970).
4. P. Bräunlich and P. Lambropoulos, *Phys. Rev. Lett.* **25**, 486 (1970).
5. P. Bräunlich, R. Hall, and P. Lambropoulos, *Phys. Rev. A* **5**, 1013 (1972).
6. M. Göppert-Mayer, *Naturwissenschaften* **17**, 932 (1929); *Ann. Phys. (Leipzig)* **9**, 273 (1931).
7. W. Kaiser and C. G. Garret, *Phys. Rev. Lett.* **7**, 229 (1961).
8. I. D. Abella, *Phys. Rev. Lett.* **9**, 453 (1962).
9. G. Breit and E. Teller, *Astrophys. J.* **91**, 215 (1940).
10. M. Lipeless, R. Novick, and N. Tolk, *Phys. Rev. Lett.* **15**, 690 (1965).
11. F. V. Bunkin and A. M. Prokhorov, *Sov. Phys. JETP* **19**, 739 (1964).
12. G. S. Voronov and N. B. Delone, *Sov. Phys. JETP* **23**, 54 (1964).
13. J. S. Bakos, *Adv. Electron. Electron Phys.* **36**, 57 (1974).
14. J. F. Ward and H. C. New, *Phys. Rev. Lett.* **19**, 556 (1967).
15. J. F. Young, G. C. Bjorklund, A. H. Kung, R. B. Miles, and S. E. Harris, *Phys. Rev. Lett.* **27**, 1551 (1971).
16. K. M. Leung, J. F. Ward, and B. J. Orr, *Phys. Rev. A* **9**, 2440 (1974).
17. S. E. Harris and D. M. Bloom, *Appl. Phys. Lett.* **24**, 229 (1974).
18. W. Zernik, *Phys. Rev.* **132**, 320 (1963); **133**, A117 (1964).
19. P. Lambropoulos, C. Kikuchi, and S. K. Osborn, *Phys. Rev.* **144**, 1081 (1966).
20. F. Shiga and S. Iamura, *Phys. Lett.* **25A**, 706 (1967).
21. C. Lecompte, G. Mainfray, C. Manus, and F. Sanchez, *Phys. Rev. Lett.* **32**, 265 (1974); *Phys. Rev. A* **11**, 1009 (1975).

22. R. A. Fox, R. M. Kogan, and E. J. Robinson, *Phys. Rev. Lett.* **25** 1416 (1971).
23. P. Lambropoulos, *Phys. Rev. Lett.* **28**, 585 (1972); **29**, 453 (1972).
24. S. Klarsfeld and A. Maquet, *Phys. Rev. Lett.* **29**, 79 (1972).
25. J. Brossel, B. Cagnac, and A. Kastler, *C.R. Acad. Sci.* **237**, 984 (1973).
26. C. Cohen-Tannoudji and A. Kastler, *Prog. Opt.* **5**, 3 (1966).
27. V. W. Hughes and L. Grabner, *Phys. Rev.* **79**, 314 (1950).
28. D. Pritchard, J. Apt, and T. W. Ducas, *Phys. Rev. Lett.* **32**, 641 (1974).
29. L. S. Vasilenko, V. P. Chebotaev, and A. V. Shishaev, *JETP Lett.* **12**, 133 (1970).
30. F. Biraben, B. Cagnac, and G. Grynberg, *Phys. Rev. Lett.* **32**, 643 (1974).
31. P. Lambropoulos, in *Advances in Atomic and Molecular Physics*, Vol. 12, Eds. D. R. Bates and B. Benderson, Academic Press, New York (1976).
32. A. Messiah, *Quantum Mechanics*, Vol. 2, Wiley, New York, p. 87 (1965).
33. L. A. Lompre, G. Mainfray, C. Manus, S. Repoux, and J. Thebault, *Phys. Rev. Lett.* **36**, 949 (1976).
34. J. E. Bayfield and P. M. Koch, *Phys. Rev. Lett.* **33**, 258 (1974).
35. M. L. Goldberger and K. M. Watson, *Collision Theory*, Wiley, New York (1964).
36. P. Lambropoulos, *Phys. Rev.* **164**, 84 (1967).
37. P. Lambropoulos, *Phys. Rev. A* **9**, 1992 (1974).
38. L. I. Schiff, *Quantum Mechanics* McGraw-Hill, New York (1955).
39. W. Louisell, *Radiation and Noise in Quantum Electronics*, McGraw-Hill, New York (1964).
40. W. L. Peticolas, R. Norris, and K. E. Rieckhoff, *J. Chem. Phys.* **42**, 4164 (1965).
41. J. Eichler, *Phys. Rev. A* **9**, 1762 (1974).
42. B. Honig and J. Jortner, *J. Chem. Phys.* **47**, 3698 (1967).
43. M. A. Kovner and O. M. Parshkow, *Opt. Spectrosc.* **32**, 591 (1972).
44. J. E. Bjorkholm and P. F. Liao, *Phys. Rev. Lett.* **33**, 128 (1974).
45. L. Vriens, *Opt. Commun.* **11**, 396 (1974).
46. P. Lambropoulos and M. R. Teague, *J. Phys. B: Atom Molec. Phys.* **9**, 587 (1976).
47. M. R. Teague and P. Lambropoulos, *Phys. Lett.* **56A**, 285 (1976).
48. V. Jacobs, *Phys. Rev. A* **4**, 939 (1971).
49. M. S. Pindzola and H. P. Kelly, *Phys. Rev. A* **11**, 1543 (1975).
50. E. J. Robinson and S. Geltman, *Phys. Rev.* **153**, 4 (1967).
51. A. Gold, in *Quantum Optics*, Ed. R. J. Glauber, Academic Press, New York (1969).
52. H. B. Bebb and A. Gold, *Phys. Rev.* **143**, 1 (1966).
53. A. Gold and H. B. Bebb, *Phys. Rev. Lett.* **14**, 60 (1965).
54. H. B. Bebb, *Phys. Rev.* **149**, 25 (1966).
55. H. B. Bebb, *Phys. Rev.* **153**, 23 (1967).
56. V. M. Morton, *Proc. Phys. Soc. London* **92**, 301 (1967).
57. A. Dalgarno and J. T. Lewis, *Proc. R. Soc. London* **A233**, 70 (1955).
58. D. C. Allison and A. Dalgarno, *Proc. Phys. Soc. London* **81**, 23 (1963).
59. A. Dalgarno, *Mon. Not. R. Astron. Soc.* **131**, 311 (1966).
60. C. Schwartz, *Ann. Phys. (N.Y.)* **2**, 169 (1959).
61. C. Schwartz and J. J. Tiemann, *Ann. Phys. (N.Y.)* **2**, 178 (1959).
62. W. Zernik, *Phys. Rev.* **135**, A51 (1964); **176**, 420 (1968).
63. W. Zernik and R. W. Klopfenstein, *J. Math. Phys.* **6**, 262 (1965).
64. G. A. Victor, *Proc. Phys. Soc.* **91**, 825 (1967).
65. G. W. F. Drake, G. A. Victor, and A. Dalgarno, *Phys. Rev.* **180**, 25 (1969).
66. M. Karplus and H. J. Kolker, *J. Chem. Phys.* **39**, 1493 (1963).
67. A. Dalgarno and G. A. Victor, *Proc. Phys. Soc.* **87**, 371 (1966).
68. V. Jacobs and J. Mizuno, *J. Phys. B* **5**, 1155 (1972).
69. J. Mizuno, *J. Phys. B* **6**, 314 (1973).
70. T. N. Chang and R. T. Poe, *Phys. Rev. A* **11**, 191 (1975).

71. B. J. Choudhury and R. P. Gupta, *Phys. Lett.* **50A**, 377 (1974).
72. Y. Gontier and M. Trahin, *Phys. Rev.* **172**, 83 (1968); *Phys. Rev. A* **4**, 1896 (1971).
73. S. I. Vetchinkin and S. V. Khristenko, *Opt. Spectrosc.* **25**, 1365 (1968).
74. S. Klarsfeld, *Phys. Lett.* **30A**, 382 (1969).
75. C. K. Au, *Phys. Lett.* **51A**, 442 (1975).
76. E. Karule, *J. Phys. B* **4**, L67 (1971).
77. B. A. Zon, N. L. Manakov, and L. P. Rapoport, *Sov. Phys. Doklady* **14**, 904 (1970).
78. B. A. Zon, N. L. Manakov, and L. P. Rapoport, *Sov. Phys. JETP* **34**, 515 (1972).
79. V. A. Davydkin, B. A. Zon, N. C. Manakov, and L. P. Rapoport, *Sov. Phys. JETP* **33**, 70 (1971).
80. N. L. Manakov, M. A. Preobragensky, and L. P. Rapoport, Proceedings of the Eleventh International Conference on Phenomena of Ionized Gases, Prague (1975), p. 23.
81. M. R. Cervanen and N. R. Isenor, *Opt. Commun.* **13**, 175 (1957).
82. J. L. Debethune, *Nuovo Cimento* **12B**, 101 (1972).
83. M. C. Teich and G. J. Wolga, *Phys. Rev. Lett.* **16**, 625 (1966).
84. P. Lambropoulos, *Phys. Rev.* **164**, 84 (1967).
85. S. N. Dixit and P. Lambropoulos, *Phys. Rev. Lett.* **40**, 111 (1977).
86. Y. R. Shen, *Phys. Rev.* **155**, 921 (1967).
87. B. R. Mollow, *Phys. Rev.* **175**, 1555 (1968).
88. G. S. Agrawal, *Phys. Rev. A* **1**, 1445 (1970).
89. F. Sanchez, *Nuovo Cimento* **27B**, 305 (1975).
90. L. Armstrong, P. Lambropoulos, and N. K. Rahman, *Phys. Rev. Lett.* **36**, 952 (1976).
91. J. Mostowski, *Phys. Lett.* **56A**, 87 (1976).
92. R. Glauber, *Phys. Rev.* **130**, 2529 (1963); **131**, 2766 (1963).
93. H. D. Simaan and R. Loudon, *J. Phys. A* **8**, 539 (1975).
94. S. Carusotto, E. Polacco, and M. Vaselly, *Lett. Nuovo Cimento* **2**, 628 (1969).
95. J. Krasinski, S. Chudzynski, W. Majewski, and M. Glodz, *Opt. Commun.* **12**, 304 (1974).
96. J. Krasinski, B. Karczewski, W. Majewski, and M. Glodz, *Opt. Commun.* **15**, 409 (1975).
97. J. P. Hermann and J. Ducuing, *Phys. Rev. A* **5**, 2557 (1972).
98. S. L. Chin, N. R. Isenor, and M. Young, *Phys. Rev.* **188**, 7 (1969); S. L. Chin, *Phys. Rev. A* **5**, 2303 (1972).
99. B. Held, G. Mainfray, C. Manus, and J. Morellec, *Phys. Rev. Lett.* **28**, 130 (1972).
100. J. S. Bakos, A. Kiss, L. Szabo, and M. Tendler, *Phys. Rev. Lett.* **39A**, 238, 317 (1972); **41A**, 163 (1972).
101. G. A. Delone, N. B. Delone, and G. K. Piskova, *Sov. Phys. JETP* **35**, 672 (1972).
102. E. H. A. Granneman and M. J. Van der Wiel, *J. Phys. B* **8**, 1617 (1975); *Rev. Sci. Instr.* **46**, 332 (1975).
103. E. H. A. Grannemann, M. Klewer, and M. J. Van der Wiel, *Ninth International Conference on Physics of Electronic and Atomic Collisions*, Seattle, Washington, Abstracts, p. 471 (1975).
104. M. R. Cervanen, R. H. Chan, and N. R. Isenor, *Can. J. Phys.* **53**, 1573 (1975).
105. Y. Flank and A. Rachman, *Phys. Lett.* **53A**, 247 (1975).
106. M. R. Teague and P. Lambropoulos, *J. Phys. B* **9**, 1251 (1976).
107. L. Spitzer and J. L. Greenstein, *Astrophys. J.* **114**, 407 (1951).
108. J. Shapiro and G. Breit, *Phys. Rev.* **113**, 179 (1959).
109. C. P. Rapoport and B. A. Zon, *Phys. Lett.* **26A**, 564 (1968).
110. B. A. Zon and L. P. Rapoport, *JETP Lett.* **7**, 70 (1968).
111. W. R. Johnson, *Phys. Rev. Lett.* **29**, 1123 (1972).
112. C. P. Rapoport, Proceedings of the Conference on the Interaction of Electrons with Strong Electromagnetic Fields, Budapest, Hungary, p. 99 (1973).

113. M. Lipeless, L. Gampel, and R. Novick, *Bull. Am. Phys. Soc.* **7**, 69 (1962).
114. R. Novick, in *Physics of the One- and Two-Electron Atom*, F. Bopp and H. Klein-poppen, North-Holland, Amsterdam (1969).
115. R. C. Elton, L. J. Palumbo, and H. R. Griem, *Phys. Rev. Lett.* **20**, 783 (1968).
116. D. O'Connell, K. J. Kollath, A. J. Duncan, and H. Kleinpoppen, *J. Phys. B* **8**, L214 (1975).
117. R. W. Schmieder and R. Marrus, *Phys. Rev. Lett.* **25**, 1092 (1972); *Phys. Rev. A* **5**, 1160 (1972).
118. W. D. Metz, *Science* **176**, 394 (1972).
119. R. S. Van Dyck, Jr., C. E. Johnson, H. A. Shugart, *Phys. Rev. A* **4**, 1327 (1971).
120. E. A. Hinds, J. E. Clendenin, and R. Novick, *Bull. Am. Phys. Soc. Ser. II* **21**, 84 (1976).
121. H. Krüger and A. Oed, *Phys. Lett.* **54A**, 251 (1975).
122. J. Ducuing, G. Hauchercorne, A. Mysyrowicz, and F. Pedere, *Phys. Lett.* **28A**, 746 (1969).
123. P. Platz, *Appl. Phys. Lett.* **17**, 537 (1970); *J. Phys. (Paris)* **32**, 773 (1971).
124. G. Fornaca, F. Giamanco, A. Giulietti, and M. Vaselli, *Lett. Nuovo Cimento* **9**, 395 (1974).
125. P. P. Sorokin and N. Braslau, *IBM J.* **8**, 177 (1964).
126. A. M. Prokhorov, *Science* **149**, 828 (1965).
127. R. L. Garwin, *IBM J.* **8**, 338 (1964).
128. A. S. Selivanenko, *Opt. Spectrosc.* **21**, 54 (1955).
129. Letokhov, *JETP Lett.* **7**, 221 (1968).
130. L. L. Hope and M. O. Vasell, *Phys. Lett.* **31A**, 256 (1970).
131. H. P. Yuen, *Appl. Phys. Lett.* **26**, 505 (1975).
132. B. P. Kirsanov and A. S. Selivanenko, *Opt. Spectrosc.* **23**, 242 (1967).
133. S. E. Harris, *Phys. Rev. Lett.* **31**, 341 (1973).
134. S. E. Harris, J. F. Young, A. H. Kung, D. M. Bloom, and G. C. Bjorklund, in *Proceedings of the Laser Spectroscopy Conference*, Eds. R. G. Bremer and A. Mooradian, Plenum, New York (1974).
135. A. H. Kung, J. F. Young, and S. E. Harris, *Appl. Phys. Lett.* **22**, 301 (1973).
136. A. H. Kung, J. F. Young, G. C. Bjorklund, and S. E. Harris, *Phys. Rev. Lett.* **29**, 985 (1972).
137. R. L. Carman, *Phys. Rev. A* **12**, 1048 (1975).
138. P. P. Sorokin, N. S. Shiren, J. R. Lankard, E. C. Hammond, and T. G. Kazyaka, *Appl. Phys. Lett.* **10**, 44 (1967).
139. R. L. Carman and W. H. Lowdermilk, *Phys. Rev. Lett.* **33**, 190 (1974).
140. G. M. Krochik and Yu. G. Khronopulo, *JETP Lett.* **21**, 274 (1975).
141. N. Bloembergen, *Am. J. Phys.* **35**, 989 (1967).
142. Y. R. Shen, *Rev. Mod. Phys.* **48**, 1 (1976).
143. R. B. Miles and S. E. Harris, *Appl. Phys. Lett.* **19**, 385 (1971).
144. R. T. Hodgson, P. P. Sorokin, and J. J. Wynne, *Phys. Rev. Lett.* **32**, 343 (1974).
145. P. P. Sorokin, J. J. Wynne, and J. R. Lankard, *IEEE J. Quantum Electron.* **QE-9**, 227 (1973).
146. J. J. Wynne, P. P. Sorokin, and J. R. Lankard, in *Proceedings of the Laser Spectroscopy Conference*, Eds. R. G. Brewer and A. Mooradian, Plenum, New York (1974).
147. D. M. Bloom, J. T. Yardley, J. F. Young, and S. E. Harris, *Appl. Phys. Lett.* **24**, 427 (1974).
148. J. Ducuing, in *Proceedings of the International School of Physics Enrico Fermi—Quantum Optics*, Ed. R. J. Glauber, Academic Press, New York (1969).
149. D. M. Bloom, G. W. Bekkers, J. F. Young, and S. E. Harris, *Appl. Phys. Lett.* **26**, 687 (1975).
150. Yu. P. Raizer, *Sov. Phys. Usp.* **8**, 650 (1966).

151. C. Cohen-Tannoudji, in *Cargese Lectures in Physics*, Ed. M. Levy, Gordon and Breach, New York (1967).

152. C. Cohen-Tannoudji and S. Haroche, *J. Phys. (Paris)* **30**, 125, 153 (1969).

153. P. Lambropoulos, *Phys. Rev. A* **9**, 1992 (1974).

154. S. Stenholm, Chapter 3 (Part A) of this work.

155. W. Behmenburg, Chapter 27 of this book.

156. C. S. Chang and P. Stehle, *Phys. Rev. Lett.* **30**, 1283 (1973).

157. P. Agostini, G. Barjot, J. F. Bonnal, G. Mainfray, C. Manus, and J. Morellec, *IEEE J. Quantum Electron.* **QE-4**, 667 (1968).

158. P. Agostini, G. Barjot, G. Mainfray, G. Manus, and J. Thebault, *Phys. Lett.* **31A**, 367 (1970); *IEEE J. Quantum Electron.* **QE-6** 782 (1970).

159. G. S. Voronov and N. B. Delone, *Sov. Phys. JETP* **23**, 54 (1966).

160. G. S. Voronov, G. A. Delone, and N. B. Delone, *Sov. Phys. JETP* **24**, 1122 (1967).

161. T. B. Bystrova, G. S. Voronov, G. A. Delone, and N. B. Delone, *Sov. Phys. JETP Lett.* **5**, 178 (1967).

162. G. Laplanche, A. Durrieu, Y. Flank, and A. Rachman, *Phys. Lett* **55A**, 13 (1975).

163. H. A. Bethe and E. E. Salpeter, *Quantum Mechanics of One- and Two-Electron Atoms*, Springer, Berlin (1957).

164. P. Lambropoulos, *Bull. Am. Phys. Soc.* (1975).

165. M. D. Levenson and N. Bloembergen, *Phys. Rev. Lett.* **32**, 645 (1974).

166. R. M. Kogan, R. A. Fox, G. T. Burnham, and E. J. Robinson, *Bull. Am. Phys. Soc.* **16**, 1411 (1971).

167. P. Agostini and P. Bensoussan, *Appl. Phys. Lett.* **24**, 216 (1974).

168. P. Agostini, P. Bensoussan, and M. Movssessian, *Phys. Lett.* **53A**, 89 (1975).

169. G. A. Delone, Second Conference on Interaction of Electronics with Strong Electromagnetic Fields, Budapest (Hungary) October 6–10, 1975.

170. Y. Gontier and M. Trahin, *Phys. Rev. A* **7**, 2069 (1973).

171. V. Jacobs, *J. Phys. B* **6**, 1461 (1973).

172. P. Lambropoulos, *Phys. Rev. Lett.* **30**, 413 (1973).

173. J. Cooper and R. N. Zare, in *Lectures in Theoretical Physics*, Vol. 11, 100, Eds. S. Geltman, K. T. Manhantahappa, and W. E. Brittin, Gordon and Breach, New York (1969).

174. J. C. Tully, R. S. Berry, and B. J. Dalton, *Phys. Rev.* **176**, 95 (1968).

175. M. Lambropoulos and R. S. Berry, *Bull. Am. Phys. Soc.* **17**, 371 (1972).

176. S. Edelstein, M. Lambropoulos, J. Duncanson, and R. S. Berry, *Phys. Rev. A* **9**, 2459 (1974).

177. Y. Gontier, N. K. Rahman, and M. Trahin, *Phys. Rev. Lett.* **34**, 779 (1975); *J. Phys. B* **8**, L179 (1975).

178. E. Arnos, S. Klarsfeld, and S. Wane, *Phys. Rev.* **7**, 1559 (1973).

179. P. J. Farago and D. W. Walker, *J. Phys. B* **6**, L280 (1973); P. S. Farago, D. W. Walker and J. S. Wykes, *J. Phys. B* **7**, 59 (1974).

180. M. Lambropoulos, unpublished.

181. P. Lambropoulos, G. Doolen, and S. P. Rountree, *Phys. Rev. Lett.* **34**, 636 (1975).

182. M. Lambropoulos, S. E. Moody, S. J. Smith, and W. C. Lineberger, *Phys. Rev. Lett.* **35** 159 (1975).

183. C. F. Burrell and H.-J. Kunze, *Phys. Rev. Lett.* **29**, 1445 (1972).

184. D. Popescu, C. B. Collins, B. W. Johnson, and I. Popescu, *Phys. Rev. A* **9**, 1182 (1974).

185. P. Esherick, J. A. Armstrong, R. F. Dreyfus, and J. J. Wynne, *Phys. Rev. Lett.* **36**, 129 (1976).

186. P. Bensoussan, *Phys. Rev. A* **11**, 1787 (1975).

187. W. E. Lamb, *Phys. Rev.* **134**, A1429 (1964).
188. H. R. Schlossberg and A. Javan, *Phys. Rev.* **150**, 267 (1966); *Phys. Rev. A* **5**, 1974 (1972).
189. W. Demtröder, "Laser High-Resolution Spectroscopy," Chapter 15 in Part II of this work.
190. R. Salomaa and S. Stenholm, *Opt. Commun.* **16**, 292 (1976).
191. B. Cagnac, G. Grynberg, and F. Biraben, *J. Phys. (Paris)* **34**, 845 (1973).
192. D. E. Roberts and E. N. Fortson, *Phys. Rev. Lett.* **31**,,1539 (1973).
193. J. E. Bjorkholm and P. F. Liao, *IEEE J. Quant. Electron.* **10**, 906 (1974).
194. B. DiBartolo, *Optical Interactions in Solids*, p. 366, Wiley, New York (1968).
195. P. L. Kelley, H. Kildal, and H. R. Schlossberg, *Chem. Phys. Lett.* **27**, 62 (1974).
196. T. W. Hänsch, K. C. Harvey, G. Meisel, and A. L. Schawlow, *Opt. Commun.* **11**, 50 (1975).
197. F. Biraben, B. Cagnac, and G. Grynberg, *Phys. Lett.* **49A**, 71 (1974).
198. K. C. Harvey, R. T. Hawkins, G. Meisel, and A. L. Schawlow, *Phys. Rev. Lett.* **34**, 1073 (1975).
199. F. Biraben, B. Cagnac, and G. Grynberg, *Phys. Lett.* **48A**, 469 (1974).
200. M. M. Salour, *Opt. Commun.* **18**, 377 (1976).
201. P. F. Liao and J. E. Bjorkholm, *Phys. Rev. Lett.* **36**, 1543 (1976).
202. H. T. Duong, S. Liberman, J. Pinard, and J. L. Vialle, *Phys. Rev. Lett.* **33**, 339 (1974).
203. A. Rosen and I. Lindgren, *Phys. Scr.* **6**, 109 (1972).
204. N. Bloembergen, M. D. Levenson, and M. M. Salour, *Phys. Rev. Lett.* **32**, 867 (1974).
205. P. F. Liao and J. E. Bjorkholm, *Phys. Rev. Lett.* **34**, 1 (1975).
206. P. F. Liao and J. E. Bjorkholm, *Opt. Commun.* **16**, 392 (1976).
207. F. Biraben, G. Grynberg, E. Giacobino, and J. Bauche, *Phys. Lett.* **56A**, 441 (1976).
208. C. D. Harper and M. D. Levenson, *Phys. Lett.* **56A**, 361 (1975).
209. A. Flusberg, T. Mossberg, and S. R. Hartmann, *Phys. Lett.* **55A**, 403 (1976).
210. B. Cagnac, *Ann. Phys. (Paris)* **9**, 223 (1975).
211. Y. Kato and B. P. Stoicheff, *J. Opt. Soc. Am.* **66**, 490 (1976); K. C. Harvey and B. P. Stoicheff, *Phys. Rev. Lett.* **38**, 537 (1977).
212. M. D. Levenson, C. D. Harper, and G. L. Eesley, to be published.
213. S. A. Lee, R. Wallenstein, and T. W. Hänsch, *Phys. Rev. Lett.* **35**, 1262 (1975).
214. T. W. Hänsch, S. A. Lee, R. Wallenstein, and C. Wieman, *Phys. Rev. Lett.* **34**, 307 (1975).
215. P. J. Mohr, *Phys. Rev. Lett.* **34**, 1050 (1975).
216. C. Wieman and T. W. Hansch, *Phys. Rev. Lett.* **36**, 1170 (1976).
217. G. Grynberg, F. Biraben, M. Bassini, and B. Cagnac, *Phys. Rev. Lett.* **37**, 283 (1976).

20

Fast-Beam (Beam-Foil) Spectroscopy

H. J. ANDRÄ

1. Introduction

Fast-beam spectroscopy (FBS) dates back to the discovery of canal rays,[1] the first technique for the production of fast ionic or atomic beams with particle velocities of the order of $v = 10^7 - 10^8 \text{ cm s}^{-1}$. The corresponding apparatus consisted of a glass discharge tube with a gas pressure of about $10^{-1} - 10^{-2}$ mbar and a potential difference of a few kilovolts between the anode and cathode. The cathode filled the whole inner tube cross section and split the tube into two compartments (see Figure 1), which were connected only by a capillary hole drilled through the cathode, the so-called canal. Because of the sharp potential drop in a discharge close to the cathode, ions were accelerated onto the cathode surface so that some of them penetrated with high velocity through this canal into the other part of the tube. On their way, these so-called canal rays experience enough gas collisions for recombination and excitation so that light-emitting rays of moderately well-defined velocity leave the rear of the cathode.

It was soon recognized that this light emission could be used for time-resolved spectroscopy.[2] In particular, Wien[3] started in 1919 a series of measurements of intensity decay versus distance from the canal exit with an apparatus schematically shown in Figure 1. The right half of the tube was specially pumped, in this case, in order to reduce gas collisions of the free decaying canal rays. The average particle velocity was determined via spectral Doppler shift when the beam was observed along its axis. With this

H. J. ANDRÄ • Fachbereich Physik der Freien Universität Berlin, 1 Berlin 33, Boltzmannstrasse 20, West Germany.

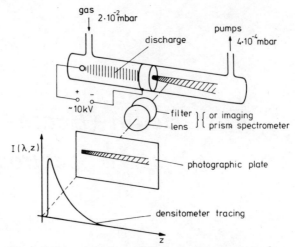

Figure 1. Schematic view of Wien's canal-ray apparatus for
atomic lifetime measurements.

velocity the time scale of the photograph or the densitometer tracing of
Figure 1 could be calibrated by the simple relation $t = x/v$ such that mean
lives of neutral and singly ionized excited states could be determined. This
technique thus represents the classical precursor of all modern fast-beam
time-resolved experiments, and most of the early results have been
assembled in a review article by Wien in 1927.[4]

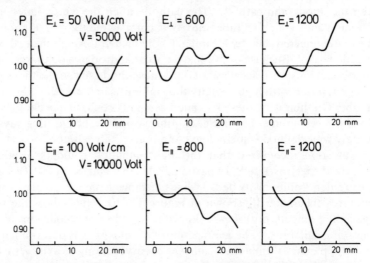

Figure 2. Stark beats in the polarization P (defined in the text) of the H_β
emission of H canal rays exposed to electric fields E parallel and
perpendicular to the rays at two discharge voltages V (velocities).

There are, however, important difficulties connected with the apparatus in Figure 1. They arise from the still poor vacuum conditions, from the coarsely defined recombination and excitation region close to the cathode surface and inside the canal, and therefore also from the only moderately well-defined acceleration. Together with the cascade problem these difficulties lead to mean lifetimes that are systematically too long compared to those obtained using other experimental techniques and thus cast some doubt on these experiments.

Even more exciting, from a modern point of view, than these early lifetime measurements was the discovery of the partial linear polarization of the light emitted from canal rays,[5,6] its use for an in-beam Hanle-effect measurement,[7,8] and for the subsequent observation of Stark-effect quantum beats[9-11] when excited HI canal rays were exposed to electric or magnetic fields. In particular these latter experiments can be considered as the first "Doppler-free" measurements of nonzero energy splittings in atomic physics. They are therefore of such fundamental importance that I reproduce in Figure 2 some of the results obtained by Walerstein[11] in 1929. He exposed hydrogen canal rays to electric fields E parallel and perpendicular to the beam axis right at the exit of the canal. At right angles to the beam and to the field he observed the linear polarization $P = I_{\parallel}/I_{\perp}$ of mainly H_{β} emission as a function of distance from the canal exit, where $I_{\parallel}$, $I_{\perp}$ are the intensities polarized parallel and perpendicular to the beam, respectively. The plotted polarization P clearly exhibited oscillations which increased in their spatial frequency with the field strength in full accord with our knowledge of Stark beats.

At this time only Kramer's[12] calculation of the Stark effect of the hydrogen fine structure in weak external fields existed. It was not fully correct but gave reasonable numbers. There was thus only qualitative support of Walerstein's interpretation of his results in Figure 2 as being due to Stark-split fine-structure frequencies. A group of experimentalists had thus developed, in essence, a high-resolution technique that permitted observation of level separations much smaller than the Doppler width of the spectral line. However, the huge number of superimposed beat frequencies in the case of hydrogen prevented them from clear-cut interpretations of their results and therefore possibly also from a full appreciation of their own discovery of a "Doppler-free" spectroscopic method.

Actually, these experiments could have started a dramatic development in atomic spectroscopy. Unfortunately, however, the interest in these measurements faded as other fields in physics became more fashionable. Historically one can only state that these efforts disappeared together with Wien's canal rays from the research scenery in the years around 1930.

It was not until 1963 that Kay[13] rediscovered Wien's classical fast-beam technique. With the use of more modern equipment, i.e., an energetically well-defined accelerator ion beam, high vacuum, and a spatially well-defined thin exciter foil, the former experimental difficulties here removed. With these refinements, fast beams offered unique conditions for time-resolved spectroscopy, a fact that was spotted by Bashkin[14] in 1964. It led him to initiate varied research efforts with this "new" light source, which became established as "beam-foil spectroscopy" (BFS).

As a spectroscopic source BFS offers time-resolved observation, nearly unperturbed emission conditions in a high vacuum environment, and a multiple-collision excitation mechanism which allows one to populate nearly any excited level in nearly any charge state of any element by choice of beam energy. In comparison to traditional sources, which are in general specially designed to meet only one of these requirements, the combination of these three basic features has made BFS a versatile tool in atomic physics.

The time resolution is so conspicuous that it is obvious why BFS is most widely used for the measurement of atomic lifetimes. This application has developed so rapidly that already more excited level lifetimes have been measured by BFS than by all other methods put together.

The spectroscopic potential of BFS was initially somewhat obscured by poor spectral resolution, resulting from the Doppler broadening of in-flight emitted radiation. Owing to the surprising observation of large numbers of previously unknown lines, however, it became more and more apparent that in BFS multiply excited levels and Rydberg levels with high angular momenta are highly populated and can be observed without perturbation. These levels are either not excited or are quenched by collisions in traditional sources. BFS has thus opened new fields of spectroscopic research, which also include Auger-electron spectroscopy of multiply excited levels, and has produced a wealth of hitherto unavailable data on atomic structure.

In addition the beam-foil source also possesses another property, namely, it can produce coherently and nonisotropically excited levels due to the short time interval and the axial symmetry of the interaction, respectively. This property not only allowed application of the known "Doppler-free" high-resolution methods of atomic physics to the fast beam but also led to the development of quantum beat spectroscopy,[15] a "new" high-resolution tool in atomic physics, which, in particular, takes full advantage of the excellent time resolution of BFS. It is being successfully used today to obtain precise fine-structure and hyperfine-structure information on previously inaccessible excited levels.

Aside from such pure beam-*foil* experiments, which easily dominate the field, any alteration of one or more of the three basic features, i.e., the unperturbed emission, the time resolution, or the excitation, may pave the way for other classes of experiments. In particular the replacement of the ion–foil interaction as means of excitation has proved to be very effective. Such replacements are, for instance, differentially pumped gas targets for improved spectral resolution and specific metastable level preparations, or crossed laser beams for the selective excitation of fast beams. They have become essential ingredients for Lamb-shift measurements in high-Z hydrogenlike ions or for lifetime measurements with unprecedented precision, respectively.

With regard to these various existing and possible modifications of BFS the description "fast-beam spectroscopy" (FBS) is certainly more representative of the subject today than is the original one if one wants to include all accelerator-based atomic physics devoted to the study of the properties of excited levels of free atoms. The overlap with the neighboring field of "collisional interactions of fast beams" is of course established by the mutual interest in understanding the excited level formation. It is sad to note, however, that in the past little attention has been paid to this problem.

An extensive literature covering the early developments up to the present state of FBS is available, and includes the proceedings of five international conferences,[16–20] a large number of review articles,[21–39] and an up-to-date book.[40] Furthermore several contributions to the present book also cover important FBS results. In order not to bore my expert colleagues with "another" listing of the numerous achievements of FBS, but on the other hand to provide newcomers with an introductory survey on the field, I derive at length in Section 2 the experimental procedures and intrinsic problems of FBS from what is known about the ion–foil interaction mechanism. With this information, backed up by the many reviews, it is only necessary to present some of the important spectroscopic and lifetime results in Section 3. Somewhat more space is given in Section 4 to the experiments related to the coherent and anisotropic excitation in FBS, which are only briefly discussed in most reviews. Also for this reason, an extra section is added discussing the laser excitation of fast beams, an area that is establishing new frontiers of precision and high resolution in FBS.

2. The Fast-Beam Spectroscopic Source

2.1. Basic Principle

The basic principle of fast-beam spectroscopy (FBS) was illustrated already in Figure 1 with Wien's apparatus. With the use of more modern equipment it has been modified to the experimental arrangement shown in

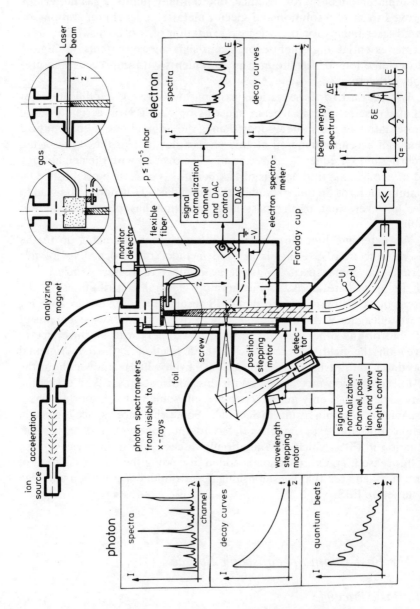

Figure 3. Experimental arrangement of a typical FBS laboratory. For details see text.

Figure 3, which is currently typical of most FBS laboratories. The ions to be studied are generated in an ion source, accelerated to energies ranging from 10 keV to several 100 MeV, and are then momentum analyzed in a magnetic field to yield monoenergetic, isotopically pure beams of 0.1–10 μA. This beam is collimated and passed through a target, which may consist of a thin self-supporting solid foil (carbon foils of 1–20 μg/cm^2 are usually used), a differentially pumped gas target of ≤ 0.1 μg/cm^2 at $p \leq$ 1 mbar, or a crossed laser beam in resonance with a transition of the ions. Because of their predominant use, we will discuss first only foil targets, unless otherwise stated. When traversing the foil the ions experience many elastic and inelastic collisions so that the beam emerges from the foil in a highly excited state, consisting generally of ions in various stages of ionization and excitation. This excitation energy is spontaneously released by the ions downstream from the foil in the form of a rich photon spectrum ranging from the optical into the x-ray region and an electron spectrum ranging from a few eV to several keV. The foil-excited beam thus represents a fast-moving spectroscopic source in a high vacuum environment of $p \leq 10^{-5}$ mbar at average velocities v' of the order of 10^8 cm/s. It is therefore ideally suited for spectral and transient studies of photon and electron emission from the excited beam.

The spectra are measured with appropriate spectrometers which accept photons or electrons from a beam section Δz at a fixed, short distance z from the foil. Modern spectrometers are equipped with photo-electric or electron detectors operating in pulse-counting mode so that spectra can be recorded by feeding the output pulses into a multiscaler whose channel sweep is synchronized to the wavelength sweep. For rapid scanning a strip chart recorder fed by a ratemeter may sometimes replace the multiscaler.

For measurements of the time-dependent emission of a specific spectral line the spectrometer wavelength is set on this line and the output intensity is recorded as a function of the distance z from the foil. This is simply achieved by mechanically moving the foil along the beam axis and feeding the detector pulses again into a multiscaler whose channel sweep is synchronized in this case to the z displacement of the foil. By observing a beam section $\Delta z = 0.01$ cm, a time resolution of the order of $\Delta t = \Delta z/v' \simeq 10^{-10}$ s can be directly achieved, easily allowing the measurement of most atomic excited level lifetimes, typically lying in the 10^{-6}–10^{-10} s range and corresponding to spatial intensity decay lengths of 100–0.01 cm behind the foil.

For the time-scale calibration of these spatially resolved measurements the average velocity v' of the foil-transmitted beam has to be determined with the highest possible accuracy.

2.2 Ion-Beam-Target Interaction

As in early traditional spectroscopy, little attention has been devoted in the past to the fast-beam excitation mechanism itself, since knowledge of the excitation mechanism is unnecessary for many applications. Consequently the present knowledge of ion-beam–foil or –gas interactions is still quite rudimentary.[41]

In recent years, however, a growing demand has developed for better charge-state identifications of spectral lines,[42,43] for an understanding of Rydberg[44–46] and multiply excited level[38] populations, and for the improvement of alignment and orientation of excited levels necessary for quantum beam measurements.[32,36] This has initiated an increasing number of attempts to gather information on the excitation mechanism from measurements of spectra, emission yields, and polarizations. These efforts have been supplemented by intense activities in the field of "atomic collisions in solids,"[20,47,48] which yield information on elastic and inelastic processes inside a foil by measurements of x-ray spectra and yields, or charge state and angular distributions of ions after passage through a foil. The presently available material can only be considered as a first step to an understanding of the ion–target interaction and is still far from providing a general model. But it may allow us to give a short summary of the current understanding of processes yielding fast excited beams.

(a) *The Target Interior.*[49] In a dilute gas electron loss and capture collisions involving ground-state ions interacting with ground-state gas atoms (molecules) lead to a dynamic charge state equilibrium for gas layers which are sufficiently thick. The collision time is long compared to excited-state relaxation times so that the charge-state distribution is basically determined by Bohr's suggestion[50] that outer electrons with orbital velocities v_e smaller than the ion velocity v_i will be lost and the maximum cross section for capture and loss will appear for orbitals $v_e \approx v_i$. As the gas density increases, the collision time becomes too short for outer electron excited states to relax so that one obtains charge-state equilibrium *and* excitation equilibrium. Under these conditions levels that cannot be excited by a single collision due to selection rules[51] may become populated by stepwise collision excitation. At even higher densities a situation is approached that applies in its extreme to energetic ions penetrating through foils where the collision time becomes short compared even to lifetimes of inner-shell excitations, of the order of 10^{-14}–10^{-15} s. At the same time outer electron orbits become ill defined owing to the time–energy uncertainty principle, and in addition they cannot continue to exist during passage through the solid because of their geometrical size.[52] However, not all of these electrons are fully lost to the moving ion since some of them may form an electron cloud which follows the ion's wake through the solid and partially screens the charge of its core.[53]

When approaching the exit surface of the foil the ions have reached a charge state and excitation equilibrium with a considerable fraction of inner-shell excitations and may be accompanied by an electron cloud in their wake. It should be noted that this internal charge-state equilibrium may be quite different from the external charge-state distribution measured at some distance behind the foil.[54] The relaxation of inner-shell excitations by autoionization and Auger processes[55] after leaving the foil may increase the average charge, whereas electron capture at the exit surface of the foil may reduce it.

(b) Foil Exit. From the observation of practically all excited levels in the fast excited beam spectra and from the former argument that outer electron orbits are ill defined and cannot exist inside the foil, one must conclude that the excited-state formation takes place at the exit surface of the foil, i.e., at its last layer including an electron density extending into the vacuum. Experimental evidence for the importance of the exit surface for the final state formation has been found in beam-foil polarization measurements with the foil normal tilted out of the beam axis (see Section 4.3).[56] The appearance of circularly polarized light components in these experiments must clearly be interpreted as a surface effect. From absolute excitation probability measurements[20,57,58] n^{-3} population dependences were found within various Rydberg series which suggest electron capture from the foil surface to the projectile.[59,60] From the comparison of an internal charge state equilibrium as determined by spectral x-ray yields with the corresponding external charge state distribution it has also been concluded that a dramatic rearrangement of the ion near the surface must take place, which may also include electron capture.[61]

Near, and at, the surface several processes are expected to contribute to the actual formation of the excited levels. Starting from a certain depth near the exit surface the dynamic excitation and charge-state equilibrium of the bulk becomes perturbed by increasing excitations of the more strongly bound electrons which survive through the last layers. The closer the ion comes to the surface the more excited states with lesser binding energies can survive until all excited states stemming from collisions with the last layer of atoms can survive into the vacuum. In these collisions direct excitation, charge exchange, and Fano–Lichten[62-64] promotional excitations may appear equally well.

Directly at the surface, or just past the surface, electron capture to the projectile may also play an important role. The electron cloud moving with the ion or in the wake of the ion supposedly supplies sufficient surplus electrons at the surface for such capture processes.[65] Depending on the

state of the ion leaving the foil, an electron may be transferred to a
nonexcited ion core or to an ion core with one or more inner-shell holes. In
the first case all excited n, l states above the core may be populated,
including those with high n, l quantum numbers. In the second case,
multiply excited states can be formed, which indeed become strongly
populated by the beam-foil interaction as verified by photon and electron
spectra. On the basis of these electron capture processes Veje[65] formulated an "independent electron model" for which he assumes a total
probability α that an electron is transferred from the foil surface to a
bound state outside the ion core and a total probability β that a vacancy
exists in the core of the projectile as it leaves the foil. By assuming α and β
to be independent functions of energy for a certain element, Veje could
reproduce with the use of basic laws of probability the relative excitation
functions within all level schemes of light elements (e.g., LiI $1s^2nl\,^2L$,
LiI $1s2snl\,^4L$, LiII $1snl\,^3L$ or LiII $1snl\,^1L$) and the charge-state distribution as a function of energy of that element. On the basis of this
success for light elements one may conclude that indeed an excited level
formation outside the ion core takes place at the foil surface independently of the ion core's history within the foil. Which detailed processes
are predominantly responsible for this behavior remains unknown for the
present.

As a consequence of the excited state formation at the surface one can
derive a formation time $\Delta t \sim 2 \times 10^{-15}$ s, if one assumes an interaction zone
of ~ 20 Å at the surface. Thus the foil excitation can be considered as an
extremely sharp excitation pulse in time which establishes a phase relation
between all levels to within 2×10^{-15} s at the exit of the foil. Excited levels
with energy separations less than 2 eV therefore become coherently
excited by the foil excitation process.

As long as the processes for the excited-state formation are not
understood, unfortunately no model for the m-sublevel population can be
developed. Only general structures of anisotropic m-sublevel populations
may be derived from symmetry arguments.[30] The fast-beam excitation
geometry singles out a unique spatial axis, the beam axis. The resulting
axial symmetry of the interaction with a perpendicular foil allows noniso-
tropic excitation (i.e., different population of $|m|$ substrates) called align-
ment, which leads to nonisotropical angular distribution of the emission
and to linear polarization of the emitted light. By tilting the foil normal
out of the beam axis the axial symmetry is reduced to reflection symmetry
on a plane containing the beam axis and the foil normal. As a result,
different populations of m substates, called orientation, may appear,
leading to the emission of circularly polarized light. In practice, however,
each level must be examined experimentally for evidence of such aniso-
tropies.

Assuming that a level was anisotropically excited at the surface, one may draw conclusions about possible anisotropy transfer processes which can take place immediately after the foil. According to the "independent-electron model" one can expect a relatively large fraction of the foil-transmitted particles to be in multiply excited states. Those which rapidly decay by autoionization or x-ray emission can contribute to the population of excited levels by cascade. Since in general anisotropy is carried away by these autoionization electrons and x-ray photons,[66] the anisotropy of such cascade populated levels can become quite low. No evidence has, however, been found as yet for such processes to be important after the foil for optically decaying levels.

(c) *Macroscopic Excited-Beam Properties.* The energy loss ΔE, the energy straggling δE, and the angular spread $\langle \theta \rangle$ of the excited beam may be called macroscopic properties in contrast to the microscopic excitations of the projectiles. They are caused by the elastic and inelastic multiple collisions inside the target and seriously affect the FBS work with respect to its time-scale calibration and spectral resolution.

With elastic collisions one associates the energy loss and momentum transfer with the recoil of target nuclei, whereas the inelastic collisions are solely connected with the energy loss due to electronic excitations and ionizations in the target atoms and projectiles. One distinguishes, therefore, in the literature[67,68] between nuclear (elastic) and electronic (inelastic) contributions to the stopping power, i.e., the (energy loss)/ (atoms × unit target area), and the energy straggling. The nuclear stopping power is thus related to the repulsive potentials of the colliding particles, which can be calculated *ab initio* with good accuracy.[69] The agreement between theoretical and experimental nuclear stopping powers is therefore quite satisfactory for all energies and all projectile—target contributions. For the electronic stopping power, which exceeds by far the nuclear stopping power in the practical region of 5–1000 keV/amu, the situation is less satisfactory.[70-72] It depends sensitively on the electronic structure of the target atoms and projectiles and exhibits, at fixed energy, strong oscillations as a function of the nuclear charges Z_1 or Z_2 of the projectile or target atom, respectively. The theoretical description of this behavior is at present still quite unsatisfactory and reproduces the experimental data only to within $\pm 20\%$. As a consequence the calculation of the energy loss from the incoming beam energy and the foil thickness in a FBS experiment is limited in accuracy by these boundaries.

Owing to the statistical nature of these energy loss collisions one obtains an average energy loss ΔE with a finite width δE, the energy straggling. In a simple model one can obtain a relation between ΔE and δE by assuming that $\Delta E = N\varepsilon$, where N is the average number of collisions (i.e., roughly the number of atomic layers of the foil) and ε is the average energy loss per collision. The order of magnitude of the energy straggling is then given by $\delta E = \varepsilon N^{1/2} = N^{-1/2}\,\Delta E$, which amounts to about 10% of ΔE for 100 atomic layers foil thickness, in good agreement with the observations. From this model one can also directly deduce that ΔE increases linearly with the foil thickness and δE with the square root of the foil thickness.

The momentum transfer in elastic collisions causes an angular deflection in each collision and gives rise to multiple angular scattering of the projectiles.[73] Originating from the same physical process as the nuclear stopping power, it can also be treated theoretically in quite satisfactory agreement with experiments.[74] The resulting angular distribution of the foil-transmitted beam is, according to Figure 5 of Ref. 74, for very thick foils, a Gaussian, and for thinner foils changing to an approximate Gaussian with more and more enhanced tails. Its half-width depends according to Figure 3 of Ref. 74 linearly on the foil (gas) thickness below $\sim 0.05\ \mu\text{g/cm}^2$; for thicker layers there is a gradual transition to a square root dependence. At 5 $\mu\text{g/cm}^2$ typical half-widths of the order of a few degrees must then be expected at around 50 keV beam energy.[75] Fortunately, however, the half-width is roughly proportional to E^{-1}, so that the beam becomes increasingly well collimated with higher beam energies.

(d) Mechanical Properties of Carbon Foils under Ion Bombardment. The stopping power and the angular scattering increase with the nuclear charge of the target material.[67-74] In combination with the easy handling of carbon foils in the laboratory this has led to the prevailing use of carbon foils in FBS. The better physical arguments for Be foils do not justify the hazards associated with the handling of this poisonous material.

Under ion bombardment the carbon foils suffer foil breakage after a certain dose of ion irradiation.[76,77] This dose is independent of the ion current density, the vacuum conditions, and the foil thickness, and seems to be related to the nuclear stopping power, which is responsible for atomic displacements in the target. This behavior may be interpreted as the production of crystal defects which lead to foil breakage when an average displacement energy per atom has been deposited in the foil.[76] The heating of foils under ion bombardment to $> 500°C$ has been claimed to increase the foil lifetime considerably.[78] This procedure has, however, not been applied successfully yet in FBS.[76]

Aside from this technical problem the foils show an even worse behavior for long-lasting time-resolved experiments: The foils thicken under ion bombardment[79] and it is sad to note that nearly all time-resolved experiments in FBS have been carried out under serious foil-thickening conditions. A careful study of this effect[77] shows that under the normally used vacuum conditions a thickness increase of 30 ng/(cm² min) can be observed with an N^+ ion current density of $1.27\,\mu A/cm^2$ at 750 keV. This rate is energy dependent and increases with lower energies. The authors interpret this effect as deposition of carbon cracked out of residual gas hydrocarbons since they have shown that it depends dramatically on the vacuum conditions and can be reduced by surrounding the foil by a cold trap at $p = 10^{-6}$ Torr until it finally changes into a low-rate foil thinning due to sputtering.[80] This sputtering is always present on both sides of the foil but is more intense in the forward beam direction owing to knock-on collisions. As a consequence one finds the foil-transmitted excited beams always slightly contaminated by excited atoms ejected out of the foil with far lower energy than the beam. In spectra measured close to the foil one therefore always observes H, C, and O lines, the intensities of which increase with the mass of the projectiles.

(e) *Alternative Excitation Methods.* Both ion-beam–foil or –gas excitations are nonselective so that cascading may severely perturb excited mean lifetime determinations. In order to circumvent this problem the foil or gas target can be replaced by a laser beam (see Section 5).[81,82] This allows the selective excitation of higher-lying levels from ionic ground states or from gas or foil prepopulated decaying levels. This technique is, however, limited by the presently available laser wavelengths.

For the application of high-resolution techniques to the fast-beam source, nonisotropic sublevel excitations are required.[32] As a new source for the production of nonisotropically oriented excited states, the ion beam solid surface interaction at grazing incidence has been developed (see Section 4.3).[83] The surface-scattered (reflected) projectiles emit spectra very similar to those of foil-transmitted beams but exhibit in addition large fractions of circular polarization. However, even less is known about the ion–surface interaction than about ion–foil interaction, and the macroscopic scattered beam properties still need to be determined.

2.3. Experimental Equipment and Procedures

The ion-beam–target interaction including the mechanical instabilities of foils is responsible for a number of difficulties and limitations of FBS. They may be partially overcome by skilled modification of the equipment for these problems as well as by the data collection and data analysis procedures.

2.3.1. Accelerators

The velocity v' of the projectiles determines the accessibility of lifetime ranges and of the average charge state after the target.

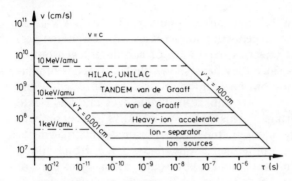

Figure 4. Practical limits of FBS for lifetime measurements with commonly used accelerators.

The measurable lifetimes τ are experimentally limited by the minimum spatial resolution of about $v'\tau = 0.00^1$ cm and the maximum useful decay length of about $v'\tau = 100$ cm. Depending on the accelerator, lifetimes may be measured from 2×10^{-12} s up to 10^{-5} s as shown in Figure 4, first introduced by Bromander[84] but slightly modified here. Accelerators covering this whole range up to the broken line in Figure 4 have been used already in FBS with total currents delivered between a few nanoamperes up to tens of microamperes.

Although the charge-state distribution after the target is somewhat related to the excited beam velocity, a detailed prediction for various projectile, target, and energy combinations is not possible. How it changes with velocity or energy must therefore be derived from actual measurements as listed in Refs. 85 and 86 and as shown in Figure 5a for a light element in a medium energy range. In particular it is important to note the large difference in average charge which may occur after gas or foil targets in Figure 5b. This is attributed to the so-called "density effect," the

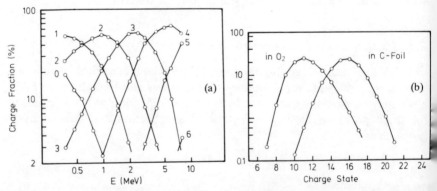

Figure 5. (a) Charge-state distribution of C ions after passing through a thin carbon foil as a function of energy. (b) Charge-state distributions of ^{79}Br after passing at 42 MeV through an O_2 gas target and a thin carbon foil—the "density effect."

explanation of which is, however, still a matter of debate.[49,55,87] The charge-state accessible in an experiment thus depends strongly on the choice of the accelerator energy and of the target. For the study of neutral atoms, heavy ion accelerators up to a few hundred keV are best suited, whereas hydrogenlike krypton[88] can be studied in the 10 MeV/amu range of heavy-ion linear accelerators. For practical work all accelerators have in common energy and ion source instabilities. After a high-quality momentum analysis of the resulting beam both effects add up at the target (behind a small aperture) to awkward ion current fluctuations which have to be corrected for in the data-collection procedures.

2.3.2. Vacuum

One of the unique features of FBS relies on the fact that once the ions have been excited by the target interaction they can reemit in a high vacuum environment. This allows one in general to neglect collisional perturbations of the excited ions by the residual gas. At the same time the excited beam density at typical beam current densities of a few 10 $\mu A/cm^2$ is only 10^5–10^6 cm^{-3} and is thus orders of magnitude lower than in most conventional spectroscopic sources. As a consequence, interionic fields and perturbations, imprisonment of resonance radiation and stimulated emission can be fully neglected. An excited level is therefore in general considered to decay in the absence of external fields fully without perturbation in FBS.

For precise time-resolved experiments, however, in particular for intensity decay measurements with spatially long decay lengths, collisional quench cross sections of the order of $\sigma = 10^{-14}$ cm^2 lead at a pressure of $p = 10^{-5}$ mbar to considerable errors in lifetime determinations. In Figure 6 the relative errors $\Delta\tau/\tau$ are plotted versus the product $p\sigma$ and the true decay length $\tau v'$. Accordingly errors of a few percent may arise for decay lengths longer than 10 cm under the above conditions. At the same time

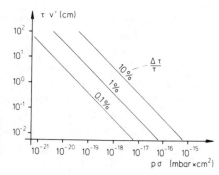

Figure 6. Lines of constant error $\Delta\tau/\tau$ [%] for a lifetime measurement of decay length $\tau v'$ at $p\sigma$ due to collisions with residual gas.

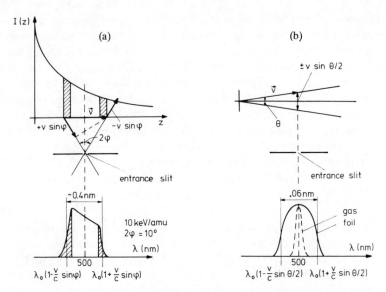

Figure 7. (a) Influence of the finite optical acceptance angle 2φ on the spectral resolution for an ideally parallel beam. (b) Influence of the beam divergence on the spectral resolution for very small φ.

collisional reexcitation of the beam and excitation of the residual gas itself may occur, which gives rise to a background emission and thus reduce the signal-to-background ratio.

In order to achieve the unprecedented condition of fully negligible collisional perturbations in FBS a vacuum system must be installed that allows pressures $p \leqslant 10^{-6}$ mbar. Furthermore the partial pressure of hydrocarbons must be kept as low as possible in order to reduce the foil thickening under ion bombardment mentioned earlier.

2.3.3. Spectrometers and Resolution

The low luminosity of the excited beam requires high-speed spectrometers for both photon and electron spectroscopy.

(*a*) *Photon Spectrometers.* For the spectral analysis of photons this requirement is met by commercial blazed grating spectrometers with focal lengths between 0.25 and 1 m from the visible (< 1000 nm) into the far vacuum ultraviolet (vuv) (> 50 nm) with $f/5.3$ to $f/11.4$ apertures. In the extreme vuv (50 nm$> \lambda > 1$ nm) grazing incidence grating spectrometers with only $f/42$ become necessary. For the x-ray region Si(Li) detectors or crystal spectrometers are being used for low and high resolution, respectively. The specially developed Doppler-tuned x-ray spectrometer will be discussed in Section 3.4. Most of the spectrometers are mounted relative to the excited beam in the so-called "side-on" geometry shown in Figure 3, which directly allows time-resolved measurements.

According to the macroscopic properties of the excited beam a spectrometer views with a finite optical acceptance angle a fast emitting beam with spatial intensity decay, angular spread θ, and energy straggling δE. In order to discuss the resulting Doppler broadening of spectral lines one may at first neglect the energy straggling and separate the influence of the finite acceptance angle and of the angular scattering as shown in Figures 7a and 7b, respectively. Because of the observation of a spatially decaying intensity one obtains, with a finite acceptance angle 2φ, an *asymmetric* Doppler profile, a fact that has only recently been carefully investigated.[89] The width is roughly determined by $\Delta\lambda = 2\lambda (v/c) \sin \varphi$ so that the resulting resolution is given by $\lambda/\Delta\lambda = c/[2(2E/M)^{1/2} \sin \varphi]$ and is thus proportional to $(E/M)^{-1/2}$. For the example in Figure 7a one obtains for $E/M = 10$ keV/amu and $2\varphi = 10°$, a resolution of $\lambda/\Delta\lambda = 1.24 \times 10^3$. This is a rather bad resolution compared to traditional sources. It becomes even worse with higher energy and the determination of a line center is complicated by the asymmetry. Improvements can at first sight only be achieved by a reduction of the acceptance angle 2φ. In particular in the visible region, numerous such attempts have been made by accepting only light emitted perpendicular to the beam axis where the loss in solid angle was compensated by observing longer beam sections. Starting by using a simple lens at focal distance from the spectrometer entrance slit,[90] taking some advantage of the azimuthal emission with an anamorphic condensing system with two cylindrical lenses,[91] and finally accomplishing full 2π-azimuthal acceptance with an axicon,[92] these systems could indeed improve the resolution up to the point where the angular scattering of Figure 7b becomes the dominant contribution to the linewidth. Resolutions of $\lambda/\Delta\lambda = 2.9 \times 10^3$ have been observed at energies of $E/M = 70$ keV/amu. These methods[93] furthermore allow a simultaneous calibration of the spectra by standard line sources when the second-order Doppler shift of typically $\Delta\lambda/\lambda = 10^{-4} - 10^{-5}$ is corrected for. The main disadvantage is, however, the loss of spatial resolution in some cases.

An ingeneous technique which preserves the spatial resolution and reduces the Doppler broadening of Figure 7a has been developed by Stoner and Leavitt[94,95] by "refocusing" the spectrometers. Consider Figure 8. Light of wavelength λ_0 from a source at rest is shown to be focused (broken line) onto the exit slit S' of a concave grating spectrometer. When the same spectrometer is illuminated from a fast beam, which emits a line at λ_0 in its rest frame, then the extreme rays in the spectrometer are shifted to the blue (B) or red (R) and follow the full lines, thus producing in the plane of the exit slit S' a broad image of the entrance slit R'-B'. The extension of the full lines leads, however, to a spatial crossover at the position S. Hence, moving the exit slit into the S position seemingly removes the Doppler width and simultaneously enhances the signal-to-background ratio. It must be stressed that the wavelength distribution passing through the slit S is still the Doppler-broadened, asymmetric distribution of Figure 7a; only the spatial imaging of the entrance slit has been improved at S.

For actual spectra recorded with a refocused spectrometer this has of course the effect of an apparent better spectral resolution, which can be used for the separation of narrow spectral structures as demonstrated in Figure 9. The normal Doppler width was $\Delta\lambda = 5.5$ Å for this case,[95] which could be reduced to "3 Å" by moving S' only "midway" to S and finally to "0.9 Å" when S' was positioned at S as determined from the relation $S - S' = d = (v/c)(\lambda/k)$, where k is the inverse linear dispersion of the spectrometer. The final limitation of "$\Delta\lambda = 0.9$ Å" is again set by the Doppler broadening stemming from the angular spread of the beam in Figure 7b.

The advantage of this refocusing method is not only the preserved spatial resolution for time-resolved measurements but also the fact that no optical imaging components are needed outside the spectrometer. Further addition of imaging optics can dramatically improve the intensity gain,[96] but in the layout of Figure 8 the method can be applied from the visible to the extreme vuv.[97] Problems do arise, however, when the center of a line is to be determined for spectroscopic purposes. Since a refocused spectrometer has lost its resolution for a line source at rest, the calibration of the instrument becomes difficult if not impossible. Actually one has to use known lines emitted from a fast beam under similar conditions as the lines to be determined, i.e., in the same spectral region, from a beam with equal velocity, and with similar spatial decay length.

The reduction of the Doppler broadening introduced by the angular spread of the beam in Figure 7b is at present for side-on geometries only

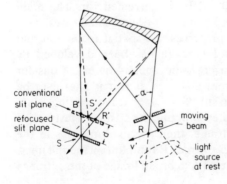

conventional slit plane
refocused slit plane
moving beam
light source at rest

Figure 8. The refocusing of a spectrometer for a fast-moving light source. For explanation see text.

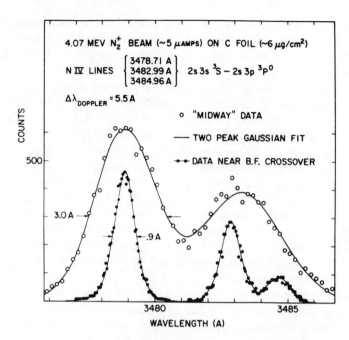

Figure 9. The effect of refocusing on the apparent spectral resolution in FBS. "Data near B.F. crossover" are taken with a fully refocused spectrometer, whereas for the "MIDWAY" data it was only partially refocused. The actual linewidth without refocusing was 5.5 Å.

possible by reducing the angular spread itself.[98,99] The resolution is given by $\lambda/\Delta\lambda = c/2(2E/M)^{1/2}\sin(\theta/2)$. Since θ is roughly proportional to $(E/M)^{-1}$ the resolution increases with $(E/M)^{1/2}$. However, not much use can be made of this dependence since the excitation function of a spectral line is significantly different from zero only over a small range of v or $(E/M)^{1/2}$. θ is also approximately proportional to the square root of the target thickness. The best results can therefore be obtained either with the thinnest mechanically stable carbon foils of $\sim 1\ \mu\mathrm{g/cm}^2$ available at present or with gas targets with thickness $\leq 0.1\ \mu\mathrm{g/cm}^2$. An improvement of the "linewidth" in Figure 9 by a factor of 8–10 could then be achieved, and indeed "linewidths" of 0.01–0.02 nm at 400–500 nm were reported ($\lambda/\Delta\lambda = 5\times 10^4$!) with the use of a refocused spectrometer in conjunction with a gas target.[100]

Another way of coping with the angular scattering is, at the cost of spatial resolution, the use of so called "end-on" geometries, where the line of sight is against the beam direction. The system that nearly eliminates both broadening effects of Figure 7 is a lens centered on the beam axis which images the target on the spectrometer entrance slit, also located on-axis, with the light acceptance cone adapted to the beam scattering cone (see Figure 6 of Ref. 93). The main beam is stopped in front of the lens by a small Faraday cup. With such a geometry the spectra are Doppler shifted and Doppler broadened by the beam velocity and by the velocity spread, respectively. The resolution $\lambda/\Delta\lambda = c/\delta v$ is then inversely proportional to the velocity spread δv and reaches with 5 $\mu g/cm^2$ foils values of the order of 10^4 when other broadening effects are neglected. Experimental results of $\lambda/\Delta\lambda \leqslant 6 \times 10^{3}$ [101] have been reported with foil targets. This represents only a slight improvement over the best "side-on" results at the cost of spatial resolution and wavelength uncertainty due to velocity uncertainty.

Most laboratories prefer, therefore, to work with side-on geometries, with resolutions ranging from 10^3 to 5×10^4 at best. It should be noted, however, that these resolutions are approximately preserved for all beam energies, i.e., for all ionization stages, since the contributions to $\lambda/\Delta\lambda$ from the finite acceptance angle $[\propto (E/M)^{-1/2}]$ and from the angular scattering $[\propto (E/M)^{1/2}]$ yield an energy-independent resolution. For the investigation of levels of highly ionized species, FBS at high energies may therefore become superior to other sources even with respect to resolution and wavelength determination[102] due to the strong perturbation by the high density and temperature necessary in these sources for the production of these levels.[103,104]

   (b) *Electron Spectrometers.* In order to obtain the largest possible transmission, the electrostatic cylindrical mirror analyzer (CMA) in Figure 10 has usually been used in FBS. It accepts electrons emitted at a mean polar angle of 42.3° in the laboratory frame over an azimuthal angle of nearly 2π. The electrons enter through a pair of entrance apertures (1) a field region between a grounded cylinder (3) and a cylinder at potential U (4) such that electrons with the correct energy are bent through the pair of exit apertures (2) onto a channeltron detector (6). Spectra of electrons emitted from a finite section of the beam can thus be measured as a function of the applied voltage U in the range from 0 to several keV electron energy. Theoretical treatments[105] of this spectrometer yield an instrumental resolution that is directly related to the total solid angle accepted from an extended source at rest. Typical instrumental resolutions for the CMA in Figure 10 are 10^3 at a solid angle $\Delta\Omega \approx 10^{-3} \times 4\pi$ or 10^2 at $\Delta\Omega \approx 10^{-2} \times 4\pi$.

   With a fast-moving beam source with angular spread and energy straggling, however, dynamic (Doppler) broadening by far exceeds the instrumental widths. At a polar angle of 42.3° the finite acceptance angle of the spectrometer, the angular spread of the beam, and the energy straggling contribute equally, and the same discussion as for photon spectroscopy applies again. Although in principle possible, no refocusing technique

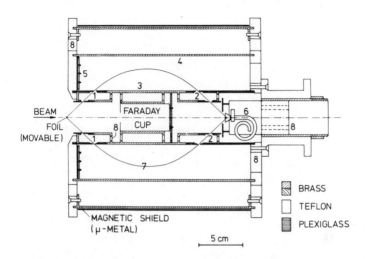

Figure 10. Layout of an electrostatic cylindrical mirror analyzer. (1) Entrance aperture. (2) Exit aperture. (3) Inner cylinder. (4) Outer cylinder. (5) Fringe plate. (6) Channeltron. (7) Electron trajectory. (8) Pumping holes.

have been employed as yet, and the experimental resolutions observed range from $E/\Delta E = 60$ with 7 $\mu g/cm^2$ foil at 32 keV/amu[106] to $E/\Delta E \approx$ 130 with 2 $\mu g/cm^2$ foil at 420 keV/amu.[107] As in photon spectroscopy the resolution can be improved by confining the polar acceptance and by use of a gas target. An experiment at very low energy of 10 keV/amu yielded $E/\Delta E = 125$ with a gas target.[108] As another possibility the parallel plate spectrometer (PPS) as indicated in Figure 3 could be used. It would accept electrons only in a small azimuthal sector, i.e., in a much smaller total solid angle than the CMA, but it would allow the measurement of the polar angular distribution of the electron emission from the beam and the measurement of energy spectra in near "end-on" geometry. As discussed for photon spectroscopy this geometry should yield in conjunction with a gas target an unprecedented resolution of $E/\Delta E > 1000$ for electrons emitted from fast beams. The PPS has, however, not been employed yet on a fast beam. Instead it has been used with great success on vapor sources at "rest" excited by energetic ion beams. Such a combination yields improved experimental resolution of the order of $E/\Delta E \approx 250$ for the observation of the same spectra[109] as obtained with gas excitation of a fast beam.[108] With a commercial ESCA-spectrometer a resolution of $E/\Delta E = 1330$ has been achieved when Ar gas was excited by 2- and 4-MeV H^+ beams.[110]

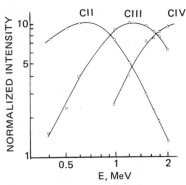

Figure 11. Relative intensity variations of C II (387.6, 392.0 nm), C III (388.7 nm), and C IV (393.5 nm) lines with C^+ beam energy after foil excitation. The intensities are normalized to 10 at their maximum.

2.3.4. Charge-State Identification

Besides the highly resolved measurement of a spectral line the charge-state determination of its emitter is of similar importance for its final classification. After Kay[111] found in 1965 that the dependence of the relative line intensity on the beam energy is similar to the energy dependence of the charge-state fraction of its emitters, the comparison of charge-state fractions and relative line intensities has become the standard method for charge-state identifications in FBS. A representative example[112] of this procedure is shown in Figure 11, which in comparison with Figure 5a allows a clear-cut identification. However, in many cases significant deviations from such good overlap of the relative dependences have been found.[42,51,113,114] According to the "independent electron model" of the final-state formation[65] these deviations can be explained for light elements and can consequently be corrected for. For heavier elements and higher charge states, however, systematic comparisons with the known lines of the spectrum become necessary in order to avoid errors of one charge at most. It should be noted that this procedure can equally well be applied to fast-beam electron spectroscopy.

Another method has also occasionally been used,[115,116] which takes advantage of the different Doppler shift of lines that are emitted from differently charged particles accelerated by an external field towards or away from the optical detection. Since it takes some time for the particles to be sufficiently accelerated for a detectable Doppler shift, this method can only be applied to longer-lived levels. Most laboratories prefer therefore the above described procedure.

2.3.5. Detectors

In the early 1960's the first FBS experiments were performed using spectrographs with photographic plates as detectors for spectroscopic a well as time-resolved studies.[21-24] Photographic plates offer in a sense the "multiplex advantage," since they work as signal integrators simultaneously for a whole spectrum or for all intensity decay curves of such a whole spectrum when the beam is imaged parallel on the entrance slit of the spectrograph. Spectra and decay curves thus stay fully unaffected by beam fluctuations. Their low sensitivity can be compensated by the use of image intensifiers[101] so that they are still an attractive alternative to modern photoelectric detectors, in particular for spectroscopic purposes, where their narrow dynamic range (< 3 decades) and poor linearity are not as important. The latter two deficiencies are, however, crucial to intensity decay measurements. Therefore photoelectric detectors have become standard equipment in all FBS laboratories. In general they are operated together with amplifiers and discriminators in pulse-counting mode and offer by this means a wide dynamic range of about 8 decades over which they are also linear, and give low noise. Cooled photomultipliers with quantum efficiencies (QE) between 5% and 30% and $\leqslant 50$ counts/s noise are used from 1000 to 105 nm. Open channeltrons with QE = 2%–20% then cover the range down to 2 nm. At even higher energies, flow-through proportional counters are utilized and Si(Li) detectors take over beyond the use of bent-crystal spectrometers. Open electron multipliers or channeltrons are used for electron detection with particle efficiencies of 60%–80%.

One thus obtains a direct-reading digital signal that can easily be processed with standard electronic equipment. In contrast to the photographic plate, however, one obtains with photoelectric detection only a point-by-point measurement. Hence, such measurements become sensitive to fluctuations of the beam, to the foil alteration, and to foil breakage, against which one has to take precautions by normalization procedures.

2.3.6. Mechanical Target Motion

Except for special requirements (e.g. laser excitation) the target is mechanically moved under controlled conditions relative to the detection region for time-resolved measurements. The target is usually mounted on a kinematically supported carriage which can slide (roll) on a lathe-bed-type track parallel to the beam axis. Also mounted in bearings on and along the track is a precision screw which can be driven from outside the target chamber with a stepping motor. It fits through a backlash-free nut mounted on the carriage such that the carriage position can be controlled step-wise with a precision and reproducibility of $< 25 \ \mu$m.

A gas target is simply mounted on this carriage with a flexible gas inlet system from the outside. For foil exchange under vacuum, special precautions are required for a reproducibility in position of $< 25 \ \mu$m when frequent foil breakage occurs. For this purpose the foil must be floated onto accurately machined target holders with ≤0.3 cm diameter beam holes, so that the foils cannot sag more than 25 μm (!), and the foil exchange mechanism (usually a rotable wheel mounted on the carriage with many foil holders at its outer periphery[117]) must operate within a tolerance of <25 μm.

Such mechanical systems have proved to work with an overall reproducibility of the *foil* position of less than 25 μm,[118] this being necessary for high time resolution quantum beat or extremely short intensity decay measurements. (See Section 3.4 for a case without foil breakage.)

2.3.7. Signal Monitors

The standard procedure in FBS is the normalization of the measured signal relative to the number of target transmitted particles as derived from the charge collected in a Faraday cup[22–24] on beam axis behind the target (see Figure 3). This cup current is well suited as a monitor for measurements at constant beam energy and is thus often used for the recording of spectra or transient decay curves, particularly in electron spectroscopy work, where the geometry of the spectrometer in Figure 10 does not allow other monitoring devices. Its quality suffers seriously, however, from foil deterioration and breakage. These drawbacks can be eliminated by normalizing the signal to the number of *excited* particles after the target for which the "white" light emission at a fixed distance from the target is a good indicator.[119–121] It may be monitored by a photomultiplier mounted either in the vacuum on the target carriage or outside the target chamber with a flexible fiber connection to the target carriage as indicated in Figure 3. Light monitors with wave length selection are rarely used because of their low count rate, which introduces additional statistical fluctuations in the signal.

For the measurement of energy-dependent excitation functions both monitor signals, i.e., the Faraday-cup current or the detected "white" light emission, are unsuitable because of their own energy dependence due to the change of the charge state distribution or of the averaged excitation probability with energy, respectively. The number of target-transmitted particles is then best derived from the number of particles entering the target, i.e., directly from the incoming ion beam current. It can be measured again with the faraday cup when the current reading is corrected for by the ratio of the cup current with and without the foil at each energy setting.[114] It is also possible to insulate the target chamber from its supports as well as from the beam-defining aperture, so that the current collected by the whole chamber corresponds to the particles entering the target.[51] It is obvious, however, that these procedures suffer again from changes in the foil quality during the recording.

2.3.8. Standard Experimental Procedure

For the recording of a spectrum or an intensity decay curve, digital signals from the spectrometer detection (SD) and from the monitor detector (MD) have to be processed in logical relation to either the stepping-motor-controlled wavelength scan or target position scan. This may be done by feeding the MD signal into a preset counter which controls the time during which SD counts are accumulated in a certain channel of a multichannel scaler (MCS) at a certain wavelength (position) setting. The recording of a whole spectrum (decay curve) may then start at λ_0 (z_0) with the accumulation of SD counts in channel 1 of the MCS until the monitor counter reaches for the first time the preset number of counts. At this instant all counting is stopped, the MCS input is advanced to channel 2, the wavelength (position) is stepped by a chosen increment $\Delta\lambda$ (λz), the preset counter is reset, and the system is started again for accumulation into channel 2 at $\lambda_0 + \Delta\lambda$ $(z_0 + \Delta z)$. This procedure is repeated until one whole scan over the desired spectrum (decay curve) is stored in the MCS. Because of the noise in both detectors, however, this normalization is still not perfect and needs further refinement. This would require the simultaneous recording of the real measuring time for each channel in a second MSC in order to correct for the detector noise contributions in the data analysis. This is, however, rarely done and one chooses instead to average over residual fluctuations by repeating and adding many of the above scans in the MCS to the final spectrum (decay curve). The electronic control system for this procedure can be set up with discrete modules or can be fully incorporated into a minicomputer (e.g., PDP 8, PDP 11, HP 2100, etc.), which may work then as an MSC and controls the stepping motors upon receiving the MD counts.

For a quantum beat measurement exactly the same procedure as for an intensity decay measurement is used except that the observed beam section Δz must be chosen to be short compared to the expected spatial period of the beats. The use of polarizers in the optical path is not absolutely necessary but may serve to improve the beat amplitude with respect to the nonbeating background.

2.3.9. Energy Determination

For time-resolved experiments the excited-beam energy must be accurately known for the time scale calibration. It may be determined either by subtracting the calculated energy loss in the target from the presumably known accelerator energy or by a direct measurement behind the target. It should be accurate to within $\pm 1\%$ for lifetime and to within $\pm 0.2\%$ for quantum beat measurements.

For the calculation of the energy loss, however, the stopping power is theoretically known only to within $\pm 20\%$ in the important energy region of 5–1000 keV/amu[70,71] and the determination of the actual foil thickness and its composition introduces further uncertainties of typically $\geqslant \pm 10\%$. Hence one obtains an uncertainty of the calculated excited beam energy which increases with the foil thickness and reaches $\pm 1\%$ at around $3 \mu g/cm^2$ for medium Z ions. Consequently one may tolerate this calculational method for lifetime measurements with thinner targets than $3 \mu g/cm^2$ at all energies, whereas thicker targets used with energies 5–100 keV/amu require an energy measurement. For higher energies this uncertainty becomes relatively less important.[72] One must also note that very often the actual accelerator energy may not be well enough known for the application of this method.

The spectral Doppler shift method of Wien[4] in "end-on" geometry is still quite attractive for the measurement of the average velocity of the excited beam.[122,123] With an assumed standard resolution of $\lambda/\Delta\lambda = 5 \times 10^3$ for this geometry and with an expected accuracy of the determination of the shifted line center to $\pm \Delta\lambda/5$ one can obtain $\pm 1\%$ energy uncertainty above 30 keV/amu and $\pm 0.2\%$ above 750 keV/amu beam energy.

The requirement of $\delta E/E \leqslant 2 \times 10^3$ for quantum beat measurements can at present only be met by an energy analysis, for which 90° electrostatic analyzers as in Figure 3 with fringing field correction have been very successful.[124] Their calibration factor, the ratio between applied voltage and true ion energy, may be deduced either from an accurate mechanical design or from a measurement with known ion energy. In the latter case an atomic clock in the form of a known quantum beat frequency[118,81] can serve for the determination of the ion energy if the accelerator voltage cannot be accurately measured.

Since all charge states leaving a target after equilibration have experienced the same history, they have also been found to possess the same energy.[118,81] The energy of neutrals can thus also be measured with an energy analyzer by measuring any one of the simultaneously appearing charged ions.

However, the multiple collisions in the target also complicate the determination of the average energy of the emitting ions, since the mean scattering angle is related to the nuclear mean energy loss.[75] Therefore beam energy spectra must be measured as a function of scattering angle in order to find the "true" average energy. A possible dependence of the excitation probability on the scattering angle has, however, never been taken into account as yet.

As an advantage of the use of energy analyzers the continuous control of the average ion energy during a time-resolved experiment must be also considered. This allows, for instance, firstly, the acceptance only of foils yielding the same average ion velocity for a given measurement, secondly, the replacement of a foil by a new one as soon as it has thickened more than a tolerable amount, and thirdly, the continuous monitoring of foil breakage. These aspects are very important for precise time-resolved experiments in general and quantum beats in particular and it is sad to note that they have been ignored in most FBS experiments to date.

2.3.10. Lifetime Determination

The actual transient measurement for a given "line" (strictly speaking for an observed wavelength interval $\Delta\lambda$) yields a recording of a spatial intensity decay curve from which the lifetime of "the" excited level has to be determined. Even if one assumes that all ions emitting into the wavelength interval $\Delta\lambda$ are decaying exponentially in time with time constant τ, this recording will be composed of constant detector noise, beam background, plus the spatial intensity decay $I(z) = I_0 \exp(-z/v\tau)$ convoluted with the Δz-detection function, the velocity distribution due to energy straggling and angular scattering, and a z-dependent detection probability $\varepsilon(z)$ which accounts for angularly scattered ions missing the viewing region of the spectrometer.

The convolution with the Δz-detection function affects the recorded intensity only up to a distance $\Delta z_b/2$ past the foil, where Δz_b is the base line width of the detection function. Thereafter it decays exponentially again[125] with the time constant τ. This effect must therefore be carefully analyzed only for decay length $v\tau \leq \Delta z_b/2$.

The influence of the velocity distribution due to both the energy straggling and the angular scattering is more serious since it simulates a faster decay for small z and a slower decay for large z. In systematic analysis of this problem it was shown for Gaussian velocity and angular distributions[126,127] with $1/e$ full widths of 0.2–$0.3v'$ and 0.2–0.3 rad, respectively, that the true decay constant may appear to be shortened by 1%–3% in the recorded intensity decay curve. This applies only to the worst FBS conditions of low-energy (~ 2.5 keV/amu) medium-Z ions excited by 8 μg/cm^2 foils. At higher energies and thinner foils this deviation from the true decay constant becomes much smaller and can be neglected compared to other sources or errors.

The detection probability $\varepsilon(z)$ is, in principle, a function with negative slope and thus always simulates a shorter decay constant. Depending on each individual experimental design, on the average scattering angle, and on the decay length to be measured, it may easily cause errors of a few percent. In addition it also affects the beam background contribution to the recorded signal. This causes further complications in the final extraction of the "true" decay constant, but it also gives a direct way of measuring $\varepsilon(z)$. Setting the spectrometer on a residual gas line directly allows the recording of $\varepsilon(z)$ plus detector noise when the same ions are passed through the same foils at the same energy as for the actual intensity decay measurement. This method has been successfully used in precision experiments[81] and allows one, in principle, to correct the recorded intensity decay curve of interest with a measured $\varepsilon(z)$ function. Hence, one may conclude that the above-

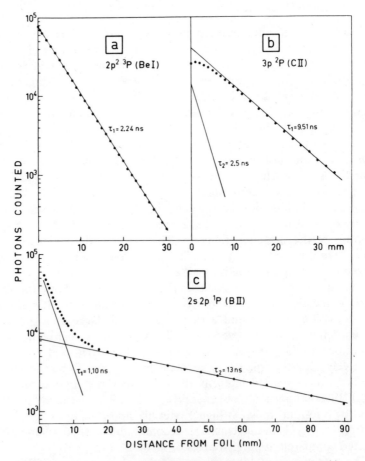

Figure 12. Three characteristic examples of decay curves encountered in FBS after foil excitation.

mentioned influences on the recorded signal can be well controlled in a *careful* experiment at a level of accuracy of $\Delta\tau/\tau \approx 0.01$ so that the recorded spatial intensity decay curve can be accepted as a true replica of the time-dependent intensity decay.

The initial assumption of an ideal exponential decay in time for the above discussion is, however, fully unrealistic.

Firstly, the poor spectral resolution in FBS often does not allow a clear separation of nearby lines so that more than one line (blend lines) may be detected in the wavelength interval $\Delta\lambda$ and therefore also more than one decay constant appears in the recorded intensity decay curve. Even with the use of refocused spectrometers and gas targets (not for short decay lengths!) one may therefore be forced to measure only well-resolved lines or to deal with several decay constants in the data analysis.

Secondly, since the collisional excitation mechanism populates essentially all excited levels of a given charge state, any level one wants to study is repopulated by cascades. Consequently all measured decay curves in collision excited FBS experiments are to a greater or lesser extent composed of several exponential decays. It is thus an enormous problem to extract the "true" lifetime of a given level from the measured decay curve and it is this problem which ultimately limits the relative accuracy of careful FBS lifetime determinations.

Three characteristic examples[34] of measured decay curves are assembled in Figure 12. The curve (a) shows a seemingly single exponential decay which is usually interpreted as "unaffected by cascades" so that its slope may directly yield the lifetime. The curve (b) is an example of a "growing-in" cascade, which appears as the difference of two exponentials. It is usually interpreted as a situation where the level of interest ($\tau = 9.5$ ns) is fed by a "single" faster decaying level ($\tau = 2.5$ ns). Curve (c), which appears as the sum of two exponentials may be regarded as the standard case in FBS. The general interpretation is the feeding of the level of interest ($\tau = 1.1$ ns) by a "single" longer-lived higher level.

According to these interpretations the customary method for extracting mean lifetimes from the measured decay curves consists of approximating the measured curve through the adjustment of the parameters I_i and τ_i in the fitting function $I(t) = \Sigma_i I_i \exp(-t/\tau_i)$ via a nonlinear least-squares fitting procedure. The physics involved clearly requires an infinite sum of exponentials to be fitted to the data. However, it is seldom that FBS decay curves can be fitted to more than three exponentials (six free parameters), and the examples in Figure 12 are very well approximated by only one or two exponentials. The reason for this fundamental restriction is the limited accuracy of each individual data point and the principal difficulty that most lifetimes involved in a multiple cascade are of the same order of magnitude. Reducing the sum of exponentials in the fitting function to 1, 2, or 3 thus introduces ambiguities that can hardly be eliminated. It is therefore useless to speak of "single" feeding levels in the interpretation of Figures 12b and 12c. Instead, the cascade contributions in these examples represent the average of many levels with one dominating the others. In view of this discussion, measured curves of the type in Figure 12a become the most difficult ones for an unambiguous lifetime determination, since they do not necessarily preclude cascades.

More sophisticated methods of data analysis, incorporation of the knowledge about the level scheme of the ion studied into the analysis, and the measurement of as many decay curves of higher-lying levels as possible in such a level scheme can help to reduce the ambiguities in the lifetime determination of a specific level. An excellent summary on the present state of the art of extracting mean lifetimes from FBS data has recently been presented by Curtis.[128] We may therefore adopt his conclusion: "In the past few years the measurement of beam-foil decay curves has been substantially improved and mean lives measured by this technique must be considered to be among the most reliable values presently available. Cascade effects, once a serious problem, are now correctly accounted for in the vast majority of cases," and we may add that the present standard is characterized by an average relative accuracy of FBS lifetimes of ± 10%.

In order to further improve this accuracy attempts have been made to reduce or eliminate the cascade problem in FBS. Among these, the selective laser excitation of the fast beam[38,81,82] has been the most successful one and has yielded the most accurate lifetime known in atomic physics as yet.[81] Because of the special excitation involved, a special chapter of this article will be devoted to this fast-beam laser (FBL) technique.

For further information on experimental techniques, equipment, and procedures the reader is referred to the conference proceedings,[16–20] the many reviews,[21–39] and in particular to the article[117] by Bashkin in Ref. 40.

2.4. Specific Features of Fast-Beam Spectroscopy

One can summarize the preceding sections with a list of specific features that characterize the fast-beam spectroscopic source:

(a) Unselective foil or gas excitation of all levels of a given charge state.
(b) Any charge state of any element may be obtained by choice of beam energy. Limitations are set by presently available energies of $\leq 10\,\text{MeV}/\text{amu}$.
(c) Collisional perturbation of the excited levels is negligible in a vacuum of $p \leqslant 2 \times 10^{-6}$ mbar.

(d) Owing to a beam density of 10^5–$10^6\,\text{cm}^{-3}$, interionic fields, imprisonment of radiation, and stimulated emission can be fully neglected.

(e) Low excited beam luminosity requires single-photon or -electron counting.

(f) Isotopically pure beams yield clean emission spectra with minor H, C, O-line contamination from foil-ejected atoms.

(g) The spectral resolution $\lambda/\Delta\lambda$ is Doppler limited with refocused spectrometers to 5×10^3 for foil and to 5×10^4 for gas targets. The corresponding limit for electron spectroscopy is presently $E/\Delta E \simeq 150$.

(h) Excellent time resolution of $\Delta t = 10^{-10}$–$10^{-11}\,\text{s}$ with a definition of $t = 0$ to within $\delta t \leqslant 10^{-14}\,\text{s}$.

(i) Time scale calibration to within $\pm 10^{-3}$ in the nanosecond range by excited-beam energy measurement.

(j) Excellent differential and integral transient detection linearity.

(k) Cascade repopulation complicates lifetime determinations.

(l) Levels are coherently excited due to the short excited-state formation time of $\sim 2 \times 10^{-15}\,\text{s}$.

(m) Aligned and oriented excited levels can result from perpendicular and tilted foil interaction, respectively.

(n) Scattering of ions from surfaces at grazing incidence yields strongly oriented excited levels.

(o) Selective laser excitation eliminates the cascade problem and conserves all the other excellent properties of FBS.

3. Levels and Lifetimes

The poor spectral resolution in FBS is compensated to a certain extent by the unperturbed emission and distinct excitation conditions. The special spectroscopic domain of FBS is thus the investigation of levels that are either weakly populated or strongly perturbed in traditional sources. These include highly excited,[42,84] multiply excited,[28,38] and metastable levels.[27] They give rise to many transitions between previously unknown levels and occur in FBS spectra of practically any element. Their spectroscopic investigation, often together with the lifetime determination of their upper levels, will be our main concern in the following sections.

In view of the articles by Beyer and Wiese in this book we devote only two short sections to Lamb-shift measurements and to the main field of FBS, the measurement of lifetimes of well-established levels.

For more detailed information on the topics of this chapter the reader is referred to recent surveys by Martinson,[129] Sellin,[130] Marrus,[88] and Wiese[131] in addition to the many special reviews.

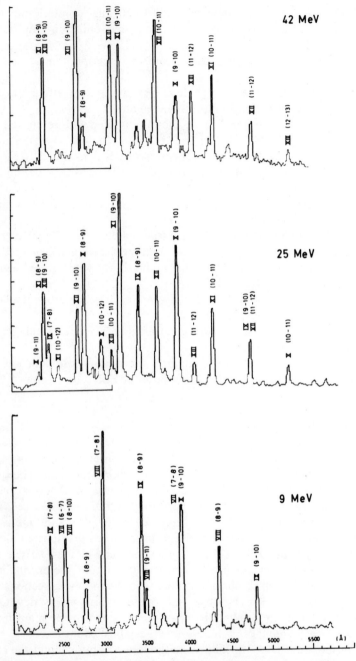

Figure 13. Examples of chlorine spectra after foil excitation at three different Cl$^+$ beam energies.

3.1. Rydberg Levels

In arbitrary ions, hydrogenic levels with high n, l quantum numbers above the ionic core are populated with surprisingly large probabilities by the beam-foil interaction as can be deduced from the assignments in the representative visible spectrum of Cl ions[113] at three different energies in Figure 13. Several authors[45,84,116] relate this finding to the unique interaction mechanism at the foil surface. They argue, in full accord with our earlier discussion, that such levels must be formed at the surface. As possible processes a direct collision with a last layer atom would not account for the large l values observed,[84] an electron capture out of the free-electron gas above the surface would not yield large enough population probabilities,[45] so that an electron capture from a last-layer atom is presently the most widely accepted explanation. The subsequent unperturbed decay after the abundant excitation turns FBS into an ideal tool for the investigation of such levels.

The absolute term values T of these Rydberg levels (measured positive from the ionization limit) can be derived for high-l nonpenetrating orbits from the relativistically corrected hydrogenic term values T_H, which must be further corrected for the dipolar and quadrupolar polarization Δ_p of the ion core.[132] The term values become

$$T = (R\zeta^2/n^2)\{1+(\alpha^2\zeta^2/n^2)[n/(l+0.5)-\tfrac{3}{4}]\}+\alpha_d R\langle r^{-4}\rangle+\alpha_q R\langle r^{-6}\rangle \quad (1)$$

where R is the Rydberg constant for the particular ion mass, ζ is the charge of the ion core, α is the fine-structure constant, α_d and α_q are the dipole and quadrupole polarizabilities of the ion core, respectively, and $\langle r^{-4}\rangle$, $\langle r^{-6}\rangle$ are expectation values calculated with hydrogenic wave functions.[133]

This so-called polarization formula thus allows the determination of α_d and α_q from measured transitions between Rydberg levels.[134,135] Their close relationship to dielectric constants or to shielding factors in hyperfine-structure studies has always attracted great theoretical interest for their calculation.[136] These efforts can now be supplemented by experimental determinations of α_d and α_q for practically any ion core via FBS studies of Rydberg levels.

The smallest polarization effects are expected for systems with one electron outside a closed ion core, i.e., for alkali-like ions. As an example we choose the Na-like level scheme of Cl VII in Figure 14, as constructed from beam-foil measurements with 1–5 MeV Cl$^+$ beams.[137] From this level scheme one can derive $\alpha_d = 0.0434a_0^3$ and $\alpha_q = 0.0478a_0^5$ (a_0 being the Bohr radius), which indeed supports the expectation that for alkali-like Rydberg levels polarization effects can nearly be neglected.

This situation changes dramatically when one electron is added to the core, i.e., for example, a Mg-like Cl VI ion. The additional $3s$ electron

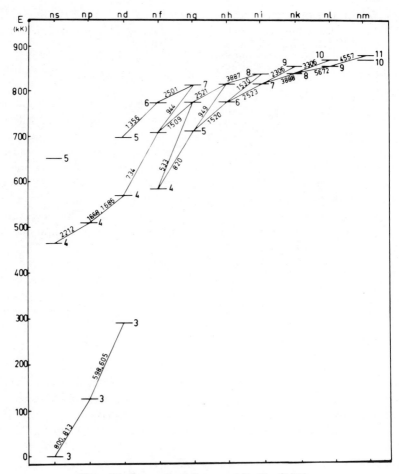

Figure 14. Energy level diagram of Na-like Cl VII.

causes large polarizabilities so that the corresponding Rydberg level scheme can be reproduced[137] with $\alpha_d = 1.62a_0^3$ and $\alpha_q = 9a_0^5$ except for levels strongly affected by possible configuration interaction.[138]

This shows that FBS studies of Rydberg levels are indeed very well suited for the determination of such polarizabilities as well as for a better understanding of the final state formation at the exit of the foil. Rydberg levels of a large number of ions have been identified already in the corresponding beam-foil spectra,[139,140] for which a comprehensive list of references may be found in Ref. 129. Their full interpretation with respect to both aspects[141] is, however, still in its infancy.

Lifetime measurements of Rydberg levels after foil excitation are unfortunately strongly affected by cascades.[84,137,138] This is, in particular, the case for the outermost transitions between $(n, l = n - 1) - (n - 1, l = n - 2)$ levels, for which one can show on the basis of a population probability proportional to ζ^2/n^3 and of theoretical calculations of the transition probabilities[142,143] that they are specifically affected by third-order cascades. Also blending by unresolved $l - l'$ transitions may superimpose further exponentials. Both effects make a proper lifetime extraction out of a measured decay curve extremely difficult. One therefore assumes at present[84,137] that the rather accurate theoretical calculations of the hydrogenic transition probabilities between Rydberg levels[144] are more reliable than the measured lifetimes[137,141,145] after foil (gas) excitation.

3.2. Multiply Excited Terms by Photon Detection

Another category of transitions is also copiously observed in beam-foil spectra as displayed by the marked lines of a Li spectrum[146] in Figure 15. These lines cannot be classified as normally known Li III $nl-n'l'$, Li II $1snl-1sn'l'$, or Li I $1s^2nl-1s^2n'l'$ transitions. Only three of them had previously been observed in a hollow-cathode discharge[147] but were left unclassified until theoretical calculations[148,149] suggested that they connect doubly excited $1snln'l'$ ^4L-quartet terms of neutral Li.

As doubly (multiply) excited terms we define terms with two (or more) electrons excited to higher principal quantum numbers than the ground-state configuration. This definition also includes the well-known displaced terms[150] which converge to a higher series limit than the first one. In this and the next section we shall discuss only multiply excited terms that lie energetically above the first ionization limit and are thus subject to radiative decay *and* autoionization. Since the autoionization process is governed

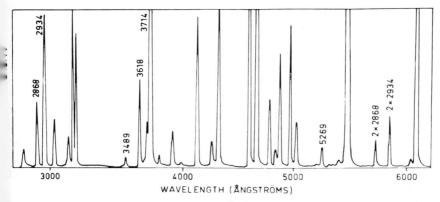

Figure 15. Spectrum of a foil-excited 56-keV $^7Li^+$ beam. The transitions between doubly excited quartet terms in Li I are indicated by wavelength.

Table 1. Selection Rules for Autoionization and Relative Transition Rates; Other Contributions from Hyperfine Interaction are Negligible

Term	Interaction	ΔL	ΔS	ΔJ	ΔF	Parity change	Relative transition rate	Range in τ (sec)
H_C	Coulomb	0	0	0	0	No	1	10^{-15}–10^{-10}
H_{SO}	Spin–orbit	0, ±1	0, ±1	0	0	No	$\alpha^4 Z^6$	10^{-9}–10^{-4}
H_{SOO}	Spin–other-orbit	0, ±1	0, ±1	0	0	No	$\alpha^4 Z^4$	10^{-8}–10^{-4}
H_{SS}	Spin–spin	0, ±1, 2	0, ±1, 2	0	0	No	$\alpha^4 Z^4$	10^{-8}–10^{-3}
H_{FC}	Fermi contact	0	0	0, ±1	0	No	$\alpha^4 Z^4 (m_e/M_p)^2$	$\geq 10^{-7}$

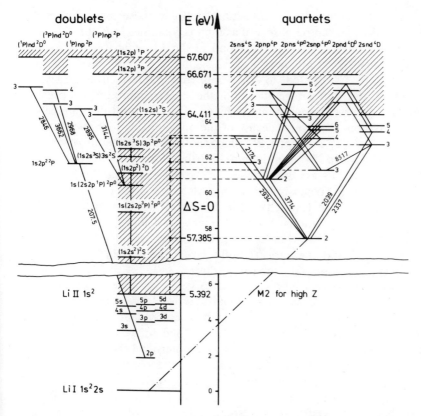

Figure 16. Presently known energy term diagram of Li I including doubly excited doublet and quartet terms. Vertical lines indicate decays via autoionization and the diagonal lines represent radiative decays. The continua are shown as hatched regions.

by the selection rules[151,152] in Table 1, situations due to the angular coupling of multiply excited terms may arise where rapid Coulomb autoionization is forbidden (metastable against Coulomb autoionization) so that radiative decays become competitive with the much weaker relativistic (magnetic) autoionization channels in Table 1. It is the observation of these radiative decays that will concern us in this section. They have been reviewed by Berry[38] and are discussed also in the article by Williams in this book.

We choose Li I as an example for the demonstration of the characteristic multiply excited term features which can be studied by FBS. In this case a whole $1snln'l'$ ^4L-quartet term scheme ($S = 3/2$) may exist in the energy range of 57.4–66.7 eV above the $1s^2 2s$ ground state, which cannot Coulomb autoionize ($\Delta S = 0$) into the adjacent doublet continuum ($S =$

1/2). As a result radiative transitions between these terms can compete successfully and can be observed as the marked lines in Figure 15. A total of 17 new lines could thus be found in the Li beam-foil spectra in the region from 200 to 600 nm.[153-158] Together with theoretical energy calculations,[149,159] limited to 15 terms and modest in accuracy, they represent the key information for the construction of the previously unknown quartet term scheme in Figure 16. This term diagram certainly needs revision, however, because of the questionable identification of the terms $1s2p4s\,^4P^0$ and $2p5s\,^4P^0$ which should Coulomb autoionize into the adjacent $1s2s\varepsilon p\,^4P^0$ continuum. Furthermore, the energy of the $1s2s3p\,^4P^0$ term stays questionable, since the 851.7-nm line could never be observed in beam-foil spectra. Also the higher series members $1s2snp\,^4P^0$ are not justified by the theoretical predictions[149,159] in their present energy positions. With this many, and more, open questions it becomes obvious that even for Li, as one of the "simplest" atoms, further spectroscopic and theoretical studies are necessary in order to finally understand the metastable quartet terms.

The selection rules of Table 1 also lead to doublet terms which are metastable against Coulomb autoionization.[148] The term $1s2p^2\,^2P$ may serve as an example. In pure LS coupling it cannot Coulomb-autoionize to $1s^2\varepsilon p\,^2P^0$ due to $\Delta\pi = 0$ or to $1s^2d\varepsilon\,^2D$ due to $\Delta L = 0$ and thus stays metastable like all $1s2pnl$, $L = l$ terms. Similarly all the doublet terms in Figure 16 drawn outside the hatched doublet continua are metastable against Coulomb autoionization such that radiative transitions between them become observable. So far the six indicated transitions in Figure 16 have been observed and correspondingly assigned in beam-foil spectra.[155,157] Two of them have the rapidly Coulomb autoionizing lower term $1s(2s2p\,^1P)^2P^0$ with large width in common and appear therefore as broad spectral lines.[157] The other type of transitions that is worth noting connects these metastable terms directly with the singly excited Li I terms. The principal transition $1s^22p^2P^0 - 1s2p^2\,^2P$ as indicated in Figure 16 could indeed be identified at 20.75 nm in a beam-foil spectrum.[155] Hence the other series members $1s^22p\,^2P^0-1s2pnp\,^2P$ or the so-called screened satellites $1s^2nl\,^2L-1s2pnl\,^2L$ to the Li II $1s^2\,^1S_0-1s2p\,^1P$ resonance transition are also expected but have not yet been observed in Li I. As a consequence of their existence, however, the lifetime of the $(1s2p\,^1P)3d\,^2D^0$ was found to be extremely short: $63(4)\,ps.$[157]

FBS allows the easy extension of such investigations along the Li I isoelectronic sequence for which previously no information was available. The quartet system of Be II,[156,160-162] B III,[163-165] C IV,[146,166] N V[167-169] has been studied with similar success as for Li I, and several lines have been observed in O VI,[169,170] F VII,[170,171] and Ne VIII.[168,169] It is interesting to note that the quartet terms stay accessible to such radiative studies for high Z owing to the scaling of the E1 transition rate with Z^4 in competition to the relativistic autoionization rates of Table 1. At the same time higher terms of the multipole expansion of the transition operator scale at even higher powers of Z. The M2 transition rate of the forbidden $1s^2 2s\,^2S-1s2s2p\,^4P^0$ decay (broken line in Figure 16) scales,[172] e.g., with Z^8, a transition which could indeed be observed in Al XI at $1563.65(90)\,eV$,[173] in S XIV at $2424\,eV$,[174] and in Cl XV at $2745.5\,eV$.[174] It thus allows the determination of the excitation energy of the $1s2s2p\,^4P^0$ term in these high-Z ions with high precision.

In the doublet system the principal allowed transition $1s^2 2p\,^2P^0-1s2p^2\,^2P$ could also be observed along the isoelectronic sequence up to O VI.[175] Owing to the scaling with Z the lifetimes of the upper terms become extremely short in this case (of the order of 1 ps) and thus require extremely good spatial resolution and accuracy of the mechanical foil drive with a step size as short as 2.5 μm for their determination.[175]

In the heavier alkali isoelectronic sequences multiply excited metastable doublet and quartet terms can also be formed that are only partly homologous to the terms in the Li sequence due to the core excitation from the $2p^6$, $3p^6$, ... subshells instead of the $1s^2$ shell. They are also subject to stronger spin–orbit interaction and therefore more accessible to autoionization electron spectroscopy (see Section 3.3), but radiative transitions should still be observable. So far only a few lines have been tentatively associated with such quartet transitions[176–178] and the satellites to the Na II resonance lines seem to stem from screened doublet transitions.[176]

In order to generalize the occurrence of radiative transitions between multiply excited terms we can say that for all atoms with one electron above closed shells (alkalilike) in their ground state, doublet and quartet terms can be formed above the first ionization limit by multiple excitations which are metastable against Coulomb autoionization. For atoms with two electrons above closed shells (alkaline-earth-like) in their ground state, multiply excited, metastable singlet, triplet, and quintet terms can be formed, and so on. Be I is thus the lightest atom where a $1s2snln'l'$ quintet term scheme may become excited with the lowest term being $1s2s2p^2\,^5P$. In the next heavier atom B I, a $1s2snln'l'n''l''$ sextet term scheme with the lowest term $1s2s2p^3\,^6S^0$ can appear for the first time, and so on. Hence one can expect a large number of radiative transitions between such terms in beam-foil spectra, of which, however, none has been observed or classified as yet, although the existence of such terms has been verified with electron spectroscopy (see Section 3.3).

The so-called displaced terms are just one special class of these generalized multiply excited terms. They leave the closed shells unaffected and occur in atoms with two or more electrons above closed shells. They

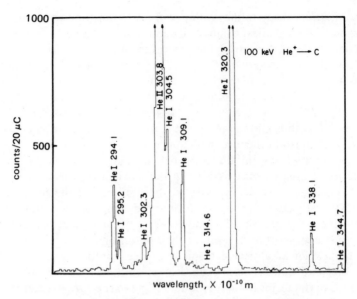

Figure 17. He I $1s\,nl\;^{1,3}L-2p\,nl\;^{1,3}L(L=l)$ satellite transitions accompanying the He II Ly_α 30.38-nm resonance line.

are a standard feature of traditional spectroscopy[150] so that we can restrict the discussion to the single example of Be I. The displaced Be I terms $2pnl\;^{1,3}L$ ($n > 2$) lie between the Be II $2s\;^2S$ and $2p\;^2P$ terms. When $L = l$, they cannot Coulomb autoionize and thus radiatively decay mainly to Be I $2snl\;^{1,3}L$ terms. Hence these transitions from displaced to "normal" Be I terms are "satellites" to the Be II $2s\;^2S-2p\;^2P$ 313.1-nm resonance line (some of them far away from it), which could clearly be observed in beam-foil spectra.[156,161]

The simplest possible atom that can exhibit multiply excited, metastable terms is of course He I (no electron above a closed shell in the ground state). It is mentioned here last because its multiply excited terms have been most intensively studied by traditional means (see Williams in this book) and FBS and have been reviewed at length,[38,129] so that we only need to sketch the main characteristics. While the traditional studies were mainly concerned with the autoionization terms,[179,180] FBS contributed most of the information on the terms metastable against Coulomb autoionization.[181–184] As before, these are the $2pnl\;^{1,3}L$ ($L = l$) terms which decay dominantly to $1snl\;^{1,3}L$ terms and thus give rise to satellites to the He II–Ly_α 30.38-nm transition as impressively demonstrated by a beam-foil spectrum[182] in Figure 17. Also a large number of transitions between the metastable terms could be identified in FBS spectra[184] and the term diagram could be constructed (see Figure 6 of the article by Williams).

For Li II the same type of $1snl$ $^{1,3}L$–$2pnl$ $^{1,3}L$ ($L = l$) satellite lines to the Li III–Ly$_\alpha$ 13.5-nm transition could be observed with the lifetimes of the upper terms varying from 10 to 54 ps.[155,183,184] For the higher iso-electronic sequence members FBS observations are sparse and limited to a few such satellite transitions to Ly$_\alpha$ lines.[166,185]

With the examples discussed FBS has proved to be an excellent tool for the photon spectroscopic study of multiply excited terms. The abundant population of these terms is consistent with the "independent-electron model" of the excited-state formation at the exit surface of the foil[65] and can thus be expected for all elements. Hence one can look forward to a wealth of new information on transitions connecting multiply excited terms from future FBS studies.

3.3. Multiply Excited Terms by Electron Detection

All terms discussed in Section 3.2, i.e., multiply excited terms above the first ionization limit and metastable against Coulomb autoionization, do autoionize in competition to possible radiative decay modes with a finite probability due to the relativistic (magnetic) interactions in Table 1 and can thus also be studied by means of electron spectroscopy. [134] In this section we call them metastable or delayed autoionizing terms in contrast to prompt autoionizing terms, by which we define multiply excited terms above the first ionization limit which predominantly decay by allowed Coulomb autoionization. These two categories of terms, as discussed also in the article by Williams in this book, differ by orders of magnitude with respect to their lifetimes and can therefore easily be distinguished in a fast-beam electron spectroscopy experiment with gas or foil excitation. Autoionization electrons of prompt decaying terms (10^{-12}–10^{-15} s) can only be detected right at the interaction region of a gas target or right after the foil exit, whereas *only* electrons from delayed autoionizing terms (10^{-5}–10^{-12} s) are detected at some distance (delay time) downstream from the interaction region.

This becomes evident from the prompt (a) and the delayed (b) Li electron spectra in Figure 18, which were recorded with the spectrometer in Figure 10 at 282 keV beam energy after foil excitation.[186] Owing to the angle of detection of 42.3° relative to the beam axis, the measured laboratory electron energy (E_{LAB}) must be transformed into the actual spec-troscopically interesting electron energy in the center-of-mass system of the emitting atom (E_{cms}), as described in a recent review[130] and else-where.[187–189] This transformation reduces the resolution in Figure 18 to only $E/\Delta E \approx 36$.

For a short interpretation of the spectra we again use the term diagram in Figure 16, where autoionization decays are indicated by vertical lines and for which the excitation energies are obtained by adding E_{cms} to the Li I ionization potential of 5.392 eV. The first (lowest) peak of the delayed spectrum can be associated with the lowest quartet term $1s2s2p\ ^4P^0$, which is also metastable against E1 radiative decay. According to Table 1 its $J = 1/2,\ 3/2$ levels can autoionize via SO and SOO interaction to the $1s^2\varepsilon p\ ^2P_{1/2,3/2}$ continuum states, whereas the $J = 5/2$ level can decay only via the weaker SS interaction to the $1s^2\varepsilon f\ ^2F_{5/2}$ continuum state. This leads to different autoionization rates for different J levels within this term,[190] a phenomenon that is generally known for relativistic interactions as differential metastability[190] and will become important for lifetime measurements later on. For spectroscopic purposes the average transition rate only is of interest and is of the order of $10^6\ s^{-1}$ for this case[190] easily allowing the observation of the resulting εp and εf electrons as the proposed line in the spectrum. Its energy of $E_{cms} = 52.1(5)$ eV directly yields the excitation energy of the $1s2s2p\ ^4P^0$ term of $E = 57.5(5)$ eV above the Li I ground state in accord with an earlier result[152] of 57.3(3) eV. The large uncertainty of the FBES result stems mainly from the uncertainty of the beam velocity, which is crucial for the absolute calibration of the energy scale in this detection geometry.

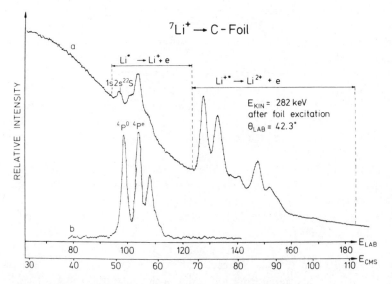

Figure 18. (a) Prompt ($t = 0$ ns) and (b) delayed ($t = 3.7$ ns) spectrum of autoionization electrons from a foil-excited Li beam.

Similarly all the other quartet terms autoionize and give rise to electron emission at similar rates but in competition to allowed E1 decay modes. One therefore expects only strong autoionization lines from the decay of the lower terms which are fed by cascades from the higher ones. The corresponding autoionization transitions are indicated in Figure 16 as broken lines. They allow us to assign the decay of the $1s2p^2\,^4P$ term to the second peak, which appears indeed at 3.33 eV above the first peak, a separation that is in good agreement with the optical $1s2s2p\,^4P^0$–$1s2p^2\,^4P$ transition energy of 3.338 eV.[147] There are probably other terms $(1s2s3p\,^4P^0, 1s2s3s\,^4S)$ contributing to the second peak, but poor resolution does not allow any further explanation of the rest of the delayed spectrum with the help of the present quartet term diagram.

No evidence has been found as yet that metastable Li I doublet terms or Li II singlet and triplet terms are contributing to the delayed electron spectrum. This is in accord with the observation of allowed E1 radiative transitions of the satellite types Li I $1s^2nl\,^2L$–$1s2npl\,^2L$ $(L = l)$ and Li II $1snl\,^{2,3}L$–$2pnl\,^{1,3}L$ $(L = l)$, respectively[154,157] which are responsible for the extremely short lifetimes of these metastable terms of the order of 1–100 ps.[154,157] At the delay time of the spectrum no such term has survived.

The prompt spectrum in Figure 18 exhibits two distinct autoionization line structures superimposed on a continuous electron background. This background is due to direct collision processes in the foil and can be subtracted. The resulting hardly resolved spectral structures up to $E_{cms} = 65$ eV can be associated with Li I promptly decaying doublet terms and above $E_{cms} = 65$ eV with Li II promptly decaying singlet and triplet terms. On the basis of earlier uv-absorption experiments[191] and theoretical predictions[192,193] a gross interpretation is possible,[186] but it would of course be more satisfactory to improve the experimental resolution for a detailed analysis.

By replacing the foil by a gas target in the detection region much of the dynamic line broadening can be eliminated. Using the same type of spectrometer a prompt spectrum with greatly improved resolution could thus be measured in Figure 19, which covers the energy range of Li I autoionization electrons.[108,194] The assignments in Figure 19 coincide with doublet terms[108] drawn within the doublet continuum of Figure 16) which promptly decay to the $1s^2\varepsilon l$ continuum states (vertical full lines). Three terms were known already,[191] so that an accurate energy determination of the others was possible with an accuracy of ± 0.05 eV.

A further improvement of resolution by a factor of 2 was achieved by exciting Li vapor at rest with a He^+ beam and by analyzing the emitted electrons with a parallel plate spectrometer.[109] The same lines as in Figure 19 appeared in the spectrum, but in addition also the $1s2s2p\,^4P^0$ decay was observed. With the improved accuracy of this experiment it was possible to determine the term energies with an accuracy of ± 0.01 eV, so that the quartet term diagram is now absolutely calibrated with this accuracy. Also the three Li II $2s^2\,^1S$, $2s2p\,^3P$, and $2s2p\,^1P$ term energies could be determined with the same accuracy in this experiment. This method of target deexcitation must therefore be considered as a successful alternative to FBES for the investigation of promptly autoionizing terms as long as atoms with a low degree of ionization are to be studied. It is, however, an open experimental question at present whether it can be applied for high charge states or not.

Along the isoelectronic sequence delayed spectra similar to the Li I spectrum have been observed[188,189,108,195,28] up to Ar^{15+}. In general the two lowest peaks can be associated with the $1s2s2p\,^4P^0$ and $1s2p^2\,^4P$ terms, of which the second one should be rapidly depleted at longer delay times owing to the allowed optical E1 transition to the $1s2s2p\,^4P^0$ term. However, starting with Be, spectrometric structures above the lowest peak in Figure 20 survive with long decay constants which are assumed not to stem from cascading. These structures could be interpreted for Be as the autoionization decay of the lowest Be I $1s2s2p^2\,^5P$ term[106] to the two Be II $1s^2 2s$ and Be II $1s^2 2p$ final atomic terms which give rise to two lines separated by the verified final-state splitting of 3.96 eV. In B spectra one

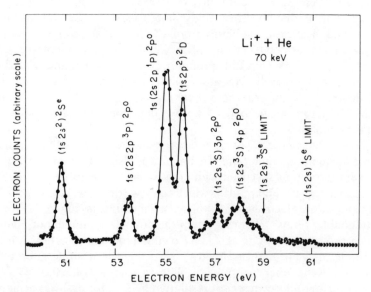

Figure 19. Prompt spectrum of Li autoionization electrons emitted from a He-gas-excited Li beam. The energies are given in c.m.s. Note the improved resolution compared to Figure 18.

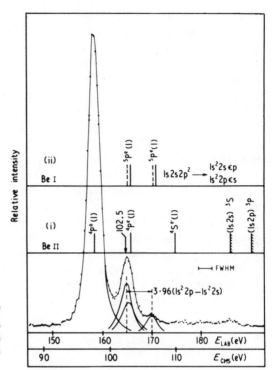

Figure 20. Delayed spectrum of Be autoionization electrons emitted at a delay of 7.4 ns from a 284-keV Be beam after foil excitation.

more peak compared to the Be spectrum survives at long decay times and can be associated with the autoionization of the lowest BI $1s2s2p^3\,{}^6S$ sextet term.[196] These are indications that metastable quintet and sextet term systems exist and that they are populated by the beam-foil inter-action.

Depending on the charge-state distribution in measurements on the higher Li isoelectronic sequence members one can interpret initially uni-dentified lines as quintet and sextet contributions.[107,188,197] It may serve as an example to mention one of the first delayed fast-beam electron spectra in Figure 21 as measured by the Oak Ridge group,[189] which initiated and established the field of FBES.[188,108,28,195] The peaks A and B could initially not be interpreted but are now assumed to stem from quintet and sextet terms.[188,107,197]

Much interest was also devoted to the lifetime measurements of the $1s2s2p\,{}^4P^0$ quartet term[188,189,108,195,28] along the isoelectronic Li I sequence. According to Table 1 the relativistic interactions scale with different powers of Z. The differential metastability of this term thus leads to very fast decays of the ${}^4P^0_{1/2,3/2}$ levels for high Z so that the decay of the ${}^4P^0_{5/2}$ level can be measured alone at longer delay times as shown in Figure

22 for Cl^{14+}. Since only the spin–spin interaction is responsible for its decay, lifetime measurements are thus a direct measure of this interaction. The results for $Z = 8$–18 have been recently reviewed by the Oak Ridge group,[198] which show a systematic deviation from the semiempirical scaling law[190] $1/\tau(^4P_{5/2}) = 1.13 \times 10^5 \ (Z - 1.75)^3 \ s^{-1}$. This deviation was found to be due to the rapid onset of an M2 radiative decay channel to the Li I $1s^2 2s$ ground state[172] which scales with $\sim Z^8$. Incorporation of this decay mode into the theory gives good agreement with the experiments somewhat closer to a $1/\tau \propto 1.13 \times 10^5 \ (Z - 1.75)^4 \ s^{-1}$ scaling. As a result of these findings attempts were made to directly observe this M2 radiation. They were successful in the x-ray region for both S^{13+} and Cl^{14+} and yielded lifetimes in excellent agreement with the electron measurements.[199]

Recently FBES has also been applied for the observation of prompt spectra of systems homologous to Li or Be^+ after gas excitation.[108] The rather complex spectra obtained can, however, be interpreted only with moderate success, mainly due to the lack of theoretical predictions and to poor resolution.

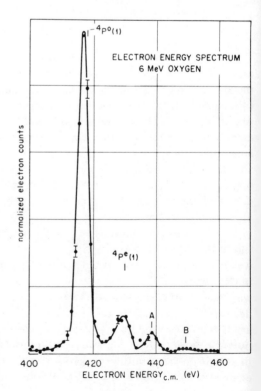

Figure 21. Features of a delayed electron spectrum from a 6-MeV oxygen beam near the lowest Li-like quartet autoionization decay.

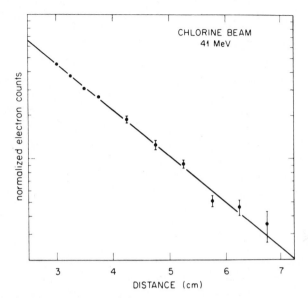

Figure 22. Intensity decay curve of electrons from the autoionization of the $(1s2s2p)^4P_{5/2}$ level of Cl^{14+}.

We may thus conclude that FBES has proved so far to be a valuable and complementary technique for the study of multiply excited terms. In particular it also allows the study of Coulomb autoionizing terms that are otherwise not accessible. However, it is also obvious from the examples presented that a greatly improved resolution is necessary in FBES before it can significantly contribute to the understanding of multiply excited term structures. The use of parallel-plate spectrometers in end-on geometry should allow such improved resolutions.

3.4. Metastable Levels in H- and He-like Ions

In the past, little experimental information was available concerning the forbidden decay modes of the $n = 2$ levels in H- and He-like ions. This was mainly due to the lack of suitable experimental techniques for the unperturbed study of these levels. The very first experiment in this field was thus reported as late as 1965 on the He II $2s\ ^2S_{1/2}$ two-photon (2E1) decay,[200–219] a decay mode that had been theoretically predicted as early as 1931.[203] Further theoretical work since 1940,[204–219] mainly motivated by astrophysical spectroscopic observations,[220–224] has clarified the decay modes of these levels as shown in Figure 23 for Ar^{17+} and Ar^{16+} as examples.

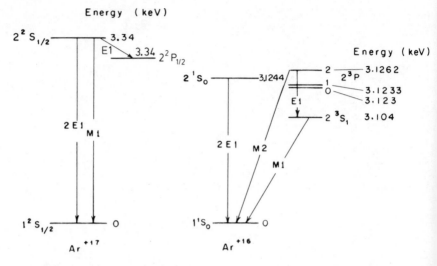

Figure 23. Decay modes of H-like and He-like $n = 2$ levels in Ar^{17+} and Ar^{16+}.

These theoretical calculations predict a scaling with high powers of Z for the "forbidden" transition probabilities, so that their use in astrophysical problems[225] requires laboratory tests of the theory on high-Z ions, for which FBS offers ideal conditions. The first such FBS experiment was reported in 1968[226] and was followed by great activity in this field at Oak Ridge, Berkeley, and Kansas State. We shall discuss only the main aspects of these experiments following the excellent reviews[27,37,88,227–229] on this field and include the latest results in a comparison between theory and experiments in Figures 25a, b.

According to Eq. (2)[207] and Eq. (3)[217] the two-photon transition probability

$$A_{2E1}(1\,^2S_{1/2}-2\,^2S_{1/2}) = 8.2283(1)\,Z^6\,s^{-1} \tag{2}$$

$$A_{M1}(1\,^2S_{1/2}-2\,^2S_{1/2}) = 2.496 \times 10^{-6}\,Z^{10}\,s^{-1} \tag{3}$$

of the $2s^2\,S_{1/2}$ level to the ground state $1s\,^2S_{1/2}$ outweighs the M1 transition probability up to $Z \approx 42$, as can be extrapolated with Figure 25a. Fully negligible in comparison to these two decay modes is the E1 decay to the $2p\,^2P_{1/2}$ level, which can be approximated by[88]

$$A_{E1}(2\,^2P_{1/2}-2\,^2S_{1/2}) = 2 \times 10^{-10}\,Z^{10}\,s^{-1} \tag{4}$$

In experiments with $Z < 45$ it is thus of interest to clearly identify the 2E1 transition, which is characterized (a) by the simultaneous emission of two photons, the energy-sum of which equals the transition energy E_0, (b) by an angular correlation $(1 + \cos^2 \theta)$ between the two photons, and (c) by a broad, symmetric energy distribution of the photons between zero and E_0. For the case of Ar^{17+} this identification was possible by observing the two photons in coincidence with two Si(Li) detectors mounted on opposite sides of the beam facing each other.[227] The pulse heights of coincident pulses could be added electronically and the pulses of the energy sum so obtained yielded a line at the expected transition energy of 3.3 keV in the spectrum in Figure 24b whereas the coincident "singles" output of each detector alone gave a continuous energy distribution up to 3.3 keV in Figure 24a, distorted of course by the energy-dependent efficiency of the detectors. In order to reduce the background from the He-like Ar^{16+} $1s^2\,^1S_0$–$1s2s\,^1S_0$ two-photon decay, which has similar transition probability,[210]

$$A_{2E1}(1\,^2S_0\text{–}2\,^1S_0) = 2 \times 8.228(Z-\sigma)^6\,s^{-1} \tag{5}$$

(σ being a screening constant) a fully stripped Ar^{18+} beam of 412 MeV was excited by a thin nonequilibrating $10\ \mu g/cm^2$ C foil to yield a ratio of $Ar^{17+}/Ar^{16+} \approx 20:1$. This allowed the use of even the noncoincident pho-

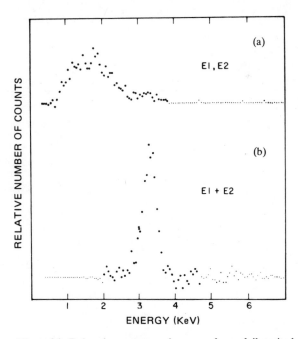

Figure 24. Delayed spectrum of x-rays from foil-excited Ar^{17+} beam. (a) Number of photons contributing to a coincident event in (b) vs. photon energy. (b) Number of coincident events vs. sum energy of the two coincident photons.

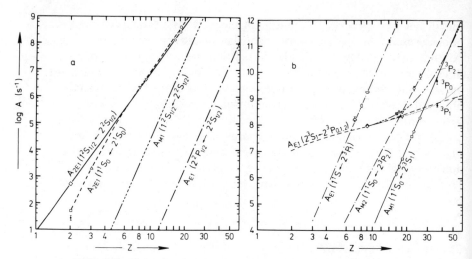

Figure 25. Comparison of theory and experiment for transition probabilities of various decay channels of H- and He-like ions. (a) For dominant 2E1 decays. (b) For dominant E1, M2, and M1 decays. References (a): $\Diamond = 232$, $\nabla = 231$, $\bigcirc = 227$, $\phi = 235$, $\square = 233$, $\triangle = 234$; (b): $\simeq = 226$, $\triangleleft = 247$, $\triangledown = 248$, $x = 249$, $\Diamond = 244$, $\blacksquare = 246$, $\triangle = 243$, $\nabla = 245$, $\Phi = 239$, $\mathbb{D} = 238$, $\square = 240$, $\bigcirc = 239$, $\bullet = 241$.

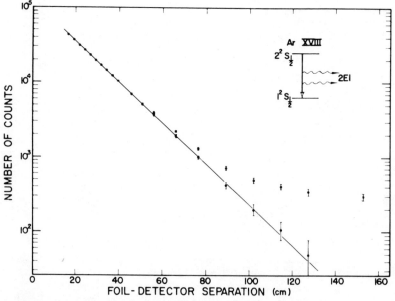

Figure 26. Intensity decay curve of the 2E1 spectrum of Ar^{17+} shown in Figure 24a.

ton pulses of Figure 24a of both detectors for the current-normalized intensity decay measurement versus distance from the foil in Figure 26. After subtraction of a constant background the lifetime $\tau(\text{Ar}^{17+} 2\,^2S_{1/2}) = 3.54(25)$ ns could be extracted from this measurement,[230] in good agreement with Eq. (2), as indicated in Figure 25a. The same measurement on S^{15+} yielded $\tau = 7.3(7)$ ns, also shown in Figure 25a to be in good agreement.

An interesting alternative for the detection of the population $N(2\,^2S_{1/2}, t)$ was employed for the same lifetime measurements on O^{7+} and F^{8+}. By letting the excited beam pass through a magnetic field region the Lorentz electric field quenched the $2\,^2S_{1/2}$ level and the subsequent Ly_α radiation could be observed by a proportional counter as a measure of $N(2\,^2S_{1/2})$. The lifetimes obtained[231] were $\tau(\text{O}^{7+}) = 453(43)$ ns and $\tau(\text{F}^{8+}) = 237(17)$ ns, also in good agreement with the theory (in Figure 25a), which is thus proved to be correct within $\pm 10\%$ from $Z = 2$, where a different type of experiment[232] also agrees with theory, up to $Z = 18$.

The two-photon decay of the He-like $1s2s\,^1S_0$-level of Ar^{16+} could also be measured in a manner essentially identical to the Ar^{17+} experiment.[227] By exciting an Ar^{14+} beam of 412 MeV by the same thin foil a ratio $\text{Ar}^{16+}/\text{Ar}^{17+} \approx 6:1$ could be obtained for this case. In spite of this ratio the measured decay curve had to be corrected for contaminations by the undistinguishable Ar^{17+} 2E1 photons. The final result is $\tau(\text{Ar}^{16+}- 2\,^1S_0) = 2.3(3)$ ns in good agreement with the theory.[219] Also shown in

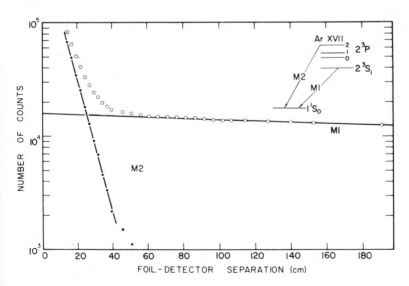

Figure 27. Intensity decay curve of the 3.1-keV transition of $\text{Ar}^{16}+$.

Figure 25a are two experimentally different measurements on $He^{(233)}$ and $Li^{+(234)}$ which prove the theory to be correct within the $\pm 15\%$ from $Z = 2$ to $Z = 18$ for this level too. One earlier result on He deviates by a factor of 2 [37]

The decay of the He-like $2\,^3S_1$ level was initially assumed also to proceed via a spin–orbit-induced two-photon decay.[204] Astrophysical observations suggested later, however, that a M1 transition is the dominant decay mode.[221–224] This finding was supported by renewed theoretical efforts, which predict an M1-transition probability[219] as shown in Figure 25b whose asymptotic Z dependence for large Z can be approximated by[215,227]

$$A_{M1}(1\,^1S_0 - 2\,^3S_1) = 1.66 \times 10^{-6} Z^{10}\,s^{-1} \tag{6}$$

In the first FBS experiments on this transition the single-photon decay could indeed be verified by observing spectral lines at the energies predicted[236] for Ar^{16+}, S^{14+}, and Si^{12+}. The subsequent lifetime measurements on these ions up to delay times of only $\sim 1/4\tau$ $(v\tau \approx 8m!)$ in Figure 27 seemed to indicate, however, a systematic deviation from the theory to shorter lifetimes,[227,231,237] which could only be accounted for when measurements up to delay times of 1.2τ revealed a superimposed faster decaying $(\tau' \approx \tau/2)$ component near the foil,[238] probably a blend line.[229] Improved measurements[238–241] over long delay times yield good agreement with the theory for ions from $Z = 16$ to $Z = 36$, as shown in Figure 25b. An experimentally different measurement on He I with $+$ 300% and -100% error bars has also been reported,[242] in qualitative agreement with the theory.

Still closer to the foil one finds in all measurements on $2\,^3S_1$ an additional fast decaying blend[227,239,243] in Figure 27 which stems from the unresolved M2 transition from the $2\,^3P_2$ level to the ground state $1\,^1S_0$. The transition probability is shown in Figure 25b and approaches a Z^8 scaling for high Z.[219] It competes with an allowed E1-transition to the $2\,^3S_1$ level with a probability[219] also shown in Figure 25b. A_{E1} is larger for $Z < 18$ and A_{M2} becomes dominant for $Z > 18$, but the difference between both is less than one order of magnitude above $Z = 12$. We therefore plot in Figure 25b only the experimental transition probabilities of the stronger decay mode as derived from the measured total transition probability after subtraction of the theoretical transition probability of the weaker decay mode. The results shown for $Z = 9^{(244)}$ and $18^{(245)}$ were obtained by E1 observation with better accuracy than the others from M2 observation.[239,243,246] Good agreement with theory[219] is found for all measured cases except for Cl^{15+}, which seems to agree with another calculation.[243]

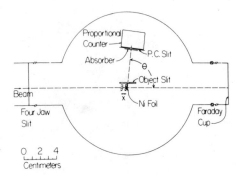

Figure 28. Schematic view of a Doppler-tuned x-ray spectrometer, specially designed for FBS.

0   2   4
Centimeters

A similar situation arises for the decay of the $2\,^3P_1$ level either via allowed E1 transition to the $2\,^3S_1$ level or via spin–orbit-induced E1 transition to the $1\,^1S_0$ ground state. The transition probability of the $1\,^1S_0$–$2\,^3P_1$ decay mode scales approximately with $\sim Z^{10}$ and becomes considerably larger than the $2\,^3S_1$–$2\,^3P_1$ transition probability for $Z \geqslant 7$.[219] The experimental results for N, O, F plotted in Figure 25b are again derived from the measured total transition probability[227,247,248] by subtracting the contribution from the weaker $2\,^3S_1$–$2\,^3P_1$ transition, whereas the results for Si and S do not require this correction.[249] The agreement with the theory is found to be quite good in all five cases.

It is worth noting that the Si and S measurements[249] could be performed with a single foil and an interferometer-controlled step size of only 0.5 μm(!) [recall Section 2.3(f)], a spatial resolution of $\Delta z = 20$ μm, and a so-called Doppler-tuned x-ray spectrometer (DTS). This spectrometer was specifically designed for FBS[250,251] and deserves therefore a short explanation. It makes full use of the Doppler effect instead of compensating for it. It consists of an assembly of detector and absorber with sharp absorption edge which can be rotated (θ) around the observation region as shown in Figure 28.[252] If one assumes that the ions emit in their rest frame a single x-ray line energetically below the absorption edge, which is, at $\theta = 0°$, Doppler shifted in the lab-frame to higher energy above the absorption edge, then the detector signal will reproduce the edge structure, as θ is varied. Several close-lying emission lines thus yield a multistep detector signal as shown in Figure 29a,[252] which after differentiation gives the x-ray line spectrum in Figure 29b. The DTS provides higher detection efficiency than a crystal spectrometer of comparable resolution and is thus advantageous for FBS work. For the lifetime measurements of the He-like Si and S $2\,^3P_1$ levels the difference of the detector counts above and below the absorption edge had to be recorded versus distance from the foil.[249,252]

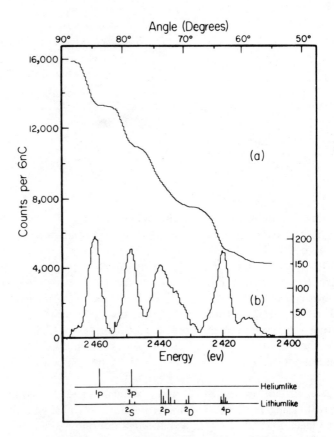

Figure 29. X-ray spectrum as observed with a Doppler-tuned spectrometer from a 60-MeV sulfur beam after foil excitation. (a) Recorded signal. (b) Derivative of the signal in (a).

The $2\,^3P_0$ level has no "forbidden" decay channel and can only decay to the $2\,^3S_1$ level by allowed E1 transition since the decay to the $1\,^1S_0$ ground state is strictly forbidden when the nuclear spin $I = 0$. Experiments on F^{7+}, Ar^{16+}, and Kr^{34+} have been performed[241,244,245] which yield agreement with the theory[219] in Figure 25b only for F and Ar, thus proving that the effect of the hyperfine interaction, which may mix $^3P_0$ with $^3P_1$, is still negligible for the case of F. The measured transition probability for Kr is lower than the theory[219] used throughout Figure 25b, which does not include the Lamb shift. The inclusion of the Lamb shift reduces the $2\,^3S_1$–$2\,^3P_0$ energy splitting ω_{10} such that the ω_{10}^3 factor in the expression for the transition probability reduces the calculated value and brings it into agreement with the measurement.[241]

In conclusion, generally excellent agreement within $\pm 10\%$ has been found between numerous experimental FBS results and the theoretical transition probabilities of the "forbidden" decay modes of H- and He-like $n = 2$ levels. The theory has thus been proved to provide reliable transition probabilities for these levels with an accuracy $\leq \pm 10\%$ for use in other fields of physics. Further FBS experiments are expected to clarify the influence of the Lamb shift and hyperfine interaction on the transition probabilities as well as to detect the M1 $1\,^2S_{1/2}$–$2\,^2S_{1/2}$ decay channel for the first time.

3.5. Lamb Shift in High-Z Ions

As outlined in detail in the article by Beyer (Chapter 12) in this work, FBS has contributed considerably to the present experimental knowledge of radiative corrections in ionic systems. It is the ease with which H-like $2s\,^2S_{1/2}$ metastable levels can be populated by fast-beam–foil or –gas interaction that has made FBS the basic tool for all H-like $n = 2$ Lamb-shift ($\mathscr{L}$) measurements beyond $Z = 3$.[253-262] The Stark-quenching technique,[253] so far the most widely used method for high Z, additionally takes advantage of the time resolution of FBS by transforming the measurement of the $\mathscr{L}$ splitting into a field and $\mathscr{L}$ dependent lifetime measurement. But also for an improved rf measurement on hydrogen[263,264] the spatial time resolution was necessary to apply the separated rf field (Ramsey) technique with a well-controlled time separation in the nanosecond range. Laser-induced "rf" experiments on high-Z ions[259-261] make use of the fast beam by Doppler-tuning the resonance signal.

The goal of all these experiments, in particular of the high-Z ones, is a test of the theoretical calculations in Table 2, which are based on a series expansion in αZ[265-267] or on closed expressions including all orders of αZ.[268-271] The required relative accuracies of $<0.5\%$ for $Z = 10$ or $<1\%$ for $Z = 20$ could not be achieved in the experiments as yet in order to distinguish between the columns of Table 2. Essentially one is facing the principal problem that the fractional width $\Gamma/\mathscr{L}$ increases with Z as shown in Figure 30. Hence, the line center has to be determined to $1/41$ and $1/29$ of the linewidth for $z = 10$ and $z = 20$, respectively, in order to obtain the required accuracy. With the laser-induced "rf" experiments the authors expect this to be possible in the near future.[259-261]

In order to circumvent this linewidth problem in hydrogenlike ions, considerations[88] and first experimental attempts have been started to use the $n = 2$ He-like metastable levels of Section 3.5 for a test of the high-Z lamb shift.[241] The $2\,^3P_0$–$2\,^3S_1$ transition is particularly well suited for such a study since the Lamb shift alters the transition energy, e.g., in Kr XXXV by $\sim 4\%$. The limitation by the natural linewidth is negligible (see Figure

Table 2. Comparison of Three Theoretical Predictions with Experimental Results of the Lamb Shift

Z	Series[265-267] expansion	Erickson[268]	Mohr[271]	FBS Experiments	Other experiments
1	1.057900	1.057910(10)	1.057867(13)	1.057893(20)[263,264]	1.05790(6)[272]
					1.05777(6)[273]
2	14.04391	14.04464(54)	14.04205(55)		14.0462(12)[274]
					14.0402(18)[275]
3	62.753	62.762(9)	62.7375(66)	63.031(327)[253]	62.765(21)[276]
6	782.87	783.68	781.99(21)	780.1(80)[254,257,258]	
8	2 200.2	2 205.2	2 196.21(92)	2 202.7(11.0)[255,257,258]	
				2 215.6(7.5)[256]	
9	3 360	3 343.1(1.6)	3 339(35)[259-261]		
10	4 869	4 889(5)	4 861.1(2.7)		
20	54 600	56 400(300)	55 116(37)		
30	207 000	234 000(3000)	223 030(320)		
40	460 000	640 000(20000)	600 000(3000)		

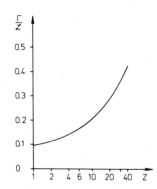

Figure 30. Natural fractional width $\Gamma/\mathscr{L}$ of $n = 2$ Lamb shifts in hydrogenic ions as function of Z.

25b), so that one expects to determine the $^3P_0-^3S_1$ 28.3-nm line center of Kr XXXV to 10^{-5}, thus yielding an accuracy for the Lamb-shift contribution of 2.5×10^{-4} in this case.[241]

3.6. Regularities of Oscillator Strengths

During the past decade FBS has become the major source for mean lives τ of atomic and ionic excited levels. The wealth of information available today requires and allows a systematic organization of the material with respect to relevant atomic variables such that a critical evaluation of the experimental and theoretical data may become possible and predictions about as yet unmeasured levels may be made simply by interpolation or extrapolation techniques. A detailed analysis of such systematic trends is presented in this book's article by Wiese and so we shall restrict ourselves to a few representative examples here.

The Z expansion of the (absorption) oscillator strength has become the most valuable scheme in systematizing the experimental data along isoelectronic sequences.[277-287] The dimensionless oscillator strength f_{ik} of a transition between two levels i (lower) and k (upper) is related to the spontaneous transition probability A_{ki} by

$$f_{ik} = (mc/8\pi^2 e^2)\lambda_{ik}^2 (g_k/g_i)A_{ki}$$
$$= 1.499\lambda_{ik}^2[\text{cm}](g_k/g_i)A_{ki}[\text{s}^{-1}] \tag{7}$$

where λ_{ik} is the transition wavelength, g_k, g_i are the statistical weights of the levels k, i, respectively, and m, c, and e are the electron rest mass, the speed of light, and the electron charge, respectively. Since the lifetime τ_k is determined by

$$\tau_k = 1/\sum_i A_{ki} \tag{8}$$

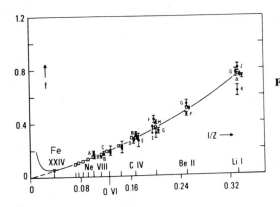

Figure 31. Oscillator strength vs. $1/Z$ for the $2s\,^2S-2p\,^2P^0$ resonance doublet of the Li sequence. Note the relativistic correction (full line) at low values of $1/Z$ to the nonrelativistic $1/Z$ expansion (broken line). For the references see Ref. 131.

a direct determination of f_{ik} by a lifetime measurement is only possible when the sum reduces to a single term, i.e., for excited levels which can only decay to a single lower level,

$$f_{ik} = 1.499\lambda_{ik}^2[\text{cm}](g_k/g_i)(1/\tau_k[\text{s}]) \tag{9}$$

In general, however, several transitions from level k to lower levels are allowed so that the f_{ik} ($f$value) of a specific transition can only be derived from the lifetime τ_k with the knowledge of the relative branching ratio $a_{ki} = A_{ki}\tau_k$, which must be determined, e.g., by absolute intensity measurements of all these allowed transitions. The f value is then given by

$$f_{ik} = 1.499\lambda_{ik}^2[\text{cm}](g_k/g_i)a_{ki}(1/\tau_k[\text{s}]) \tag{10}$$

According to perturbation theory the f value of a specific transition $i \to k$ can be expanded in terms of inverse powers of $Z^{(131,277-287)}$:

$$f = f_0 = f_1 Z^{-1} + f_2 Z^{-2} + \cdots = \sum_{n=0}^{\infty} f_n Z^{-n} \tag{11}$$

and is valid in this form for a whole isoelectronic sequence except near the neutral end where strong perturbations by configuration interaction may occur.[282]

Its asymptotic behavior for large Z approaches in the nonrelativistic limit hydrogenic f values, i.e., $f_0 = 0$ for $\Delta n = 0$ transitions and $f_0 \neq 0$ for $\Delta n \neq 0$ transitions. However, it has recently been shown[131,285,286] that relativistic corrections cause considerable deviations from this behavior for high Z as shown for the well-established trend of the Li $2s\,^2S-2p\,^2P$ sequence in Figure 31. The experimental points are all FBS data obtained between 1965 and 1973 except for a Hanle-effect measurement on Li I They agree with the Z-expansion calculation as well as with the more elaborate calculation[279] close to the neutral end of the sequence. Excep

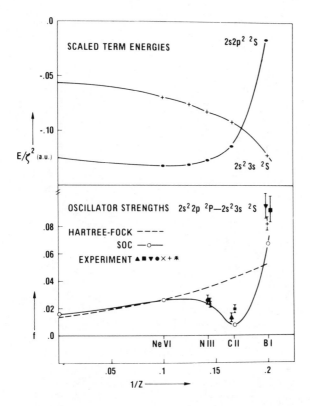

Figure 32. Term energies (in a.u.) and oscillator strengths
for the boron isoelectronic sequence. $\zeta = Z - N + 1$,
with N = number of electrons. For the experimental
values see Ref. 285.

for the value K (1967) they exhibit a consistency within $\pm 10\%$, which is
typical for FBS data.

The great value of such graphical presentation of the experimental
material along well-founded trends is thus clearly demonstrated. It indeed
allows in this case the extrapolation of f values of the higher members of
the sequence. Unfortunately, no experimental data exist as yet to verify the
relativistic effects for $Z > 20$.

Besides the "basic" trend (in Figure 31) which is typical for the
majority of all established cases (nonzero asymptotic behavior
included),[282] considerable irregularities are observed in situations where
either a cancellation in the transition integral or strong configuration
interaction occurs. An impressive example of the latter mechanism is the

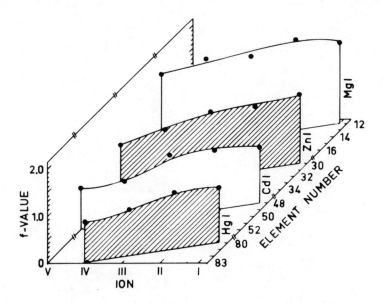

Figure 33. Oscillator strengths for the lowest-lying $ns\ ^1S-np\ ^1P$ transitions in the Mg I, Al II, Si III, P IV, S V homologous sequences.

BI $2s^2 2p\ ^2P-2s^2 3s\ ^2S$ isoelectronic sequence in Figure 32, where close to C II a crossing of the term energies (upper part of Figure 32) of the $2s^2 3s$ and $2s2p^2$ configurations, i.e., also a strong mixing of the wave functions occurs. This produces the pronounced dip in the f value curve in Figure 32, which is well established by the experimental points and a "superposition of configurations" (SOC) calculation.[288] The experimental investigation of such noticeable irregularities with the help of FBS thus allows a direct study of correlation effects in atomic systems.

Regularities of f values can also be found for homologous sequences, which are, however, theoretically corroborated only in an approximate way.[280] We present therefore in Figure 33 only an attempt to establish such trends on the basis of experimental data for the $ns\ ^1S-np\ ^1P$ transitions in the Mg I, Al II, Si III, P IV, S V homologous sequences.[289] This example clearly shows that such regularities may also be useful for the approximate determination of unknown f values by interpolation techniques.

Also f-value regularities within a spectral series are well known.[280,290,291] They may be used in systematizing experimental FBS lifetime data and may help to reveal series perturbations (cancellations) in previously inaccessible series.

4. Coherence, Alignment, Orientation: "Doppler-Free" Methods

For this section one can adopt many of the concepts outlined in the articles by Dodd and Series, Happer and Gupta, Blum, and Beyer elsewhere in this book; these are referred to as $|D+S|$, $|H+G|$, $|BL|$, and $|BE|$, respectively. Further reviews by Fano and Macek,[30] Andrä,[32] Haroche,[292] and Macek and Burns[66] will also be helpful for a deeper understanding of this particular field.

4.1. Quantum Interference—An Introduction

Among all the interrelated coherence effects in fluorescence experiments like the Hanle effect [zero-field level crossing (ZFLC)], high-field level crossing (HFLC), or quantum beats (QB) in atomic physics and the time-integral or time-differential perturbed angular correlation techniques in nuclear physics, the QB interference phenomenon is conceptually the

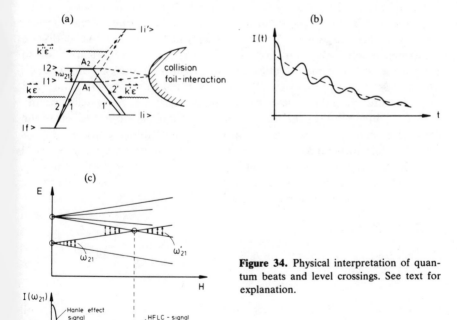

Figure 34. Physical interpretation of quantum beats and level crossings. See text for explanation.

simplest and most basic one from which all the other methods can be derived. Instead of presenting its quantum electrodynamic derivation[292,293] we choose its physical interpretation and show in Figure 34a a four-level system with two nearly degenerate excited levels 1 and 2 separated by an energy interval $\hbar\omega_{21}$, for which we use as a definition throughout this chapter $\omega_{ij} = \omega_i - \omega_j$. We assume that an ensemble of such atoms is irradiated by a broadband light pulse (Fourier-limited bandwidth suffices) of average frequency $\omega_{(\bar{2}\bar{1})i}$ $[\omega_{\bar{2}\bar{1}} = (\omega_1 + \omega_2)/2]$ and length δt and we observe the remitted (scattered) light intensity as a function of time with a detection time window Δt. At first we approximate δt and Δt by δ functions centered at $t_0 = 0$ and t, respectively, and we can then describe this light-scattering process by probability amplitudes A_1 and A_2:

$$A_1 = \text{const} \langle f, t, \mathbf{k}\varepsilon| V_{\mathbf{k}\varepsilon}|1, t, 0 \rangle\, e^{(-i\omega_1 - \gamma_1/2)t} \langle 1, 0, 0| V_{\mathbf{k'}\varepsilon'}|i, 0, \mathbf{k'}\varepsilon' \rangle \qquad (12)$$

$$A_2 = \text{const} \langle f, t, \mathbf{k}\varepsilon| V_{\mathbf{k}\varepsilon}|2, t, 0 > e^{(-i\omega_2 - \gamma_2/2)t} \langle 2, 0, 0| V_{\mathbf{k},\varepsilon'}|i, 0, k'\varepsilon' \rangle \qquad (13)$$

for the two different paths of scattering through levels $|1\rangle$ and $|2\rangle$ in Figure 34a with the notation |atomic state, time, field state⟩, where the time evolution in both levels $|1\rangle$ and $|2\rangle$ is written out explicitly. *If* these two paths are *in principle indistinguishable* a general postulate of quantum mechanics requires that A_1 and A_2 must be added and their sum must be squared in order to obtain the probability $P(f, t, \mathbf{k}\varepsilon)$ for detecting a photon $\mathbf{k}\varepsilon$ and finding the atom in its final state $|f\rangle$ at time t. The two paths are indistinguishable if the transitions 1′, 2′ and 1, 2 in Figure 34a are allowed for photons of the same direction and polarization in the excitation $(\mathbf{k'}, \varepsilon')$ and detection $(\mathbf{k}, \varepsilon)$ processes, respectively, if the transitions 1′, 2′ and 1, 2 are energetically indistinguishable owing to the time–energy uncertainty relation, which requires then $\Delta t, \delta t \ll 2/\omega_{21}$, and if the atoms are left in the same final state $|f\rangle$. These are practically the same interference conditions that we all know from Young's double-slit experiment, except that we deal here in the time–energy domain. As a result of these basic quantum interference considerations we obtain for $P(f, t, \mathbf{k}\varepsilon)$, which is equivalent to the scattered light intensity $I(t, \mathbf{k}\varepsilon)$,

$$P(f, t, \mathbf{k}\varepsilon) \propto I(t, \mathbf{k}\varepsilon)$$

$$\propto \left| \underbrace{\sum_{n=1,2} \langle f, t, \mathbf{k}\varepsilon| V_{\mathbf{k}\varepsilon}|n, t, 0 \rangle \langle n, 0, 0| V_{\mathbf{k'}\varepsilon'}|i, 0, \mathbf{k'}\varepsilon' \rangle}_{B_n}\, e^{(-i\omega_n - \gamma_n/2)t} \right|^2 \qquad (14)$$

which can be simplified by the conditions

$$B_n = B_n^*, \qquad \gamma_1 = \gamma_2 = \gamma \qquad (14a)$$

to the final result

$$I(t, \mathbf{k}\varepsilon) \propto [B_1^2 + B_2^2 + 2B_1 B_2 \cos \omega_{21} t]\, e^{-\gamma t} \qquad (15)$$

Hence, we expect to observe an exponential decay with a superimposed quantum beat in Figure 34b, the frequency of which directly represents the excited level separation. We thus have available an experimental technique that allows the measurement of fine- and hyperfine-structure splittings without applying any external field, and for which the Doppler effect is reduced by a factor of $\omega_{(\overline{21})f}/\omega_{21} \approx 10^5$, since it is the emission probability of *each single* atom that oscillates.

If one chooses not to use time-resolved but time-integral detection, i.e., Δt is spread out from $t = 0$ to ∞, then one obtains the scattered light intensity as a function of the level separation ω_{21}:

$$I(\omega_{21}, \mathbf{k\varepsilon}) = \int_0^\infty I(t, \mathbf{k\varepsilon}) \, dt \propto \frac{2B_1 B_2 \gamma}{\gamma^2 + \omega_{21}^2} \tag{16}$$

It thus establishes the connection between QBs and the Hanle effect as well as the HFLC (see |H+G|) experiments, where ω_{21} is varied via Zeeman effect in response to an external magnetic field variation as schematically shown in Figure 34c.

This short sketch of quantum interference of course relies not only on the pulsed light excitation. It equally well applies to photon-($\mathbf{k''\varepsilon''}$)–photon-($\mathbf{k\varepsilon}$) coincidence experiments or to collisional (foil) excitation of the two levels 1, 2 as indicated in Figure 34a as long as the condition of principally indistinguishable paths is guaranteed.

Equation (14) may also be written

$$I(t, \mathbf{k\varepsilon}) \propto |\langle f, t, \mathbf{k\varepsilon} | V_{\mathbf{k\varepsilon}} | \Psi_i(t) \rangle|^2 \tag{17}$$

with

$$|\Psi_i(t)\rangle = \sum_{n=1,2} |n, t, 0\rangle\langle n, 0, 0| \, V_{\mathbf{k'\varepsilon'}} |i, 0, \mathbf{k'\varepsilon'}\rangle \, e^{-i\omega_n t - \gamma_n t/2} \tag{18}$$

$$|\Psi_i(t)\rangle = \sum_{n=1,2} C_{ni} |n, t, 0\rangle \, e^{-i\omega_n t - \gamma_n t/2}$$

[see Eq. (3) in |D+S|]. In the eigenstate representation $|\Psi_i(t)\rangle$ is called a coherent superposition state. Hence one can say that interference or QBs will occur if such a superposition state is produced and can decay to the same final atomic and field state. Equivalently the density matrix of such an excited superposition state

$$\rho_i(t) = \begin{pmatrix} C_1 C_1^* \, e^{-\gamma_1 t} & C_1 C_2^* \, e^{-i\omega_{21} t - (\gamma_1 + \gamma_2)t/2} \\ C_1^* C_2 \, e^{+i\omega_{21} t - (\gamma_1 + \gamma_2)t/2} & C_2 C_2^* \, e^{-\gamma_2 t} \end{pmatrix} \tag{19}$$

exhibits off-diagonal matrix elements, so that one can say equally well: Coherence exists and interference occurs if off-diagonal excited-state density matrix elements in the eigenstate representation[294] are generated by the excitation process.

The general method of preparing such superposition states (off-diagonal density matrix elements) is the application of a sudden perturbation to the system which breaks the symmetry of the preexisting Hamiltonian. This includes not only the pulsed (collision) excitation, but also the sudden change of external fields (see the discussion of this point in $|D + S|$) or the sudden change of the atomic environment when a fast atom is leaving a foil surface.[32]

So far we have assumed δ functions for the excitation and detection time windows and we shall continue to do so in order to simplify all theoretical expressions for the time-dependent emitted light intensities. For actual experimental situations, however, these time windows have finite widths δt and Δt, respectively. As a consequence one has to convolute the amplitudes A_1 and A_2 in Eqs. (12) and (13) with the excitation pulse shape of the field amplitude [see Eq. (9) in $|D + S|$] and the intensity in Eq. (15) with the detection function in order to obtain the real signal. Assuming both to be square shaped with widths δt and Δt, respectively, normalized heights of 1, and δt, $\Delta t \ll \gamma^{-1}$, one obtains

$$I(k\varepsilon, t) \propto \left[B_1^2 + B_2^2 + 2B_1 B_2 \frac{\sin \omega_{21} \delta t/2}{\omega_{21} \delta t/2} \frac{\sin \omega_{21} \Delta t/2}{\omega_{21} \Delta t/2} \cos \omega_{21} t \right] e^{-\gamma t}$$

(20)

The amplitude of the oscillating term, the beat amplitude, is thus reduced and becomes zero for $\delta t = 2\pi/\omega_{21}$ or $\Delta t = 2\pi/\omega_{21}$. In other words, δt and Δt determine the frequency resolution of the experiment. Since very often the frequencies of a QB experiment are determined by Fourier transform techniques, the reduction of the beat amplitude is best shown by the frequency response curve together with the Fourier transform

$$I(\omega) = \int_0^\infty \frac{\sin \omega_{21} \delta/2}{\omega_{21} \delta/2} \frac{\sin \omega_{21} \Delta t/2}{\omega_{21} \Delta t/2} \cos \omega_{21} t \exp(-i\omega t - \gamma t) \, dt \quad (21)$$

of purely the damped oscillating term of Eq. (20) in Figure 35, where $2B_1 B_2$ is normalized to 1 and δt is assumed to be short compared to Δt. It is interesting to note that the frequency response curve crosses zero at $\omega = 2\pi/\Delta t$, but would allow the measurement of frequencies beyond $\omega = 2\pi/\Delta t$ with negative sign and strongly reduced amplitude up to $4\pi/\Delta t$ and so on. However, no use has been made yet of this behavior for the measurement of high frequencies. Also interesting to note is the linewidth $\Delta \omega = 2(3^{1/2})\gamma$ of the Fourier transform[81,118] of *intensity* beats, which is thus a factor of $3^{1/2}$ wider than the corresponding double-resonance signals. Because of the simultaneous measurement of γ and ω in a QB experiment, however, one can reduce this width considerably by analyzing only the undamped oscillation up to delay times t where the beat signal is still exceeding background and noise contributions.[81,295]

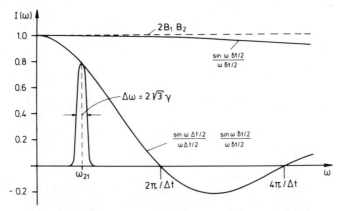

Figure 35. Frequency response function of quantum beat amplitudes on the frequency ω for excitation and detection time windows $\delta t = 0.1 \cdot \Delta t$ and Δt, respectively. Schematic Fourier transform of beats at ω_{21}.

After these introductory remarks on the fundamental principles of the QB phenomenon by use of a model level system and its experimental implications, we must discuss further conditions that are necessary for the actual observation of QBs. They can be derived from the description of realistic cases for which we choose Stark-beat experiments.

4.2. Stark Beats

A realistic excited low-level system can be approximated by the hydrogen $n = 2$ Lamb-shift splitting $2s_{1/2}-2p_{1/2}$. We neglect the existence of the $2p_{3/2}$ level, of the hyperfine structure, of the m sublevels and can use then the simple notation given in Figure 36a. Owing to the parity selection rule for the usual atomic E1 transitions only the $|2p\rangle$ state can decay to the $|1s\rangle$ ground state. First we assume that only the $|2s\rangle$ state is initially populated, which can be experimentally easily achieved by waiting after a beam-foil (gas) excitation until the $|2p\rangle$ state has decayed away completely. This system would stay like this forever (2E1 and M1 transitions neglected), no coherence exists, and no E1 decay to the ground state is possible. If we now enter suddenly, at $t = 0$, a region with electric field F_0 as shown in Figure 36b such that the sudden perturbation can be applied, then we have to expand the initial field-free eigenstates $\{a\}$ with respect to the new eigenstates in the field $\{c\}$. Since only the $2s$ level was populated a particularly simple expression including the time evolution for $t > 0$ is obtained:

$$I(t) \propto \left| \sum_{n=1,2} \langle f|V|n\rangle \, e^{(-i\omega_n - \gamma_n/2)t} \langle n|2s\rangle \right|^2 \tag{22}$$

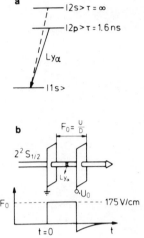

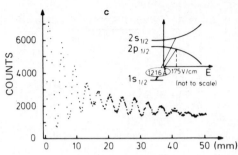

Figure 36. Stark beats in H I Ly_α observation. (a) Simplified level scheme. (b) Field electrodes, electric field F_0, and direction of detection. (c) Field-dependent level splitting and observed quantum beats.

The new eigenstates in the field are linear combinations of the field-free ones

$$|1\rangle = \beta_{ss}|2s\rangle + \beta_{ps}|2p\rangle, \qquad \langle 1|2s\rangle = \beta_{ss}^*$$
$$|2\rangle = \beta_{sp}|2s\rangle + \beta_{pp}|2p\rangle, \qquad \langle 2|2s\rangle = \beta_{sp}^* \tag{23}$$

so that both can decay to the *same final* $|1s\rangle$ ground state owing to their $|2p\rangle$ admixtures.

We have thus obtained a situation where a sudden perturbation, which breaks the symmetry of the pre-existing field-free Hamiltonian, has produced a coherent superposition state in the field-eigenstate representation, which gives rise to a special QB pattern[32,296] in Figure 36c according to the expression

$$I(t) \propto A\,e^{-\gamma_1 t} + B\,e^{-\gamma_2 t} + C\,e^{-(\gamma_1/2 + \gamma_2/2)t}\cos(\omega_1 - \omega_2)t \tag{24}$$

where all constants depend on the field strength. The additional modulation with beat amplitude minima at 17 and 51 mm from the field entrance is stemming from the hyperfine structure which we neglected in our description.

An improved version of this experiment with better time resolution Δt allowed also the observation of the $2s_{1/2}$–$2p_{3/2}$ splitting in the field,[297,298] but the ultimate goal of determining the Lamb shift to high accuracy as a fraction of the fine-structure splitting was not achieved.

If the exciting foil is placed in the plane of the left electrode in Figure 36b both the $|2s\rangle$ and $|2p\rangle$ states are simultaneously but *incoherently* excited when a H^+ beam is passing through the foil. (The recently observed partial s–p coherence[229–301] is neglected here.) Without a field one has then a diagonal density matrix in the field-free eigenstate representation, where p_1 and p_2 give the populations of the $2s$ and $2p$ states, respectively. In order to obtain

$$\rho\{a\} = \begin{array}{c} 2s \\ 2p \end{array}\!\!\begin{array}{cc} 2s & 2p \\ \left(\begin{array}{cc} p_1 & 0 \\ 0 & p_2 \end{array}\right) \end{array} \xrightarrow[\text{field}]{U\rho\{a\}U^{-1}=\rho\{c\}} \rho\{c\} = \frac{1}{2}\begin{array}{c} \\ \\ \end{array}\!\!\begin{array}{cc} 1 & 2 \\ \left(\begin{array}{cc} p'_{1_r} & x \\ x & p'_2 \end{array}\right) \end{array} \tag{25}$$

the density matrix in the field, one assumes that the field is suddenly switched on right after the $2s$- and $2p$-state formation. This sudden perturbation determines the transformation matrix U. As a result, off-diagonal density matrix elements in the field eigenstate representation occur if and only if $\rho\{a\}$ was not proportional to the unit matrix, i.e., $p_1 \neq p_2$! Hence QBs become observable only if $p_1 \neq p_2$. The experimental result in Figure 21 of $|\text{BE}|$ exhibits a large enough difference of p_1 and p_2 for a clean observation of the field-induced Ly_α Stark beats.[302] It also reveals the increase of the beat frequency with applied electric field according to the increase of the level splitting as indicated in Figure 36c. It is important to note that for this particular case coherence between different parity states is established by the external field, which leads to an intensity pulsation of the spatially isotropic Ly_α emission.[32]

When the neglected s–p coherence,[299–301] i.e., a nonzero off-diagonal matrix element σ_{sp} of the excitation density matrix $\rho\{a\}$, is taken into account a phase shift of the QBs at $t = 0$ with respect to 0 or π of Eqs. (15) and (24) may appear. It was shown[299] that σ_{sp} can be determined from the residual beat signal of the difference of the QBs in an axial field parallel or antiparallel to the beam axis. Measurements indeed established the existence of a nonzero σ_{sp},[300] and a thorough analysis including the full $n = 2 - l$, m_l basis yielded the complete excitation density matrix[301] at 210 keV beam energy:

$$\rho\{a\} = \begin{array}{cc} \begin{array}{c} l \rightarrow \\ \downarrow \\ m_l \rightarrow \\ \downarrow \\ 0\ 0 \\ 1\ 1 \\ 1\ 0 \\ 1\ -1 \end{array} & \begin{array}{cccc} 0 & 1 & 1 & 1 \\ 0 & 1 & 0 & -1 \\ \left(\begin{array}{cccc} 0.56 & 0 & 0.22\,e^{i2.06} & 0 \\ 0 & 0.15 & 0 & 0 \\ 0.22\,e^{-i2.06} & 0 & 0.14 & 0 \\ 0 & 0 & 0 & 0.15 \end{array}\right) \end{array} \end{array} \tag{26}$$

For a complete description of the hydrogen $n = 2$ Stark beats after foil excitation the whole uncoupled basis $\{a'\}$ including electronic spin s and nuclear spin I has to be set up with both s and I assumed to be unaffected by the foil interaction:

$$\rho\{a'\} = \rho\{a\} \otimes \frac{1}{2s+1} \otimes \frac{1}{2I+1} \tag{27}$$

The "sudden" transformation of $\rho\{a'\}$ into the field eigenstate representation yields then a 16×16 matrix, which gives rise to a large number of
beat frequencies [in a pure axial electric field, "only" 12 arise, owing to the
(m)-degeneracy]. It is this large number of frequencies that make the
analysis of Stark-beat experiments on higher n states quite difficult. For
instance Walerstein[11] dealt in the historic experiments of Figure 2 with
$32|m|$ substates of $n = 4$, and Bickel and Bashkin[303,304] dealt in their
rediscovery of the Stark beats in 1965 with $64|m|$ substates of $n = 8$ in He II.
In both cases an interpretation of the data was hopeless. However, such
Stark-beat experiments clearly demonstrated the great advantages of FBS
for such high time resolution studies. They attracted a lot of attention from
many experimentalists,[305-319,302] ultimately leading to the development of
FBS zero-field quantum beat experiments.[14,134]

4.3. Excitation Symmetry: Foils, Tilted Foils, Inclined Surfaces

So far we have only discussed QBs involving different parity states
which can be interpreted as time-dependent pulsations of the transition
rate of the atom. In the sections to come we concentrate on coherence
phenomena between equal parity states which become observable only
when the nonisotropic radiation pattern of the atoms periodically changes
its shape (see $|D+S|$) relative to a given quantization axis, thus yielding
again intensity oscillations for fixed directions of detection. Hence an initial
nonisotropic excitation must take place in order to make interference
phenomena observable at all.

The theory of perturbed angular correlations[320-322] is most appropriate for their description and allows in its extension to the Liouville
formalism[323] (see Sections 5.3 and 5.4 in $|BL|$) the elegant derivation of
most equations necessary for the understanding of the experiments. We
start with Eq. (108) of $|BL|$ and rewrite it for the detection of $k\varepsilon$ photons

$$I_{k\varepsilon}(t) = (D_{k\varepsilon}^+ | e^{-(i/\hbar)\hat{H}t} | \rho(0)) \tag{28}$$

where the detection operator $D_{k\varepsilon}$ is defined by

$$D_{k\varepsilon} = g(k\varepsilon) \cdot V_{k\varepsilon}^+ \cdot V_{k\varepsilon} \tag{29}$$

with the detection efficiency $g(k\varepsilon)$ and the interaction operators $V_{k\varepsilon}$. By
expanding Eq. (28) in terms of an irreducible tensor base (here related to
the orbital angular momentum standard base) we obtain Eq. (109) of $|BL|$
with the modified notation $T(jj')_{kq} = U_q^k(jj')$:

$$I_{k\varepsilon}(t) = \sum_{\substack{kk' \\ qq'}} (D_{k\varepsilon}^+ | U_q^k(LL))(U_q^k(LL) | e^{-(i/\hbar)\hat{H}t} | U_{q'}^{k'}(LL))(U_{q'}^{k'}(LL) | \rho(0))$$

$$\tag{30}$$

Table 3. Components ϕ_q^k of the Polarization Density Tensor of Eq. (34) for Linearly and Circularly Polarized Light, with the Geometry Defined in Figure 37

Component	Linearly polarized	Positive helicity σ^+	Negative helicity σ^-
ϕ_0^0	$(\frac{1}{3})^{1/2}$	$(\frac{1}{3})^{1/2}$	$(\frac{1}{3})^{1/2}$
ϕ_0^1	0	$-(\frac{1}{2})^{1/2}\cos\vartheta$	$+(\frac{1}{2})^{1/2}\cos\vartheta$
$\phi_{\pm 1}^1$	0	$\pm\frac{1}{2}\sin\vartheta\, e^{\pm i\varphi}$	$\pm\frac{1}{2}\sin\vartheta\, e^{\pm i\varphi}$
ϕ_0^2	$(\frac{1}{6})^{1/2}[3\cos^2\beta(\cos^2\vartheta-1)+1]$	$\dfrac{1}{2(6)^{1/2}}(3\cos^2\vartheta-1)$	$\dfrac{1}{2(6)^{1/2}}(3\cos^2\vartheta-1)$
$\phi_{\pm 1}^2$	$\mp\frac{1}{2}e^{\pm i\varphi}[\cos^2\beta\sin 2\vartheta\pm i\sin\vartheta\sin 2\beta]$	$\mp\frac{1}{4}\sin 2\vartheta\, e^{\pm i\varphi}$	$\pm\frac{1}{4}\sin 2\vartheta\, e^{\pm i\varphi}$
$\phi_{\pm 2}^2$	$\frac{1}{2}e^{\pm 2i\varphi}[\sin^2\beta-\cos^2\vartheta\cos^2\beta\mp i\cos\vartheta\sin 2\beta]$	$\frac{1}{4}\sin^2\vartheta\, e^{\pm 2i\varphi}$	$\frac{1}{4}\sin^2\vartheta\, e^{\pm 2i\varphi}$

We abbreviate this equation by

$$I_{\mathbf{k}\varepsilon}(t)\propto \sum_{\substack{kk'\\qq'}} {}^{LL'}A_q^k(\mathbf{k}\varepsilon)\times G_{qq'}^{kk'}(t)\times {}^{LL'}\rho_{q'}^{k'}(0) \tag{31}$$

with the excitation tensor components defined by

$$ {}^{LL'}\rho_{q'}^{k'}(0)=\sum_{LM,L'M'}\rho_{LM,L'M'}(-1)^{L-M}\begin{pmatrix} L & k' & L' \\ -M & q' & M' \end{pmatrix}(2k'+1)^{1/2} \tag{32}$$

and the detection tensor components defined by

$$ {}^{LL'}A_q^k(\mathbf{k}\varepsilon)\propto(-1)^{1+L+L_0}\begin{Bmatrix} 1 & 1 & k \\ L & L & L_0 \end{Bmatrix}|\langle L\|r\|L_0\rangle|^2\phi_q^k(\mathbf{k}\varepsilon) \tag{33}$$

and the perturbation factor $G_{qq'}^{kk'}(t)$ describing the actual time evolution of the system, which must be calculated for each particular experiment. In Eq. (33) one may replace L, L_0 by J, J_0 when a fine-structure multiplet is spectrally resolved. The polarization density tensor components $\phi_q^k(\mathbf{k}\varepsilon)$ in Eq. (33) are defined for E1 transitions by

$$\phi_q^k(\mathbf{k}\varepsilon)=\sum_{q_1 q_2} e_{q_1}(e_{q_2})^*(-1)^{1-q_1}\begin{pmatrix} 1 & k & 1 \\ -q_1 & q & q_2 \end{pmatrix}(2k+1)^{1/2} \tag{34}$$

They fully describe the angular and polarization dependence of the photon emission and are tabulated in Table 3 with the direction of emission $\mathbf{k}$ and the linear polarization ε defined in Figure 37 and the circular polarization determined by $\varepsilon(\sigma\pm)=(1/2)^{1/2}(\varepsilon^{(1)}\pm i\varepsilon^{(2)})$. The table clearly shows that the ϕ_0^0 component represents the isotropic emission, whereas ϕ_q^1 and ϕ_q^2 components are responsible for the nonisotropic circularly and linearly polarized light emission, respectively. By anticipating the condition $G_{qq'}^{kk'}(t)=G_{qq'}^{kk'}(t)\,\delta_{kk'}$ for all experiments to be discussed here, it becomes thus necessary to excite nonzero ρ_q^1-orientation or ρ_q^2-alignment tensor

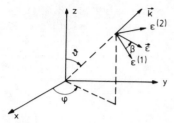

Figure 37. Geometry of detection and definition of the polarization vector ε.

components in order to make interference phenomena of equal-parity states observable.

A particularly simple situation arises for light excitation from level L_i to L. The ρ_q^k excitation tensor components can then be set equal to the A_q^k detection tensor components with appropriate changes:

$$^{LL}\rho_q^k(0) = {^{LL}}B_q^k(\mathbf{k'\varepsilon'}) \propto (-1)^{1+L+L_i}\begin{Bmatrix} L & 1 & k \\ L & L & L_i \end{Bmatrix}|\langle L\|r\|L_i\rangle|^2[\phi_q^k(\mathbf{k'\varepsilon'})]^* \quad (35)$$

The direction and polarization of the incoming light thus fully determines the excitation tensor components with the limitation on k set by the multipolarity of the $E1$ transition to $k \le 2$.

For gas, foil, or surface excitation no adequate theory exists as yet. One therefore fully relies on symmetry arguments in order to predict at least which ρ_q^k components can become nonzero for certain excitation geometries[324] (see Section 4.2 of |BL|). However, based on the argument that the interaction time in these geometries is short compared to fs or hfs periods, one can assume that initially a pure Coulomb interaction takes place[325] which can align or orient only the orbital angular momentum and leaves the electronic and nuclear spins S, I isotropic. This assumption has been confirmed by the foil and inclined-surface experiments which have been analyzed with respect to its validity.[326,83] The $^{LL}\rho_q^k(0)$ tensor components of the uncoupled orbital angular momentum base are thus the relevant quantities to be discussed and compared in experiments, since they represent directly the result of the interaction mechanism.

For the following sections it is therefore of interest to express the emitted light intensity in terms of these $^{LL}\rho_q^k$ excitation tensor components. For singlet transitions $L \to L_0$ one obtains from Eq. (31) with $G_{qq'}^{kk'}(t) = e^{-\gamma_L t}\delta_{kk'}\delta_{qq'}$

$$I_{\mathbf{k}\varepsilon}(t) \propto \sum_{kq}(-1)^{1+L+L_0}\begin{Bmatrix} 1 & 1 & k \\ L & L & L_0 \end{Bmatrix}|\langle L\|r\|L_0\rangle|^2\phi_q^k(\mathbf{k},\varepsilon){^{LL}}\rho_q^k e^{-\gamma_L t} \quad (36)$$

In order to obtain the intensity of a resolved multiplet transition $J \to J_0$ in LS-coupling one projects $\rho(0) = (1_S/2S+1) \otimes \rho_L$ on an uncoupled tensor base, transforms into the coupled $\{SLJM_J\}$ base, and calculates the intensity in this coupled base for which the detection tensor components of Eq. (33) can be used after replacing L, L_0 by J, J_0, respectively:

$$I_{\mathbf{k}\varepsilon}(t) \propto (D_J^+(\mathbf{k}\varepsilon)|e^{-(i/\hbar)\hat{H}t}|1_{S}\rho_L) 1/(2S+1) \tag{37}$$

$$I_{\mathbf{k}\varepsilon}(t) \propto \sum_{\substack{K'Kk'k \\ Q'Qq'q}} (D_{\mathbf{k}\varepsilon}^+ | U_{Q'}^{K'}(JJ))(U_{Q'}^{K'}(JJ)|e^{-(i/\hbar)\hat{H}t}|U_Q^K(JJ))$$

$$\times (U_Q^K(SL)J(SL)J|U_{q'}^{k'}(SS)U_q^k(LL))$$

$$\times (U_{q'}^{k'}(SS)U_q^k(LL)|1_S\rho_L(0)) \frac{1}{(2S+1)} \tag{37a}$$

With[373]

$$|1_j) = \sum_j |U_0^0(jj))(2j+1)^{1/2} \tag{38}$$

and with Eq. (33) and $G_{QQ'}^{KK'}(t) = e^{-\gamma_J t}\delta_{KK'}\delta_{QQ'}$ this becomes

$$I_{\mathbf{k}\varepsilon}(t) \propto \sum_{kqKQ} {}^{JJ}A_Q^K(\mathbf{k}\varepsilon) e^{-\gamma_J t}(U_Q^K(LS)J(LS)J|U_0^0(SS)U_q^k(LL))^{LL}\rho_q^k(0)\frac{1}{2S+1} \tag{39}$$

With[30]

$$(U_Q^K(j_1 j_2)j(j_1'j_2')j'|U_0^0(j_1 j_1')U_q^k(j_2 j_2'))$$

$$= \delta_{Kk}\delta_{Qq}[(2j+1)(2j'+1)(2k+1)]^{1/2}\begin{Bmatrix} j_1 & j_2 & j \\ j_1' & j_2' & j' \\ 0 & k & K \end{Bmatrix} \tag{40}$$

one finally obtains

$$I_{\mathbf{k}\varepsilon}(t) \propto \sum_{kq} (-1)^{1+J+J_0}\begin{Bmatrix} 1 & 1 & k \\ J & J & J_0 \end{Bmatrix} |\langle J\|r\|J_0\rangle|^2 \phi_q^k(\mathbf{k}\varepsilon) e^{-\gamma_J t}$$

$$\times \frac{(2J+1)^2}{2S+1}(-1)^{L+J+S+k}\begin{Bmatrix} J & J & k \\ L & L & S \end{Bmatrix} {}^{LL}\rho_q^k(0) \tag{41}$$

For the axially symmetric gas collisional and perpendicular foil (foil normal parallel to the beam) excitation[324] one can deduce from Section 4.3 of $|BL|$ that only k = even and q = 0 components with respect to the beam as quantization axis can become nonzero. For E1 transitions it is thus only the existence of a nonzero ${}^{LL}\rho_0^2$ alignment tensor component that will permit the observation of interference effects for this excitation geometry with the detection of linearly polarized or unpolarized light.

When the foil normal is tilted out of the beam direction the symmetry of the excitation process is lowered to reflection symmetry on a plane

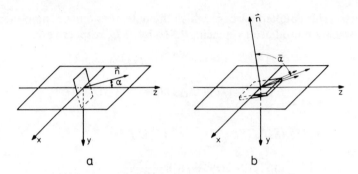

Figure 38. (a) Tilted-foil and (b) inclined-surface geometry.

containing the beam axis and the foil normal. If we choose this to be the $y-z$ plane (not $x-z$ as in $|BL|$) in Figure 38a, it implies the condition

$$\rho_q^k = (-1)^k \rho_{-q}^k \qquad (42)$$

and allows the following nonzero ρ_q^k components:

$$\rho_0^0, \qquad \rho_1^1 = -\rho_{-1}^1, \qquad \rho_0^2, \qquad \rho_{+1}^2 = \rho_{-1}^2, \qquad \rho_{+2}^2 = \rho_{-2}^2 \qquad (43)$$

Hence, interference effects may become observable with this geometry owing to the existence of nonzero $^{LL}\rho_q^2$ alignment or $^{LL}\rho_{\pm1}^1$ orientation tensor components not only in linearly but also in circularly polarized light detection, respectively.

Reflection symmetry on the $y-z$ plane also holds for the excitation of surface scattered ions when an ion beam is impinging along the z axis on an inclined surface at grazing incidence[327] as shown in Figure 38b. Hence, the same conclusions on the $^{LL}\rho_q^k$ components can be drawn for this geometry as for the tilted-foil excitation.

The lack of an appropriate theory for the various collisional inter-action mechanisms requires each level to be examined experimentally for evidence of nonzero $^{LL}\rho_q^2$ or $^{LL}\rho_q^1$ tensor components. This information may be either directly obtained from polarization measurements[325,328] or from zero-field QB, Zeeman-beat, Hanle-effect, or HFLC experiments.

In zero-field QB and Zeeman-beat experiments the relative beat amplitudes plus their phases, and in Hanle effect or HFLC experiments the relative signal strengths plus the signal shapes, indirectly allow the deter-mination of some of the $^{LL}\rho_q^k$ components of Eq. (43) for which the formulas will be derived later on.

In polarization measurements one generally determines the state of polarization of the light emitted by an atomic ensemble in terms of normalized Stokes parameters M/I, C/I, and S/I, where the Stokes parameters[329] I, M, C, and S can be expressed in terms of linearly polarized $I(\beta)$ (β defined in Figure 37) and circularly polarized $I(\sigma\pm)$ ($\sigma\pm = \pm$ helicity) intensity components of the emitted light:

$$I = I(0) + I(90) = I(\sigma^-) + (\sigma^+)$$

$$M = I(0) - I(90)$$

$$C = I(45) - I(135) \tag{44}$$

$$S = I(\sigma^-) - I(\sigma^+)$$

Through Eqs. (36) and (41) these Stokes parameters are directly related to the $^{LL}\rho_q^k$-tensor components. By a measurement of the normalized Stokes parameters at two different emission angles ϑ in the x–z plane[330] one can thus in principle determine all five independent $^{LL}\rho_q^k$ components of Eq. (43). In practice, however, one often limits the measurement simply to M/I (fractional linear polarization) or S/I (fractional circular polarization) when searching for alignment or orientation, respectively. In order to compare the different measured quantities of these experiments one uses the normalized $^{LL}\rho_q^k$ components of the lowest symmetry in Eq. (43) to define alignment parameters A_q^k.

$$A_0^2 = \frac{^{LL}\rho_0^2}{^{LL}\rho_0^0}, \qquad A_0^{col}(0) = \begin{Bmatrix} 1 & 1 & 2 \\ L & L & L \end{Bmatrix} [6(3L+1)]^{1/2} A_0^2 \tag{45a}$$

$$A_1^2 = \frac{^{LL}\rho_1^2}{^{LL}\rho_0^0}, \qquad A_{1-}^{col}(0) = \begin{Bmatrix} 1 & 1 & 2 \\ L & L & L \end{Bmatrix} 2[2L+1]^{1/2} A_1^2 \tag{45b}$$

$$A_2^2 = \frac{^{LL}\rho_2^2}{^{LL}\rho_0^0}, \qquad A_{2+}^{col}(0) = \begin{Bmatrix} 1 & 1 & 2 \\ L & L & L \end{Bmatrix} 2[2L+1]^{1/2} A_2^2 \tag{45c}$$

and an orientation parameter O_1^1 with respect to the beam as quantization axis.

$$O_1^1 = \frac{^{LL}\rho_{+1}^1}{^{LL}\rho_0^0}, \qquad O_{1+}^{col}(0) = -\begin{Bmatrix} 1 & 1 & 1 \\ L & L & L \end{Bmatrix} 2[L(L+1)(2L+1)]^{1/2} O_1^1 \tag{45d}$$

They are directly related to the often-used normalized collisional alignment A^{col} and orientation O^{col} parameters[30,66] as defined in Eqs. (45a)–(45d) for our coordinate systems in Figure 38 with the y–z plane as reflection symmetry plane. In order to compare them with those defined in the also often-used coordinate system with the x–z plane as reflection symmetry plane,[66] they can be transformed as follows:

$$A_0^{col}(xz) = A_0^{col}(yz), \qquad A_{1+}^{col}(xz) = -iA_{1-}^{col}(yz)$$

$$A_{2+}^{col}(xz) = -A_{2+}^{col}(yz), \qquad O_{1-}^{col}(xz) = O_{1+}^{col}(yz) \tag{46}$$

The first definitive experimental indication of a nonzero A_0^2 alignment after perpendicular foil excitation was observed in a zero-field QB experiment on He I $n = 3$ and H I $n = 3, 4$ fine-structure (fs) splitting,[15] which yields after later refinements for He I 3^3P $A_0^2 = -0.2(A_0^{col} = -0.14)$ at 300 keV beam energy. Since then alignment determinations after foil excitation were performed on numerous levels, mostly of $Z \leq 18$ elements, and scattered over a wide energy range.[331-345] Macek and Burns[66] have recently discussed most of these results in terms of the collision parameters. For *non*hydrogenlike systems, A_0^2 is always found to be negative, which means that the orbital angular momenta are preferentially produced in a disk perpendicular to the beam direction $(M_L = 0)$. However, no other systematic trend has been found as yet and the values of A_0^2 scatter between 0 and -0.2, with some levels exhibiting no alignment at all [recall the consequences of the independent-electron model in Section 2.2(b)]. For hydrogenlike systems the most thorough study has been performed on $H n = 2$ by QB and polarization measurements[337,344,346] from 12 to 1600 keV. A_0^2 is measured positive for most of this energy range with a maximum of $A_0^2 \approx 0.32$, but it changes sign to become negative below 45 keV. For $n = 3$ similarly a positive A_0^2 was found above 225 keV[336] and more recently also for H-like O VIII $n = 9, 10, 12$; $l = n = 1$ levels as well as for the hydrogenic O VII $n = 8, l = 7$ level at 36 MeV.[345] One may thus conclude that hydrogenlike and hydrogenic levels, which can be strongly affected by surface field Stark effect, are the only ones that exhibit positive A_0^2 for most energies, whereas all other levels exhibit negative A_0^2. Except for H $n = 2$ the energy variation of the alignment has been measured only over rather narrow energy regions and has been generally found to be small.

Gas excitation of energetic ions leads to a more complex alignment behavior characterized by a strong energy dependence with pronounced maxima and minima possible in the lower-energy range and a smooth asymptotic behavior for high energies.[335,343,347-350] For a discussion of some aspects of these phenomena the reader is referred to the article by Kleinpoppen and Scharmann in Chapter 8 of this work. In actual FBS experiments one can take advantage of these alignment fluctuations by searching for maxima of the order of $|A_0^2| \approx 0.3$ via careful adjustment of the beam energy.

The tilted-foil geometry stands out for the possible appearance of partially circularly polarized light emission in the x direction due to the allowed $^{LL}\rho_{+1}^1$-tensor components. The first evidence for such circular polarization was found by Berry *et al.*[330] in 1974 on the He I $2s$ ^1S-3p ^1P 501.6-nm transition at tilt angles α up to 45°. They measured the normalized Stokes parameters at $\vartheta = 90°$ and 53° and could thus determine the four $^{LL}\rho_q^k/^{LL}\rho_0^0$ ratios of Eq. (45), from which the collision parameters

POLARIZATION, %

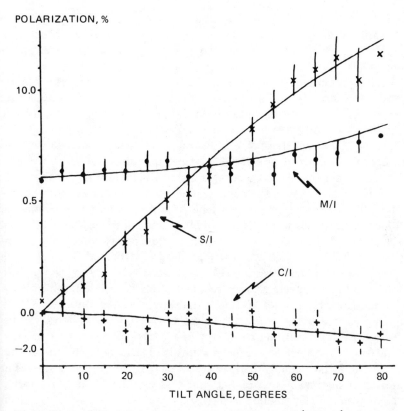

Figure 39. Normalized Stokes parameters for the Ne III $3s'$ $^1D-3p'$ ^1F 286.6-nm transition as a function of the tilt angle α at $E = 1$ MeV. The data are fitted to the theory by Lombardi in Ref. 357.

could be deduced. In particular they obtained at $\alpha = 45°$ $S/I = 0.105(10)$, $^{LL}\rho_1^1/^{LL}\rho_0^0 = +0.07(1)$, and $O_{1+}^{col}(yz) = -0.057(7)$. With these values they clearly established the influence of the surface on the foil interaction mechanism since the reflection symmetry of the tilted-foil geometry becomes effective only when the foil surface is involved in the final excited-state formation because the interaction stays axially symmetric inside the bulk of the tilted foil.

Further measurements with improved accuracies on He I, O II, Ne II, III, and Ar II lines fully confirmed this first observation at various beam energies.[351-355,56] The sign of $S/I(^{LL}\rho_1^1/^{LL}\rho_0^0)$ indicated for all cases that the initial orbital angular momentum of the excited term was oriented and pointing in the $(-x)$ direction. It was also found that S/I always increases with tilt angle α approximately proportionally to $\sin \alpha$, for which Figure 39 is a representative example. Furthermore, the total fractional polarization

$f_p = (M^2 + C^2 + S^2)^{1/2}/I$ was observed to increase with tilt angle. The tilted-foil geometry may therefore be of practical importance for the application of high-resolution techniques on levels that show only small anisotropy at $\alpha = 0°$.

In theoretical interpretations of this so-called tilted-foil effect either surface electric fields or the atomic locations in the final layer of the surface were assumed to be responsible for the effect.[356-361] The results of these treatments give different angular dependences on the tilt angle of M/I, C/I, and S/I. Besides those results that have to be fully discarded owing to $\sin 2\alpha$ or $\tan \alpha$ dependences for S/I, even Lombardi's result[357] of a $\sin \alpha$ dependence cannot be accepted as a conclusive model since the author concedes himself that many dependences may be obtained with appropriate sets of parameters. At least his model of coherent excitation of different parity levels with subsequent linear and quadratic Stark effect in a surface electric field does not contradict the observations. In a recent similar calculation Band[362] also obtains agreement with the results in Figure 39 with an appropriate set of parameters.

Even without having an accepted theoretical explanation available for the tilted-foil effect, one can easily imagine that the direction of the outgoing excited particles relative to the surface normal is important for the size of S/I. According to this picture a surface-scattered ion beam should also exhibit oriented excited states in close analogy to a tilted-foil experiment. In particular, when the scattered particles leave the surface at a grazing angle, maximum orientation can be expected owing to the $\sin \alpha$ dependence of the orientation in tilted-foil experiments.

In first experiments He^+, Ar^+, Ne^+, and Kr^+ ion beams with energies up to 300 keV were scattered from mechanically polished graphite, Ta, or Cu surfaces at grazing angles of incidence of $1°-1.5°$.[327] In all cases large values of S/I in the line emission of surface-scattered ions or neutralized atoms could be observed which by far exceeded the S/I values of tilted-foil experiments. In a subsequent, more careful study S/I was measured over the whole Ar II spectrum from 3000 to 5500 Å when Ar^+ ions of 300 keV

Table 4

Ar II multiplet	J_0	J	λ (nm)	$(S/I)_{exp}$	$(S/I)_{theor}$
$4s'\,^2D-4p'\,^2F$	5/2	7/2	460.9	0.76(4)	0.66
	5/2	5/2	463.7	0.17(7)	0.21
	3/2	5/2	459.0	0.72(4)	0.69
$4s\,^4P-4p\,^4P^0$	5/2	5/2	480.6	0.15(3)	0.18
	5/2	3/2	473.6	-0.14(4)	-0.17
	3/2	5/2	500.9	0.50(10)	0.60
	3/2	1/2	489.8	0.10(3)	0.14

energy were impinging on a mechanically polished Cu surface at a grazing angle of incidence of $1°$.[83,363] In order to keep the surface clean a liquid N_2 fed cold shield surrounded the target at an overall vacuum pressure of 2×10^{-6} mbar. The experiment yielded a S/I_{av}, averaged over all significant Ar II lines in this spectral region of $S/I_{av} \approx 0.4$. However, large variations of S/I from line to line with occasionally even negative values were found as shown in Table 4.

An explanation of this behavior can be based on a pure orbital angular momentum orientation by the short ($<10^{-13}$ s) ion–surface Coulomb interaction. When the ion is leaving the interaction region the corresponding isotropic spin S has to be coupled to the oriented $\langle \mathbf{L} \rangle$ in order to form all allowed fs eigenstates $(LS)J$ which then can decay radiatively within a multiplet to lower states $(L_0 S)J_0$. Under such circumstances S/I for observation along the x axis in Figure 38b can be derived from Eq. (41) in terms of the $^{LL}\rho_q^k$ components:

$$
\left(\frac{S}{I} \right)_{\text{theor}} = {}^{LL}\rho_1^1 \begin{Bmatrix} 1 & 1 & 1 \\ J & J & J_0 \end{Bmatrix} \begin{Bmatrix} J & J & 1 \\ L & L & S \end{Bmatrix} \left({}^{LL}\rho_0^0 \left(\frac{1}{3} \right)^{1/2} \begin{Bmatrix} 1 & 1 & 0 \\ J & J & J_0 \end{Bmatrix} \begin{Bmatrix} J & J & 0 \\ L & L & S \end{Bmatrix} \right.
$$

$$
+ \left. \left[\frac{1}{2} {}^{LL}\rho_2^2 - \left(\frac{1}{24} \right)^{1/2} {}^{LL}\rho_0^2 \right] \begin{Bmatrix} 1 & 1 & 2 \\ J & J & J_0 \end{Bmatrix} \begin{Bmatrix} J & J & 2 \\ L & L & S \end{Bmatrix} \right)^{-1} \tag{47}
$$

The four unknown $^{LL}\rho_q^k$ components in this expression could be adjusted such that all measured S/I_{exp} values in Table 4 are best reproduced by S/I_{theor} with a single set:

$$
{}^{LL}\rho_1^1 / {}^{LL}\rho_0^0 = +0.5, \qquad {}^{LL}\rho_0^2 / {}^{LL}\rho_0^0 = 0.2, \qquad {}^{LL}\rho_2^2 / {}^{LL}\rho_0^0 = 0.15 \tag{48}
$$

The good agreement in Table 4 allows us to deduce from $^{LL}\rho_0^0 = 0.5$ that a unique orbital angular momentum $\langle L_x \rangle < 0$ is generated for all excited levels in Table 4. This result of a pure orbital angular momentum orientation by the ion–surface interaction has been further confirmed by measurements on other Ar II, Ne II, Kr II, N II, C II, O II multiplets at energies between 50 and 600 keV. In all cases one finds the orbital angular momentum strongly oriented and pointing in the $(-x)$ direction. The energy dependence within the quoted energy region was small, with a trend to lower S/I values for lower energies. A rapid drop of S/I was observed with increasing angle of incidence, whereas a nearly constant S/I was obtained when measured as a function of the averaged angle $90-\bar{\alpha}$ (see Figure 38b) of the outgoing excited particles, which was, however, observable only up to $90-\bar{\alpha} \approx 6°$ because of intensity reasons.[364]

For the interpretation of this strong orientation effect one has available at present only the theoretical considerations for the tilted-foil effect and a simple model proposed by Schröder and Kupfer[365] which gives the physical picture for the so-called torque model. If one assumes a hypo-

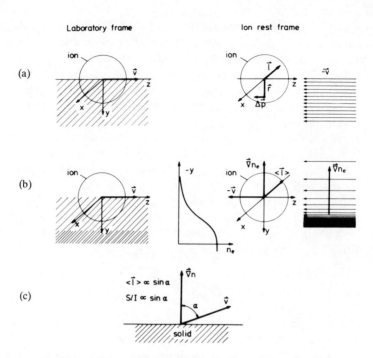

Figure 40. Physical model for the direction and angular dependence of the large orientations observed in inclined-surface and tilted-foil experiments. See text for explanations.

thetical surface with sharp boundary in Figure 40a with an ion moving parallel to it with velocity $\mathbf{v}$ then the ion experiences in its rest frame a beam of electrons (atoms) with $-\mathbf{v}$ which fills out only the space below the $x-z$ plane $(+y)$. When an electron is captured out of this beam by the ion, it possesses a momentum $\Delta\mathbf{p}$ which yields, together with $\mathbf{r}$, an orbital angular momentum pointing in the $(-x)$ direction. (A momentum transfer $\Delta\mathbf{p}$ from beam electrons to ionic electrons gives the same result.) Because of the limitation of the beam to the lower $(+y)$ hemisphere of the ion, the averaging over all possible $\Delta\mathbf{p}$ and $\mathbf{r}$ leads to an effective $\langle l_x \rangle < 0$.

For a more realistic surface in Figure 40b the solid–vacuum discontinuity may be approximated by an exponentially decaying electron density n_e, which can be expressed as a density gradient ∇n_e. In the ion rest frame this gradient ∇n_e determines the ratio of electrons passing the ion in the lower $(+y)$ hemisphere $(\langle l_x \rangle < 0)$ to those passing in the upper $(-y)$ hemisphere $(\langle l_x \rangle > 0!)$ and thus determines a unique direction for the interaction process as well as the size of the effective orientation. Hence, the orientation of the orbital angular momentum can be expressed by the vector product $[-\mathbf{v} \times \nabla n_e] \propto \langle \mathbf{l} \rangle$ in the ion rest frame which corresponds to $[\nabla n_e \times \mathbf{v}] \propto \langle \mathbf{l} \rangle$ in the laboratory frame.

Even without knowing the details of the interaction mechanism this formula will dominate any mechanism we may think of. It thus directly explains the large orientations obtained from ion–surface interactions at grazing incidence as well as the $(\sin \alpha)$ dependence $S/I \propto |v| |\nabla n_e| \sin \alpha$ of S/I in tilted-foil experiments, if we assume the same basic mechanism in both cases.

One should note that the same physical picture also applies to the production of aligned excited states in axially symmetric excitation geometry. In the rest frame of the ion, the beam of electrons with $-\mathbf{v}$ then fills the space above and below the x–z plane. Assuming the same types of momentum transfer $\Delta \mathbf{p}$ (capture or collisional transfer), one obtains the angular momenta preferentially arranged in a disk in the x–y plane ($M_L = 0$) as actually observed for all cases except for hydrogenic levels.

4.4. Zeeman Beats and Hanle Effect

Zeeman beats are conceptually the simplest quantum beats and can easily be visualized with Hanle's early picture of a precessing classical dipole.[366] The Liouvillian for a situation with the z axis as field and quantization axis becomes

$$\hat{H} = \hbar \omega_L \hat{J}_z - i\hbar\gamma \qquad \hbar\omega_L = g\mu_B B_z \tag{49}$$

with ω_L being the Larmor frequency and with $\hat{J}_z$ being the z component of the angular momentum superoperator, which satisfies the eigenvalue equation

$$\hat{J}_z |U_q^k(JJ)\rangle = q|U_q^k(JJ)\rangle \tag{50}$$

One thus readily obtains for the perturbation factor in this geometry

$$G_{qq'}^{kk'}(t) = e^{-i\omega_L qt - \gamma t} \delta_{kk'} \delta_{qq'} \tag{51}$$

which has to be inserted in Eqs. (36) or (41) for singlet or resolved fine-structure multiplet transitions, respectively.

For gas or perpendicular foil excitation the axial symmetry allows only $q = 0$ tensor components to be nonzero. Hence, with the field parallel to the beam the oscillating term in Eq. (51) drops out and consequently no beats can be observed. One therefore has to apply a perpendicular field with respect to the beam axis and describe the system with this field axis as a new quantization axis. The ρ_q^k components in this new coordinate system are obtained by a passive $\pi/2$ rotation:

$$\overline{\rho_q^k} = \rho_0^k D_{0q}^k(0, \pi/2, 0): \qquad \overline{\rho_0^0} = \rho_0^0 \quad \overline{\rho_0^2} = -\tfrac{1}{2}\rho_0^2, \quad \overline{\rho_{\pm}^2} = \tfrac{3}{8}\rho_0^2 \tag{52}$$

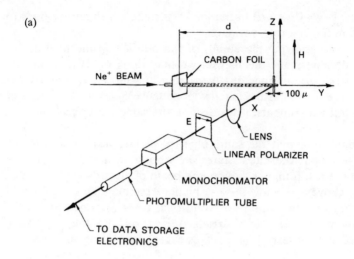

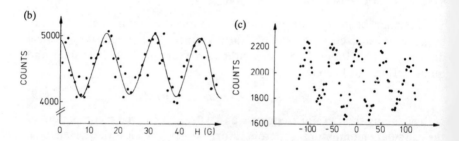

Figure 41. (a) Geometry for the observation of Zeeman beats at fixed distance d from the exciting foil by variation of the magnetic field H. (b) Zeeman beats in the Ne II $3d^4D$–$4f^4D$ 422.0-nm emission as observed with the geometry (a); $d = 3.59$ cm. (c) Zeeman beats in the Ar II $4s'^2D_{5/2}$–$4p'^2F_{7/2}$ 460.9-nm emission as observed with a tilted foil and the magnetic field parallel to the beam; $d = 3.59$ cm.

of which the $\rho_{\pm 2}^2$ components are responsible for the observable Zeeman beats with intensity

$$I(t) \propto A[1 + B \cos 2\omega_L t]\, e^{-\gamma t} \qquad (53)$$

We choose as an example the geometry used by Liu *et al.*[331] in Figure 41a. Instead of using t as a variable at fixed ω_L according to Eq. (53) they kept t fixed by observing a small section of the beam at a fixed distance d downstream from the exciting foil and varied ω_L by changing the magnetic field strength. Consequently they could observe in the Ne II $3d^4D$–

$4f^4D$ 422.0-nm transition Zeeman beats as a function of field strength in Figure 41b. The main interest in such measurements is the determination of atomic g values,[29,332,341,343,367-370] which require, however, relative accuracies of better than 10^{-3} in order to be of significance to atomic structure calculations. Because of the lack of a careful velocity analysis most of these measurements failed, however, to reach this level of accuracy.

Furthermore, one has to pay attention to the influence of cascades on the measured Zeeman beat frequencies, a problem that has been thoroughly discussed by Dufay[371] and others.[372,373] for a resolved fine-structure multiplet cascade $(SL_2)J_2 \to (SL_1)J_1 \to (SL_0)J_0$ one obtains with the use of Eq. (I.13) of Ducloy and Dumont[374] and with the transient convolution in Eq. (12) of Dufay[371]

$$I_{k\varepsilon}(t) \propto \sum_{kq} (-1)^{1+J_1+J_0} \begin{Bmatrix} 1 & 1 & k \\ J_1 & J_1 & J_0 \end{Bmatrix} \phi_q^k(k\varepsilon)^{J_1 J_1}\rho_q^k(t) \tag{54}$$

with

$$^{J_1 J_1}\rho_q^k(t) = (-1)^{L_1+J_1+S+k}\frac{2J+1}{2S+1}\begin{Bmatrix} J_1 & J_1 & k \\ L_1 & L_1 & S \end{Bmatrix} {}^{L_1 L_1}\overline{\rho_q^k}(0)\, e^{-i\omega_1 qt - \gamma_1 t}$$

$$+ (-1)^{1+J_1+S+L_2+2J_2}\frac{(2J_2+1)^2}{(2S+1)}\begin{Bmatrix} J_2 & J_2 & k \\ L_2 & L_2 & S \end{Bmatrix}$$

$$\times \begin{Bmatrix} J_2 & J_2 & k \\ J_1 & J_1 & 1 \end{Bmatrix} \gamma_2 {}^{L_2 L_2}\overline{\rho_q^k}(0)$$

$$\times \frac{e^{-i\omega_1 qt - \gamma_1 t} - e^{-i\omega_1 qt - \gamma_1 t}}{(\gamma_1 - \gamma_2) + i(\omega_1 q - \omega_2 q)} \tag{55}$$

where ω_1 and ω_2 are the Larmor frequencies of the feeding and the actually measured level, respectively. The first term in this equation represents the direct excitation of level J_1 and the second term describes the cascading from level J_2 to level J_1 which essentially allows the observation of the Zeeman beat frequency of level J_2 in the lower $J_1 \to J_0$ transition in cosine and sine contributions. As an illustration, we show for the cascade in Figure 42 only the oscillatory terms of Eq. (55) in Figure 42a under the following assumptions: $\omega_1 = 13.33\ s^{-1}$ $\omega_2 = 12.00\ s^{-1}$, $^{L_1}\rho_2^2(0) = 5^{L_2 L_2}\rho_2^2(0)$. The cascade contribution with clearly different frequency from the beats from the direct excitation becomes significant at about $\tau_1 = 1/\gamma_1$. For accurate g-value determinations one must therefore measure at as short delay times as possible, but even then an analysis in terms of more than one frequency is necessary.

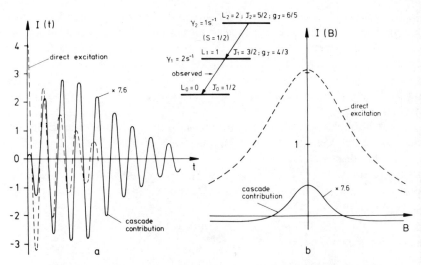

Figure 42. The influence of a direct cascade on (a) the Zeeman beats and (b) the Hanle effect of the lower transition shown in the level scheme indicated. The intermediate level is assumed to be populated five times as much as the upper feeding level. See text for details.

With tilted-foil excitation the set of ρ_q^k components of Eq. (43) has been observed to be nonzero. According to Eq. (51) one can thus expect to observe Zeeman beats with a field parallel to the beam. In particular, with circularly polarized light detection in the x direction one can observe

$$I(t) \propto A[1 + B \cos \omega_L t + C \cos 2\omega_L t] e^{-\gamma t} \qquad (56)$$

due to the $\rho_{\pm 1}^1$ and $\rho_{\pm 2}^2$ contributions, respectively. The experimental example in Figure 41c on the Ar II $4s'\,^2D_{5/2}$–$4p'F_{7/2}$ 460.9-nm transition clearly shows these Zeeman beats as measured with the geometry of Figure 41a with circular polarizer and the field parallel to the beam.[352] With A, B, and C expressed in terms of the $^{LL}\rho_q^k$ components one can use Eq. (56) to determine these $^{LL}\rho_q^k$ components from the experimental data. No further use has been made as yet of this special tilted-foil geometry.

As shown with Eq. (16) the time integration of Eqs. (53) and (56) yields the well-known Hanle-effect signals (see $|D+S|$ and $|H+G|$). Except for one experiment,[352] all Hanle-effect measurements in FBS have been performed with axially symmetric gas or foil excitation conditions.[375, 378,343,370] We thus insert $\omega_{21} = 2\omega_L$ in Eq. (16) and obtain with the full width ΔB at half-maximum of the resulting Hanle-effect signal the lifetime $\tau = \hbar/g\mu_B\,\Delta B$ of the excited level with the commonly known g value. However, the fast beam does in general impose an observationally limited time integration up to T instead of infinity.[314,375] For gas excitation the situation becomes even more complicated since the observation time is limited to ΔT and the atoms reaching the observation region may have been excited everywhere between the beam entrance into the gas target and the observation region itself. As a result one deals with broadened Hanle signals with sidebands, and a careful evaluation of the theoretical signal shape is necessary in order to determine lifetimes from FBS Hanle signals.[375]

As an example we show in Figure 43 the first FBS Hanle-effect lifetime determination on the He 3^1P level and with H_2-gas excitation where M/I is plotted as P as a function of the magnetic field strength.[375] The distortion of the ideal Lorentzian signal shape is negligible in this case owing to the short lifetime of the He 3^1P level, which could thus be determined by a fit of a Lorentzian to the data. Quite a number of lifetimes have been determined in this way by foil and gas excitation with a general claim of accuracy of the order of 5%–10%.

The FBS Hanle effect is, as a time integral method, in principle velocity independent to first order and can be applied with an extended gas target even for heavy particles without suffering from foil breakage. However, it is of course subject to cascade distortions in the same way as the Zeeman beats are. The affected signal shape can be obtained by time-integration of Eq. (55). As an illustration we use the same cascade under the same assumptions as for the discussion of the cascade influence on Zeeman beats in Figure 43. Time integration of the beats in Figure 43a from $t = 0$ to infinity directly yields the Hanle-effect signals in Figure 43b. Again the contributions from direct excitation of level J_1 and from cascade repopulation are separately shown, such that the sum of both would be expected in an actual experiment. The cascade contribution clearly distorts the pure Lorentzian of the direct excitation in such a way that if the sum were interpreted as a Lorentzian a longer lifetime $\tau > \tau_1 = 1/\gamma_1$ (besides a bad fit) would be obtained. A fit with the time-integrated equation (55) is

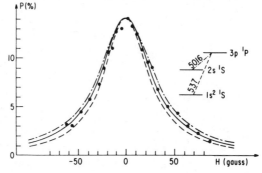

Figure 43. Hanle-effect signal in the He I $2s^1S–3p^1P$ 501.6-nm emission from a gas-excited He beam. Circles: experimental data. Solid line: the best fit by a Lorentzian curve with $\tau = 1.8$ ns. Dashed line and dot-dashed line: Lorentzian curves with $\tau = 2.0$ and 1.6 ns, respectively.

thus necessary at least in order to carefully extract lifetimes from collisional excited Hanle-effect signals.[379]

4.5. Zero-Field Quantum Beats

As pointed out in Section 4.1, coherent superposition states can be generally prepared by a sudden perturbation of the system which breaks the symmetry of the preexisting Hamiltonian. In a foil excitation experiment the Coulomb interacting foil environment, which acts in the uncoupled base $\{LM, SM_S, IM_I\}$ only on the orbital angular momentum and leaves S and I isotropic (if LS coupling holds), represents the preexisting Hamiltonian which is suddenly changed at the exit of the foil to the field-free Hamiltonian of the atom (as the "perturbation"). One thus has, at the foil exit, the excitation density matrix in the uncoupled base representation

$$\rho(0) = \frac{1_S}{2S+1} \otimes \frac{1_I}{2I+1} \otimes \rho_L(0) \tag{57}$$

which must be suddenly transformed into the field free (coupled) eigenbase in order to conveniently describe the evolution of the atom in the field-free environment. Since fs or hfs QBs are observed principally in unresolved multiplets, one averages in the detection over nuclear and electronic spin variables so that the detection operator acts on the orbital angular momentum in the uncoupled base. For the light intensity we can thus write in LS coupling

$$I_{\mathbf{k}\varepsilon}(t) \propto \langle 1_I 1_S D_L^+(\mathbf{k}\varepsilon) | \, e^{-(i/\hbar)\hat{H}t} \, |1_I 1_S \rho_L(0) \rangle \frac{1}{(2I+1)(2S+1)} \tag{58}$$

With the use of Eq. (38) and the expansion in terms of an irreducible tensor base one obtains a modified Eq. (30) with the perturbation factor replaced by

$$G_{qq'}^{kk'}(t) = \langle U_0^0(II)U_0^0(SS)U_q^k(LL) | \, e^{-(i/\hbar)\hat{H}t} \, |U_0^0(II)U_0^0(SS)U_{q'}^{k'}(LL) \rangle \tag{59}$$

$$= \sum_{\substack{J_1 J_2 \\ F_1 F_2}} \frac{(2J_1+1)(2J_2+1)(2F_1+1)(2F_2+1)}{(2S+1)(2I+1)} \begin{Bmatrix} J_1 & J_2 & k \\ L & L & S \end{Bmatrix}^2 \begin{Bmatrix} F_1 & F_2 & k \\ J_2 & J_1 & I \end{Bmatrix}$$

$$\times \exp\left[-i(\omega_{J_1F_1} - \omega_{J_2F_2})t - \gamma t\right]\delta_{kk'}\delta_{qq'} \tag{60}$$

where use has been made of Eq. (106) of |BL| and of Eq. (40). [See also the derivation of Eq. (41).] Equation (31) with Eqs. (33) and (60) thus fully describes zero-field fs and hfs QB experiments in LS coupling with J as good quantum number after foil or light excitation.[32,324,380-382] For $I = 0$ this perturbation factor reduces to an expression similar to Eq. (101) o (a)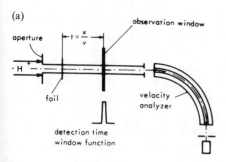

(b)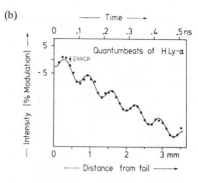

Figure 44. (a) Geometry for the observation of zero-field QBs after foil excitation. (b) Zero-field H I $2p\ ^2P_{1/2}-^2P_{3/2}$ fs QBs at 10969 MHz observed in H Lyα emission without polarization. The solid line is a fit to the data. (c) Schematic view of the transient change of the angular distribution of the total Ly$_\alpha$ emission in the zero-field QB experiment of (b). Initial $M_L = \pm 1$ state population is assumed.

(c)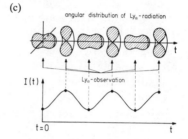

$|BL|$ for zero-field fs QBs. The triangular rules $\Delta(J_1J_2k)$ and $\Delta(F_1F_2k)$ for the $6j$ symbols clearly show that $k \neq 0$ is required for the observation of QBs, in full accord with our remarks at the beginning of Section 4.3 (see also $|D + S|$).

Although the basic formulas had been given already by Breit in 1933[383] and were intelligibly reinterpreted by Franken[384] in 1961 and by Macek in 1969,[385] there existed a number of technical reasons that prevented experimentalists from performing such zero-field QB experiments. Zero-field fs- or hfs-splittings of typically ≥ 100 MHz require extremely short light excitation pulses and high time resolution and linearity in the detection, which were not available in atomic physics until the discovery of fast-beam spectroscopy. It was therefore the privilege of FBS to introduce zero-field QB spectroscopy in 1970 as a new tool in atomic physics.[15,314] According to Section 2.4, FBS offers ideal conditions for such measurements as soon as $^{LL}\rho_q^k$ components with $k \neq 0$, i.e., alignment or orientation, are obtained by the excitation mechanism.

As a demonstration we choose a measurement of the H I $2p\ ^2P$ fs splitting (the hfs can be neglected!) observed in Ly$_\alpha$ radiation after foil excitation[337] with the setup sketched in Figure 44a. Since it is rather troublesome to measure polarization at a wavelength of 121.6 nm one simply measures the unpolarized Ly$_a$ intensity through a slit system which defines a detection function of 0.3 mm halfwidth. At 270 keV beam energy this corresponds to a time resolution of ~40 ps and easily allows the resolution of the expected 10,969 GHz fs frequency in Figure 44b. In this particular case, the fs splitting was of course known beforehand to a high degree of accuracy, and therefore the result in Figure 44b served for the determination of the alignment from the beat amplitude as produced by the foil interaction. It is instructive to use this simple example for a schematic view of the angular distribution of emitted radiation in a zero-field QB experiment. Owing to the axially symmetric beam-foil interaction a nonisotropic but axially symmetric radiation pattern occurs at $t = 0$. Without external fields it is only subject to the isotropic fs (spin–orbit) interaction and must therefore stay axially symmetric for all times with the dynamic changes schematically sketched in Figure 44c together with the corresponding light emission perpendicular to the beam axis.[32]

Numerous fs zero-field QB measurements have been performed during the past 7 years, and they were mostly devoted to a better understanding of the QB phenomenon itself, to alignment measurements, and to the experimental improvement of the method.[15,314,316,326,336,337,342,382,386,388,389] Only a few results in the He, Li isoelectronic sequences yielded new information on atomic fine structures.[118,338,387,390,391]

As a particularly interesting example for a hfs QB experiment we discuss a measurement of the ^{14}N IV $2s3p$ ^3P structure with perpendicular- and tilted-foil excitation.[56] According to the level structure in Figure 45f one expects, with perpendicular-foil excitation ($k = 0, 2$), only the beat frequencies 2ω, 3ω, and 5ω since the 1ω beat between the $F = 1$ and $F = 0$ levels is not allowed owing to the $\Delta(F_1F_2k)$ rule in Eq. (50) when $k = 2$. A measurement in linearly polarized light detection shows indeed in Figure 45a only the 2ω and 3ω frequencies, with the 5ω frequency not resolved. With the foil tilted to $\alpha = 45°$ the same result is obtained in linearly polarized light detection with nearly doubled signal in Figure 45b, the fourier transform of which in Figure 45c clearly shows the absence of the 1ω beat frequency. It should, however, occur in circularly polarized light detection due to the nonzero $^{LL}\rho^1_{\pm 1}$ components in a tilted-foil excitation. Figure 45d indeed shows the dominant contribution of the 1ω beat in a plot of S/I versus distance from the foil as verified by the fourier transform in Figure 45e and in excellent agreement with the prediction. It should be noted, however, that plotting of S/I leads in general to a rather complicated quantum beat pattern since according to Eq. (47) oscillating terms will appear not only in the numerator but also in the denominator.

Particularly in light atoms one often finds a situation with the hfs being of the same order of magnitude as the fs. As an example we take the fs–hfs of the ^{7}Li II $1s2p$ ^3P term in Figure 46. Anticipating the experimental

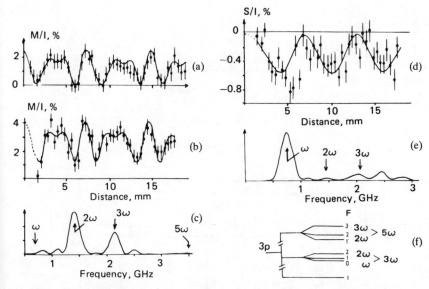

Figure 45. Zero-field QBs in the ^{14}N IV $3s^3S$–$3p\ ^3P$ 348.0-nm emission. (a) Foil at $\alpha = 0°$ and M/I detection. (b) Foil at $\alpha = 45°$ and M/I detection. (c) Fourier transform of (b). (d) Foil at $\alpha = 45°$ and S/I detection. (e) Fourier transforms of (d). (f) Schematic hfs levels of ^{14}N IV $3p\ ^3P$.

result one notes indeed that frequencies 5 or 6 are of the same order of magnitude as the fs splittings of ~ 42 GHz and ~ 83 GHz. As a consequence J is not a good quantum number anymore and the eigenstates are given by $|\alpha\rangle = \sum_J C_{\alpha J}|IJF\rangle$. The corresponding zero-field QB perturbation factor is then

$$G_{qq}^{kk}(t) = \sum_{\alpha_1 F_1 \alpha_2 F_2} \frac{(2F_1+1)(2F_2+1)}{(2S+1)(2I+1)} e^{-i(\omega_{\alpha_1}-\omega_{\alpha_2})t - \gamma t}$$

$$\times \left| \sum_{J_1 J_2} (-1)^{J_1+J_2} C_{\alpha_1 J_1} C_{\alpha_2 J_2}^* \begin{Bmatrix} J_1 & J_2 & k \\ L & L & S \end{Bmatrix} \right.$$

$$\times \left. \begin{Bmatrix} F_1 & F_2 & k \\ J_2 & J_1 & I \end{Bmatrix} [(2J_1+1)(2J_2+1)]^{1/2} \right|^2 \tag{61}$$

In general such situations lead to a complicated level scheme and thus also to a confusing QB pattern for which Figure 46 may stand here as an excellent example, measured with a perpendicular foil.[118] The time resolution is about 40 ps at 300 keV ^{7}Li$^+$ beam energy and the time scale is calibrated to 1.5×10^{-3} accuracy. This quality allows the extraction of six clearly resolved frequencies from the fourier transform with an accuracy of 3×10^{-3} to 5×10^{-3}, which are in excellent agreement with a recent cal-

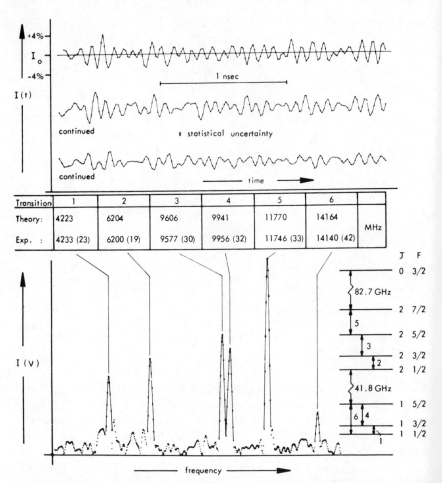

Transition	1	2	3	4	5	6	
Theory:	4223	6204	9606	9941	11770	14164	MHz
Exp. :	4233 (23)	6200 (19)	9577 (30)	9956 (32)	11746 (33)	14140 (42)	

Figure 46. ^{7}Li II $2p\ ^{3}P$ (fs) hfs measurement with the zero-field QB method after foil excitation. The upper part shows the beat pattern in the $2s^{3}S-2p\ ^{3}P$ 548.5-nm emission after subtraction of the nonoscillating intensity and continued twice. The full line connects the data points only. The lower part shows the fourier transform of these beats with the zero point of the frequency being suppressed. In the center six experimentally determined frequencies as indicated in the level diagram are compared with the theory by Jette et al., Phys. Rev. A **9**, 2337 (1974).

culation. Furthermore the beat amplitudes are in full accord with Eq. (61) if a frequency-response correction due to the finite width of the detection function is applied to the measured beat amplitudes in Figure 46.

These two examples of zero-field QB hyperfine-structure measurements after foil excitation clearly demonstrate the great potential of this method, and it has already been successfully applied to various terms in various charge states of light elements[392–399,389,391] which are not accessible by other known techniques. Unfortunately, however, the alignment is often small and imposes a serious limitation on the use of the zero-field QB method with foil excitation. As shown in Figures 39 and 45 the tilting of the foil may somewhat improve this situation for the $^{LL}\rho_q^2$ components, but the large increase of the $^{LL}\rho_{\pm1}^1$ components with tilt angle α does not yield an improvement on the beat amplitudes.

In order to circumvent this fundamental problem use can be made of the large orientations obtained with ion-beam–surface interaction at grazing incidence (IBSIGI).[83] As discussed by Fano and Macek[30] this large electronic orientation is partly transfered to the nucleus via hf interaction such that the surface-scattered ion beam will become nuclear spin polarized. This beam can then be used in a perpendicular-foil zero-field QB experiment with circularly polarized light detection of the hfs QBs, where the initial anisotropy at the foil exit is supplied by the nucleus and not by the foil-excitation mechanism.[400]

The transfer of the electronic orientation to the nucleus can be derived from

$$^{II}\rho_q^k(t) = \left(U_q^k(II) 1_S \cdot 1_L \middle| e^{-(i/\hbar)\hat{H}t} \middle| 1_I 1_S \rho_L \right) \frac{1}{(2S+1)(2I+1)} \tag{62}$$

which yields after expansion in terms of irreducible tensor operators and averaging over all QBs

$$^{II}\rho_q^k = \sum_{J,F} (-1)^{J+L+S+k} \frac{(2J+1)^2(2F+1)^2}{(2S+1)(2I+1)} \begin{Bmatrix} J & J & k \\ L & L & S \end{Bmatrix}$$

$$\times \begin{Bmatrix} F & F & k \\ J & J & I \end{Bmatrix} \begin{Bmatrix} F & F & k \\ I & I & J \end{Bmatrix} {}^{LL}\rho_q^k \tag{63}$$

From this equation one can deduce an orientation transfer to the nucleus between 10% and 32% depending on the oriented electronic term and the nuclear spin. If one assumes that the subsequent perpendicular-foil excitation leads to an isotropically excited electronic state one can give the intensity for a resolved $J \to J_0$ transition of a foil-excited nuclear-spin-polarized ion beam according to

$$I_{\mathbf{k}\varepsilon}(t) \propto \left(1_I D_J^+(\mathbf{k}\varepsilon) \middle| e^{-(i/\hbar)\hat{H}t} \middle| \rho_I 1_J \right) \frac{1}{(2J+1)} \tag{64}$$

For the description of an actual experiment with a $^{14}N^+$ beam with $I = 1$, one uses a quantization axis along the x axis in Figure 38b and neglects all $^{LL}\rho_q^k$ components of the IBSIGI except $^{LL}\rho_0^0$ and $^{LL}\rho_0^1$. After expansion of Eq. (64) in terms of irreducible tensor operators one obtains for S/I after *perpendicular* foil excitation and simultaneous observation of several fs multiplet components along the $(+x)$ direction

$$\frac{S}{I}(t) \propto \left(\frac{^{II}\rho_0^1}{^{II}\rho_0^0}\right) a_J \left[\sum_{J,F} (2F+1)^2 \begin{Bmatrix} 1 & 1 & 0 \\ J & J & J_0 \end{Bmatrix} \begin{Bmatrix} F & F & 0 \\ I & I & J \end{Bmatrix} \begin{Bmatrix} F & F & 0 \\ J & J & I \end{Bmatrix} \right]^{-1}$$

$$\times \sum_J \left[\sum_F (2F+1)^2 \begin{Bmatrix} 1 & 1 & 1 \\ J & J & J_0 \end{Bmatrix} \begin{Bmatrix} F & F & 1 \\ I & I & J \end{Bmatrix} \begin{Bmatrix} F & F & 1 \\ J & J & I \end{Bmatrix}\right.$$

$$+ 2 \sum_{F_1 < F_2} (-1)^{F_1+F_2+2J+2I} (2F_1+1)(2F_2+1)$$

$$\left. \times \begin{Bmatrix} 1 & 1 & 1 \\ J & J & J_0 \end{Bmatrix} \begin{Bmatrix} F_1 & F_2 & 1 \\ J & J & I \end{Bmatrix} \begin{Bmatrix} F_1 & F_2 & 1 \\ I & I & J \end{Bmatrix} \cos{(\omega_{JF_1} - \omega_{JF_2})t} \right]$$

$$(65)$$

where the summation over I includes all transitions of the $(LS)J \to (L_0 S)J_0$ multiplets which are detected simultaneously with detection efficiency a_J. This formula applies to the observation of the four ^{14}N II $2s^2 2p3s\ ^3P$– $2s^2 2p3p\ ^3D$ multiplet components with shorter wavelengths in an experi-

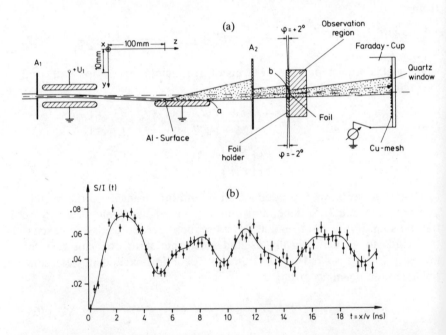

Figure 47. (a) Experimental arrangement for the observation of zero-field QBs in S/I detection after nuclear spin polarization of a 300-keV ^{14}N$^+$ beam via IBSIGI on the Al surface shown and passage through a perpendicular foil. The y axis is stretched by a factor of 4 compared to the z axis. The beam path is sketched with an approximate scattered beam distribution. (b) Zero-field QBs in S/I detection of the four ^{14}N II $2p3s\ ^3P$–$2p3p\ ^3D$ multiplet components with wavelengths 566.7, 567.6, 568.0, and 568.6 nm.

ment with a 300-keV $^{14}N^+$ ion beam scattered off an aluminum surface at $\sim 0.5°$, which then passes through a nearly perpendicular foil as sketched in Figure 47a. The result in Figure 47b clearly shows the feasibility of such experiments where the electronic orientation is obviously zero at $t = 0$, so that Figure 47b actually represents the dynamic orientation transfer from the nucleus to the initially isotropic electronic level of the ion. These QBs are in full accord with Eq. (65) and allow the determination of the hf interaction of the ^{14}N II $2s^2 2p3p\ ^3D$ term.[400]

Once a nuclear-spin-polarized ion beam is obtained by IBSIGI this method no longer requires alignment or orientation in the direct electronic foil excitation process as formerly necessary in all fast-beam QB experiments. Hence, a wide area is opened up for systematic hfs studies of all charge states and of all those configurations (by zero-field QB or HFLC techniques) that can be observed with circularly polarized light detection.

Fortunately, zero-field QBs are much less distorted by cascade repopulation than Zeeman beats or Hanle-effect signals. Following the analysis of Dufay[371] with the use of Eq. (I.47) of Ducloy and Dumont[374] one obtains for a fs cascade $(SL_2)J_2J_2' \rightarrow (SL_1)J_1J_1' \rightarrow (SL_0)$

$$I_{k\varepsilon}(t) \propto \sum_{\substack{kq \\ J_1J_1'}} (-1)^{-1-L_0+J_1+S+k+J_1+S+k} \frac{(2J_1+1)(2J_1'+1)}{(2S+1)} \begin{Bmatrix} 1 & 1 & k \\ L_1 & L_1 & L_0 \end{Bmatrix}$$

$$\times \begin{Bmatrix} J_1 & J_1' & k \\ L_1 & L_1 & S \end{Bmatrix} \phi_q^k(k\varepsilon)\,{}^{J_1J_1'}\rho_q^k(t) \quad (66)$$

with

$${}^{J_1J_1'}\rho_q^k(t) = (-1)^{L_1+J_1+S+k} \begin{Bmatrix} J_1 & J_1' & k \\ L_1 & L_1 & S \end{Bmatrix} {}^{L_1L_1}\rho_q^k(0)\exp(-i\omega_{J_1J_1'}t - \gamma_1 t)$$

$$+ \sum_{J_2J_2'} \gamma_2(2L_2+1)(2J_2+1)(2J_2'+1)$$

$$\times(-1)^{L_2+J_2+J_2'+J_1+S+1} \begin{Bmatrix} J_2 & J_2' & k \\ J_1' & J_1 & 1 \end{Bmatrix} \begin{Bmatrix} J_1' & J_2' & 1 \\ L_2 & L_1 & S \end{Bmatrix}$$

$$\times \begin{Bmatrix} J_1 & J_2 & 1 \\ L_2 & L_1 & S \end{Bmatrix} \begin{Bmatrix} J_2' & J_2 & k \\ L_2 & L_2 & S \end{Bmatrix} {}^{L_2L_2}\rho_q^k(0)$$

$$\times \frac{\exp(-i\omega_{J_2J_2'}t - \gamma_2 t) - \exp(-i\omega_{J_1J_1'}t - \gamma_1 t)}{(\gamma_1 - \gamma_2) + i(\omega_{J_1J_1'} - \dot{\omega}_{J_2J_2'})} \quad (67)$$

The first part in Eq. (67) represents again the direct excitation of the (SL_1) term and the second part describes the coherent, dynamic alignment (orientation) transfer via the cascade from the (SL_2) term to the (SL_1) term. The latter leads to cosine and sine contributions at J_1J_1' and J_2J_2' fs frequencies in the observation of the principally unresolved $(SL_1) \rightarrow (SL_0)$

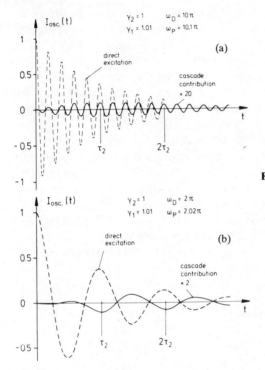

Figure 48. The influence of a direct cascade on zero-field QBs. A term diagram as in Figure 42 is chosen with unresolved fs. The fs beat frequency ω_D and γ_2 of the feeding term are assumed to be nearly equal to ω_P and γ_1 of the term to be studied, which is assumed to be populated five times as much as the feeding term at equal alignment A_0^2. (a) The oscillating intensity for $\omega_D > 2\pi\gamma_1$. (b) The oscillating intensity for $\omega_D = 2\pi\gamma_1$.

transition. However, these contributions will be small and negligible as long as the denominator in Eq. (67) is large, i.e., when $\gamma_1 \neq \gamma_2$ and $\omega_{J_1J_i} \neq \omega_{J_2J_2'}$. Since this is normally the case in zero-field QB experiments one can in general neglect cascade contributions. In order to prove this statement we choose the cascade in Figure 42 for a fs QB experiment under the following assumptions: $\gamma_2 = 1s^{-1}$, $\gamma_2 = 2s^{-1}$, $\omega_D = \omega_{3/2,5/2}(^2D) = 10\pi \, s^{-1}$, $\omega_p = \omega_{1/2,3/2}(^2P) = 50\pi \, s^{-1}$, $^{L_1L_1}\rho_0^2(0) = 5^{L_2L_2}\rho_0^2(0) = 1$, and consider only the oscillatory contributions. As a result one obtains indeed negligible cascade contributions as compared to the first direct excitation term:

$$I_{\text{osc}}(t) \propto \cos \omega_P t \, e^{-\gamma_1 t} - (+5 \times 10^{-6} \cos \omega_P t - 6 \times 10^{-4} \sin \omega_P t) \, e^{-\gamma_1 t}$$

$$- (1.9 \times 10^{-5} \cos \omega_D t - 7.5 \times 10^{-5} \sin \omega_D t) \, e^{-\gamma_2 t} \qquad (68)$$

Significant cascade contributions to zero-field QB experiments can thus be expected only when accidentally close coincidences occur between upper $\omega_{J_2J_i}$ and lower $\omega_{J_1J_i}$ beat frequencies and between γ_1 and γ_2. For the illustration of this situation we choose the same example as before with the special assumptions shown in Figure 48. The result in Figure 48a for a high beat frequency shows a cascade contribution of only 2% at $t = 2\tau_2$ and is thus negligible for a determination of ω_p to within 10^{-3} accuracy. Problems may thus only occur for the case of Figure 48b with a low beat frequency of $\omega_p \approx 1/\tau_1$. The cascade contribution reaches a level of about 20% at $t = 2\tau_2$ and must therefore be considered in a data analysis. However, this is an extremely unsuitable example, which can hardly even be measured with good accuracy in the absence of cascades.

Hence, one can conclude that cascades do not affect the determination of fs or hfs splittings with the zero-field QB technique at a level of accuracy of 10^{-3}.

4.6. High-Field Level Crossing

As an alternative to zero-field QB measurements one may consider the fast-beam HFLC method. As a time integral method it has the great advantage of being velocity independent and should therefore allow better accuracies than the zero-field QB technique when no velocity analysis is performed. However, the HFLC signal is in principle smaller than the QB signal since according to Figure 34c only two out of $(2S + 1)(2L + 1)(2I + 1)$ substates of a given term are contributing to the signal, whereas all the others are responsible for the large background. Hence, the alignment produced by foil or gas excitation also imposes (besides the beam bending in the necessary large fields) serious limitations on the applicability of the fast-beam HFLC technique.

For the theoretical description we refer to $|H + G|$ and show here only one of the rare applications of fast-beam HFLC for the determination of the Be II $5f^2F$ fs splitting in Figure 49. The figure shows the $(7/2, -7/2 - 5/2, -3/2) - \Delta m = 2$ crossing with a signal of only $\sim 0.16\%$, which clearly demonstrates the difficulties involved in such measurements.[343]

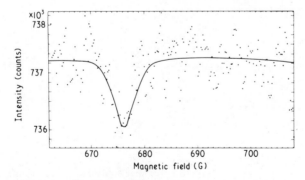

Figure 49. High-field $(j = 7/2, \quad m = -7/2) - (j = 5/2, \quad m = -3/2) \, \Delta m = 2$ level crossing for the determination of the Be II $5f^2F$ fs with gas excitation of a fast Be^+ beam.

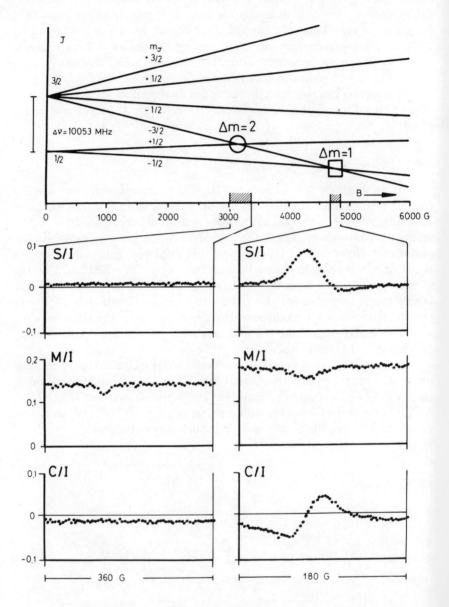

Figure 50. High-field $\Delta m = 1$ and 2 level crossings in the S/I, M/I, and C/I detection of the Li I $2s\ ^2S-2p\ ^2P$ 670.7-nm emission after IBSIGI in a magnetic field parallel to the incident beam direction.

A considerable improvement of the signal can be expected by inclined-surface excitation, which yields large orientations and avoids beam bending by applying the magnetic field parallel to the surface-scattered beam direction. First test measurements on the known high-field structure of the Li I $2p\,^2P$ term[401] demonstrate the feasibility of this new method in Figure 50, where measurements of S/I, M/I, and C/I are plotted versus field strength. As expected, one observes predominantly $\Delta m = 1$ crossings in S/I and C/I measurements and $\Delta m = 2$ in M/I measurements. The signal size is sufficient for future applications of this method, but a careful analysis of the obvious signal distortions is necessary before unknown structures can be measured.

4.7. Fast-Beam rf Experiments

Practically all experiments of this type so far have been devoted to hydrogenlike levels. Their discussion is fully covered by the article of Beyer |BE| in this work.

5. Selective Laser Excitation

According to the specific features of FBS in Section 2.4 any means of selective excitation of a single level (term) of a single charge state of the fast-moving beam would yield unprecedented experimental conditions for its transient spectroscopic study. As such means, selective electron or light excitation can be considered, of which light excitation is favored by a factor of $\sim 10^6$ owing to the atomic excitation cross section of $\sigma \approx 10^{-10}$ cm^2 for visible light at resonance[402,403] compared to $\sigma \leqslant 10^{-16}$ cm^2 for electrons near threshold.[404] With this large cross section for resonant light excitation, estimates for an actual fast-beam excitation show[405,406] that a photon flux of $\phi \approx 10^{16}$ photons/s mm^2 (mrad)^2 GHz is necessary in order to obtain a detectable fluorescence signal of the order of 10^4 photon counts s^{-1} in the typical setup of Figure 3 with an overall detection efficiency of $\varepsilon \approx 10^{-5}$. Such high photon fluxes are presently available only from lasers, which leads us to restrict the discussion on selective excitation of fast beams in this chapter to resonant laser excitation.

5.1. Precision Lifetimes

The most obvious application of the unprecedented properties of FBS combined with selective laser excitation is the cascade-free measurement of lifetimes of excited levels with high accuracy. Since most of the errors of the other known techniques for lifetime measurements[407,408] are strongly

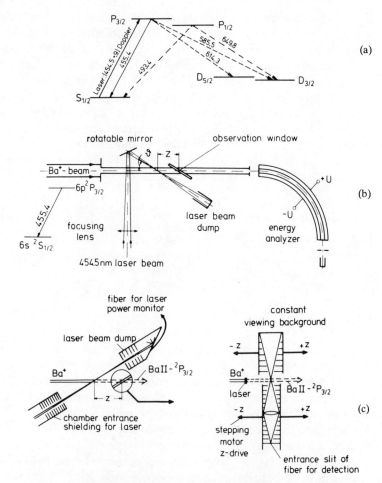

Figure 51. (a) Diagram of the five lowest Ba II energy levels. (b) Schematic view of the experimental setup used for laser excitation of a fast Ba$^+$ beam with a Doppler-tuned fixed-frequency argon-ion laser. The observation window is moved along the beam axis relative to the fixed excitation region. (c) Schematic design of the laser stray light suppressors for the laser beam and the detection optics.

reduced (if not eliminated) with this fast-beam laser (FBL) method, relative accuracies below $\Delta\tau/\tau = 0.01$ can be achieved for the first time in atomic physics.

The basic features of such experiments may be best described by way of an example of the precise lifetime determination of the ^{138}Ba II $6p\,^2P_{3/2}$ level for which the experimental setup is shown in Figure 51. According to the level scheme this level can be excited from the ground state with a cw argon-ion laser line at 454.5 nm Doppler-tuned to the resonance line at 455.4 nm by letting the laser intersect the Ba^+-ion beam at an angle ϑ based on the relation

$$\lambda_{laser} = \lambda_{atom}\left(1 - \frac{v}{c}\cos\vartheta\right)\bigg/\left(1 - \frac{v^2}{c^2}\right)^{1/2} \tag{69}$$

In order to achieve good time resolution ($\delta t \approx 0.5$ ns) the laser beam is focused by a lens via a mirror onto the ion beam and the observation window is also tilted to the same angle ϑ. The resonance can be tuned in either by rotation of this mirror or by velocity variation of the ion beam, of which the latter method allows even the elimination of this mirror when the laser beam enters the target chamber approximately at the correct angle ϑ.

The excitation probability P_{12} for such a geometry must be calculated for an effective broad laser bandwidth $\sigma[s^{-1}]$. This effective absorption linewidth takes into account the laser width, time uncertainty, and Doppler broadening due to ion and laser beam divergences and energy spread of the ion beam [see Eq. (73) in Section 5.3]. With an interaction time $T = 0.5$ ns, a power density $I_0 = 300$ mW/7 mm$^2 \times$ c, and an estimated Gaussian width $\sigma = 4$ GHz one can still use perturbation theory and obtains with

$$P_{12}(6^2S_{1/2} \to 6^2P_{3/2}) = \int_0^\infty \frac{I_0\lambda^3 A_{12}}{4(2\pi)^{3/2}\sigma\hbar} e^{-(\omega-\omega_{12})^2/2\sigma^2} \frac{\sin^2(\omega-\omega_{12})T/2}{(\omega-\omega_{12})^2} d\omega \tag{70}$$

$P_{12} = 1.6\%$ as excitation probability for each Ba^+ ion passing through the laser beam. Hence, with a beam of 5 μA Ba^+ one expects 5×10^{11} ions per second to be excited, and the count rate for a precision experiment will be thus high enough even for detection geometries with low efficiency.

The design of such detection geometries crucially depends on the laser stray light in the target chamber, which can be rather high when the laser beam is reflected off mirror or laser dump surfaces as in Figure 51b. A high-quality measurement (i.e., low background) was therefore only possible after elimination of this mirror, installation of stray-light baffles as shown in Figure 51c, and by detecting the resonance line spectrally resolved and 0.9 nm away from the Doppler-shifted laser line through a spectrometer. The connection from the movable detection optics to the spectrometer was a flexible fiber. With this geometry an overall detection efficiency of less than 10^{-6} was achieved, which allowed the observation of a maximum fluorescence intensity of $\sim 10^4$ counts s^{-1} at a detected laser stray light of <1 count s^{-1}. The laser stray light can be further greatly reduced when the laser is not stopped in a dump inside the chamber and can instead leave the target chamber through a Brewster window. This improves the detection efficiency up to $\sim 10^{-3}$ by replacing the lens, fiber, and spectrometer simply by a slit optics, and interference filter, and a

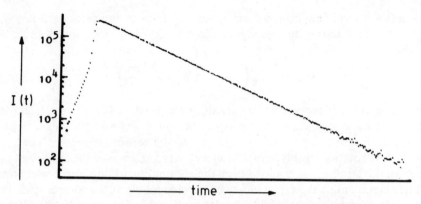

Figure 52. Cascade-free intensity decay curve of the ^{138}Ba II $6s$ $^2S_{1/2}$–$6p$ $^2P_{3/2}$ 455.4-nm emission after laser excitation as measured with the experiment setup shown in Figure 51b.

multiplier inside the target chamber.[295] At reduced excitation probability with similar laser power, signals of 10^5 counts s^{-1} could be observed at a maximum laser stray light of <20 counts s^{-1}.

The actual experiment with the setup in Figure 51c was performed as described in Section 2.3.8 normalized to the beam current and with simultaneous recording of the laser power in order to correct the data for laser fluctuations. A result is shown in Figure 52. The data were corrected for dead times, the laser fluctuations, and laser stray light. The beam background and the multiplier noise was subtracted after separate measurements of each contribution in order to obtain a pure single exponential for a least-squares-fit analysis. The statistical reproducibility of the lifetimes so obtained was 10^{-3}. However, a number of systematic errors, as discussed in Sections 2.3.2, 2.3.6, 2.3.9, and 2.3.10 have to be even more carefully determined or estimated than in normal beam-foil experiments. Furthermore, results from modified experimental arrangements (e.g., Figures 51b and 51c) have to be compared before one can obtain proof of the unprecedented precision of $\pm0.25\%$ that can be achieved with this FBL technique for the determination of lifetimes. A detailed description of the procedures is given in Ref. 81, from which we quote the final result: $\tau($Ba II $6p$ $^2P_{3/2}) = 6.312 \pm 0.016$ ns. The great improvement over other methods becomes obvious when comparison is made with the most accurate published Hanle-effect result of $\tau = 6.27 \pm 0.25$ ns[409] of this level.

After the first successful FBL lifetime measurement it was suggested of course[406] that this method should be extended to excitation from metastable levels and neutral ground states prepopulated by ion-beam–gas collisions, and to the use of cw dye lasers or even pulsed dye lasers besides

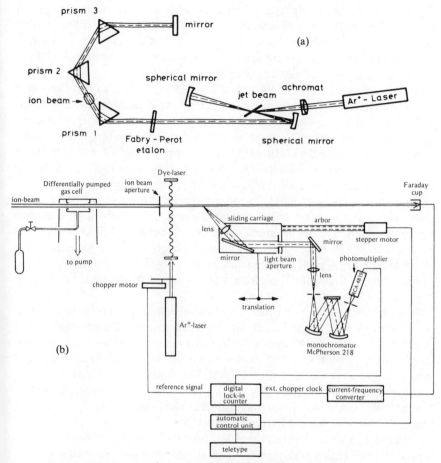

Figure 53. (a) Schematic drawing of the dye-laser resonator for intracavity excitation of fast beams with the experimental arrangement shown in (b) for the prepopulation of neutral ground states or metastable levels by a gas target far upstream ($\sim 1~\mu$s) from the laser interaction region. Later on a velocity analysis was incorporated in this setup.

fixed-frequency Doppler-tuned cw lasers. Since 1973 the development is thus characterized by the successive employment of these various excitation regimes, of which we will mention a cw dye laser and a pulsed dye laser experiment.

With the high detection efficiency of $\varepsilon \approx 10^{-3}$ mentioned above, cw dye laser outputs of the order of a few mW/mm^2 (mrad)^2 GHz are sufficient for precise lifetime determinations and allow, owing to their tunability, the very convenient excitation at $\vartheta = 90°$. In order to increase the cw dye laser power density, an intracavity excitation of the ion beam in

Figure 53a has been employed[410] which yields 5–10 W intracavity power with rhodamine 6 G at a bandwidth of ~0.05 nm (~50 GHz in the visible). At an effective absorption linewidth of 4 GHz this corresponds to a useful laser power density of $I_0 \approx 500 \, \text{mW/mm}^2 \, c$, which is sufficient to excite the Ba II $6p \, ^2P_{3/2}$ level from the metastable Ba II $5d \, ^2D_{3/2}$ level with a transition probability $A \approx 0.04 \, A(6 \, ^2P_{3/2} - 6 \, ^2S_{1/2})$ at 585.4 nm wavelength according to the level scheme in Figure 51a. This metastable level is already populated to about 5% by the ion source so that the experimental arrangement in Figure 53b could be used without the gas target to measure $\tau(\text{Ba II } 6p \, ^2P_{3/2}) = 6.28 \pm 0.06 \, \text{ns}$ by detecting the 455.4-nm resonance radiation far away in wavelength from the stray laser light. In order to eliminate background contributions the laser was chopped in this experiment at a frequency of 82 Hz, and only the difference signal (laser-on)– (laser-off) was recorded. With the gas target 0.6 m in front of the laser interaction region the experiment could be run equally well, since a similar population of the metastable level could be obtained by gas collisions, while the $6p \, ^2P_{3/2}$ population including long-lived cascades had decayed after a time of flight of ~1 μs from the gas target to the laser interaction region.

With this type of experiment, i.e., the prepopulation of ground states or metastables by a gas target far upstream from the laser interaction region or of metastables directly by the ion source, numerous lifetimes have been measured with cw dye lasers and Doppler-tuned fixed-frequency lasers as listed in Table 5.[81,82,410–417] The limits of application of this technique are, however, clearly set by the wavelength range of cw dye lasers from ~430 to ~800 nm and the accidental close coincidences between atomic and fixed frequency laser lines.

For the study of the resonance levels of most neutral or singly ionized atoms uv laser action is required, which is presently not yet available from frequency-doubled cw dye lasers with sufficient power. Except for accidental close coincidences between resonance lines and fixed-frequency uv laser lines one is therefore presently dependent on the use of frequency-doubled pulsed nitrogen laser pumped dye lasers. The main problem with such pulsed lasers is the fantastically small duty cycle of the order of 5×10^{-6}, whereas the power available of a few hundred watts usually suffices to saturate the atomic beam during the laser pulses of 10 ns typical length. Estimates for a 1-μA ion beam yield then a maximum signal of ~6 photons per laser pulse at a detection efficiency of 10^{-3} and a detection time window of 0.1 τ, which at a repetition rate of 500 s⁻¹ would yield a large enough signal of 3000 counts s⁻¹ for a precision experiment. Unfortunately, however, the detection electronics can process only one count per laser pulse in photon-counting mode. The effective signal must therefore be limited to about 0.2 counts per laser pulse in order to reduce

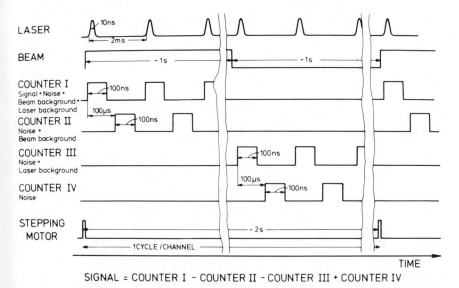

Figure 54. Laser, atomic beam, detection, and stepping motor timing in an experiment using a pulsed dye laser for obtaining the pure fluorescence signal of the beam according to the formula on the bottom of the figure. For use with cw lasers the "laser-on" periods are set equal to the counter time windows of ~0.5 s.

the necessary dead-time correction $n_0 = -r \ln(1 - n/r)$ to about 10%, where r is the laser repetition rate and n, n_0 are the measured and real photon rates. In order to obtain an accurate lifetime result at such low count rates of $100 \, \mathrm{s}^{-1}$ one must carefully subtract all background contributions during the experiment. To this end a gating sequence of the ion beam and the detection channel takes place at each position setting of the general FBS experimental procedure in Section 2.3.8, as indicated in Figure 54. The detection channel is thus only opened for periods of 100 ns and output pulses are directed into different counters I–IV for the combinations laser-on–beam-on, laser-off–beam-on, laser-on–beam-off, and laser-off–beam-off, respectively. Hence, data are accumulated simultaneously in four memory blocks of the multiscaler from which the pure signal can be extracted after completion of the experiment. This procedure has turned out to be very effective not only for pulsed laser but also for cw laser experiments with gate windows of ~0.5 s.

Using this pulsed dye laser excitation and gated detection technique,[418] measurements without frequency doubling have been successfully performed on the resonance levels of Ba II and Sr II as tests for future frequency-doubling experiments. A careful analysis and investigation of systematic errors of these results yielded lifetimes of these levels of 1%

Table 5. List of FBL Lifetime Measurements Performed So Far

Year	Level	τ (ns)	Beam energy (keV)	Velocity analysis	Excitation from initial level	Excitation from wavelength (nm)	Decay to final level	Decay to wavelength (nm)	Type of laser	ϑ	Level preparation	Ref.
1972	Ba II $6p\,^2P_{3/2}$	6.25 ± 0.06	336	no	$6s\,^2S_{1/2}$	$454.5+0.9$	$6s\,^2S_{1/2}$	455.4	argon ion	30	ion source	405
1973	Ba II $6p\,^2P_{3/2}$	6.21 ± 0.06	336	no	$6s\,^2S_{1/2}$	$454.5+0.9$	$6s\,^2S_{1/2}$	455.4	argon ion	30	ion source	405
1974	Ba II $6p\,^2P_{3/2}$	6.32 ± 0.06	336	yes	$6s\,^2S_{1/2}$	$454.5+0.9$	$6s\,^2S_{1/2}$	455.4	argon ion	30	ion source	36
1974	Ba II $6p\,^2P_{3/2}$	6.28 ± 0.06	240	yes	$5d\,^2D_{3/2}$	585.4	$6s\,^2S_{1/2}$	454.8	cw-dye	90	ion source	410
1974	Ne I $3p\,[\tfrac{5}{2}]_2$	19.6 ± 0.2	52	yes	$3s\,[\tfrac{3}{2}]_2$	594.5	$3s\,[\tfrac{3}{2}]_1$	608.6	cw dye	90	gas target	410
1975	Ba II $6p\,^2P_{3/2}$	6.3 ± 0.2	45	no	$5d\,^2D_{5/2}$	614.1	$6s\,^2S_{1/2}$	455.4	cw dye	90	ion source	411
1975	Ba II $6p\,^2P_{1/2}$	8.0 ± 0.3	45	no	$5d\,^2D_{3/2}$	649.8	$6s\,^2S_{1/2}$	493.4	cw dye	90	ion source	411
1975	Ba II $6p\,^2P_{3/2}$	6.312 ± 0.016	336	yes	$6s\,^2S_{1/2}$	$454.5+0.9$	$6s\,^2S_{1/2}$	455.4	argon ion	30	ion source	81
1975	Rb II $5p\,^3[\tfrac{2}{2}]_1$	8.04 ± 0.08	310	yes	$5s\,^3[\tfrac{2}{2}]_2$	$476.5+1.1$	$5s\,^3[\tfrac{2}{2}]_1$	515.2	argon ion	33.8	gas target	412
1975	Mg I $4s\,^3S_1$	9.68 ± 0.06	260	yes	$3p\,^3P_0$	$514.5+2.2$	$3p\,^3P_2$	518.3	argon ion	27.7	gas target	81
1975	Sr I $5d\,^3D_3$	16.29 ± 0.24	100	yes	$5p\,^3P_2$	$496.5-0.28$	$5p\,^3P_2$	496.2	argon ion	111.4	gas target	413
1975	Sr I $5d\,^3D_2$	16.34 ± 0.13	150	yes	$5p\,^3P_2$	$496.5+0.29$	$5p\,^3P_1$	487.2	argon ion	72.5	gas target	413
1975	Sr I $5d\,^3D_1$	16.49 ± 0.10	150	yes	$5p\,^3P_1$	$488.0-0.38$	$5p\,^3P_0$	483.2	argon ion	114.0	gas target	413
1975	Sr I $5p^2\,^3P_2$	7.89 ± 0.05	210	yes	$5p\,^3P_1$	$472.7-0.46$	$5p\,^3P_2$	481.2	argon ion	115.3	gas target	413
1975	Sr I $4f\,^1F_3$	34.15 ± 0.38	235	yes	$4d\,^1D_2$	$514.5+1.1$	$4d\,^1D_2$	515.6	argon ion	29.4	gas target	413
1975	Na I $3p\,^2P_{3/2}$	16.1 ± 0.20	100	yes	$3s\,^2S_{1/2}$	589.0	$3s\,^2S_{1/2}$	587.4	cw dye	90	gas target	82
1975	Na I $3p\,^2P_{1/2}$	16.3 ± 0.16	100	yes	$3s\,^2S_{1/2}$	589.6	$3s\,^2S_{1/2}$	588.0	cw dye	90	gas target	82
1975	Ne I $3p\,[\tfrac{3}{2}]_2$	19.7 ± 0.3	?	yes	$3s\,[\tfrac{3}{2}]_2$	594.5	$3s\,[\tfrac{3}{2}]_2$	607.9	cw dye	90	gas target	82
1975	Ne I $3p\,[\tfrac{3}{2}]_2$	19.5 ± 0.3	?	yes	$3s'[\tfrac{1}{2}]_2$	614.3	$3s'[\tfrac{1}{2}]_1$	690.5	cw dye	90	gas target	82
1975	Li II $2p\,^3P$	37.1 ± 0.4	200	yes	$2s\,^3S$	548.5	$2s\,^3S$	544.5	cw dye	90	gas target	82

Year	Species	Value		no/yes						Method		Source	Ref
1975	Sc II $3d4pz\ {}^1P_1$	9.2 ± 0.5	212	no	$501.7 + 1.4$	$b\ {}^1D_2$	353.6	$a\ {}^1D_2$	argon ion	30	ion source	414
1976	Sc II $3d4pz\ {}^3F_2$	6.2 ± 0.2	45	yes	$363.8 + 0.4$	$a\ {}^3D_1$	441.5	$a\ {}^3F_2$	argon ion	22.7	ion source	415
1977	La II $y\ {}^3F_4$	455 ± 11	30	no	579.8	$a\ {}^3F_4$	626.2	$a\ {}^3D_3$	cw dye	90	ion source	416
1977	La II $y\ {}^3F_3$	430 ± 10	30	no	580.6	$a\ {}^3F_3$	639.1	$a\ {}^3D_2$	cw dye	90	ion source	416
1977	La II $y\ {}^3F_2$	511 ± 13	30	no	580.8	$a\ {}^3F_2$	652.7	$a\ {}^3D_1$	cw dye	90	ion source	416
1977	La II $z\ {}^1D_2$	573 ± 21	50	no	588.1	$a\ {}^3D_1$	?	?	cw dye	90	ion source	417
1977	La II $z\ {}^3D_1$	43.8 ± 1.0	50	no	530.4	$a\ {}^3D_2$	?	?	cw dye	90	ion source	417
1977	la II $z\ {}^3D_2$	51.1 ± 1.6	50	no	530.4	$a\ {}^3D_3$	?	?	cw dye	90	ion source	417
1977	La II $z\ {}^3D_3$	67.8 ± 1.7	50	no	464.5	$a\ {}^3F_3$	?	?	cw dye	90	ion source	417
1977	Ba II $6p\,{}^2P_{3/2}$	6.31 ± 0.07	300	yes	455.4	$6s\,{}^2S_{1/2}$	455.4	$6s\,{}^2S_{1/2}$	pulsed dye	90	ion source	418
1977	Ba II $6p\,{}^2P_{1/2}$	7.92 ± 0.08	300	yes	493.4	$6s\,{}^2S_{1/2}$	493.4	$6s\,{}^2S_{1/2}$	pulsed dye	90	ion source	418
1977	Sr II $5p\,{}^2P_{3/2}$	6.69 ± 0.07	300	yes	407.7	$5s\,{}^2S_{1/2}$	407.7	$5s\,{}^2S_{1/2}$	pulsed dye	90	ion source	418
1977	Sr II $5p\,{}^2P_{1/2}$	7.47 ± 0.07	300	yes	421.5	$5s\,{}^2S_{1/2}$	421.5	$5s\,{}^2S_{1/2}$	pulsed dye	90	ion source	418
1977	Ba II $6p\,{}^2P_{3/2}$	6.31 ± 0.05	254	yes	$454.5 + 0.9$	$6s\,{}^2S_{1/2}$	455.4	$6s\,{}^2S_{1/2}$	argon ion	0	ion source	426

accuracy,[419] as shown with all other FBL lifetime results in Table 5. First measurements with frequency doubling yielded data of similar quality with incomplete data analysis as yet.[420] It should be noted that pulsed laser excitation of atoms at rest or at thermal velocities with electronic time resolution in the detection is only seemingly a competitive method since the excellent detection linearity and time-scale calibration of the FBL technique with pulsed-laser excitation guarantees a better ultimate accuracy.

As another excitation regime the two-step laser excitation, for which saturating pulsed dye lasers are particularly well suited, will further extend the applicability of the FBL method. The main breakthrough for the FBL method is, however, expected from the feasibility of laser excitation from prepopulated short-lived levels k to higher-lying levels n, i.e., from a modified two-step excitation (MTSE), where the first step is accomplished by gas or foil excitation only a few nanoseconds before the laser interaction takes place.[410] The basic experimental problem of this MTSE technique is the large background of the $k-n$ transition from direct excitation of and cascades through level n on which one finds superimposed the actual signal due to population transfer by the laser from level k to level n. It was shown for ideal conditions, i.e., no laser stray light and no residual gas, that the difference signal (laser-on) − (laser-off) represents a single exponential with the decay constant γ_n of the upper level n [410,128]:

$$N_n^1(t) - N_n^0(t) = [N_n^1(0) - N_n^0(0)]\, e^{-\gamma_n t} = N_n^d(t) = N_n^d(0)\, e^{-\gamma_n t} \qquad (71)$$

where N_n^1, N_n^0 are the populations of level n with laser on and off, respectively, after the laser interaction, which ends at $t = 0$. In reality, however, collisions with the residual gas may cause considerable deviations from a single exponential of $N_n^d(t)$ when the collisional repopulation from k to n is markedly different from the average collisional repopulation of level n. In this case one subtracts different beam backgrounds and obtains therefore a distorted signal $N_n^d(t)$. This effect becomes particularly troublesome when a gas target is used a few millimeters upstream from the laser interaction region, whereas foil preexcitation with good vacuum conditions (see Figure 6) should allow it to be neglected.

At present two groups [421,422] are working on this MTSE technique and have achieved values of $N_n^d(0)/N_n^0(0) \approx 0.2$–$0.3$ in excitations from short-lived (<27 ns) levels, but no completed data analysis for a lifetime determination has been reported yet. A very weak argon-ion laser induced change of population (too small for a lifetime measurement) of the $10M$ level from the short-lived $9L$ level in hydrogenic ^{19}F IX has also recently been achieved.[423] It should be noted that for the above large population transfers Eq. (70) is not valid any more. One must then take the coherence properties of the broadband exciting laser light explicitly into account[424,425] when solving the Bloch equations for such a system[426] in order to accurately predict such large transfer probabilities. (This is of course necessary also for all pulsed dye laser FBL experiments.)

As an interesting alternative to all the preceding crossed-beam experiments the superimposed-beam geometry has recently been applied for a FBL lifetime measurement,[427] which may be particularly well suited for MTSE experiments. For the example of Ba II 6p $^2P_{3/2}$ an argon ion laser beam at 454.5045 (air) wavelength was unidirectionally superimposed on a Ba$^+$ beam of 253.65 keV, which was according to Eq. (69) 2.5 keV above resonance. By passing both beams through a triple of narrow-spaced electrodes (the two outer ones grounded and the central one on potential $+ U$) it was possible to slow the ion beam down into resonance and to reaccelerate it again out of resonance over a total distance of only 3.4 mm. This resulted in a sufficiently sharp cutoff of the excitation region for an accurate lifetime determination with normal side-on detection. The velocity was simultaneously determined by a different set of electrodes further downstream, which allowed the accurate measurement of the resonance fluorescence as a function of the applied potential. The advantages of this superimposed-beam technique for lifetime measurements are, first, a possible longer beam–laser interaction time with sharp cutoff to yield higher excitation probabilities and, secondly, its easy combination with a foil or a gas target for MTSE experiments.

5.2. Quantum Beats

One of the early motivations for the introduction of FBL experiments was the goal of achieving maximum possible alignment (or orientation) of excited levels with polarized laser light for precise fs, hfs, and g-value quantum beat experiments.

As an example we chose the hfs of the ^{137}Ba II 6p $^2P_{3/2}$ level. This case is of particular interest since it will give us in Section 5.3 an estimate of the spectral resolution that can be obtained with laser fluorescence spectroscopy on fast beams. Considering the level scheme in Figure 55 we find the ground state split by 8.04 GHz,[428] whereas the upper states have a maximum separation of 659 MHz only. When Doppler-tuning the laser resonance by changing the angle ϑ in Figure 51b and recording the 455.4-nm fluorescence emission, a partially resolved double-peak structure as shown in Figure 55 is observed. It clearly indicates that selective excitation from either one of the ground-state hyperfine components $F = 1$ or $F = 2$ is achieved, giving an estimate on the resonance linewidth of ~ 5 GHz, or a resolution of about $\lambda/\Delta\lambda = 150000$. As a consequence of the selection rules one has then only coherent excitation of either the upper three or the lower three hfs components of the excited level and expects

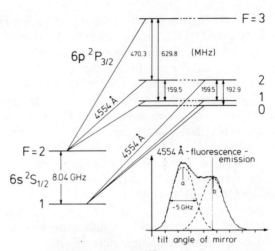

Figure 55. Level scheme of ^{137}Ba II indicating the selective laser excitation from $F = 2$ ground state (a) to the upper three and from $F = 1$ ground state (b) to the lower three hfs components of the excited $6p^2P_{3/2}$ level. (a) and (b) correspond to the two angular settings of ϑ as indicated in the lower part of the figure.

therefore only beats among the upper three or the lower three hfs components, depending on which angular setting (a or b in Figure 55) of ϑ has been chosen. This is exactly what has been observed in the experiment: In the upper part of Figure 56 two low-frequency beats from the lower three hfs components appear and in the lower part one low-frequency plus two high-frequency beats appear from the upper three hfs components, each time with their Fourier transforms inserted. Recalling the discussion on linewidths of quantum beat ($\Delta\nu = 3^{1/2}\gamma/\pi = 88$ MHz) or double resonance ($\Delta\nu = \gamma/\pi = 50$ MHz) experiments in Section 4.1, it is important to note that the frequencies ω_{21} and ω_{20} are actually resolved here beyond the natural width. This was possible owing to the simultaneous measurement of the frequencies and the decay up to 7τ with good enough statistics, so that the data could be divided by the measured exponential to yield undamped quantum beats up to 7τ, which by fourier transformation result in the good resolution shown. Such a procedure fully relies on the excellent detection linearity of FB experiments, which makes FBL QB experiments superior to pulsed-laser-excited QBs detected with presently known electronic time resolution.

The QBs in Figure 56 were observed in unpolarized light along a z axis perpendicular to the beam after excitation along an x axis also perpendicular to the beam with linear polarization vector parallel to the z

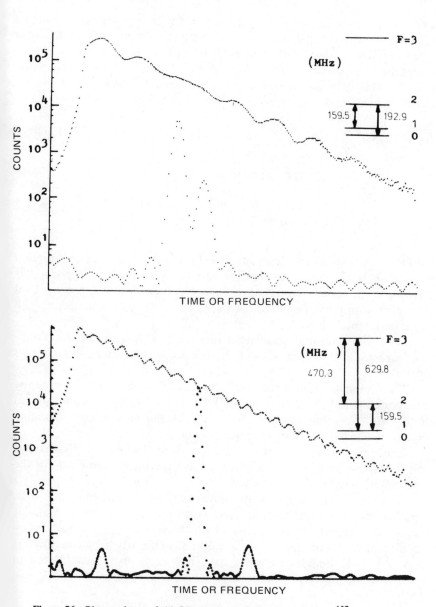

Figure 56. Observed zero-field QBs in the 455.4-nm emission of ^{137}Ba II and their Fourier transforms as calculated after dividing the data by an exponential and then subtracting the constant background. The inserted level schemes indicate the measured beat frequencies. Note that the time and frequency scales are different in the top and bottom part of the figure.

axis. As one would expect for broadband (nonselective) excitation one seems to observe at first sight only intensity minima at $t = 0$. However, a more detailed analysis with a direct least-squares fit to the data clearly shows that the beat frequency $\omega(F_1 = 1 - F_2 = 2)$ starts out with a maximum at $t = 0$. This is in full accord with the theoretical description of the hfs quantum beats under selective excitation given by the intensity formula

$$I_{k\varepsilon}(t) = \frac{1}{N} e^{-\gamma t} \sum_{kq} (-1)^{k+2J+J_0-I+\bar{F}_0} (2\bar{F}_0+1)\Phi_q^k(k\varepsilon) \begin{Bmatrix} 1 & 1 & k \\ J & J & J_0 \end{Bmatrix} [\overline{\Phi_q^k(k'\varepsilon')}]^*$$

$$\times \sum_{F_1 F_2} (-1)^{F_1+F_2}(2F_1+1)(2F_2+1)$$

$$\times \begin{Bmatrix} 1 & 1 & k \\ F_1 & F_2 & F_0 \end{Bmatrix} \begin{Bmatrix} F_1 & J & I \\ J_0 & F_0 & I \end{Bmatrix} \begin{Bmatrix} F_2 & J & I \\ J_0 & F_0 & I \end{Bmatrix} \begin{Bmatrix} J & F_1 & I \\ F_2 & J & k \end{Bmatrix} e^{-i\omega_{F_1 F_2} t}$$

$$(72)$$

for the excitation from a $\overline{I_0 F_0}$ state to $I(F_1, F_2)$ states and the observation of the $J-J_0$ transition, which was first derived by Macek.[429] Applying this formula to the specific case studied here, one indeed obtains the ω_{12} beat frequency with opposite phase compared to all the others in accord with the experimental result.

From the measured frequencies one can readily deduce the hf coupling constants,[430] which show a significant improvement over earlier work.[431]

In a similar experiment the ^{23}Na I $3p$ $^2P_{3/2}$ hfs could be recently remeasured with FBL QBs up to 9τ (!).[295] This allowed the full resolution of this hfs[81] for the first time in zero field and yielded corresponding improvements particularly of the quadrupole coupling constant.

These measurements clearly demonstrate that whenever FBL excitation is possible excellent conditions for QB experiments are given for the determination of fs, hfs, or g values.[432]

One should add at this point another category of Doppler-free FBL experiments, where the laser is used as an E1 "rf" source for double-resonance experiments on high-Z hydrogenic ions[423] for the determination of the Lamb shift[259-261] as outlined in the article by Beyer in this book. Also M1 double-resonance experiments are being considered in nonhydrogenic highly stripped ions.[433]

5.3. High-Resolution Fluorescence Spectroscopy

The linewidth of the Doppler-tuned, partially resolved resonance curve in Figure 55 is to a large extent due to the bandwidth of the free-running argon–ion laser used. When switching to single-mode operation this contribution becomes negligible and the residual linewidth is then approximately given by the simple addition of the contributions from velocity straggling of the ion beam $\Delta v/v$, the ion-beam divergence $\Delta\varphi$, the laser divergence $\Delta\vartheta$, and time-frequency uncertainty $v \sin \vartheta/\pi \Delta x$, where Δx is the geometrical diameter of the laser beam, and $\beta = v/c$ [36]:

$$\Delta\nu = \nu\beta \left| \frac{\Delta v}{v}\cos\vartheta + (\Delta\varphi + \Delta\vartheta)\sin\vartheta \right| + \frac{v \sin\vartheta}{\pi \Delta x} \qquad (73)$$

At 335 keV, ^{137}Ba$^+$ energy a considerably improved linewidth of $\Delta\nu = 1.64$ GHz could be observed[36] in accord with Eq. (73) when $\Delta v/v \approx 5 \times 10^{-4}$, $\Delta\varphi \approx 10^{-3}$, $\Delta\vartheta \approx 5 \times 10^{-4}$ (no focusing lens!), and $\Delta x = 2$ mm are estimated. This corresponds to a resolution of $\nu/\Delta\nu \approx 4 \times 10^5$ and represents thus a great improvement over general fast-beam resolutions.

It is interesting to note that the last three terms of Eq. (73) can be eliminated for $\vartheta \to 0$, i.e., in end-on geometry or parallel superimposed beams (SB) geometry. Simultaneously the interaction time T of Eq. (70) becomes correspondingly long as the beams overlap, so that a large excitation probability at low laser power (a few mW) may be obtained. In such favorable conditions, the resolution becomes $\nu/\Delta\nu = c/\Delta v$ and simply depends on the velocity uncertainty of the ion beam, and the resonance may be simply Dopper tuned by energy variation of the ion beam.

In the first application of this SB geometry the uv 363.79-nm argon-ion laser line was used to excite high Rydberg levels ($n = 40$–55) in H I from the H I $2s$ level with energy tuning of the resonances.[434] Since this experiment was not particularly designed for high resolution, "only" $\nu/\Delta\nu \approx 2.5 \times 10^5$ was achieved at 9.2 keV beam energy with non-single-mode operation of the laser.

It is obvious, however, that with single-mode lasers the resolution depends mainly on the energy spread ΔU_0 of the ion source and the stability of the acceleration voltage U. Under the assumption of an ideally stable U it was shown that a velocity bunching effect occurs[435] as a result of the acceleration such that the relative velocity uncertainty reduces according to $\Delta v/v = U_0/2U$, where ΔU_0 is the initial energy uncertainty at the ion source. Hence, by acceleration to high energies the influence of the velocity spread of the source on the resolution can be strongly reduced so that high-resolution fluorescence spectroscopy down to the natural linewidths should become possible.

An attempt has been made to make use of this velocity bunching by exciting the ^{131}Xe II $[5p^4\,^3P]\,6p\,^4P^0_{5/2}$ level from the metastable $5d\,^4F_{7/2}$ level (prepopulated in the ion source as in former FBL experiments[410,414–417]) with a single-mode cw dye laser at 605 nm after acceleration of ^{131}Xe$^+$ to 20 keV and by monitoring the $6s\,^4P_{5/2}$–$6p\,^4P^0_{5/2}$ 529-nm transition.[436] With the experimental arrangement shown in

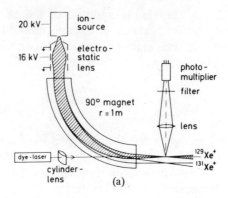

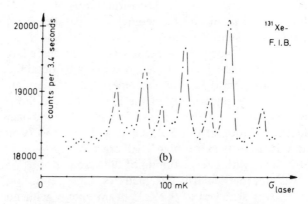

Figure 57. (a) Experimental arrangement for high-resolution fluorescence spectroscopy in SB geometry with a scanning dye laser. (b) Hyperfine structure of the ^{131}Xe II 605-nm as observed with the setup in (a). Further details see text.

Figure 57a the HFS of the 605 nm transition in ^{131}Xe II in Figure 57b could be resolved by tuning the laser through the resonances, since energy tuning was not possible with this setup. The high resolution of $\nu/\Delta\nu = 3.7 \times 10^6$ however, did not correspond at all to the expected higher resolution with velocity bunching. This was probably due to instabilities of U. This measurement thus simply represents the verification of Eq. (73) with $\vartheta = 0$ and $\Delta v/v$ determined by the energy stability of the whole system.

A similar experiment is described in Figure 24 of the article by Kluge in Chapter 17 of this work, where Na$^+$ was accelerated to 5 keV, then neutralized and excited by a tuned cw single-mode laser to yield 30 MHz linewidth or a resolution of $\nu/\Delta\nu = 2.5 \times 10^7$ (!) after subtraction of the

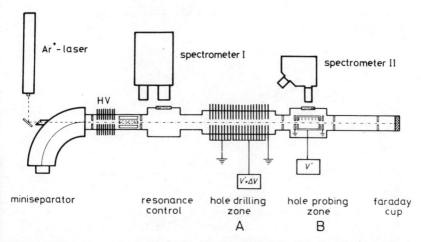

ministeparator resonance control — hole drilling zone A — hole probing zone B — faraday cup

Figure 58. Experimental arrangement for Doppler-tuned "in-flight Lamb-dip spectroscopy." See text for explanation.

natural width. This corresponds approximately to the expected resolution with the velocity bunching effect.

Essentially the basic principle of the latter two experiments is very similar to the Doppler-free saturation spectroscopy techniques,[437] where in general atoms within a narrow velocity interval Δv are prepared out of the thermal velocity distribution by a first laser beam and are then probed by a second laser beam. Here the narrow velocity interval Δv is prepared by electromagnetic ion optical means, which is then probed by a laser beam. A very elegant combination of both techniques was first introduced

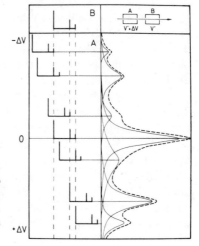

Figure 59. Reconstruction of the Lamb-dip structure of the ^{137}Ba II $6s^2 S_{1/2}(F = 2)$–$6p$ $^2P_{3/2}(F' = 1, 2, 3)$ transitions by shifting the three holes burned in the region A through the three equivalent probing transitions in region B by variation of ΔV.

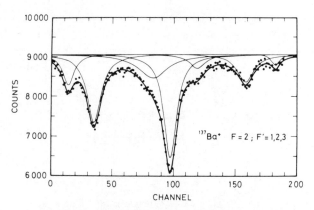

Figure 60. Experimental result of a Doppler-tuned "in-flight Lamb-dip" measurement of the ^{137}Ba II $6p\,^2P_{3/2}$ hfs with the setup of Figure 58. The result can be fully explained with the reconstruction in Figure 59. The left–right asymmetry of the data is explained in Ref. 440.

by Dufay *et al.*[438] An improved version[439] of their experiment for the measurement of the ^{137}Ba II $6p\,^2P_{3/2}$ hfs is shown in Figure 58. A single-mode argon-ion laser beam at 454.5 nm was superimposed on a high-energy Ba^1 beam. With the acceleration voltage HV the beam energy was Doppler-tuned through the resonance at ~251.5 keV (with a width of ~1.2 keV for ^{138}Ba$^+$ owing to instabilities of HV) and was then kept fixed above the resonance at about 254 keV as controlled by spectrometer I, which accepts the resonance fluorescence at 455.4 nm. In zone A the beam was decelerated (Doppler switched) into resonance by the potential V' such that according to the level scheme of ^{138}Ba$^+$ in Figure 51a population was pumped out of the ground state through the $6p\,^2P_{3/2}$ level into the $5d\,^2D_{3/2,5/2}$ levels for a very narrow velocity class of ions in resonance with the laser, i.e., a very narrow hole was burned into the broad velocity distribution of the ground-state ions. After reacceleration out of zone A the beam was again decelerated by a potential V' into zone B in order to probe this hole by observing a minimum (Lamb dip) in the resonance fluorescence detected through spectrometer II when ΔV in zone A was zero. By variation of ΔV the hole could thus be scanned with a dispersion of 2.6 MHz/V to yield for ^{138}Ba$^+$ a power-broadened (~3 mW) dip-width of ~110 MHz as compared to the natural width of 50 MHz.[440]

When ^{137}Ba$^+$ was used simultaneously, three holes, corresponding to the $F' = 1, 2, 3$ *or* 0, 1, 2, states of the $6p\,^2P_{3/2}$ level, respectively, could be burned into the $F = 2$ or $F = 1$ ground-state velocity distributions separated by 3.08 kV (see Figure 55). By choosing $F = 2$ the corresponding three holes were probed in zone B with the three equivalent transitions to yield the central fluorescence dip in Figure 59 for $\Delta V = 0$. When ΔV was swept the three holes shifted across the three probing transitions as indicated in Figure 59, which explains the whole dip structure observed in Figure 60.[440] From such results the authors[439] could extract the hfs splittings of the ^{135,137}Ba II $6p\ ^2P_{3/2}$ level in excellent agreement with the FBL QB results[430] and with about half the accuracy of the QB experiment.

The great advantage of this "in-flight Lamb-dip spectroscopy" technique is not only the ease with which the dip structure can be scanned by variation of a rather small voltage at a few MHz/V but also its important independence of laser and HV instabilities.

Its application will be, however, more limited than the above-mentioned velocity-bunching experiment since it relies on the three-level situation and can in principle only approach half the resolution. The combination of its beautiful Doppler-switching technique with velocity-bunched ion beams may therefore be an attractive tool for the future.

For completeness one should add that high-resolution Lamb-dip spectroscopy on fast beams had initially been suggested for other angles of incidence than $\vartheta = 0°$.[441] A geometry with $\vartheta = 90°$ could indeed be successfully used for a new measurement of the relativistic Doppler shift on 5–50 keV neutralized Ne beams.[442] In a recent refinement of this experiment advantage could be taken of the high time resolution of the FBL technique for the observation of optical "Ramsey" fringes in saturated absorption with separated laser fields, i.e., with transient successive, phase-correlated laser pulses[443] in the rest frame of the atoms. Since this technique suffers less from power broadening than normal Lamb-dip spectroscopy, improvements in ultimate spectral resolution can be expected from it.

Acknowledgments

I am grateful to E. Kupfer and H. Schröder for helpful discussions and suggestions and to all members of the Berlin FBS group for the stimulating collaboration during the past years. I wish to thank those authors and journals who gave permission to reproduce or adapt figures, and Mrs. Sedatis for drawing the remaining figures. The photographic work by Mr. Lemma, the typing of the manuscript by Mrs. Kline and Mrs. Lange, and the reading of part of the manuscript by W. Brewer is also greatly appreciated. This work has been supported by the "Sonderforschungsbereich 161 der Deutschen Forschungsgemeinschaft."

References and Notes

1. E. Goldstein, *Sitzungsber. Berlin Akad.*, 25 July (1868); *Wied. Ann. Phys. Chem.* **38** (1898).
2. J. Stark, *Ann. Phys. (Leipzig)* **49**, 731 (1916).
3. W. Wien, *Ann. Phys. (Leipzig)* **60**, 597 (1919).
4. W. Wien, in *Handbuch der Experimental-Physik*, Eds. W. Wien and F. Harms, p. 431, Akademische Verlagsgesellschaft, Leipzig (1927).
5. J. Stark and H. Kirschbaum, *Ann. Phys. (Leipzig)* **43**, 991 (1914).
6. J. Stark and H. Luneland, *Ann. Phys. (Leipzig)* **46**, 68 (1915).
7. H. Rausch von Traubenberg, *Naturwissenschaft* **10**, 791 (1922).
8. H. Rausch von Traubenberg and S. Levy, *Z. Phys.* **44**, 549 (1927).
9. K. L. Hertel, *Phys. Rev.* **29**, 848 (1927).
10. E. Rupp, *Ann. Phys. (Leipzig)* **85**, 515 (1928).
11. I. Walerstein, *Phys. Rev.* **33**, 800 (1929).
12. H. A. Kramers, *Z. Phys.* **3**, 199 (1920).
13. L. Kay, *Phys. Lett.* **5**, 36 (1963).
14. S. Bashkin, *Nucl. Instr. Meth.* **28**, 88 (1964).
15. H. J. Andrä, *Phys. Rev. Lett.* **25**, 325 (1970).
16. *Beam-Foil Spectroscopy*, Vols. I and II, Proceedings of the First International Conference in Tucson, Arizona, USA, Ed. S. Bashkin, Gordon and Breach, New York (1968).
17. "Beam-Foil Spectroscopy," Proceedings of the Second International Conference in Lysekil, Sweden, Eds. I. Martinson, J. Bromander, and H. G. Berry, *Nucl. Instr. Meth.* **90** (1970).
18. Proceedings of the Second European Conference on "Beam-Foil Spectroscopy and Connected Topics," held at L'Institut de Physique Nucléaire de l'Université de Lyon, France, 20–21 July (1971).
19. "Beam-Foil Spectroscopy," Proceedings of the Third International Conference in Tucson, Arizona, USA, Ed. S. Bashkin, *Nucl. Instr. Meth.* **110** (1973).
20. *Beam-Foil Spectroscopy*, Proceedings of the Fourth International Conference in Gatlinburg, USA, Vol. 1, 2, Eds. I. A. Sellin and D. J. Pegg, Plenum Press, New York (1976).
21. S. Bashkin, *Science* **148**, 1047 (1965).
22. W. S. Bickel, *Appl. Opt.* **6**, 1309 (1967).
23. S. Bashkin, *Appl. Opt.* **7**, 2341 (1968).
24. W. S. Bickel, *Appl. Opt.* **7**, 2367 (1968).
25. S. Bashkin, in *Atomic Physics 2*, Ed. P. G. Sandars, p. 43, Plenum Press, London (1971).
26. W. S. Bickel, *Surf. Sci.* **37**, 971 (1973).
27. R. Marrus, in *Atomic Physics 3*, Eds. S. J. Smith and G. K. Walters, p. 291, Plenum Press, New York (1973).
28. D. J. Pegg, P. M. Griffin, I. A. Sellin, W. W. Smith, and B. Donally, in *Atomic Physics 3*, Eds. S. J. Smith and G. K. Walters, p. 327, Plenum Press, New York (1973).
29. D. A. Church and C. H. Liu, *Physica* **67**, 90 (1973).
30. U. Fano and J. Macek, *Rev. Mod. Phys.* **45**, 553 (1973).
31. S. Bashkin, in *Progress in Optics*, Vol. XII, Ed. E. Wolf, p. 289, North-Holland, Amsterdam (1974).
32. H. J. Andrä, *Phys. Scr.* **9**, 257 (1974).
33. I. Martinson, *Phys. Scr.* **9**, 281 (1974).
34. I. Martinson and A. Gaupp, *Phys. Rep.* **15C**, 113 (1974).

35. I. Martinson, in *Nuclear Spectroscopy and Reactions*, Part C, VIII, p. 467, Academic Press, New York (1974).
36. H. J. Andrä, in *Atomic Physics 4*, Eds. G. zu Putlitz, E. W. Weber, and A. Winnacker, p. 635, Plenum Press, New York (1975).
37. P. Richard, in *Atomic Inner-Shell Processes*, Ed. B. Crasemann, p. 135, Academic Press, New York (1975).
38. H. G. Berry, *Phys. Scr.* **12**, (1975).
39. F. M. Pipkin, in *Atomic Physics 4*, Eds. G. zu Putlitz, E. W. Weber, and A. Winnacker, p. 119, Plenum Press, New York (1975).
40. *Beam-Foil Spectroscopy*, Ed. S. Bashkin, Topics in Current Physics, Vol. 1, Springer-Verlag, Heidelberg (1976).
41. W. S. Bickel, *Nucl. Instr. Meth.* **110**, 327 (1973).
42. M. Dufay, *Nucl. Instr. Meth.* **90**, 15 (1970).
43. G. W. Carriveau and S. Bashkin, *Nucl. Instr. Meth.* **90**, 203 (1970).
44. M. Dufay, A. Denis, and J. Desesquelles, *Nucl. Instr. Meth.* **90**, 85 (1970).
45. W. N. Lennard and C. L. Cocke, *Nucl. Instr. Meth.* **110**, 137 (1973).
46. F. Hopkins, J. Sokolov, and P. von Brentano, in *Beam-Foil Spectroscopy*, Vol. 2, Eds. I. A. Sellin and D. J. Pegg, p. 553, Plenum Press, New York (1976).
47. *Atomic Collisions in Solids*, Vols. 1 and 2, Eds. S. Datz, B. R. Appleton, and C. D. Moak, Plenum Press, New York (1975).
48. *Atomic Collisions in Solids*, Eds. F. W. Saris and W. F. van der Weg, North-Holland, Amsterdam (1976).
49. S. Datz, *Nucl. Instr. Meth.* **132**, 7 (1976).
50. N. Bohr, *Phys. Rev.* **59**, 270 (1941).
51. N. Andersen, W. S. Bickel, R. Boleu, K. Jensen, and E. Veje, *Phys. Scr.* **3**, 255 (1971).
52. J. D. Garcia, *Nucl. Instr. Meth.* **110**, 245 (1973).
53. W. Brandt, in *Atomic Collisions in Solids*, Vol. 1, Eds. S. Datz, B. R. Appleton, and C. D. Moak, p. 261, Plenum Press, New York (1975).
54. I. S. Dmitriev, V. S. Nikolaev, and Ya. A. Teplova, *Phys. Lett.* **27A**, 122 (1968).
55. H. D. Betz, *Rev. Mod. Phys.* **44**, 465 (1972).
56. H. G. Berry, in *Beam-Foil Spectroscopy*, Vol. 2, Eds. I. A. Sellin and D. J. Pegg, p. 755, Plenum Press, New York (1976).
57. J. Davidson, *Phys. Rev. A* **12**, 1350 (1975).
58. H. H. Bukow, H. von Buttlar, G. Heine, and M. Reinke, in *Beam-Foil Spectroscopy*, Vol. 1, Eds. I. A. Sellin and D. J. Pegg, p. 263, Plenum Press, New York (1976).
59. D. R. Bates and A. Dalgarno, *Proc. Phys. Soc. London A* **66**, 972 (1953).
60. J. R. Hiskes, *Phys. Rev.* **180**, 146 (1969).
61. R. J. Fortner and J. D. Garcia, *Phys. Rev. A* **12**, 856 (1975).
62. U. Fano and W. Lichten, *Phys. Rev. Lett.* **14**, 627 (1965).
63. M. Barat and W. Lichten, *Phys. Rev. A* **6**, 211 (1972).
64. W. Lichten, in *Atomic Physics 4*, Eds. G. zu Putlitz, E. W. Weber, and A. Winnacker, p. 249, Plenum Press, New York (1975).
65. E. Veje, *Phys. Rev. A* **14**, 2077 (1976).
66. J. Macek and D. Burns, in *Beam-Foil Spectroscopy*, Vol. 1, Ed. S. Bashkin, p. 235, Springer-Verlag, Heidelberg (1976).
67. J. Lindhard and M. Scharff, *Phys. Rev.* **124**, 128 (1961).
68. J. Lindhard, M. Scharff, and H. E. Schiøtt, *Mat. Fys. Medd. Dan. Vid. Selsk.* **33**, 2 (1963).
69. W. D. Wilson, L. G. Haggmark, and J. P. Biersack, *Phys. Rev. B* **15**, 2458 (1977), and references quoted therein.
70. W. Pietsch, U. Hauser, and W. Neuwirth, *Nucl. Instr. Meth.* **132**, 79 (1976), and references quoted therein.

71. D. J. Land and J. G. Brennan, *Nucl. Instr. Meth.* **132**, 89 (1976).
72. L. C. Northcliffe and R. F. Schilling, *Nucl. Data Tables A* **7**, 233 (1970).
73. L. Meyer, *Phys. Stat. Sol.* **44**, 253 (1971).
74. P. Sigmund and K. B. Winterbon, *Nucl. Instr. Meth.* **119**, 541 (1974).
75. G. Högberg, H. Nordén, and H. G. Berry, *Nucl. Instr. Meth.* **90**, 283 (1970).
76. P. Dobberstein and L. Henke, *Nucl. Instr. Meth.* **119**, 611 (1974).
77. P. D. Dumont, A. E. Livingston, Y. Baudinet-Robinet, G. Weber, and L. Quaglia, *Phys. Scr.* **13**, 122 (1976).
78. J. L. Yntema, *Nucl. Instr. Meth.* **113**, 605 (1973).
79. W. S. Bickel and R. Buchta, *Phys. Scr.* **9**, 148 (1974).
80. P. Sigmund, *Phys. Rev.* **184**, 383 (1969).
81. H. J. Andrä, in *Beam-Foil Spectroscopy*, Vol. 2, Eds. I. A. Sellin and D. J. Pegg, p. 835, Plenum Press, New York (1976).
82. H. Harde, in *Beam-Foil Spectroscopy*, Vol. 2, Eds. I. A. Sellin and D. J. Pegg, p. 859, Plenum Press, New York (1976).
83. H. J. Andrä, R. Fröhling, H. J. Plöhn, and J. D. Silver, *Phys. Rev. Lett.* **37**, 1212 (1976).
84. J. Bromander, *Nucl. Instr. Meth.* **110**, 11 (1973).
85. A. B. Wittkower and H. D. Betz, *At. Data* **5**, 113 (1973).
86. J. B. Marion and F. C. Young, *Nuclear Reaction Analysis*, North-Holland, Amsterdam (1968).
87. H. D. Betz, *Nucl. Instr. Meth.* **132**, 19 (1976).
88. R. Marrus, in *Beam-Foil Spectroscopy*, Ed. S. Bashkin, p. 209, Springer-Verlag, Heidelberg (1976).
89. M.-W. Chang and W. S. Bickel, *J. Opt. Soc. Am.* **65**, 1376 (1975).
90. G. W. Carriveau, M. H. Doobov, H. J. Hay, and C. J. Sofield, *Nucl. Instr. Meth.* **99**, 439 (1972).
91. L. Kay and B. Lightfoot, *Nucl. Instr. Meth.* **90**, 289 (1970).
92. J. O. Stoner, Jr., *Appl. Opt.* **9**, 53 (1970).
93. G. S. Bakken and J. A. Jordan, Jr., *Nucl. Instr. Meth.* **90**, 181 (1970).
94. J. O. Stoner, Jr., and J. A. Leavitt, *Appl. Phys. Lett.* **18**, 368 (1971).
95. J. O. Stoner, Jr., and J. A. Leavitt, *Appl. Phys. Lett.* **18**, 477 (1971).
96. K.-E. Bergkvist, in *Beam-Foil Spectroscopy*, Vol. 2, Eds. I. A. Sellin and D. J. Pegg, p. 719, Plenum Press, New York (1976).
97. J. A. Leavitt, J. W. Robson, and J. O. Stoner, Jr., *Nucl. Instr. Meth.* **110**, 423 (1973).
98. J. O. Stoner, Jr., and L. J. Radziemski, Jr., *Nucl. Instr. Meth.* **90**, 275 (1970).
99. J. O. Stoner, Jr., and J. A. Leavitt, *Opt. Acta* **20**, 435 (1973).
100. J. O. Stoner, Jr., and L. J. Radziemski, Jr., *Appl. Phys. Lett.* **21**, 105 (1972).
101. G. S. Bakken, A. C. Conrad, and J. A. Jordan, Jr., *J. Phys. B* **2**, 1378 (1969).
102. See, e.g., D. J. Pegg, P. M. Griffin, H. H. Haselton, R. Laubert, J. R. Mowat, R. S. Thoe R. S. Peterson, and I. A. Sellin, *Phys. Rev. A* **10**, 745 (1974).
103. N. J. Peacock, in *Beam-Foil Spectroscopy*, Vol. 2, Eds. I. A. Sellin and D. J. Pegg, p 925, Plenum Press, New York (1976).
104. H. R. Griem, *Plasma Spectroscopy*, McGraw-Hill, New York (1964).
105. J. S. Risley, *Rev. Sci. Instr.* **43**, 95 (1972), and references quoted therein.
106. R. Bruch, G. Paul, and J. Andrä, *J. Phys. B* **8**, L253 (1975).
107. J. P. Forrester, R. S. Haselton, P. M. Griffin, D. J. Pegg, H. H. Haselton, K. H. Liao, I A. Sellin, J. R. Mowat, and R. S. Thoe, in *Beam-Foil Spectroscopy*, Vol. 1, Eds. I. A Sellin and D. J. Pegg, p. 451, Plenum Press, New York (1976).
108. D. J. Pegg, in *Beam-Foil Spectroscopy*, Vol. 1, Eds. I. A. Sellin and D. J. Pegg, p. 419 Plenum Press, New York (1976).
109. P. Ziem, R. Bruch, and N. Stolterfoht, *J. Phys. B* **8**, L480 (1975).
110. D. Schneider, K. Roberts, B. M. Johnson, J. Whitenton, and C. F. Moore, in *Beam-Fo*

Spectroscopy, Vol. 2, Eds. I. A. Sellin and D. J. Pegg, p. 615, Plenum Press, New York (1976).
111. L. Kay, *Proc. Phys. Soc. London* **85**, 163 (1965).
112. M. C. Poulizac, M. Druetta, and P. Ceyzeriat, *J. Quant. Spectr. Rad. Trans.* **11**, 1087 (1971).
113. R. Hallin, J. Lindkog, A. Marelius, J. Phil, and R. Sjödin, *Phys. Scr.* **8**, 209 (1973).
114. B. Dynefors, I. Martinson, and E. Veje, *Phys. Scr.* **12**, 58 (1975); **13**, 308 (1976).
115. G. W. Carriveau and S. Bashkin, *Nucl. Instr. Meth.* **90**, 203 (1970).
116. W. N. Lennard, R. M. Mills, and W. Whaling, *Phys. Rev. A* **6**, 884 (1972).
117. S. Bashkin, in *Beam-Foil Spectroscopy*, Ed. S. Bashkin, p. 5, Springer-Verlag, Heidelberg (1976).
118. W. Wittmann, Dissertation, Free University of Berlin.
119. P. L. Smith, W. Whaling, and D. L. Mickey, *Nucl. Instr. Meth.* **90**, 47 (1970).
120. J. H. Brand, C. L. Cocke, B. Curnutte, and C. Swenson, *Nucl. Instr. Meth.* **90**, 63 (1970).
121. D. J. Pegg, E. L. Chupp, and L. W. Dotchin, *Nucl. Instr. Meth.* **90**, 71 (1970).
122. E. H. Pinnington, *Nucl. Instr. Meth.* **90**, 93 (1970).
123. B. L. Cardon, J. A. Leavitt, M. W. Chang, and S. Bashkin, in *Beam-Foil Spectroscopy*, Vol. 1, Eds. I. A. Sellin and D. J. Pegg, p. 251, Plenum Press, New York (1976).
124. S. K. Allison, S. P. Frankel, T. A. Hall, J. H. Montague, A. H. Morrish, and S. D. Warshaw, *Rev. Sci. Instr.* **20**, 735 (1949), and references quoted therein.
125. W. S. Bickel and A. S. Goodman, *Phys. Rev.* **148**, 1 (1966).
126. L. J. Curtis, H. G. Berry, and J. Bromander, *Phys. Scr.* **2**, 216 (1970).
127. L. Kay, *Phys. Scr.* **5**, 139 (1972).
128. L. J. Curtis, in *Beam-Foil Spectroscopy*, Ed. S. Bashkin, p. 163, Springer-Verlag, Heidelberg (1976).
129. I. Martinson, in *Beam-Foil Spectroscopy*, Ed. S. Bashkin, p. 33, Springer-Verlag, Heidelberg (1976).
130. I. A. Sellin, in *Beam-Foil Spectroscopy*, Ed. S. Bashkin, p. 265, Springer-Verlag, Heidelberg (1976).
131. W. Wiese, in *Beam-Foil Spectroscopy*, Ed. S. Bashkin, p. 147, Springer-Verlag, Heidelberg (1976).
132. B. Edlén, in *Handbuch der Physik*, Vol. 27, Ed. S. Flügge, p. 80, Springer, Berlin (1964).
133. K. Bockasten, *Phys. Rev. A* **9**, 1087 (1974).
134. P. Vogel, *Nucl. Instr. Meth.* **110**, 241 (1973).
135. J. W. Swenson and B. Edlén, *Phys. Scr.* **2**, 235 (1974).
136. A. Dalgarno, *Adv. Phys.* **11**, 281 (1962).
137. S. Bashkin, J. Bromander, J. A. Leavitt, and I. Martinson, *Phys. Scr.* **8**, 285 (1973).
138. S. N. Bhardwaj, H. G. Berry, and T. Mossberg, *Phys. Scr.* **9**, 331 (1974).
139. L. C. McIntyre, J. D. Silver, and N. A. Jelley, in *Beam-Foil Spectroscopy*, Vol. 1, Eds. I. A. Sellin and D. J. Pegg, p. 331, Plenum Press, New York (1976).
140. J. P. Buchet, A. Denis, J. Desesquelles, M. Druetta, and J. L. Subtil, in *Beam-Foil Spectroscopy*, Vol. 1, Eds. I. A. Sellin and D. J. Pegg, p. 355, Plenum Press, New York (1976).
141. H. G. Berry and C. H. Batson, in *Beam-Foil Spectroscopy*, Vol. 1, Eds. I. A. Sellin and D. J. Pegg, p. 367, Plenum Press, New York (1976).
142. W. Gordon, *Ann. Phys. (Leipzig)* **2**, 1031 (1929).
143. Formula (3) in Ref. 137 derived by J. D. Garcia.
144. L. C. Green, R. P. Rush, and C. D. Chandler, *Astrophys. J., Suppl. Ser.* **3**, 37 (1957).
145. S. Bashkin, in *Beam-Foil Spectroscopy*, Vol. 1, Eds. I. A. Sellin and D. J. Pegg, p. 129, Plenum Press, New York (1976).
146. I. Martinson, *Nucl. Instr. Meth.* **90**, 81 (1970).

147. G. Herzberg and H. R. Moore, *Can. J. Phys.* **37**, 1293 (1959).
148. J. D. Garcia and J. E. Mack, *Phys. Rev.* **138**, A987 (1965).
149. E. Holøien and S. Geltman, *Phys. Rev.* **153**, 81 (1967).
150. H. G. Kuhn, *Atomic Spectra*, Longmans, London (1968).
151. I. A. Sellin, *Nucl. Instr. Meth.* **110**, 477 (1973).
152. P. Feldman and R. Novick, *Phys. Rev.* **160**, 143 (1967).
153. W. S. Bickel, I. Bergström, R. Buchta, L. Lundin, and I. Martinson, *Phys. Rev.* **178**, 118 (1969).
154. J. P. Buchet, A. Denis, J. Desesquelles, and M. Dufay, *Phys. Lett.* **A28**, 529 (1969).
155. J. P. Buchet, M. C. Buchet-Poulizac, H. G. Berry, and G. W. F. Drake, *Phys. Rev. A* **7**, 922 (1973).
156. H. G. Berry, J. Bromander, I. Martinson, and R. Buchta, *Phys. Scr.* **3**, 63 (1971).
157. H. G. Berry, E. H. Pinnington, and J. C. Subtil, *J. Opt. Soc. Am.* **62**, 767 (1972).
158. F. Gaillard, M. L. Gaillard, J. Desesquelles, and M. Dufay, *Compt. Rend.* **269**, 420 (1969).
159. A. W. Weiss, unpublished, but results appear in most of the references 153–158.
160. T. Andersen, K. A. Jessen, and G. Sørensen, *Phys. Rev.* **188**, 76 (1969).
161. S. Hontzeas, I. Martinson, P. Erman, and R. Buchta, *Phys. Scr.* **6**, 55 (1972); *Nucl. Instr. Meth.* **110**, 51 (1973).
162. J. Bromander, O. Poulsen, and J. L. Subtil, *Phys. Scr.* **7**, 283 (1973).
163. I. Martinson, W. S. Bickel, and A. Ölme, *J. Opt. Soc. Am.* **60**, 1213 (1970).
164. H. G. Berry and J. L. Subtil, *Phys. Scr.* **9**, 217 (1973).
165. K. X. To, E. J. Knystautas, R. Drouin, and H. G. Berry, in *Beam-Foil Spectroscopy*, Vol. 1, Eds. I. A. Sellin and D. J. Pegg, p. 385, Plenum Press, New York (1976).
166. H. G. Berry, M. C. Buchet-Poulizac, and J. P. Buchet, *J. Opt. Soc. Am.* **63**, 240 (1973).
167. J. P. Buchet and M. C. Buchet-Poulizac, *J. Opt. Soc. Am.* **64**, 1011 (1974).
168. E. J. Knystautas and R. Drouin, in *Beam-Foil Spectroscopy*, Vol. 1, Eds. I. A. Sellin and D. J. Pegg, p. 393, Plenum Press, New York (1976).
169. J. P. Buchet, A. Denis, J. Desesquelles, M. Druetta, and J. L. Subtil, in *Beam-Foil Spectroscopy*, Vol. 1, Eds. I. A. Sellin and Dr. J. Pegg, p. 355, Plenum Press, New York (1976).
170. D. J. Pegg, P. M. Griffin, H. H. Haselton, R. Laubert, J. R. Mowat, R. S. Thoe, R. S. Peterson, and I. A. Sellin, *Phys. Rev. A* **10**, 745 (1974).
171. D. J. Irwin and R. Drouin, in *Beam-Foil Spectroscopy*, Vol. 1, Eds. I. A. Sellin and D. J. Pegg, p. 347, Plenum Press, New York (1976).
172. K. T. Cheng, C. P. Lin, and W. R. Johnson, *Phys. Lett.* **48A**, 437 (1974).
173. J. R. Mowat, K. W. Jones, and B. M. Johnson, *Phys. Rev. A* **14**, 1109 (1976).
174. C. L. Cocke, B. Curnutte, and R. Randall, *Phys. Rev. A* **9**, 1823 (1974).
175. E. J. Knystautas and R. Drouin, in *Beam-Foil Spectroscopy*, Vol. 1, Eds. I. A. Sellin and D. J. Pegg, p. 377, Plenum Press, New York (1976).
176. H. G. Berry, R. Hallin, R. Sjödin, and M. Gaillard, *Phys. Lett.* **50A**, 191 (1974).
177. L. Lundin, B. Engman, J. Hilke, and I. Martinson, *Phys. Scr.* **8**, 274 (1973).
178. B. Emmoth, M. Braun, J. Bromander, and I. Martinson, *Phys. Scr.* **12**, 75 (1975).
179. G. J. Schulz, *Rev. Mod. Phys.* **45**, 378 (1973).
180. R. P. Madden and K. Codling, *Astrophys. J.* **141**, 364 (1965).
181. H. G. Berry, I. Martinson, L. J. Curtis, and L. Lundin, *Phys. Rev. A* **3**, 1934 (1971).
182. E. J. Knystautas and R. Drouin, *Nucl. Instr. Meth.* **110**, 95 (1973).
183. H. G. Berry, J. Desesquelles, and M. Dufay, *Phys. Rev. A* **6**, 600 (1972).
184. H. G. Berry, J. Desesquelles, and M. Dufay, *Nucl. Instr. Meth.* **110**, 43 (1973).
185. D. L. Matthews, W. J. Braithwaite, H. W. Wolter, and C. F. Moore, *Phys. Rev. A* **8**, 1397 (1973).

186. R. Bruch, G. Paul, H. J. Andrä, and L. Lipsky, *Phys. Rev. A* **12**, 1808 (1975).
187. R. Bruch, Dissertation, Free University of Berlin.
188. D. J. Pegg, I. A. Sellin, R. Peterson, J. R. Mowat, W. W. Smith, M. D. Brown, and J. R. MacDonald, *Phys. Rev. A* **8**, 1350 (1973).
189. I. A. Sellin, *Nucl. Instr. Meth.* **110**, 477 (1973).
190. M. Levitt, R. Novick, and P. D. Feldman, *Phys. Rev. A* **3**, 130 (1971).
191. D. L. Ederer, L. Lucatorto, and R. P. Madden, *Phys. Rev. Lett.* **25**, 1537 (1970).
192. V. V. Balashov, S. J. Grishanova, H. M. Kruglowa, and V. Senashenko, *Opt. Spectrosc.* **28**, 466 (1970).
193. R. H. Perrott and A. L. Stewart, *J. Phys. B* **1**, 381, 1226 (1968).
194. D. J. Pegg, H. H. Haselton, R. S. Thoe, P. M. Griffin, M. D. Brown, and I. A. Sellin, *Phys. Rev. A* **12**, 1330 (1975).
195. B. Donnally, W. W. Smith, D. J. Pegg, M. Brown, and I. A. Sellin, *Phys. Rev. A* **4**, 122 (1971).
196. R. Bruch. J. Andrä, and G. Paul, in *Beam-Foil Spectroscopy*, Vol. 1, Eds. I. A. Sellin and D. J. Pegg, p. 437, Plenum Press, New York (1976).
197. B. R. Junker and J. N. Bardsley, *Phys. Rev. A* **8**, 1345 (1973).
198. H. H. Haselton, R. S. Thoe, J. R. Mowat, P. M. Griffin, D. J. Pegg, and I. A. Sellin, *Phys. Rev. A* **11**, 468 (1975).
199. C. L. Cocke, B. Curnutte, and R. Randall, *Phys. Rev. A* **9**, 1823 (1974).
200. M. Lipeles, R. Novick, and N. Tolk, *Phys. Rev. Lett.* **15**, 690 (1965).
201. C. J. Artura, N. Tolk, and R. Novick, *Astrophys. J.* **157**, L181 (1969).
202. R. Novick, in *Physics of the One- and Two-Electron Atoms*, Eds. F. Bopp and H. Kleinpoppen, p. 269, North-Holland, Amsterdam (1969).
203. M. Geoppert-Mayer, *Ann. Phys.* (*Leipzig*) **9**, 273 (1931).
204. G. Breit and E. Teller, *Astrophys. J.* **91**, 215 (1940).
205. L. Spitzer and L. L. Greenstein, *Astrophys. J.* **114**, 407 (1951).
206. J. Shapiro and G. Breit, *Phys. Rev.* **113**, 179 (1959).
207. S. Klarsfeld, *Phys. Lett.* **30A**, 382 (1969).
208. M. Mizushima, *Phys. Rev.* **134**, A883 (1964).
209. R. H. Garstang, *Astrophys. J.* **148**, 579 (1967).
210. G. W. F. Drake, G. A. Victor, and A. Dalgarno, *Phys. Rev.* **180**, 25 (1969).
211. H. R. Griem, *Astrophys. J.* **156**, L103 (1969).
212. G. W. F. Drake and A. Dalgarno, *Astrophys. J.* **157**, 459 (1969).
213. G. W. F. Drake, *Astrophys. J.* **158**, 1199 (1969).
214. G. Feinberg and J. Sucher, *Phys. Rev. Lett.* **26**, 681 (1971).
215. G. W. F. Drake, *Phys. Rev. A* **3**, 908 (1971).
216. W. R. Johnson, *Phys. Rev. Lett.* **29**, 1123 (1972).
217. G. W. F. Drake, *Phys. Rev. A* **5**, 1979 (1972).
218. W. R. Johnson and C. Lin, *Phys. Rev. A* **9**, 1486 (1974).
219. C. D. Lin, W. R. Johnson, and A. Dalgarno, *Phys. Rev. A* **15**, 154 (1977).
220. R. C. Elton, L. J. Palumbo, and H. R. Griem, *Phys. Rev. Lett.* **20**, 783 (1968).
221. W. M. Neupert and M. Swartz, *Astrophys. J.* **160**, L189 (1970).
222. G. A. Doschek, J. F. Meekins, R. W. Kreplin, T. A. Chubb, and H. Freidman, *Astrophys. J.* **164**, 165 (1971).
223. A. B. C. Walker, Jr., and H. R. Rugge, *Astron. Astrophys.* **5**, 4 (1970), and references therein.
224. A. H. Gabriel and C. Jordan, *Nature* **221**, 947 (1969).
225. G. R. Blumenthal, G. W. F. Drake, and W. H. Tucker, *Astrophys. J.* **172**, 205 (1972).
226. I. A. Sellin, B. L. Donnally, and C. Y. Fan, *Phys. Rev. Lett.* **21**, 717 (1968).
227. R. Marrus and R. W. Schmieder, *Phys. Rev. A* **5**, 1160 (1972).

228. R. Marrus, *Nucl. Instr. Meth.* **110**, 333 (1973).
229. C. L. Cocke, in *Beam-Foil Spectroscopy*, Vol. 1, Eds. I. A. Sellin and D. J. Pegg, p. 283, Plenum Press, New York (1976).
230. R. W. Schmieder and R. Marrus, *Phys. Rev. Lett.* **25**, 1245 (1970).
231. C. L. Cocke, B. Curnutte, J. R. MacDonald, J. A. Bednar, and R. Marrus, *Phys. Rev. A* **9**, 2242 (1974).
232. M. H. Prior, *Phys. Rev. Lett.* **29**, 611 (1972).
233. R. S. Van Dyck, Jr., C. E. Johnson, and H. A. Shugart, *Phys. Rev. A* **4**, 1327 (1971).
234. M. H. Prior and H. A. Shugart, *Phys. Rev. Lett.* **27**, 902 (1971).
235. A. S. Pearl, *Phys. Rev. Lett.* **24**, 703 (1970).
236. R. Marrus and R. W. Schmieder, *Phys. Lett.* **32A**, 431 (1970).
237. R. W. Schmieder and R. Marrus, *Phys. Rev. Lett.* **25**, 1245 (1970).
238. J. A. Bednar, C. L. Cocke, B. Curnutte, and R. Randall, *Phys. Rev. A* **11**, 460 (1975).
239. H. Gould, R. Marrus, and P. Mohr, *Phys. Rev. Lett.* **33**, 676 (1974).
240. H. Gould, R. Marrus, and R. W. Schmieder, *Phys. Rev. Lett.* **31**, 504 (1973).
241. H. Gould and R. Marrus, in *Beam-Foil Spectroscopy*, Vol. 1, Eds. I. A. Sellin and D. J. Pegg, p. 305, Plenum Press, New York (1976).
242. H. W. Moos and J. R. Woodworth, *Phys. Rev. Lett.* **30**, 775 (1973).
243. C. L. Cocke, B. Curnutte, J. R. MacDonald, and R. Randall, *Phys. Rev. A* **9**, 57 (1974).
244. J. R. Mowat, P. M. Griffin, H. H. Haselton, R. Laubert, D. J. Pegg, R. S. Peterson, I. A. Sellin, and R. S. Thoe, *Phys. Rev. A* **11**, 2198 (1975).
245. W. A. Davis and R. Marrus, in *Beam-Foil Spectroscopy*, Vol. 1, Eds. I. A. Sellin and D. J. Pegg, p. 317, Plenum Press, New York (1976).
246. C. L. Cocke, B. Curnutte, and R. Randall, *Phys. Rev. A* **9**, 1823 (1974).
247. I. A. Sellin, M. Brown, W. W. Smith, and B. Donally, *Phys. Rev. A* **2**, 1189 (1970).
248. J. R. Mowat, I. A. Sellin, R. S. Peterson, D. J. Pegg, M. D. Brown, and J. R. MacDonald, *Phys. Rev. A* **8**, 145 (1973).
249. S. L. Varghese, C. L. Cocke, and B. Curnutte, *Phys. Rev. A* **14**, 1729 (1976).
250. R. W. Schmieder and R. Marrus, *Nucl. Instr. Meth.* **110**, 459 (1973).
251. R. W. Schmieder, *Rev. Sci. Instr.* **45**, 687 (1974).
252. S. L. Varghese, C. L. Cocke, B. Curnutte, and R. R. Randall, in *Beam-Foil Spectroscopy*, Vol. 1, Eds. I. A. Sellin and D. J. Pegg, p. 299, Plenum Press, New York (1976).
253. C. Y. Fan, M. Garcia-Munoz, and I. A. Sellin, *Phys. Rev.* **161**, 6 (1967).
254. D. E. Murnick, M. Leventhal, and H. W. Kugel, *Phys. Rev. Lett.* **27**, 1625 (1971).
255. M. Leventhal, D. E. Murnick, and H. W. Kugel, *Phys. Rev. Lett.* **28**, 1609 (1972).
256. G. P. Lawrence, C. Y. Fan, and S. Bashkin, *Phys. Rev. Lett.* **28**, 1613 (1972).
257. H. W. Kugel, M. Leventhal, and D. E. Murnick, *Phys. Rev. A* **6**, 1306 (1972).
258. M. Leventhal, *Nucl. Instr. Meth.* **110**, 343 (1973).
259. H. W. Kugel, M. Leventhal, D. E. Murnick, C. K. N. Patel, and O. R. Wood, II, *Phys. Rev. Lett.* **35**, 647 (1975).
260. D. E. Murnick, in *Beam-Foil Spectroscopy*, Vol. 2, Eds. I. A. Sellin and D. J. Pegg, p. 815, Plenum Press, New York (1976).
261. H. W. Kugel, M. Leventhal, D. E. Murnick, C. K. N. Patel, and O. R. Wood, II, in *Abstracts of the Fifth International Conference on Atomic Physics*, Berkeley 1976, p. 208.
262. H. Gould and R. Marrus, II, in *Abstracts of the Fifth International Conference on Atomic Physics*, Berkeley 1976, p. 207.
263. S. R. Lundeen and F. M. Pipkin, *Phys. Rev. Lett.* **34**, 1368 (1975).
264. F. M. Pipkin, in *Atomic Physics 4*, Eds. G. zu Putlitz, E. W. Weber, and A. Winnacker, p. 119, Plenum Press, New York (1975).
265. T. Applequist and S. J. Brodsky, *Phys. Rev. Lett.* **24**, 562 (1970).

266. S. J. Brodsky and S. D. Drell, *Ann. Rev. Nucl. Sci.* **20**, 147 (1070).
267. B. E. Lautrup, A. Peterman, and E. de Rafael, *Phys. Rep.* **3**, 193 (1972), and references quoted therein.
268. G. W. Erickson, *Phys. Rev. Lett.* **27**, 780 (1971).
269. P. J. Mohr, *Ann. Phys. (N.Y.)* **88**, 26, 52 (1974).
270. P. J. Mohr, *Phys. Rev. Lett.* **34**, 1050 (1975).
271. P. J. Mohr, in *Beam-Foil Spectroscopy*, Vol. 1, Eds. I. A. Sellin and D. J. Pegg, p. 89, Plenum Press, New York (1976).
272. R. Robiscoe and T. Shyn, *Phys. Rev. Lett.* **24**, 559 (1970).
273. S. Triebwasser, E. S. Dayhoff, and W. E. Lamb, Jr., *Phys. Rev.* **89**, 98 (1953).
274. M. A. Narasimham and R. L. Strombotne, *Phys. Rev. A* **4**, 14 (1971).
275. E. Lipworth and R. Novick, *Phys. Rev.* **108**, 1434 (1957).
276. M. Leventhal, *Phys. Rev. A* **11**, 427 (1975).
277. D. Layzer, *Ann. Phys. (N.Y.)* **8**, 271 (1959); *Int. J. Quantum Chem.* **1**, 45 (1967).
278. M. Cohen and A. Dalgarno, *Proc. R. Soc. London A* **280**, 258 (1964).
279. A. Weiss, *Astrophys. J.* **138**, 1262 (1963).
280. W. L. Wiese and A. W. Weiss, *Phys. Rev.* **175**, 50 (1968).
281. A. W. Weiss, *Nucl. Instr. Meth.* **90**, 121 (1970).
282. A. W. Smith and W. L. Wiese, *Astrophys. J. Suppl. Ser. No. 196* **23**, 103 (1971).
283. A. Dalgarno, *Nucl. Instr. Meth.* **110**, 183 (1973).
284. M. W. Smith, G. A. Martin, and W. L. Wiese, *Nucl. Instr. Meth.* **110**, 219 (1973).
285. A. W. Weiss, in *Beam-Foil Spectroscopy*, Vol. 1, Eds. I. A. Sellin and D. J. Pegg, p. 51, Plenum Press, New York (1976).
286. G. A. Martin and W. L. Wiese, *Phys. Rev. A* **13**, 699 (1976).
287. E. Hylleraas, *Z. Phys.* **65**, 209 (1930).
288. A. Weiss, in *Advances in Atomic and Molecular Physics*, Vol. 9, Eds. D. R. Bates and I. Estermann, p. 1, Academic Press, New York (1973).
289. T. Andersen, A. Kirkegård Nielsen, and G. Sørensen, *Nucl. Instr. Meth.* **110**, 143 (1973).
290. A. Fillipov and V. K. Prokof'ev, *Z. Phys.* **56**, 458 (1929).
291. A. N. Fillipov, *Z. Phys.* **69**, 526 (1931).
292. S. Haroche, in *High Resolution Laser Spectroscopy*, Ed. K. Shimoda, Topics in Applied Physics, Vol. 13, p. 253, Springer-Verlag, Berlin (1976).
293. W. W. Chow, M. O. Scully, and J. O. Stoner, Jr., *Phys. Rev. A* **11**, 1380 (1975).
294. M. O. Scully, in *Atomic Physics*, Eds. B. Bederson, V. W. Cohen, and F. M. J. Pichanik, p. 81, Plenum Press, New York (1969).
295. Th. Krist, P. Kuske, A. Gaupp, W. Wittmann, and H. J. Andrä, *Phys. Lett.* **61A**, 94 (1977).
296. A. Gaupp, Diploma Thesis, Freie Universität Berlin.
297. G. W. F. Drake and A. van Wijngaarden, in *Beam-Foil Spectroscopy*, Vol. 2, Eds. I. A. Sellin and D. J. Pegg, p. 749, Plenum Press, New York (1976).
298. A. van Wijngaarden, E. Goh, G. W. F. Drake, and P. S. Farago, *J. Phys. B* **9**, 2017 (1976).
299. I. G. Eck, *Phys. Rev. Lett.* **31**, 270 (1973).
300. I. A. Sellin, J. R. Mowat, R. S. Peterson, P. M. Griffin, R. Laubert, and H. H. Haselton, *Phys. Rev. Lett.* **31**, 1335 (1973).
301. A. Gaupp, H. J. Andrä, and J. Macek, *Phys. Rev. Lett.* **32**, 268 (1974).
302. H. J. Andrä, *Phys. Rev. A* **2**, 2200 (1970).
303. S. Bashkin, W. S. Bickel, D. Fink, and R. K. Wangsness, *Phys. Rev. Lett.* **15**, 284 (1965).
304. W. S. Bickel and S. Bashkin, *Phys. Rev.* **162**, 12 (1967).
305. S. Bashkin and G. Beauchemin, *Can. J. Phys.* **44**, 1603 (1966).
306. W. S. Bickel, *J. Opt. Soc. Am.* **58**, 713 (1968).

307. I. A. Sellin, C. D. Moack, P. M. Griffin, and J. A. Biggerstaff, *Phys. Rev.* **184**, 56 (1969).
308. I. A. Sellin, C. D. Moack, P. M. Griffin, and J. A. Biggerstaff, *Phys. Rev.* **188**, 217 (1969).
309. P. M. Griffin, J. A. Biggerstaff, I. A. Sellin, and C. D. Moack, in *Physics of One- and Two-Electron Atom*, Eds. F. Bopp and H. Kleinpoppen, p. 387, North-Holland, Amsterdam (1969).
310. O. A. Keller and R. T. Robiscoe, *Phys. Rev.* **188**, 82 (1969).
311. I. A. Sellin, P. M. Griffin, and J. A. Biggerstaff, *Phys. Rev. A* **1**, 1553 (1970).
312. Yu. P. Sokolov, *JETP Lett.* **11**, 359 (1970).
313. I. A. Sellin, *Nucl. Instr. Meth.* **90**, 329 (1970).
314. H. J. Andrä, *Nucl. Instr. Meth.* **90**, 301 (1970).
315. P. H. Heckmann, *Z. Phys.* **250**, 42 (1972).
316. H. J. Andrä, P. Dobberstein, A. Gaupp, and W. Wittmann, *Nucl. Instr. Meth.* **110**, 301 (1973).
317. M. J. Alguard and G. W. Drake, *Nucl. Instr. Meth.* **110**, 311 (1973).
318. E. H. Pinnington, H. G. Berry, J. Desesquelles, and J. L. Subtil, *Nucl. Instr. Meth.* **110**, 315 (1973).
319. M. J. Alguard and C. W. Drake, *Phys. Rev. A* **8**, 27 (1973).
320. H. Frauenfelder and R. M. Steffen, in *Alpha-, Beta- and Gamma-Ray Spectroscopy*, Vol. 2, Ed. K. Siegbahn, p. 997, North-Holland, Amsterdam (1968).
321. R. M. Steffen and H. Frauenfelder, in *Perturbed Angular Correlations*, Eds. E. Karlsson, E. Matthias, and K. Siegbahn, p. 1, North-Holland, Amsterdam (1964).
322. R. M. Steffen, in *Angular Correlation in Nuclear Disintegration*, Eds. H. van Kruyten and B. van Nooijen, p. 2, Rotterdam University Press (1971).
323. H. Gabriel and J. Bosse, in *Angular Correlations in Nuclear Disintegration*, Eds. H. van Kruyten and B. van Nooijen, p. 394, Rotterdam University Press (1971).
324. D. G. Ellis, *J. Opt. Soc. Am.* **63**, 1232 (1973).
325. I. Percival and M. Seaton, *Phil. Trans. R. Soc. London, Ser. A* **25**, 113 (1958); *Colloq. Int. CNRS* **162**, 21 (1967).
326. D. J. Burns and W. H. Hancock, *J. Opt. Soc. Am.* **63**, 241 (1973).
327. H. J. Andrä, *Phys. Lett.* **54A**, 315 (1975).
328. H. Kleinpoppen, in *Physics of One- and Two-Electron Atoms*, Eds. F. Bopp and H. Kleinpoppen, p. 612, North-Holland, Amsterdam (1969).
329. M. Born and E. Wolf, *Principles of Optics*, Pergamon Press, New York (1970).
330. H. G. Berry, L. J. Curtis, D. G. Ellis, and R. M. Schectman, *Phys. Rev. Lett.* **32**, 751 (1974).
331. C. H. Liu, S. Bashkin, W. S. Bickel, and T. Hadeishi, *Phys. Rev. Lett.* **26**, 222 (1971).
332. D. A. Church and C. H. Liu, *Phys. Rev. A* **5**, 1031 (1972).
333. T. Andersen, O. Poulsen, and G. Sørensen, in Proceedings of the Second European Conference on Beam-Foil Spectroscopy and Connected Topics held at L'Institut de Physique Nucléaire de l' Université de Lyon, France, 20 to 21 July (1971).
334. H. G. Berry and J. L. Subtil, in Proceedings of the Second European Conference on Beam-Foil Spectroscopy and Connected Topics, held at L'Institut de Physique Nucléaire de l'Université de Lyon, France, 20 to 21 July (1971).
335. M. Carré, J. Desesquelles, M. Dufay, and M. L. Gaillard, in Proceedings of the Second European Conference on Beam-Foil Spectroscopy and Connected Topics held at L'Institut de Physique Nucléaire de l'Université de Lyon, France, 20–21 July (1971).
336. D. J. Lynch, C. W. Drake, M. J. Alguard, and C. E. Fairchild, *Phys. Rev. Lett.* **26**, 1211 (1971).
337. P. Dobberstein, H. J. Andrä, W. Wittmann, and H. H. Bukow, *Z. Phys.* **257**, 272 (1972).
338. O. Poulsen and J. L. Subtil, *Phys. Rev. A* **8**, 1181 (1973).

339. J. Yellin, T. Hadeishi, and M. C. Michel, *Phys. Rev. Lett.* **30**, 417 (1973).
340. J. Yellin, T. Hadeishi, and M. C. Michel, *Phys. Rev. Lett.* **30**, 1286 (1973).
341. M. Druetta and A. Denis, *Nucl. Instr. Meth.* **110**, 291 (1973).
342. A. Denis, J. Desesquelles, M. Druetta, and M. Dufay, in *Beam-Foil Spectroscopy*, Vol. 2, Eds. I. A. Sellin and D. J. Pegg, p. 799, Plenum Press, New York (1976).
343. O. Poulsen, T. Andersen, and N. J. Skouboe, *J. Phys. B* **8**, 1393 (1975).
344. H. Winter and H. H. Buckow, *Z. Phys. A* **277**, 27 (1976).
345. L. J. Curtis, R. Hallin, J. Lindskog, J. Pihl, and H. G. Berry, *Phys. Lett.* **60A**, 297 (1977).
346. H. Winter, private communication.
347. T. Andersen, A. Kirkegård Nielsen, and K. J. Olsen, *Phys. Rev. Lett.* **31**, 739 (1973).
348. T. Andersen, A. Kirkegård Nielsen, and K. J. Olsen, *Phys. Rev. A* **10**, 2174 (1974).
349. N. Andersen, T. Andersen, and K. Jensen, *J. Phys. B* **9**, 1373 (1976).
350. N. H. Tolk, J. C. Tully, C. W. White, J. Kraus, A. A. Monge, D. L. Simms, M. F. Robbins, S. H. Neff, and W. Lichten, *Phys. Rev. A* **13**, 969 (1976).
351. D. A. Church, W. Kolbe, M. C. Michel, and T. Hadeishi, *Phys. Rev. Lett.* **33**, 565 (1974).
352. C. H. Liu, S. Bashkin, and D. A. Church, *Phys. Rev. Lett.* **33**, 993 (1974).
353. H. G. Berry, S. N. Bhardwaj, L. J. Curtis, and R. M. Schectman, *Phys. Lett.* **50A**, 59 (1974).
354. H. G. Berry, L. J. Curtis, and R. M. Schectman, *Phys. Rev. Lett.* **34**, 509 (1975).
355. J. D. Silver and L. C. McIntyre, Jr., in *Beam-Foil Spectroscopy*, Vol. 2, Eds. I. A. Sellin and D. J. Pegg, p. 773, Plenum Press, New York (1976).
356. T. G. Eck, *Phys. Rev. Lett.* **33**, 1055 (1974).
357. M. Lombardi, *Phys. Rev. Lett.* **35**, 1172 (1975).
358. R. M. Herman, in *Beam-Foil Spectroscopy*, Vol. 2, Eds. I. A. Sellin and D. J. Pegg, p. 809, Plenum Press, New York (1976).
359. Y. B. Band, *Phys. Rev. Lett.* **35**, 1272 (1975).
360. R. M. Herman, *Phys. Rev. Lett.* **35**, 1626 (1975).
361. E. L. Lewis and J. D. Silver, *J. Phys. B* **8**, 2697 (1975).
362. Y. B. Band, *Phys. Rev. A* **13**, 2061 (1976).
363. H. J. Andrä, R. Fröhling, and H. J. Plöhn, in *Inelastic Ion-Surface Collisions*, Eds. N. H. Tolk, J. C. Tully, W. Heiland, and C. W. White, Academic Press, New York (1977).
364. R. Fröhling, Diploma Thesis, Free University of Berlin (1978). See also Reference 363.
365. H. Schröder and E. Kupfer, *Z. Phys. A* **279**, 13 (1976).
366. W. Hanle, *Z. Phys.* **30**, 93 (1924).
367. C. H. Liu and D. A. Church, *Phys. Lett.* **35A**, 407 (1971).
368. C. H. Liu, M. Druetta, and D. A. Church, *Phys. Lett.* **38A**, 49 (1972).
369. D. A. Church and C. H. Liu, *Nucl. Instr. Meth.* **110**, 267 (1973).
370. M. Gaillard, M. Carré, H. G. Berry, and M. Lombardi, *Nucl. Instr. Meth.* **110**, 273 (1973).
371. M. Dufay, *Nucl. Instr. Meth.* **110**, 79 (1973).
372. J. Macek, *Phys. Rev. A* **1**, 618 (1970).
373. R. K. Wangsness, *Phys. Rev. A* **4**, 1275 (1971).
374. M. Ducloy and M. Dumont, *J. Phys. (Paris)* **31**, 419 (1970).
375. M. Carré, J. Desesquelles, M. Dufay, and M. L. Gaillard, *Phys. Rev. Lett.* **27**, 1407 (1971).
376. D. A. Church, M. Druetta, and C. H. Liu, *Phys. Rev. Lett.* **27**, 1763 (1971).
377. M. Carré, M. Gaillard, and J. Desesquelles, *Nucl. Instr. Meth.* **110**, 295 (1973).
378. C. H. Liu, R. B. Gardiner, and D. A. Church, *Phys. Lett.* **43A**, 165 (1973).
379. M. Carré, Thesis, University of Lyon (1975).
380. J. Macek and D. J. Jaecks, *Phys. Rev. A* **4**, 2288 (1971).

381. H. G. Berry, J. L. Subtil, and M. Carré, *J. Phys.* (*Paris*) **33**, 947 (1972).
382. J. Bosse and H. Gabriel, *Z. Phys.* **266**, 283 (1974).
383. G. Breit, *Rev. Mod. Phys.* **4**, 504 (1932); **5**, 91 (1933).
384. P. A. Franken, *Phys. Rev.* **121**, 508 (1961).
385. J. Macek, *Phys. Rev. Lett.* **23**, 1 (1969).
386. D. J. Burns and W. H. Hancock, *Phys. Rev. Lett.* **27**, 370 (1971).
387. W. Wittmann, K. Tillmann, H. J. Andrä, and P. Dobberstein, *Z. Phys.* **257**, 279 (1972).
388. H. G. Berry and J. L. Subtil, *Phys. Rev. Lett.* **27**, 1103 (1971).
389. W. Wittmann, K. Tillmann, and H. J. Andrä, *Nucl. Instr. Meth.* **110**, 305 (1973).
390. G. Astner, L. J. Curtis, L. Liljeby, S. Mannervik, and I. Martinson, *J. Phys. B* **9**, L345 (1976).
391. A. Gaupp, M. Dufay, and J. L. Subtil, *J. Phys. B* **9**, 2365 (1976).
392. K. Tillmann, H. J. Andrä, and W. Wittmann, *Phys. Rev. Lett.* **30**, 155 (1973).
393. H. G. Berry, J. L. Subtil, E. H. Pinnington, H. J. Andrä, W. Wittmann, and A. Gaupp, *Phys. Rev. A* **7**, 1609 (1973).
394. H. G. Berry, E. H. Pinnington, and J. L. Subtil, *Phys. Rev. A* **10**, 1065 (1974).
395. O. Poulsen and J. L. Subtil, *J. Phys. B* **7**, 31 (1974).
396. J. D. Silver, J. Desesquelles, and M. L. Gaillard, *J. Phys. B* **8**, L219 (1975).
397. J. L. Subtil, P. Ceyzériat, A. Denis, and J. Desesquelles, *J. Phys.* (*Paris*) **37**, 1299 (1976).
398. J. L. Subtil, P. Ceyzériat, J. Desesquelles, and M. Druetta, in *Beam-Foil Spectroscopy*, Vol. 2, Eds. I. A. Sellin and D. J. Pegg, p. 791, Plenum Press, New York (1976).
399. J. L. Subtil, Thesis, University of Lyon (1977).
400. H. J. Andrä, H. J. Plöhn, A. Gaupp, and R. Fröhling, *Z. Phys. A* **281**, 15 (1977).
401. H. J. Plöhn and H. J. Andrä, to be published.
402. A. C. G. Mitchell and M. W. Zemansky, *Resonance Radiation and Excited Atoms*, MacMillan, London (1934).
403. W. Heitler, *Quantum Theory of Radiation*, Clarendon Press, Oxford (1949).
404. B. L. Moiseiwitsch and S. J. Smith, *Rev. Mod. Phys.* **40**, 238 (1968).
405. H. J. Andrä, A. Gaupp, K. Tillmann, and W. Wittmann, *Nucl. Instr. Meth.* **110**, 453 (1973).
406. H. J. Andrä, A. Gaupp, and W. Wittmann, *Phys. Rev. Lett.* **31**, 501 (1973).
407. A. Corney, *Adv. Electron. Electron Phys.* **29**, 115 (1969).
408. R. E. Imhof and F. H. Read, *Rep. Progr. Phys.* **40**, 1 (1977).
409. A. Gallagher, *Phys. Rev.* **157**, 24 (1967).
410. H. Harde and G. Guthörlein, *Phys. Rev. A* **10**, 1488 (1974).
411. A. Arnesen, A. Bengtson, R. Hallin, S. Kandela, T. Noreland, and R. Lidholt, *Phys. Lett.* **53A**, 459 (1975).
412. M. Gaillard, H. J. Andrä, A. Gaupp, W. Wittmann, H. J. Plöhn, and J. O. Stoner, Jr., *Phys. Rev. A* **12**, 987 (1975).
413. H. J. Andrä, H. J. Plöhn, W. Wittmann, A. Gaupp, J. O. Stoner, Jr., and M. Gaillard, *J. Opt. Soc. Am.* **65**, 1410 (1975).
414. J. O. Stoner, Jr., L. Klynning, I. Martinson, B. Engman, and L. Liljeby, in *Beam-Foil Spectroscopy*, Vol. 2, Eds. D. A. Sellin and D. J. Pegg, p. 873, Plenum Press, New York (1976).
415. A. Arnesen, A. Bengtson, L. J. Curtis, R. Hallin, C. Nordling, and T. Noreland, *Phys. Lett.* **56A**, 355 (1976).
416. A. Arnesen, A. Bengtson, R. Hallin, and T. Noreland, *J. Phys. B* **10**, 565 (1977).
417. A. Arnesen, A. Bengtson, R. Hallin, J. Lindskog, C. Nordling, and T. Noreland, Uppsala University Institute of Physics report No. 952 (1977).

418. M. Gaillard, H. J. Plöhn, H. J. Andrä, D. Kaiser, and H. H. Schulz, in *Beam-Foil Spectroscopy*, Vol. 2, Eds. I. A. Sellin and D. J. Pegg, p. 853, Plenum Press, New York (1976).
419. P. Kuske, N. Kirchner, and D. Kaiser, *Phys. Lett.* **64A**, 377 (1978).
420. N. Kirchner and P. Kuske, private communication.
421. D. Schulze-Hagenest, H. Harde, W. Brand, and W. Demtröder, *Z. Phys.* **A282**, 149 (1977).
422. L. Winkowski, Diploma Thesis, Free University of Berlin (1978).
423. J. D. Silver, N. A. Jelley, and L. C. McIntyre, in Abstracts of the Fifth International Conference on Atomic Physics, p. 205, Berkeley (1976).
424. P. Avan and C. Cohen-Tannoadji, *J. Phys. B* **10**, 155 (1977).
425. H. J. Kimble and L. Mandel, *Phys. Rev. A* **15**, 689 (1977).
426. P. Kuske, Diploma Thesis, Free University of Berlin, Germany (1978).
427. H. Winter and M. Gaillard, *Z. Phys.* **A281**, 311 (1977).
428. F. v. Sichart, H. J. Stöckmann, H. Ackermann, and G. zu Putlitz, *Z. Phys.* **236**, 97 (1970).
429. J. H. Macek, private communication (1974).
430. M. Kraus, Diploma Thesis, Free University of Berlin, Germany; results published in *Beam-Foil Spectroscopy*, Vol. 2, Eds. I. A. Sellin and D. J. Pegg, p. 835, Plenum Press, New York (1976).
431. W. Becker, W. Fischer, and H. Hühnermann, *Z. Phys.* **216**, 142 (1968).
432. O. Poulsen and P. S. Ramanujam, *Phys. Rev. A* **14**, 1463 (1976).
433. H. J. Andrä, J. Macek, J. D. Silver, N. Jelley, and L. C. McIntyre, in *Beam-Foil Spectroscopy*, Vol. 2, Eds. I. A. Sellin and D. J. Pegg, p. 877, Plenum Press, New York (1976).
434. P. M. Koch, L. D. Gardner, and J. E. Bayfield, in *Beam-Foil Spectroscopy*, Vol. 2, p. 829, Plenum Press, New York (1976).
435. S. L. Kaufman, *Opt. Commun.* **17**, 309 (1976).
436. Th. Meier, H. Hühnermann, and H. Wagner, *Opt. Comm.* **20**, 397 (1977).
437. V. S. Letokhov, in *High Resolution Laser Spectroscopy*, Ed. K. Shimoda, Topics in Applied Physics, Vol. 13, p. 95, Springer-Verlag, Berlin (1976).
438. M. Dufay, M. Carré, M. L. Gaillard, G. Meunier, H. Winter, and Z. Zgainski, *Phys. Rev. Lett.* **37**, 1678 (1976).
439. H. Winter and M. Gaillard, *J. Phys. B* **10**, 2739 (1977).
440. F. Beguin, M. L. Gaillard, H. Winter, and G. Meunier, *J. Phys.* (*Paris*) **38**, 1185 (1977).
441. W. Chow, A. D. Maio, and M. O. Scully, *Nucl. Instr. Meth.* **110**, 469 (1973).
442. J. J. Snyder and J. L. Hall, in *Lecture Notes in Physics*, Vol. 43, Eds. S. Haroche, J. C. Pebay-Peyroula, T. W. Hänsch, and S. E. Harris, p. 6, Springer-Verlag, Berlin (1975).
443. J. C. Bergquist, S. A. Lee, and J. L. Hall, *Phys. Rev. Lett.* **38**, 159 (1977).

21
Stark Effect

K. J. Kollath and M. C. Standage

1. Introduction

At the turn of this century the splitting and shift of spectral lines emitted by atoms in an electric field were considered too small to be detected in a laboratory experiment.[1] It therefore came as a surprise when in 1913 Stark[2] and LoSurdo[3] independently discovered considerable splittings of the Balmer lines of hydrogen in electric fields of the order of 10^5 V cm^{-1}. Stark applied electric fields to canal rays which traveled through a condenser with a gap small enough to avoid discharges. Measuring the applied voltage for a known condenser gap, Stark could observe a splitting of the Balmer lines that was linear in the electric field strength. A more detailed analysis shows that the exceptionally close energy gap between states of different parity in the hydrogen spectrum is responsible for this very large and linear effect in the Balmer lines, and that most other atoms have the small second-order Stark effect originally expected.

The linear Stark effect of hydrogen gave considerable support to the then new atomic theory of Bohr, when Epstein and Schwarzschild independently calculated the effect using quantized orbits. The first application of the quantum mechanical perturbation theory by Schrödinger was also devoted to the Stark effect and in first order the linear effect was derived again. The second- and third-order contributions were even more important because the old theory and the quantum-mechanical calculation gave different results, the latter being in better agreement with experiments.[4]

In the subsequent development of atomic physics the Stark effect has not received the same attention as its magnetic counterpart, the Zeeman

---

K. J. Kollath • Universität Münster, Münster, Germany.   M. C. Standage • Griffith University, Brisbane, Australia.

effect. The reason for this is the fact that in most atomic states moderate magnetic fields cause a first-order Zeeman effect, which is well understood theoretically, so that fine-structure (fs) and hyperfine-structure (hfs) intervals may be measured in comparison with the known Zeeman splitting. Such comparison is utilized in a number of precision methods like optical magnetic double resonance, level-crossing, and anticrossing techniques. The Stark effect, however, cannot be used so conveniently since it requires very high electric fields for the first excited states and is theoretically more complicated, the second-order shift being calculated from an infinite sum of radial matrix elements.

On the other hand, the connection of the Stark effect with the radial matrix elements provides a powerful test for the quality of approximate wave functions and oscillator strength calculations. The importance of the Stark effect is intuitively clear considering that the electric field affects the electron charge distribution directly. One of the effects of this perturbation is the appearance of spectral lines that are forbidden in the unperturbed atomic system,[5,6] like $n_1 {}^2S_{1/2}-n_2 {}^2S_{1/2}$ transitions. Their existence can be explained by the mixture of excited states with different orbital quantum numbers L by the electric field and has been used successfully in the assignment of complex spectra.

More recently, the progress made in experimental as well as theoretical methods has focused new interest on the Stark effect. On the experimental side, several methods achieving very-high-energy resolution have been developed and methods of investigating highly excited states showing large polarizabilities have been introduced. In both cases the Stark effects produced by electric fields available in the laboratory can be measured. On the theoretical side, methods of dealing with the infinite sum of matrix elements and with electron correlation in many-electron atoms have been improved.[7]

Apart from its fundamental role in atomic physics the Stark effect has been of importance in the determination of fundamental constants[8] and in the measurements of the hydrogen Lamb shift.[9] It has also been used extensively in line-broadening theory and has been applied to plasma diagnostics. More recently much interest has been attracted by the dynamical Stark effect, that is, the response of atomic systems to alternating electric fields of high and even optical frequencies. This topic is described in Chapter 3 of this book. In the present chapter we shall concentrate on uniform constant electric fields applied to atomic systems.

In the following theoretical section we shall outline the theory of the Stark effect using the perturbation approach. Included is a complete treatment, with Lamb shift and hfs, of the Stark effect for the hydrogen atom in the level $n = 2$.

In the third section we shall review a selection of experimental techniques involving the Stark effect and describe some of the recent experiments. Reviews of earlier work may be found in the articles of Bonch-Bruevich and Khodovoi[10] and Buckingham.[11] The latter article is of particular value for the reader interested in a more detailed treatment of the Stark effect in molecules.*

2. Theory

A comparison of the field strengths available in the laboratory of the order of 10^6 V cm^{-1} with the atomic Coulomb field of the order of

$$E = \frac{1}{4\pi\varepsilon_0} \frac{e}{a_0^2} = 5 \times 10^9 \text{ V cm}^{-1} \tag{1}$$

shows that the external electric field can be considered as a small perturbation to the internal field.

According to general nondegenerate perturbation theory[12] the eigenvalues W and the eigenstates $|\psi\rangle$ which obey the time-independent Schrödinger equation

$$(H_0 + H')|\psi\rangle = W|\omega\rangle \tag{2}$$

may be written up to second order as

$$W = E_m + \langle m|H'|m\rangle + \sum_n' \frac{|\langle m|H'|n\rangle|^2}{E_m - E_n} \tag{3}$$

$$|\psi\rangle = |m\rangle + \sum_k' |k\rangle \frac{\langle k|H'|m\rangle}{E_m - E_k}$$

$$+ \sum_k' |k\rangle \left[\sum_n' \frac{\langle k|H'|n\rangle\langle n|H'|m\rangle}{(E_m - E_k)(E_m - E_n)} - \frac{\langle k|H'|m\rangle\langle m|H'|m\rangle}{(E_m - E_k)^2} \right]$$

Here the primes denote the omission of the term $k = m$ or $n = m$ from the sum and

$$H_0|m\rangle = E_m|m\rangle \tag{4}$$

describes the unperturbed atom. The perturbation Hamiltonian H' for the

---

* Since completion of this review a book on the Stark effect has been published by N. Ryde.[11a]

external electric field can be written in the dipole approximation as

$$H' = -\mathbf{P} \cdot \mathbf{E} \tag{5}$$

where $\mathbf{P} = -\Sigma_i e \, \mathbf{r}_i$ is the atomic electric dipole operator.*

The Hamiltonian of an unperturbed atomic or molecular system is invariant against inversion of the coordinate system, denoted by the operator P. This means that H_0 and P commute and therefore that the eigenfunctions $|m\rangle$ of H_0 are also eigenfunctions of P. Since $p^2 = 1$, the eigenvalues of P can only be $p = \pm 1$ and the eigenfunctions are said to have $\{^{\text{even}}_{\text{odd}}\}$ parity if $P|m\rangle = \pm |m\rangle$. For example, the eigenstates of an electron in a central potential have even parity if the orbital quantum number l is even and odd parity if l is odd, since the angular part of its eigenfunction, the spherical harmonics, have the property

$$PY^l_m(\theta, \phi) = Y^l_m(\pi - \theta, \pi + \phi) = (-1)^l \, Y^l_m(\theta, \phi) \tag{6}$$

while the radial and the spin part of the total eigenfunction remain unchanged by the inversion.

The perturbation Hamiltonian H', however, does not commute with P,

$$PH' = +eP \sum_i (\mathbf{r}_i \cdot \mathbf{E}) = -e \sum_i (\mathbf{r}_i \cdot \mathbf{E})P = -H'P \tag{7}$$

but the anticommutator $H'P + PH'$ of the two operators vanishes. It follows that the matrix elements for the operator H' between eigenstates of P are zero, unless the eigenstates belong to eigenvalues of P with opposite sign. Then matrix elements for H' only exist between states of opposite parity, and the expectation values of H' vanish.

Since H' does not commute with P the eigenstates ψ of the total Hamiltonian H have no definite parity, and for the single electron in a spherical potential perturbed by an electric field the eigenfunctions ψ must be combinations of unperturbed eigenfunctions with different l values.

From the above it follows that there is no first-order Stark effect except in the case when states of opposite parity are degenerate. In the general perturbation treatment of such degenerate states, the matrix of the perturbation Hamiltonian H' is the first diagonalized within the degenerate manifold of states and the set of independent linear combinations of the degenerate eigenfunctions is determined which form the proper zero-order eigenfunctions for the perturbed system. If eigenfunctions with different parity belong to the same group of degenerate states the electric field perturbation can produce a linear Stark effect.

In the following we shall consider in some detail the Stark effect of the hydrogen atom, where in certain approximations we find a linear Stark

---

* In this chapter $e = 1.6 \times 10^{-19}$ C and the charge of the electron $q_e = -e$.

effect. Hydrogen also serves as an instructive example for topics like field ionization, mixing of states, and change of lifetime. In Section 2.2 we turn our attention to nonhydrogenic atoms.

2.1. Stark Effect of Hydrogen

2.1.1. Stark Effect of Hydrogen Neglecting Fine Structure

(a) *Energy Calculations by Perturbation Theory.* If fine structure (fs), Lamb shift, and hyperfine structure (hfs) are neglected, the eigenstates of a hydrogenic system are specified by the principal quantum number n, the orbital quantum number l, and the quantum number of its projection on the quantization axis m_l: $|nlm_l\rangle$. Taking $n = 2$ as an example, there are an s state with even parity, $|200\rangle$, and three p states with odd parity, $|21-1\rangle$, $|210\rangle$, $|211\rangle$, all degenerate in this approximation.

Choosing $\mathbf{E}$ in the direction of the quantization axis, for a single electron H' becomes $H' = ezE$. Because of the axial symmetry of the problem, $[l_z, H'] = 0$ and the only nonvanishing matrix elements are $\langle 200|H'|210\rangle$ and $\langle 210|H'|200\rangle$, assuming electric fields small enough ($E < 10^5$ V/cm) that the calculation can be restricted to $n = 2$. Using the known wave functions, the secular equation for the $n = 2$ manifold is

$$
\begin{vmatrix}
-W^1 & -3ea_0E & 0 & 0 \\
-3ea_0E & -W^1 & 0 & 0 \\
0 & 0 & -W^1 & 0 \\
0 & 0 & 0 & -W^1
\end{vmatrix} = 0
\tag{8}
$$

There are four roots $W^1_{1\cdots4} = 0, 0, \pm(3ea_0E)$, with the corresponding eigenfunctions $|\Psi_1\rangle = |211\rangle$ and $|\Psi_2\rangle = |2\,1-1\rangle$ as before, and $|\Psi_{3,4}\rangle = 2^{-1/2}$ $(|200\rangle \pm |210\rangle)$. As is to be expected, the states belonging to the eigenvalues $\pm(3ea_0E)$ no longer have definite parity.

In this approximation there is a level shift linear in the electric field strength E which partly removes the fourfold degeneracy. The hydrogen atom in the $n = 2$ state behaves as though it has a permanent electric dipole moment of magnitude $3ea_0$ which can be oriented parallel, antiparallel, or perpendicular to the external field.

The original Stark-effect calculations by Schrödinger[13] worked with the same approximation, i.e., neglected fs, Lamb shift, and hfs but applied a parabolic coordinate system to match the axial symmetry of the problem:

$$
\xi = r + z, \qquad \eta = r - z, \qquad \phi = \arctan(y/x)
\tag{9}
$$

In parabolic coordinates not only the Schrödinger equation for the unperturbed atom can be separated* but also the equation including the Stark perturbation. For low electric fields the resulting separated differential equations may be solved by a perturbation procedure.[15]

In first order the shift from the degenerate level

$$W^0 = -RZ^2/n^2 \tag{10}$$

is

$$W^1 = \tfrac{3}{2}(n/Z)ea_0E(n_1 - n_2) \tag{11}$$

where the Rydberg energy is $R = me^4/2\hbar^2(4\pi\varepsilon_0)^2$ and n_1, n_2 are the parabolic quantum numbers with the condition

$$0 \leq n_1,\ n_2 \leq (n-1)$$

For the above example, $n = 2$ and $n_1 = 0, 1$; $n_2 = 0, 1$ so that Eq. (11) results in the same energies W_i^1 as found from Eq. (8).

In second order the perturbation procedure results in

$$W^2 = -\frac{(4\pi\varepsilon_0)}{16}a_0^3E^2\left(\frac{n}{Z}\right)^4[17n^2 - 3(n_1 - n_2)^2 - 9m_l^2 + 19] \tag{12}$$

where $n = n_1 + n_2 + m_1 + 1$. Equation (12) describes a quadratic effect producing a lowering of all energy levels and, since the effect is stronger in higher n, resulting in a red shift of all transition lines.

In third order one obtains

$$W^3 = \frac{3}{32e}(4\pi\varepsilon_0)^2a_0^5E^3\left(\frac{n}{Z}\right)^7(n_1 - n_2)[23n^2 - (n_1 - n_2)^2 + 11m_l^2 + 39] \tag{13}$$

a result also derived by Tsai[16] using pure operator techniques.

For still higher fields ($E > 10^6$ V cm^{-1} for the Balmer-α line) the WKB method was applied to the separated differential equations. In the over-

---

* This property is a result of the $O(4)$ symmetry of the hydrogen problem when spin and relativistic motion are neglected. The invariance of the Hamiltonian under rotation and parity operation for any spherical potential $V(r)$, the $O(3)$ symmetry which is equivalent to the conservation of angular momentum, leads to the degeneracy with respect to the directional quantum number m and allows separation in the spherical coordinate system. The additional symmetry of the Coulomb potential, equivalent to the conservation of the Runge–Lenz vector, corresponds to the additional degeneracy with respect to the orbital quantum number l. This higher symmetry allows the basis set for the Hilbert subspace of eigenfunctions belonging to the same energy to be chosen in two different physically significant ways. The two sets correspond to the set of ψ_{nlm_l} obtained by separating the Schrödinger equation in spherical coordinates, and to the set of Stark states $\psi_{n\,n_1\,n_2}$ obtained from separation in parabolic coordinates. Hughes[14] has shown that the transformation between the two sets may be expressed in Clebsch–Gordon coefficients.

lapping range of electric fields the results are in agreement with the perturbation theory.

The results of the experimental work, performed almost exclusively on the Balmer lines, are in good agreement with the calculations up to field strengths when the spectral lines are quenched by field ionization.

(b) *Field Ionization.* The field ionization can take place because the potential of the hydrogen atom perturbed by an electric field

$$V = -Z/r + Ez \tag{14}$$

forms a potential barrier in the direction of the anode, whose height and width depend on the external electric field. Classically one expects that the electron in a particular state leaves the atom when the term energy, $W_n = -Z^2 R/n^2$ exceeds the maximum of the barrier

$$V_{max} = -2(EZ)^{1/2} \tag{15}$$

The critical field for ionization of a particular level n is

$$E = \frac{Z^3 R^2}{4} \frac{1}{n^4} \tag{16}$$

so that highly excited states are ionized by smaller electric fields than states with lower energy.

In quantum mechanics one finds that none of the previously bound states is stationary any more but all have a finite lifetime, since the electron can eventually tunnel through the barrier leaving the atom ionized.

A more detailed analysis by Lanczos[17] confirmed that excited states with high principal quantum number n are ionized in lower fields than low-n states. However, within a group of Stark states with the same n the low-energy states ($n_2 > n_1$) are quenched in lower electric fields. This at first surprising result was confirmed by experiments which showed quenching of the red Stark components at smaller fields than quenching of the violet components.[18] The effect can be understood from the charge distribution of Stark states.[15] For $n_1 > n_2$ the electron is mostly on the cathode side of the atom avoiding the barrier, while for $n_1 < n_2$ the electron is close to the area of low potential where the tunneling takes place.

Since the early treatment by Lanczos the problem of the hydrogen atom perturbed by the electric field has received much attention and the Stark shifts and ionization probabilities have been calculated by various methods. The finite ionization probability means that a stationary level of the unperturbed hydrogen atom is turned into a narrow band of continuum by the electric field. The problem of calculating the center of this band and its width is quite similar to the problem of quasibound states in scattering theory.

The most recent treatment is by Damburg and Kolosov,[19] who described a numerical method based on modified parabolic coordinates. In the nonrelativistic case this method is applicable to arbitrary quantum numbers. For comparison with earlier calculations, the authors listed the transition probability for the Stark state $n = 10$, $n_1 = 0$, $n_2 = 9$, $m = 0$, which is $7 \times 10^6 \, \text{s}^{-1}$ to $1.3 \times 10^{12} \, \text{s}^{-1}$ for the range of electric fields from $4.6 \times 10^4 \, \text{V/cm}$ to $6.7 \times 10^4 \, \text{V/cm}$.

As may be seen from the calculated values, the lifetime changes rapidly near the critical field strengths. This is so, since the barrier penetration probability depends exponentially on the barrier dimensions, which are determined by the external field. Near the critical field the ionization probability changes from very small values to almost unity over a small range of the external field strength.

The field ionization has been utilized in the detection of highly excited states in fast atomic hydrogen beams leading to a very good detection efficiency. From theory it can be estimated that the hydrogen ions produced from states belonging to different n may be separated experimentally only up to $n = 7$. For higher n the Stark states of neighboring n with large $|n_1 - n_2|$ have similar energies at the electric field required for ionization. Nevertheless Riviere et al.[20] and Bayfield[21] succeeded in resolving highly excited states between $n = 9$–20 and $n = 13$–28, respectively, in fast-beam experiments. The better resolution obtained in these experiments may be explained by the preferential population of states with low angular momentum l by the charge exchange collisions used for excitation. In this way the overlapping Stark states with large $|n_1 - n_2|$ were less strongly populated.

The field ionization has also been applied to alkali atoms,[22,22a] and the critical field for ionization of ns states from $n = 26$ to 37 has been studied using two-step laser excitation.[23] The laser provided sufficient resolution to distinguish different n manifolds, and since for the alkali atoms states of different l within the same n are well separated in energy, the ns and nd states could be populated separately by the two-step excitation. Using the effective quantum number n^*, the $(n^*)^4$ dependence of Eq. (16) was confirmed. The absolute value for the critical field must be modified to include the Stark shift of the levels under investigation.

More recently the tunneling rates have been measured for the excited states of sodium for the principal quantum numbers $n = 12$–14.[24] The rates found for sodium are in good agreement with calculations by Bailey et al.[25] and Damburg and Kolosov[19] for hydrogen.

Field ionization also found considerable interest in plasma physics for application in fusion reactors using the injection of highly excited neutral beams.[26] For use in plasma physics and also in laser and solid-state physics, Fauchier and Dow[27] calculated energy levels of the hydrogenic system also for the range of electric fields larger than the internal Coulomb field.

2.1.2. Stark Effect of Hydrogen Including Fine Structure

So far we have looked at electric fields producing Stark shifts much larger than the fine structure. For small electric fields of the order of 100 V cm^{-1}, the relativistic fine structure may no longer be neglected in the calculations of the Stark effect.

In the Pauli approximation to the Dirac equation the eigenstates are characterized by the quantum numbers $|njm\rangle$, where $\mathbf{j}$ is the total angular momentum obtained by adding orbital $\mathbf{l}$ and spin $\mathbf{s}$ momenta $\mathbf{j} = \mathbf{l} + \mathbf{s}$, and m is the projection quantum number for $\mathbf{j}$.

For calculations of the energy levels in combined electric and magnetic fields we are interested in electric fields with arbitrary direction relative to the quantization axis. The perturbation Hamiltonian* H' may then be written using the spherical tensor notation

$$H' = e \sum_{\mu=-1}^{1} (-1)^{\mu} E_{-\mu} r C_{\mu} \tag{17}$$

where

$$C_{\mu}^{1} = \left(\tfrac{4}{3}\pi\right)^{1/2} Y_{\mu}^{1}(\theta, \phi)$$

$$E_0 = E_z, \qquad E_{\pm 1} = \mp 2^{-1/2}(E_x \pm iE_y)$$

The matrix elements of H' are

$$\langle n'j'm'|H'|njm\rangle = \sum_{\mu=-1}^{1} (-1)^{\mu} E_{-\mu} \langle n'j'm'|rC_{\mu}^{1}|njm\rangle \tag{18}$$

and application of standard angular momentum techniques leads to

$$\langle n'j'm'|H'|njm\rangle = e \sum_{\mu=-1}^{1} (-1)^{\mu} E_{\mu} (-1)^{j'-m'+j+3/2}$$

$$\times [(2j+1)(2j'+1)(2l+1)(2l'+1)]^{1/2} \langle n'l'|r|nl\rangle$$

$$\times \begin{pmatrix} j' & 1 & j \\ -m' & \mu & m \end{pmatrix} \begin{pmatrix} l' & 1 & l \\ 0 & 0 & 0 \end{pmatrix} \begin{Bmatrix} l' & j' & s \\ j & l & 1 \end{Bmatrix} \tag{19}$$

This formulation contains the well-known selection rules for electric dipole transitions

$$\Delta j = \pm 1, 0 \quad \text{and} \quad \Delta l = \pm 1$$

---

* H' of Eq. (5) is a good approximation in the relativistic case for low Z.[28]

with $\mu = \Delta m = 0$ for the field parallel to the quantization axis and $\mu = \Delta m = \pm 1$ for a perpendicular field.

(a) *Calculation for $n = 2$.* If we again consider the $n = 2$ level of hydrogen, according to the Dirac–Pauli theory there are four degenerate $P_{3/2}$ states and a group of four degenerate $S_{1/2}$, $P_{1/2}$ states. Since the latter group contains states of different parity, a linear Stark effect is expected from degenerate perturbation theory. For **E** parallel to the quantization axis the matrix elements of H' are

$$e\langle 2l'\tfrac{1}{2}m | r_0 E_0 | 2l\tfrac{1}{2}m \rangle = -2me\, 3^{1/2} a_0 E_0 \qquad (20)$$

For the two values $m = \pm 1/2$ one obtains submatrices which have identical secular equations

$$\begin{vmatrix} -W^1 & e\,3^{1/2}a_0 E_0 \\ e\,3^{1/2}a_0 E_0 & -W^1 \end{vmatrix} = 0 \qquad (21)$$

leading to the roots $W^1_{1,2} = \pm e\,3^{1/2}a_0 E_0$ and the eigenfunctions

$$|\psi_{1,2}\rangle = 2^{-1/2}(|2l'\tfrac{1}{2}m\rangle \pm |2l\tfrac{1}{2}m\rangle)$$

The degeneracy is only partially removed as states with $\pm m$ remain degenerate and the energies correspond to an electric dipole of magnitude $e\,3^{1/2}a_0$ parallel or antiparallel to the field.

However, because of the quantum electrodynamical corrections to the Dirac theory, the $S_{1/2}$ and $P_{1/2}$ states of $n = 2$ are not degenerate but are split by the Lamb shift S of about 1058 MHz. Therefore, if the zero energy is chosen midway between the states, the matrix reads

$$\begin{vmatrix} +S/2 - W^1 & e\,3^{1/2}a_0 E_0 \\ e\,3^{1/2}a_0 E_0 & -S/2 - W^1 \end{vmatrix} = 0 \qquad (22)$$

with the eigenvalues $W^1_1 = W^1_+ = +[(S/2)^2 + (e\,3^{1/2}a_0 E_0)^2]^{1/2}$ and $W^1_2 = W^1_- = -W^1_+$. The corresponding eigenstates are combinations of the zero-field eigenstates

$$|\psi_\pm\rangle = C_{1\pm}|20\tfrac{1}{2}m\rangle + C_{2\pm}|21\tfrac{1}{2}m\rangle$$

with coefficients depending on the external field:

$$C^2_{1\pm} = \frac{(W^1_\pm + S/2)^2}{(e\,3^{1/2}a_0 E_0)^2 + (W^1_\pm + S/2)^2}$$

$$C^2_{2\pm} = \frac{(e\,3^{1/2}a_0 E_0)^2}{(e\,3^{1/2}a_0 E_0)^2 + (W^1_\pm + S/2)^2} \qquad (23)$$

It follows that even for hydrogen for low electric fields there is no linear Stark effect, as can be shown by expanding

$$W_{\pm}^{1} \approx \pm \left[\frac{S}{2} + \frac{1}{2} \frac{(e3^{1/2}a_0)^2}{S/2} E_0^2 \right] \tag{24}$$

Only for fields producing Stark shifts much larger than the Lamb shift, may its contribution to the eigenvalue be neglected, and the Stark effect becomes linear to good approximation. For such fields, however, the Stark effect is of the order of the fs, and other fs states of the same n must be considered as well, leading again to a nonlinear behavior. The Stark effect in this region, where the Stark shifts are comparable to the fs splittings, has been calculated first by Lüders[29] by numerically diagonalizing the corresponding matrix with the electric field parallel to the quantization direction. At the high-field end of his range, where the Stark splitting is larger than the fs, his calculations join to the theory given in Section 1.1.

A complete treatment of the $n = 2$ Stark effect should include the hfs as well. Then the states are characterized by quantum numbers $|nljFm_F\rangle$ and the Stark-effect matrix elements are

$$\langle n'l'j'F'M_F'|H'|nljFM_F\rangle = e \sum_{\mu=-1}^{1} (-1)^{j'+F'+j+F+1-M_F'}$$

$$+ [(2F+1)(2F'+1)(2j+1)(2j'+1)(2l'+1)(2l+1)]^{1/2}$$

$$+ \begin{pmatrix} l' & 1 & l \\ 0 & 0 & 0 \end{pmatrix} \begin{pmatrix} F' & 1 & F \\ -M_F' & \mu & M_F \end{pmatrix}$$

$$\times \begin{Bmatrix} j' & F' & \frac{1}{2} \\ F & j & 1 \end{Bmatrix} \begin{Bmatrix} l' & j' & \frac{1}{2} \\ j & l & 1 \end{Bmatrix} \langle n'l'|r|nl\rangle \tag{25}$$

The results of numerical diagonalization of the $n = 2$ matrix are shown in Figure 1 using Schrödinger radial eigenfunctions, which are a very good approximation for low Z.[30] The hf splitting is not resolved on the scale shown.

(b) *Lifetime of the Metastable State.* The Stark effect not only shifts the energy levels of atomic states, it also affects the lifetime of excited states since it mixes states that have different transition probabilities in absence of the field. This effect is of particular interest for metastable states whose very long lifetime may be reduced considerably by electric fields.

In the level $n = 2$ of hydrogen, for example, the $2^2S_{1/2}$ state is metastable in the absence of an electric field. The parity rule does not allow electric dipole transitions of the $2S$ into the $1S$ ground state and the lifetime of $\tau = 1/8$ s is determined by the two-photon decay.[31,31a] The $2P$

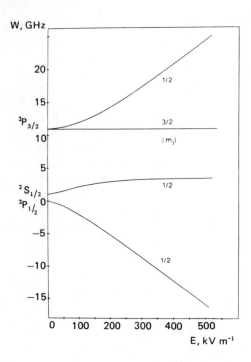

Figure 1. Low-field Stark effect of hydrogen, $n = 2$.

states, on the other hand, decay rapidly with an allowed single-dipole transition to the ground state. Their lifetime is only $\tau_p = 1.6$ ns.

It is clear that the mixture of $2P$ and $2S$ states must lead to a considerable change in the lifetime of the state, which in zero field was the $2S$.[15] Since the two states are well separated in zero field and the interval increases with the Stark effect, the states $|\psi_\pm\rangle$ decay separately with lifetimes $\tau_\pm(E)$ which depend on the electric field. The transition probability of $|\psi_\pm\rangle$ to the ground state is essentially that of the $|2P\rangle$ content, so that the lifetimes are

$$\tau_\pm(E) = \tau_p C_{2\pm}^{-2} \tag{26}$$

For the $2S$ state this means that the lifetime τ_+ for small electric fields decreases quadratically with the field strength. A more detailed study of the lifetime including hfs and radiative widths was made by Heckmann.[32] His aim was to show that amplitude modulations of quantum beats observed by Andrä[33] in Ly_α radiation from beam-foil excitation originated from the hfs.

Since the coefficients $C_{i\pm}$ depend on the zero-field interval between the interacting states, a measurement of the lifetime of hydrogenic metastables in electric fields can be used for Lamb-shift determinations. (See Chapter 12 of this work.)

The polarization of the electric-field-induced radiation also depends on the zero-field intervals and can be used for Lamb-shift measurements. For a practical application to hydrogenlike ions the anisotropy of the quench radiation, which is equivalent to the polarization, is even more promising.[34]

The very simple approximation of using only P states belonging to the same value $n = 2$ works well for hydrogenic systems only. For other atoms, for example for the 2^1S_0 helium, all the ^1P states of the discrete spectrum and of the continuum must be taken into account.[35]

2.2. Stark Effect for Nonhydrogenic Atoms

For nonhydrogenic atoms, the eigenstates of the unperturbed Hamiltonian H_0 have definite parity so that the first-order perturbation term is zero and the energy shift of the state $|m\rangle$ is given to second order by

$$\Delta E(m) = \sum_n' \frac{|\langle m|H'|n\rangle|^2}{E_m - E_n} \qquad (27)$$

Only the even terms in the expansion of $\Delta E(m)$ are nonzero for non-hydrogenic atoms, the ratio of these successive terms being of order $|H'/(E_m - E_n)|^2$.

The Stark shift can be related to the optical oscillator strengths, which also depend on the modulus squared of the electric dipole matrix elements. If the electric field is directed along the axis of quantization z, the connection between the oscillator strengths f_{mn} and the electric dipole matrix elements of $\langle m|P_z|n\rangle$ is

$$f_{mn} = \frac{8\pi^2 m}{h e^2}(E_m - E_n)|\langle m|P_z|n\rangle|^2 \qquad (28)$$

where m is the electronic mass and h is Planck's constant. Combining Eq. (27) and Eq. (28) shows the connection between the Stark shift and the atomic oscillator strengths:

$$\Delta E(m) = \frac{h e^2 E_z^2}{8\pi^2 m}\sum_n' \frac{f_{mn}}{(E_m - E_n)^2} \qquad (29)$$

Experimental or theoretical estimates of the oscillator strength may be used to calculate the Stark shift. Usually only a few of the most significant transitions are included even though $\sum_n'$ is in general an infinite sum over the discrete atomic energy levels and the continuum. The errors caused by truncation of $\sum_n'$ have been discussed for Rb and Cs by Khadjavi et al.[36]

A straightforward expansion of Eq. (27) for an eigenstate $|nJm\rangle$ of the unperturbed atomic Hamiltonian H_0 gives[37]

$$\Delta E(Jm) = (A + Bm^2)E_z^2$$

where

$$A = \sum_{n'} \left[\frac{|(nJ\|P_z\|n'J+1)|^2 (J+1)^2}{E_{nJ} - E_{n'J+1}} + \frac{|(nJ\|P_z\|n'J-1)|^2 J^2}{E_{nJ} - E_{n'J-1}} \right] \quad (30)$$

and

$$B = \sum_{n'} \left[\frac{|(nJ\|P_z\|n'J)|^2}{E_{nJ} - E_{n'J}} - \frac{|(nJ\|P_z\|n'J+1)|^2}{E_{nJ} - E_{n'J+1}} - \frac{|(nJ\|P_z\|n'J-1)|^2}{E_{nJ} - E_{n'J-1}} \right]$$

n, J, and m are, respectively, the principal, angular momentum, and magnetic quantum numbers of the $|nJm\rangle$ eigenstate and the terms $(nJ\|P_z\|n'J)$ are reduced matrix elements of the electric dipole operator. Thus, the Stark shifts of the sublevels of an eigenstate of an atom characterized by angular momentum J depends on two terms: AE_z^2, which produces a shift common to all sublevels, and $Bm^2 E_z^2$, which causes differential shifts by lifting the degeneracy between sublevels of different $|m|$.

More recently, the use of irreducible tensor operators[36,38,39] has led to a generalization of the theory outlined above and to an elucidation of the symmetry properties of the Stark effect. In this approach, the energy shift of the eigenstate $|m\rangle$ is written in terms of an effective Hamiltonian

$$H'_E = \sum_{n}{}' \frac{H'|n\rangle\langle n|H'}{E_m - E_n} \quad (31)$$

The Stark shift is then formally equivalent to a first-order perturbation

$$E(m) = \langle m|H'_E|m \rangle \quad (32)$$

Consideration of possible off-diagonal elements $\langle m'|H'_E|m\rangle$ shows that since the effective Hamiltonian is composed of products of the electric dipole operator H', clearly H'_E is diagonal in the nuclear and electronic spin quantum numbers I and S. Furthermore, because the level shifts due to the electric field are almost always small compared with the atomic fine-structure separations, the matrix elements of H'_E which are off-diagonal in n, L, and J will be very small and can be neglected.

The use of an effective first-order Stark Hamiltonian is advantageous in many cases of interest where the Stark shifts are comparable with the hyperfine structure. The energy eigenvalues within a given level J are then calculated by diagonalization of the sum of the two perturbation Hamiltonians $H = H'_E + H_{\text{hfs}}$.

The general symmetry properties of the effective Hamiltonian can be seen by rewriting it in the form of a multipole operator expansion,[39] which gives

$$H'_E = \sum_{L} K_L P^L (\hat{\mathbf{E}} \cdot \hat{\mathbf{J}}) \quad (33)$$

The operators $P^L(\hat{\mathbf{E}} \cdot \hat{\mathbf{J}})$ can be shown to be scalar products of a spherical harmonic $Y^L(\hat{\mathbf{E}})$ and a spherical harmonic operator $Y^L(\hat{\mathbf{J}})$, such that

$$P^L(\hat{\mathbf{E}} \cdot \hat{\mathbf{J}}) = [4\pi/(2L+1)] Y^L\hat{\mathbf{E}}) \cdot Y^L(\hat{\mathbf{J}}) \qquad (34)$$

where $\hat{\mathbf{E}}$ and $\hat{\mathbf{J}}$ are, respectively, unit vectors denoting the direction of the electric field and the atomic angular momentum. Combining Eq. (33) and Eq. (34) gives

$$H'_E = \sum_L b_L Y^L(\hat{\mathbf{E}}) \cdot Y^L(\hat{\mathbf{J}}) \qquad (35)$$

This reformulation of the effective Hamiltonian separates specific details concerning the atomic structure (such as the energy-level spacings and the electric dipole matrix elements) contained in the terms b_L from the general features of the coupling of the atomic angular momentum to the electric field which are described by the scalar product terms $Y^L(\hat{\mathbf{E}}) \cdot Y^L(\hat{\mathbf{J}})$. Two features of the general form of Eq. (35) emerge from consideration of the quadratic dependence of H'_E on the electric field. Since the components of the electric field are rank-1 tensors, it follows that $Y^L(\hat{\mathbf{E}})$ cannot have rank greater than 2. In addition, because the effective Hamiltonian is invariant to the inversion operation $\mathbf{E} \to -\mathbf{E}$, whereas the spherical harmonic transforms as $Y^L_M(\hat{\mathbf{E}}) \to Y^L_M(-\hat{\mathbf{E}}) = (-1)^L Y^L_M(\hat{\mathbf{E}})$, it follows that L has to be even, Equation (35) reduces to

$$H'_E = b_0 Y^0(\hat{\mathbf{E}}) \cdot Y^0(\hat{\mathbf{J}}) + b_2 Y^2(\hat{\mathbf{E}}) \cdot Y^2(\hat{\mathbf{J}}) \qquad (36)$$

Thus, the effective Hamiltonian consists of only a monopolar term ($L = 0$) and a quadrupolar term ($L = 2$).

The detailed expansion of H'_E may be carried out by writing the spherical harmonic operator $Y^L_M(\hat{\mathbf{J}})$ as[40]

$$Y^L_M(\hat{\mathbf{J}}) = \sum_m |Jm\rangle\langle J, m-M|(2L+1)^{1/2}(-1)^{m-J}\begin{pmatrix} J & J & L \\ m & M-m & -M \end{pmatrix} \qquad (37)$$

The spherical harmonic $Y^L_M(\hat{\mathbf{E}})$ is constructed in an analogous way from spherical components $\hat{\mathbf{E}}_\mu = \mathbf{E}_\mu/E$ of the electric field unit vector $\hat{\mathbf{E}}$:

$$Y^L_M(\hat{\mathbf{E}}) = \sum_\mu \hat{\mathbf{E}}_\mu \hat{\mathbf{E}}_{M-\mu}(2L+1)^{1/2}(-1)^{M+1}\begin{pmatrix} 1 & 1 & L \\ \mu & M-\mu & -M \end{pmatrix} \qquad (38)$$

The corresponding inverse relations are given in Ref. 40. Expansion of Eq. (31) in terms of spherical basis components E_μ and P_ν, where P_ν are spherical components of the electric dipole unit vector, gives

$$H_E' = E^2 \sum \frac{(-1)^{\mu+\nu}|Jm\rangle\langle Jm|\hat{\mathbf{E}}_{-\mu}P_\mu|J'm'\rangle\langle J'm'|\hat{\mathbf{E}}_{-\nu}P_\nu|Jm''\rangle\langle Jm''|}{E_{nJ}-E_{n'J'}}$$

(39)

where the identity operator $1 = \Sigma_m |Jm\rangle\langle Jm|$ has been used in the expansion. The summation is over the quantum n', J', m, m', m'', μ, and ν. Taking the electric field in the z direction considerably simplifies the expansion of H_E'. Setting $\mu = \nu = 0$ and combining Eqs. (37)–(39) gives

$$H_E' = E^2 \sum (-1)^{J-m+1}|(nJ\|P_0\|n'J')|^2 \begin{pmatrix} J & 1 & J' \\ -m & 0 & m \end{pmatrix}\begin{pmatrix} J' & 1 & J \\ m & 0 & -m \end{pmatrix}$$

$$\times [(2L+1)(2L'+1)]^{1/2}\begin{pmatrix} 1 & 1 & L \\ 0 & 0 & 0 \end{pmatrix}\begin{pmatrix} J & J & L' \\ m & -m & 0 \end{pmatrix}Y_0^L(\hat{\mathbf{E}})Y_0^{L'}(\hat{\mathbf{J}})$$

(40)

where the matrix elements of the electric dipole operator have been written as[41]

$$\langle nJm|P_0|n'J'm'\rangle = (-1)^{J-m}\begin{pmatrix} J & 1 & J' \\ -m & 0 & m' \end{pmatrix}(nJ\|P_0\|n'J')$$

(41)

and the summation is over n', J', m, L, and L'. Using the identities which connect products of $3-j$ symbols to $6-j$ symbols,[41] Eq. (40) reduces to

$$H_E' = E^2 \sum_L K_L' Y_0^L(\hat{\mathbf{E}})Y_0^L(\hat{\mathbf{J}})$$

(42)

where

$$K_L' = \sum_{n'J'} \frac{(-1)^{1+J+J'}}{E_{nJ}-E_{n'J'}}\begin{Bmatrix} J & J & L \\ 1 & 1 & J' \end{Bmatrix}|(nJ\|P_0\|n'J')|^2$$

(43)

Comparison of Eqs. (42) and (35) shows that both equations have the same multipole form, while Eq. (43) shows the explicit dependence of the multipole expansion coefficients on the detailed atomic structure.

The classical form of the interaction between an atom and an electric field is described by the induced dipole moment $\mathbf{a} \cdot \hat{\mathbf{E}}$, where $\mathbf{a}$ is the classical polarizability tensor of the atom. The interaction energy is given by $-\frac{1}{2}\mathbf{E} \cdot \mathbf{a} \cdot \mathbf{E}$. Since the effective Hamiltonian can be expressed as the sum of a monopolar and a quadrupolar term, corresponding scalar and tensor (or quadrupolar) polarizabilities, α_0 and α_2 are defined in the following way. The Stark shift of the state $|nJm\rangle$ is written as

$$\langle nJm|H_E'|nJm\rangle = -\frac{1}{2}\alpha_0 E^2 - \frac{1}{2}\alpha_2 \frac{3m^2-J(J+1)}{J(2J-1)}E^2$$

(44)

where a comparison of Eq. (44) with Eq. (42) and (43) shows that

$$\alpha_0 = \frac{-2/3}{2J+1} \sum_{n'J'} \frac{|(nJ\|P_0\|n'J')|^2}{E_{nJ} - E_{n'J'}} \tag{45}$$

$$\alpha_2 = 2\left[\frac{10J(2J-1)}{3(2J+3)(J+1)(2J+1)}\right]^{1/2} \sum_{n'J'} (-1)^{J+J'+1}$$

$$\times \begin{Bmatrix} J & J & 2 \\ 1 & 1 & J' \end{Bmatrix} \frac{|(nJ\|P_0\|n'J')|^2}{E_{nJ} - E_{n'J'}}$$

In the case of atoms with nonzero nuclear spin, an eigenstate of the atom is represented by $|nIJFM\rangle$, where the total angular momentum is $\mathbf{F} = \mathbf{I} + \mathbf{J}$. By making use of the fact that the state $|nIJFM\rangle$ can be written as a sum of products of the electronic states $|nJm\rangle$ and the nuclear spin states $|IM_I\rangle$ such that

$$|nIJFM\rangle = \sum_m (2F+1)^{1/2}(-1)^{J-I-M} \begin{pmatrix} I & J & F \\ M-m & m & -m \end{pmatrix} |nJm\rangle|I, M-m\rangle \tag{46}$$

and using the orthogonality relations between 3-j symbols,[41] the effective Hamiltonian for this case H_E'' reduces to

$$H_E'' = \sum \frac{\mathbf{E}\cdot\mathbf{P}|n'J'm'\rangle\langle n'J'm'|\mathbf{E}^*\cdot\mathbf{P}^*}{E_{nJ} - E_{n'J'}} \left(1 + \frac{\Delta E_{\text{hfs}}}{\Delta E_{\text{opt}}}\right) \tag{47}$$

where the sum is over $n'J'$ and m'.

The small correction term $\Delta E_{\text{hfs}}/\Delta E_{\text{opt}}$, which arises because the energy denominator terms in Eq. (47) include the hyperfine splittings, can be neglected since the optical energy intervals are about 10^4 cm^{-1}, whereas the hyperfine splittings are about 10^{-1} cm^{-1}. Thus, provided the small correction term is neglected, the effective Hamiltonian is the same for either the $I = 0$ or the $I \neq 0$ cases. The first-order matrix element between the states $|nIJFM\rangle \equiv |FM\rangle$ and $|nIJF'M'\rangle \equiv |F'M'\rangle$ can be rewritten with the aid of Eq. (46), to give

$$\langle F'M'|H_E'|FM\rangle =$$

$$\sum_{mm'} [(2F+1)(2F'+1)]^{1/2}(-1)^{2J-2I-M-M'} \begin{pmatrix} I & J & F \\ M-m & m & -m \end{pmatrix}$$

$$\times \begin{pmatrix} I & J & F' \\ m'-m & m' & -m' \end{pmatrix} \langle I, M'-m'|I, M-m\rangle\langle nJm'|H_E'|nJm\rangle \tag{48}$$

Since the nuclear spin states are orthogonal and because the electric field has been chosen parallel to the z axis, H_E' is diagonal in both m and M and

Eq. (48) reduces to

$$\langle F'M|H'_E|FM\rangle = \sum_m [(2F+1)(2F'+1)]^{1/2}(-1)^{2(J-I-M)}$$

$$\times \begin{pmatrix} I & J & F \\ M-m & m & -m \end{pmatrix} \begin{pmatrix} I & J & F' \\ M-m & m & -m \end{pmatrix}$$

$$\times \langle nJm|H'_E|nJm\rangle \tag{49}$$

Whether H'_E is also diagonal in F is dependent on the relative size of the electric field shifts compared with the hyperfine structure. For the alkali metal atoms, electric-field-induced shifts that are comparable with the hyperfine structure are experimentally realizable, whereas, for elements such as mercury and cadmium, this is not the case.

To illustrate the effect nuclear spin has on the Stark effect consider a $J = 1$ eigenstate of ^{199}Hg which has a nuclear spin of $I = 1/2$. The total angular momentum quantum number F can take on the value of $1/2$ and $3/2$ and the hyperfine Stark splittings are

$$\Delta E(\tfrac{3}{2}, M) = -\tfrac{1}{2}\alpha_0 E^2 + \tfrac{1}{2}\alpha_2 E^2(\tfrac{5}{4} - M^2)$$

$$\Delta E(\tfrac{1}{2}, M) = -\tfrac{1}{2}\alpha_0 E^2 \tag{50}$$

Thus, for the $I \neq 0$ case, the Stark effect has a M^2 dependence, as compared with m^2 for the $I = 0$ case.

3. Experimental Methods

The original methods of measuring the Stark effect have been applied for a considerable time.

In the canal-ray method of Stark, ions are extracted from a dc discharge through a small hole in the cathode and the light emitted after recombination is observed. An electric field is applied parallel or perpendicular to the beam in a condenser with a narrow gap to avoid discharges. This method has been recently used to measure the Stark shifts in the H_β and H_γ line of hydrogen with an accuracy better than 1%.[42] Improved resolution allowed measurements of the linear effect down to 1.7 kV cm^{-1}, mainly limited by Doppler broadening, contradicting the controversial measurements of Steubing and Junge.[43] Their results, which did not show any Stark shift below fields of 2 kV cm^{-1}, were contradicted before by Rother[44] using a "Drehfeld" method somewhat similar to atomic-beam radio-frequency experiments.

In this "Drehfeld" method canal rays were first subjected to a very high electric field, which ionized the low-energy Stark states of a level n

The resulting asymmetric population among the Stark states was detected by measuring the ratio of the intensities of the weaker red-shifted components to the violet-shifted components of the line. Then the canal ray transversed a region where a smaller, nonionizing electric field perpendicular to the canal ray was suddenly reversed. In this region some of the atoms experienced a nonadiabatic transition which repopulated the previously ionized Stark states producing a new intensity ratio between red- and violet-shifted compoents. The experimental transition probability, obtained by comparison of the intensity ratios in the high- and low-field regions, was compared with the calculated probability for such nonadiabatic transitions based on the known field distribution and the velocity of the canal rays and by assuming the theoretical Stark splitting. The experiment confirmed the linear Stark effect for fields as low as $1.5\,\mathrm{kV\,cm}^{-1}$ to an accuracy of about 10% and clearly disagreed with the results of Steubing and Junge.

The second classic method, the Lo Surdo cell, has also found a number of applications. In particular, it allows the investigation of the cathode material in the discharge evaporated by cathode sputtering. Since the electric field strength cannot be measured directly the Lo Surdo cell is unsually calibrated against hydrogen.

In the following we discuss a number of high-resolution spectroscopic methods such as level crossings, anticrossings, and pulsed fields. The application of maser and laser techniques to Stark-effect measurements is also reviewed.

3.1. Level Crossings

In the level-crossing technique the resonance fluorescence of atoms in external fields is observed. Under certain geometrical conditions a change in the intensity or polarization of the fluorescent light may be observed when certain sublevels are tuned to degeneracy by the external field. The first observation of this effect was made by Hanle[45] in the special case of degeneracy in zero magnetic field; the method was extended to degeneracy in finite magnetic fields by Franken[46] and transferred to the case of degeneracy produced by an electric field by Khadjavi et al.[47] The intensity of the resonance fluorescence associated with the level crossing of excited substates $|\mu\rangle$ and $|\mu'\rangle$ which decay radiatively to substates $|m\rangle$ and $|m'\rangle$ of the ground state is described by the Breit-Franken[46] formula:

$$I = C \sum_{\mu\mu'} \frac{f_{\mu m}f_{m\mu'}g_{\mu'm'}g_{m'\mu}}{\Gamma + iw_{\mu\mu'}} \qquad (51)$$

Here Γ is the excited-state radiative decay constant and $w_{\mu\mu'} = (E_\mu - E_{\mu'})/h$ the splitting frequency of the excited substates which depends on the

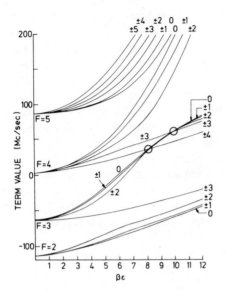

Figure 2. Stark shifts in the ^{133}Cs $7\,^2P_{3/2}$ hfs levels. The scalar Stark shift common to all levels shown has been suppressed. (After Khadjavi et al.[36])

external field; $f_{\mu m} = \langle \mu | \hat{\mathbf{f}} \cdot \mathbf{P} | m \rangle$ and $g_{m'\mu} = \langle m' | \hat{\mathbf{g}} \cdot \mathbf{P} | \mu \rangle$ are matrix elements describing the excitation and emission processes; $\hat{\mathbf{f}}$ and $\hat{\mathbf{g}}$ are unit vectors which represent the polarization of the exciting radiation and the polarization selected by the analyzer, respectively; the factor C is proportional to the intensity of the exciting light, which is assumed to be constant over the manifold of level-crossing substrates.

Since the observation of level crossings requires coherence in the excitation process between the crossing levels, these levels must have the same parity and in practice belong to the same fs or hfs manifold. Therefore, only differential Stark shifts between states of different m may be measured and only the tensor polarizability is derived from level-crossing experiments. With the level-crossing technique measurements of Stark shifts may be made by pure electric field crossings if the zero-field fs or hfs splitting is known or they may be made in parallel magnetic and electric fields, when the Stark shifts are effectively measured in terms of the known Zeeman effect.

3.1.1. Pure Electric Field Level Crossings

The pure electric field level-crossing technique was applied by Khadjavi et al.[36] to Rb and Cs and by Schmieder et al.[48] to K. The Stark effect is sufficiently large in the alkalis that crossings can occur between different hyperfine levels (see Figure 2). The Hamiltonian for the excited state of a atom with hyperfine structure in an electric field is

$$H = H_{hfs} + H'_E \tag{52}$$

where

$$H_{hfs} = A\mathbf{I}\cdot\mathbf{J} + B\frac{3(\mathbf{I}\cdot\mathbf{J})^2 + \frac{3}{2}(\mathbf{I}\cdot\mathbf{J}) - I(I+1)J(J+1)}{2I(2I-1)2J(2J-1)}$$

The parameters A and B are the hyperfine-structure constants. Diagonalization of Eq. (52) yields the energy eigenvalues E_μ and $E_{\mu'}$ which are inserted in the expression for the level-crossing curve, Eq. (51). The value of the atomic polarizability that yields the best fit of the level-crossing curve to the experimental data is then determined.

More recently the pure electric field level-crossing techniques has been used by Bhaskar and Lurio[49] to measure the tensor polarizability of the 2^1P state of atomic helium. Since the $2^1P - 1^1S$ transition is in the vacuum ultraviolet (58.4 nm), the lack of efficient polarizers at such wavelengths leads to considerable experimental difficulties. The following method was used to overcome these problems. Figure 3 shows the energy-level scheme for helium. An atomic beam of 2^1S_0 metastables was produced by electron bombardment of ground-state helium. The 2^1S_0 atoms are excited to the 2^1P state by linearly polarized 2058.2-nm resonance

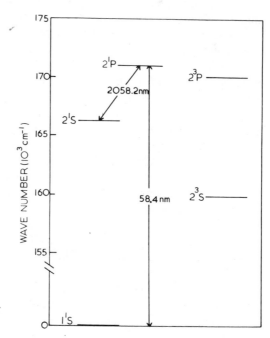

Figure 3. Energy-level scheme for helium.

radiation and the unpolarized 58.4-nm fluorescent radiation from the spontaneous decay to the ground state was observed with a channel electron multiplier. Quenching of some of the 2^1S_0 atoms through mixing of the 2^1S_0 state with the 2^1P_1 state via the electric field leads to the emission of 60.1-nm "quench" photons. Because of the lack of spectral filters in the vacuum ultraviolet, the observed signal consists of "quench" photons and fluorescent (58.4 nm) photons which are detected with about the same efficiency by the channel electron multiplier.

The electric field dependence of the level-crossing signal is not simply given by the Breit–Franken formula because the electric field quenching rate depends on the square of the electric field. Bhaskar and Lurio show how by comparing the ratio of intensities for suitably chosen polarizer angles the quenching contribution to the level-crossing signal can be eliminated. The measured value was $\alpha_2 = (55.8 \pm 3.4) \text{kHz}(\text{kV/cm})^{-2}$ compared with an estimated value of 52.8 kHz $(\text{kV/cm})^{-2}$ obtained using experimental oscillator strengths. These values disagree with a theoretical value of 59.8 kHz $(\text{kV/cm})^{-2}$ obtained by Deutsch et al.[50]

3.1.2. Level Crossings in Combined Electric and Magnetic Fields

In this type of experiment, a level-crossing is generated by the magnetic field and the shift of its magnetic field position by the electric field is measured, or a group of substates which is split by the electric field are made degenerate again by the magnetic field.

The former method has been applied to a measurement of the low-field Stark effect in the level $n = 2$ of hydrogen.[51] Level crossings between the states $2^2P_{3/2}$, $m_j = -3/2$ and $2^2P_{1/2}$, $m_j = 1/2$ were observed as intensity changes of about 10% in the Ly_α resonance fluorescence from an atomic beam of hydrogen.

Because of the narrow fine structure in the hydrogen spectrum, for low electric fields the Stark effect can be fully described by considering the $n = 2$ manifold of states only. In this approximation the $P_{3/2}$, $m_j = \pm 3/2$ state are not affected by the electric field and the shift of the level-crossing position is entirely due to the $P_{1/2}$ Stark effect. The shift can be calculated by numerical diagonalization of the $n = 2$ matrix of 16 hyperfine states for combined Zeeman and Stark-effect perturbations.

Figure 4 shows the result of the calculations using the correct values for the electric field matrix elements (solid line) and matrix element reduced by 2% (broken line). The figure shows that the experimental results are in agreement within the margin of about 2%.

Level crossings in parallel magnetic and electric fields have also been applied in the Stark-effect measurements of the even isotopes of Hg and Cd.[36] In principle, the zero-field level-crossing (the electric field analog of

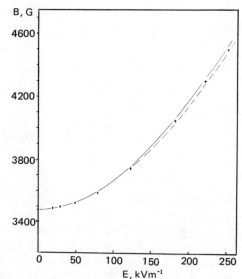

Figure 4. Position of the level crossing between $2^2P_{3/2}$ $m_j = -3/2$ and $2^2P_{1/2}$ $m_j = 1/2$ versus electric field strength. Experimental points[51] in comparison with theory (solid line) and theory reduced by 2% (broken line).

the Hanle effect) could be used to determine the Stark splitting provided the effective lifetime of the crossing substates is known. However, because the uv resonance lines of Hg and Cd are sensitive to radiation trapping and the shape of the zero-field level-crossing curve is altered by stray magnetic fields, parallel magnetic and electric fields were used to obtain a level-crossing at finite electric field. In the odd isotopes of Hg and Cd, hyperfine structure is present, but, unlike the alkalis, the experimentally realizable Stark shifts are small compared to the hyperfine splittings so that substates having different F values cannot be made to cross.

The Hamiltonian for the n^3P_1 states of Cd and Hg is given by

$$H = H_B + H'_E$$

where

$$H_B = g_J\mu_0 J_z B \tag{53}$$

g_J is the Lande g value, μ_0 is the Bohr magneton, and B is the magnetic field. For a fixed electric field three level crossings occur: one at zero magnetic field and two when $g_J\mu_0 B_C \pm \frac{3}{2}\alpha_2 E^2 = 0$ (see Figure 5), where B_C is the crossing field value. Since the magnetic field and the g_J values are known, the Zeeman effect provides a measure of the Stark splitting.

One source of systematic error that can occur in such experiments is that anticrossing effects[52] can arise if the electric and magnetic fields are not parallel. The angle ϕ that can be tolerated between the electric and magnetic field directions can be estimated by requiring that the Zeeman

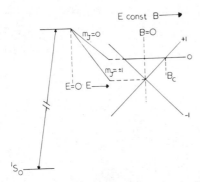

Figure 5. Level crossings for sublevels of $^3P_1$ states of Cd and Hg in parallel electric and magnetic fields.

frequency shift associated with the transverse component of the magnetic field $(B_C \sin \phi)$ is smaller than the radiative width Γ, which gives $\Gamma = 1/\tau > (g_J \mu_0 B_C \sin \phi)/h$. Thus, the longer the lifetime of the excited states, the more collinear the fields have to be. For a typical electric field of 50 kV cm^{-1}, the crossing value of the magnetic field is approximately 3 G for both Hg and Cd. For the Hg 6^3P state ($\tau = 117$ ns), the angle that can be tolerated is $\phi \simeq 10°$, but, for the Cd 5^3P state ($\tau = 2.4 \ \mu$s) the angle is $\phi \simeq 0.5°$, which demands precise alignment of the fields for Cd measurements.

Table 1 shows a comparison of experimental and theoretical values obtained for α_2 by Khadjavi et al.[36] and Schmieder et al.[48] The theoretical values were obtained by these authors using the Bates and Damgaard method in which the outer electron is assumed to move in a Coulombic potential. The approximation works well for the alkali atoms that have one valence electron; however, for heavy atoms such as Cd and Hg with two valence electrons, not surprisingly, the approximation works less well.

It is of interest to compare the experimental tensor polarizability for Cd shown in Table 1 with measurements obtained by Aleksandrov and Khromov[53] using the method of beats. These authors obtained a value of $\alpha_2 = (1.68 \pm 0.06) \text{ kHz (kV/cm}^{-2})$ using a pure electric field method and $\alpha_2 = (1.34 \pm 0.07) \text{ kHz (kV/cm}^{-2})$ using the parallel electric and magnetic

Table 1. Experimental and Theoretical Tensor Polarizabilities for ^{39}K, ^{85}Rb, ^{133}C and the Even Isotopes of Cd and Hg in Units of MHz $(\text{kV/cm})^{-2}$

	State				
α_2	^{39}K$(4^2P_{3/2})$	^{85}Rb$(5^2P_{3/2})$	^{133}Cs$(6^2P_{3/2})$	Cd$(5^3P_1)^a$	Hg$(6^3P_1)^a$
Expt.	-0.26 ± 0.04	-0.52 ± 0.02	-1.08 ± 0.04	1.70 ± 0.07	1.57 ± 0.06
Theory	-0.26	-0.49	-1.05	1.3	0.92

^{a} kHz $(\text{kV/cm})^{-2}$.

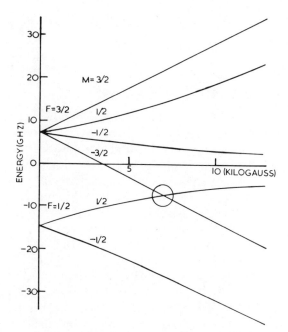

Figure 6. Level crossing in the 6^3P_1 state of ^{199}Hg.

fields. A possible explanation for the discrepancy may lie in the high degree of collinearity of the magnetic and electric fields required for Cd measurements, or perhaps the presence of stray fields.

Kaul and Latshaw[54] have performed a measurement of the Stark shift in the Hg 6^3P_1 state in which the shift of a high-field magnetic level-crossing in the odd isotope of mercury (^{199}Hg) was measured when an additional electric field was applied parallel to the magnetic field. For the Hg 6^3P_1 state, a level-crossing occurs between the $|F = \frac{3}{2}, M = -\frac{3}{2}\rangle$ and "F" $= \frac{1}{2}, M = \frac{1}{2}\rangle$ substates (see Figure 6) at a magnetic field of about 7 kG, a field high enough to cause partial decoupling of the electronic and nuclear angular momenta. The total angular momentum quantum number F is no longer a good quantum number and the $|M_J M_I\rangle$ representation has to be used. At the level-crossing, the crossing substates are composed of a mixture of $\|M_J M_I\rangle$ substates:

$$|F = \tfrac{3}{2}, M = -\tfrac{3}{2}\rangle = |M_J = -1, M_I = -\tfrac{1}{2}\rangle \tag{54}$$

and

$$|\text{``}F\text{''} = \tfrac{1}{2}, M = \tfrac{1}{2}\rangle = \left(\tfrac{2}{3}\right)^{1/2}|M_J = 0, M_I = \tfrac{1}{2}\rangle - \left(\tfrac{1}{3}\right)^{1/2}|M_J = +1, M_I = -\tfrac{1}{2}\rangle$$

The energy shifts of the crossing substates in the presence of the electric field is given by the matrix element $\langle FM|H'_E|FM\rangle$. The energy shifts are

$$\Delta E(F = \tfrac{3}{2}, M = -\tfrac{3}{2}) = -\tfrac{1}{2}(\alpha_0 + \alpha_2)E^2$$

and   (55)

$$\Delta E("F" = \tfrac{1}{2}, M = \tfrac{1}{2}) = -\tfrac{1}{2}(\alpha_0 - \alpha_2)E^2$$

The effect of the electric field is to shift the $F = \tfrac{3}{2}$ level further downward in energy than the "F" $= \tfrac{1}{2}$ level by an amount $\alpha_2 E^2$, thereby causing the level-crossing to occur at a lower magnetic field. Knowledge of the Zeeman slopes (dE/dB) in the vicinity of the level-crossing and the shift in the magnetic field of the level-crossing allows the differential Stark shift to be determined. The value that was obtained for the tensor polarizability of the Hg 6^3P_1 state was $\alpha_2 = (1.58 \pm 0.02)$ kHz $(kV/cm)^{-2}$.

The results for the 6^3P_1 state were used by Chantepie for his measurement of the Stark effect in the $6^{1,3}D_J$ states of ^{199}Hg.[55]

Other Stark experiments for atoms with two valence electrons using the level-crossing method with parallel magnetic and electric fields have been done for the lowest P terms of the even isotopes of Ca, Sr, Ba by Kreutzträger and von Oppen.[56] Since the g_J value for these terms was not known with much better accuracy, the authors stated the experimental results for the ratio $\gamma = \alpha_2/g_J$. They also quoted the ratios of the results for different atoms, e.g., $\gamma(Sr)/\gamma(Ca)$. This ratio, which is independent of the absolute determination of the electric field strength with its relatively high error contribution, can be measured with better accuracy than $\gamma(Sr)$ itself.

3.2. Level Anticrossings

If an electric field perpendicular to the magnetic field is applied, the crossings of Zeeman branches turn into anticrossings,[52] and this technique may be used to investigate levels not accessible to level-crossing observation. The pure anticrossing technique utilizes population differences between the states under investigation very much like the radio-frequency resonance methods. Instead of tuning the Zeeman states until their distance matches the rf, for electric-field-induced anticrossings the levels are tuned to cross and a dc electric field perpendicular to the magnetic field is applied. Because of the close proximity of the two levels, a small perturbation by the electric field is sufficient to mix the states, and the corresponding equilization of the previously different populations may be observed as intensity changes in the spectral lines emitted by either of the levels.

Compared to the rf methods, which can easily produce electric dipole transitions with $\Delta l = 1$ but require very high rf fields to produce $\Delta l >$ multiquantum transitions, the anticrossing method produces relatively easily anticrossings between states with $\Delta l > 1$. As for the rf methods, the anticrossing signals suffer from saturation, line broadening, and shift from the applied dc field.* In fact, it is sometimes difficult experimentally to reduce stray electric fields to such a level to observe $\Delta l = 1$ anticrossings. For $\Delta l = 2$ or 3 anticrossings, however, a limited range of electric fields may be observed and the shift of the anticrossing position caused by the relative Stark shift between the anticrossing levels can be measured.

This method has been applied by Beyer et al.[58] to $S-D$ and $S-F$ anticrossings in the level $n = 4$ of He$^+$. Differential population of the Zeeman states was produced by electron bombardment of He in a sealed glass system at a pressure of 0.01 Torr. The anticrossings were detected as intensity changes in the 486.6 nm ($n = 4 \rightarrow n = 3$) transition line as the magnetic field was swept through the resonant magnetic field position, and the shift of this position was measured for various electric fields perpendicular to the magnetic field.

Since He$^+$ is a hydrogenic system, the observed Stark shifts could again be compared with theoretical values obtained by numerical diagonalization of the combined Stark- and Zeeman-effect matrices. In the electric field range from 0 to 200 V cm^{-1} experimental results and theory agree in most cases within the experimental error of about 10%.

Similar Stark-effect investigations have been done in the helium atom using electric-field-induced anticrossings, singlet–triplet anticrossings, and electric-field-induced singlet–triplet anticrossings.[59] The three types of anticrossings differ in the perturbation producing the signal, which is the external electric field, the internal singlet–triplet interaction due to part of the spin–orbit Hamiltonian, and the combination of external and internal perturbations, respectively. The third type is the most interesting for Stark-effect investigations, since it allows one to observe separately the Stark shift of a single anticrossing between a particular Zeeman branch of the n^1D and a particular branch of the n^3D levels.

In the experiments the shift of an anticrossing was observed as function of the applied electric field and the result was compared with the theoretical shift, which was calculated from second-order perturbation theory according to Eq. (27). Instead of the infinite sum only the contributions from P and F states of the same principal quantum number n were calculated and hydrogenic eigenfunctions were used to evaluate the dipole matrix elements. Good agreement was found within the experimental errors of about 10% for the principal quantum numbers $n = 5, 6, 7$.

In the case of the pure singlet–triplet and the pure electric-field-induced anticrossings, the comparison is more complicated because several

---

The signals for electric-field-induced anticrossings have been calculated for a three-level hydrogenic system by Glass-Maujean and Descoubes[57] using the density-matrix formalism.

anticrossings between Zeeman branches of the n^1D with the corresponding n^3D or nL branches were observed together. Although the Stark shift of a single signal component can be calculated as before, the relative weight of a component within the unresolved signal is not known, so that the Stark shift of the observed signal cannot be calculated accurately.

In the case of the electric-field-induced anticrossings the relative weight is particularly important since the Stark shifts of the single components differ in sign and in magnitude. Therefore, no agreement between experimental and theoretical shift, calculated by assuming equal relative weight for all components, has been found.

For the pure singlet–triplet anticrossings, where the Stark shift of the different components differ only little and where all anticrossing components are in saturation so that the relative weights can be estimated with some certainty, good agreement between theory and experiment was found for $n = 7$ and 8. For $n = 9$ and 10 the results differ by a factor of 2 and 5, respectively, since these levels are more sensitive to small stray fields in the experimental setup, so that the results are less reliable.

3.3. The Pulsed-Field Method

The pulsed-field method of Sandle et al.[60] involves the observation of transients in radiation resonantly scattered by a vapor following the sudden application of a pulsed electric field. See the article by Dodd and Series, Chapter 14 of this work.

The general form of the transients is given by

$$I(t) = C \sum_{mm'} \langle m|F|m'\rangle\langle m'|G|m\rangle\{(\Gamma + iw^+_{mm'})^{-1} + (\Gamma + iw^-_{mm'})^{-1}$$

$$+ (\Gamma + iw^+_{mm'})^{-1}\} \exp\left[-(\Gamma + iw^+_{mm'})t\right] \qquad (56)$$

where $I(t)$ is the fluorescent intensity emitted at time t. The field pulse is assumed to commence at $t = 0$. F is the excitation operator

$$F = \sum_{\mu} \hat{\mathbf{f}} \cdot \mathbf{P}|\mu\rangle\langle\mu|\hat{\mathbf{f}}^* \cdot \mathbf{P}^* \qquad (57)$$

where $\hat{\mathbf{f}}$ is the polarization unit vector of the exciting radiation and the $|\mu$ are substates of the ground state. G, the emission operator, is analogous to the excitation operator F except that unit vector $\hat{\mathbf{g}}$, representing the polarization of the fluorescent radiation selected by the analyzer, replace $\hat{\mathbf{f}}$. The matrix elements of F and G are taken between the excited substate $|m\rangle$ and $|m'\rangle$. Γ is the excited-state decay constant and $w^-_{mm'}$ and $w^+_{mm'}$ refer, respectively, to the values of the splitting frequencies between the excited substates $|m\rangle$ and $|m'\rangle$ before, and after, the field pulse is applied.

The time-independent part of Eq. (57) is the familiar Breit–Franken formula Eq. (51) for the steady scattered fluorescent intensity. The time-dependent terms describe the transient behavior, which under suitable conditions takes the form of a damped oscillation beginning at $t = 0$. The modulation frequency can be any of the excited-state splitting frequencies $w^+_{mm'}$, but in practice the range of frequencies that can be observed has an upper limit determined by either the response time of the detection equipment or the rise time of the field pulse. The lower limit is set by the radiative width of the state.

Sandle $et\ al.$[60] used the pulsed electric field method to measure the magnitude and sign of the tensor polarizability of the 6^3P_1 state of ^{198}Hg and ^{199}Hg. A schematic diagram of the experimental arrangement is shown in Figure 7. The scattered fluorescent radiation was analyzed for either linear or circular polarization, the latter being required to determine the sign of α_2. The electric field, in the range 25–100 kV cm^{-1}, was provided by a pulsing unit which produced voltage pulses with a rise time of about 2 ns and a duration of about 100 ns. The rise time of the field pulse has to be short compared to any characteristic evolution time (e.g., the lifetime) of the excited state. A single-photon counting system consisting of nuclear timing equipment and a pulse-height analyzer was used to measure the time delay between the switching on of the electric field pulse and the

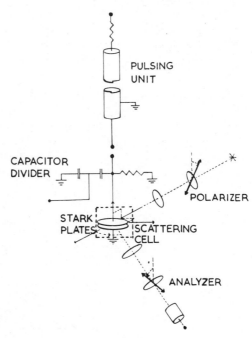

Figure 7. Schematic diagram of the pulsed electric field experiment of Sandle $et\ al.$[60]

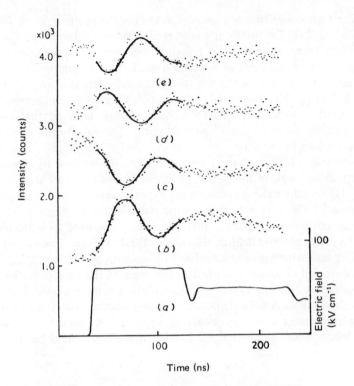

Figure 8. (a) Electric field pulse. Experimental data for ^{198}Hg. (b) Crossed analyzer. (c) Parallel analyzer. (d) LHC analyzer. (e) RHC analyzer. Curves are displaced to show relative phases.

detection of a photon. Typical experimental data are presented in Figure 8, which shows the fluctuations in the scattered fluorescent intensity as a function of the time after the field pulse is switched on.

Experimental values obtained were $\alpha_2(^{198}\text{Hg}) = +(1.58 \pm 0.02) \text{ kHz}$ $(\text{kV/cm})^{-2}$ and $\alpha_2(^{199}\text{Hg}) = +(1.01 \pm 0.03) \text{ kHz} (\text{kV/cm})^{-2}$. The ratio of the ^{199}Hg to ^{198}Hg polarizabilities of 0.64 ± 0.04 agrees with the theoretical ratio [see Eq. (50)] of 0.67 to within experimental error. The ^{198}Hg values

Table 2. Electric Dipole Polarizabilities α_0 of the Alkali Ground States (10^{-24} cm^3)

Li	Na	K	Rb	Cs	Reference
—	24.4 ± 1.7	45.2 ± 3.2	48.7 ± 3.4	63.3 ± 4.6	61
24.3 ± 0.5	23.6 ± 0.5	43.4 ± 0.9	47.3 ± 0.9	59.6 ± 1.2	62
24.74	22.33	42.97	45.49	61.19	63
24.6	23.8	43.2	48.2	61.0	64

Table 3. Polarizabilities (10^{-24} cm^3) of Metastable Levels of He ($^3S_1$) and Other Noble Gases ($^3P_2$)

Polarizability	He	Ne	Ar	Kr	Xe	Reference
α_0		27.8	47.9	50.7	63.6	62
		27.8	48.1	53.5	62.5	67^a
	46.767					66
α_2		−1.1	−3.2	−3.9	−6.1	62
	$(0.51 \pm 0.03) \times 10^{-3}$	−0.963	−2.95	−3.89	−6.03	65
		±0.04	±0.15	±0.21	±0.39	
	$(0.51 \pm 0.02) \times 10^{-3}$					68

^{a} Coulomb approximation method.

agree with recent measurements made by Khadjavi et al.[36] and Kaul and Latshaw.[54] It is worth noting that the tensor polarizability of the Hg 6^3P_1 state has now been measured by four different experimental techniques. Blamont[60a] made the first measurements in 1957 using the double-resonance method. He obtained a value of $\alpha_2 = (1.42 \pm 0.03)$ kHz (kV/cm)$^{-2}$, which is in disagreement with the more recent measurements.

3.4. Atomic-Beam Experiments

Several atomic-beam methods have been applied to Stark-effect measurements. An example for the electric analog to the Stern–Gerlach experiment is the determination of the alkali ground-state polarizabilities, by Hall and Zorn.[61]

A well-collimated beam of alkali atoms is deflected in an inhomogeneous electric field and the deflection is measured by a hot-wire detector. In their experiments Hall and Zorn had to use careful velocity selection since the calculation of the polarizability from the measured deflection depends on the knowledge of the velocity of the atoms.

Another atomic-beam method, the $E-H$ gradient balance method, is[62] independent of the atomic velocity. In this method congruent inhomogeneous electric and magnetic fields are applied across an atomic beam in a configuration where the ratio of field to field gradient is constant over the beam cross section. If the atoms in the beam have a polarizability α and a magnetic moment μ, electric and magnetic forces may be balanced so that

$$\alpha E \frac{dE}{dz} = \mu \frac{dH}{dz} \tag{58}$$

For known μ, E and H may be measured to obtain α. The results for the

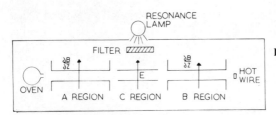

Figure 9. Schematic diagram of atomic-beam apparatus for studying Stark shifts associated with the D lines of the alkaline metals.

ground states of the alkalis are shown in Table 2 together with theoretical values of Sternheimer[63] (perturbation calculation) and of Adelman and Szabo[64] (Coulomb-like approximation).

The method has also been applied to the metastable $^3P_2$ states of the noble-gas atoms. The differential Stark splittings of these states and of He had been measured previously by Player and Sandars[65] by atomic-beam resonance measurements, observing the quadratic field dependence of the $M_J = 0 \rightarrow 1$ transitions. In Table 3 the results of Player and Sandars are listed together with the results obtained by the $E-H$ gradient balance method. The latter were normalized to the scalar polarizability of the $^3S_1$, $M_J = 1$ state of helium known from an accurate variational calculation,[66] thus avoiding systematic errors in the experimental determination of the fields.

Also listed is the result for helium of Ramsey and Petrasso,[68] who measured the quadratic shift between the $M_J = 1$ and $M_J = 0$ Zeeman levels. The very small shift, which should be zero in the LS coupling scheme, was $\Delta = (1.27 \pm 0.04)\, 10^{-6}\, \mathrm{Hz}/(\mathrm{V}/\mathrm{cm})^2$. The corresponding tensor polarizability of $\alpha_2 = (3.41 \pm 0.11)\, 10^{-3}\, a_0^3$ is in good agreement with a third-order calculation of α_2 by Sandars including the noncontact spin–spin interaction between the two electrons.

A further atomic-beam method has been applied by Marrus *et al.*[69] using a Rabi-type atomic-beam technique to measure the Stark effect in the D lines of ^{85}Rb and ^{133}Cs. A schematic diagram of the apparatus is shown in Figure 9. A modified atomic-beam apparatus was used with flop-in magnetic geometry. The C region was fitted with a pair of electric field plates and the gap between the plates was illuminated by a spectral lamp filled with the appropriate isotope. A spectral filter was used to isolate the D_1 and D_2 lines.

The basic method employed for measuring Stark shifts involves the comparison of shifts of atomic absorption lines in an electric field with the $S_{1/2}$ ground-state hfs. The excited-state hfs is about an order of magnitude smaller and is ignored in these experiments. The atomic-beam apparatus refocuses atoms that undergo the transition $M_J = +\frac{1}{2} \leftrightarrow M_J = -\frac{1}{2}$ in the C region. At zero electric field, atoms of the same isotopic species are contained in the lamp absorb radiation from the lamp. In the subsequent

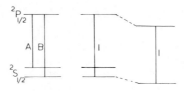

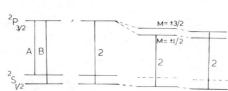

Figure 10. Schematic diagram of energy levels. The lamp emits the lines A and B. At zero electric field the absorption lines 1 and 2 overlap with the emission line B. Signals are observed at electric fields for which lines 1 and 2 coincide with line A.

decay, half the atoms undergo spin-flip and contribute to the flop-in signal. As the electric field is increased from zero, the ^2P and ^2S levels are displaced downwards, reducing the transition energy (see Figure 10). When the electric field is sufficiently large that the absorption lines are shifted by an amount equal to the ground-state hfs, a new overlap of absorption and emission lines occurs and another maximum in the flop-in signal is observed. One maximum results for the $^2P_{1/2}$ state, two for the $^2P_{3/2}$ state because the electric field lifts the degeneracy between the $M_J = \pm\frac{3}{2}$ and $M_J = \pm\frac{1}{2}$ sublevels. To obtain ^2P polarizabilities, the ground-state polarizability has to be known. The measurements of Salop et al.[70] were used in these experiments.

Table 4. Excited-State Polarizabilities (10^{-24} cm^3) Obtained by Marrus et al.[69,71]

Atom and level	$\alpha(n^2P_{1/2}, M_F = \pm1/2)$	$\alpha(n^2P_{3/2}, M_F = \pm1/2)$	$\alpha(n^2P_{3/2}, M_F = \pm3/2)$
^{23}Na($n = 3$)	51 ± 5	67 ± 8	38 ± 4
	51^a	63^a	39^a
^{39}K($n = 4$)	87 ± 13	114 ± 16	68 ± 10
	93^a	109^a	80^a
	94 ± 13^c	121 ± 16^c	75 ± 10^c
^{85}Rb($n = 5$)	112 ± 17	148 ± 23	102 ± 15
	116^a	151^a	108^a
	119 ± 17^c	155 ± 23^c	109 ± 15^c
^{133}Cs($n = 6$)	187 ± 29	273 ± 42	196 ± 30
	192^a	246^a	191^a
	187^b	273^b	200^b
	194 ± 29^c	280 ± 42^c	203 ± 30^c

^{a} Theoretical values based on the Coulomb approximation of Bates and Damgaard.[73]
^{b} Theoretical values of Stone.[74]
^{c} Values derived by Molof et al.[62] using their experimental result for the ground-state polarizability.

Table 5. Quadratic Stark Splittings of the Zeeman Substrates $(I+\frac{1}{2}, -I-\frac{1}{2})-(I+\frac{1}{2}, -I-\frac{1}{2})-(I+\frac{1}{2}, -I+\frac{1}{2})$ [$\times10^{-10}$ Hz (V cm^{-1})$^{-2}$] and $(I+\frac{1}{2}, 0)-(I-\frac{1}{2}, 0)$ [$\times10^{-6}$ Hz (V cm^{-1})$^{-2}$] of the Ground States of Alkali atoms

Transition	^{7}Li	^{23}Na	^{39}K	^{85}Rb	^{87}Rb	^{133}Cs	Reference
$(I+\frac{1}{2}, -I-\frac{1}{2})$ ↔ $(I+\frac{1}{2}, -I+\frac{1}{2})$		-11.2 ± 1.0	-4.8 ± 0.3	-22.8 ± 1.2	-104.4 ± 6.7	-137.2 ± 7.9	76
		$+5.2\pm1.8$	$+2.2\pm4.0$	-15.6 ± 2.0	-92.4 ± 10.9	-126.8 ± 8.7	76
		-9.7 ± 1.9	-5.0 ± 1.5				76
		-13	-7	-39	-150	-155	76
$(I+\frac{1}{2}, 0)$ ↔ $(I-\frac{1}{2}, 0)$	-0.061 ± 0.002	0.124 ± 0.004	-0.071 ± 0.002	-0.546 ± 0.012	-1.23 ± 0.03	-2.25 ± 0.05	77
	-0.078	-0.152	-0.099	-0.737	-1.66	-3.78	77
						-2.248	78

Marrus and Yellin[71] have modified the technique to measure the Stark effect in the D lines ^{39}K. Because the ground-state hfs of ^{39}K is smaller than the Doppler width in a ^{39}K spectral lamp, a ^{39}K atomic-beam absorption filter was used to obtain narrow absorption lines in the broad lamp emission. At zero electric field, when the absorption lines of atoms in the atomic-beam overlap with the absorption lines of atoms in the absorption filter, a minimum in the flop-in signal was observed. When the shift of the absorption lines by the electric field caused further overlaps, additional minima in the flop-in signal were observed.

The same atomic-beam technique has been used by Duong and Picque[72] to measure the polarizabilities of the $3^2P_{1/2}$ and $3^2P_{3/2}$ levels of Na.

Table 4 shows experimental polarizabilities obtained in these experiments using the ground-state polarizabilities of Salop[70] and those of Molof.[62] Theoretical calculation of polarizabilities are also shown.

The quadratic Stark splittings between the Zeeman substates ($F = I + \frac{1}{2}$, $M_F = -I + \frac{1}{2}$) and ($I + \frac{1}{2}, -I - \frac{1}{2}$) in the ground states of alkali atoms have been measured by Carrico et al.[75] and Stein et al.[76] using an atomic-beam technique. Measurements on the Stark splittings of the $(I + \frac{1}{2}, 0) \leftrightarrow (I - \frac{1}{2}, 0)$ substates have been measured by Mowatt.[77]

The experiments were performed on an atomic-beam magnetic resonance apparatus. In the C region, the atomic beam passes between electric field plates which are placed between the rf loops of a Ramsey double-hairpin structure. The transition between the substates is induced by tuning the rf loops. Large ac (frequency ω) and dc voltages were applied to the field plates. The resulting detector signal contained modulations at ω (dc $\neq 0$) and 2ω (dc = 0), which were recorded using phase-sensitive detection. The results obtained from these experiments are shown in Table 5. In general, the experimental shifts are smaller in magnitude than the theoretical shifts except in the case of ^{133}Cs, where the agreement between them is excellent for the $(I = \frac{1}{2}, 0 \leftrightarrow (I = \frac{1}{2}, 0)$ transition.

Experiments using fast-beam methods to measure the Stark effect and the coherent excitation of states with different orbital quantum numbers are discussed in the article by Andrä, Chapter 20 of this work.

3.5. Maser and Laser Stark-Effect Experiments

Gibbons and Ramsey[79] have measured the changes in the hyperfine separation of the ground state of atomic hydrogen in an electric field. A hydrogen maser, which operates on the ground-state hyperfine transition, was modified by the addition of a second storage bulb placed in an electric field. During the radiative lifetime of the atoms, a few tenths of a second, some atoms passed from the maser cavity into the electric field region and back again. These atoms radiated at a frequency that was determined by their average hyperfine separation during their lifetime. The relationship between a change in the hyperfine separations of atoms in the electric field region and the corresponding change in the maser output frequency was determined experimentally by applying magnetic fields of different strengths to the second storage bulb. The change in the maser output frequency was measured as a function of the Zeeman shifts in the second storage bulb, providing a calibration of Stark-induced shifts.

The change in the $\Delta M_F = 0$ transition frequency was $\delta_\nu(\text{Stark}) = -\beta E^2$; $\beta = (7.9 \pm 1.2) \times 10^{-4} \text{ Hz} (\text{kV cm}^{-1})^{-2}$. This result is in good agreement with a previous measurement[80] and with theory.

Lasers have been used in Stark-effect investigations applying their outstanding properties of high intensity for optical excitation, the high resolution for selective population and for accurate measurement of Stark shifts, and the short pulse duration for quantum beat experiments.

The Stark effect in highly excited states of the alkali has been recently studied by several investigators using stepwise excitation techniques. Hogervorst and Svanberg[81] and Belin et al.[82] have measured differential Stark shifts in the D states of ^{39}K, ^{85}Rb, and ^{133}Cs using level-crossing spectroscopy. An atomic beam was excited in a two-step process. A powerful rf lamp was used for the first step from the ground state to the first excited P state. In the second step, intense radiation from the laser excited atoms to the D state. Level-crossing experiments using parallel electric and magnetic fields were carried out on the 6, 7, and $8\,^2D_{3/2,5/2}$ levels of ^{85}Rb and the 8, 9, and $10\,^2D_{3/2,5/2}$ levels of ^{133}Cs. Even though the laser radiation was not the "white" light source required for the strict validity of the Breit–Franken formula, the presence of radiation trapping in the D lines, the small D level hyperfine splittings, and the laser frequency drifts caused the excitation to appear "white." Another requirement for the strict validity of the Breit–Franken formula is that the sublevels of the P state be incoherently and equally populated. Excitation by an unpolarized rf lamp and radiation trapping in the D lines ensured that this requirement was satisfied.

In this experiment coherent excitation of magnetic sublevels of the intermediate P state is produced by the σ component of the unpolarized rf lamp. This coherence can be transferred to the D states at the zero-field crossing and contributes to the level-crossing signal. However, because the radiation width of the D state is generally much smaller than the lower-lying P state, the signal structure is dominated by the D-state contribution. Also the experimental geometry was chosen to separate the coherence transfer and the level-crossing contribution to the signal.

Resolved level-crossing signals were observed in the $^2D_{3/2}$ levels and in the high-field measurements ($\sim$3 kV/cm) on the $^2D_{5/2}$ levels. Low-field measurements were carried out to determine the hyperfine-structure magnetic dipole interaction constant A. In the low-field experiments on the $^2D_{5/2}$ levels, the level crossings were poorly resolved and the signal was analyzed using computer fits of the Breit–Franken formula. Experimental values for K, Rb, and Cs are shown in Table 6. Note the good agreement between experimental and theoretical values of the tensor polarizability. The theoretical results were obtained using the Bates and Damgaard Coulomb approximation.

Fredriksson and Svanberg[83] have used stepwise excitation with an rf lamp and a single-mode dye laser to measure the scalar Stark shifts for highly excited S and D states of ^{133}Cs. In these experiments, the energy levels were scanned by the Stark effect until excitation occurred for a set laser frequency. Experimental and theoretical values for the scalar polarizability of the $9^2D_{5/2}$, $10^2S_{1/2}$, and $11^2S_{1/2}$ levels are included in Table 6, which also shows the results of the most recent experiments by Belin et al.[84]

Table 6. Highly-Excited-State Scalar and Tensor Polarizabilities [MHz $(kV/cm)^{-2}$] for K, Rb, and Cs

Element	Level	α_0 (expt.)	α_0 (theory)	$|\alpha_2$ (expt.)$|$	α_2 (theory)
^{39}K	$5^2D_{3/2}$			9.6 ± 0.5	-9.6
	$5^2D_{5/2}$			13.4 ± 0.7	-13.5
	$6^2D_{3/2}$			33.8 ± 1.7	-32.5
	$6^2D_{5/2}$			47.6 ± 2.4	-45.3
^{85}Rb	$6^2D_{3/2}$			0.105 ± 0.007	-0.150
	$6^2D_{5/2}$			0.94 ± 0.05	0.845
	$7^2D_{2/2}$			5.0 ± 0.3	4.6
	7^2D_{52}			11.9 ± 0.6	11.1
	$8^2D_{3/2}$			27.0 ± 1.4	27.6
	$8^2D_{5/2}$			56.9 ± 3.0	52.4
	$9^2D_{5/2}$			180.3 ± 9.0	
^{133}Cs	$8^2D_{3/2}$			82.5 ± 4.0	83.9
	$8^2D_{5/2}$			182.0 ± 10.0	168
	$9^2D_{3/2}$			313.0 ± 15.0	295
	$9^2D_{5/2}$	-509.0 ± 25.0	-440	660.0 ± 35.0	592
	$10^2D_{3/2}$			840.0 ± 40.0	848
	$10^2D_{5/2}$			1770.0 ± 90.0	1705
	$10^2S_{1/2}$	123.0 ± 6.0	118		
	$11^2S_{1/2}$	322.0 ± 16.0	300		
	$13^2D_{5/2}$			19 ± 1^a	20^a
	$14^2D_{5/2}$			37 ± 2^a	38^a
	$15^2D_{5/2}$			70 ± 4^a	70^a
	$16^2D_{5/2}$			120 ± 6^a	124^a
	$17^2D_{5/2}$			199 ± 10^a	213^a
	$18^2D_{5/2}$			323 ± 16^a	348^a

^{a} GHz/$(kV/cm)^{-2}$.

The scalar polarizabilities of even more highly excited states of cesium-$40 \leq n \leq 60$ have been measured by van Raan et al.[85] High-lying P states were produced by photoexcitation in atomic beam using the frequency-doubled light of a dye laser. The subsequent deflection of the beam in an inhomogeneous electric field, averaged over the unresolved Stark states, is a measure for the scalar polarizability α_0.

From Eq. (45) the n dependence of the polarizability for hydrogenic systems can be estimated considering that the reduced dipole matrix elements scale according to n^2 while the fine-structure splitting is proportional to n^{-3}, so that $\alpha \sim n^7$. The results by van Raan et al. between $n = 40$ and $n = 60$ are consistent with the n^7 behavior.

A discussion of the Stark effect in highly excited states of sodium may be found in the article by Kleppner, Chapter 16 of this book.

The high-resolution capability of lasers has also been utilized in the measurement of the polarizabilities of the $5s\,^2S_{1/2}$ and $4d\,^2D_{3/2,5/2}$ states of Na by Harvey et al.[86] The states were populated by two-photon excitation of counterpropagating laser beams to avoid Doppler broadening, and the uv fluorescence of the excited states was monitored. In Figure 11 the fluorescence intensity is compared for zero electric field and for an electric field of 2.5 kV/cm, showing the shift and splitting by the Stark effect. The results are given in Table 7 in comparison with calculations

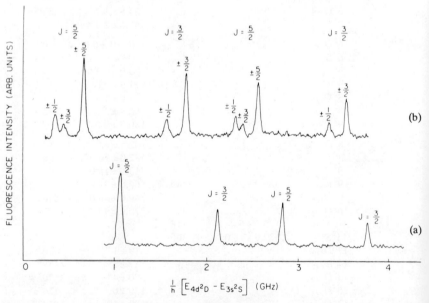

Figure 11. Stark effect of the $3s\,^2S$–$4d\,^2D$ transition of sodium. The two-photon fluorescence is shown for (a) $E = 0$ and (b) $E = 2.5$ kV/cm.[86]

Table 7. Polarizabilities of Sodium by Harvey et al.[86]

	State		
Polarizability	$4^2D_{3/2}$	$4^2D_{5/2}$	$5^2S_{1/2}$
α_0			
experiment	155.3 ± 1.7	156.1 ± 1.3	5.2 ± 0.3
theory	152.02	151.97	5.81
α_0			
experiment	-38.5 ± 0.7	-53.2 ± 0.5	
theory	-35.64	-51.06	

according to the Coulomb approximation, which are in close agreement.

The short duration of the light output from a pulsed laser can be utilized for excitation of quantum beats. Stark-effect measurements based on this method were done by Fabre and Haroche,[87] who observed Stark shifts in the fine-structure beat frequencies for highly excited nD states of sodium ($n = 10, 11, 12$).

Similarly Hese et al.[88] have remeasured the tensor polarizability of the $6s\,6p\ ^1P_1$ of barium by observing quantum beats. The advantage of this method lies in the direct measurement of the Stark splitting without comparison with the Zeeman effect, although in this particular case the accuracy of the result $\alpha_2 = (-10.79 \pm 0.29)\,\text{kHz}/(\text{kV}/\text{cm})^2$ is not yet quite as good as that of the level-crossing experiment.[56]

3.6. Summary

The Stark-effect investigations reviewed in the previous sections are listed in Table 8, ordered by element and state. In the table reference is made to the section where the investigation is mentioned, which is equivalent to a listing of the method applied in the experiment.

As is obvious from the number of recent publications quoted in the preceding sections, the modern experimental and theoretical methods allow one to tackle the Stark-effect problem more and more successfully. This is true even for more complicated atomic systems, where the Stark-effect investigation provides valuable additional information.

In many cases the investigations may be done as before by comparing the Stark-effect to the Zeeman-effect, to hyperfine- or fine-structure splittings; in other cases the Stark-effect may be measured directly. It can be hoped that the quantum beat methods will be applicable to atomic or molecular systems where no other data are available for comparison.

Table 8. Listing of Elements and States for Which Stark-Effect Experiments Are Reviewed in This Chapter

Element	State/line	Chapter section	Reference
H	Ground	3.5	79, 80
	$n = 2$	3.1.2	51
	H_β, H_γ	3	42–44
He	2^3S_1	3.4	65, 66, 68
	2^1P_1	3.1.1	49, 50
	$n^{1,3}D, n = 5{-}10$	3.2	59
He$^+$	$n = 4$	3.2	58
Li	Ground	3.4	62–64, 77
Ne	m^3P_2	3.4	62, 65, 67
Na	Ground	3.4	61–64, 70, 75–77
	$3^2P_{1/2,3/2}$	3.4	69, 71–73
	$4^2D_{3/2,5/2}$	3.5	86
	$5^2S_{1/2}$	3.5	86
	$nD, n = 10, 11, 12$	3.5	87
Ar	m^3P_2	3.4	62, 65, 67
K	Ground	3.4	61–64, 70, 75–77
	D line	3.4	71
	$4^2P_{1/2,3/2}$	3.1; 3.4	48, 69, 71, 73
	$n^2D_{3/2,5/2}, n = 5, 6$	3.5	81, 82
Ca	4^1P_1	3.1.2	56
Kr	m^3P_2	3.4	62, 65, 67
Rb	Ground	3.4	61–64, 70, 75–77
	D lines	3.4	69
	$5^2P_{1/2,3/2}$	3.1.1; 3.4	36, 69, 71, 73
	$n^2D_{3/2,5/2}, n = 6{-}9$	3.5	81–84
Sr	5^1P_1	3.1.2	56
Cd	5^3P_1	3.1.2	36, 53
Xe	m^3P_2	3.4	62, 65, 67
Cs	Ground	3.4	61–64, 70, 75–77, 78
	D lines	3.4	69
	$6^2P_{1/2,3/2}$	3.1.1; 3.4	36, 69, 71, 73, 74
	$n^2P, 40 \leqslant n \leqslant 60$	3.5	85
	$n^2D_{3/2}, n = 8{-}10$	3.5	81–84
	$n^2D_{5/2}, n = 8{-}10, 13{-}18$	3.5	81–84
	$n^2S_{1/2}, n = 10, 11$	3.5	81–84
Ba	6^1P_1	3.1.2; 3.5	56, 88
	6^3P_1	3.1.2	56
Hg	6^3P_1	3.1.2; 3.3	36, 54, 60, 60a
	$6^{1,3}D_J$	3.1.2	55

The recently developed very accurate methods for detection of Stark shifts will certainly find applications in plasma diagnostics. The construction of an instrument capable of measuring small electric fields should be possible on the basis of the high-resolution laser experiment[89] where a field of 10 V/cm in a Wood discharge was detected or using very highly excited states in a beat experiment[87] for even smaller electric fields.

In summary we can say that with the increase of work being done in the field of the Stark-effect considerable progress can be expected for the near future.

References

1. W. Voigt, *Ann. Phys.* (Leipzig) **69**, 297 (1899); **4**, 197 (1901).
2. J. Stark, *Sitzungsber. Akad. Wiss. Berlin* **47**, 932 (1913).
3. A. Lo Surdo, *Atti R. Accad. Naz. Lincei* **22**, Part 2, 664 (1913).
4. E. U. Condon and G. H. Shortley, *The Theory of Atomic Spectra*, Cambridge University Press, Cambridge (1963).
5. A. M. Yancharina and I. I. Muravev, *Opt. Spectrosc.* **39**, 127 (1975).
6. B. Bouchiat, *J. Phys. (Paris)* **37**, L79 (1976).
7. H. J. Werner and W. Meyer, *Phys. Rev. A* **13**, 13 (1976).
8. J. A. Blackman and G. W. Series, *J. Phys. B* **6**, 1090 (1973).
9. T. W. Shyn, W. L. Williams, R. T. Robiscoe, and T. Rebane, *Phys. Rev. Lett.* **22**, 1273 (1969).
10. A. M. Bonch-Bruevich and V. A. Khodovoi, *Sov. Phys. Usp.* **10**, 637 (1967).
11. A. D. Buckingham, in *Physical Chemistry Series I*, Vol. 3, Ed. D. A. Ramsey, Medical and Technical Press, Lancaster, England (1972).
11a. N. Ryde, *Atoms and Molecules in Electric Fields*, Almquist and Wiksell, Stockholm (1976).
12. L. I. Schiff, *Quantum Mechanics*, McGraw-Hill, New York (1968).
13. E. Schrödinger, *Ann. Phys. (Leipzig)* **80**, 437, 457 (1926).
14. J. W. B. Hughes, *Proc. Phys. Soc. London* **91**, 810 (1967).
15. H. A. Bethe and E. E. Salpeter, *Quantum Mechanics of One- and Two-Electron Atoms*, Springer Verlag, Berlin (1957),
16. W. Tsai, *Phys. Rev. A* **9**, 1081 (1974).
17. C. Lanczos, *Z. Phys.* **68**, 204 (1931).
18. H. Rausch von Traubenberg, *Z. Phys.* **71**, 291 (1931).
19. R. J. Damburg and V. V. Kolosov, *J. Phys. B* **9**, 3149 (1976).
20. A. C. Riviere, in *Methods of Nuclear Physics*, Eds. B. Bederson and W. L. Fite, Academic Press, New York (1968).
21. J. E. Bayfield, G. A. Khayrallah, and P. M. Koch, *Phys. Rev. A* **9**, 209 (1974).
22. M. G. Littmann, M. L. Zimmermann, T. W. Ducas, R. R. Freeman, and D. Kleppner, *Phys. Rev. Lett.* **36**, 788 (1976).
22a. C. Fabre, P. Goy, and S. Haroche, *J. Phys. B* **10** L183 (1977).
23. T. W. Ducas, M. G. Littman, R. R. Freeman, and D. Kleppner, *Phys. Rev. Lett.* **35**, 366 (1975).
24. M. G. Littman, M. L. Zimmermann, and D. Kleppner, *Phys. Rev. Lett.* **37**, 486 (1976).
25. D. S. Bailey, J. R. Hiskes, and A. C. Riviere, *Nucl. Fusion* **5**, 41 (1965).
26. R. F. Post, T. K. Fowler, J. Killeen, and A. A. Mirin, *Phys. Rev. Lett.* **31**, 280 (1973).

27. J. Fauchier and J. D. Dow, *Phys. Rev. A* **9**, 98 (1974).
28. J. E. Cordle, *J. Phys. B* **7**, 1284 (1974).
29. G. Lüders, *Ann. Phys. (Leipzig)* (6) **8**, 301 (1951); *Z. Naturforsch.* **5a**, 608 (1950).
30. R. G. Kulkarni, N. V. V. J., Swamy, and E. Chaffin, *Phys. Rev. A* **7**, 27 (1973).
31. D. O'Connell, K. J. Kollath, A. J. Duncan, and H. Kleinpoppen, *J. Phys. B* **8**, L214 (1975).
31a. H. Krüger and A. Oed, *Phys. Lett.* **54A**, 251 (1975).
32. P. H. Heckmann, *Z. Phys.* **250**, 42 (1972).
33. H. J. Andrä, *Phys. Rev. A* **2**, 2200 (1971).
34. G. W. F. Drake, P. S. Farago, and A. van Wijngaarden, *Phys. Rev. A* **11**, 1621 (1975).
35. C. E. Johnson, *Phys. Rev. A* **7**, 872 (1973).
36. A. Khadjavi, A. Lurio, and W. Happer, *Phys. Rev.* **167**, 128 (1968).
37. E. U. Condon, *Phys. Rev.* **43**, 648 (1933).
38. J. R. P. Angel and P. G. H. Sandars, *Proc. R. Soc. London A* **305**, 125 (1968).
39. R. W. Schmieder, *Am. J. Phys.* **40**, 297 (1972).
40. W. Happer and E. B. Saloman, *Phys. Rev.* **160**, 23 (1967).
41. A. Messiah, *Quantum Mechanics*, North Holland, Amsterdam (1969).
42. R. Gebauer and H. Selhofer, *Acta Phys. Austriaca* **31**, 8 (1970).
43. W. Steubing and W. Junge, *Ann. Phys. (Leipzig)* (6) **5**, 108 (1949).
44. H. Rother, *Ann. Phys. (Leipzig)* (6) **17**, 185 (1956).
45. W. Hanle, *Z. Phys.* **30**, 93 (1924).
46. P. A. Franken, *Phys. Rev.* **121**, 508 (1961).
47. A. Khadjavi, W. Happer, and A. Lurio, *Phys. Rev. Lett.* **17**, 463 (1966).
48. R. W. Schmieder, A. Lurio, and W. Happer, *Phys. Rev. A* **3**, 1209 (1971).
49. N. D. Bhaskar and A. Lurio, *Phys. Rev.* **10**, 1685 (1974).
50. C. Deutsch, H. W. Drawin, and L. Herman, *Phys. Rev. A* **3**, 1879 (1971).
51. K. J. Kollath and H. Kleinpoppen, *Phys. Rev. A* **10**, 1519 (1974).
52. H. Wieder and T. G. Eck, *Phys. Rev.* **153**, 103 (1967).
53. E. B. Aleksandrov and V. V. Khromov, *Opt. Spectrosc.* **18**, 313 (1965).
54. R. D. Kaul and W. S. Latshaw, *J. Opt. Soc. Am.* **62**, 615 (1972).
55. M. Chantepie, *J. Phys. Lett. (Paris)* **35**, L173 (1974).
56. A. Kreutzträger and G. von Oppen, *Z. Phys.* **265**, 421 (1973); A. Kreutzträger, G. von Oppen, and W. Wefel, *Phys. Lett.* **49A**, 241 (1974).
57. M. Glass-Maujean and J. P. Descoubes, *Opt. Commun.* **4**, 345 (1972).
58. H. J. Beyer, H. Kleinpoppen, and J. M. Woolsey, *J. Phys. B* **6**, 1849 (1973).
59. H. J. Beyer and K. J. Kollath, *J. Phys. B* **10**, L5 (1977).
60. W. J. Sandle, M. C. Standage, and D. M. Warrington, *J. Phys. B* **8**, 1293 (1975).
60a. J. E. Blamont, *Ann. Phys. (Paris)* **2**, 35 (1957).
61. W. D. Hall and J. C. Zorn, *Phys. Rev. A.* **10**, 1141 (1974).
62. R. W. Molof, H. L. Schwartz, T. M. Miller, and B. Bederson, *Phys. Rev.* **10**, 1131 (1974).
63. R. M. Sternheimer, *Phys. Rev.* **183**, 112 (1969).
64. S. A. Adelmann and A. Szabo, *Phys. Rev. Lett.* **28**, 1427 (1972).
65. M. A. Player and P. G. H. Sandars, *Phys. Lett.* **30A**, 475 (1969).
66. K. T. Chung and R. P. Hurst, *Phys. Rev.* **152**, 35 (1966).
67. E. J. Robinson, J. Levine, and B. Bederson, *Phys. Rev.* **146**, 95 (1966).
68. A. T. Ramsey and R. Petrasso, *Phys. Rev. Lett.* **23**, 1478 (1969).
69. R. Marrus and D. McColm, *Phys. Rev. Lett.* **15**, 813 (1965); R. Marrus, D. McColm, and J. Yellin, *Phys. Rev.* **147**, 55 (1966); R. Marrus, E. Wang, and J. Yellin, *Phys. Rev. Lett.* **19**, 1 (1967).
70. A. Salop, E. Pollack, and B. Bederson, *Phys. Rev.* **124**, 1431 (1961).
71. R. Marrus and J. Yellin, *Phys. Rev.* **177**, 127 (1969).
72. H. T. Duong and J. L. Picque, *J. Phys. (Paris)* **33**, 513 (1972).

73. D. R. Bates and A. Damgaard, *Phil. Trans. R. Soc. London A* **242**, 101 (1949).
74. P. M. Stone, *Phys. Rev.* **127**, 1151 (1962).
75. J. P. Carrico, A. Adler, M. R. Baker, S. Legowski, E. Lipworth, P. G. H. Sandars, T. S. Stein, and M. C. Weisskopf, *Phys. Rev.* **170**, 64 (1968).
76. T. S. Stein, J. P. Carrico, E. Lipworth, and M. C. Weisskopf, *Phys. Rev.* **2**, 1093 (1970).
77. J. R. Mowatt, *Phys. Rev.* **5**, 1059 (1972).
78. J. D. Feichtner, M. E. Hoover, and M. Mizushima, *Phys. Rev.* **137**, 702 (1965).
79. P. C. Gibbons and N. F. Ramsey, *Phys. Rev. A* **5**, 73 (1972).
80. E. N. Fortson, D. Kleppner, and N. F. Ramsey, *Phys. Rev. Lett.* **13**, 22 (1964).
81. W. Hogervorst, and S. Svanberg, *Phys. Scr.* **12**, 67 (1975).
82. G. Belin, L. Holmgren, I. Lindgren, and S. Svanberg, *Phys. Scr.* **12**, 287 (1975).
83. K. Fredriksson and S. Svanberg, *Phys. Lett.* **53A**, 461 (1975).
84. G. Belin, L. Holmgren, and S. Svanberg, *Phys. Scr.* **13**, 315 (1976).
85. A. F. J. van Raan, G. Baum, and W. Raith, *J. Phys. B* **9**, L349 (1976).
86. K. C. Harvey, R. T. Hawkins, G. Meisel, and A. L. Schawlow, *Phys. Lett.* **34**, 1073 (1975).
87. C. Fabre and S. Haroche, *Opt. Commun.* **15**, 254 (1975).
88. A. Hese, A. Renn, and H. S. Schweda, to be published.
89. C. Wiemann and T. W. Hänsch, *Phys. Rev. Lett.* **36**, 1170 (1976).

22

Stored Ion Spectroscopy

Hans A. Schuessler

1. Introduction

Since the development of the three-dimensional quadrupole trap[1,2] and its introduction into atomic physics in the late 1950s, most applications have relied on the charge-to-mass selective storage feature of this device. It is therefore being used as a mass spectrometer[3] but, in spectroscopy, also as an ideal tool to suspend charged particles, such as ions, ion molecules, and electrons, in ultrahigh vacuum for long periods of time. This last feature is the subject matter of this chapter. The long storage time, combined with the almost complete isolation from environmental perturbations, leads to the high intrinsic accuracy and precision inherent in the spectroscopy of stored ions. The state of the art of three-dimensional quadrupole ion traps and of their application to high-resolution spectroscopy will be described. This field was pioneered by H. G. Dehmelt, who also reviewed[4] the subject in 1967 and 1969. Several new developments have evolved since then and will be discussed here.

Section 2 gives an overview of the operating techniques of quadrupole ion traps. Section 3 describes the use of quadrupole ion traps in spectroscopy. Four techniques are illustrated, the ion storage exchange collision method, state selection by selective quenching, optical pumping and laser spectroscopy of stored ions, and alignment of ions by state selective photodissociation. Section 4 describes the precision obtained so far and the limitations of the techniques. Recent experiments to reduce the main sources which broaden and shift the magnetic resonance signals of stored ions, namely, spin exchange collisions, charge transfer collisions, and the second-order Doppler effect, are discussed.

HANS A. SCHUESSLER • Department of Physics, Texas A&M University, College Station, Texas 77843.

Section 5 deals with some of the results of precision measurements obtained so far.

2. The Containment of Isolated Ions

2.1. The Trap Geometry

In order to confine a particle to a finite volume, a three-dimensional potential well must be generated which a particle of sufficiently low energy cannot leave. When trapped in a harmonic well, the oscillation frequency of the particle is a simple harmonic motion of frequency ω. It is possible to generate a good approximation of such a harmonic potential well with the radio-frequency (rf) quadrupole arrangement to be described here and to achieve in this way three-dimensional electrodynamic containment by strong focusing.

An efficient focusing system requires a purely radial field. The restoring force acting on a particle must be proportional to the distance from the center. In a rf quadrupole, an average focusing force is produced that is proportional to the distance from the center.

Figure 1 shows the main electrodes of a quadrupole trap. Such a storage device consists basically of three electrodes: The central ring electrode, which is a hyperboloid of revolution symmetric about the z axis, and the two end cap electrodes, which are complementary hyperboloids. Figure 2a shows a completely assembled ion trap and Figure 2b a disassembled one. Slots are cut into the central ring electrode to focus a laser or atomic beam into the trapping region for spectroscopy. An electron gun is located on each side of the two end cap electrodes. The electrons enter the trapping region through an array of holes a tenth of a millimeter in diameter and produce ions from the background gas by electron impact. Microwave energy is fed to the trap through a waveguide to induce magnetic resonance transitions.

The trap electrodes are placed to within one thousandth of an inch of their correct position. This is done by means of precision ground glass spacers. The surface of the molybdenum electrodes is highly polished to minimize the effect of local inhomogeneities on the trapping field. For many applications, where trapping times of a few seconds are sufficient, the mechanical tolerance requirements of the assembly are less stringent than stated and some traps have even been built with cylindrical ring electrodes.[5]

A discussion of the storage properties of cylindrical traps[6] was carried out, but there is only a limited amount of experimental data available for comparison with the well-known features of standard quadrupole traps.

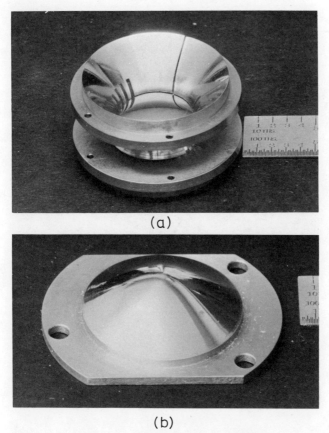

(a)

(b)

Figure 1. Electrodes of a rf quadrupole trap. (a) Central ring electrode. (b) One of the two end cap electrodes. The complementary hyperboloids of revolution are made of molybdenum.

Even though the potential function of the cylindrical trap has a saddle point at the origin and the deviations from the ideal field geometry are therefore generally not evident near the center, they will become noticeable when resonance effects occur. For such effects, the action of a weak disturbance accumulates in time and ion orbits become quickly unstable at otherwise stable operating parameters.

The fields required for electrodynamical containment of ions in an rf quadrupole trap are produced by applying suitable rf and dc potentials to the ring and the two end cap electrodes. This is shown in Figure 3 for a symmetrical ion trap characterized by

$$r_0^2 = 2z_0^2$$

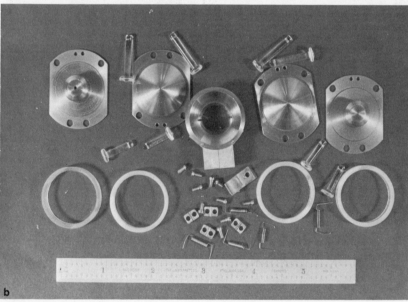

Figure 2. Complete ion trap. (a) Assembled trap. A glass-filled waveguide is butted against the central ring electrode to induce magnetic dipole transitions on stored ions. The electrodes are separated by glass spacers. The two outermost electrodes are focusing electrodes for the electron gun located on each side of the end cap electrodes. (b) Disassembled trap showing the various components.

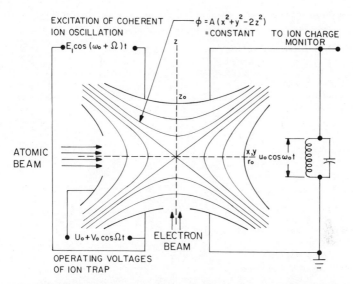

Figure 3. Cross section of the quadrupole ion trap showing the potential distribution of the trapping voltage $U_0 - V_0 \cos \Omega t$. The ion oscillation is excited by a homogeneous rf field at $\Omega + \omega_0$. The voltage $u_0 \cos \omega_0 t$ induced into the tank circuit by the cooperative ion motion is detected as an ion number signal. The atomic beam and electron beam windows are indicated.

Here r_0 is the minimum radius of the ring electrode and $2z_0$ is the minimum separation of the end cap electrodes. Typical experimental parameters for the storage and manipulation of, for instance, $^3\text{He}^+$ ions are listed in Table 1. The average well depths generated are only a fraction of the applied voltages, but well depths of a few electron volts are sufficient to store even relatively hot ions. In a rf quadrupole trap, storage times are long compared to the coherence times of most spectroscopic experiments. Measured storage times range from many seconds for light ions to hours for heavy ions.[7]

2.2. Description of the Ion Motion

The theory of the ion motion in a three-dimensional rf quadrupole trap defined by

$$(z/z_0)^2 - (r/r_0)^2 = \pm 1*$$

has been treated in detail.[1,8] Only the most important relations are summarized here. When driving voltages of amount $U = U_0 - V_0 \cos \Omega t$ are

---

* The plus sign describes the surface of the two end cap electrodes and the minus sign gives the form of the ring electrode.

Table 1. Typical Experimental Parameters for Storage of ^{3}He$^+$ Ions[14]

Axial dimension	$z_0 = 25.5$ mm
Radial dimension	$r_0 = 36.0$ mm
dc bias	$U_0 = 0$ V
ac trapping amplitude	$V_0 = 137$ V peak to zero
Frequency of trapping rf	$\Omega = 2\pi \times 1$ Mhz
Detection frequency	$\omega_0 = 2\pi \times 134$ kHz
Axial oscillation frequency	$\omega_z = 2\pi \times 123$ kHz
Axial well depth	$D_z = 6.1$ V
Radial well depth	$D_r = 3.05$ V
Stored charge	$q = 8 \times 10^6 e$
Electron current	$i_i = 10$ mA
Electron acceleration voltage	$U_e = 500$ V
Electron pulse duration	$t_e = 80$ msec
Coherent excitation amplitude	$V_{\Omega + \omega} = 3$ V
Duration of excitation pulse	$T_{\Omega + \omega} = 40$ msec
Excitation frequency	$\omega_0 + \Omega = 2\pi \times 1134$ kHz
Detection sweep amplitude	$U_s = 0$ to -3.6 V
Total detection sweep duration	$t_s = 32$ msec
Duration of traversal through ω_z resonance	$t_m = 4$ msec
Capacity between end caps and ring electrode	$C = 34$ pF
Quality factor of detection circuit	$Q_t = 50$
Quality factor of stored ions	$Q_i = 100$
Ion storage time without Cs beam	$Ti = 25$ sec
Ion storage time with Cs beam	$T_0 = 0.4$ sec
^{3}He density	$N_{He} = 2 \times 10^8$ atoms/cm^3
Cs density at site of trap	$N_{Cs} = 1 \times 10^8$ atoms/cm^3
Ultimate vacuum	$p = 1 \times 10^{-10}$ Torr
Ratio of ion-number signal to rms fluctuations of ion-number signal in presence of Cs beam at end of interaction interval (800 msec) comparing consecutive signals	$S/N_f = 400:1$
Ratio of ion-number signal to rms fluctuations without Cs beam	$S/N_{f0} = 2000:1$
Ratio of ion-number signal to rms thermal noise	$S/N_t = 8000:1$

applied to the trap electrodes, a rotationally symmetric potential

$$\phi(r, z) = (U/4z_0^2)(r^2 - 2z^2)$$

is generated. The field at any point inside the structure is given by

$$E_z = (U/z_0^2)z \quad \text{and} \quad E_r = -(U/2z_0^2)r$$

The equations of motion for particles of a charge-to-mass ratio e/M are then

$$d^2z/dt^2 - (e/M)(U_0/z_0^2)z + (e/M)(V_0/z_0^2)z \cos \Omega t = 0$$

$$d^2r/dt^2 + (e/M)(U_0/2z_0^2)r - (e/M)(V_0/2z_0^2)r \cos \Omega t = 0$$

These Mathieu-type equations can be written in canonical form as

$$d^2 u_i / d\xi^2 + (a_i - 2q_{i,} \cos 2\xi) u_i = 0, \qquad i = r, z$$

where

$$\xi = \Omega t/2, \qquad a_z = -2a_r = -4eU_0 / Mz_0^2 \Omega^2, \qquad q_z = -2q_r = -2eV_0 / Mz_0^2 \Omega^2$$

The solutions of the Mathieu equations are either stable or unstable depending on the values of a_i and q_i. In an experiment, U_0, V_0, and Ω have to be suitably chosen so that a_i and q_i lie inside the stability diagram for both the r and z directions simultaneously. The solution for stable ion motion can be written as a power series

$$U_i(\xi) = A \sum_{n=-\infty}^{+\infty} C_{2n} \cos (2n + \beta_i)\xi + B \sum_{n=-\infty}^{+\infty} C_{2n} \sin (2n + \beta_i)\xi$$

Here A and B are constants of integration that depend on the initial conditions $u_i(0)$ and $\dot{u}_i(0)$ when the ion is formed inside the trap. The expansion coefficients C_{2n} and β_i are functions of a_i and q_i only. According to the equation for stable ion motion, the frequencies are

$$\omega_n = (n \pm \beta/2)\Omega$$

The fundamental frequencies

$$\omega_{r0} = \beta_r \Omega/2 \quad \text{and} \quad \omega_{z0} = \beta_z \Omega/2$$

are dominant for a_i, $q_i \ll 1$.

In addition to having a stable orbit, an ion must also avoid hitting the electrodes. It has been shown[2] that, for $a_z = 0$ and $\beta_z = 0.5$, this is the case for half of all ions which are formed evenly distributed over radius and time in the $z = 0$ plane.

Ideally the maximum energy eD that an ion can have is determined by the trap dimensions r_0 and z_0. Neglecting higher-order terms it holds that

$$eD = (M\omega_{r0}^2 r_0^2 + M\omega_{z0}^2 z_0^2)/2$$

which yields

$$eD = (m\Omega^2/8)(\beta_r^2 r_0^2 + \beta_z^2 z_0^2)$$

However, even in ultrahigh vacuum, collisions quickly equalize the ion temperature in the r and z directions, and the maximum ion energy is more realistically given by the depth of the potential well in the r direction alone.

Other treatments of the ion motion[4] are based on the electric pseudopotential[9]

$$\psi(\bar{r}, \bar{z}) = (e/4m\Omega^2)E_0^2(\bar{r}, \bar{z})$$

where E_0 is the amplitude of the electric field and $\bar{r}$ and $\bar{z}$ are the coordinates of the guiding center about which the forced micromotion of the ion with amplitude $\zeta = \zeta_0 \cos \Omega t$ occurs. This theory applies in the quasistationary limit when $|\dot{z}| \ll \zeta_0 \Omega$.

Computer simulations by numerical integration[10,11] of the Mathieu equations have been carried out for a variety of initial conditions, such as the phase of the rf field at ion formation, initial position, and initial kinetic energy of the ion. More recently matrix methods,[12,13] based on phase space dynamics, were used for efficient calculations of a multitude of different ion orbits.

2.3. Measurement of the Ion Number Signal

An ion number signal can be measured in a variety of ways. In the absorption method,[2] the damping of a weakly driven detection tank circuit is observed when the ion motion frequency is swept through resonance. This technique requires high-amplitude stability of the driving frequency ω_0. The emission method[14] avoids this difficulty by measuring the emitted energy of the coherently excited ion cloud. Normally, the motional macromotion frequency ω_z is set slightly below the detection frequency ω_0. To sweep ω_z through resonance, a linear negative-going dc sweep is applied to the ring electrode. Simultaneously with the detection sweep, another pulse at the sideband frequency $\Omega + \omega_0$ coherently excites the ion macromotion when $\omega_z = \omega_0$, finally driving the ions against the electrodes. Hereby, an emission signal voltage is induced, which, after narrow-banded amplification, is rectified and displayed on a scope.

In the bolometric technique,[14,15] the voltage induced without excitation of the ion motion is observed. This technique relies on the incoherent ion motion and is nondestructive. In still another method,[16] the ions are extracted from the trap with an ion lens and are counted by a particle multiplier.

The effects of the space charge of stored ions and their use to determine the total stored ion charge are discussed next. The space charge of the ions slightly modifies the well depth of the trap. However, an exact treatment of the problem is complex and would involve the knowledge of individual ion orbits.

In order to derive a value of the correct order of magnitude, we assume a spherical charge distribution and zero dc bias voltage. A more detailed analysis, using the actual ellipsoidal ion cloud and an additional dc voltage, is given elsewhere.[4] If a uniform charge distribution of density $\tilde{n}$ in a trapping region of radius z_0 is assumed, the space-charge effect is equivalent to an additional dc bias voltage U_i. The space charge is defocusing in all directions, which means that the virtual bias voltage U_i has a different sign for the r and z directions. With this approximation, U_i

follows from the Poisson equation as

$$U_i = \bar{n}z_0^2/4\varepsilon_0$$

where ε_0 is the permittivity constant. Using the smallest characteristic dimension $z_0 = 25$ mm of the trap structure, described in Table 1, the virtual bias voltage is

$$U_i = 2.8 \times 10^{-6}\bar{n}$$

The procedure for measuring the total ion charge is outlined in the following.[14]

The trap parameters were chosen to store both $^3He^+$ and a small number of H_2^+ test ions. Then the H_2^+ ion-number signal was detected in the conventional way by observing the damping of a tuned detection circuit at $\omega_z(H_2^+)$, when the fundamental ion oscillation frequency was varied through resonance. In addition, all $^3He^+$ ions could be driven out of the trap by a strong rf pulse at $\Omega + \omega_z(^3He^+)$. The amplitude of this pulse was adjusted before the actual measurement by making the $^3He^+$ ion signal disappear. If now the $^3He^+$ ions were completely thrown out in every other detection cycle, the observed H_2^+ signal appeared space-charge-shifted during one cycle and unshifted in the following cycle. From the size of the shift, when interpreted as an additional dc bias voltage, the $^3He^+$ ion number is derived. With respect to the change in D_z the measured shift was equivalent to an applied dc voltage $U_i = 400$ mV, or, according to the above, was caused by a space charge of density $\bar{n} = 1.4 \times 10^5$ ions/cm^3. The total number of stored ions follows to $n = 8 \times 10^6$, if the assumption of a uniform charge distribution over the trapping region of radius $z_0 = 25$ mm is taken into account. Care was taken to obtain a large amplitude of the H_2^+ oscillation for the purpose of averaging the space charge of the whole $^3He^+$ cloud.

3. Reorientation Spectroscopy of Stored Ions

3.1. The Ion Storage Exchange Collision Method

This general method was developed for the spectroscopy of ions in ^2S and ^2P ground states and can be applied to both positive and negative ions. It requires no ion resonance radiation and relies on spin-dependent collision processes for producing and detecting polarized ions. So far, the technique has been applied to the measurement of the hyperfine-structure splittings of the hydrogenlike $^3He^+$ ion and of the H_2^+ ion.[17]

Figure 4 shows the apparatus used in the $^3He^+$ experiment.[14] $^3He^+$ ions are created inside the ion trap by pulsed electron bombardment of the

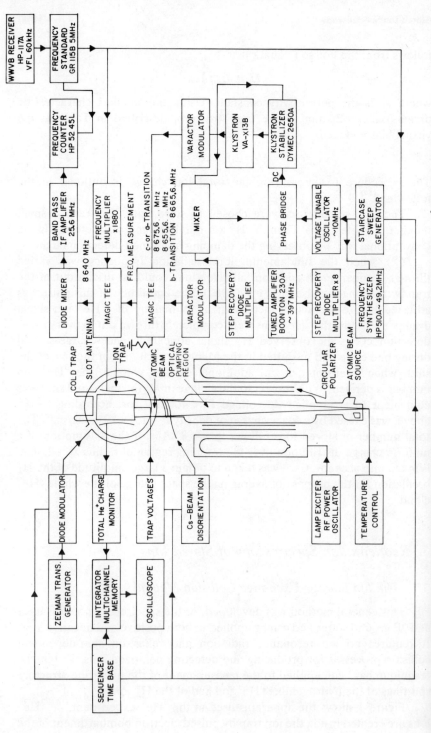

Figure 1. Detailed block diagram of the ion storage exchange collision apparatus for the ^{3}He$^+$ hfs experiment.

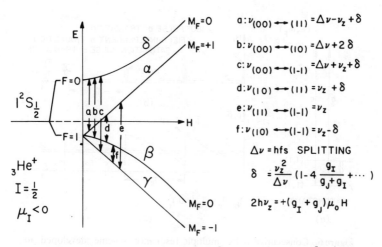

Figure 5. Breit–Rabi diagram of the $^2S_{1/2}$ ground state of $^3He^+$.

background ^3He gas. A weak beam of cesium atoms passes through the trap by a number of slots cut into the ring electrode. The entire apparatus is submerged in a homogeneous magnetic field, produced by a pair of 150-cm-diameter Helmholtz coils. The cesium atomic beam is polarized by optical pumping and transfers polarization to the stored ions by spin exchange collisions. After leaving the rf quadrupole arrangement, the cesium beam is collected on a cold trap. The circularly polarized pumping light is effectively focused into the beam region by cylindrical mirrors. Microwave power is fed into the trap by a $\lambda/2$ slot antenna cut into one end cap electrode.

Figure 5 shows the hyperfine-structure (hfs) Zeeman levels of $^3He^+$ in a Breit–Rabi diagram. All possible transitions are indicated and were observed in the experiment. Of particular interest was the field-independent transition $(0, 0) \leftrightarrow (1, 0)$, which is labeled b and had a linewidth of 10 Hz. This transition alone does not change the ion polarization. However, it was made directly observable by a consecutive-pulse multiple-resonance scheme.

The principle of the scheme is illustrated in Figure 6. A diagram of the participating hfs levels, the sequence of pulses, the development of the instantaneous polarization p and of the average polarization $\bar{p}$, as a function of time, are depicted. The scheme consists of applying consecutively three rf pulses. The first pulse, centered at the d transition resonance frequency, is short against the spin exchange time. The pulse width was adjusted to cause the populations of the α and β states to be inverted at the end of each pulse. This pulse is a 180° pulse since the reversal of the occupation numbers is analogous to a 180° precession of the macroscopic

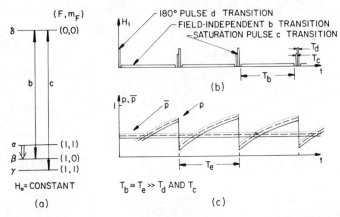

Figure 6. Consecutive-pulse multiple-resonance scheme developed to make the b transition directly observable. (a) Diagram of involved hfs levels. (b) Sequence of b, c, d pulses. T_b, T_c, T_d are respectively, the duration of the b, c, and d transition pulses. H_1 is the amplitude of the applied rf field. (c) Time evolution of the instantaneous ion polarization p and the average ion polarization $\bar{p}$. Three cases are shown schematically. The dotted curves represent the case in which only the d pulse was applied. The dashed curves illustrate the case in which the d and b transition pulses were applied consecutively. The solid curves show the complete pulse scheme, when the d, b, and c transition pulses were applied in sequence.

polarization at angular frequency ω_1 in a two-level system. The pulse height H_1 and the width T_d fulfill the relation

$$\omega_1 T_d = \gamma H_1 T_d = \pi, \qquad T_d \ll T_e$$

The third pulse saturates the c transition. The field-independent b transition of interest is induced by the central pulse. The duration of the outer $\Delta m_F = \pm 1$ pulses is very short compared to that of the central pulse, which is applied over one spin exchange period in a weak atomic beam. Typical pulse widths are 0.2 msec for the 180° pulse at the d transition, 3.2 msec for the c transition saturation pulse, and 76.6 msec for the central transition pulse.

For the purpose of explanation, the ideal case in which atomic beam and ion polarization are both initially unity is discussed. The process is described in three steps. For each step the evolution of both the instantaneous polarization p and the average polarization $\bar{p}$ is shown.

First, only the 180° pulse is applied. If no magnetic dipole transition occur during the time of the second pulse, for instance, when the frequency is off-resonance, the spin-exchange process tends to pump the ions back into the initial α state until the next 180° pulse interchanges the population numbers of the α and β states. This situation is indicated by dotted lines. The net effect is a decrease in the average ion polarization $\bar{p}$ and is shown by the horizontal line in the graph of the time evolution of p in Figure 6.

In the next step, the frequency of the b transition is on-resonance during the second pulse. There is now a certain probability that ions are sorted out into the δ state and are not available in the following 180° pulse. Therefore, both the instantaneous and the average polarizations decrease further as shown by the dashed lines. At this stage a b transition resonance signal may already be detected.

If, in addition, the c transition is fixed at resonance and saturated during the third pulse period, and if the b transition is also at resonance, some ions that have gone from the β to the δ state may end up in the γ state, which has a negative contribution to the polarization. Thus, the effect of the added c saturation pulse is to reinforce the b transition signal. This situation of the complete consecutive-pulse multiple-resonance scheme is indicated by solid lines. In a measurement, the frequency of the central b transition pulse was varied through resonance.

A description of the various collision processes that occur when the polarized cesium atomic beam passes through the trap follows. In a first reaction, polarization is transferred to the stored $^3He^+$ ions by spin exchange collisions. In this process the $^3He^+$ ions become polarized, whereas the polarization of the cesium atoms is lost. Neglecting at this point the relatively slow ion loss processes, the rate equations[18] for the population numbers of the $^3He^+$ ion, for arbitrary electron spin polarization P of the cesium atoms, are

$$\dot{n}_\alpha = -2(1-P)n_\alpha + (1+P)n_\beta + (1+P)n_\gamma$$

$$\dot{n}_\beta = +(1-P)n_\alpha - 3n_\beta + (1+P)n_\gamma + n_\delta$$

$$\dot{n}_\gamma = +(1-P)n_\beta - 2(1+P)n_\gamma + (1-P)n_\delta$$

$$\dot{n}_\delta = +(1-P)n_\alpha + n_\beta + (1+P)n_\gamma - 3n_\delta$$

The time is in units of $4T_e$ with T_e being the average spin exchange time.

These equations are solved under steady-state conditions with disorienting rf fields either present or absent. Without rf fields and with polarizing and relaxing effects in equilibrium, we have $\dot{n}_\alpha = \dot{n}_\beta = \dot{n}_\gamma = \dot{n}_\delta = 0$, and unnormalized occupation numbers may be found to be $n_\alpha = (1+P)/(1-P)$, $n_\gamma = (1-P)/(1+P)$, $n_\delta = n_\beta = 1$. According to this, the equilibrium ion polarization $p = (n_\alpha - n_\gamma)/n$ is equal to the atomic polarization P. Here the total ion number $n = n_\alpha + n_\beta + n_\gamma + n_\delta$.

The last step in a magnetic resonance experiment is to monitor the changes in the ion polarization, when a resonant rf field is applied. This was achieved by a suitably chosen charge transfer reaction by which ions are slowly lost from the trap. The fact that the spin exchange cross section is an order of magnitude larger than the charge transfer cross section ensures that the collision partners are highly polarized before the charge transfer becomes effective. The charge transfer process is spin dependent since it proceeds along the following near resonant channels:

$$Cs(6s\,{}^2S_{1/2}) + He^+(1s\,{}^2S_{1/2}) \to \begin{cases} He^*(1s2p\,{}^1P_1) + Cs^+({}^1S_0) - 0.52\text{ eV} \\ He^*(1s2p\,{}^3P_{2,1,0}) + Cs^+({}^1S_0) - 0.27\text{ eV} \\ He^*(1s2s\,{}^1S_0) + Cs^+({}^1S_0) + 0.08\text{ eV} \\ He^*(1s2s\,{}^3S_1) + Cs^+({}^1S_0) + 0.88\text{ eV} \end{cases}$$

A large number of channels of a different nature is available at the same time:

$$Cs(6s\,{}^2S_{1/2}) + He^+(1s\,{}^2S_{1/2}) \to He(1s^2\,{}^1S_0) + Cs^{+*}[5p^5({}^2P)nl] + \Delta E_{n,l}$$

where $n = 6$–10, and $l = s$, p, and d. In this reaction an electron is transferred from the closed $5p^6$ shell of cesium to the $1s$ shell of helium. Orbital overlap requires a close collision with a small cross section, but a large number of near resonant channels and other channels are present that will all contribute to the singlet cross section Q_1.

It has been shown[19] that total spin angular momentum is conserved in this reaction. Therefore, collisions in which the symmetry of the spin state of the total system is even are treated independently of those for which the spin state is odd. Emphasizing the spin-dependent aspects of the charge transfer process, the following specific reactions occur: A singlet channel,

$$Cs{\uparrow} + He^+{\downarrow} \to He^*({\uparrow}{\downarrow} - {\downarrow}{\uparrow}) + Cs^+ + \Delta E_1$$

leads to the singlet excited states described. Two triplet channels,

$$Cs{\uparrow} + He^+{\uparrow} \to He^*{\uparrow}{\uparrow} + Cs^+ + \Delta E_3$$

and

$$Cs{\uparrow} + He^+{\downarrow} \to He^*({\uparrow}{\downarrow} + {\downarrow}{\uparrow}) + Cs^+ + \Delta E_3$$

lead to the triplet excited states described. The arrows represent the orientation of the electron spin along the external field, just before the collision for the left-hand side, and just after the collision for the right-hand side of the equation. Depending on the relative spin orientation of the collision partners, the charge transfer then proceeds with different cross sections dominantly along either the singlet or the triplet channels.

The time evolution of the ion number n of polarized ${}^3He^+$ under the influence of charge exchange collisions with the polarized cesium beam is described by charge exchange rate equations in terms of the spin polarizations $\langle s_z \rangle/s$, which are labeled p for ions and P for atoms. The change of the stored ion number n is given by

$$\frac{dn}{dt} = -\left(\frac{n}{T_0}\right)\left[1 - pP\left(\frac{\Delta Q}{4\bar{Q}}\right)\right]$$

Here T_0 is the average ion lifetime for a given unpolarized atomic beam,

$$\Delta Q = Q_1 - Q_3, \text{ and } \bar{Q} = \tfrac{1}{4}(Q_1 + 3Q_3).$$

In the experiment, a polarization signal $S(p, P)$, defined as the relative difference in ion numbers remaining in the trap after a time t with collision partners either polarized or unpolarized, was measured and is

$$S(p, P) = [n(p, P) - n(0, 0)]/n(0, 0)$$

Assuming only statistical fluctuations, the polarization signal $S[p(t), P(t)]$ has an optimum value of

$$S^*(p, P) = \tfrac{1}{2}pP(\Delta Q/\bar{Q}), \qquad S^* \ll 1$$

for $t = 2T_0$. The polarization signal disappears when either p or P are destroyed. P is easily destroyed by applying the Zeeman frequencies of the cesium ground state. The $^3\text{He}^+$ ions become completely disoriented by a saturating rf field at the double quantum transition frequency with $p = 0$; they become only partly disoriented when the other hfs resonance frequencies are applied and $p > 0$.

In order to observe a hfs transition, a rf field is applied to the polarized stored ion sample. When the rf field is swept through resonance, the polarization changes and, owing to the spin-dependent charge transfer, with it the number of stored ions left in the trap after a fixed interaction time. A resonance curve is obtained by counting the number of ions for each value of the frequency.

Figure 7 shows the measuring sequence. The number of stored ions is plotted against time. At the beginning of each observation cycle the trap is cleared of ions remaining from the previous cycle. Then new $^3\text{He}^+$ ions are formed by a short electron pulse from the He^3 background gas. With the polarized cesium beam on all the time, the ions are polarized by spin exchange in a short time and are subsequently slowly neutralized by spin-dependent charge exchange. If the polarization is destroyed by a resonant rf field, the ion number decays faster, thus giving rise to a magnetic resonance signal. At the end of an observation cycle, the ion number is measured by the emission detection scheme described in Section 2.3 and is fed into a signal-averaging system. Then the process starts all over again and the whole sequence is repeated at fixed intervals of one second until a sufficiently high signal-to-noise ratio is obtained.

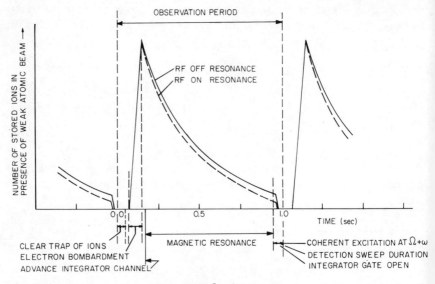

Figure 7. Time evolution of the stored $^3\mathrm{He}^+$ ions during the periodic sequence of manipulating pulses.

All possible hfs transitions in the ground state of $^3\mathrm{He}^+$ were observed by applying suitably oriented rf fields at resonance with the various hfs splittings. The field-dependent transitions showed linewidths of 2 kHz, due to residual field inhomogeneities and hum. It was, however, possible to reduce the linewidth of the almost field-independent transition to 10 Hz. Using the Breit–Rabi formula and the value of the double quantum transition for field calibration, the hyperfine-structure splitting $\Delta\nu_{1s}$ of $^3\mathrm{He}^+$ was obtained and is

$$\Delta\nu_{1s} = 8665.649867(10)\,\mathrm{MHz}$$

In a subsequent similar experiment, the ion storage exchange collision technique was extended to the H_2^+ ion.[17] The Zeeman splitting of the rotational $K = 0$ ground state having $^2\Sigma_g^+$ electronic symmetry was observed. Magnetic resonance transitions $\Delta F = 0$, $\Delta m_F = \pm 1$ were measured in H_2^+ and compared with double quantum transitions $\Delta F = 0$, $\Delta m_F = \pm 2$ in the ground state of $^3\mathrm{He}^+$. Since both ion species are stored practically in the same volume, field sampling errors are avoided. Again a polarized cesium atomic beam was used to polarize the H_2^+ ions by spin exchange and to detect changes in the polarization by spin-dependent charge transfer collisions. The charge transfer collisions proceed along several near-resonant singlet and triplet channels to the $n = 2$ excited states of the neutral H_2 molecule.

3.2. State Selection by Selective Quenching of Stored Ions

A recent precision experiment[20] investigated the $2s\,^2S_{1/2}$ excited state of the $^3He^+$ ion, complementing the measurements of the hfs in the ground state of $^3He^+$ just described. Metastable $^3He^+$ ions were created by pulsed electron bombardment and stored in an electrostatic ion trap.[21] The state-selection and resonance-detection schemes are the timelike analog of an earlier ion beam experiment in the same state. They consist of applying three rf pulses consecutively. During the A period, rf transitions are induced from the $2s\,^2S_{1/2}$, $F = 1$ state to the $2p\,^2P_{1/2}$, $F = 0, 1$ states at the Lamb-shift frequency. The $2p\,^2P_{1/2}$ state decays immediately to the ground state emitting 304-Å Ly_α radiation. After the A pulse, an excess of ions exists in the $2s\,^2S_{1/2}$, $F = 0$ state. During the C period, the hfs splitting is measured by stepwise varying a rf field through resonance with the $F = 0$ to $F = 1$ hfs splitting. When the Lamb-shift frequency is reapplied during the B period, the number of newly induced Ly_α photons is detected, producing in this way a resonance signal. A linewidth of about 1 kHz was obtained when storing the ions for a few msec. The latest result of the measurements is

$$\Delta\nu_{2s} = 1083.3549825(76)\,\text{MHz}$$

The error is largely due to corrections, involving a possible Stark shift, and effects due to magnetic field inhomogeneities.

3.3. Optical Pumping and Laser Spectroscopy of Stored Ions

As for atoms, resonance light can also be used to produce polarization of stored ions. Population inversion of the ground-state hfs levels of $^{199}Hg^+$ was achieved by intensity pumping with an accidentally coincident $^{202}Hg^+$ transition. Hereby the 1942-Å line of a $^{202}Hg^+$ lamp excites only one of the hfs components, namely, the transition from the $6s\,^2S_{1/2}$, $F = 1$ state to the $6p\,^2P_{1/2}$, $F = 1$ state of $^{199}Hg^+$. In the resonant absorption and subsequent emission processes, the ions are accumulated in the $6s\,^2S_{1/2}$, $F = 0$ ground state of $^{199}Hg^+$. The field-independent $\Delta F = \pm 1$, $\Delta m_F = 0$ transition is then induced by shining suitably oriented 40-GHz radiation into the quadrupole trap with a high-gain horn. The transition is monitored by the increase in the resonance fluorescence. By extrapolating to zero magnetic field, the hfs splitting of $^{199}Hg^+$ is found to be[22]

$$\Delta\nu_{6s} = 40\,507.348050(50)\,\text{MHz}$$

The experiment is being continued with a new apparatus and an improvement of more than an order of magnitude in the signal-to-noise ratio has been reported.[23]

The wavelength tunability of dye lasers is of particular value for the excitation of resonance fluorescence of stored ions. This has been demonstrated[7] for Ba^+ ions, where the resonance transition from the $6s\,^2S_{1/2}$ ground state to the $6p\,^2P_{1/2}$ state is at 4934 Å. With a nitrogen-laser-pumped dye laser, a signal-to-noise ratio of the resonance fluorescence, corresponding to an ion number signal of more than 600:1, was obtained with laser repetition rates of 20 Hz and integration times of 1 s. This ratio is comparable to an ion number signal obtained by the coherent excitation method (see Table 1). So far, the Doppler width of the fluorescence line was too large to allow clear resolution of the isotope shifts and the hfs components. However, after cooling the ions, this structure will be readily observable.

A second optical detection scheme[24] for monitoring stored Ba^+ ions used a high-power hollow-cathode lamp and ions stored in a cylindrical trap. 4555-Å light excites ions from the $6s\,^2S_{1/2}$ ground state to the $6p\,^2P_{3/2}$ state which decays not only back to the ground state but also to the metastable $5d^9\,6s^2\,^2D_{5/2}$ state emitting light at 6143 Å. Since detection can then be effected at a wavelength that is different from excitation, the scheme will be particularly useful when saturating laser excitation is applied and the intensity of the directly scattered light is high.

3.4. Alignment of Ions by Selective Photodissociation

In this method[5,16,25,26] H_2^+ ions are produced in an rf quadrupole ion trap by electron bombardment of H_2 gas. Photodissociation by a high-intensity mercury arc lamp can then be effected from the $v \geqslant 4$ vibrational levels of the $^2\Sigma_g^+$ electronic bound state to the vibrational continuum of the $^2\Sigma_u^+$ repulsive electronic state. The pulsating electric dipole moment[27] produced during such a transition lies in the direction of the internuclear axis. This causes the dissociation rate to be largest when the electric light vector is parallel to the internuclear axis and to be smallest when the electric light vector is perpendicular to the internuclear axis. For linearly polarized light, the dissociation rates of individual Zeeman levels with different magnetic quantum numbers $|m_F|$ can differ by as much as a factor of 2. As a consequence of these orientation-dependent dissociation rates, an alignment in the ion ensemble builds up. The various hfs Zeeman levels acquire a nonthermal distribution independent of the sign of m_F. If a rf field is applied at resonance between two hfs Zeeman levels, ions will be continuously transferred from the longer-living level to the other, resulting in an increased rate of photodissociation. Counting the H^+ photodissociation products as a function of the frequency yields a resonance signal. The most recent version of this experiment[28] realized an interesting extension of a consecutive-pulse multiple-resonance scheme to molecular ions. Two adiabatic passages and one 180° pulse were used to exchange the populations of pairs of hfs Zeeman levels. The 180° pulse measures the transition of interest. The sequence is chosen so that with the 180° pulse on resonance the populations of hfs Zeeman levels with greatly differing photodissociation rates are exchanged, but when the resonance condition is not met there is practically no change. Using the $|F_2, F\rangle$ representation, hfs transitions were observed with high accuracy between the $|1/2, 3/2\rangle$ and $|1/2, 1/2\rangle$ states for three vibrational states and are

$$\Delta\nu_4 = 15.371407(2) \text{ MHz}$$

$$\Delta\nu_5 = 14.381513(2) \text{ MHz}$$

$$\Delta\nu_6 = 13.413460(2) \text{ MHz}$$

The linewidth of the field-independent resonance transitions was less than 5 Hz.

4. Limitations of Stored Ion Spectroscopy

4.1. Line Shifts and Widths of Resonances of Stored Ions

The sources for frequency shifts and widths of magnetic resonance signals on stored ions have been analyzed[14] in detail and only the most important ones will be discussed here. Some of the effects are similar to the ones in other storage devices, such as the hydrogen maser, the hydrogen storage beam tube, and even the rubidium cell standard. However, the magnitude of the various effects is quite different. Also, additional effects occur that are specific to stored ions. These are caused by the trapping fields and the periodic ion motion.

The accuracy of an ion storage experiment is presently limited by the insufficient knowledge of shifts due to the second-order Doppler effect, spin exchange collisions, and charge transfer collisions.

The first-order Doppler effect does not, for all practical purposes, broaden or shift the line but leads to a modification of the hfs spectrum, which will be discussed in Section 4.3.

Because of the second-order Doppler effect, the transition frequency of ions with velocity v is shifted by

$$\frac{\nu - \nu_0}{\nu_0} = -\frac{1}{2}\frac{v^2}{c^2} = -\frac{\langle E \rangle}{E_0}$$

where $\langle E \rangle$ is the average kinetic energy of an ion in the trap and $E_0 = Mc^2$ is the rest energy.

This effect amounted to 4 Hz for $^3\text{He}^+$ and 0.06 Hz for $^{199}\text{Hg}^+$ at an energy of a few electron volts. It seems possible to reduce the second-order

Doppler contribution considerably by radiatively cooling the ions to room temperature or lower. In an experiment with protons,[29] the disordered ion motion was coupled to an external tank circuit where the ion energy was dissipated into heat. However, for heavy ions this method seems to be difficult since the time constant for cooling t_{0z}, given by

$$t_{0z} = M(2z_0)^2/q^2 R$$

is rather long. Here M and q are the mass and charge of the ion and R is the resistance of the external tank circuit. For a trap with a storage volume of 1 cm^3 the time constants for cooling range from milliseconds for electrons to hundreds of seconds for heavy ions.

Other means to reduce the average ion temperature are possible, such as using collisions of the stored ions with inert light buffer gas atoms at room temperature or, in an ion storage exchange collision experiment, with the cesium beam. Such viscous drag cooling is very effective for heavy ions, when a light gas such as helium is employed. In preliminary experiments the possibility of viscous drag cooling has been shown to be feasible and one-dimensional model calculations have been performed.[30,31]

A novel method of motional sideband cooling has also been proposed.[32] It is based on the modulation of radiation scattered by stored ions. The spectrum of a coherently excited ion consists of a carrier at ν_0 and symmetric sidebands spaced around it at multiples n of the ion oscillation frequencies ν_i. In a suitably chosen case, an ion is excited by radiation of energy $h(\nu_0 - n\nu_i)$ but will reradiate symmetrically at all sideband frequencies or with an average energy $h\nu_0$. The energy excess $nh\nu_i$ of this process can only come from the oscillatory motion of an ion in the trap. In this way the motional energy is reduced and the ion is left closer to the center of the trap.

The second-order Doppler shift is

$$(\nu - \nu_0)/\nu_0 = -1.4 \times 10^{-13} T/m$$

as a function of the absolute temperature T and atomic mass m of the ion. If the ions were cooled to room temperature, the effect for $^3\mathrm{He}^+$ would be of magnitude 10^{-11} and, for the heavier Hg^+ ion, of 10^{-13}.

Another important process affecting the precision in an ion storage experiment is spin exchange. An estimate of the relative spin exchange shift f follows from $f = \frac{1}{4} N(4/3)\pi r_{SE}^3$, where r_{SE} is the radius of the fractional spin exchange volume defined by $\pi r_{SE}^2 = 2Q_{SE}$. As a numerical example for the spin exchange shift take $N = 10^8$ particles/cm^3 and $2Q_{SE} = 2.8 \times 10^{-14}$ cm^2. The fractional spin exchange shift f is then $f \approx 10^{-13}$.

Charge transfer collisions will also broaden and shift the hfs transition in a way similar to spin exchange. Assuming that nuclear polarization is partially conserved between collisions, the modification of the hfs splitting for resonant charge transfer is of comparable size as for spin exchange, but is much smaller in nonresonant charge transfer collisions.

Viscous drag cooling of paramagnetic ions by a light diatomic buffer gas will cause the well-known foreign gas pressure shifts,[33] but contrary to optical pumping experiments on atoms, efficient cooling of stored ions requires only partial pressures of about 10^{-5} Torr and these shifts are a few millihertz or smaller.

Frequency shifts due to dispersive effects occurring for atoms in a radiation field[34] may play a role when laser pumping of stored ions is used. For monochromatic excitation such an energy shift ΔE is in lowest order

$$\Delta E = (k - k_0)V_{ab}^2/[(k - k_0)^2 + (\Gamma/2)^2]$$

Here V_{ab} is the matrix element of the electric dipole operator, k the wave vector of the laser radiation, k_0 the wave vector of the atomic transition, and Γ the damping constant of the excited state. Additional light shifts[35,36] have been studied for atoms and need also be considered for stored ions. The total light shift can amount to a few hertz but is controllable and in a precision experiment must be reduced by decreasing the light intensity and the amount of detuning.

A frequency shift by cavity pulling is practically negligible. The narrow resonance linewidth obtainable for stored ions is electrically equivalent to an oscillator having a high quality factor Q_R. For $Q_R = 1 \times 10^{10}$ and a cavity quality factor $Q_c = 200$, the "pulling factor" is $(Q_c/Q_R)^2 = 4 \times 10^{-16}$.

The effects that are peculiar to the ion storage tube are contributions due to the electric fields used for the containment of the ions and the electric and magnetic fields associated with the ion motion itself.

In lowest order the Stark effect shifts the hfs frequency of an ion by a fractional amount dE^2. For hydrogenlike atoms[37] $d = 16a_0^4/Z^6e^2$. The size of the shift for singly charged ions is of the order of 10^{-17} and is clearly negligible. The near absence of shifts due to the containing fields is one of the major advantages of the ion storage technique.

The rf magnetic fields, caused by the displacement currents produced by the trapping fields, are a few milligauss and the motional magnetic fields of the oscillating ions are even smaller. All such magnetic fields are further reduced for cool ions, which will be concentrated in the field-free center of the quadrupole. The shifts due to these rf magnetic fields have not been observed since they are of the order of a few millihertz or smaller. This value follows from the second-order magnetic field dependence $\Delta \nu_m$ of a field-insensitive transition, which for instance for $^{199}Hg^+$ is given by $\Delta \nu_m/H^2 = 89$ Hz/G^2, where H is the magnetic field.

In order to further develop trapped ion spectroscopy, the following experiments were undertaken to minimize and control the two main sources of uncertainty, namely, broadening and shifts due to collisions and the second-order Doppler effect.

4.2. Pulsed Spin Precession Reorientation Method

The main collisional shift is caused by spin exchange, a process which is well understood for atom–atom collisions but not for ion–atom collisions. The pulsed spin precession reorientation method[38] was developed to study spin exchange collisions between free ions and atoms. It was applied to stored $^3He^+$ ions, which interacted with a beam of polarized cesium atoms. The atomic beam and the ion storage parts of the apparatus were similar to the ones described in Section 3.1.

For the purpose of explanation, the ideal case of complete polarization with $P = 1$ is assumed. If we assume in addition that the ions have no hyperfine structure, then their electronic spins are all parallel to H_0. This simplified case is easy to discuss. Consider that the constant magnetic field H_0 points in the z direction and a weak rf field of frequency ω is applied in the x–y plane. When the rf field is at resonance with the Zeeman splitting of the ions, the total electronic ion spin is turned out of the z direction. If the H_1 field were applied continuously, the total ion spin would rotate periodically from a parallel to an antiparallel direction and back with respect to H_0. This is possible since, in an ion trap, the ions are ideally isolated from each other and the phase memory time is long.

However, in the pulsed spin precession reorientation scheme, the ions are exposed to a sequence of short resonant rf pulses and not to a continuous rf field. For each pulse, the length T_d and height H_1 are adjusted to produce an inversion of the electronic spin polarization. The pulse parameters then fulfill the conditions $\gamma H_1 T_d = \pi$, $T_d < T_e$, where $\gamma =$

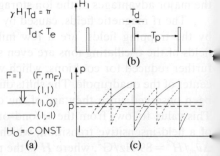

Figure 8. Pulsed spin precession reorientation method. (a) Diagram of the hfs Zeeman levels of interest. The magnetic dipole transition, which inverts the occupation number of the (1,1) and (1,0) levels, is indicated. (b) Sequence of the inverting pulses. The frequency of the pulsed rf field is at resonance with one of the hfs Zeeman splittings. T_d and T_p are, respectively, the duration of the inverting pulses and their separations. (c) Time evolution of the instantaneous ion polarization p (sawtoothlike lines) and of the average ion polarization $\bar{p}$ (horizontal lines).

$\mu_B g_S/h$ is the gyromagnetic ratio. The last condition means that the resonant rf pulses, which invert the polarization, are strong enough so that the rotation of the electronic spin is fast against the spin exchange pumping of the polarized cesium beam. The polarized cesium beam passes continuously through the trapping region and partly reorients the inverted electronic spins of the ions during the long rf free periods T_p between the short inverting pulses. The average polarization was measured as a function of the pulse separation T_p. Such average polarizations are depicted in Figure 8 for the case of $^3\text{He}^+$, including hyperfine structure, by horizontal lines for two different values of T_p. The solid line represents the case in which the pulse separation was equal to the spin exchange time T_e. The dashed line represents the case with T_p half as large. The rate equations for spin exchange were solved for each hfs level taking into account that the inversion of the electronic spins by the rf pulses is achieved instantaneously.

The first step in the experiment was to adjust the inverting pulses. For a pulse of proper length and height the average polarization was at a minimum. Then the effect of spin exchange collisions between $^3\text{He}^+$ and the polarized cesium beam was investigated by observing the average polarization for different intervals T_p between the inverting pulses. From such measurements the spin exchange cross section for the He^+–Cs system was derived and is

$$Q_e = (1.4 \pm 0.8) \times 10^{-14} \, \text{cm}^2$$

This ion–atom spin exchange cross section is of similar size as in atom–atom spin exchange, showing that polarization effects due to the ion charge are small. It should be noted that, when measuring the time evolution of the ion polarization, a geometrical factor describing the spatial ion distribution in the trap does not enter into the evaluation of the cross section.

4.3. Motional Sideband Spectra

Simultaneously with the various cooling concepts to reduce the ion temperature, also methods to measure this temperature are being developed. If the ion energy and the energy distribution are known, the second-order Doppler shift can be evaluated and accounted for. The following discussion describes the various efforts undertaken to determine the ion temperature.

First-order Doppler shifts cancel to the extent that the ions do not have a net translational velocity[14] due to motional averaging. However, sidebands occur in the magnetic resonance signals[39,40] and are caused by the first-order Doppler effect and by the nonuniformity of the rf field used to induce the hfs transition on trapped ions. A measurement of the relative amplitudes of the sidebands yields sufficient information to determine the energy distribution of the ions. A short description of the effects that generate the sidebands follows.

When an ion with a hfs frequency ν_0 in its rest frame is oscillating with a characteristic ion oscillation frequency ν_i in the confined space inside the ion trap, the wave absorbed or emitted by the oscillating ion is both frequency and amplitude modulated. The two modulations are discussed separately considering the extreme cases of a traveling and a standing rf wave inside the ion trap. The actual case is usually a mixture of both.

First, we assume that only the ion but not the radiation is trapped. When an oscillating ion absorbs or emits a rf photon, the absorption or emission frequency of the ion is changed. The change in the frequency is proportional to the component of the ion velocity in the direction of the radiation process and is described by the classic formula of the Doppler effect. Because of the periodic ion motion, the Doppler effect produces a frequency modulation of the main magnetic resonance peak. A set of lines, spaced by integer multiples of the characteristic ion oscillation frequencies from the main hfs line, results. The relative intensity of the sidebands follows from a Fourier series and depends on the ion velocity or, for a fixed ion oscillation frequency, on the size r of an ion orbit relative to the hfs wavelength λ_0. For $r > \lambda_0$, the ion velocity is large so that the periodic frequency deviation Δf is large compared with the modulation frequency $f_m = \nu_i$. In this situation, where relatively hot ions oscillate through almost

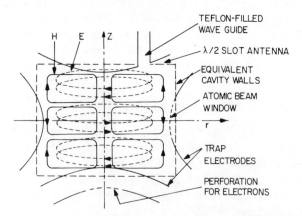

Figure 9. Cross section of the rf quadrupole trap. Substituting equivalent cylindrical surfaces for the rotationally symmetric hyperbolic surfaces, the TE_{013} field distribution is depicted. The radius of the ring electrode was $r_0 = 3.6$ cm and about equal to the wavelength of the hfs transitions studied.

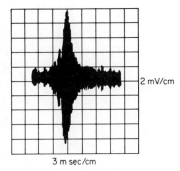

Figure 10. Thermal ion-number signal. Without excitation of the ion motion, the induced voltage in a fixed tuned circuit is observed, when sweeping the macromotion frequency through resonance. Experimental parameters: $V_0 = 137\ V_{peak}$, $U_0 = 0$ V, $U_s = -2.5$ V, $i_e = 10$ mA, $T_e = 80$ ms, sensitivity 2 mV/cm, amplification 2000, total sweep period 32 ms.

the whole volume of the trap, strong sidebands of high order are observed. When the ions are cooled to a point where the ion velocity is small, then $\Delta f < f_m$, and the relative strength of the sidebands decreases.

Next, we assume that both the ion and the radiation are trapped. Figure 9 shows for this case, in a schematic drawing of the trap, the distribution of the rf field. When the ion motion occurs over regions z and r that are large compared to the half-period variation of the field, the nodes of the resonant rf field lead to a strong amplitude modulation of the induced magnetic dipole transitions. When the ion energy is sufficiently decreased, the orbit of an ion is small compared to the half-period variation of the field, so that the resonant rf field seen by the ions is almost uniform and only a small amplitude modulation and small sidebands result. Magnetic-field-dependent sidebands[41] that are due to the Larmor precession of the ions in the external magnetic field have also been observed. Their separation from the central resonance peak was varied by inducing the transition in different external magnetic fields.

Considering that trapped ions are in thermal equilibrium with all degrees of freedom of the ion motions, the average ion energy can be obtained by measuring the thermal noise of stored ions induced in a cold tank circuit. Such a signal is shown for trapped $^3He^+$ ions in Figure 10. The bolometric technique is nondestructive for the total number of ions, which must be known. It is also advantageous in such a measurement of the ion temperature when the number of trapped ions is small, so that the thermal noise dominates the statistical Poisson noise.

In the case of laser excitation of stored ions, the Doppler width of the resonant radiation has been measured.[7] Varying the well depth of the trap by changing the operating parameters showed that the average ion energy is a constant fraction of the well depth.

So far, all determinations of the ion temperature and energy distribution are rather crude and more work is needed to accurately determine these parameters.

5. Some Results of Precision Measurements of Stored Ions

The hfs splittings measured for the hydrogenlike $^3He^+$ ion can be compared with accurate calculations and, therefore, serve as a sensitive test for nuclear structure and quantum electrodynamical theories. Considering the electron and nucleus as structureless point dipoles, the following extended Fermi formula[42,43] holds for S states of hydrogenlike atoms:

$$\Delta \nu_F = \frac{8}{3} Z^3 \alpha^2 cR_\infty n^{-3} \frac{\mu_n}{\mu_0} \left(1 + \frac{1}{2I}\right) \left[1 + \frac{m}{M}\right]^{-3} \left[1 + \frac{\alpha}{2\pi} - 0.328 \frac{\alpha^2}{\pi^2}\right]$$

$$\times [1 + B(n)(Z\alpha)^2] \left[1 - \alpha^2 Z\left(\frac{5}{2} - \ln 2\right)\right]$$

$$\times \left[1 + r(\alpha^3; n, Z) + r\left(\alpha^2 \frac{m}{M}; n, Z\right)\right]$$

The first term in square brackets is the reduced mass correction, the second is the anomalous electron magnetic moment contribution, the third is the relativistic Breit correction with $B(1) = 3/2$ for the $n = 1$ ground state and $B(2) = 17/8$ for the $n = 2$ first excited state, the fourth is the radiative correction of Kroll and Pollock, and the fifth contains higher-order radiative corrections.[44,45] Using $\alpha^{-1} = 137.035987(29)$,[46] $R_\infty = 109737.3143(10)\,cm^{-1}$,[47] $\mu_n/\mu_0 = 1.1587414(9) \times 10^{-3}$,[48] and $m/M = 0.000181954(1)$,[49] we calculate the values of the ground-state hfs splitting of $^3He^+$ to be

$$\Delta \nu_F(1s) = 8667.61(5)\,MHz$$

and of the first excited state to be

$$\Delta \nu_F(2s) = 1083.60(1)\,MHz$$

For comparison with the experiment, the hfs anomaly defined by

$$\Delta \nu_{exp}/\Delta \nu_F = 1 + \delta$$

is introduced. Here $\Delta \nu_{exp}$ and $\Delta \nu_F$ are for the same atomic state. It is well known that δ is independent of n and indentical in the 1s and 2s states up to effects of order $(Zr_{nucleus}/a_0)^2$ or 10^{-8}. With the values of the experiments discussed in Sections 3.1 and 3.2 we obtain

$$\delta = -226(5)\,ppm$$

for both the 1s and 2s states, demonstrating excellent agreement between the two ion storage experiments.

Another quantity of interest is the ratio of the hfs splitting[45] in the $2s$ and $1s$ states defined as

$$8\Delta\nu_{2s}/\Delta\nu_{1s} = 1 + R$$

In this ratio the nuclear structure effects drop out, making an investigation of the state-dependent radiative correction derived from quantum electrodynamics possible. Extending the latest theoretical values to include also nuclear recoil,

$$R_{\text{theory}} = 137.300(9)\,\text{ppm}$$

The largest contribution to R is due to the well-established relativistic Breit correction

$$R_B = (5/8)z^2\alpha^2 = 133.128\,\text{ppm}$$

Combining the results of the two ion storage experiments, the experimental value is

$$R_{\text{exp}} = 137.323(7)\,\text{ppm}$$

To obtain the purely quantum electrodynamic contributions, we subtract the Breit term in both cases and obtain

$$R_{\text{theory}}^{\text{QED}} = 4.173(9)\,\text{ppm}, \qquad R_{\text{exp}}^{\text{QED}} = 4.195(7)\,\text{ppm}$$

The agreement is to about 0.5%. A comparison of the results of the ion storage experiments on $^3\text{He}^+$ with those of the corresponding precision measurements on hydrogen and deuterium by maser and optical pumping techniques is made in Table 2.

Table 2. Comparison of the Measured hfs Splitting[a] in the $1s$ and $2s$ States of H, D, and $^3\text{He}^+$

	H	D	$^3\text{He}^+$
$1s$	1420.405751768(2)[b]	327.38435251(5)[c]	8665.649867(10)[d]
$2s$	177.556842(10)[e]	40.924439(20)[f]	1083.3549825(76)[g]
R_{exp}	34.49(6)	34.09(50)	137.323(7)
R_{theory}	34.45(2)[h]	34.53(2)[h]	137.300(9)
δ	-10	157	$-226(5)$

[a] hfs splittings in Mhz, R and δ in ppm.
[b] C. Audoin and J. Vanier, *J. Phys. E, Sci. Instr.* **9**, 697 (1976).
[c] D. J. Larson, P. A. Valberg, and N. F. Ramsey, *Phys. Rev. Lett.* **23**, 1369 (1969).
[d] H. A. Schuessler, E. N. Fortson, and H. G. Dehmelt, *Phys. Rev.* **187**, 5 (1969).
[e] J. Gruenebaum and P. Kusch, Columbia Radiation Laboratory Quarterly Report, September 15 (1960).
[f] J. W. Heberle, H. A. Reich, and P. Kusch, *Phys. Rev.* **104**, 1585 (1956).
[g] M. H. Prior and E. C. Wang, *Phys. Rev. Lett.* **35**, 29 (1975).
[h] M. M. Sternheim, *Phys. Rev.* **130**, 211 (1963).

The hfs measurements on the H_2^+ ion are of fundamental importance for molecular theories since H_2^+ is the simplest of all molecules. The Schrödinger equation can be exactly solved in the approximation of stationary nuclei and electronic wave functions have been calculated with high accuracy.[50,51] The hfs interaction, however, is only reasonably well established and theoretical values for the hfs splitting differ considerably.[52,53] In this situation, existing theories were tested by the measurement of suitably chosen hfs transitions on stored molecular ions.

The hyperfine structure of H_2^+ is calculated using the following Hamiltonian for the hfs interaction:

$$H = b\mathbf{I} \cdot \mathbf{S} + cI_z S_z + d\mathbf{S} \cdot \mathbf{K} + f\mathbf{I} \cdot \mathbf{K}$$

Theoretically the coupling constants b, c, d, f can only be approximately determined and depend on the average internuclear separation. $\mathbf{S}$, S_z, $\mathbf{I}$, and I_z are the electron spin angular momentum operator, its z component, and the nuclear spin angular momentum operator and its z component, respectively, and $\mathbf{K}$ is the rotational momentum operator of the two nuclei. The presently known data are listed in Table 3.

Table 3. Transition Frequencies and Hamiltonian Coefficients[a] of H_2^+ for Vibrational States $v = 4\text{–}8$, and Rotational States $K = 1$ and $K = 2$[b]

v	Frequencies between states $\lvert F_2 F\rangle \leftrightarrow \lvert F_2' F'\rangle$					
	$\lvert\frac{3}{2},\frac{3}{2}\rangle-\lvert\frac{3}{2},\frac{5}{2}\rangle$	$\lvert\frac{3}{2},\frac{3}{2}\rangle-\lvert\frac{3}{2},\frac{1}{2}\rangle$	$\lvert\frac{1}{2},\frac{3}{2}\rangle-\lvert\frac{1}{2},\frac{1}{2}\rangle$	$\lvert\frac{3}{2},\frac{5}{2}\rangle-\lvert\frac{1}{2},\frac{3}{2}\rangle$	$\lvert\frac{3}{2},\frac{3}{2}\rangle-\lvert\frac{1}{2},\frac{3}{2}\rangle$	$\lvert\frac{5}{2}\rangle-\lvert\frac{3}{2}\rangle$
4	5.721	74.027	15.371 15.371407(2)[c]	1270.550	1276.271	81.121
5	5.258	68.933	14.381 14.381513(2)[c]	1243.251	1248.509	75.601
6	4.817	63.989	13.413 13.413460(2)[c]	1218.154	1222.971	70.231
7	4.395	59.164	12.461	1195.156	1199.551	64.977
8	3.989	54.425	11.517	1174.169	1178.159	59.804

	Coefficients				
	b	c	d	f	d
4	804.065	98.034	32.636	0.038	32.448
5	788.846	91.180	30.421	0.036	30.240
6	775.006	84.540	28.266	0.034	28.092
7	762.494	78.074	26.156	0.032	25.991
8	751.271	71.733	24.080	0.030	23.922
		$K = 1$			$K = 2$

[a] K. B. Jefferts, *Phys. Rev. Lett.* **23**, 1476 (1969).
[b] All values are given in MHz and a uniform error of ± 1.5 kHz was assigned.
[c] S. C. Menasian, Thesis, University of Washington (1973).

6. Conclusion

Spectroscopy of stored ions has a wide range of possible applications. It can be applied to charged atomic systems with preferably paramagnetic ground states. A number of collisional reactions is being used for polarization and detection of polarization changes, such as, for instance, spin exchange, spin-dependent charge transfer, spin-dependent excitation transfer, ion–molecule reactions, electron and photon impact ionization, and state-selective photodissociation. However, so far, only a handful of precision experiments on stored ions have been carried out. It appears that there exists considerable potential for improvement of the technique. The experiments described in Sections 4.2 and 4.3 above show the way in which future efforts may have to proceed.

New methods for stored ions will employ ultra-narrow-banded dye lasers and precision techniques already proven for atoms, such as, for instance, saturated absorption and suitable two-photon spectroscopy. Experiments on a single stored ion[54,55] promise an important breakthrough even though an experimental demonstration has yet to be reported. The following elegant scheme has been envisioned.[56] Parallel dye laser beams at 4936 Å and 6499 Å are focused upon a single Ba^+ ion stored in a miniature quadrupole trap. The ion is then excited back and forth between the $6s\,^2S_{1/2}$ ground state and the $5d^9\,6s^2\,^2D_{3/2}$ metastable state emitting fluorescent light at the two wavelengths.

The Ba^+ ion is also a good candidate[57] for a two-photon absorption experiment. The transition of interest from the $5d^{10}\,6s\,^2S_{1/2}$ state to the $5d^9\,6s^2\,^2D_{5/2}$ state occurs at 5630 Å and has the necessary long-lived upper state for narrow resonances.

Precision experiments on stored ions have already provided sensitive tests for accurate quantum electrodynamical and nuclear structure theories of the hfs of hydrogenlike systems.

Technical applications of stored ion methods exist and have been discussed with respect to their potential use as primary frequency standards.[58] Of all existing and proposed devices for primary frequency standards, ion storage tubes exhibit the highest atomic line Q values $\nu/\Delta\nu$ approaching now 10^{10}. The shortcomings of stored ion experiments, namely, low signal-to-noise ratios and a relatively large second-order Doppler shift, have to be minimized in future experiments.

Acknowledgments

This work was supported in part by the U.S. National Bureau of Standards, the National Science Foundation, the Center for Energy and

Mineral Resources of Texas A & M University, the Robert A. Welch Foundation of Texas, and the Research Corporation of New York.

References

1. W. Paul, O. Osberghaus, and E. Fischer, *Forschungsber. Wirtsch. Verkehrsminist. Nordrhein-Westfalen* No. 415 (1958).
2. E. Fischer, *Z. Phys.* **156**, 1 (1959).
3. W. Paul, H. P. Reinhardt, and U. von Zahn, *Z. Phys.* **152**, 143 (1958).
4. H. G. Dehmelt, *Advan. At. Mol. Phys.* **3**, 53 (1967); **4**, 109 (1969).
5. C. B. Richardson, K. B. Jefferts, and H. G. Dehmelt, *Phys. Rev.* **165**, 80 (1968).
6. M. N. Benilan and C. Audoin, *Int. J. Mass Spectrom. Ion Phys.* **11**, 421 (1973).
7. R. Ifflaender and G. Werth, *Metrologia* **13**, 167 (1977).
8. R. F. Wuerker, H. Shelton, and R. V. Langmuir, *J. Appl. Phys.* **30**, 342 (1959).
9. P. L. Kapitsa, *Zh. Eksperim. i Teor. Fiz.* **21**, 588 (1951).
10. P. H. Dawson and N. R. Whetten, *J. Vac. Sci. Technol.* **5**, 1 (1968).
11. P. H. Dawson and N. R. Whetten, *Int. J. Mass Spectrom. Ion Phys.* **2**, 45 (1969).
12. M. Baril and A. Septier, *Rev. Phys. Appl.* **9**, 525 (1974).
13. P. H. Dawson and C. Lambert, *Int. J. Mass Spectrom. Ion Phys.* **16**, 269 (1975).
14. H. A. Schuessler, E. N. Fortson, and H. G. Dehmelt, *Phys. Rev.* **187**, 5 (1969).
15. H. G. Dehmelt and F. L. Walls, *Phys. Rev. Lett.* **21**, 127 (1968).
16. K. B. Jefferts, *Phys. Rev. Lett.* **20**, 39 (1968).
17. H. A. Schuessler, *Bull. Am. Phys. Soc.* **13**, 1674 (1968).
18. F. G. Major and H. G. Dehmelt, *Phys. Rev.* **170**, 91 (1968).
19. H. A. Schuessler, *Metrologia* **13**, 109 (1977).
20. M. H. Prior and E. C. Wang, *Phys. Rev. Lett.* **35**, 29 (1975).
21. K. H. Kingdon, *Phys. Rev.* **121**, 408 (1923).
22. F. G. Major and G. Werth, *Phys. Rev. Lett.* **30**, 1155 (1973).
23. M. D. McGuire, R. Petsch, and G. Werth, in Abstracts of the Fifth International Conference on Atomic Physics, p. 407 (1977).
24. J. L. Duchene, C. Audoin, and J. P. Schermann, *C.R. Acad. Sci.* 24 Mai (1976).
25. H. G. Dehmelt and K. B. Jefferts, *Phys. Rev.* **125**, 1318 (1962).
26. K. B. Jefferts, *Phys. Rev. Lett.* **23**, 1476 (1969).
27. R. S. Mulliken, *J. Chem. Phys.* **7**, 20 (1939).
28. S. C. Menasian, Thesis, University of Washington (1973).
29. D. Church and H. G. Dehmelt, *J. Appl. Phys.* **40**, 3421 (1969).
30. J. André and J. P. Schermann, *Phys. Lett.* **45A**, 139 (1973).
31. J. André, *J. Phys. (Paris)* **37**, 719 (1976).
32. H. Dehmelt, *Nature (London)* **262**, 777 (1976).
33. M. Arditi and R. R. Carver, *Phys. Rev.* **112**, 449 (1958).
34. J. P. Barrat and C. Cohen-Tannoudji, *J. Phys. Radium* **22**, 329, 443 (1961).
35. B. S. Mathur, H. Tang, and W. Happer, *Phys. Rev.* **171**, 11 (1968).
36. S. Yeh and P. Stehle, *Phys. Rev. A* **15**, 213 (1977).
37. C. Schwartz, *Ann. Phys. (N.Y.)* **2**, 156 (1959).
38. H. A. Schuessler, *Phys. Lett.* **30A**, 350 (1969).
39. H. A. Schuessler, *Appl. Phys. Lett.* **18**, 117 (1971).
40. F. G. Major and J. L. Duchene, *J. Phys. (Paris)* **36**, 953 (1975).
41. H. A. Schuessler, *Bull. Am. Phys. Soc.* **16**, 532 (1971).
42. H. A. Bethe and E. E. Salpeter, in *Quantum Mechanics of One and Two-Electron Atoms*, p. 110, Springer-Verlag, Berlin (1957).

43. D. Greenberg and H. M. Foley, *Phys. Rev.* **120**, 1684 (1960).
44. D. E. Zwanziger, *Phys. Rev.* **121**, 1128 (1961).
45. M. M. Sternheim, *Phys. Rev.* **130**, 211 (1963).
46. P. T. Olson and E. R. Williams, Proceedings of the Fifth Conference on Precision Electromagnetic Measurements, Paris (1975).
47. T. W. Haensch, *Phys. Rev. Lett.* **32**, 1336 (1974).
48. W. L. Williams and V. W. Hughes, *Phys. Rev.* **185**, 1251 (1969).
49. A. H. Wapstra and N. B. Gove, *Nucl. Data Tables* **9**, 267 (1971).
50. D. R. Bates, K. Ledsham, and A. L. Stewart, *Phil. Trans. R. Soc.* (*London*) **246**, 215 (1953).
51. D. M. Bishop, *Mol. Phys.* **28**, 1397 (1974).
52. W. B. Somerville, *Mon. Not. R. Astron. Soc.* **139**, 163 (1968).
53. S. K. Luke, *Astrophys. J.* **156**, 761 (1969).
54. H. G. Dehmelt, *Bull. Am. Phys. Soc.* **18**, 1521 (1973).
55. D. J. Wineland, R. E. Drullinger, and F. C. Walls, *Phys. Rev. Lett.* **40**, 1639 (1978).
56. H. G. Dehmelt and P. Toschek, *Bull. Am. Phys. Soc.* **20**, 61 (1975).
57. P. L. Bender, J. L. Hall, R. H. Garstang, F. M. J. Pichanick, W. W. Smith, R. L. Barger, and J. B. West, *Bull. Am. Phys. Soc.* **21**, 599 (1976).
58. H. A. Schuessler, *Metrologia,* **7**, 103 (1971).

23

The Spectroscopy of Atomic Compound States

J. F. WILLIAMS

1. Introduction

This brief report on the spectroscopy of compound states of atoms presents a discussion of their energy levels, widths, spectral classification, and modes of decay. Generally, mention is not made of intensities or oscillator strengths as reliable estimates are frequently not available because of the difficulties of absolute calibration of apparatus. Reference to theoretical treatments of the subject is made only to clarify the nature of the levels. Diagrams and tables of energy levels are given. The comprehensive review by Schultz[1] in 1973 considered atomic compound states primarily as short-lived, excited states of negative ions which arose either from the temporary capture of electrons in collision with neutral atoms or from the excitation of fast negative ions in various collision processes with gaseous targets. The interesting feature of these negative ion states is that they overlap an adjacent continuum of states of the neutral atom with a consequent interaction of configurations which strongly influences their nature. A more general concept of a compound state includes all states whose nature is determined by configuration interaction effects rather than by an independent particle description and so, in the widest sense, includes nearly all atomic states. The following section discusses some general features of atomic states and then narrows the contents of the paper to atomic states that lie in a continuum.

J. F. WILLIAMS • Department of Pure and Applied Physics, Queen's University of Belfast, BT7 1NN, Northern Ireland.

State Description

The setting for this paper is developed from a simple picture of the bound stationary states of an atom. In terms of a zero-order approximation,[2] each energy state may be described by an independent-particle model with a single configuration in which the electrons are assigned to single-particle, hydrogenlike orbitals. The total wave function is given by an antisymmetrized product of orbitals, $\phi_i(nlms)$, which are populated according to the Pauli exclusion principle. The orbitals satisfy the self-consistent field equation $H\phi_i = E\phi_i$, where the one-particle Hamiltonian has the form $H = \Sigma_i f_i + \Sigma_{i,j} g_{ij}$ and the $f_i = -\Delta_i^2 - 2Z/r_i$ and $g_{ij} = 2/r_{ij}$. The f_i operate on the coordinates of only one electron while the g_{ij} operate on the coordinates of pairs of electrons and so include the averaged affects of the interelectron interactions represented by the Coulomb and exchange potentials of all the other electrons in the atom. The most elaborate and well-known procedure of this type is the Hartree–Fock self-consistent field method.[3] The traditional analyses of atomic optical spectra have been made from this basis, that is, a single configuration may be assigned to each energy level. It forms the practical method of labeling spectral lines.

The theoretical positions of the term values arising from a given electron configuration are determined by diagonalizing the combined electrostatic and spin–orbit interaction energy matrices.[2] For the simplest configuration, theory predicts certain interval ratios which may be compared directly with experiment. For other cases where there are more term values than radial integrals, it has become accepted practice to treat the latter as adjustable parameters and their values are determined to give the best least-squares fit to the experimental spectrum. Traditionally this approach has been used to analyze photon emission and absorption spectra and it now also forms the basis for assignment of simple scattered and ejected electron spectral lines into series.

As early as 1930[4] it was recognized that the assignment of precise configurations to term levels was an approximation. By considering the nondiagonal elements between configurations of the same parity, the energies of the terms may be displaced from the position where they would be found in the absence of such interaction and the wave function of each term becomes a linear combination of wave functions associated with the interacting configuration. Then the wave function may be approximated by

$$\psi = a_0\phi_0 + \sum_i a_i\phi_i \qquad (1)$$

where ϕ_0 is a reference configuration and ϕ_i are the correlation configurations. The many theoretical papers on configuration interaction, or correlation, effects are reviewed by Jucys[5] for ground states, by Weiss[6] for excited states, and by Burke[7] for photon and electron scattering processes. Some conceptual understanding of the diversity of the effects of configuration interactions upon spectra is obtained from the work of Nicolaides,[8] who also gives references to recent theoretical work.

Many examples of configuration interaction effects, mainly for low-lying states of neutral atoms, are given in the above papers. Experimentally these effects are revealed mainly as line perturbations, asymmetries, and intensity peculiarities. They are most readily studied by considering isoelectronic spectra. These perturbations may be either weak or strong. In the former case the matrix elements coupling the perturbing configuration to a series are localized to several lines, while in the latter case the perturber is strongly coupled to all members of a series. An example[6] of a strongly perturbed series is the aluminium ion $3snf\ ^3F$ series perturbed by the $3p3d\ ^3F$ state which lies between the $6f$ and $7f$ lines. The perturbation results in an asymmetric distribution of oscillator strengths with a peak above the perturbing state such that the envelope of the distribution resembles the autoionization line shapes to be discussed later. A frequently quoted example[9] is found in the magnesium isoelectronic series for the perturbations between the ^1D and ^3D terms of configurations $1s^2 2s^2 2p^6 3s3d$ and with the ^1D term with the doubly excited $3p^2$ configuration. In Al II and Si III the $3p^2\ (^1D)$ level lies below the two $3s3d$ levels but in Mg I the $3p^2(^1D)$ lies above the first ionization limit. These excited states are sufficiently mixed that any single configuration has little meaning. Other examples are the $2s^2$ and $2p^2$ mixing in the ground state of neutral beryllium[10] and the perturbation of the $3s^2 3p$ and $3s^2 nd(^2D)$ series by the $3s3p^2(^2D)$ state of the aluminum sequence[11] which have been studied in optical spectra. Generally such effects are well understood for the lighter atoms. Heavier atoms are discussed elsewhere in this book. In spite of these, and other, demonstrated correlation effects it is still widely assumed in the literature that it makes sense to label a state by a single configuration, even if only for ease of discussion.

Recent interest in compound states concerns cases where, in Eq. (1), the reference configuration overlaps a Rydberg or continuum series. This report will now be generally confined to such cases.

Spectroscopic studies are no longer limited to observations of photons but include the use of electrons, metastable atoms, and ions as either probe or observed particles. The role of these particles in studying doubly excited compound states has been discussed by Fano[12-15] and can be inferred from Figure 1,[13] which shows some transition processes for the helium atom. For the simplest case of such two-electron atoms (H⁻, He, Li⁺, etc.) there is a set of configurations $(1s, nl)$, in an independent-particle model, in which one electron is always in the lowest $(1s)$ orbital and an infinite number of sets of configurations $(n'l', nl)$ in which both electrons are in

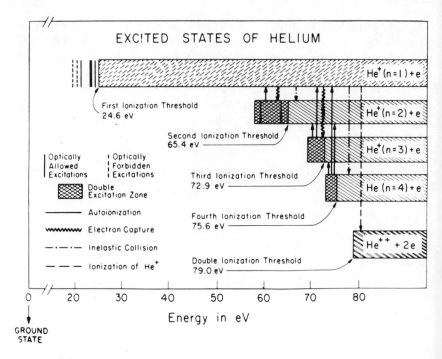

Figure 1. Diagram showing several channels in the spectrum of helium and their inter-connection via autoionization processes. Also shown are other interchannel transition processes which are implicitly related to autoionization. (Figure taken from Fano, [13] Figure 1.)

excited orbitals. For each orbital $(n'l')$ there are an infinite number of series of Rydberg states converging to a higher ionization limit and each series overlaps at least one continuum. In the neighborhood of the energy of a given excited state the true wave function is then a superposition of the multiexcited and continuum configurations of the form of Eq. (1). The strength of the interaction determines the finite lifetime of the compound state which has been labeled as a' short-lived, unstable, metastable, autoionizing, autodetaching, autoemitting, or autoejecting state depending upon the collision process under study. In nearly all cases their quasi-stationary nature has produced structure in an absorption or emission spectrum and it has been this structure that has initially aroused experimental interest. If the state is observed in inner shell ionization or excitation of the neutral or negative systems, they have been referred to as Auger, autoionizing, or autodetaching states.

The above considerations are readily generalized for multielectron atoms for which there are four different types of compound states: (a) multiply excited states with two, or more, simultaneously excited states, for example, the $2s^2 2p$ state of He^-; (b) a subshell electron may be excited to an outer orbital, for example, the $3s3p^6 4p$ state of argon; (c) an inner-shell vacancy, for example, the $2s3s^2 3p^6$ state of argon, may autoionize to the $2s^2 3s^2 3p^4$ state in an Auger transition; (d) a rearrangement of core electrons for example, the $(2p^3\ {}^4S)2p\ {}^3P$ state of oxygen can be excited to the $(2p^3\ {}^2D)\ 4s\ {}^3D$ which is above the first ionization potential. Detailed descriptions of the many aspects of such compound states have been reviewed from a theoretical viewpoint by Burke,[7,16] Smith,[17] Fano,[13,15] and Fano and Cooper[18] and from an experimental viewpoint by Schulz,[1] Ehrhardt,[19] Melhorn,[20] Rudd and Smith,[21] and Krause.[22]

The choice of incident (or probe) particle should be determined primarily by the type of coupling required in order to study a given aspect of the atomic state. This principle is trivial for electronic excitation, which simply requires electrons, photons, or atoms to provide electromagnetic coupling. The incident particle is then further selected by the strength of the coupling (cross section) and the selection rules[21] for each particle. For the multiply excited levels which have excitation energies above an ionization limit, state deexcitation may occur either via autoionization or radiation decay. The autoionization rate R is given simply as $R = (4\pi/h)|\langle k|Q|i\rangle|^2 \delta(E_i - E_k)$ where the electrostatic operator $Q = \Sigma\, e^2/r_{ik}$. Typical autoionization lifetimes of 10^{-15}–10^{-13} sec are obtained for the Coulomb interaction selection rules of $\Delta J = 0$, $\Delta L = 0$, and $\Delta S = 0$ for LS coupling and no parity change. If there are no available continuum states then autoionization via magnetic interactions, such as spin–orbit, spin–other-orbit, and spin–spin interactions are possible with lifetimes of the order of R times α^{-4}, or longer, where α is the fine-structure constant. Alternatively photon emission, with lifetimes of the order of 10^{-11}–10^{-8} s for uv to visible radiation is possible. The dipole radiative decay rate, A_{ik}, is given by the well-known formula $A_{ik} = \frac{4}{3}\alpha w^2/e^2 |\langle k|Q|i\rangle|^2$, where the dipole operator Q is equal to $\Sigma\, r_{ik}$. The absorption oscillator strength, f, is equal to $1.5\lambda^2(g_i/g_k)A_{ik}$, where g_i and g_k are the statistical weights of the upper and lower states, respectively.

2. Experimental Considerations

The method of preparation of compound states is simple by either electron,[23] photon,[24] neutral atom,[25] or positive[26] or negative[27] ion interaction with an atom, which may be either in gaseous[23] or solid[28] form. The method of observation is either to study the absorption in the photon[24] or electron[23] beam or to observe the scattered electrons,[29] the ejected electrons,[28,30-33] radiated photons,[34] metastable atoms,[35] ions[36] or to detect, in coincidence, the scattered and ejected electrons.[37]

Table 1. Summary of Spectroscopic Methods for Studying Atomic Compound States

Method	Incident particle	Best resolution	Target	Collected particle	Reference
1. Total scattering	Electron	0.04 eV	Atomic H	Electron (transmitted)	Sanche and Schulz[23]
2. Photon absorption	Photon	Optical	Rare gases	Photon (transmitted)	Madden and Codling[38]
3. Elastic scattering	Electron	0.06 eV	Atomic H	Electron	McGowan[39]
4. Inelastic scattering	Electron	0.03 eV	He, Ne, Ar	Electron (scattered)	Kuyatt et al.,[40] Ehrhardt[19]
5. Ejected electron detection	He	~0.5 eV	He	Electron (ejected)	Berry[42]
	H^+, H_2^+, He^+	0.25 eV	He	Electron (ejected)	Rudd[44]
	H^-, O^-, Cl^-, Br^-	0.06 eV	Rare gases	Electron (ejected)	Edwards et al.[43]
	Electron	0.012 eV	Rare gases	Electron (ejected)	Read et al.[30]
	Li	0.5 eV	Carbon foil	Electron (ejected)	Andra et al.,[28] Sellin[45]
	He^+		Carbon foil	Photon	Berry et al.[46]
6. Deexcitation photon	Electron	—	He, Ar	Photon	Heddle et al.,[34] Heidemann et al.[47]
	Electron	0.45 eV	Ba^+, Ng^+	Photon	Dunn et al.[48]
7. Ion/atom detection	Electron	0.12 eV	Rare gases	Ion	Marmet et al.[36]
	Electron	—	He^+	Ion He^{++}	Daly and Powell[41]
	Electron	0.05 eV	Rare gases	Metastable atom	Pichanick and Simpson[49]
8. (e, 2e) coincidence	Electron	0.6 eV	He	Scattered and ejected electrons	Weigold et al.[37]

A summary of these spectroscopic methods with other examples[38-49] is given in Table 1. Apparatus details are given in the references. The choice of the incident and observed particles is subject also to various experimental considerations. The great advantage of narrow linewidth of a photon probe[50,51] is counterbalanced in the visible and uv wavelength range by the limited range of a tunable dye laser and in the far uv range by the relative difficulties of working with synchrotron radiation. The observation of photons is more difficult than electrons or ions because of the limited wavelength response and polarization dependence of the photon detector and the large size of monochromators. Photons are a selective probe for dipole transitions and are useful for studying states whose separations and widths are less than the best $(10^{-2}$ eV) electron energy resolution.

Incident heavy particles permit the study of compound states of highly charged ions in highly excited states[38] and of negative ions such as O^-, C^-, Cl^-, and even H^{2-}, which are not readily studied by temporary electron attachment to the appropriate parent ion or atom. Their main disadvantages are that all states are populated (in contrast to the photon probe) and that the positions and widths of the observed spectral lines are strongly influenced by kinematic effects, for example,[27] the 9–6 eV (^1S)H^- line can be shifted at least from 0.5 to 30 eV laboratory energy depending upon the ion energy and observation angle.

The most sensitive method of negative ion compound state detection is still the transmission method[1] for incident electrons in which the absorption at zero scattering angle is measured. The high sensitivity results from the fact that electrons scattered throughout 4π solid angle appear as absorption from the incident beam. This high sensitivity is gained, however, at the loss of information derived from differential scattering measurements about the symmetry of the compound state.

Since the review of Schulz[1] in 1973, refinements[30] have been made in the methods involving detection of the ejected electrons. The energies of the electrons ejected from an autoionizing state are independent of the incident particle energy. This implies that the incident (electron) beam can be formed from a high-intensity gun, rather than a monochromator, because good energy resolution is required only in the analyzer. By operating the gun and analyzer in a "constant-energy-loss mode,"[30] in which the incident electron energy is scanned synchronously with the analyzer voltage, it is possible to separate the ejected electron spectra from the simpler energy-loss spectra. This ambiguity does not arise for heavy particle impact on gases, which has its own difficulty of interpretation arising from the Doppler-shifted spectra of electrons from the incident atoms relative to the spectra from the target atoms.

3. Interpretation of Spectra

3.1. Threshold Spectra

For incident electron energies near an autoionizing state energy considerable care must be taken in deducing the state energy from the measured energies of either the scattered or ejected electrons. The scattering process can be represented by

$$e(E_0) + A \rightarrow (A^-)^{**} \rightarrow A^{**}(E_c) + e(E_s)$$

$$\downarrow$$

$$A^+(E_i) + e(E_e) + e(E_s)$$

Energy conservation gives $E_c = E_0 - E_s = E_i + E_e$. It has been found that the ejected electrons[30] do not appear at the expected threshold of $E_0 = E_c$ but at a higher energy $E_0 = E_c + \Delta E$ while the scattered electrons[52] appear at a lower energy $E_0 = E_c - \Delta E$. As E_0 increases above the threshold energy, ΔE becomes progressively smaller. The effect is largest for shorter lifetime states. This behavior has been explained[53-58] by a classical post-collision Coulomb interaction and resulting energy exchange between the slower electron and the faster scattered electron. If the autoionizing state has a lifetime τ, the slow electron of velocity v is ejected at a distance $r = v\tau$ producing an energy shift $\Delta E = (v\tau)^{-1}$. For example, for the $2s^2$, ^1S state of helium this explanation gives $E \approx 1$ eV at threshold, in agreement with the observed value of 0.9 eV. The effect has been studied for the states in helium and for the states in neon.

An exciting interpretation of ejected electron spectra has been reported recently by Morgenstern et al.[59] for He$^+$ on He collision. Structure in the electron spectrum from autoionizing states has been attributed to interference effects arising from the coherent excitation of several states in only one atom and also from the coherent excitation of such states in both the target atom and the incident ion. They propose that for electron impact excitation of autoionizing states, the former mechanism can explain the postcollision interaction observations. Their evidence as shown in Figure 2 strongly suggests their interpretation is correct so that further manifestations of such interference effects may be discernable in electron spectra and a closer examination of previous results may be required.

3.2. Line Profiles

The fundamental quantitative descriptions of the autoionizing line profile in terms of quantum theory were given by Fano[66] and Cooper.[19] The measurement of line profiles was discussed by Fano and others.[67,68]

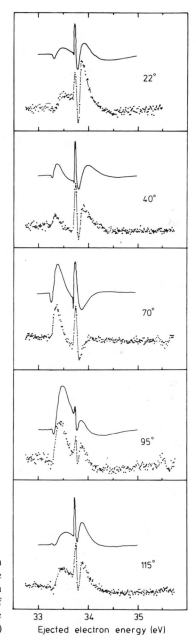

Figure 2. A comparison of the ejected electron spectra from helium obtained by the measurements of Hicks *et al.*[57] (bottom curves) and semiclassical calculations of Morgenstern *et al.*[59] (upper curves). (Figure taken from Morgenstern *et al.*,[59] Figure 2.)

The original formulations concerned photon absorption profiles; however, the measurements now include electron and ion profiles.

The interpretations of observations of the energy and scattering angle of either the ejected electron or the scattered electron are based on the fact that the interference between the direct and compound state scattering amplitudes depends on the electron momentum. Measurements of the momentum of only one of the electrons do not specify, that is, integrates over, the momentum of the undetected electron. The advantages of specifying the momentum of both electrons were realized several years ago by Ehrhardt *et al.*[60] for nonresonant scattering. However, only recently have Balashov *et al.*[61] pointed out from theory the advantages of studying autoionizing states by a coincidence method and the first measurements been made by Weigold *et al.*[37]

Balashov *et al.* used a plane-wave Born approximation to derive the sixfold (coincidence) differential cross section in the form

$$\frac{d^6\sigma}{dk_1\,dk_2}=f(k^1,K)+[a(k^1,K)\varepsilon+b(k^1,K)](1+\varepsilon^2)^{-1} \tag{2}$$

where

$$f(k^1,K)=c|\tau(k^1,K)|^2$$

$$a(k^1,K)=d\,\mathrm{Re}\,\{\tau^*(k^1,K)\tau^L(k^1,K)[q(K)-i]\}$$

$$b(k^1,K)=c\{|\tau^L(k^1,K)|^2[q^2(K+1)]+2\,\mathrm{Im}\,[\tau^*(k^1,K)\tau^L(k^1,K)][q(K)-i]\}$$

$$c=4a_0^2k/(k_0K^4)$$

and $q(K)$ is the Fano profile parameter given by $(q^2-1)/2q=b/a$; $\varepsilon = 2(E-E_r)/\Gamma$, E_r and Γ = energy and width of the resonance, dk and dk^1 are the momenta of the scattered and ejected electrons, respectively, $k_1 = k_0-k$ is the momentum transfer from the incident electron, a_0 is the Bohr radius, and $T(k^1,K)$ is the transition matrix element, which is expressed as the sum of the amplitudes $\tau(k^1,K)$ and $\tau^L(k^1,K)$ for the direct and autoionizing processes, respectively.

The terms of Eq. (2) have a simple interpretation: $f(k^1,K)$ describes the cross section of the direct ionization process, $a(k^1,K)$ describes the asymmetry of the resonance, and $b(k^1,K)$ describes the resonance part of the total cross section. It is noted that the parameters Γ and E_r relate only to the dynamics of the target atom and are independent of the mode of excitation. That is, values determined from photon, electron, and heavy-particle impact experiments should be comparable. This is not true for other parameters. The amplitudes τ and τ^L depend upon the interference terms.

Equation (1) can be integrated to yield expressions in the well-known Fano form for either the scattered or ejected electron distributions; however, in the latter case the terms of the expression are complicated combinations of all the interfering amplitudes corresponding to the various excitation multipoles of the atom. Studies of these distributions have begun[62,63] to produce values of E_r and Γ for autoionizing states in helium. By formulating the profile parameter $q(K)$ as a function of momentum transfer K, Balashov *et al.*[61,64] have calculated the parameter $q(K)$ for the first three ^1P autoionizing transitions in He. They show that the q value of -2.8 ± 0.25 determined for optical experiments,[38] which determine the original Fano q parameter, is consistent with the value of -1.87 determined from 500-eV inelastic electron scattering[65] and with the value of -1.75 ± 0.15 determined by Melhorn[31] for 4000-eV electron inelastic scattering because of the dependence of $q(K)$ upon the momentum transfer studied in each experiment. The approximate constancy of $q(K)$ for $K^2 < 0.5$ a.u. is the reason the interpretation of such ejected electron profiles is possible at all using the Fano equation with $q(K)$ in the optical limit. Balashov[64] provides graphs of the variation of $q(K)$ for $0 < K^2 < 10$ a.u. for $E_0 = 400$ and 4000 eV.

The first coincidence measurements have been made recently by Weigold *et al.*[37] for the unresolved $2s2p(^1D)$ and $2p^2(^1P)$ states of helium as shown in Figure 3. The interference between the resonant and background scattering is clearly seen. The resonant scattering is consistent with the prediction[61] that it should be symmetric about the momentum transfer axis and in the backward direction (of $-K$) should be larger than the direct scattering. Figure 3 also shows the direct and resonant cross-section values determined from the measurements. Emphasis has been given to this initial experiment because of its promise and indications for future work in specifying all the kinematical parameters of the scattering process. As indicated by Weigold *et al.*,[37] when the direct scattering cross section is negligible only the first term of the $b(k^1, K)$ factor of Eq. (2) remains and this factor contains the sum over the magnetic substates. Then it should be possible to determine the scattering amplitudes, rather than cross sections, for the magnetic substates. This is an area of considerable interest in which significant developments can be expected.

The profile of the negative ion autoionizing states is simply analyzed in the scattering channel below the first inelastic threshold, where the differential cross section takes on a simple form, in an expansion of the direct and exchange scattering phase shifts,

$$\frac{d\sigma}{d\theta} = |f(\theta)|^2 + |g(\theta)|^2 \tag{3}$$

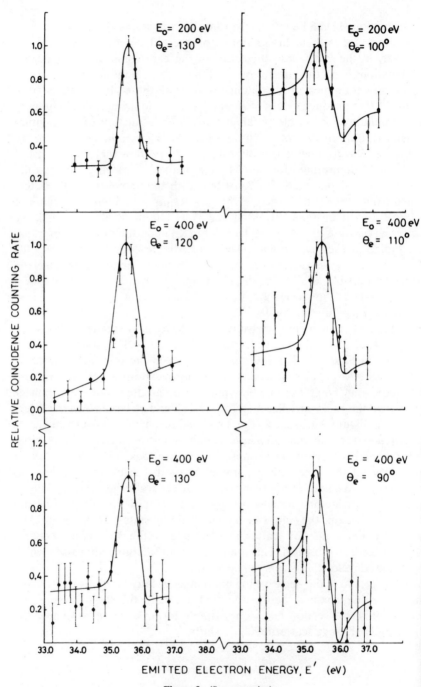

Figure 3. (See opposite)

where

$$f(\theta) = (2ik)^{-1}\left\{\sum_{j,l}(l+1)\{[\exp(2i\delta_{l+\frac{1}{2}})-1]+l[\exp(2i\delta_{l-1/2})-1]\}\right.$$
$$\left.\times P_l(\cos\theta) - P_{l'}(\cos\theta) + (j'+1/2)\exp(2i\delta_{j'})P_{l'}(\cos\theta)\right\}$$

$$g(\theta) = (2ik)^{-1}\left\{\sum_{j,l}[\exp(2i\delta_{l+1/2})-\exp(2i\delta_{l-1/2})]P_{l'}(\cos\theta)\right.$$
$$\left.+(j'-l')\exp(2i\delta_{j'})P_{l'}(\cos\theta)\right\}$$

where $\delta_{j=l\pm1/2}$ are the phase shifts corresponding to scattering for a given l value and alternative spin states; P_l and $P_{l'}$ are Legendre functions, $\delta_{j'}$ is the resonant phase shift, which increases by π radians over the width of the resonance and in one of the phases for an isolated resonance. Equation (3) can be rearranged[35] in the standard Fano form to allow an analysis of the scattering data to yield the resonant state parameters of E_0, Γ, q, and ρ^2. The advantage of this approach is that the symmetry of the state is determined. This information is not available from other experimental methods.

3.3. Line Series

Most of the work on compound states may be described as providing further examples of the classification of lines into Rydberg series of the appropriate charged or uncharged atom. The guidelines for interpreting the experimental spectra were provided by Fano and Cooper,[18] who showed the following for photon absorption measurements:

(i) The line profiles should be approximately equal for all lines of a Rydberg series.

(ii) The linewidths should vary from line to line approximately as the inverse cube of the effective principal quantum number n^*.

(iii) The line intensities generally decrease with increasing n^*.

(iv) The energy levels of a Rydberg series are represented by $E_n = E_\infty - I_H Z^2/(n-\sigma)^2$, where Z is the charge of final state configuration, E_∞ is the energy of the series limit, which can usually be obtained from optical data, and σ is the quantum defect, a very slowly varying function of E_n.

The use of these guidelines will be illustrated in the following section.

Figure 3. The coincident detection of the scattered and ejected electrons for electron impact excitation of the (unresolved) $(2s2p)^1P$ and $(2p^2)^1D$ states of helium. The relative coincidence counting rates are shown as a function of the energy and angle of the ejected electron for incident electron energies of 200 and 400 eV and a scattering angle of 10°. (Figure taken from Weigold et al.,[37] Figure 1.)

4. Atomic Data

4.1. Atomic Hydrogen

Some attention[69,70] has been given to the possibility of the existence of single excited, autoionizing states $(1snl)$ of H^- in order to explain the presence of diffuse absorption bands at 6180, 4890, 4760, and 4330 Å in interstellar spectra. Spence and Inokuti[71] argue that such states are highly improbable on both experimental and theoretical grounds. Their high-sensitivity electron transmission experiment showed that if any structure were present in the region of 0.5–9.5 eV then its maximum intensity was less than a few percent of that of the lowest doubly excited $(2l2l')$ H^- states below 10.2 eV and considerably less than the excited $(3l3l')$ states below 12.1 eV. Their assembled theoretical arguments, based on accurate wave function calculations, variationally calculated phase shifts for electron scattering, and the fact that the spectral distribution of the dipole oscillator strength (photoabsorption cross section) is consistent with all known properties of the H^- ion, all indicate that bound singly excited states of H^- are highly improbable.

Considerably more attention has been given to the doubly excited states. Detailed angular distribution studies,[72] coupled with a phase shift analysis method, have clarified earlier experimental understanding[39,73–75] of the $n = 2$ ^1S, ^3P, and ^1D Feshbach resonances. The earlier data were difficult to interpret because of the prevailing state-of-the-art techniques which limited the signal-to-noise ratio as well as the energy (0.050 eV) and angular (5°) resolutions of the electron beams and analyzers. The ^1S, ^3P, and ^1D nature of the resonances at 9.557, 9.735, and 10.122 eV ± 0.010 eV, respectively, has been shown in the latest data.[72] The energies and widths of these states are in agreement, within the experimental errors, with those values determined by the methods of transmission total cross section[76,77] and H^- ion impact[43] as shown in Table 2(a). Most of the theoretical work on resonances has been concerned with these states in H^- that are used as a test of the validity of approximations prior to their extension to more complicated atoms. Details of the theoretical work are given elsewhere[16,27,78]; but calculated values of the state widths and energies, representative of various methods,[79–85] are given in Table 2(b). The latest theoretical work is discussed in Ref. 80. Risley et al.[43] have given a fairly complete compilation of experimental and theoretical results up to 1974. As pointed out by Bain et al.,[86] the calculated energies should be converted to electron volts using the rydberg for infinite mass ($R_\infty = 13.60583$ eV) for comparison with electron impact experimental energies and by using the rydberg for reduced mass ($R_m = 13.60398$ eV) for comparison with photoabsorption resonance energies. This correction

Table 2.

(a) The Experimental Values for the Energies (in eV) and Widths (in eV) of Lower Resonant States of H^- [a]

State	Electron impact				Ion impact,
	Sanche and Burrow[76]	McGowan[39]	Spence[77]	Williams[72]	Risley et al.[43]
(i) Energies					
$2s^2(^1S)$	9.558(10)	9.56(1)		9.557(10)	9.59(3)
$2s2p(^3P)$	9.738(10)	9.71(3)		9.735(10)	9.76(3)
$2p^2(^1D)$	10.128(10)	10.130(5)			10.18(3)
$2s2p(^1P)$		10.21(2)		10.21(2)	
$3s^2(^1S)$		11.65(3)		11.74(8)	
$3p^2(^1D)$		11.77(2)	11.86(2)	11.85(8)	11.86(4)
$3s3p(^1P)$		11.89(2)		11.94(8)	
$3s4s(^1S)$				12.05(8)	
(ii) Widths					
$2s^2(^1S)$		0.043(5)		0.045(5)	
$2s2p(^3P)$	0.0056(5)	0.009		0.0060(5)	
$2p^2(^1D)$	0.0073(20)				
$2s2p(^1P)$				0.014(5)	

(b) Theoretical Values of Energies (E_r), in eV, and Decay Widths (Γ), in eV, of H^- Resonances

State	Energy	Width	Method	Reference
$(2s^2)^1S$	9.554	0.0411	Feshbach projection operator	Chung and Chen[79]
	9.557	0.0476	Feshbach projection operator	Bhatia and Temkin[80]
	9.557	—	Complex rotation	Doolen et al.[81]
	9.557	0.0472	Kohn variational	Shimamura[82]
	9.557	0.0559	Stabilization	Bhatia[83]
	9.560	0.0475	Close coupling	Burke et al.[84]
	9.571	0.0492		Nesbet and Lyons[85]
$(2s2p)^3P$	9.738	0.0063	Feshbach projection operator	Bhatia and Temkin[80]
	9.740	0.0049	Stabilization	Bhatia[83]
	9.7417	0.0059	Close coupling	Burke et al.[84]
$(2p^2)^1D$	10.124	0.010	Projection operator	Bhatia and Temkin[80]
	10.127	0.0088	Close coupling	Burke et al.[84]

[a] The numbers in parentheses are the probable errors in the last significant digits.

ncreases the values listed by Risley et al. by about $+0.008$ eV. There is
agreement between experiment and theory within the experimental errors,
n describing both the state energies and widths. In order to differentiate
between the various theoretical calculations, the experimental errors need

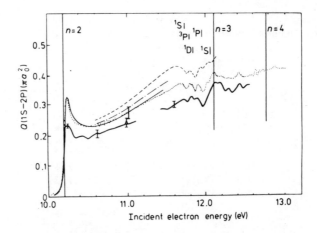

Figure 4. The $2P$ state total excitation cross section is shown
as a function of incident electron energy. Feshbach-type
resonances are seen below, the $n = 3$ and 4 levels of the
neutral atom and a well-defined ^1P shape resonance is
shown just above the $n = 2$ level. The figure is taken from
Williams.[72] The dotted and full curves with error bars,
are photon emission data while other curves are close-
coupling calculated values explained in Ref. 72.

to be reduced by an order of magnitude, which does not appear to be
probable in the near future.

The description of the $n = 3$ states of H⁻ below 12.1 eV is less clear
because of the similar experimental limitation that the best electron energy
resolution is not snall enough compared with the separation of the states. A
total cross-section transmission measurement[77] identified only a single ^1D
peak; no structure has been observed in elastic scattering, while at least
four states have been seen in the $2S$ and $2P$ inelastic decay channels[87,88]
by observing the prompt 1216-Å photons from the $2P$ state, the quenched
$2S$ state photons, and the 10.2 eV energy loss electrons. These data are
shown in Table 2(a). The $2P$ state observations are shown in Figure 4. A
clear interpretation of the Feshbach-type resonant structure of the $n = 3$
states is not possible; however, probable assignments are indicated in
Table 2(a). The assignments are deduced by comparison with the isoelectronic He atom spectra and with theory.

The narrow ^1P shape resonance just above the $n = 2$ level has been
studied[87,88] in both the $2S$ and $2P$ decay channels by observation of the
1216-Å photons. An energy of 10.22 ± 0.02 eV and a width of $0.014 \pm$
0.005 eV are in agreement with theoretical predictions.[16] This resonance
is interesting because it is the simplest example of a shape resonance for

which the binding is provided by the angular momentum barrier and the short-range nuclear attraction. Because the resonance lies close to the $n = 2$ level the angular momentum barrier is wide, with the result that the state width is small. A similar resonance just above the $n = 3$ level is perhaps indicated in experimental data.[87,88]

The predicted $^3P^e$ state at 10.10 eV[89,90] is of interest because it is the lowest even-parity state for which decay via optical emission is allowed. Selection rules forbid decay via autoionization provided LS coupling dominates. It has not been identified in experiments, although the iso-electronic He state at 59.7 eV has been seen via its optical decay to $1snp(^3P)$ states in beam-foil experiments.[46] Optical transitions from odd-parity ^1P states to the $1s2s(^1S)$ ground state have been predicted by Macek[91] from the $1s2p(^1P)$ state (1130 Å) and by Oberoi[92] from the five $(n = 3)$ ^1P states (113–1094 Å) and the five $(n = 4)$ ^1P states (999–980 Å). A recent plasma experiment,[93] specially designed to search for the $(n = 2)$ ^1P optical decay, did not find any of the above lines.

4.2. Inert Gas Atoms

4.2.1. Neutral Atom States

The variety of types of excitation that lead to the observed discrete structure will be considered initially from the point of view of the electron configurations, rather than on an atom-by-atom basis. The inert gases generally have a configuration of the type $(n-1)p^6 (n-1)d^{10} ns^2 np^6 {}^1S_0$. A detailed analysis of the possible autoionizing transitions observed in the prototype photon absorption measurements of Madden and Codling[94] has been given in a number of papers.[95–97] The transitions are produced by the excitation of either one or two electrons from successively more tightly bound shells. The one-electron transitions that they observed, in order of increasing excitation energy, were as follows:

(i)   An outer p electron excitation of the type $np^6 {}^1S_0 \to np^5 ms {}^1P_1$ or $np^5 md {}^1P_1$. These states lie between the $^2P_{3/2}$ and $^2P_{1/2}$ limits and interact with the $p^5 {}^2P_{3/2} \eta s$ and ηd continua.

(ii) An inner s electron is excited $ns^2 np^6 {}^1S_0 \to nsnp^6 mp {}^1P_1$, the upper state lying 10–20 eV above the $np^5 {}^2P$ ionization limits.

(iii) For Kr and Xe an inner d electron can be excited in transitions $(n-1)d^{10} ns^2 np^6 {}^1S_0 \to (n-1)d^9 ns^2 np^6 mp {}^1P_1$.

(iv) In Xe an inner p electron can be excited to an outer ms orbit.

The two-electron excitations, in order of increasing energy, are of the following types: (i) both electrons from the outer p shell $ns^2 np^6 {}^1S_0 \to ns^2 np^4 ml\,m'l'$; (ii) one s and one p electron; and (iii) in Kr and Xe, one d

and one p electron. These transitions concern only those upper states that can be reached by dipole radiation from a $^1S_0$ ground state. For neon, where LS coupling generally prevails, the upper state is a $^1P_1$ state, but for the heavier inert gases LS coupling does not generally prevail, so that all final states having $J = 1$ can be populated. Generally only in the elements of low atomic number have other methods, mainly the beam-foil method, enabled identification of terms of doubly excited states which cannot be reached by dipole excitation from the ground state. So it is to the optical absorption measurements that one must turn in order to develop the basic spectral features.

Only the helium and neon spectra have been analyzed in detail.[95] From the resonance parameters Γ, q, and ρ^2 of the lower states in neon and the derived oscillator strengths, it was established (i) that the onset of the autoionizing states contributes only in a minor way to the total continuum oscillator strength, (ii) the double-electron excitations interact with a smaller fraction of the accessible continua than the single-electron excitations, and (iii) intensity sharing and line perturbations occurred between the double-electron excitation spectra with configurations $2s^2 2p^4(^3P)3s(^2P)np(^1P_0^1)$ and $2s^2 2p^4(1D)\ 3s(^2D)np(^1P_1^0)$, that is with $2s^2 2p^4(^3P)$ and $2s^2 2p^4(^1D)$ grandparent states. These features are characteristic of configuration interaction between series. Subsequent studies of other atomic systems have tended to analyze their data along similar lines. Detailed discussions of the higher-lying states are given elsewhere.[96]

(a) *Helium.* The doubly excited $^1P_1^0$ states are only a small number of the infinity of Rydberg series of doubly excited states of helium having symmetries $^{1,3}S^e$, $^{1,3}P^{0,3}$, $^{1,3}D^{0,e}$, and so on, which converge to the excited states of the ion. They can be discussed in two groups according to whether they can or cannot autoionize under the Coulomb selection rules of no change in parity, L, S, or J. The second group is of the type $2p\ nl^{1,3}L$, where $l = L > 1$, 1, $n > 2$. The problem of classification of these states is seen by considering how the number of terms increases as n increases. For the $2l2l'$ configurations there are six terms, while for the $2l3l'$ configurations there are 20 terms, which all lie within several volts of one another. Considerable progress has been made in studying the two groups of states below the $n = 2$ state of He^+ at 65.4 eV.

The autoionizing states have been observed in electron[32,56,99] and ion[44,63,98] impact experiments. Figure 5 is a typical spectrum from low-energy electron impact studies[56] showing the ejected electron spectrum from the 61–65 eV states. The autoionizing states were excited at a constant 10 eV above threshold so that there are no ambiguities of interpretation arising from threshold effects. Table 3(a) gives values for all the observed structures as well as various calculated values. The four lowest states (not shown in Figure 5) are well identified. Through configuration

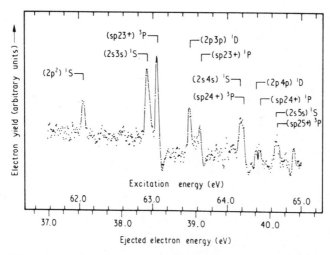

Figure 5. An ejected electron spectrum at an angle of 110° in the region between 61 and 65 eV excitation energy in helium. The doubly excited autoionizing states were excited at a constant 10 eV above threshold by electron impact. (Figure taken from Hicks and Comer,[56] Figure 5.)

interaction, the $^{1,3}P^0$ terms become linear combinations of $2snp$, $2pns$, and $2pnd$ configurations of the type $(2snp \pm 2pns)$. The + states are strongly autoionizing whereas the − states have longer lifetimes. They have been studied theoretically[80,100–106] using a variety of methods as shown in Table 3(a). The latest, and probably the most accurate, values calculated by Bhatia and Temkin[80] using the projection-operator formalism, indicate that there is agreement between theory and experiment but that the accuracy of both should be improved. Similarly, the $^{1,3}S$ terms are combinations of the $2sns$ and $2pnp$ configurations[101,102] and the $^{1,3}D$ terms are combinations of the $2pnp$, $2snd$, and $2pnf$ configurations.[107] The state identifications of Table 3 can only be regarded as probable for other than the first six states because the small separations of the levels are comparable to the experimental resolutions and the lines predicted by theory[107] are unidentified in experiments, for example the ^3P at 63.260 eV, the ^1P at 62.760 eV, the ^1D at 63.874 eV, and the ^3S states.

The nonautoionizing states have been detected only by dipole radiative decay in beam-foil experiments.[46,108–110] Figure 6 shows an energy-level diagram of the associated doubly excited states and their probable configurations. Lines have been observed around 3000 Å and identified as originating from transitions between doubly excited levels, whereas the 300-Å lines arise from transitions from doubly excited to singly excited states. Several general features are discernable.[109] The autoionization

Table 3.
(a) Energies of Autoionizing Levels in Helium^a

State	Experiment (E_r)						Theory (E_r)			
	Hicks and Comer[56]	Madden and Codling[38]	Rudd[44]	Oda et al.[32]	Siegbahn et al.[99]	Bordenave-Montesquieu et al.[63]	Burke and Taylor[104]	Bhatia et al.[102]	Bhatia and Temkin[80]	O'Malley and Geltman[103]
$(2s^2)^1 S$	57.82(4)		57.82(5)	57.90(5)	57.95(3)	57.86(6)	57.842	57.817	57.844	57.832
$(2s2p)^3 P$	58.30(3)		58.34(5)	58.30(5)		58.36(6)	58.317	58.298	58.321	58.300
$(2p^2)^1 P$	59.89(3)		60.0	59.90(5)	59.86(2)		59.911	59.902	59.915	
$(2s2p)^1 P$	60.130	60.130(15)	60.1	60.13	60.12(1)	60.06(6)	60.149	60.143	60.145	60.186
$(2p^2)^1 S$	62.06(3)		62.15(5)	62.00(5)		62.14(6)	62.134	62.062	62.091	62.160
$(2s3s)^3 S$		62.756(10)						62.759		
$(2s3s)^1 S$	62.94(3)		62.95(5)	62.80(5)	62.94(2)	62.96(6)	62.975	62.953	62.962	62.959
$(2p23+)^3 P$	63.07(3)		63.08(5)					63.097	63.107	63.145
$(2p3p)^1 D$	63.50(3)				63.50(2)			63.515	63.526	
$(sp23+)^1 P$	63.65(3)	63.653(7)	63.65	63.50(5)	63.65(2)	63.65		63.667	63.661	63.712
$(2p3p)^3 S$										
$(2s4s)^1 S$	64.18(3)		64.22(5)		64.22(3)	64.25(6)		64.182	64.101	64.314
$(sp24+)^3 P$	64.23(3)		64.22(5)							64.498
$(2p4p)^1 D$	64.39(3)				64.38(3)			64.403	64.415	
$(sp24+)^1 P$	64.45(3)	64.462(7)	64.46(5)	64.40(5)	64.45(2)	64.53(6)				
$(2s5s)^1 S$	64.67(4)		64.71(5)		64.70(2)					
$(sp25+)^3 P$	64.69(4)		64.71(5)							

Table 3 (continued)

(b) Widths (in eV) of Autoionizing Levels in Helium

State	Hicks and Comer[56]	Madden and Coding[38]	Burke and Taylor[104]	Bhatia et al.[106]	Bhatia and Temkin[80]
$(2s^2)^1P$	0.138 ± 0.015		0.124		0.125
$(2s2p)^3P$	< 0.015		0.0090	0.0084	0.0089
$(2p^2)^1D$	0.072 ± 0.018		0.0662	0.0729	0.0729
$(2s2p)^1P$	0.042 ± 0.018	0.038 ± 0.004	0.0388	0.0374	0.0363
$(2s3s)^1S$	0.041 ± 0.010		0.0363		0.0387

[a] All energies are quoted in electron volts. The figures in parentheses denote the expected error in the last digit of the experimentally measured values.

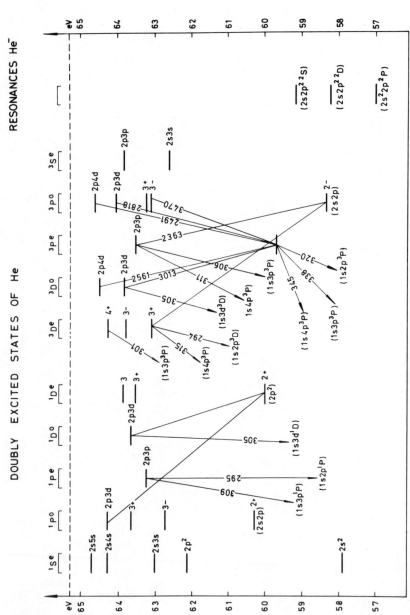

Figure 6. Energy-level diagram for the doubly excited neutral atom and the triply excited negative ion of helium. Those states are shown and are indicated with the transition line and wavelength. All other states (except

rates for many of the triplet states were sufficiently slow that many more radiative transitions originated from triplet than from singlet states. Lines that terminate on strongly autoionizing states, such as the $2P^2\ {}^1D^2$, are considerably broadened by the very short lifetime of the lower state.

Figure 6 also shows the strongly autoionizing states, discussed in the previous paragraph. It appears that nearly all of the expected states have been correctly identified in either the autoionizing or radiative decay channels. The notable exception is the $2s3s\ {}^3S^e$ state at about 62.5 eV, which is still missing from all experimental identifications.

Extensions of these spectral observations on the radiative decays for the lowest doubly excited states have been made for ten members of the helium I isoelectronic sequences. Buchet et al.[111] studied the beam-foil spectrum in the vicinity of 150 Å and observed transitions from $2p^3({}^3P)$, $2p3p({}^3P,\ {}^3D)$, and $2p3d({}^3D)$ doubly excited Li II with lifetimes of about 3×10^{-11} sec. A variety of other techniques have contributed to the data shown in Table 3. Full details of the works and their references are given elsewhere.[109]

(b) Neon to Xenon. Notable additions to the catalog of singly excited autoionizing states that lie between the ${}^2P_{3/2}$ and ${}^2P_{1/2}$ ionization limits have been made by the experiments of Stebbings et al.[112] Atoms in a metastable state are photoexcited by a tunable laser to autoionizing states that are not accessible by dipole transitions from the ground state. The autoionizing state is detected by the resulting ion production. In argon the $3p^5({}^2P_{1/2})np'$ and nf' ($9 < n < 26$) levels and in krypton the $4p^5({}^2P_{1/2})np'$ and nf' ($8 < n < 25$) levels have been observed. The krypton data are shown in Figure 7. These levels did not show detectable (less than 6 cm^{-1}) perturbations so that unambiguous assignments were made from a quantum defect extrapolation of the known lower terms of each series.

The multiply excited states of the heavier inert-gas atoms have not received the same attention as helium. The complexity of the spectra is indicated by noting, firstly the increase in the number of levels in just the photon absorption spectra in passing from He to Ne and, secondly, from Figure 6, how the helium spectrum has evolved beyond the photon absorption spectra. The basic reason for this lack of data arises from the increased experimental difficulty and lack of electron energy resolution for studying regions with a greater density of states. Only in neon has a more detailed spectrum accumulated from electron impact,[113–116] ion impact,[117] and theory.[118] The states of configuration $2P^43snl({}^1S,\ {}^{1,3}P)$ with grandparents $2p^4({}^1S,\ {}^3P,\ {}^1D)$ and $2s2p^6nl({}^{1,3}S,\ {}^{1,3}P)$ which in turn act as parent states of the negative ion resonances are shown later in Figure 12.

4.2.2. Negative Ion States. The resonant states of He$^-$ below the first ionization threshold at 24.56 eV were discussed in detail by Schulz.[1] One outstanding problem concerns the many structures between 19.45 and

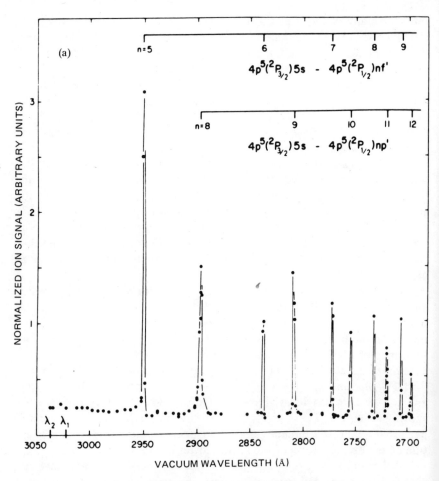

Figure 7. (a) The autoionizing $4p^5(^2P_{1/2})\,np^1$ and nf^1 states of krypton excited by laser impact upon the $5s$ metastable state. (b) (See opposite.) The associated term diagram for part (a). (Figure taken from Dunning and Stebbings,[112] Figure 3.)

20.4 eV reported by Golden and Zecca[119] but as yet unconfirmed by other workers. Golden et al.,[120] in a separate but similar electron-transmission-type experiment, obtained further measurements to support the earlier work; however, Andrick et al.[121] argue, from the basis of a phase shift analysis of elastic angular distribution measurements, that if resonances exist at 19.54 and 19.72 eV then their width must be of the order of 10^{-5} eV rather than the claimed 10^{-3} eV width. An electron energy resolution considerably better than the present 0.030 eV is required for a definitive measurement.

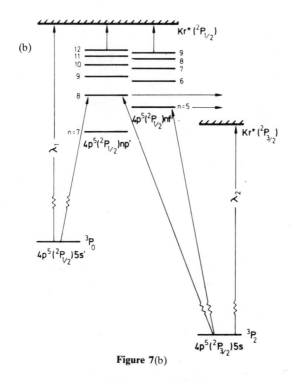

Figure 7(b)

An analysis of the helium resonance series has been made by Heddle.[34] Mainly from measurements of optical excitation functions[122] for the 3,4,5,6 ^1S and 3,4,5 ^3S states, series of the type $1sns^2(^2S)$ and $1snsnp(^2P)$ and possibly $1s3p^2(^2D)$ and $1s3p^2(^2S)$ have been identified as indicated in Figure 8. The classification is based on the identification of the lowest members of the series from electron elastic scattering data and on a quantum defect extrapolation. Interpretation of electron impact data is difficult in this energy region because of threshold effects and closely spaced neutral atom states. The study of the negative ion series would appear to be ideally suited to the technique of tunable laser photon absorption by a low-lying resonant state.

The two lowest doubly excited states of He, the $2s^2(^1S^e)$ at 57.84 eV and $2s2p(^3P^0)$ at 58.32 eV, each support a temporary negative ion resonant state at 57.16 and 58.25 eV which have been classified as $2s^22p(^2P^0)$ and $2s2p^2(^2D^e)$ Feshback resonances,[123] respectively. These negative ion states have been seen in electron-transmission-type experiments[23,40,124] and their decay modes have been identified as (i) single-electron emission[113] into various excited states (2^3S, 2^1S, and 2^1P at least) and (ii) two-electron emission to produce ground-state helium ions.[125,126]

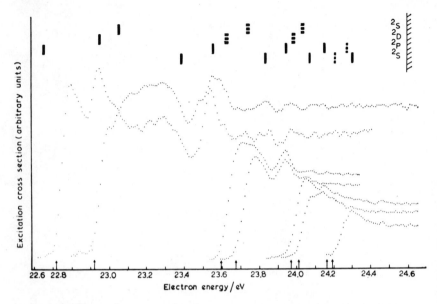

Figure 8. ^2S, ^2P, and ^2D resonance series in helium are shown along the top of the figure above the positions as observed in the excitation functions of the 3^3S, 3^1S, 4^3S, 4^1S, 5^3S, 5^1S, and 6^3S states (curves from left to right) in decay to the 2^3P or 2^1P states. (Figure taken from Heddle,[34] Figure 8.)

A third He⁻ state has recently been seen by Spence[52] using a modulated, potential well technique which is essentially similar to the usual trapped electron method. This feature at 59.0 eV (with a width of 0.4 eV) has been attributed to the configuration $2s2p^2(^2S^e)$ on the basis of arguments by Fano and Cooper[123] and close-coupling calculations by Ormonde et al.[127] This resonance has been identified through its single electron decay to the $2s2p(^3P)$ state, where it is seen as a small peak just above the ^3P threshold. It also probably decays to the $1s2s(^3S)$ state but it has not been seen in the ground-state decay channel.[57] Higher-lying resonances may have been seen in He⁺ decay channels[126] and in optical excitation functions.[47] Some of the observed structure may also be interpreted in terms of doubly excited neutral atoms. Table 4 indicates the energies of the observed structures and probable dominant configurations of nearby He⁻ states predicted[127] by close-coupling theory. The uncertainty of the configuration assignment for these triply excited three-electron states increases as one moves up in energy and the density of states increases. For example, in an eigenvalue-type calculation,[128] the lowest ^2P resonance wave function contains 73% $2s^2p$ configuration with the remainder composed of $2p^3$ and $2s2pnd$ configurations and the lowest ^2D resonance wave function contains 86% $2s2p^2$ plus $2s^2nd$ and $2p^2nd$

Table 4. The Energies (in eV) of Triply Excited Resonant States of He⁻

Dominant configuration	Theory[127]	Experiment		
		Heidemann et al.[47]	Grissom et al.[125]	Quemener et al.[126]
$2s^22p(^2P)$			57.21	57.15
$2s2p^2(^2D)$			58.31	58.23
$2s2p^2(^2S)$	59.4	59.7(?)	59.42	
$2s^23p(^2P)$	59.14		58.79(?)	
$2s2p3p(^2D)$	60.17		60.41	
$2s2p3p(^2S)$	60.35			
$2s3p^2(2D)$	62.95	62.4(?)	61.93	
$2s3p^2(^2S)$	63.32			
$2s3s3p(^2P)$	63.45			

configurations. The study of this spectral region is continuing with considerable vigor.

Below the first ionization potential of the heavier inert gases several interesting studies of negative ion resonances have been made since the review by Schulz[1] in 1973. Phase-shift analysis studies[129] of the elastic scattering in argon as shown in Figure 9 have confirmed the lowest resonances to be $3s^23p^54s^2\,^2P_{3/2}$ and $^2P_{1/2}$ at 11.07 and 11.24 eV with widths of about 0.004 and 0.04 eV and in krypton to be $4s^24p^55s^2\,^2P_{3/2}$ and $^2P_{1/2}$ at 9.47 and 10.11 eV with widths of 0.008 and 0.07 eV, respectively. Confirmations of the argon and krypton data were provided in similar phase-shift analysis experiments by Williams and Willis[130] and Swanson et al.[35]

In krypton, Swanson et al.[35] identified a series of sharp resonances between 10.5 and 12.0 eV to have a probable configuration

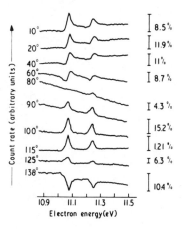

Figure 9. The angular dependence of the $^2P_{3/2}$ and $^2P_{1/2}$ resonances of the negative argon ion below the first excited state of the neutral atom measured by electron elastic scattering with an energy resolution of 0.03 eV. (Figure from Weingartshofer et al.,[129] Figure 2.)

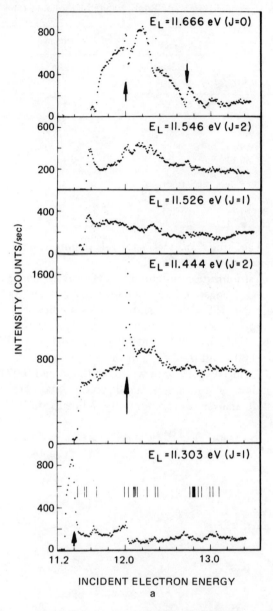

Figure 10. (a) The excitation function of the $4p^5(^2P_{3/2})5p$ ($J = 0, 1, 2, 3$) states of krypton measured by electron energy loss spectroscopy with a resolution of 0.030 eV. The levels of

(*continued opposite*)

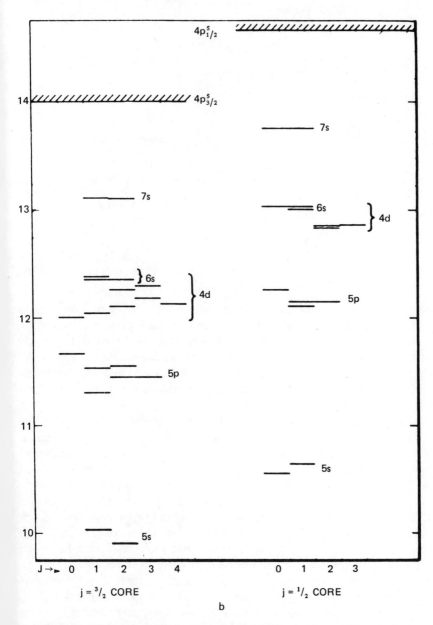

the neutral atom are shown in the lowest section of the figure and the probably negative ion resonance levels are indicated by arrows. (b) The relevant energy-level diagram for krypton. (Figure from N. Swanson et al.,[35] Figure 7.)

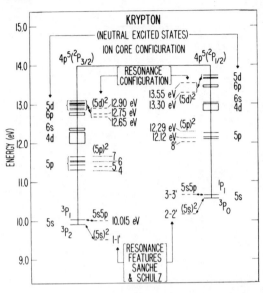

Figure 11. Energy-level diagram of singly excited parent neutral atom states and negative ion resonant states in krypton. The measured resonance energies by Sanche and Schulz show a one-to-one correspondence with nearby neutral atom states. (Figure from Spence and Noguchi,[131] Figure 6.)

$4p^5 5s4d(^2P_{3/2,1/2})$. Their studied decay modes were to the four $4p^5_{3/2,1/2}5s$ and six $4p^5_{3/2}5p$ states, and Figure 10 shows the excitation function for the latter states for $J = 0–3$. The energy-level diagram for the $J = 0, 1, 2$, levels of the $5p$ and $4d$ levels of the neutral atom shown in the bottom of the figure illustrates the complexity of trying to study such closely spaced levels with the current state-of-the-art electron energy resolution of about 0.030 eV.

State identification was asserted by comparison with isoelectronic Rb I. Spence and Noguchi[131] have also studied this resonance region. They identify the same resonances to be the type $4p^5(^2P_{3/2,1/2})5s5p$ and $5p^2$ rather than $5s4d$ or $5s6s$. Their arguments are based on the general guideline that seems to be followed in most negative ion resonance spectra, namely, that the binding energy of an additional electron to a parent Rydberg state is likely to be small or zero when the excited electron has different n and l from the parent. Figure 11 shows a probable energy-level diagram of neutral excited parent states and their associated Feshbach resonances as identified by Spence and Noguchi from an analysis of the data of Sanche and Schulz.[23] Unfortunately neither experiment has sufficient information to resolve the resonance identification problem.

Above the first ionization potential of the inert gases, only Carette *et al.*[29] for neon have made a detailed study of resonances. Figure 12 shows their probable identification of both shape and Feshbach resonances together with their parent states. The decay modes via one-electron emission to the excited states $3s$, $3s'$, $3p'$, $4s'$, and $3d$ as well as to the ground state of the neutral atom were observed. A main point of the identification rests, in turn, upon the correct identification, mainly in optical absorption

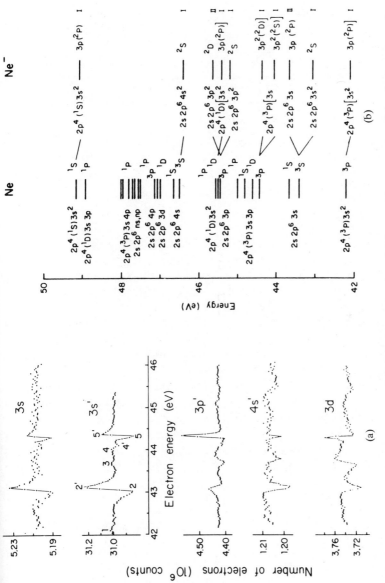

Figure 12. (a) The excitation functions of the $3s$, $3s'$, $3p'$, $4s'$, and $3d$ states of neon measured by electron energy loss spectroscopy with an electron energy resolution of 0.03 eV. (b) The energy diagram of the autoionizing states of neon and their associated negative ion resonant states. The Roman numerals on the right indicate the resonance type as I (Feshbach) or II (shape). (Figure from Roy et al.,[29] Figures 2 and 4.)

Table 5. Excited Levels of the Neutral Krypton Atom and Related Negative Ion
States between 22 and 32 eV

	Neutral atom		Negative ion	
Configuration	Valin and Marmet[133] (± 0.05 eV)		Configuration	Valin and Marmet[133]
(a) Doubly excited				
$(^3P_2)5s^2$	22.83		$(^3P_2)5s^25p$	22.68
$(^3P_{0,1})5s^2$	23.39			
$(^3P_2)5s5p$	24.13		$(^3P_2)5s5p^2$	24.03
$(^1D)5s^2$	24.56		$(^1D)5s^25p(?)$	25.88
$(^1D)5s5p(^3P)$	26.00			
$(^1D)5s4d(^3D)$	26.35			
$(^1D)5s4d(^1D)$	26.77			
$(^1D)5s4d(^1P)$	26.95			
(b) Singly excited				
$(4s4p^65s(^1S)$	23.83			
$4s4p^64d(^3D)$	25.41		$4s4p^64d^2$	24.92

experiments, of the parent states and upon the guidelines of Taylor *et al.*[132] that when the parent atom has many terms with the same electronic configuration, the reference for the determination of the resonance type is the energy position of the lowest term. The literature contains details of the various decay modes and energies of the resonant states,[29] the influence of selection rules on possible state configurations of the parent states[116] and threshold effects upon state energies.[116]

Recently the Laval group[133] have studied the krypton spectrum between 22 and 32 eV. They have identified new lines of the neutral atom of the type $4s4p^6nl$ and $4s^24p^4nl$, $n'l'$ and several related negative ion states as shown in Table 5. Their work deserves mention because of the empirical techniques used to sort out a complicated spectrum in which very few levels have been positively identified. Unfortunately it is generally not possible to unambiguously assign term values to many of the resonant states because it is frequently not clear whether the order of the parent excited states is the same as the ordering of the associated resonances. In other words, more attention should be given to the coupling in both the parent and negative ion states.

4.3. Alkali Atoms

4.3.1. Neutral Atom States

Early work on doubly excited lithium spectra was confined to the study of $1s2s2p(^4P)$ metastable states near 57.3 eV in electron impact experiments.[134] Their study remains one of the best in the literature on the properties of such states, including the selection rules. The early photon absorption measurements[135] identified states in the region from 50 to 70 eV arising from transitions $(1s^2 2s)^2 S_{1/2}$ to $1s2snp$ for $n = 2$–10 and to $1s2pns$ for $n = 2$–11, that is one inner s electron excitations and simultaneous inner s and outer s electron excitations. Subsequent experiments recording ejected electron spectra from fast Li^+ beams incident upon a foil[28] and fast H^+ and He^+ beams incident upon Li vapor[137] have extended the spectra up to 160 eV with transitions of the type $(1s^2 2s)^2 S_{1/2}$ to $(2snl)^1 S$, $^1 P$, $^3 P$ of Li^+ with double K shell vacancy. Transitions involving quartet states were also seen in the beam-foil data that are reported elsewhere in this book. Figure 13 shows the 50–60 eV region of the lithium vapor spectra with 0.12 eV resolution. The $^4 P$ state at 52 eV is strongly excited with He^+ but not with H^+. The difference between the He^+ and H^+ impact spectra of Li reflects the ability of the He^+ interactions to allow exchange and spin-flip processes to occur. Numerical values of the state energies are given in Table 6. The lower states are well identified as $1s(2s2p\ ^3P)^2 P$, $1s(2s2p\ ^1P)^2 P$ and the $1s2s\ ^3Snp$ series converging to the $1s2s\ ^3S$ limit of Li^+. Close-coupling calculations[137,28] have assisted the identification of these, and higher, states above the $1s2s\ ^3S$ limit. They

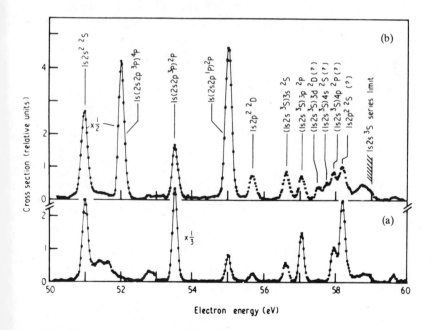

Figure 13. The ejected electron spectra of lithium from (a) 100 keV H^+ and (b) 200 keV He^+ ions incident upon a lithium vapor target. (Figure taken from Ziem et al.,[136] Figure 1.)

show that the appropriate zero-order coupling is $1s(2snp^{1,3}P)^2P$ or $1s(2pns^{1,3}P)^2P$ rather than $(1s2s^{1,3}S)np^2P$ or $(1s2p^{1,3}P)ns^2P$ for the lowest states. These doublet levels of the type $(1s2ln'l')^2L$ converge to four series limits $(1s2s)^{1,3}S$ and $(1s2p)^{1,3}P$ and there is strong mixing of the $(1s2s)^1Snl$ and $(1s2p)^3Pnl$ configurations. The odd-parity $^2P^0$ states of configuration $(1s2s)^1Sns$ show a good example of the effects of perturbation by the $2(^3P)3s$ level which has been pushed down between the $n=6$ and 7 states below the 2^3S threshold by the strong interaction with the $n=3$ states.

Recent studies using the beam-foil technique[138-140] have observed the radiation decay of doubly excited states of Li II and triply excited states of Li I. The positions and radiative lifetimes (of the order of 2×10^{-11} sec) have been calculated for the $(2p3p+2s3d)^3D$ states. The $(-)^3D$ was found to autoionize but the $(+)^3D$ decayed by dipole radiation to the $2p^2(^3P)$, $2p3p(^3P)$ radiating states of Li^+. The weak 155-Å radiation from the $2p^2{}^3P$ state to the $1s3p{}^3P$ was also observed. Also in Li II, the 1036-Å transition between the doubly excited states $2p3d{}^3D \rightarrow 2p^2{}^3P$ was observed. Triply excited states of Li I of configurations $2s^22p$ and

Table 6. Energies (in eV) of Singly and Doubly Excited Autoionizing States of Li I and Li II as Determined from Ejected Electron Spectra from a Li Vapor Target under H^+ and He^+ Ion Impact[a]

	State energy (eV)		
Configuration	Ziem et al.[136] ± 0.010	Ederer et al.[136]	Theory
Li I			
$1s2s^2(^2S)$	56.352		56.51
$1s2s2p(^4P^0)$	57.385		57.47
$1s(2s2p^3p)^2P$	58.912	58.912	58.96
$1s(2s2p^1p)^2P$	60.397	60.398	60.60
$1s2p^2(^2D)$	61.065		61.45
$1s2s(^3S)3s(^2S)$	61.995		
$1s2s(^3S)3p(^2P)$	62.425	62.421	62.46
$1s2s(^3S)3d(^2D)$	62.98		62.91
$1s2s(^3S)4s(^2S)$	63.17		
$1s2s(^3S)4p(^2P)$	63.35	63.358	63.36
$1s2s(^3S)4d(^2D)$	63.58		63.56
Li II			
$2s^2(^1S)$	151.65		151.72
$2s2p(^3P)$	152.41		152.47
$2s2p(^1P)$	155.70		155.94

[a] Data taken mainly from Ziem et al.[136] See Ref. 136 for details of the theoretical values and other experimental values. The assignments of the last four singly excited states is uncertain.

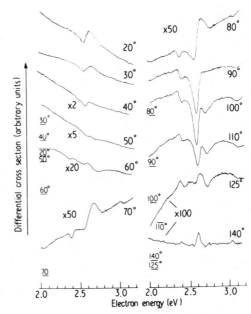

Figure 14. Elastic angular distributions for electron scattering from potassium. The energy range covers the second, third, and fourth inelastic thresholds (the 5^2S at 2.61 eV, 3^3D 2.67 eV, and the 5^2P at 3.06 eV). The zero lines for each trace are indicated by short bars on the lefthand side of each trace for the corresponding angles. (Figure from Eyb,[143] Figure 3.)

$2s2p^2$ at about 80 eV have been identified through their decay via autoionization to the $(1s2s)^{1,3}S$; $^{1,3}P$, nl continua. The beam-foil technique permits the production of these states in large enough numbers for these states to be seen in emission.

4.3.2. Negative Ion States

There are no measurements of negative ion resonances in Li or Na although they have been predicted as mentioned by Schulz.[1] Low-energy electron elastic scattering measurements in sodium[141] and potassium[142] have revealed structure near the lowest excited state, but threshold phenomena (cusps) seem to explain the observations. Recently Eyb[143] identified potassium resonances at 2.4, 2.6, and 2.68 eV in the s, d, and p waves from the angular dependence of elastically scattered electrons as indicated in Figure 14. The resonances are of the Feshbach type associated with the second, third, and fourth lowest excited states.

A tunable laser photodetachment measurement from alkali negative ions by a JILA group[144] has revealed extremely narrow and strong resonant states near the lowest excited ^2P levels of the neutral alkalis. The accuracy of the laser wavelength and the resolution (0.15 Å) allows resonance widths to be measured to two orders of magnitude smaller than can be measured from electron impact studies. The best example of their work is the Cs^- photodetachment cross section for the process $Cs^-(^1S)+$

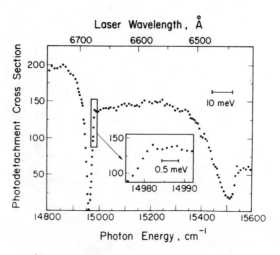

Figure 15. Photodetachment cross section for Cs⁻ showing the resonant state about 0.003 eV below the onset of the Cs($6^2P_{3/2,1/2}$) states. (Figure from Patterson *et al.*,[144] Figure 2.)

$hv \rightarrow \text{Cs}(^2S_{1/2}, {}^2P_{1/2,3/2}) + e$ as shown in Figure 15. The opening of the first inelastic channel ($^2P_{1/2}$) at about $14{,}982 \text{ cm}^{-1}$ (6675 Å) is shown in the inset, and about 3 meV below this threshold the strong Fano profile is attributed to a doubly excited Cs⁻ state. The Rb⁻ state is similar but the resonance is narrower with a width of 0.150 meV. The minima can be explained if, as in the other alkalis, a resonance exists in the ^1P partial wave about 3 meV below threshold which leads to a window-type profile ($q = 0$) in the photodetachment cross section. The resonances have been tentatively identified as of the type $np(n+1)s$. Experiments are in progress to look for resonances associated with higher excited states.

4.4. Atomic Oxygen

The knowledge of O⁻ resonant states has considerably improved with recent experimental[145,146] and theoretical[147,148] studies.

Figure 16 is an energy-level diagram showing the locations of the observed resonant O⁻ states, together with their parent (O) and grand-parent (O⁺) states. The combined uncertainties of experimental (0.05 eV) and theoretical (0.11 eV) values reasonably locates these states. Their most probable configurations have been indicated by theory using the method of configuration interaction, in which a given state is assumed to be adequately represented by a superposition of two particle configurations coupled in an *LS* scheme to a fixed positive-ion core. The dominant configurations are $1s^2 2s^2 2p^3(^4S, {}^2D, {}^2P)\{3s^2; 3s3p; 3p^2; 4s^2\}^{2S+1}L$ and all appear to be Feshbach-type resonances. A shape resonance of width 1 eV appearing in a close-coupling calculation in the ^2P partial wave near 11.5 eV is not readily identified with any of the experimental resonances.

The measured energies of these states are listed in Table 7. The appearance of states in the electron and O^- ion experiments illustrates the selection rules. Photodetachment of ground-state O^- ions can only excite the ^2P state at 9.50 eV; electron impact can excite all doublet and quartet states and O^- ion impact will excite only doublet states with atomic targets if spin exchange scattering is negligible. Thus the identification of the structure at 10.63 eV as a ^6P state, for example, must await confirmation, and it is not clear why the 9.50-eV resonance is not seen in the electron scattering experiment yet appears strongly in ion impact experiments. All of the previous experimental measurements were made with O (or O^-)

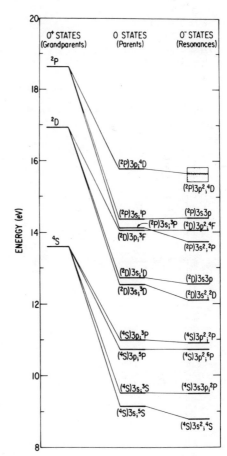

Figure 16. Energy-level diagram of O^+ grandparent, O parent, and O^- resonant states. The dominant O^- states have configurations $1s^2, 2s^2 2p^3$ $(^4S, {}^2D, {}^2P) \, 3s^2; 3s3p; 3p^2; 4s^2 \, {}^{2S+1}L.$ (Figure from Spence,[146] Figure 4.)

Table 7. Comparison of Experimental and Theoretical O^- Resonance Energies and Electron Affinities[a]

Grandparent (parent) resonance configurations	Resonance energies			Electron affinity	
	Theory, Matese[147]	Experiment, Spence[146]	Experiment, Edwards and Cunningham[25]	Matese[147]	Spence[146]
$^4S(3s^5S)3s\,^4S$	8.69	8.78	—	0.456	0.366
$^4S(3s^5S)3p\,^6P$	9.14	—	—	0.008	—
$^4S(3s^3S)3p\,^2P$	9.50	—	9.50	0.018	—
$^4S(3p^5P)3p\,^6P$	10.63	10.73	—	0.016	0.010
$^4S(3p^3P)3p\,^2P$	10.88	10.90	10.87	0.107	0.089
$^4S(4s^5S)4s\,^4S$	11.63	—	—	0.205	—
$^2D(3s^3D)3s\,^2D$	12.05	12.10	12.12	0.484	0.441
$^2D(3s^1D)3p$	—	12.55	—	—	0.178
$^2P(3s^3P)3s\,^2P$	13.65	13.71	13.71	0.473	0.410
$^2D(3p^3F)3p\,^4F$	14.09	14.05	14.06	0.006	0.048
$^2P(3s^1P)3p$	—	14.40	—	—	0
$^2D(4s^3D)4s\,^2D$	14.95	—	—	0.221	—
$^2P(3p^3D)3p\,^4D$	15.75	15.65	—	0.022	0.130
$^2P(4s^3P)4s\,^2P$	16.65	—	—	0.219	—

[a] All energies are in electron volts and are referred to the ground state of atomic oxygen.

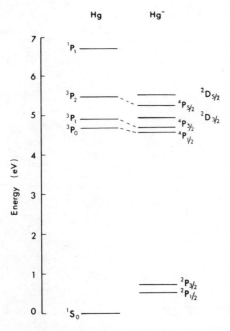

Figure 17. Energy-level diagram of Hg and Hg⁻. (Figure from Heddle,[149] Figure 1.)

beams which may have contained excited atoms (or ions) and did not make angular studies of the scattered electrons to assist state identification.

4.5. Mercury

Low-energy negative ion resonances in mercury have been studied because of their possible use as a source of low-energy spin-polarized electrons. Recently Heddle[149] has analyzed the experimental data[40,150–152] on the basis of comparison of energy levels with the Tl I, Pb II, and Bi III isoelectronic sequence, the importance of magnetic interactions in the excitation of quartet terms and consideration of the decay channels studied in each experiment. The energies of the terms of the resulting $6s^2 6p$ and $6s 6p^2$ configurations are shown in Figure 17. It is seen that the ^4P resonances are bound to their respective ^3P parents by 0.12 $(J = \frac{1}{2})$, 0.21 $(J = \frac{3}{2})$ and 0.23 $(J = \frac{5}{2})$ eV.

Acknowledgments

The author is grateful to his colleagues Dr. R. Browning and Dr. A. Crowe for helpful discussions. This article was completed in August, 1976.

References

1. G. J. Schulz, *Rev. Mod. Phys.* **45**, 378 (1973).
2. J. C. Slater, *Quantum Theory of Atomic Structure*, Vols. 1 and 2, McGaw-Hill, New York (1960).
3. D. R. Hartree, *The Calculation of Atomic Structures*, Wiley, New York (1957).
4. E. U. Condon and G. H. Shortley, *The Theory of Atomic Spectra*, Cambridge University Press, New York (1935).
5. A. P. Jucys, *Adv. Chem. Phys.* **14**, 191 (1969).
6. A. W. Weiss, *Adv. At. Mol. Phys.* **9**, 1 (1973).
7. P. G. Burke, in *Electron and Photon Interactions with Atoms*, Eds. H. Kleinpoppen and M. R. C. McDowell, pp. 1–25, Plenum Press, New York (1976).
8. C. A. Nicolaides and D. R. Beck, *J. Phys. B* **9**, L259 (1976); C. A. Nicolaides, *Phys. Rev. A* **6**, 2078 (1972).
9. R. N. Zare, *J. Chem. Phys.* **45**, 1966 (1966); **47**, 3561 (1967).
10. H. Odabasi, *J. Opt. Soc. Am.* **59**, 583 (1969).
11. C. Froese-Fischer, *J. Quant. Spect. Radiat. Transfer*, **8**, 755 (1968); A. W. Weiss, *Phys. Rev. A* **9**, 1524 (1974).
12. U. Fano and C. D. Lin, in *Atomic Physics*, Vol. 4, Eds. G. Zu Putliz, E. W. Weber, and A. Winnacker, pp. 47–70, Plenum Press, New York (1975).
13. U. Fano, in *Atomic Physics*, Vol. 1, Eds. B. Bederson, V. W. Cohen, and F. M. J. Pichanick, pp. 209–225, Plenum Press, New York (1969).

14. U. Fano, in *Ninth International Conference on the Physics of Electronic and Atomic Collisions*, Eds. J. S. Risley and R. Geballe, pp. 27–39, University of Washington Press, Seattle, Invited Papers (1975).

15. U. Fano and C. D. Lin, in *Eighth International Conference on the Physics of Electronic and Atomic Collisions*, Eds. B. C. Cobic and M. V. Kurepa, pp. 229–238, Institute of Physics, Belgrade (1973).

16. P. G. Burke, *Adv. Phys.* **14**, 521 (1965); *Adv. At. Mol. Phys.* **4**, 173 (1968).

17. K. Smith, in *Physics of One and Two Electron Atoms*, Eds. F. Bopp and H. Kleinpoppen, pp. 559–573, North-Holland Publishing Company, Amsterdam (1969).

18. U. Fano and J. W. Cooper, *Rev. Mod. Phys.* **40**, 441 (1968).

19. H. Ehrhardt, in *Physics of One and Two Electron Atoms*, Eds. F. Bopp and H. Kleinpoppen, pp. 598–611, North-Holland Publishing Company, Amsterdam (1969).

20. W. Melhorn, in *Ninth International Conference on the Physics of Electronic and Atomic Collisions*, Eds. J. S. Risley and R. Geballe, pp. 756–765, University of Washington Press, Seattle (1976).

21. M. E. Rudd and K. Smith, *Phys. Rev.* **169**, 79 (1968).

22. M. O. Krause, in *Atomic Innershell Processes*, Ed. B. Crasemann, Chap. 6, Academic Press, New York (1975).

23. L. Sanche and G. J. Schulz, *Phys. Rev. A* **6**, 69 (1972); **5**, 1672 (1972).

24. D. L. Ederer, T. B. Lucatorto, E. G. Saloman, R. P. Madden, M. Manalis, and J. Sugar, in *Electron and Photon Interactions with Atoms*, Eds. H. Kleinpoppen and M. R. C. McDowell, pp. 69–81, Plenum Press, New York (1976).

25. A. K. Edwards and D. L. Cunningham, *Phys. Rev. A* **8**, 168 (1973).

26. M. E. Rudd and J. Macek, *Case Stud. At. Phys.* **3**, 47 (1974).

27. J. S. Risley, A. K. Edwards, and R. Geballe, *Phys. Rev. A* **9**, 115 (1974).

28. R. Bruch, G. Paul, J. Andra, and L. Lipsky, *Phys. Rev. A* **12**, 1808 (1975).

29. D. Roy, A. Delage, and J. D. Carette, *Phys. Rev. A* **12**, 45 (1975).

30. J. Comer and F. H. Read, *J. Electron Spectrosc. Relat. Phenom.* **1**, 3 (1972).

31. W. Melhorn, *Phys. Lett.* **21**, 155 (1966),

32. N. Oda, F. Nishimura, and S. Tahira, *Phys. Rev. Lett.* **24**, 42 (1970).

33. H. Suzuki, A. Koniski, M. Yamamoto, and K. Wakiya, *J. Phys. Soc. Japan* **28**, 534 (1970).

34. D. W. O. Heddle, in *Electron and Photon Interactions with Atoms*, Eds. H. Kleinpoppen and M. R. C. McDowell, pp. 671–678, Plenum Press, New York (1976); *Contemp. Phys.* **17**, 443 (1976).

35. N. Swanson, J. W. Cooper, and C. E. Kuyatt, *Phys. Rev. A* **8**, 1825 (1973); D. Andrick, *Adv. At. Mol. Phys.* **9**, 207 (1973).

36. E. Bolduc, J. J. Quemener, and P. Marmet, *Can. J. Phys.* **49**, 3095 (1971); *J. Chem. Phys.* **57**, 1957 (1972).

37. E. Weigold, A. Ugabe, and P. J. O. Teubner, *Phys. Rev. Lett.* **35**, 209 (1975).

38. R. P. Madden and K. Codling, *Astrophys. J.* **141**, 364 (1965).

39. J. W. McGowan, *Phys. Rev.* **156**, 165 (1967).

40. C. E. Kuyatt, J. A. Simpson, and J. A. Mielczarek, *Phys. Rev. A* **138**, 385 (1965).

41. N. R. Daly and R. E. Powell, *Phys. Rev. Lett.* **19**, 1165 (1967).

42. H. W. Berry, *Phys. Rev.* **121**, 1714 (1961); **127**, 1634 (1962).

43. A. K. Edwards, J. S. Risley, and R. Geballe, *Phys. Rev. A* **3**, 583 (1971); **10**, 220 (1974).

44. M. E. Rudd, *Phys. Rev. Lett.* **13**, 503 (1964).

45. I. A. Sellin, *Nucl. Instrum. Methods* **110**, 477 (1973).

46. H. G. Berry, I. Martinson, L. J. Curtis, and L. Lundin, *Phys. Rev. A* **3**, 1934 (1971).

47. H. G. M. Heidemann, W. van Dalfsen, and C. Smit, *Physica* **51**, 215 (1971); *J. Phys. B* **7**, L493 (1974).

48. D. H. Crandall, R. A. Phaneuf, and G. H. Dunn, *Phys. Rev. A* **11**, 1223 (1975).
49. F. M. J. Pichanick and J. A. Simpson, *Phys. Rev.* **168**, 64 (1968).
50. W. R. S. Garton, *Adv. At. Mol. Phys.* **2**, 93 (1966); G. V. Marr, *Photoionization Processes in Gases*, Academic Press, New York (1967).
51. T. W. Hansch, in *Dye Lasers*, Topics in Applied Physics, Vol. 1, Ed. F. P. Schafer, Chap. 5, Springer-Verlag, New York (1974); D. J. Bradley, P. Ewart, J. V. Nicholas, J. R. D. Shaw, and D. G. Thompson, *Phys. Rev. Lett.* **31**, 263 (1973); J. L. Carlsten, T. J. McIlrath, and W. H. Parkinson, *J. Phys. B* **8**, 38 (1975).
52. D. Spence, *Phys. Rev. A* **12**, 2353 (1975).
53. F. H. Read, in *Ninth International Conference on the Physics of Electronic and Atomic Collisions*, Eds. J. S. Risley and R. Geballe, University of Washington Press, Seattle (1975).
54. R. B. Barker and H. W. Berry, *Phys. Rev.* **151**, 14 (1966).
55. C. Bottcher and K. R. Schneider, *J. Phys. B* **9**, 911 (1976).
56. P. J. Hicks and J. Comer, *J. Phys. B* **8**, 1866 (1975).
57. P. J. Hicks, S. Cvejanovic, J. Comer, F. H. Read, and J. M. Sharp, *Vacuum* **24**, 573 (1974).
58. A. J. Smith, P. J. Hicks, F. H. Read, S. Cvejanovic, G. C. M. King, J. Comer, and J. M. Sharp, *J. Phys. B* **7**, L496 (1974).
59. R. Morgenstern, A. Niehaus, and U. Thielmann, *Phys. Rev. Lett.* **37**, 199 (1976); *J. Phys. B* **9**, L363 (1976).
60. H. Ehrhardt, K. H. Hesselbacher, K. Jung, E. Schubert, and K. Willmann, *J. Phys. B* **7**, 69 (1974), and references therein.
61. V. V. Balashov, S. S. Lipovetsky, and V. S. Senashenko, *Phys. Lett.* **39A**, 103 (1972); *Zh. Eksp. Teor. Fiz.* **63**, 1622 (1972); *Sov. Phys. JETP* **36**, 858 (1973).
62. F. Gelebart, R. J. Tweed, and J. Peresse, *J. Phys. B* **7**, L174 (1974).
63. A. Bordenave-Montesquieu, A. Gleizes, M. Rodiere, and P. Benoit-Cattin, *J. Phys. B* **6**, 1997 (1973).
64. V. V. Balashov, S. S. Lipovetsky, V. S. Senashenko, A. V. Pavlichenkov, and A. N. Polyndov, *Opt. Spectrosc.* **32**, 4 (1972).
65. E. N. Lassettre and S. Silverman, *J. Chem. Phys.* **40**, 1265 (1964).
66. U. Fano, *Phys. Rev.* **124**, 1866 (1961).
67. B. W. Shore, *J. Opt. Soc. Am.* **57**, 881 (1976).
68. J. Comer and F. H. Read, *J. Phys. E* **5**, 211 (1972).
69. W. van Rensbergen, *J. Quant. Spectrosc. Radiat. Transfer* **11**, 1125 (1971); **12**, 1105 (1972).
70. M. Rudkjobing, *J. Quant. Spectrosc. Radiat. Transfer* **13**, 1479 (1973).
71. D. Spence and M. Inokuti, *J. Quant. Spectrosc. Radiat. Transfer* **14**, 953 (1974).
72. J. F. Williams, in *Electron and Photon Interactions with Atoms*, Eds. H. Kleinpoppen and M. R. C. McDowell, pp. 309–338, Plenum Press, New York (1976).
73. J. W. McGowan, E. M. Clarke, and E. K. Curley, *Phys. Rev. Lett.* **15**, 917 (1965); **17**, 66(E) (1966).
74. G. J. Schulz, *Phys. Rev. Lett.* **13**, 583 (1964).
75. H. Kleinpoppen and V. Raible, *Phys. Lett.* **18**, 24 (1965).
76. L. Sanche and P. D. Burrow, *Phys. Rev. Lett.* **29**, 1639 (1972).
77. D. Spence, *J. Phys. B* **8**, L42 (1975).
78. J. C. Y. Chen, *Nucl. Instrum. Methods* **90**, 237 (1970).
79. K. T. Chung and J. C. Y. Chen, *Phys. Rev. A* **6**, 686 (1972).
80. A. K. Bhatia and A. Temkin, *Phys. Rev. A* **11**, 2018 (1975); **8**, 2184 (1973).
81. G. D. Doolen, J. Nuttall, and R. W. Stagat, *Phys. Rev. A* **10**, 1612 (1974).
82. I. Shimamura, *J. Phys. Soc. Japan* **31**, 852 (1971).
83. A. K. Bhatia, *Phys. Rev. A* **9**, 9 (1974); **10**, 729 (1974).

84. P. G. Burke, in *Invited Papers of the Fifth International Conference on the Physics of Electronic and Atomic Collisions, 1967*, pp. 128–139, University of Colorado Press, Boulder, Colorado (1968).

85. R. K. Nesbet and J. D. Lyons, *Phys. Rev. A* **4**, 1812 (1971).

86. R. A. Bain, J. N. Bardsley, B. R. Junker, and C. V. Sukumer, *J. Phys. B* **7**, 2189 (1974).

87. J. F. Williams, *J. Phys. B* **9**, 1519 (1976).

88. J. W. McGowan, J. F. Williams, and E. K. Curley, *Phys. Rev.* **180**, 132 (1969).

89. V. L. Jacobs, A. K. Bhatia, and A. Temkin, *Astrophys. J.* **191**, 785 (1974).

90. G. W. Drake, *Astrophys. J.* **184**, 145 (1973).

91. J. Macek, *Proc. Phys. Soc. (London)* **92**, 365 (1967).

92. R. S. Oberoi, *J. Phys. B* **5**, 1120 (1972).

93. W. R. Ott, J. Slater, J. Cooper, and G. Gieres, *Phys. Rev. A* **12**, 2009 (1975).

94. R. P. Madden and K. Codling, *J. Opt. Soc. Am.* **54**, 268 (1964).

95. K. Codling, R. P. Madden, and D. L. Ederer, *Phys. Rev.* **155**, 27 (1967).

96. R. P. Madden and K. Codling, in *Autoionization*, Ed. A. Temkin, pp. 129–151, Mono Book Corporation, Baltimore (1966).

97. R. P. Madden and K. Codling, *Phys. Rev. Lett.* **10**, 516 (1963).

98. M. E. Rudd, *Phys. Rev. Lett.* **15**, 580 (1965).

99. K. Siegbahn *et al.*, *ESCA Applied to Free Molecules*, North Holland, Amsterdam (1969).

100. J. W. Cooper, U. Fano, and F. Prats, *Phys. Rev. Lett.* **10**, 518 (1963).

101. P. E. Burke and D. D. McVicar, *Proc. Phys. Soc. (London)* **86**, 989 (1965).

102. L. Lipsky and A. Russek, *Phys. Rev.* **142**, 59 (1969).

103. T. F. O'Malley and S. Geltman, *Phys. Rev. A* **137**, 1344 (1965).

104. P. G. Burke and A. J. Taylor, *Proc. Phys. Soc. (London)* **88**, 549 (1966).

105. A. K. Bhatia, A. Temkin, and J. F. Perkins, *Phys. Rev.* **153**, 177 (1967); **182**, 15 (1969); *A* **6**, 120 (1972).

106. A. K. Bhatia, P. G. Burke, and A. Temkin, *Phys. Rev. A* **8**, 21 (1973); **10**, 459 (1974).

107. J. W. Cooper, S. Ormonde, C. H. Humphrey, and P. G. Burke, *Proc. Phys. Soc. (London)* **91**, 285 (1967).

108. J. L. Tech and J. F. Ward, *Phys. Rev. Lett.* **27**, 367 (1971).

109. H. G. Berry, J. Desesquelles, and M. Dufay, *Phys. Rev. A* **6**, 600 (1972); *Nucl. Instrum. Methods* **110**, 43 (1973).

110. E. J. Knystautas and R. Drouin, *Nucl. Instrum. Methods* **110**, 95 (1973).

111. J. P. Buchet, M. C. Buchet-Poulizac, H. G. Berry, and G. W. F. Drake, *Phys. Rev. A* **7**, 922 (1973).

112. R. F. Stebbings and F. B. Dunning, *Phys. Rev. A* **8**, 665 (1973); **9**, 2378 (1974).

113. J. A. Simpson, G. E. Menendez, and S. R. Mielczarek, *Phys. Rev. A* **139**, 1039 (1965).

114. S. Tahira, F. Nishimura, and N. Oda, *J. Phys. B* **6**, 2306 (1973).

115. E. Bolduc and P. Marmet, *Can. J. Phys.* **51**, 2108 (1973).

116. J. M. Sharp, J. Comer, and P. J. Hicks, *J. Phys. B* **8**, 2512 (1975).

117. A. K. Edwards and M. E. Rudd, *Phys. Rev.* **170**, 140 (1968).

118. L. A. Parcell, J. Langlois, and J. M. Sichel, *Chem. Phys. Lett.* **25**, 390 (1974).

119. D. E. Golden and A. Zecca, *Phys. Rev. A* **1**, 241 (1970).

120. D. E. Golden, F. D. Schowengerdt, and J. Macek, *J. Phys. B* **7**, 478 (1974).

121. D. Andrick and L. Langhans, *J. Phys. B* **8**, 1245 (1975).

122. D. W. O. Heddle, R. G. W. Keesing, and J. M. Kurepa, *Proc. R. Soc. (London) A* **334**, 135 (1973); **337**, 435 (1974).

123. U. Fano and J. W. Cooper, *Phys. Rev. A* **137**, 1364 (1965); **138**, 400 (1965).

124. P. D. Burrow and G. J. Schulz, *Phys. Rev. Lett.* **22**, 1271 (1969).

125. J. T. Grisson, R. N. Compton, and W. R. Garrett, *Phys. Lett.* **30A**, 117 (1969).

126. J. J. Quemener, C. Paquet, and P. Marmet, *Phys. Rev. A* **4**, 494 (1971).

127. S. Ormonde, F. Kets, and H. G. M. Heideman, *Phys. Lett.* **50A**, 147 (1974).
128. C. A. Nicolaides, *Phys. Rev. A* **6**, 2078 (1972).
129. A. Weingartshofer, K. Willman, and E. M. Clarke, *J. Phys. B* **7**, 79 (1974).
130. J. F. Williams and B. A. Willis, *J. Phys. B* **8**, 1641 (1975).
131. D. Spence and T. Noguchi, *J. Chem. Phys.* **63**, 505 (1975).
132. H. S. Taylor, G. V. Nazaroff, and A. Golebiewski, *J. Chem. Phys.* **45**, 2872 (1966).
133. M. Valin and P. Marmet, *J. Phys. B* **8**, 2953 (1975).
134. P. Feldman and R. Novick, *Phys. Rev.* **160**, 143 (1967).
135. D. L. Ederer, T. Lucatorto, and R. P. Madden, *Phys. Rev. Lett.* **25**, 1537 (1970).
136. P. Ziem, R. Bruch, and N. Stolterfoht, *J. Phys. B* **8**, L480 (1975).
137. J. W. Cooper, M. J. Conneely, K. Smith, and S. Ormonde, *Phys. Rev. Lett.* **25**, 1540 (1970).
138. M. Ahmad and L. Lipsky, *Phys. Rev. A* **12**, 1176 (1975).
139. I. Martinson, *Phys. Scr.* **9**, 281 (1974).
140. I. Martinson and A. Gaupp. *Phys. Rep.* **15C**, 113 (1974).
141. D. Andrick, M. Eyb, and H. Hoffmann, *J. Phys. B* **5**, L15 (1972).
142. M. Eyb and H. Hoffmann, *J. Phys. B* **8**, 1095 (1975).
143. M. Eyb, *J. Phys. B* **9**, 101 (1976).
144. T. A. Patterson, H. Hotop, A. Kasdan, D. W. Norcross, and W. C. Lineberger, *Phys. Rev. Lett.* **32**, 189 (1974).
145. D. Spence and W. A. Chupka, *Phys. Rev. A*, **10**, 71 (1974).
146. D. Spence, *Phys. Rev. A* **12**, 721 (1972).
147. J. J. Matese, *Phys. Rev. A* **10**, 454 (1975).
148. J. J. Matese, S. P. Rowntree, and R. J. W. Henry, *Phys. Rev. A* **7**, 846 (1973).
149. D. W. O. Heddle, *J. Phys. B* **8**, L33 (1975).
150. T. W. Ottley and H. Kleinpoppen, *J. Phys. B* **8**, 621 (1975).
151. M. Duweke, N. Kirchner, E. Reichert, and E. Staudt, *J. Phys. B* **6**, L208 (1973).
152. I. P. Zapesochnyi and O. B. Skpenik, *Sov. Phys. JETP* **23**, 592 (1966).

24

Optical Oscillator Strengths by Electron Impact Spectroscopy

W. R. NEWELL

1. Introduction

The measurement of an optical oscillator strength (OOS) for an atomic transition has "classically" been done using the standard methods of absorption spectroscopy,[1] i.e., beam attenuation, and of anomalous dispersion and emission techniques. All of these methods involve the production of light sources and/or the detection of emitted photons coupled with the experimental difficulties[2] incurred, especially at shorter wavelengths. Consequently these classical methods have been applied mostly to transitions in the visible region of the spectrum. Recent developments,[3] however, have extended the convenient usable range of such laboratory light sources to less than 600 Å; but for shorter wavelengths the requirement of a synchrotron radiation source becomes necessary, with its attendant cost and size.

The advent of electron spectroscopy has made available a further method for the study of atomic transitions, i.e., discrete excitations, autoionizing transitions, and ionization phenomena. The production of continuously variable high-resolution electron beams provides a convenient and inexpensive probe to excite selectively atomic transitions. In this chapter we will confine the discussion to experiments which have been performed to make measurements on atomic oscillator strengths for both discrete and continuum transitions. The continuously variable and wide range of momentum transfer, the constant detection efficiency afforded by the use of electrons, and the low gas densities employed make electron

W. R. NEWELL • Department of Physics and Astronomy, University College London, Gower Street, London WC1E 6BT.

Table 1. Relative Probabilities for Optical Excitation, P_{op}, and Electron Impact Excitation, P_{ele}

Transition	Selection	P_{op}	P_{ele}
1^1s–2^1p	Dipole-allowed	1	1
$1's$–$2's$	Parity-forbidden	10^{-5}–10^{-8}	0.18
1^1s–2^3p	Spin-forbidden	10^{-10}	0.2

spectroscopy a very attractive mode of investigation. In addition to the large dynamical range obtained in these types of experiments, an advantage of electron excitation is that it is not restricted by the optical selection rules, hence parity-violating and spin-forbidden transitions are easily detected and studied. The relative probabilities (P) for photon excitation, P_{op}, and electron impact excitation, P_{ele}, for various transitions in helium gas are given in Table 1.[4] With P_{op} set equal to P_{ele} for the dipole-allowed transition, the relative ease with which "forbidden" transitions can be excited is clearly demonstrated.

For more about oscillator strengths, see Chapter 23 (Wiese) of this work.

2. Apparatus

Since in electron-energy-loss spectroscopy the incident electron is acting as a probe of the atomic energy levels, its energy definition needs to be less than the energy separations normally encountered between different atomic states. Consequently, in order to obtain an energy-loss spectrum, the primary electron beam must be highly monochromatic in energy.

In a standard electrostatically focused electron gun using a tungsten filament, the energy spread in the electron-beam energy can be as high as 500 meV due to the thermal spread in the emitted electron velocities— assuming that the detrimental electron optical aberation effects are minimized. By selecting an indirectly heated dispenser cathode,[5] which operates at a lower temperature, this spread in energy can be reduced to ~200 meV. Still this value is too high for useful energy-loss work to be performed, and ideally a value of 20 meV should be sought. To achieve this order-of-magnitude reduction it is necessary to use a velocity-dispersive element which selects only that fraction of the electron gun current which falls within the required electron energy spread. Predominantly two types of electrostatic analyzer have been used: 127° cylindrical and 180° hemispherical. There are definite differences in the operational properties of each type (i.e., dispersion, abberation, transmission efficiency, energy resolution) due to the different geometries employed, but normally personal preference determines which type is used. A complete electron-energy-loss spectrometer,[6] which used 180° hemispherical analyzers, is

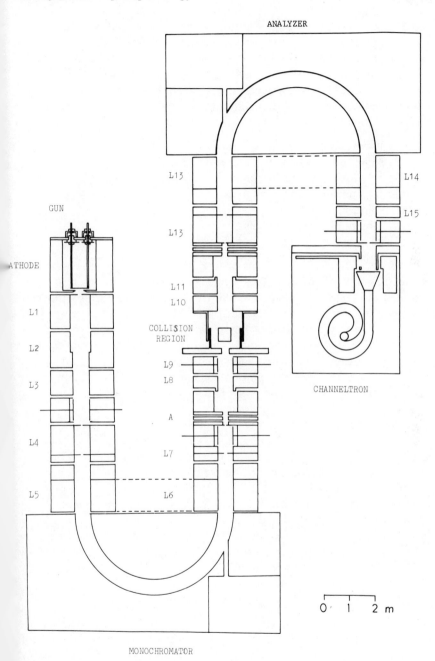

Figure 1. A typical double hemispherical energy-loss electron spectrometer.[6] The instrument is shown in the zero-angle position with the atom beam perpendicular to the plane of the page.

shown in Figure 1. The gun produces a well-focused electron beam (10^{-7} A) of spot size 0.3 mm in the entrance plane of the 180° monochromator. A fraction of the dispersed beam, determined by the virtually focused aperture A, is selected and accelerated by the lenses L6, L7, L8, L9 from the 5-eV analyzing energy to the required experimental energy. Good velocity selection is dependent on several factors which should be optimized to give

> (1) low beam divergence of the gun
> (2) small spot size of electron beam
> (3) low analyzing velocity in the hemispherical analyzers
> (4) large mean radius of analyzer
> (5) stable electric fields
> (6) minimum magnetic fields
> (7) good mechanical alignment
> (8) uniform surface effects

Neglect of any of these factors can result in poor energy resolution. Detailed discussion of these effects is given in the literature by different authors.[7-9] After scattering, the electron beam is retarded by lenses L10, L11, L12, L13 to 5 eV before being analyzed in a second 180° monochromator. The transmitted electrons are detected with a channel electron multiplier. Scattered electrons which loose energy in causing an atomic transition have the required amount of energy restored to them by an applied ramp voltage to lens L11. This enables them to be transmitted through the analyzer and to be detected.

The majority of the data presented in this article was obtained using apparatus similar to that in Figure 1. The performance of an electron spectrometer is sensitive to small changes of the factors listed above, but normally an overall instrumental energy resolution of 60 meV is easily obtained. To a limited extent time-of-flight methods[10,11] are now being used in velocity analysis, but these are restrictive over the usable energy range. The retarded potential difference method of Fox[12] is also limited in its application to electron-energy-loss spectroscopy. Other forms of analyzers employed in complementary investigations are described by Rudd,[13] to which the reader is referred for detailed comparisons of the different devices. Details of the electron optics of the electrostatic lenses are given by Pierce,[14] Harting and Read,[15] and Septier.[16]

3. Theory

The concept of oscillator strength was introduced in the 19th century to describe the electrical and optical properties of dilute matter. In pre-quantum mechanics the classical oscillator controlled the response of atomic electrons to weak perturbations such as electromagnetic fields. The oscillator strength was defined classically as the number of electrons free to oscillate at a given frequency, but with the advent of quantum mechanics the interpretation placed on the oscillator strength changed due to the statistical distribution of electron positions. Hence each response frequency (v_s) of the atomic system had associated with it a fractional oscillator strength (f_s), and the sum over all frequencies and oscillator strengths gives the total number of electrons (N) in the atom:

$$\sum_s \sum_v f_s = N$$

In order to determine the OOS from an electron scattering experiment we must establish the relationship between scattering theory and resonance absorption measurements. A recent review of the theory of inelastic scattering of fast charged particles by atoms has been given by Inokuti,[17] and optical oscillator determinations are discussed in Chapter 25. For convenience only a very brief outline of the pertinent theory will be given here.

The equivalence of electron impact excitation and response absorption becomes a good approximation if the scattering system before and after the collision can be described by the Born approximation.[18] Within the range of applicability of the first Born approximation the differential cross section $d\sigma$ for the inelastic scattering of an electron by an atom can be written as

$$\frac{d\sigma}{d\Omega} = \frac{1}{2}\frac{k_m}{k_0}\frac{|\varepsilon_{0m}(K)|^2}{K^4} \tag{1}$$

where k_0 is the incident electron momentum, k_m the scattered electron momentum, and $\mathbf{K}$ $(=\mathbf{k}_0 = \mathbf{k}_m)$ is the momentum transferred in the collision. The matrix element for the transition from state $|0\rangle$ to state $|m\rangle$ is denoted by

$$\varepsilon_{0m}(K) = \langle 0|\exp(i\mathbf{K}\cdot\mathbf{r})|m\rangle \tag{2}$$

and with the z axis normally taken as coincident with $\mathbf{K}$ the exponential becomes $\exp(iKz)$. Atomic units are used throughout this discussion. Now using the concept of a generalized oscillator strength as defined by Bethe[19] and expressed by

$$f_{0m}(K) = (E_0 - E_m)K^{-2}|\varepsilon_{0m}(K)|^2 \tag{3}$$

the differential cross section can be written as

$$\frac{d\sigma}{d\Omega} = \frac{2}{E_0 - E_m}\frac{k_m}{k_0}K^{-2}f(K) \tag{4}$$

where E_0 is the initial incident electron energy and E_m the final electron energy. Hence if the differential cross section per unit solid angle can be measured experimentally, the generalized oscillator strength (GOS) can be determined as a function of K^2. The relationship of the GOS to the OOS can be demonstrated by expanding the matrix element ε_{0m} of Eq. (3) in a power series

$$\varepsilon = \sum_l \sum_s K^l \frac{i^l}{l!} \langle 0|(\mathbf{q} \cdot \mathbf{r}_s)^l|m \rangle \tag{5}$$

where $\mathbf{q}$ is the unit vector in the direction of $\mathbf{K}$ and Σ_s is the sum over the atomic electrons. Neglecting terms higher than K^4, we get from Eq. (5)

$$|\varepsilon|^2 = K^2 \varepsilon_1 + K^4 (\tfrac{1}{4} \varepsilon_2^2 - \tfrac{1}{3} \varepsilon_3 \varepsilon_1) + O(K^6) \tag{6}$$

where ε_1 represents the dipole matrix element and ε_m, with $m > 1$, represents higher-order matrix elements. From the dynamics of the system the momentum transfer vector $\mathbf{K} = \mathbf{k}_0 - \mathbf{k}_m$ can be written as

$$K^2 = k_0^2 + k_m^2 - 2k_0 k_m \cos \theta \tag{7}$$

where θ is the angle between the incident $\mathbf{k}_0$ and scattered, $\mathbf{k}_m$ momentum vectors. Recalling that $E_0 = \tfrac{1}{2} k_0^0$ and $E_m = \tfrac{1}{2} k_n^2$, we can rewrite Eq. (7) as

$$K^2 - 4\bar{E} - 4\bar{E}(1 - W^2/4\bar{E}^2)^{1/2} \cos \theta \tag{8}$$

where $\bar{E} = \tfrac{1}{2}(E_0 + E_m)$ and $W = (E - E_m)$. This expression, when expanded, will converge rapidly for the case where $E_m \approx E_0$, i.e., conditions where the energy loss encountered by the primary beam is small compared with the energy of the primary beam. By expanding Eq. (8) and retaining only the first terms, the GOS becomes

$$f_{0m} \simeq 2W[\varepsilon_1^2 + 8\bar{E}(\sin^2 \tfrac{1}{2}\theta + W^2/16\bar{E}^2)(\tfrac{1}{4}\varepsilon_2^2 - \tfrac{1}{3}\varepsilon_3 \varepsilon_1)] \tag{9}$$

In the limit of zero momentum transfer ($K^2 \to 0$), or equivalently high-incident-electron energy ($E_0 \to \infty$) and zero-angle scattering, the GOS, Eq. (3), will converge to the OOS. Consequently, the OOS can be determined by the scattering of high-energy monochromatic electrons and measurements of the optically related quantities such as L (absorption) coefficient and A coefficient obtained from the relationships

$$f_0 = \frac{2.303}{2\pi^2 N} \int L \, dw$$

$$A_{0m} = 8 \times 10^9 \left(\frac{\nu}{Ry}\right) f_{0m}$$

$$\tag{10}$$

When, due to the symmetry properties of the transition involved, the dipole matrix element is zero, the GOS becomes

$$f_{0m} = \frac{1}{2} W \left\{ 8\bar{E} \left[\sin^2 \frac{\theta}{2} + \left(\frac{W}{4\bar{E}} \right)^2 \right] \right\} \varepsilon_2^2 \tag{11}$$

and a plot of f_m vs. K^2 (the expression in square brackets) will yield the quadrupole matrix element ε_2, where

$$\varepsilon_2^2 = \left| \langle 0 | \sum_s (\mathbf{r}_s \cdot \mathbf{q})^2 | m \rangle \right|^2 \tag{12}$$

In high-energy electron excitation, i.e., when small K^2 values (almost zero) are involved, we will expect an excitation intensity for quadrupole and dipole transitions similar to that observed in photon absorption, but due to the nature of electron excitation, with its variable range of momentum transfers, the differential cross sections for dipole [Eq. (13a)] and quadrupole [Eq. (13b)] transitions have different momentum dependences, which are given by

$$\left(\frac{d\sigma}{d\Omega} \right)_d = 8 \frac{k_m}{k_0} \frac{\bar{E}}{W^2} |\varepsilon_1|^2 \tag{13a}$$

$$\left(\frac{d\sigma}{d\Omega} \right)_q = \frac{k_m}{k_0} |\varepsilon_2|^2 \tag{13b}$$

for zero-angle scattering. Hence, at lower incident electron energies, i.e., those closer to the threshold for the transition, the quadrupole transition is preferentially excited since E_0 ($= \frac{1}{2} k_0^2$) is small, therefore providing a relatively easy method for obtaining the matrix element involved. Garstang[20] has generalized Eq. (9), and the GOS for nondipole transitions can be written as

$$f_{0m}(K) = C(E_m - E_0)K^2 \left(J_0 \left\| \sum_s r_s^2 \right\| J_m \right) \delta(J_0 J_m)$$

$$+ c(g_n/g_0)K^2 A_{0m}(E_n - E_0)^{-1} + O(k^4) \tag{14}$$

where g_0 and g_m are the statistical weights of the lower and upper levels, respectively. Equation (14) separately gives, from the same experiment, information on $S \rightarrow S$ and $S \rightarrow D$ parity-forbidden transitions. This enables the reduced matrix elements, $(J_0 \| \sum_s r_s^2 \| J_m)$ and quadrupole transition probabilities to be determined from measurement of the GOS as a function of K^2. In the case of an unclassified state the nature of the GOS dependence can be used to determine the symmetry of the state.[21] Higher-order matrix elements associated with octopole transitions have been measured in potassium,[22] and those show a K^4 dependence on the GOS.

In the case of singlet to triplet excitations the generalized oscillator strength f_T is given by

$$f_T = 4c^2 \tfrac{1}{2} W K^2 k_0^{-4} |\varepsilon_{ex}|^2 \tag{15}$$

due to the modifications necessary[23,24] in the exchange matrix element, ε_{ex}.

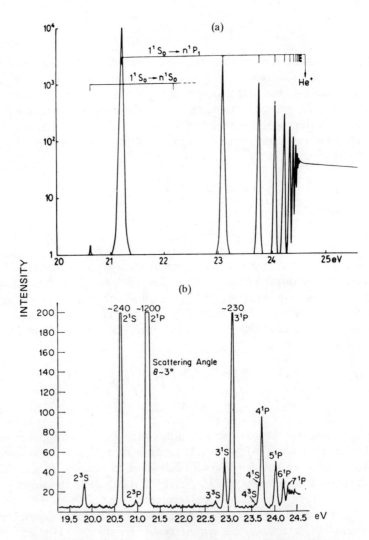

Figure 2. Electron impact spectra of helium obtained (a) at 2500 eV by Boersch *et al.*[65] and (b) at 50 eV by Lassettre *et al.*[66]

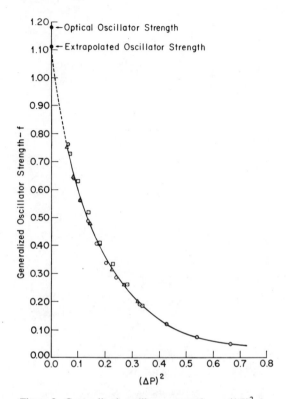

Figure 3. Generalized oscillator strength vs. $(\Delta P)^2$ for the 6^1S_0–6^1P_1 transition in mercury. Initial kinetic energies are: $\odot$, 300 3V; $\triangle$, 400 eV; $\square$, 500 eV. The solid curve is a least-squares line, and the experimental points do not depart from the line by more than experimental error.

Figures 2a and 2b show two different energy-loss spectra of helium taken at vastly different energies. The relative intensities of the various types of transitions excited at the different energies and angles are clearly demonstrated.

4. Normalization Procedures

Excluding direct absolute measurements of the differential cross section the GOSs of different transitions in the same atom must be made absolute before any optical properties can be deduced from them. Several approaches to this problem have been made by different workers.

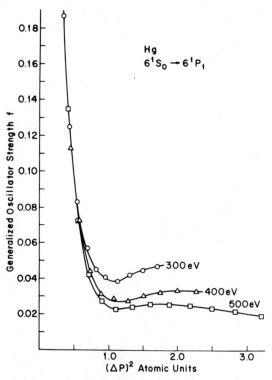

Figure 4. Generalized oscillator strength curves, f vs. $(\Delta P)^2$, for the 6^1S_0–6^1P_1 transition in mercury. Note that f is a function of both $(\Delta P)^2$ and E, which indicates deviations from the Born approximation.

(a) Simultaneous measurement of the elastic and inelastic differential cross sections over the same energy and angular range[25] allows a direct normalization to independent measured absolute elastic scattering cross sections. This method has been used by Skerbele and Lassettre.[26,27] (See Figure 3, where the relative GOS of the $6S$–$6P$ in mercury is made absolute by normalizing to the absolute elastic scattering data of Bromberg.[28])

(b) From obtained electron energy-loss spectra one transition of the spectrum, normally, the resonance, can be used to place the other transitions on an absolute scale by normalization to a calculated GOS curve. This procedure has been used in sodium for the $3S \rightarrow nL$ ($n > 3$) transitions,[30] where the $3S$–$3p$ transition has been normalized to a theoretical calculation by Moores.[18]

(c) In a similar manner the relative GOS for one, normally, the first, transition of a series can be made absolute by extrapolation (see section below) to a known optical oscillator strength. This will yield the OOSs for the higher transitions in the series which are less readily obtained by direct optical measurement.

(d) Utilization of the Thomas–Kuhn sum rule

$$\sum_m f_{0m}(K) = N \qquad (16)$$

has been demonstrated by Hertel and Ross[29] in the alkali metal spectra over the range of discrete spectral lines. In this method allowance must be made for the continuum component of the oscillator strength[31] and downward-forbidden Pauli transitions.[32]

5. Validity of the Born Approximation

It should be recalled that the GOS is a quantity which is defined within the framework of the first Born approximation (FBA). Consequently, the GOS will exhibit certain properties depending on the region of K^2 in which the experiment is performed. In the limit of $K^2 \to 0$ the GOS will extrapolate to the OOS (Figure 3), and a valid Born region can be demonstrated experimentally by determining when the GOS becomes independent of the incident electron energy. For the $6S$–$6P$ transition in mercury this independence of electron energy onsets at $K^2 = 0.7$ a.u. (see Figure 4). The deviation from the Born approximation will increase with increasing K^2 and is in general larger, at a given K^2, for symmetry-forbidden transitions than for optically allowed transitions.[25]

6. Extrapolation Procedure

Assuming the GOS data for the various transitions in the atom under investigation are made absolute either by direct experimental measurement or by a normalizing procedure, the GOS curves must be extrapolated to $K^2 = 0$, an experimentally inaccessible region, to obtain the OOS. The question arises whether the experimental data have been obtained in a valid Born region and even whether that is necessary for a correct interpretation of the data. Lassettre et al.[25] have shown that the optical selection rules are still obeyed at $K^2 = 0$ whether the Born is rigorously satisfied or not in the experimental region. In this nonphysical limit they derive the theorem

$$\lim_{K_{0m} \to 0} K_{0m} f_{0m}(K) = \lim_{K_{0m} \to 0} K_{0m} f_{0m}^B(K) \qquad (17)$$

where f is the exact nonrelativistic scattering amplitude and f^B is the corresponding Born amplitude. Bonham[33] has considered the expansion of $K^2|f(K)|^2$ about $K^2 = 0$ and confirms that the differential cross section can be regarded as a function of K^2 down to relatively low energies. The definition of a GOS by Eq. (3) is independent of the validity of the FBA, but in the extrapolation procedure the curve will always traverse a valid Born region. It is, however, advantageous to perform the experiment in a K^2 region where the FBA adequately describes the scattering process.

The extrapolation of experimental data is facilitated by an analytical expression derived by Lassettre[34]:

$$f(K) = f_0(1+x)^{-6}\left[1 + \sum_{v=1}^{\infty} c_v\left(\frac{x}{1+x}\right)^v\right] \qquad (18)$$

where f_0 is the OOS, $x = (Ka/\alpha)^2$, and $\alpha = (I/R)^{1/2} + [(I-E)^{1/2}/R]$ is given in terms of the ionization potential (I) and excitation energy (E). The coefficients C_v determine the angular dependence of the differential cross section and hence the K dependence of $f(K)$. The functional form of Eq. (18) is not without theoretical justification which depends on the singular points in the complex plane and asymptotic representation of the GOS at large K. As an example of this technique, the absolute GOS curve in Figure 3 for the transition Hg $(6S-6P)$ is extrapolated to $K^2 = 0$ using Eq. (18); this yields an OOS of 1.11, which is in good agreement with the value 1.18 obtained by Lurio[35] using the Hanle effect. The method does provide a reliable procedure for determining optical data from electron scattering experiments.

7. Summary of Results for Discrete Transitions

As indicative of the optical data obtained from electron scattering experiments we will consider the metal atoms in group I (lithium, sodium, potassium, rubidium, cesium), and group IIb (zinc, cadmium, mercury) of the periodic table. For a summary of work on gaseous atoms see the review by Inokuti.[17] The difficulty of working with metal vapors in high-resolution

Table 2. Optical Oscillator Strengths
in Mercury

Transition	OOS
$6^1S_0-6^3P_1$	0.0285 ± 0.0035
$6^1S_0-6p^1\,{}^3P$	0.704 ± 0.070
$6^1S_0-7p^1\,{}^1P$	0.067 ± 0.009

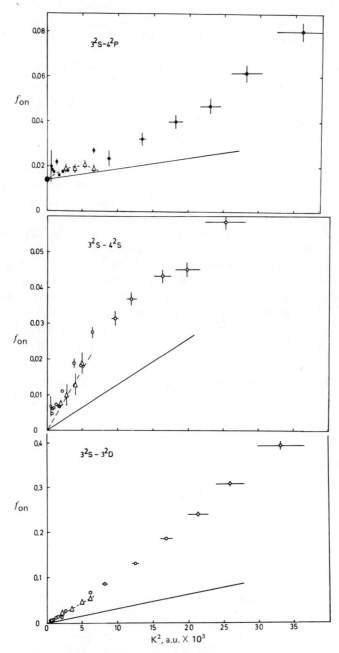

Figure 5. GOS plots for the 3S–3P, 4S, and 3D transitions in sodium.[30] Circles, Shuttleworth *et al.*[30] Triangles (joined by dashed line), Hertel and Ross.[29] Solid lines, Born amplitudes.

Table 3. Sodium Data

Transition	Experiment[a]	Theory
$S-P$	$f_0 = 1.60 \ 10^{-2}$	$1.37 \ 10^{-2b}$
$S-D$	$A = 581 \ s^{-1}$	$612 \ s^{-1c}$
$S-S$	$R = 43$ a.u.	—

[a] Shuttleworth *et al.* data.[30]   [b] Weishert and Dalgarno.[37]
[c] Ali.[38]

experiments accounts for the sparse amount of data available in the literature compared with gaseous atoms. This difficulty will nevertheless reinforce the applicability of the method as a useful tool in determining optical parameters from electron scattering experiments.

Using the absolute measurements of the $6^1S_0-6^1P_1$ transition in mercury, Lassettre has obtained, by normalization, the OOSs for other transitions in mercury as given in Table 2. The $6^1S_0-6^3P_1$ OOS determination agrees well with Lurio's[35] optical measurement and Garstang's[36] calculated value of 0.0278. The other two transitions in Table 2 are autoionizing excitations at 11 and 13 eV, for which no optical measurements have been made. Optical experiments in this wavelength range (~ 1000 Å) are difficult due to suitability of radiation source, whereas

Table 4. OOS for Some Discrete Excitations in Metal Atoms

Element	Transitions	OOS (electron scattering)	OOS (optical measurement)
Lithium	$2S-3P$	0.008^h	—
Sodium	$3S-4P$	0.016^a	0.014^b
	$3S-5P$	0.004^a	0.002^b
Potassium	$4S-5P$	0.008^c	$0.868(-2)^b$
	$4S-6P$	$0.2(-3)^c$	$0.846(-3)^b$
Rubidium	$5S-6P$	0.013^d	0.0134^b
	$5S-7P$	$0.2(-2)^d$	$0.246(-2)^b$
Cesium	$6S-7P$	$0.15(-1)^d$	$0.146(-1)^b$
	$6S-8P$	$0.18(-2)^d$	$0.284(-2)^b$
Zinc	$4S-5P$	0.05^e	—
	$4S-6P$	0.008^e	—
Cadmium	$5S-6P$	0.02^f	0.01^g
	$5S-7P$	0.004^f	—
	$5S-8P$	0.002^f	—

[a] Shuttleworth *et al.*[30]   [e] Newell and Ross.[42]
[b] Marr and Greek[32]   [f] Newell *et al.*[42]
[c] Hertel and Ross.[39]   [g] Webb and Messenger.[43]
[d] Hertel and Ross.[40]   [h] Burgess *et al.*[44]

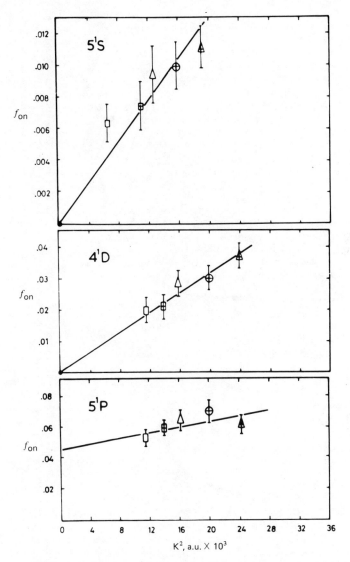

Figure 6. GOS plots for the $4S$–$5S$, $4D$, and $5P$ transitions in cadmium.[41]

electron scattering at 500 eV over an angular range 2°–7°, as in this particular experiment, is comparatively easy and more versatile.

A systematic study of the GOS of various transitions in sodium atoms by Hertel and Ross[29] and Shuttleworth *et al.*[30] yield optical data for the different types of transitions involved. The GOS curves (Figure 5), for the

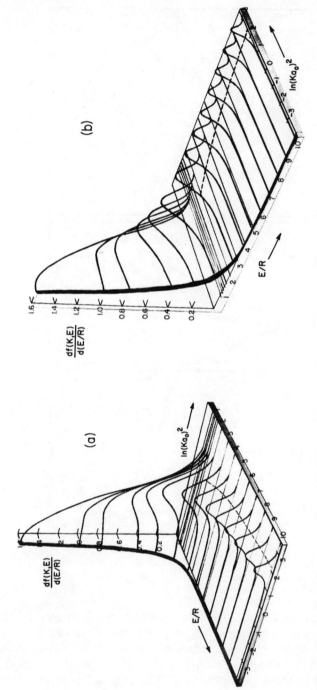

Figure 7. Photographs of a plastic model of the Bethe surface for atomic hydrogen. The horizontal axes for E/R and $\ln (Ka_0)^2$ define the base plane. The vertical axis represents $R\,df(K, E)/dE$. The 14 plates are placed at $E/R = 3/4,\ 8/9,\ 1,\ 5/4,\ 3/2,\ 2,\ 3,\ 4,\ 5,\ 6,\ 7,\ 8,\ 9,$ and 10. The first two plates represent the discrete spectrum, in which case the vertical scale corresponds to $\frac{1}{2}n^3 f_n(K)$, n being the principal quantum number. While the model shows the major portion $-4 \le \ln (Ka_0)^2 \le 3.6$, the surface indefinitely extends as a plateau toward smaller $\ln (Ka_0)^2$. The dotted curve on the base plane shows the location $(Ka_0)^2 = E/R$ of the Bethe ridge, which is the main feature for large E/R. View (a) shows the spreading of the Bethe ridge with decreasing E/R. The optical region $(Ka_0)^2 \ll 1$ develops conspicuously for small E/R. View (b) shows in front a cut at $\ln (Ka_0)^2 = -4$, a curve that closely approximates the photoabsorption spectrum $R\,df/dE$.

optically allowed $S-P$ transition, the quadrupole $S-D$ transition, and the parity-forbidden $S-S$ transition yield, respectively, an OOS, an A coefficient, and a reduced matrix element (see Section 3). From the intercepts and slopes of the GOS plots the data in Table 3 are derived. Good agreement is obtained with known theoretical calculations of these parameters. In the particular case of OOS Table 4 gives a summary of the atomic transitions of the alkali and group IIB atoms for which OOS have been determined by electron scattering. Comparison with existing optical measurements where available are good. Cadmium GOS plots for $4S-5S$, $4D$, and $5P$ transitions are shown in Figure 6.

8. The Bethe Surface

The graphs of f_{0m} vs. K^2 (Figures 3, 4, and 7) are sections of a more general plot of f_{0m} vs. K^2 vs. E; this three-dimensional plot constitutes a Bethe surface for that particular atom. The Bethe surface contains all the information for the scattering of charged particles of that particular atom within the applicability of the Born approximation. This includes not only discrete excitations but also transitions to continuum and autoionizing states. A calculated plot[17] of the Bethe surface for atomic hydrogen (Figure 7) will indicate different areas of collision physics. At small K^2 the surface essentially represents a region where photoabsorption is dominant, i.e., $df_{0m}/dE \simeq df_0/dE$. As the energy increases the theoretically minimum value of $K^2(= E^2/2E_0)$ increases and the correspondence between electron scattering and photoabsorption cross sections becomes less definite. The Bethe surface shows two distinct regions of physical interest: (i) E and K^2 small, where the dipole properties of the atom govern the scattering (soft region) and the sensitivity to the electronic structure of the atom is preserved. (ii) E and K^2 large, where the Bethe ridge appears centered around the line $K^2 = E$. In this region momentum transfers are much larger than the electron binding energies, and essentially the scattering is represented by the binary encounter model.[45]

An investigation of this surface for helium at electron impact energies of 25 keV and over the angular range 1°–10° has been completed by Wellenstein et al.[46] (Figure 8). The data in this case were normalized using the sum rule, Eq. (16), which was extended to account for continuum transitions. The shape of this surface will determine the region of Born validity and its use for obtaining OOS. Consider a slice of the Bethe surface at one energy E. A plot of f_{0m} vs. K^2 can give quantitative information on the maxima and minima measured by the experiment. The GOS can be written as[47]

$$f_{0m} = (E_0 - E)K^2 \left| \int_0^\infty P_m(r) j_\lambda(Kr) P_0(r)\, dr \right|^2$$

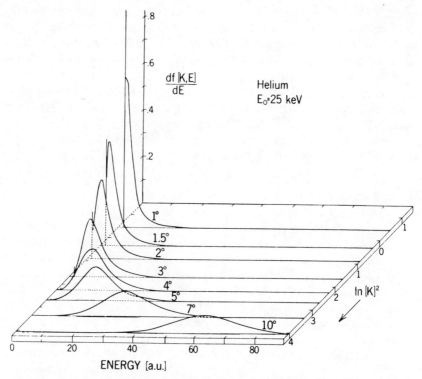

Figure 8. Bethe surface of helium obtained with 25-keV electrons. Experimental cross sections were converted to generalized oscillator strengths, and the data were placed on an absolute scale using the Bethe sum rule.

when the initial $|0\rangle$ and final states $|m\rangle$ can be represented by Hartree–Fock orbitals, where $j_\lambda(Kr)$ is a spherical Bessel function with λ determined by the orbital quantum numbers. The sign of the expression in the modulus depends on both the wave functions used and the value of K. If the overlap integral changes sign as K varies, we will obtain a zero in the GOS curve. Such minima and maxima have been observed, and they represent troughs and crests of the Bethe surface. The related optical phenomena of a Cooper minimum in photoabsorption is determined by electron scattering if the trough reaches the $K = 0$ plane.[31,48] Conversely the shape of f vs. K^2 can be used to determine the nature of the wave functions[49] of the excited state $|m\rangle$.

Figure 9 shows an f_{0m} vs. K^2 plot for the $5S$–$5P$ transition in cadmium with a minimum at $K^2 = 0.8$ a.u. and subsequent maxima at 1.5 a.u. In this particular slice of the Bethe surface the GOS does not touch the $K = 0$ plane. Several regions of Figure 9 are subject to test, i.e., the value of the

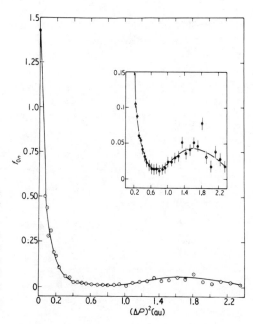

Figure 9. Generalized oscillator strength f_{0n} against the square of the momentum change, $(\Delta P)^2$, for the transition 5^1S-5^1P in cadmium vapor at an incident electron energy of 255 eV. The data have been normalized to the optical oscillator strength at $(\Delta P)^2 = 0$. The curve is fitted to Eq. (18) for $(\Delta P)^2 < 0.3$ a.u.

OOS at $K^2 = 0$ and the positions and intensities of minima and maxima. A systematic study of the resonance transitions in zinc, cadmium,[50] and mercury[51] gives reasonable agreement with theory, Table 5.

The disagreement between theory and experiment is probably due to insufficient configurations being used in the wave functions or a failure of the approximation being used to describe the scattering process. The observation of nonzero minima in optical absorption has been attributed to the departure from the applicability of central field wave functions[48] to describe the states. Hence, having measured f_{0n} as a function of K^2, we can use the extrapolation procedure to determine the OOS. The experimental determination of the positions of the minima and maxima could be used to

Table 5. Positions of Extrema in GOSs for Zn, Cd, and Hg

Atom	OOS Expt.	OOS Theor.	Minimum K^2 Expt.	Minimum K^2 Theor.	Maximum K^2 Expt.	Maximum K^2 Theor.
Zn	1.46^a	2.02^d	1.0^e	3.8^d	1.5^e	5.6^d
Cd	1.42^a	2.39^d	0.8^c	2.7^d	1.7^e	3.9^d
Hg	1.18^a	3.03^d	0.9^f	2.4^d	2.1^f	3.5^d

^{a} Lurio.[35]
^{b} Lurio and Novick.[63]
^{c} Lurio et al.[64]
^{d} Kim (see Ref. 50).
^{e} Newell and Ross.[50]
^{f} Eitel et al.[51]

put the data on an absolute scale although there are no reported examples of this in the literature.

9. Continuum Distribution of Oscillator Strength

In addition to determining the optical properties of discrete transitions by electron scattering, which is the analog of photoabsorption, we can also measure the continuum distribution of oscillator strength, which is the analog of photoionization. Above the ionization threshold the oscillator strength is expressed as $df(K)/dE$ and related to the photoabsorption cross section σ by

$$\frac{df}{dE} = \frac{me}{\pi e^2 h} \sigma(W_s) \tag{19}$$

Expanding df/dE in a power series[52] yields

$$\frac{df}{dE}(K) = \frac{df}{dE} + f^1 K^2 + \frac{1}{2} f^2 K^4 \tag{20}$$

where df/dE is the OOS for the dipole approximation, and f^1 and f^2 are higher-order terms. It is possible by suitable selection of the experimental parameters, i.e., small K^2 values, to minimize the magnitude of the higher terms in Eq. (20) and hence obtain a direct measurement of the OOS. This

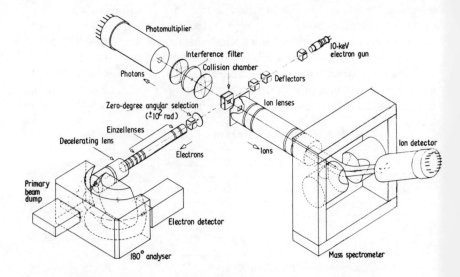

Figure 10. Schematic diagram of apparatus as used by Backx et al.[55]

Table 6. Dipole Oscillator Strengths f_n^0, and the First Derivative of the Generalized Oscillator Strengths at $K = 0$, f_n^1, for Discrete Transitions in He

State	f_n^0		f_n^1	
	Theory	Expt.[d]	Theory	Expt.[d]
2^1S	—	—	$+0.0836^d$	$+0.070$
2^1P	0.27616^a	0.276	-0.4502^d	-0.36
3^1P	0.0734^a	0.073	-0.0926^d	-0.078
$\Sigma_0^{1P} f_n^{(0)}$	0.427^c	0.421	—	—

[a] Schiff and Pekeris.[56]
[b] Kim and Inokuti.[60]
[c] Bell and Kingston,[61] Pekeris,[62] Schiff and Pekeris.[56]
[d] Backx et al.[55]·

situation is analogous to the discussion in Section 7 on discrete excitation. Experimentally it has been demonstrated that good agreement is obtained with the optically derived data for the noble gases[53] for energy transfers up to several hundred eV with incident electron energies of 10 keV. This method is unsuitable for small incident energies and energy losses since a large amount of extrapolation is required to obtain the OOS. Further, the extrapolation curve may pass through a K^2 region where the oscillator strength could vary rapidly.[27] Boersch et al.[54] have attempted, by using extreme collimation, to arrange an experiment such that K was independent of the small observation angle. Unfortunately, the data did not produce the correct shape of the helium continuum. The problems encountered in this region are circumvented by the use of coincidence detection of electrons and ions.[55]

In an electron–ion coincidence experiment the energy-loss electron is detected in coincidence with the ion produced, and in addition the photo-decay of an excited ion can also be isolated by a triple-coincidence measurement. The apparatus used by Van der Wiel[55] is shown in Figure 10. When operated in the electron–ion mode the experiment is analogous to photoionization mass spectrometry, and in the triple-coincidence mode, where the partial oscillator strengths of excited states are measured, we have the analog of a photofluorescence experiment.

The nondipole terms, previously neglected in earlier experiments, can now be included due to improved apparatus design. The differential cross section becomes

$$d\sigma(KE) = \frac{2}{E} \frac{k_m}{k_0} \left[\frac{1}{K^2} f^0(E) + f^1(E) \right] \qquad (21)$$

and with careful measurements relative oscillator strength distribution can be obtained. The measurements of the dipole oscillator strength and the

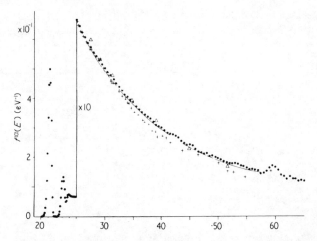

Figure 11. Optical oscillator strengths of helium: •, Backx *et al.*[55]; +, Samson[58]; △, Lowry *et al.*[57]; and ———, Burke and McVicar.[67]

first derivative of the GOS at $K = 0$ for some discrete transitions in helium are given in Table 6. The data are absolute through normalization to the $1^1s–2^1p$ value of Schiff and Pekeris.[56] The continuum distribution of OOS is obtained for helium (Figure 11). Excellent agreement is obtained with the optical photoabsorption work of Lowry *et al.*[57] and Samson.[58] Utilization of this method to obtain continuum OOS for larger energy transfers is relatively inexpensive in relation to using a synchrotron radiation source.

A direct test of the Thomas–Kuhn sum rule, Eq. (16), can be made using the continuum and discrete oscillator strength distributes. For helium we expect

$$\int_0^\infty f^0(E)\, dE = 2 \tag{22}$$

and the data of Van der Wiel yield a value of 2.06. At $K = 0$ the derivative $f'(E)$ of the GOS should obey the sum rule[17]

$$\int_0^\infty f^1(E)\, dE = 0 \tag{23}$$

whereas the experimental data yield -0.134. The nonzero value is probably due to the influence of the $(2S–2P)$He resonance in the continuum.

The orientation of the momentum transfer vector, **K**, in an $(e, 2e)$ coincidence experiment (where the ejected electron is detected in coincidence with the energy-loss electron) will experimentally provide a condition where the ejected electron can be monitored independently of the customary p dependence factor encountered in photoelectron ejection.

The measured coincidence count rate in such an experiment can be expressed as

$$dI_{coinc} = \frac{2}{E}\frac{k_1}{k_0}K^{-2}\frac{df}{dE}\left[1+\frac{1}{2}\beta(\varepsilon)(3\cos^2\psi - 1)\right]d\tau \qquad (24)$$

where ψ is the angle between the ejected electron vector and $\mathbf{K}$, and β is functionally dependent on the energy ε of the ejected electron. For small momentum transfers Eq. (24) can be written as[59]

$$I_{coinc} = I_0[1 + C_\alpha(E)\beta(\varepsilon)] \qquad (25)$$

where α is the angle between the incident and ejected electron directions, E is the energy loss of the primary electron, and $C_\alpha(E)$ is an apparatus function for the particular spectrometer used. I_0 is proportional to the OOS. In the particular case of $\alpha = 54.7°$, i.e., the magic angle of photoelectron spectroscopy,

$$I_{coinc} \approx I \qquad (26)$$

since $C_\alpha(E)$ is zero for all values of E,[59] and consequently the coincidence intensity is independent of $\beta(\varepsilon)$ at all energy-loss values. This simulation of photoelectron spectroscopy enables β independent branching ratios and partial OOS to be measured over a variable range of "photon energies" with relative ease. Such coincident techniques will no doubt provide extremely interesting optical data otherwise obtainable only by using synchrotron radiation sources.

10. Conclusion

Electron impact spectroscopy has advanced considerably over the last ten years. It can easily and readily provide accurate optical data not only for discrete atomic transitions but for continuum excitation processes as well. As an effective probe of atomic structure and properties it has definitely established itself.

References

1. A. C. G. Mitchell and N. W. Zemansky, *Resonance Radiation and Excited Atoms*, Cambridge (1934).
2. J. A. R. Samson, *Advances in Atomic and Molecular Physics*, Vol. 2, p. 177, Academic Press, New York (1966).
3. W. R. S. Garton, J. P. Connerade, M. W. D. Mansfield, and J. E. G. Wheaton, *Appl. Opt.* **8**, 919 (1969).
4. A. Skerbele, M. A. Dillon, and E. N. Lassettre, *J. Chem. Phys.* **46**, 4161 (1967).

5. G. A. Haas, *Methods Exp. Phys.* **4A**, 25 (1967).
6. T. Shuttleworth, Thesis, University College London (1977); T. Shuttleworth, W. R. Newell, and A. C. H. Smith, *J. Phys. B*, **10**, 1641 (1977).
7. P. Marmet and L. Kerwin, *Can. J. Phys.* **38**, 787 (1960).
8. C. E. Kuyatt and J. A. Simpson, *Rev. Sci. Instrum.* **38**, 193 (1966).
9. J. A. Simpson, *Rev. Sci. Instrum.* **35**, 1698 (1964).
10. J. E. Lano and W. Raith, *Phys. Rev. Lett.* **30**, 193 (1973).
11. D. G. Wilden, P. J. Hicks, and J. Cromer, *J. Phys. B* **9**, 1959 (1976).
12. R. E. Fox, W. M. Hickam, D. J. Grove, and A. S. Kjelda, *Rev. Sci. Instrum.* **26**, 1101 (1951).
13. M. F. Rudd, *Low Energy Electron Spectrometry*, Chap. 2, Wiley, New York (1972).
14. J. R. Pierce, *Theory and Design of Electron Beams*, Van Nostrand, New York (1954).
15. E. Harting and F. Read, *Electron Optics*, Elsevier, Amsterdam (1976).
16. A. Septier, *Focussing of Charged Particles*, Vols. 1 and 2. Academic Press, New York (1971).
17. M. Inokuti, *Rev. Mod. Phys.* **43**, 297 (1971).
18. H. S. W. Massey, E. H. S. Burhop, and H. B. Gilbody, *Electronic and Ionic Impact Phenomena*, Vol. 1, p. 426, Oxford (1969).
19. H. A. Bethe, *Ann. Phys. Leipzig* **5**, 325 (1930).
20. R. H. Garstang, *J. Chem. Phys.* **44**, 1308 (1966).
21. M. Kraus and Mielczarek, *J. Chem. Phys.* **51**, 5241 (1969).
22. I. V. Hertel and K. J. Ross, *J. Chem. Phys.* **50**, 536 (1969).
23. M. R. H. Rudge, *Proc. Phys. Soc. London* **86**, 763 (1965).
24. V. I. Ockur, *Sov. Phys.-JEPT* **18**, 503 (1964).
25. E. M. Lassettre, A. Skerbele, and M. A. Dillon, *J. Chem. Phys.* **50**, 1829 (1969).
26. A. Skerbele and E. N. Lassettre, *J. Chem. Phys.* **52**, 2708 (1969).
27. A. Skerbele and E. N. Lassettre, *VII I.C.P.E.A.C.*, North-Holland, Amsterdam (1971).
28. J. P. Bromberg, *J. Chem. Phys.* **51**, 4117 (1969).
29. I. V. Hertel and K. J. Ross, *J. Phys. B* **1**, 697 (1968).
30. T. Shuttleworth, W. R. Newell, and A. C. H. Smith, *J. Phys. B* **10**, 3307 (1977).
31. V. Fano and J. W. Cooper, *Rev. Mod. Phys.* **40**, 441 (1968).
32. G. V. Marr and D. M. Greek, *Proc. R. Soc. London A* **304**, 245 (1968).
33. R. A. Bonham, *J. Chem. Phys.* **56**, 762 (1972).
34. E. N. Lassettre, *J. Chem. Phys.* **43**, 4479 (1965).
35. A. Lurio, *Phys. Rev.* **140**, A1505 (1965).
36. R. H. Garstang, *J. Opt. Soc. Am.* **52**, 845 (1962).
37. J. C. Weishert and A. Dalgarno, *Chem. Phys. Lett.* **9**, 517 (1971).
38. M. A. Ali, *J. Quant. Spectrosc. Radiat. Transfer* **11** (1971).
39. I. V. Hertel and K. J. Ross, *J. Phys. B* **2**, 285 (1969b).
40. I. V. Hertel and K. J. Ross, *J. Phys. B* **2**, 484 (1969a).
41. W. R. Newell and K. J. Ross, *J. Phys. B* **5**, 701 (1972b).
42. W. R. Newell, K. J. Ross, and J. B. P. Wickes, *J. Phys. B* **4**, 684 (1971).
43. H. W. Webb and H. A. Messenger, *Phys. Rev.* **66**, 77 (1944).
44. D. E. Burgess, M. A. Hender, T. Shuttleworth, and A. C. H. Smith, *VII I.C.P.E.A.C.* **96** (1971).
45. L. Vriens, *Case Studies in Atomic Collision Physics*, Vol. I, Eds. E. W. McDaniel and M. R. C. McDowell, North-Holland, Amsterdam (1969).
46. H. F. Wellenstein, R. A. Bonham, and R. C. Ulsh, *Phys. Rev. A* **8**, 304 (1973).
47. Y. K. Kim, M. Inokuti, G. E. Chamberlain, and S. R. Mielczarek, *Phys. Rev. Lett.* **21**, 1146 (1968).
48. M. J. Seaton, *Proc. R. Soc. London* **A208**, 418 (1951).

49. R. A. Bonham and M. Fink, *High Energy Electron Scattering*, Van Nostrand Reinhold, New York (1969).
50. W. R. Newell and K. J. Ross, *J. Phys. B* **5**, 2304 (1972a).
51. W. Eitel, F. Hanne, and J. Kessler, *VII I.C.P.E.A.C.* Eds. L. M. Branscombe *et al.* North-Holland, Amsterdam (1971).
52. M. Inokuti and R. L. Platzman, *IV I.C.P.E.A.C.* Science Bookcrafter, New York (1965).
53. M. J. Van der Wiel and G. Wietes, *Physica* **53**, 225 (1971).
54. H. Boersch, J. Geiger, and B. Schröder, *V I.C.P.E.A.C.*, Leningrad, (1967).
55. C. Backx, R. R. Tol, G. P. Wight, and M. J. Van der Wiel, *J. Phys. B* **18**, 2050 (1975).
56. B. Schiff and C. L. Pekeris, *Phys. Rev.* **134**, A638 (1964).
57. J. F. Lowry, D. H. Tomboulian, and D. L. Ederer, *Phys. Rev.* **1317**, A1054 (1965).
58. J. A. R. Samson, *Advances in Atomic and Molecular Physics*, Vol. 2, p. 177, Academic Press, New York (1966).
59. A. Hammett, W. Stoll, G. Branton, C. Brian, and M. J. Van der Wiel, *J. Phys. B* **9**, 945 (1967).
60. Y. K. Kim and M. Inokuti, *Phys. Rev.* **175**, 176 (1968).
61. K. L. Bell and A. E. Kingston, *Proc. Phys. Soc. London* **90**, 31 (1967).
62. C. R. Pekeris, *Phys. Rev.* **155**, 1216 (1959).
63. A. Lurio and R. Novick, *Phys. Rev.* **137**, A608 (1964).
64. A. Lurio, R. L. de Zafru, and R. J. Goshen, *Phys. Rev.* **134**, A1198 (1964).
65. H. Boersch, J. Geiger, and B. Schröder, *Physics of One and Two Electron Atoms*, Eds. F. Bopp and H. Kleinpoppen, North-Holland, Amsterdam (1969).
66. E. N. Lassettre, A. Skerbele, M. A. Dillon, and K. J. Ross, *J. Chem. Phys.* **48**, 5066 (1968).
67. P. G. Burke and D. D. McVicar, *Proc. Phys. Soc. London*, **86**, 989 (1965).

25

Atomic Transition Probabilities and Lifetimes

W. L. WIESE

1. Introduction

The determination of atomic transition probabilities and lifetimes is an active research area, as may be seen from the following three indicators: (a) the number of research papers since the publication of the first general bibliography[1] in 1962 has grown from 650 to nearly 3000; (b) the data known in 1962 covered roughly 10^5 spectral lines, a number which has increased by an order of magnitude, to roughly 10^6; (c) the accuracy of the numerical data has significantly improved and is now, on average, typically in the range from 10% to 25% versus roughly 25%–50% 15 years ago. The numbers show that the improvement in accuracy lags behind the other two advances, and the relatively large uncertainties point to the continuing difficulties in performing accurate determinations, a central subject which will come up many times throughout this chapter. A further indication for the steady growth and development is that several approaches which dominate the field now did not even exist in the early 1960s or had not been developed to a practical state for large-scale production of high-quality data.

The principal new *experimental* developments in recent years have been

- (i) Beam-foil spectroscopy
- (ii) The utilization of stabilized arcs for heavy element studies
- (iii) The combination of two complementary techniques to determine relative values and an absolute scale in a two-step process
- (iv) Advanced lifetime and transition probability measurement techniques [such as lifetime techniques involving selective (e.g.,

W. L. WIESE • National Bureau of Standards, Washington, D.C.

tunable dye laser) excitation, the atomic beam absorption technique, the zero-field level-crossing technique, stabilized arc emission measurements].

The principal new *theoretical* developments are

  (i)   Detailed treatments of electron correlation effects by multiconfigurational approximations
  (ii) Introduction of rigorous upper and lower error bounds
  (iii) Treatment of relativistic effects
  (iv) High-volume calculations
  (v)   Regularities and systematic trends.

The techniques, from which the great majority of atomic transition probability and lifetime data are derived, are the following: Experimentally, transition probabilities are directly determined through (a) emission, (b) absorption, and (c) anomalous dispersion (hook) techniques; atomic lifetimes are *directly* determined (and thus transition probabilties indirectly) by (d) beam-foil spectroscopy, (e) the delayed-coincidence technique, and (f) the Hanle-effect (or zero-field level-crossing) technique.

Theoretically, (a) self-consistent field (Hartree–Fock) calculations in various levels of refinement, (b) nuclear charge expansion techniques, and (c) semiempirical calculations are mostly applied. In addition, regularity studies have yielded many new data.

The literature contains a number of fairly detailed reviews and descriptions of the major experimental and theoretical methods to determine transition probabilities and lifetimes. Quoted below is a list of authors who have recently reviewed specific methods:

  (1) The emission technique: Foster,[2] Neumann,[3] Wiese[4]
  (2) The absorption technique: Foster,[2] Neumann,[3] Wiese[4]
  (3) The anomalous dispersion ("hook") technique: Foster,[2] Neumann,[3] Wiese,[4] Huber,[5] Penkin[6]
  (4) Calculational methods: Crossley,[7] Layzer and Garstang,[8] Hibbert[9] (see also Hibbert, Chapter 1 of this work)
  (5) Analysis of systematic trends and regularities: Wiese and Weiss,[10] Wiese[11]
  (6) Atomic lifetime measurements:
       (a) General reviews: Foster,[2] Wiese,[4] Corney,[12] Ziock[13]
       (b) Special reviews: delayed coincidence technique: see Fowler Chapter 26 of this work; beam-foil spectroscopy: Curtis,[14] Cocke,[15] and Andrä (Chapter 20 of this work).

This summary of descriptive reviews is not meant to be complete [for a more exhaustive listing one should consult the NBS bibliography on atomic transition probabilities (Reference 16, Section 1)]; however, the cited articles contain detailed descriptions of the currently used techniques.

Comprehensive tables of critically evaluated data have been compiled by the NBS data center on atomic transition probabilities. The tabulations, comprising a total of about 11,000 lines, presently cover the elements H through Mn ($Z = 25$) for all stages of ionization.[17-21] The evaluation of data for the allowed lines of Fe, Co, and Ni is in progress, while forbidden-line data on these three elements have already been published.[19]

The major techniques are well established, and few basic changes have occurred since the above-cited recent reviews were written. Aside from mainly technological advances—for example, the use of the tunable dye laser as a new selective excitation source in lifetime experiments—these reviews still describe approximately the current status of techniques in this field, and an updated descriptive review of the methods appears to be hardly justified. In tune with the title of this book, it seems thus more appropriate to review the significant progress in this field during the last 10-15 years on a rather general level and describe briefly some of the important recent advances, especially for techniques not discussed elsewhere in this book. Thus the various major techniques listed earlier will be described only to the extent necessary for the discussion that follows, and the emphasis in this review will be put on recent advances in the determination of transition probabilities and lifetimes—a task which has not been undertaken on a general level for quite a few years.

With the literature in this field now approaching 3000 papers, it is extremely difficult to present a balanced review of all significant new developments. The choice of quoted references is often rather arbitrary and should be seen from the point of view that these are just some typical examples representative of a number of similar papers. On the other hand, it was attempted to include most of those papers that have had more than average impact on the field or on applications. Purposely, techniques which have had little recent exposure have been given more attention, and approaches discussed by other authors in this book are only briefly mentioned. Special techniques, which have yielded very few though usually quite accurate data, will generally not be discussed because of their small impact on the field. This review does not address molecular data, and the discussion on forbidden lines is limited to a short section at the end of the chapter.

Major needs for atomic transition probabilties, or in some cases for atomic lifetimes, exist in the following fields. (1) *Astrophysics*: The main application arises in the determination of stellar element abundances, where transition probabilities are the key atomic quantity. For example, the new data on elements of the Fe group have had a big impact on solar models: Changes of nearly a factor of 10 resulted in the respective element abundances at the solar photosphere due to equally large changes in the atomic transition probabilities and have also caused smaller corrections in many other solar quantities.[22,23] (2) *Space physics*: With the advent of extensive observations from satellites and rockets, space astronomers have become major users of transition probability data, primarily for the interpretation of the far-ultraviolet and soft x-ray spectral line emission from highly ionized species in the solar corona.[24] (3) *Upper atmosphere physics* (aeronomy): For the study of upper atmosphere processes, such as the absorption of solar radiation by atomic oxygen and nitrogen, accurate transition probability data for the atmospheric gases are needed.[25] (4) *Plasma physics, gaseous discharges*: For the diagnostics of plasmas as well as studies of their equilibrium states, especially the transition probabilities of stable gases are of interest. Of particular importance has been argon, which is widely used as a test gas and has found applications as a radiation and temperature standard. Thus, the prominent lines of Ar I and II have been the subject of more than 100 experimental and theoretical studies, listed in Ref. 16. Nevertheless, after all this work the transition probabilities of a number of prominent argon lines are still in appreciable doubt [see the discussion below in Section 3.1.1 and Refs. 26 and 27]. (5) *Thermonuclear fusion research*: Some of the most urgent needs for transition probabilities have arisen in thermonuclear fusion work on magnetically confined plasmas, primarily the Tokamak devices. In the very hot plasmas, minute heavy-element impurities from highly stripped ions radiate large amounts of energy away and thus contribute appreciably to plasma cooling. To analyze and model these energy-loss problems, data for highly stripped ions of wall materials like Cr, Fe, Ni, Mo, and W are needed.[28] (6) *Isotope separation by lasers*: A very promising new approach to isotope separation is the utilization of stepwise laser excitation and photoionization. The choice of various possible transitions and atomic levels depends heavily on their transition probabilities and respective lifetimes. Special efforts to provide these data are directed toward the uranium case.[29] (7) *Development of laser systems*: Atomic transition probabilities and the radiative decay rates of atomic levels are very important parameters needed to assess the potential of a system as a laser, since population inversions can be achieved only if some basic relationships and inequalities are satisfied among these quantities.[30]

2. Basic Concepts

The concepts of atomic transition probabilities and lifetimes of excited atomic states were introduced by Einstein[31] in 1916 by considering the spontaneous emission of photons in analogy to radioactive decay processes. If per unit volume element N_k atoms are in an excited quantum state k, then the number of spontaneous emissions per second into a lower quantum state i is

$$A_{ki}N_k \tag{1}$$

where thus A_{ki} is introduced as the probability, per unit time, that spontaneous emission takes place.

The radiative lifetime of an excited atomic state k follows from the consideration that this state decays radiatively, in the absence of absorption and induced emission, into a number of lower states i, so that

$$\frac{dN_k}{dt} = -N_k \sum_i A_{ki} \tag{2}$$

Integration results in

$$N_k(t) = N_{k,0} \exp\left[-\left(\sum_i A_{ki}\right)t\right] \tag{3}$$

where $N_{k,0}$ is the initial population of state k at time $t = 0$. The mean lifetime τ_k of this state, i.e., the time in which $N_k(t)$ decays to $1/e$ of its original value, thus amounts to

$$\tau_k = 1/\sum_i A_{ki} \tag{4}$$

This relation connects τ and A, and the sum reduces to a single term for low-lying quantum states, from which only one allowed transition can take place, usually to the ground state.

In addition to the atomic transition probability for spontaneous emission and the atomic lifetime, two other related quantities have found widespread use in the literature, the (absorption) oscillator strength f_{ik} (often simply called "f value") and the line strength S.

The oscillator strength, even though it is linked to classical theory, has continued to this day to be a very useful quantity. By its very nature as a "fraction" (of classical oscillators compared to the atomic number density) it has allowed the generation of sum rules, of which especially the Wigner–Kirkwood partial sum rule has become of considerable practical importance. Furthermore, for a given transition along an isoelectronic sequence the f value possesses valuable scaling properties which have been extensively used in the study of systematic trends and regularities (which will be discussed later in Section 5.1).

The atomic line strength S is widely used in the theoretical literature. It represents the square of the quantum-mechanical electric dipole moment and is a symmetrical quantity, i.e., $S = S_{ik} = S_{ki}$.

The relationships between the three equivalent quantities A_{ki}, f_{ik}, and S are given in Table 1. Oscillator strengths for the prominent lines of a spectrum, particularly the resonance lines, are generally in the range $1 >$

Table 1. Conversion Factors[a]

	A_{ki} (s^{-1})	f_{ik}	S (a.u.)
$A_{ki} =$	1	$\dfrac{6.670 \times 10^{13}}{\lambda^2} \dfrac{g_i}{g_k}$	$\dfrac{2.026 \times 10^5}{g_k \lambda^3}$
$f_{ik} =$	$1.499 \times 10^{-14} \lambda^2 \dfrac{g_k}{g_i}$	1	$\dfrac{30.38}{g_i \lambda}$
S (a.u.) $=$	$4.936 \times 10^{-16} g_k \lambda^3$	$3.292 \times 10^{-2} g_i \lambda$	1

[a] The line strength is as usual given in atomic units, A_{ki} is in s^{-1}, and f is dimensionless. The wavelength λ is in nm, and the statistical weight g_n of level n is related to the total angular momentum (or inner) quantum number J_n by $g_n = 2J_n + 1$. Further details and relationships between "multiplet" and "line" data are found in Ref. 17.

$f_{ik} > 10^{-2}$. The A_{ki}, on the other hand, span a range that covers many orders of magnitude. From the relations given in the table, it is clear that A_{ki} values for spectral lines of very short wavelengths λ_{ik}—i.e., lines in the extreme vacuum uv or x-ray regions—are very large, up to 10^{14} s^{-1} and beyond, while infrared lines have small A_{ki} values, as low as 10^3 s^{-1}. For lines in the optical spectrum (400–800 nm) A_{ki} values are usually in the range 10^5–10^8 s^{-1}.

The foregoing remarks and numerical estimates concerned allowed or electric dipole (E1) transitions. Higher-order radiation, i.e., electric multipole [quadrupole (E2), etc.] and magnetic multipole [magnetic dipole (M1), magnetic quadrupole (M2), etc.] transitions—commonly called forbidden transitions—are usually by many orders of magnitude smaller in their transition probabilities and do not need to be considered in many applications.

These transition probabilities scale, however, very strongly with nuclear charge so that for very highly ionized spectra they may compete with (or sometimes even dominate) the E1 transition probabilities with respect to the decay rates of excited levels. They are of considerable interest in astrophysics, especially for the highly ionized species of the solar corona and some of the recent major developments concerning these transitions will be briefly discussed in Section 7.

3. Transition Probability Measurements

3.1. The Emission Technique

The emission technique has been traditionally the largest supplier of experimental transition probability data and remains in this position today

The most significant recent advances in this method have been made in the further development of emission sources, which now come much closer to the ideal requirement of emission from a homogeneous, stable, and well-analyzed volume of plasma. Measurements with these new generation plasma sources have had an enormous impact on applications, especially astrophysical ones. These important developments have occurred largely unreviewed and will thus be discussed here in some detail. The emission technique has been applied in three methodically quite different variations, which are best discussed separately.

3.1.1. Absolute Values

Utilizing relation (1), one obtains for the emission coefficient ε_{ik} of an atomic transition of frequency ν_{ik}, emitted from a unit plasma volume into unit solid angle,

$$\varepsilon_{ik} = \frac{1}{4\pi} h\nu_{ik} A_{ki} N_k \qquad (5)$$

and the radiative intensity of a spectral line I_{ik} emitted from a plasma of length l along the line of observation is given by

$$I_{ik} = \int_0^l \varepsilon_{ik}\, dl \qquad (6)$$

In the general application of the emission technique, A_{ki} is obtained from Eq. (5) on an absolute scale through the determination of I_{ik} and N_k. Most of the difficulties and problems center around the determination of N_k. With advanced emission sources, i.e., stabilized and shock tubes, the experimental conditions are usually chosen such that the plasma is in a state of local thermodynamic equilibrium (LTE). Then N_k may be obtained through the Boltzmann factor[2,32,33]

$$N_k = N_a \frac{g_k}{U(T)} \exp\left(-\frac{E_k}{k_B T}\right) \qquad (7)$$

N_a denotes the total number density of the species a; T is the temperature; g_k is the statistical weight of atomic state k, which is related to the total angular momentum quantum number J (j for one-electron spectra) by $g_k = 2J_k + 1$; E_k denotes the excitation energy for level k; k_B is the Boltzmann constant; and $U(T)$ is the atomic partition function. The latter is given by

$$U(T) = \sum_{k=0}^{k^*} g_k \exp\left(\frac{-E_k}{k_B T}\right) \qquad (8)$$

where the summation is carried out up to an energy level E_{k^*} which is determined by high-density effects in plasmas.[32,33]

The determination of N_k is thus reduced to the determination of the temperature and the total number density of the species. For LTE conditions, the two quantities T and N_a are related by a set of equilibrium and conservation equations.[2,3,32,33] For stabilized arc sources, which are operated at known pressure, usually atmospheric, these relations permit the complete determination of the state of a plasma[2,3,33] from the measurement of only one of the two quantities above (or any other related one) in the case of a one-element plasma. For two-element arc plasmas, because of demixing effects,[33] two independent measurements are needed, and so on. Similarly, conservation equations for shock tubes permit the analysis of the plasma state if the initial conditions and the shock velocity are known.[2,32] For the determination of the plasma temperature or a particle density numerous diagnostic techniques have been developed,[3,32,33] the majority of which are of spectroscopic nature. But also nonspectroscopic techniques, especially laser interferometry and laser scattering,[34] have gained importance.

The important advance in the emission method has been the development of refined emission sources. Wall-stabilized arcs have become especially valuable for the following reasons: (a) The source is very stable, and its physical conditions may be accurately analyzed and varied over a range of densities and temperatures. (b) The arc source is readily adaptable to the measurement of all those elements which can be introduced as gases or gaseous compounds, both for neutral and singly ionized species. (c) For observations along the arc axis, the plasma is homogeneous in the line of sight, so that Eq. (6) may be simplified to $I_{ik} = \varepsilon_{ik}l$. Side-on observations of the cylindrically symmetrical arc plasma may be analyzed by the application of the Abel inversion process,[32,33] from which the intensity data for a set of homogeneous concentric plasma rings of width Δl may be derived.

The most advanced work with the full emission technique has been addressed to the prominent lines of neutral and singly ionized argon in the visible spectrum. Argon as an economical inert gas exhibits many desireable properties in high-current arc discharges, which have not only generated interest in various technological applications, but also make it suitable as a standard of temperature[27] in the 10,000–20,000 K range and of vacuum ultraviolet[35] radiation. These applications have thus generated strong needs for accurate Ar I and II transition probabilities and have at the same time made argon an excellent test case for studying in detail the accuracy, reliability, and limitations of the emission technique.

The NBS tabulation of argon transition probabilities[18] in 1969 contained a rather extensive review of the then available material. For the prominent 4s–5p transitions of Ar I two separate absolute scales, differing by about 30%, were supported by several emission experiments on each side, a rather puzzling situation. (Theoretical data for these 4s–5p transitions are unreliable because of severe cancellation in the transition integral.) After critical evaluation of all data, the scale supporting the higher A values was chosen for the NBS compilation on the basis of several considerations. One of these was the fact that this scale was most consistent with lifetime data, another was the argument that the authors arriving at the lower scale apparently did not include the contributions from the far-line wings in their intensity measurements. They approximated the total integrated spectral line intensity

$$I_{ik} = \int_{-\infty}^{+\infty} I(\lambda)\, d\lambda \qquad (9)$$

by

$$I_{ik} \approx \int_{-\Delta\lambda}^{+\Delta\lambda} I(\lambda)\, d\lambda \qquad (9a)$$

i.e., they extended the integration only over a range $2\,\Delta\lambda$ fairly close to the line center and thus lost the intensity in the far wings which may be taken into account by using, for example, theoretical line wing expressions. It can be readily estimated that failure to include such corrections reduces I_{ik}, and thus A_{ki}, by 20% or more.[36]

Since this was only one possible explanation, the f-value scale for the $4s$–$5p$ lines could not be considered a settled issue in 1969. Numerous new emission experiments since then have attempted to solve the discrepancy. The two most advanced emission studies are probably the very careful wall-stabilized arc experiments by Nubbemeyer[26] and Preston,[27] in which the Ar plasma has been diagnosed in several ways to overdetermine the plasma parameters and thus to achieve the highest possible accuracy. In both investigations the critical factors of the emission method have been

Table 2. Listing of Error Sources Considered by Nubbemeyer[26] and Preston[27] in Their High-Precision Argon Emission Experiments with Wall-Stabilized Arcs

1. Plasma length determination
2. Correction for line-wing loss
3. Intensity calibration
4. Self-absorption correction
5. Scattered light effects
6. Diffraction losses
7. Inhomogeneities in observed plasma column
8. Transmission of arc enclosure window
9. Detector linearity
10. High-density plasma effects
11. Boundary-layer effects (for end-on measurements)
12. Absolute pressure in the plasma and pressure variations during run
13. Spatial averaging effects in observed plasma volume
14. Uncertainties in atomic data used for diagnostics

Table 3. Transition Probabilities (in $10^5\,\text{s}^{-1}$) for Three Prominent Argon Lines From Recent Emission Measurements

Author	Ar I, 430.0 nm $4s[\frac{3}{2}]^0-5p[\frac{5}{2}]$	Ar I, 714.7 nm $4s[\frac{3}{2}]^0-4p'[\frac{3}{2}]$	Ar II, 480.6 nm $4s\,^4P_{5/2}-4p\,^4P^0_{5/2}$
Scholz and Anderson[37] (1968)	3.31	—	—
van Houwelingen and Kruithof[38] (1971)	3.03(±8.5%)	—	860(±13%)
Garz[39] (1973)	3.90	—	—
Nodwell, Meyer, and Jacobson[40] (1970)	—	12(±15.5%)	—
Pichler and Vuinovic[41] (1972)	—	9.5	—
Ranson and Chapelle[42] (1974)	2.77(±30%)	—	—
Shumaker and Popenoe[43] (1972)	—	5.57	786
Nubbemeyer[26] (1976)	3.91(±5%)	6.51(±5%)	1020(±7%)
Preston[27] (1977)	3.72(±11%)	—	749(±12%)
	3.69(±11%)	—	711(±12%)
NBS compilation[18] (1969)	3.94(±25%)	6.5(±25%)	790(±25%)

addressed in great detail and numerous sources of uncertainties have been taken into account. These are compiled in Table 2 to provide an impression of the complexity of the measurements. Total errors, estimated to be at the 5%–7% level in Nubbemeyer's work and at the 11%–12% level in Preston's work, constitute a significant achievement for this technique, where accuracies for absolute values better than 25% are rare. Both authors conclude that the nonconsideration of line-wing intensities in earlier experiments, as indicated above, represents indeed a major contribution to the reported discrepancies. In Table 3, all recent emission measurements[26,27,39–43] are compared for representative argon lines: the 430-nm line (of the $4s$–$5p$ transition array) and the 714.7-nm line ($4s$–$4p$ array) of Ar I, and the 480.6-nm line (a $4s$–$4p$ transition) of Ar II. The comparison covers the recent emission data which have appeared after publication of the NBS compilation in 1969. It is seen that for the 430-nm line of the $4s$–$5p$ array two groups of data continue to exist, either indicating an $A' \approx 3 \times 10^5\,\text{s}^{-1}$ or $A'' \approx 4 \times 10^5\,\text{s}^{-1}$. The 1969 NBS scale is in all cases in agreement with the two high-precision data sets of Nubbemeyer and Preston within the mutually estimated error limits. However, the two authors disagree appreciably for the Ar II 480.6-nm line, which indicates the presence of still undetected or underestimated error sources. Thus, the argon situation continues to be a puzzle and a challenge.

3.1.2. Relative Measurements

The determination of the plasma temperature T is often much easier to accomplish than that of the species density N_a. This is particularly the case for arc plasmas containing several elements, where the determination of the various particle densities becomes a very complex measurement problem and yields sizable uncertainties.[3,33] It is then advantageous to treat N_a as an unknown constant and to measure transition probability data within an atomic species only on a common *relative* scale. With the greatly increased capabilities to determine atomic lifetimes (Section 6), the absolute scale may usually be more accurately and conveniently obtained from the application of a suitable lifetime technique which could involve just one of the lines measured on the relative scale. Indeed, this combination of emission and lifetime measurement techniques has emerged as a very effective new approach to obtain large amounts of accurate absolute data and has thus been often applied in recent investigations on heavier elements, as will be seen from references given below. This combining of data has been also frequently practiced in the most recent NBS critical tabulations of atomic oscillator strengths for Fe-group elements.[20,21]

It should be emphasized that the measurement of only *one* lifetime (or one absolute transition probability) is sufficient to normalize all relative emission data for a given species. This is in contrast to the "branching-ratio" technique, to be discussed later, where each group of emission lines originating from a given atomic level must be separately normalized. To obtain the relative populations of the various atomic states according to Eq. (7), the main requirement is the establishment of a common excitation temperature. This requirement of "partial LTE," which has to extend only over the excited atomic states involved, is less severe than the requirement of complete LTE down to the ground state.[32,33] Much lower electron densities, which are easier to realize in practice, are sufficient to establish an excitation temperature, so that relative sets of transition probabilities may be obtained under a much wider range of plasma conditions than absolute data.

If relative values are expressed in terms of an arbitrarily chosen reference line r, transition probability ratios are obtained from Eqs. (5)–(7) as

$$\frac{A_x}{A_r} = \frac{\lambda_x g_{k,r} I_x}{\lambda_r g_{k,x} I_r} \exp\left(\frac{E_{k,x} - E_{k,r}}{k_B T}\right) \qquad (10)$$

Since the upper levels E_k of most lines occur within a narrow energy range, where the difference $|E_{k,x} - E_{k,r}|$ is at most a few times $k_B T$, such sets of relative data are rather insensitive to temperature uncertainties and may be obtained with fairly high accuracy, of the order of 10%.

While the operation of wall-stabilized arc sources with stable gases, including gaseous compounds such as CO_2 or SO_2, has not presented any significant technical difficulties, many technical problems are encountered in attempts to generate stable arc plasmas containing significant

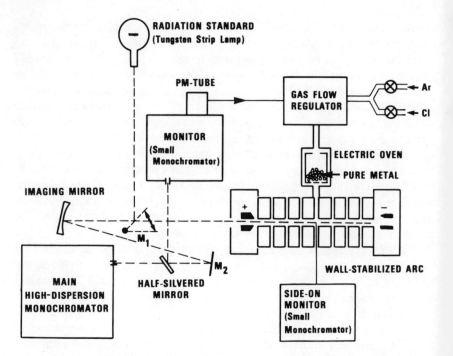

Figure 1. Schematic diagram of a wall-stabilized arc experiment to determine relative oscillator strengths of metallic elements. M_1 and M_2 are plane mirrors, with M_1 rotating around a pivot point to permit spectral radiance calibrations.

amounts of heavier elements. Only a few of these elements have compounds with vapor pressures at room temperature sufficiently high to make them suitable as substantial admixtures to an arc plasma. Thus special measurement techniques had to be developed, and stabilized arc sources had to be appreciably modified in order to study heavier elements, especially metals. The successful development of this arc technology has been the major accomplishment in this field in recent years. A very successful approach[44] is sketched in Figure 1. The metal under investigation is heated in a small oven adjacent to the arc chamber, and the latter is also heated to the same temperature to avoid condensation problems. Argon gas with an admixture of chlorine flows through the oven, and at temperatures of about 600 K a sufficient amount of gaseous metal–chloride is produced there and flows into the arc chamber. The metal atoms and ions leave the arc plasma very soon by diffusion and condense irregularly on the walls, so that the plasma is inherently unstable, and fluctuations as well as long-term drifts in the spectral signals occur. An important component for the stabilization of the plasma conditions is therefore an added feedback system (Figure 1) which monitors and controls the metal concentration in the arc. This consists of a small monochromator which continuously monitors the strength of a typical metal line and a gas flow regulator which instantaneously responds to changes in the monitor signal. Deviations from a preset signal level automatically adjust the gas flow until the correct signal is reestablished.

3.1.3. The Branching Ratio Technique

This greatly simplified version of the emission technique has become very useful in conjunction with the availability of lifetimes for higher excited atomic states, due mainly to beam-foil spectroscopy. "Branches" are defined as groups of emission lines originating from a common atomic level, as illustrated in Figure 2. If one introduces a "photon" intensity $D_{ik} = I_{ik}/h\nu_{ik}$, one obtains from Eqs. (5) and (6) for the branches from a common level k

$$D_{k1}:D_{k2}:D_{k3}:\cdots = A_{k1}:A_{k2}:A_{k3}:\cdots \tag{11}$$

Consequently, one obtains relative transition probabilities for all the "branch" lines simply by measuring their relative photon intensities. These may be converted into absolute data if the lifetime of the atomic level k is known. Selecting an arbitrary transition A_{kj} from the branches, it follows from Eq. (4) that

$$\tau_k = \frac{1}{\sum_i A_{ki}} = \frac{1}{A_{kj}\sum_i(A_{ki}/A_{kj})} \tag{12}$$

and

$$A_{kj} = \frac{1}{\tau_k}\frac{A_{kj}}{\sum_i A_{ki}} = \frac{1}{\tau_k}\frac{D_{kj}}{\sum_i D_{ki}} \tag{13}$$

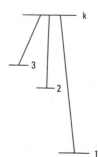

Figure 2. Branching ratios from an excited atomic level k.

Whaling and co-workers[45-50] have utilized this two-step technique to obtain numerous absolute transition probabilities for lines of Fe-group elements. They generated the emission spectra in special hollow cathode sources, where the cathode is usually a cylinder made of the metal under study. Relative photon intensities were measured of all the branch lines from a number of atomic levels. The lifetimes of each of these atomic levels were then determined by the beam-foil technique (which will be discussed later) to convert the relative intensities into absolute transition probabilities. The photoelectric total-line-intensity measurements have been quite accurate, within an uncertainty of a few percent when the light source was stable and the signal not very weak. To monitor the stability of the source and to correct for fluctuations, the signal of one of the branch lines is continuously recorded, while the others are scanned.[50]

3.1.4. Recent Results for Fe-Group Elements

After the technical problems were overcome to operate steady-state arc sources reliably in gases with admixtures of heavy elements, many measurements were carried out on the astrophysically important Fe-group elements. Large sets of mostly relative transition probabilities have been obtained by the emission technique, i.e., either by relative emission or branching ratio measurements, for most of the first and second spectra. In the following, only the principal data sources are listed: For Fe I, Bridges and Kornblith[44] performed wall-stabilized arc measurements for about 500 lines; May, Richter, and Wichelmann[51] carried out measurements with a flow-stabilized arc for another 1000 lines; for Ti I, Klemt[52] measured 139 lines with a wall-stabilized arc; Roberts et al.[53] carried out measurements on 171 Ti I, II, and III lines (mainly Ti II) with a wall-stabilized arc; and Wolnik and Berthel[54] determined absolute f values for 137 Ti I and II lines with a shock tube. For Mn I, Woodgate[55] measured 436 lines with a vortex-stabilized arc; for Ni I and II, Bell et al.,[56] Goly et al.,[57] and Heise[58] measured a total of about 200 lines with wall-stabilized arcs; for Co I and II, Roig and Miller[59] determined 150 relative transition probabilities with a shock tube; for Cr I, 136 lines were determined on an absolute scale by Wolnik et al.[60,61]; and for Cr II, 100 lines were measured with a wall-stabilized arc by Musielok and Wujec.[62]

The branching ratio technique was applied by Whaling and co-workers to Ti I (103 lines)[49]; and Ni I (97 lines).[48,50] Roberts et al.[63] utilized it for 375 lines of Ti II, 75 lines of V I and 423 lines of V II (employing a flow-stabilized arc).

In toto, this represents by far the largest number of f values recently measured with an advanced technique. Its impact, especially on astrophysics, has been enormous. The solar abundances of most Fe-group elements were substantially revised, by factors of as much as 10, largely as a result of these new accurate f value data.[22,23,46-49,51] Thus the importance of this emission work equals or surpasses that of other recent experimental results, but has received relatively little attention.

3.2. The Absorption Technique

3.2.1. The Basic Technique

Total line intensities absorbed by a hot column of vapor are measured and are related to the atomic oscillator strength in a way quite similar to the emission case. Radiation from a continuum source of incident intensity $I_0(\lambda)$ passes through a homogeneous column of gas heated in a furnace. The spectral intensity $I(\lambda)$ received by a photoelectric detector (or a photographic plate) placed behind a spectrometer is given by

$$I(\lambda) = I_0 \exp[-k(\lambda)l] + E(\lambda, T) \tag{14}$$

where $k(\lambda)$ is the atomic absorption coefficient and l the path length traversed in the heated vapor column; $E(\lambda, T)$ accounts for any emission from the furnace and is under normal operating conditions a small correction term, which is less than 1%.

Neglecting the emission term $E(\lambda, T)$ and considering only weak absorption lines, i.e., those with $k(\lambda)l \ll 1$, one obtains

$$I(\lambda) = I_0 \left\{ 1 - k(\lambda)l + \frac{[k(\lambda)l]^2}{2!} - \cdots \right\} \tag{15}$$

or, neglecting higher-order terms,

$$\frac{I_0(\lambda) - I(\lambda)}{I_0(\lambda)} \approx k(\lambda)l \tag{16}$$

The reduced total intensity absorbed by a weak line, in the literature often called the "equivalent width" W_L, thus follows as

$$W_L = \int_{-\infty}^{+\infty} \frac{I_0(\lambda) - I(\lambda)}{I_0(\lambda)} \, d\lambda = l \int_{-\infty}^{+\infty} k(\lambda) \, d\lambda \tag{17}$$

[The "equivalent width" is the width of a fictitious line of the same strength (or area) as the actual line, which is thought of as being completely absorbing over the spectral width W and completely nonabsorbing elsewhere.] The useful range of the absorption method may be extended by the application of the "curve-of-growth" technique (see, e.g., Ref. 2), according to which one may work with somewhat stronger absorption lines where the condition $k(\lambda)l \ll 1$ is not well fulfilled.

In terms of atomic parameters, equivalent expressions for W_L [or $k(\lambda)$, respectively] have been obtained from classical electron theory[64] as well as from Einstein's quantum description of emission and absorption processes[31] mentioned earlier. The general result is

$$W_L = \frac{\pi e^2}{m_e c^2} \lambda^2 N_i f_{ik} l \tag{18}$$

(In absorption experiments, it is customary to employ the absorption oscillator strength f_{ik} rather than a transition probability for absorption.) Equation (18) shows that the determination of f_{ik} requires (a) the measurement of the equivalent width [i.e., of a line intensity absorbed from a continuous background intensity I_0 as indicated in the integral on the left-hand side of Eq. (17)], (b) the measurement of the length l of the absorbing column of gas, and (c) the determination of the number density of absorbing atoms N_i in the atomic state i. The measurements of W_L and l may be performed with high precision; the difficult task is the determination of N_i. With furnaces normally in thermal equilibrium (but see Ref. 65 for a case of substantial deviations at low inert-gas-filling pressures), N_i may be expressed by the Boltzmann factor [Eq. (7)]. This substitution converts the determination of N_i into that of the total density N_a of the species a and the furnace temperature T. The determination of the furnace temperature has been accurately accomplished with optical pyrometry, but the accurate determination of N_a is often a difficult undertaking. Since N_a is a constant for species a, the determination of T suffices, however, for the determination of relative oscillator strengths within this species. No temperature measurement is necessary if all measured transitions start from the same lower state, for example, from the ground state. For the ground state, N_i equals N_a to a very good approximation. In this case, a temperature determination can be used to obtain absolute f values, if reliable vapor pressure (P) data for the element studied are known, since from the equation of state for an ideal gas,

$$N_a = P/k_B T \tag{19}$$

the required species density N_a is immediately obtained.

3.2.2. Precision Measurements

Recent absorption measurements have been concerned mainly with elements of the Fe-group. From these experiments, performed with either furnaces or shock tubes, especially the work of Blackwell and co-workers[65-67] is noteworthy. This work, underway since about 1970, has as its main goal the very precise determination of relative f values for lines of

Table 4. A Comparison of Oscillator Strength Results Obtained from the Absorption Technique and Other Recent Precision Measurements for Two Resonance Multiplets of Cr I

Multiplet	λ (nm)	Bieniewski[68] (absorption)	Becker, Bucka, and Schmidt[69] (Hanle)	Marek and Richter[70] (phase shift)	Marek[71] (delayed coincidence)	Cocke, Curnutte, and Brand[72] (beam foil)
$3d^5 4s\ a^7S - 3d^5 4p\ z^7P^0$	425.435	0.106 ± 0.011	0.111 ± 0.003	0.111 ± 0.010	0.110 ± 0.009	0.100 ± 0.01
	427.480	0.082 ± 0.008	0.0849 ± 0.002	0.084 ± 0.007	—	—
	428.972	0.059 ± 0.006	0.0616 ± 0.002	0.065 ± 0.003	—	—
$3d^5 4s\ a^7S - 3d^4 4s4p\ y^7P^0$	357.869	0.34 ± 0.04	0.356 ± 0.010	0.402 ± 0.046	—	—
	359.349	0.28 ± 0.03	0.271 ± 0.008	0.319 ± 0.036	—	—
	360.533	0.21 ± 0.02	0.220 ± 0.007	0.244 ± 0.021	—	—

the astrophysically important Fe I spectrum. Very high measurement precisions has been achieved (a) by using a very stable furnace of isothermal temperature along its complete length of about 120 cm, (b) by applying sophisticated low-noise photoelectric spectral intensity recording techniques, and (c) by using two identical very-high-resolution spectrometers for the simultaneous scanning of pairs of lines. Furthermore, the f value ratios of various line pairs are measured, and the pairs have been chosen such that many overlapping internal links are established. By this over-determination of data the internal consistency of the photometric measurements could be optimized; typically, adjustments of only 1%–2% were necessary to obtain complete consistency. Overall uncertainties in the relative f values, which span a range of nearly six orders of magnitude, were estimated to be about 5% or less and were shown to be mainly due to uncertainties in the spectral photometry.

Similarly advanced instrumentation was employed by Bieniewski[68] to determine the f values of several lines of Cr I. He used sealed quartz cells, either about 4 cm or 50 cm long, containing pure chromium crystallites, which were placed in a 2-m-long alumina ceramic tube. By appropriate heating a very stable isothermal central part of about 55 cm length could be obtained. Intensity measurements were performed by applying photographic photometry. The range of f values covered was smaller than that of Blackwell and co-workers, and the relative uncertainties are somewhat higher, close to 10%. Bieniewski also performed absolute f value measurements by employing recently published accurate vapor pressure data in conjunction with precise temperature measurements, according to Eq. (19). His absolute f values, estimated to be uncertain to 14%–15%, agree impressively with other precision measurements,[69–72] as seen from Table 4.

3.2.3. The Atomic-Beam Technique

The atomic-beam technique was developed about 30 years ago as a new approach to determine accurate *absolute* data by absorption measurements. It is applied to the special case of resonance lines and has been prominently involved in the determination of the accurate absolute f value scales for a number of heavier elements. Recent applications involved the spectra of Ti I,[73,74] V I,[75] Fe I,[76] and Cu I.[76]

An improved version of an atomic-beam apparatus[77] is shown in Figure 3. All of the major components of the instrumentation are contained in a vacuum vessel. The material under study is evaporated in a crucible, which is placed inside an electrically heated tubular furnace. The vapor emanates from a small hole at the top of the otherwise closed crucible and forms a diffuse atomic beam whose solid angle is defined by an

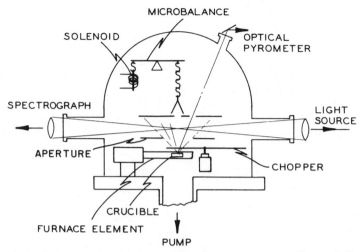

Figure 3. Improved atomic beam absorption apparatus according to Bell and Tubbs.[77]

aperture. The atoms hit the cone-shaped pan of an automated micro-balance and are deposited there. Light from a continuum source traverses the beam, as shown in the figure, and undergoes selective spectral absorption. The radiation is then focused on the slit of a spectrometer, where it is photoelectrically recorded while spectral scans are made. The concentration of absorbing atoms in the light path, and thus the required atomic number density, may be derived from the rate of increase in the weight of the pan under the assumption that all atoms hitting it stick to it. Measurements of this rate, of the temperature of the crucible (to determine the thermal velocity of the beam atoms), and of the absorption intensity are required for this technique. Furthermore, use is made of the fact that for resonance lines $N_i \approx N_a$ holds to a very good approximation. the detailed analysis of the determination of N_a from the weight-increase measurement, which is fairly complex, has been worked out by Bell et al.[78] The automated microbalance applied in the improved atomic-beam apparatus shown in the figure allows a continuous recording of the rate of deposit and thus permits the direct monitoring of any variation in N_a during the spectral scans.

   A major experimental difficulty in all absorption experiments is the maintenance of constant evaporation of heavy elements. This becomes especially critical in the production of atomic beams. For example, Bell et al.[74] encountered considerable difficulties controlling the evaporation rate of the highly reactive titanium metal. A further problem arose from the strong gettering properties of titanium, which traps residual gas together with the titanium atoms on the microbalance and therefore leads to systematic uncertainties in the number-density determination. Reinke,[73] in earlier work on titanium, as well as Mie and Richter,[75] in a study of vanadium resonance lines, used an electron gun to heat a titanium and vanadium rod, respectively, in order to evaporate the metal and produce an atomic beam. Bell et al.[74] point out, however, that this may cause the electron beam to excite a fraction of the atoms into higher states which are then primarily left in metastable levels and thus to reduce the assumed number of atoms in the ground state.

3.3. The Anomalous Dispersion or "Hook" Technique

In the two previously discussed techniques, atomic oscillator strengths have been derived by utilizing relations between the f values and total line intensities. In this technique, a relation between the f value and the index of refraction n near an absorbing wavelength λ_0, i.e., at the edge of an absorption line, is applied. The principal measurement is that of a change in n in the region of anomalous dispersion. The "hook" technique devised by Rozhdestvenskii[79] in 1912 is an elegant and convenient way to accomplish this.

The quantum-mechanical expression for the index of refraction n in the vicinity of an isolated absorption line at wavelength λ_0 is[80]

$$n - 1 = \frac{e^2 N_i f_{ik}}{4\pi \, m_e c^2} \frac{\lambda_0^3}{\lambda - \lambda_0} \left(1 - \frac{N_k g_i}{N_i g_k} \right) \qquad (20)$$

In Rozhdestvenskii's hook technique the absorbing gas is placed into one arm of an interferometer of the Mach–Zehnder or Jamin type. By the use of a compensating plate in the other arm, a very large phase difference of order 10^6 is introduced between the two different light paths. This leads to the formation of two characteristic hooks symmetric to the center of an absorption line. As can be readily shown,[2,4-7] at the positions of the hooks one obtains a special condition for $dn/d\lambda$ which is readily expressed in terms of some measurable quantities of the interferometer. On the other hand, $dn/d\lambda$ is also obtained by differentiation of Eq. (20) with respect to λ, so that $N_i f_{ik}$ is essentially determined by the precisely measurable hook distance $2(\lambda_{\text{hook}} - \lambda_0)$.

Hook separations can easily be measured to an accuracy of a few percent, and uncertainties in the interferometer constant are negligible. Line-broadening effects due to pressure and thermal Doppler broadening have been estimated[81,82] to introduce at most a few percent uncertainty into the hook separation. The term in parentheses in Eq. (20) is usually very small, below 1%, and can be neglected. However, if need be, N_k may be readily correlated to N_i by use of the Boltzmann factor, Eq. (7). In that case the temperature of the absorption tube must be measured, for which, as in the absorption technique, standard optical pyrometry may be used. Temperature measurements are not necessary if only relative f values for resonance lines are measured, i.e., if only the product $N_i f_{ik} \approx N_a f_{ik}$ is determined. But a temperature measurement is needed if N_i pertains to states other than the ground state, in order to be able to utilize the Boltzmann factors. Finally, again as in the absorption technique, absolute f values may be determined for resonance lines from known vapor pressure data by applying the equation of state [Eq. (19)] and measuring T.

The most recent applications have been addressed to a number of heavy elements. The group established by Rozhdestvenskii has continued to be active to this day. Penkin and Komarovskii[83] have, e.g., recently determined relative f values for the rare earth spectra Nd I, Sm I, Eu I, Gd I, Dy I, Tm I, and Yb I. Other recent work has been performed by Parkinson, Huber, and colleagues on the spectra of Fe I,[81,82] Cr I,[84] and Sc I,[85] and by Miyazaki et al.[86] on Cd I. A novel feature of the Sc experiment is the application of a heat-pipe furnace to provide a stable column of hot Sc vapor.

The hook method has, compared to the absorption and emission intensity measurements, the basic advantage that a wavelength distance rather than an intensity is measured. Such factors as possible self-absorption in the line center or line-wing intensity corrections do not enter into the analysis, and no spectral intensity calibrations, another source of uncertainties in the emission method, are involved.

The hook technique has, however, a smaller dynamic range than either emission or absorption intensity measurements. Weak lines are difficult to measure, since the hooks are not clearly formed. In some recent experiments hook and absorption techniques are therefore simultaneously applied in a complementary fashion. For example, an absorption tube setup has been used by Huber and Sandemann[84] to determine stronger lines of Cr I by the hook technique by placing the absorption furnace into one arm of a Mach–Zehnder interferometer; for weaker lines the absorption technique is applied, using the furnace only. Some overlap in the data is used to check the consistency between the two techniques, which was found to be excellent.

In Figure 4 some high-precision data on Fe I are compared which have been obtained from the hook,[82] as well as the emission[44] and the absorption techniques.[67] The three data sets are relative and have been normalized to essentially the same value for the principal Fe I resonance line at 372.0 nm. It is seen that the agreement is indeed very good. It must be cautioned, however, that the investigated lines all originate from states within 0.12 eV of the ground state, so that the data are not sensitive to possible temperature errors.

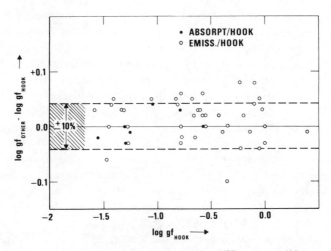

Figure 4. Comparison of precision "hook,"[82] emission,[44] and absorption[67] data for ultraviolet Fe I lines.

4. Transition Probability Calculations

A detailed description of recent advances in atomic transition probability calculations is provided by Hibbert in Chapter 1 of this work. In view of his contribution, this section on theoretical work on atomic transition probabilities will be very brief and confined to (a) some comments on the impact of recent theoretical advances on the overall progress of the field and (b) comparisons of some typical theoretical results with experimental data to assess the accuracy of the calculations. In the introduction, a listing of principal theoretical developments during the last decade was already provided. Each of these will now be taken up in turn with some remarks along the lines of (a) and (b).

4.1. Advanced Calculations

Many accurate theoretical data have resulted from the development and application of theoretical techniques which take into account inter-electron interactions for those—usually rather prominent—transitions where the atomic electrons, including the active one, penetrate each other strongly. These methods all employ some form of a multi-configurational expansion of the wave functions and are known as the superposition-of-configurations (SOC) approach, the multiconfigurational Hartree–Fock (MCHF) model, the nuclear charge expansion method, the non-closed-shell–many-electron-theory (NCMET), etc.

Table 5. Illustration of Configuration Interaction Effects for Some C I Transitions, where Lower- and Upper-State Wave Functions Are Subject to Strong Interelectronic Interactions[a]

| Transition | λ (nm) | Type | Hartree–Fock (single configuration) | Multiconfigurational approximation (SOC) including all interacting configurations through | | | | Other multiconfiguration theory | | Experiment | |
				n = 2	n ≤ 3	n ≤ 4	n ≤ 5	NCMET	Mallow and Bagus	Emission	Lifetime
2s²2p² ³P–2s2p³ ³D⁰	156.1	l	0.286[b]	0.204[b]	0.131[b]	0.122[b]	0.102[b]	0.077[c]	0.080[d]	0.091[e]	0.169[f]
		v	0.332	0.432	0.175	0.149	0.117	0.088[i]	0.122[d]	—	0.076[g], 0.082[h]
2s²2p² ³P–2s2p³ ³P⁰	132.9	l	0.202	0.260	0.131	0.121	0.097	0.092[c]	—	0.039[e]	—
		v	0.171	0.120	0.161	0.136	0.105	0.038[i], 0.059[i]	—	—	—
2s²2p² ³P–2s²2p3s ³P⁰	165.7	l	0.075	0.075	0.105	0.108	0.108	0.116[i]	—	0.17[e]	0.137[f], 0.133[g]
		v	0.094	0.094	0.123	0.124	0.123	0.146[i]	—	0.13[k]	0.14[h]

[a] Results from the single-configuration Hartree–Fock approach are compared with multiconfigurational approximations and experimental data. Also shown are the changes in the multiconfigurational results, as increasingly larger expansions of the wave functions are used to include all configurations up to the indicated principal quantum number n. It is seen that this sequence of approximations gradually converges to the experimental results. Whenever calculated, the oscillator strength data are given in both the dipole length (l) and velocity (v) forms.
[b] Reference 89. [c] Reference 90. [d] Reference 91. [e] Reference 87. [f] Reference 92. [g] Reference 93.
[h] Reference 94. [i] Reference 95. [j] Reference 96. [k] Reference 97.

Applications have been addressed mainly to prominent transitions between lower excited states of light atoms and ions, which include many in-shell ($\Delta n = 0$) transitions involving "shell-equivalent" electrons. Theoretical investigations of excited-state correlation effects were first stimulated in the early 1960's by drastic discrepancies between advanced emission experiments[88,89] and conventional Hartree–Fock calculations. The latter are based on the independent particle concept and thus do not account for correlation.

Subsequently, the advent of beam-foil spectroscopy produced another major stimulus for this theoretical work by providing a wealth of experimental comparison material, mostly on light ions. To illustrate the improvements which resulted from the advanced theoretical approaches, a comparison is given in Table 5 between single-configuration Hartree–Fock results[89] and multiconfigurational data for the case of three C I multiplets.[89–91,95,96] The table clearly shows pronounced differences between the two types of theoretical data. Also seen is the greatly improved (but still not satisfactory) agreement with the experimental comparison data[87,92–94,97] as the wave functions are expanded to include increasing numbers of interacting configurations. The detailed theoretical material (SOC) is taken from the work of Weiss,[89] who has for the ground state $2s^2 2p^2\ {}^3P$, for example, successively included all the configurations containing the $n = 2$ orbitals ($2s$, $2p$; two configurations), then added the $n = 3$ orbitals ($3s$, $3p$, $3d$) for a total for 21 configurations and the $n = 4$ orbitals ($4s$, $4p$, $4d$, $4f$) for a total of 39 configurations, etc.

Due to the large amount of such theoretical work mainly on lighter atoms and their ions (see Ref. 16), and numerous beam-foil lifetime data for comparison, a vast improvement in the f value data situation for prominent transitions between low-lying levels has resulted during the last decade, with many data estimated to be accurate to about 10% or better (see also below).

4.2. Error Bounds

Until quite recently, an assessment of the accuracy of theoretical data has been virtually impossible. Weinhold[98] developed in 1970 a scheme for establishing rigorous upper and lower error bounds to theoretical data, and subsequent applications have been undertaken by him and Anderson,[99] Sims et al.,[100] and Sims and Whitten[101] on some He-, Li-, and Be-sequence transitions, respectively. These results are very valuable in as much as they allow for the first time a theoretical assessment of the quality of experimental data and a theoretical check on the correctness of experimental error estimates. This is just the reverse of the usual situation, where the accuracy of the theoretical data is checked by comparison with reliable

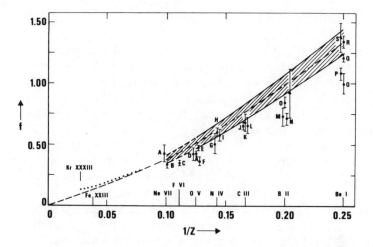

Figure 5. Theoretical upper and lower error bounds, enclosing the shaded area, and f-value data for the $2s^2\,{}^1S$–$2s2p\,{}^1P^0$ transition of the Be sequence: – –multiconfiguration Hartree–Fock calculations[101,120,121]; · · · · multiconfiguration relativistic calculations[123]; ● beam-foil data[94,102–119]; ▲ phase-shift lifetime experiment.[93] It is seen that 9 of the 22 experimental data points—with the authors' estimated error bars included—fall outside the shaded area.

experimental results. Figure 5 serves as an illustration for the usefulness of theoretical error bounds. The f value data for the Be-sequence resonance transition $2s^2\,{}^1S$–$2s2p\,{}^1P^0$ are plotted versus the inverse nuclear charge, and the shaded area indicates the range within which the correct f values must lie according to the rigorous error bounds.[101] It is seen that the majority of the rather numerous experimental data[93,94,102–119] overlap with the shaded area within the estimated uncertainties indicated by the error bars. However, 9 of the 22 data points are fully outside the shaded area. One must conclude that in these cases the experimental errors were insufficiently accounted for or additional undetected systematic errors were present. For comparison, the results of several superposition-of-configurations calculations[101,120,121] (which are practically identical) are shown as a single broken line in the shaded area.

Unfortunately, the extension of the valuable error-bound scheme to more complex atomic systems appears to be out of practical reach because of the need to use extremely accurate wave functions to obtain narrow bounds. As soon as the wave functions are only slightly less accurate, the error bounds become very wide, and are thus of little value for assessments of other data.

4.3. Relativistic Effects

Relativistic effects in transition probabilities are expected to become increasingly important with larger nuclear charge Z along an isoelectronic sequence, since the remaining bound electrons of a highly stripped ion, which move in the field of a large nuclear charge, approach relativistic speeds. The question therefore arises at what Z these effects take hold and how they manifest themselves. Aside from the traditional intermediate coupling effect, it is of great interest to find out how relativistic effects affect the line strength directly. This is also of high practical importance in connection with f-value needs for highly ionized species in controlled thermonuclear fusion research and solar corona physics. A number of relativistic calculations have been carried out during the past years for the resonance lines and for some other prominent transitions of highly stripped ions of the He, Li, Be, Na, Mg, Ar, Cu, and Zn sequences.[122-131] The results show that the range of Z, where oscillator strengths begin to get modified by relativistic effects, varies widely. For some transitions, relativistic effects set in for ions of rather low charge ($\approx +15$), while for others these effects take hold only at much higher charge states, in the range above +40. Since more pronounced relativistic modifications occur in the transition energies, while changes in the dipole moment develop slowly,

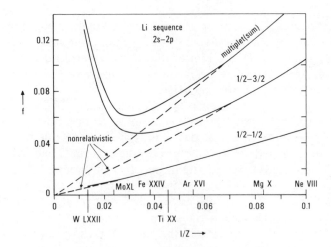

Figure 6. Relativistic oscillator strengths (solid lines) for the $2s_{1/2}-2p_{1/2}$ and $2s_{1/2}-2p_{3/2}$ transitions ($\Delta n = 0$) of very highly ionized Li-like species, according to the calculations by Kim and Desclaux.[122] The broken lines are nonrelativistic (interpolated) data[164] for comparison.

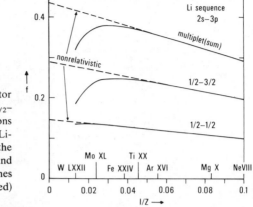

Figure 7. Relativistic oscillator strengths (solid lines) for the $2s_{1/2}$–$3p_{1/2}$ and $2s_{1/2}$–$3p_{3/2}$ transitions ($\Delta n = 1$) of very highly ionized Li-like species according to the calculations by Kim and Desclaux.[122] The broken lines are nonrelativistic (interpolated) data[164] for comparison.

the f values for $\Delta n = 0$ transitions with their small energy differences are often most strongly affected. For two lines of the Li sequence of the types $\Delta n = 0$ and $\Delta n \neq 0$, which appear to be fairly typical for the light elements, the situation is illustrated in Figures 6 and 7. It is seen that the divergence of relativistic f values from the nonrelativistic ones is at first rather slow, but that ultimately the differences become quite drastic. Younger and Weiss[132] have calculated the relativistic effects for a number of transitions of hydrogenic ions. By comparing their hydrogenic ratios of relativistic to nonrelativistic line strengths with those that have been obtained from the above-cited work on Li,[122] Be,[123] and Ar,[124] one finds that these are practically identical. This appears to be a good indication that hydrogenic corrections may be used for other one-electron-type transitions with little loss in accuracy.

4.4. High-Volume Calculations

Calculations of semiempirical nature, starting with the venerable Coulomb approximation by Bates and Damgaard[133] developed in 1949, are capable of yielding large numbers of oscillator strengths with reasonable effort. Thus, comprehensive numerical tables of Bates–Damgaard data for about 3000 lines in the visible and ultraviolet[32] and for about 17,000 infrared lines[134] have been published. The recent work by Kurucz and Peytremann[135] on 1,760,000 oscillator strengths, of which the large majority is on lines of neutral and singly ionized spectra of Fe-group elements, stands out as a most monumental effort. For their calculations, Kurucz and Peytremann utilized observed energy levels to determine by a least-squares-fit procedure scaled Thomas–Fermi–Dirac wave functions and Slater parameters. They published a table of 265,587 lines,[135]

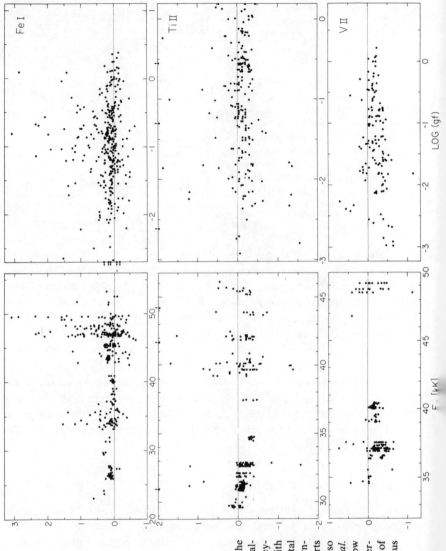

Figure 8. Comparison of the semiempirical f values calculated by Kurucz and Peytremann[135] (KP) with several sets of experimental data: by Bridges and Kornblith[44] for Fe I, by Roberts et al.[53,63] for Ti II (see also Ref. 20), and by Roberts et al.[63] for V II.[63] The plots show $\log gf_{expt} - \log gf_{KP}$ either versus the upper energy level of the transition or versus $\log gf_{expt}$.

addressed especially to line identifications and spectral synthesis in solar and stellar spectra. Kurucz and Peytremann acknowledge that it is very difficult to estimate the errors of their calculated f values. The only practical way to assess the accuracy of their data appears to be through extensive comparisons with those experimental data which can be considered to be fairly accurate. Such comparisons have now been carried out for several spectra by Younger et al.[21] and Smith[136] on a comprehensive scale. Smith's main graphical data comparison for the three spectra Fe I, Ti II, and V II is reproduced in Figure 8, and is typical for these comparisons, which involve several thousand lines altogether. Clearly, the Kurucz and Peytremann semiempirical data are subject to large and apparently randomly occurring errors. The only exception are the transitions from the ground state to the lowest excited levels, where, not unexpectedly, the data obtained from this semiempirical method are reasonably reliable. Because of these uncertainties, the recent NBS data compilations on Fe-group elements[20,21] do not contain these data except for Sc II, V IV, and Cr V, all ions of comparatively simple atomic structure.

5. Regularities in Atomic Oscillator Strengths

On the basis of quantum-mechanical considerations, several types of regularities are expected to occur in atomic oscillator strengths, of which the principal ones will be discussed below.

5.1. Systematic Trend of a Given Transition Along an Isoelectronic Sequence

This is the most important trend in terms of its practical impact. Perturbation theory yields the result[10] that the f value for a given transition along an isoelectronic sequence varies with the nuclear charge Z as

$$f = f_0 + f_1 Z^{-1} + f_2 Z^{-2} + \cdots \qquad (21)$$

where the constant term f_0 is a hydrogenic f value. For Z approaching infinity, or $1/Z \to 0$, the power series reduces to this term, which is one of two things: (a) It vanishes for all transitions in which the principal quantum number does not change, because for hydrogen all such energy levels are degenerate, that is, upper and lower levels coincide. Thus for an important class of transitions, including many principal resonance lines, e.g., the $2s$–$2p$ transition of the Li sequence, etc., the f value tends to zero as Z becomes large. However, at large Z relativistic effects set in and will ultimately change this trend completely. (See, for example, the trends illustrated in Figure 6.) (b) For all other transitions which involve a change in principal quantum number ($\Delta n \neq 0$), f_0 has readily calculable nonzero values,[10] and for one-electron systems, like the alkalis, f_0 is strictly the hydrogen value. Thus in these cases, too, a limiting value for f is established in the limit $1/Z \to 0$. While again relativistic effects at large Z values will start to cause increasing deviations from the nonrelativistically established trend, the knowledge of f_0 is nevertheless very useful for establishing the direction of the systematic trend at lower and medium Z, i.e., for the range of spectra near the important "neutral" end of an isoelectronic sequence. (Figure 7 provides an illustration for the usefulness of nonrelativistic extrapolations to $1/Z \to 0$.)

Isoelectronic trends for several hundred prominent transitions, mostly of lighter element sequences, have become well-established. A wealth of data has made this possible, of which the majority has originated from various theoretical methods and from beam-foil spectroscopy. The latter is, among the experimental techniques, most suited to follow a given transition for several ions of an isoelectronic sequence. A trend analysis is best done graphically by plotting the f value versus $1/Z$. Such a plot is not only suggested by the form of Eq. (21), but also has the advantages to compress an entire isoelectronic sequence into a manageable format, to emphasize the most important range, and conveniently to let f tend to the known or calculable (nonrelativistic) f_0 for large Z. Figure 5 may already serve as one example for a $\Delta n = 0$ transition which shows a smooth, monotonic dependence of f with $1/Z$, with f tending to zero for $1/Z \to 0$. According to Ref. 123, relativistic corrections start to exceed 10% only for $1/Z < 0.05$ and effect the trend from there to $1/Z \to 0$.

Another type of trend curve that is quite frequently encountered exhibits a maximum, usually at lower Z. This maximum can normally be correlated with configuration interaction effects. The boron sequence transition $2s^2 2p\ ^2P^0 - 2s 2p^2\ ^2D$, illustrated in Figure 9, provides an extensively studied example of this type (Refs. 90, 92–94, 96, 104, 106–109, 112, 115, 118, 119, 137–154). Weiss[137] concluded from large-scale superposition-of-configuration (SOC) calculations that the interaction of the $2s 2p^2$ state with configurations of the type $2s^2 nd$ ($n = 3, 4, \ldots$) is important for the lowest stages of ionization and causes appreciable cancellation there, thus causing the f value to be small, especially for the neutral spectrum. But as Z increases, $2s 2p^2$ rapidly separates from the $2s^2 nd$ terms, and these configuration-interaction effects strongly diminish. Another appreciable configuration mixing, that of the lower state $2s^2 2p\ ^2P^0$ with $2p^3\ ^2P^0$, remains approximately constant along the sequence. As the overall result, a maximum in the f value is expected at some low Z, and then the trend curve is expected to tend gradually to zero for this $\Delta n = 0$ transition, until at small $(1/Z)$ values relativistic effects will finally alter this trend (the region of the broken line).

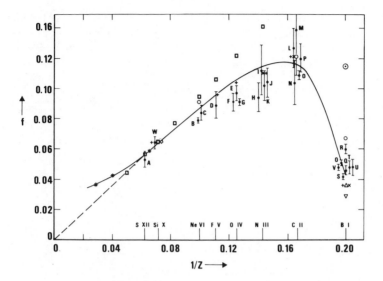

Figure 9. Oscillator strength versus $1/Z$ for the boron sequence transition $2s^2 2p\ ^2P^0-2s2p^2\ ^2D$. The data sources are: ● experimental (mostly beam foil, with error flags according to the authors' estimates): A, Ref. 138; B, 140; C, 139; D, 104; E, 106; F, 107; G, 141; H, 108, I, 109; J, 142; K, 143; L, 112; M, 92; N, 144; O, 94; P, 145; Q, 115; R, 118; S, 119; T, 93; U, 146; V, 147; W, 153. Theoretical (various multiconfigurational approaches): ◯, Ref. 137; △, 96; ▽, 149; ×, 90; +, 150; ⊙, 148; □, 152; ✳, 151; ◇, 154. The solid line is the estimated best fit to the data.

Minima in systematic trend curves are much less frequently encountered, and are usually caused by severe cancelation of positive and negative contributions to the transition integral. The various types of systematic trend curves and the correlations between the occurrence of maxima or minima and their underlying physical causes have been discussed in detail in the literature.[10,11]

5.2. Homologous Atoms

The oscillator strengths for strong comparable transitions of homologous atoms are expected to be similar on account of the analogous outer electron structure of such elements. Thus, all alkalis, for example, would be expected to exhibit similar f values for the leading lines in a Rydberg series, and it is, indeed, well known that the principal resonance lines for the alkalis have f values close to unity. However, one has to consider that as the elements within a chemical family become heavier, the outer atomic structure gets modified due to the presence of unfilled electron shells. In this

Table 6. f-Value Regularities in Homologous Atoms. Comparisons of f Values for the Resonance Lines of "Alkalis" and "Alkaline Earths"

"Alkali"	$ns\,^2S-np\,^2P^0$ Transition			"Alkaline earth"	$ns^2\,^1S-nsnp\,^1P^0$ Transition	
	n	f Value			n	f Value
Li	2	0.753^a		Be	2	1.37^c
Na	3	0.982^b		Mg	3	1.81^b
K	4	1.02^b		Ca	4	1.75^b	
Cu	4		0.64^d	Zn	4		1.46^i
Rb	5	0.98^e		Sr	5	1.99^d	
Ag	5		0.75^f	Cd	5		1.42^k
Cs	6	1.14^g		Ba	6	1.65^j	
Au	6		0.62^h	Hg	6		1.18^l

^{a} Reference 17.	^{d} Reference 155.	^{g} Reference 158. ^{j} Reference 161.
^{b} Reference 18.	^{e} Reference 156.	^{h} Reference 159. ^{k} Reference 162.
^{c} Reference 121.	^{f} Reference 157.	^{i} Reference 160. ^{l} Reference 163.

sense copper, with a completely filled M shell, has an atomic structure completely analogous to the lighter alkalis, while for potassium the $3d$ shell is still unfilled. Generally, the systematic behavior for homologous atoms is not expected to be as closely adhered to as for the case of a given transition along an isoelectronic sequence where the electron configuration remains the same and only a scaling of the nuclear charge occurs. As an illustration of this regularity, accurate f-value data[17,18,120,155–163] for the "alkali" and "alkaline earth" resonance lines are shown in Table 6, selected mainly from recent Hanle lifetime measurements. According to the Kuhn–Thomas–Reiche f-sum rule, the f sums for transitions out of the ground states are one and two for these one- and two-electron systems, respectively, and it is observed that usually about three-fourths of the f-sum is contained in the principal resonance line. It is also seen that this regularity is of a more approximate nature than the isoelectronic sequence trend, but still useful for estimates.

5.3. Regularities in Spectral Series

It has long been established that for all spectral series of hydrogen or hydrogenlike species the oscillator strengths diminish as n^{-3}, so that

$$n^3 \times f = \text{const} \tag{22}$$

Since other atoms approach a hydrogenlike energy-level structure for large principal quantum numbers, it is to be expected that f values for the high members of any spectral series will follow a similar behavior. This prediction requires a modification if quantum defects are not small, because in

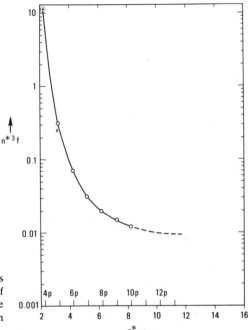

Figure 10. Systematic trend of f values within the $4s$–np spectral series of KI. Plotted is $n^{*3}f$ versus the effective principal quantum number n^*.[18]

this case n should be replaced by the effective principal quantum number.[10] An example for such a systematic trend for a spectral series is provided in Figure 10 (taken from Ref. 18).

5.4. Oscillator Strength Distributions in Spectral Series Along Isoelectronic Sequences

The systematic trend analyses discussed up to this point tie together either (a) the ions of an isoelectronic sequence for a selected transition, (b) the lines of a spectral series for some selected ion, or (c) the corresponding transitions in homologous atoms; but these various cases have remained as yet unrelated. Regularity studies may be carried a step further by inter-relating individual systematic trends through combining (a) and (b).[164] This combination permits the checking of f-value data with regard to two fundamental constraints: adherence to f-sum rules and adherence to the requirement of continuity of f value across the spectral series limit.

For some light-element sequences the oscillator strength material has become so plentiful that such generalized systematic trend studies are possible. In a recent study on the Li sequence,[164] a very high internal consistency in the resulting data set was achieved by fitting the input data

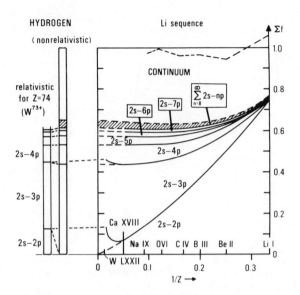

Figure 11. Oscillator strength distribution for the $2s-p$ series of the lithium isoelectronic sequence.[164] Plotted are the partial and total f sums for various ions, which are seen to be very close to unity as required by the Wigner–Kirkwood sum rule. For comparison the nonrelativistic hydrogenic f-value distribution is shown on the left as well as the relativistic distribution for $Z = 74$,[132] for which also the relativistic results for the Li-like ion by Kim and Desclaux[122] for $f(2s-2p)$ and the partial sum $[f(2s-2p)+f(2s-3p)]$ are inserted. The agreement between the oscillator strength distributions for very highly charged hydrogenic and Li-like ions is practically exact.

(a) to smooth isoelectronic sequence trends, (b) to smooth spectral series trends, and by forcing them to adhere (c) to the requirement of continuity of the f value across a spectral series limit and (d) to the Wigner–Kirkwood f-sum rule.

As an example from the above-quoted work, Figure 11 illustrates the f-value distribution in the principal spectral series $2s-p$ and its gradual variation along the isoelectronic sequence. The f-value sums have been built up from individual trend graphs for the various transitions $2s-2p$, $2s-3p$, $2s-4p, \ldots$ vs. $1/Z$ along the Li sequence and from spectral series plots $2s-p$ for the various ions of the sequence, i.e., Li I, Be II, B III, etc. Adjustments in the data were made to achieve adherence to the condition of a smooth transition from line-to-continuum f values. When the f sums for the various ions were obtained, they were found to agree with the

Wigner–Kirkwood f-sum rule (which requires $\Sigma f = 1$ for the s–p series) within 5%. The f sums are indicated by the broken line in Figure 11 and are so close to unity that no further adjustments in the data were made. Relativistic effects are included for the very highly ionized species. The f-value distributions for hydrogenlike ions, which should be asymptotically approached, are included for comparison on the left side of the graph for the nonrelativistic case as well as for the relativistic case at $Z = 74$.

6. Lifetimes of Excited Atomic States

6.1. General Remarks

For the great majority of applications, transition probabilities are the atomic quantity required, while atomic lifetimes per se are rarely needed, and their main use occurs through the relation $\tau_k = 1/\Sigma_i A_{ki}$ [Eq. (4)]. Because of the form of this relation, lifetimes can be utilized for transition probability determinations in two ways:

(1) A direct correlation can be made for such low-lying atomic states where only one particular radiative decay channel $k \to i$ exists or dominates the others, because in these cases Eq. (4) reduces to $\tau_k \approx A_{ki}^{-1}$.

(2) Lifetime measurements may provide an absolute scale for relative transition probability measurements in all other cases, where several A values contribute significantly to τ.

In this latter respect, the emission, absorption, and hook methods and the lifetime techniques are often complementary. While the former allow with ease and high accuracy the determination of the product $(N_k A_{ki})$ or $(N_i f_{ki})$ for a large number of lines [see Eqs. (5), (18), (20)], the determination of the number density N_k or N_i—usually related to the total species density N_a by the Boltzmann factor [Eq. (7)]—is difficult to perform accurately and is often best left undone. Instead, the normalization of such relative A values to the absolute scale is much more reliably accomplished by lifetime measurements. Indeed, the branching ratio technique (Section 3.1.3) has been only applied in situations where its relative emission data could be readily converted with beam-foil lifetime results to absolute values.

The great appeal of the lifetime measurement techniques is that they are conceptually quite straightforward, in contrast to the earlier-discussed methods for the direct determinations of f values. No measurements of temperatures and densities and no assumptions of thermodynamic equilibrium, etc., come into play; all such potentially serious sources of systematic error are avoided. Because of this attractive simplicity and the often resulting high accuracy of the data, lifetime measurements have been the most dynamic area in transition probability measurements, with a very steep growth rate in the last 20 years and, along the way, developments of several entirely new techniques as well as major technological advances. The following quite different techniques have emerged:

(a)   The beam-foil (as well as beam-gas) technique, which is currently the most widely applied lifetime technique. Some recent advances are studies of very highly stripped ions and measurements of extremely short lifetimes ($\approx 10^{-14}$ s).

(b)   The delayed-coincidence technique, where tunable dye lasers are increasingly applied for excitation.

(c)   The Hanle-effect or zero-field level-crossing technique, also an active area.

(d)   The phase-shift technique, practically inactive at the moment (with the exception of a recent experiment by Marek and Richter[70]).

(e)   The natural linewidth, Lamb dip, and other special techniques.

Lifetime techniques have been the subject of especially thorough reviews. First, one must mention the very comprehensive review by Corney.[12] Beam-foil lifetime measurements have been repeatedly discussed in detail, the most recent reviews being those by Curtis[14] and Cocke.[15] Furthermore, descriptions of the delayed-coincidence technique, of some aspects of beam-foil spectroscopy, and of zero-field level-crossing experiments are contained elsewhere in this book.

Thus descriptions of the above-listed techniques will be held here to a bare minimum; instead the recent progress will be reviewed, and a continuing problem as well as the major advances will be highlighted.

6.2. The Cascading Problem

Equation (2) describes the idealized case of the depletion of a selectively excited atomic level k by purely spontaneous radiative decay. In real experiments, one has to consider that the measurements are made on a large assembly of atoms which are in the gaseous or plasma state. Generally, a number of competing processes may occur then that contribute to, and thus alter, the over-all decay rate. The main processes are (a) de- and repopulation of the level by collisions, (b) repopulation by absorption of radiation (often called "radiation imprisonment or trapping"), and (c) repopulation by radiative cascading from higher-lying atomic states which may all be simultaneously populated, depending on the nature of the excitation process. Thus, normally Eq. (2) must be written in a more general form as

$$\frac{dN_k}{dt} = -N_k(t) \sum_{i=1}^{k-1} A_{ki} + Q + B + \sum_{u=k+1}^{\infty} N_u(t) A_{uk} \tag{23}$$

where Q represents a collision and B an absorption term, accounting for processes (a) and (b) above. Both are strongly density-dependent, and for most experiments they can be reduced to insignificance by operating at sufficiently low densities. For those lifetime measurements where Q is significant, it is usually a negative term, i.e., the dominant effect is the shortening of the lifetime by inelastic collisions. On the other hand, radiation imprisonment (term B) lengthens the lifetime. Both effects have been readily taken care of by variation of the density and extrapolation to zero-density conditions. Examples of density-dependent data analyses are found in Refs. 30 and 165–168 and will not be discussed further.

A much more serious problem is that posed by radiative cascading from higher atomic levels for the case of nonselective excitation. This process is expressed by the last term in Eq. (23), and u denotes atomic levels above k. Nonselective excitation is widely used in current lifetime work. It is inherent to the beam-foil technique and is applied in most delayed-coincidence and phase-shift measurements, where electron beams are operated at energies far above threshold. While it is a very general excitation scheme, allowing, in principle at least, lifetime studies of any excited atomic level, it has the serious drawback that the repopulation of level k by radiative cascading from higher levels u during the decay lengthens the lifetime. Cascading effects obviously vary with the atomic structure of the investigated ion, the location of the particular level studied, as well as the excitation conditions. Their significance is thus expected to vary widely, and each particular case requires separate attention.

Cascading has been recognized as a major problem in the beam-foil and electron-beam excitation techniques, and many studies have been devoted to its solution. The general effects on beam-foil data are most instructively illustrated by discussing theoretically simulated radiative decay curves. Several such simulation studies, in which beam-foil decays were constructed from purely theoretical data, have been recently addressed to cases where significant differences between beam-foil lifetimes and reliable theoretical data were found.[169–171]

For beam-foil conditions, i.e., for extremely low densities of order 10^7 ions per cm^3, one may neglect Q and B in Eq. (23), to be left with the cascade term, so that the population of level k is determined by the differential equation

$$\frac{dN_k(t)}{dt} = -N_k(t)\alpha_k + \sum_{u=k+1}^{\infty} N_u(t)A_{uk} \qquad (23')$$

where

$$\alpha_k = \sum_{i=1}^{k-1} A_{ki} = \tau_k^{-1}$$

is the decay constant (or inverse mean life) of level k. Analogous equations must be written for each of the cascading states $u = k + 1, k + 2, k + 3, \ldots$ in order to obtain their populations, resulting in a system of coupled first-order differential equations. Curtis[172] has developed a convenient solution for $N_k(t)$ by using a diagrammatic technique. For the purpose of illustrating the main features, this solution is presented here for the very simplified case of only two cascading states $k + 1$ and $k + 2$ feeding into the primary state k (this already illustrates the essential features):

$$
N_k(t) = \left[N_k(0) + \frac{N_{k+1}(0)A_{k+1,k}}{\alpha_{k+1} - \alpha_k} + \frac{N_{k+2}(0)A_{k+2,k}}{\alpha_{k+2} - \alpha_k} \right.
$$

$$
\left. + \frac{N_{k+2}(0)A_{k+2,k+1}A_{k+1,k}}{(\alpha_{k+2} - \alpha_k)(\alpha_{k+1} - \alpha_k)} \right] e^{-\alpha_k t}
$$

$$
+ \left[\frac{N_{k+1}(0)A_{k+1,k}}{\alpha_k - \alpha_{k+1}} + \frac{N_{k+2}(0)A_{k+2,k+1}A_{k+1,k}}{(\alpha_{k+2} - \alpha_{k+1})(\alpha_k - \alpha_{k+1})} \right] e^{-\alpha_{k+1} t}
$$

$$
+ \left[\frac{N_{k+2}(0)A_{k+2,k}}{\alpha_k - \alpha_{k+2}} + \frac{N_{k+2}(0)A_{k+2,k+1}A_{k+1,k}}{(\alpha_{k+1} - \alpha_{k+2})(\alpha_k - \alpha_{k+2})} \right] e^{-\alpha_{k+2} t} \qquad (24)
$$

$N_k(t)$ is thus obtained as a sum of exponential terms which contain the inverse lifetimes of all three states $k, k + 1, k + 2$ in the exponents. For the numerical analysis the relevant transition probabilities and the initial populations must be known. With regard to the latter, only the distribution over states, but not their absolute values, is necessary to obtain the time dependence of $N_k(t)$. Furthermore, due to spectroscopic selection rules, some of the cascade paths are actually "forbidden" for electric dipole transitions and are only allowed for magnetic dipole or electric quadrupole transitions which have normally much smaller A values. When both the $k + 1$ and $k + 2$ states cascade by electric dipole (E1) transitions directly into k, their orbital angular momentum quantum numbers l must be either equal or differ by 2. According to the $\Delta l = \pm 1$ selection rule, transitions from $k + 2$ to $k + 1$ are therefore electric-dipole-forbidden, which means that $A_{k+2,k+1}$ should usually be negligibly small. On the other hand, when state $k + 2$ cascades first by E1 radiation into $k + 1$, which then cascades into k, the transition $k + 2 \rightarrow k$ is E1 forbidden ($\Delta l = 0, \pm 2$) and $A_{k+2,k}$ is thus very small.

As a numerical example, the $4s$–$4p$ transition of Kr VIII will be now discussed, which is the resonance line of the Cu sequence. For this transition, a discrepancy of about 40% exists between a theoretical value obtained from multiconfiguration calculations[173] and several beam-foil experiment data.[174–176] The beam-foil decay curves for the $4p$ levels published by Druetta and Buchet[174] appear to indicate a two-exponential

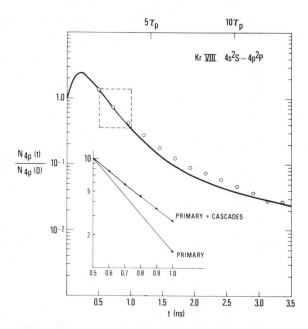

Figure 12. Semilog plots of the radiative decay for the $4p$
level of Kr VIII, obtained (a) from the beam-foil
experiment of Druetta and Buchet[174] (open circles)
and (b) from a theoretical simulation with contributions
from 30 cascade levels. The simulated curve producing
this close fit is based on a $(2l+1)n^{-3}$ distribution for
the initial populations. The insert, which represents the
important initial part of the decay (boxed region,
enlarged), demonstrates that the simulation comprising
"primary and cascades" produces an almost linear
decay just as the primary itself, but with a significantly
different slope.

decay with two well-separated decay components, the slow component
being about 13 times longer than the fast one. This decay was recently
theoretically simulated for beam-foil conditions.[171] For the initial populations of the atomic states after leaving the foil a $(2l+1)n^{-3}$ distribution was
selected, since it very closely reproduces Druetta and Buchet's experimental
decay curve and is also supported by a number of recent beam-foil studies, as
discussed, e.g., in Ref. 171. With respect to the required transition probability data, the above-cited elaborate configuration-interaction calculations by Froese Fischer[173] were used (which include relativistic corrections) for the primary $4p$ level. Approximate transition probabilities
were calculated using the frozen-core Hartree–Fock method for all cascade

Table 7. Summary of Lifetime Data (in ns) for the Kr VIII. $4p$ Level [a]

Theory (multiconfiguration approximation)	Beam-foil experiments (averaged for the $4p_{1/2}$ and $4p_{3/2}$ levels)	Simulation study (two-exponential fit)
0.26[b]	0.39 ± 0.05[d]	$0.35(2.8)$[e]
	$0.36 \pm 0.04(4.7)$[f]	
	$0.35 \pm 0.015(4.8)$[g]	

[a] When the data have been subjected to a two-exponential fit, the numbers for the long-lived cascade are given in parentheses.
[b] Used as point of departure in simulation study (Ref. 171).
[c] Reference 173.   [d] Reference 176.   [e] Reference 171.
[f] Reference 174.   [g] Reference 175.

transitions involving levels with $4 \le n \le 7$ and $0 \le l \le 3$ (l = orbital angular momentum quantum number). Furthermore, for the especially important decay branch $\cdots -9l-8k-7i-6h-5g-4f-4d-4p$, states up to $n = 20$ were included, and for the higher transitions scaled hydrogenic data were used. Thus, a sum of exponentials containing 30 different atomic decay constants α_u—i.e., contributions from 30 cascade levels—was employed to construct the temporal variation in N_k. The resultant simulated decay is compared with the beam-foil data of Druetta and Buchet[174] in Figure 12. It is seen that the theoretically constructed curve reproduces the experimental data quite well, provided a constant background count, as extrapolated from the experiment, is subtracted from the measurements. However, as the insert in the figure illustrates, the slope for the calculated primary $4p$ lifetime used in constructing the curve is quite different. It is of interest to note that the simulated decay containing 30 cascade levels still produces an approximately straight line and thus effectively masks the multiexponential character of the curve. Druetta and Buchet[174] subjected their experimental data to a frequently applied cascade analysis, a two-exponential fit, since the experimental decay appeared to be basically represented by a short- (primary) and a long-lived (cascade) component. Their analysis produced the following lifetimes: $\tau_{DB}(4p) = 0.36$ ns, $\tau_{DB}(\text{casc.}) = 4.7$ ns. For the simulated beam-foil curve the analogous analysis yields: $\tau_{sim.}(4p) = 0.34$ ns, $\tau_{sim.}(\text{casc.}) = 2.6$ ns. This close agreement implies again only that the simulated curve approximates the experimental one well, while the theoretical lifetime for the $4p$ level used in the simulation study is in both cases missed by 30%. The over-all data situation on the lifetime for this level is summarized in Table 7. One must conclude that for this transition it is impossible to extract accurately the primary lifetime by a two- (or three-) exponential cascade analysis, the fundamental reason being that *numerous* higher excited atomic states contribute significantly to the decay curve.

From the above-cited simulation study, some more general conclusions could be drawn, too. It is evident from Eq. (24) and further generalizations thereof that strong cascading effects are expected when (a) the populations of at least some states N_{k+1}, N_{k+2}, etc., are comparable to N_k, and when (b) the decay constants of some cascading levels α_{k+1}, α_{k+2}, etc., are comparable to α_k (this usually implies also a number of fairly strong coefficients $A_{k+1,k}$, etc.). For beam-foil experiments, with an approximate n^{-3} level population dependence, condition (a) is mainly encountered for primary levels with large n. There the population decrease for the cascading levels involving quantum numbers $n+1$, $n+2$, etc., is not nearly so steep as for small n. Furthermore, condition (b) is also most likely to occur for primary levels with large n values, because the differences in the decay constants between successive n levels generally diminish with increasing n. A special, but very important, case are the primary levels in the same shell with the ground state which can decay only via $\Delta n = 0$ transitions. (Such lifetime measurements usually allow direct conversions to f values.) The transition probabilities for these $\Delta n = 0$ transitions—normally quite large for neutral atoms—grow only slowly along an isoelectronic sequence with the first power of the nuclear charge Z, while the probabilities for the $\Delta n \neq 0$ cascading transitions grow with Z^4. As Z increases, the lifetimes of important cascading states therefore become comparable to the primary lifetime at some point in the sequence. Combining requirements (a) and (b), one must expect the most significant and most difficult to recognize cascading effects in beam-foil experiments for high-lying levels (say, $n \geq 4$) and for excited levels in the same shell with the ground state, especially for ions. Both these conditions point to ions of heavier elements—large Z and n—as most likely to be affected. On the other hand, for lifetimes of lower levels of light elements which are *not* in the same shell as the ground state, minimal cascading effects are anticipated.

These general expectations appear to be indeed borne out by many beam-foil results. For example, Astner et al.[177] reported a practically perfect single exponential decay for the He I $3p$ level and a lifetime measurement in excellent agreement with theory. On the other hand, Bashkin[178] has reviewed several examples of high-lying levels in lighter elements ($n = 4 - 7$), which all show appreciable cascading, as would be expected. As examples for levels involving $\Delta n = 0$ transitions, the resonance lines of alkalis are an interesting example: One observes apparently very little cascading for the upper level of the resonance transition $2s$–$2p$ of the Li sequence and excellent agreement with theory,[164] since one encounters very few instances of comparable cascade lifetimes of neighboring n states ($n = 3, 4$) along the sequence. For the homologous Na sequence $3s$–$3p$ transition, Crossley et al.[170] note a drastic change in the ratios of lifetimes of cascades to primary level along the sequence: for Na I, all cascading states have longer lifetimes than the $3p$ primary level; for Ar VIII, however, the $3p$ lifetime has become an order of magnitude longer than some close-lying cascades and comparable to some other not too distant cascade levels, which makes the extraction of an accurate lifetime much more difficult. Indeed, substantial discrepancies between beam-foil results and advanced calculations occur.[170] Proceeding to the homologous $4s$–$4p$ transition of the Cu sequence, both the large Z values along the sequence and the larger n of the primary level increase the likelihood of many significantly populated cascade levels of comparable lifetimes along the sequence. Indeed, the above-discussed simulation study on the $4p$ level and the comparison with beam-foil decay curves (Figure 12) make observed discrepancies of 30% with advanced calculations entirely plausible due to the inability to extract a precise $4p$ lifetime.

The cascade problem has been up to now discussed essentially in the framework of beam-foil experiments, in view of the large amount of beam-foil data and the many studies on this problem.[14] However, cascading is of course also encountered in all other lifetime measurements using nonthreshold electron excitation of atoms. The studies of Bennett and Kindlmann[165] on atomic lifetimes of noble gases with the delayed-coincidence technique provide an instructive example. To investigate the magnitude of cascading effects, Bennett and Kindlmann compared the decay curves for the $2p_2$ and $2p_8$ levels of Ne I (Paschen notation), which were obtained by monoenergetic electron-beam excitation, with beam

Table 8. Comparison of Lifetime Data (in ns) for Two Ne I Levels, Taken Either Very Close to or Well Above (10 eV) the Threshold Energy from the Work of Bennett and Kindlmann[165] and Klose[166]

Level (Paschen notation)	Threshold (one-exp.)	Bennett and Kindlmann			Klose, 10 eV above threshold, two-exp. fit
		11 eV above threshold			
		One-exp. fit	Two-exp. fit	Three-exp. fit	
$2p_2$	18.7 ± 0.3	59.2 ± 1.3	28.2 ± 0.3	25.2 ± 0.5	16.3 ± 0.6
			154.6 ± 1.8	90.3 ± 10	—
				284 ± 56	—
$2p_8$	19.7 ± 0.2	52.3 ± 1.2	29.0 ± 0.1	27.4 ± 0.4	24.3 ± 0.8
			158.8 ± 1.4	97.6 ± 15	—
				249 ± 50	—

energies of 0.1 and 11 eV above threshold energy. They assumed that for the latter case several exponential decay components would be significant and thus performed least-squares fits of their data to sums of exponential terms. The nonselective excitation at 11 eV above threshold indeed yielded decay curves that contained at least four exponential terms, while the threshold data exhibit only a single exponential decay. Results obtained from fits to one, two, and three exponential terms are collected in Table 8, and it is seen that for nonselective excitation the "primary" decay, i.e., the fastest decay rate, converges gradually to the value obtained from threshold excitation. But the convergence is slow; even with a three-exponential fit, the difference still amounts to 35%. Klose,[166] on the other hand, whose data are included for comparison, has obtained primary decay rates of 16.3 ns for the p_2 and 24.3 ns for the p_8 level for electron-beam excitation at about 10 eV above threshold. These data, which are either below or within 23% of Bennett and Kindlmann's threshold value, were analyzed by using a fit to two exponentials and background. The uncertainties quoted by Bennett and Kindlmann clearly emphasize the extraordinary sensitivity of the exponents and amplitudes to very small variations in the data, a point which has long been known in numerical analysis.[179]

6.3. Advances in Lifetime Techniques

6.3.1. Beam-Foil Experiments

In a typical setup, an ion beam is accelerated in a van de Graff accelerator, runs through a mass-analyzing magnet, and then passes a thin carbon foil. Interaction with the foil atoms raises a considerable proportion of the ions into excited states; a luminous beam emerges. From measurements of the intensity decay as a function of distance from the foil and knowledge of the beam velocity, atomic mean lives are directly derived.

Because of the appealing simplicity of this setup and the high versatility of ion-beam accelerators with regard to the choice of elements and states of ionization via the acceleration energy, beam-foil lifetime measurements have become quite popular and numerous results covering many different elements and ions have become available. Current summaries of this very active field are found in Refs. 14 and 15.

Of the recent significant advances, especially the attainment of very high charge states should be mentioned. With high-energy tandem accelerators, operated at 50–90 MeV, observations of H- and He-like S and Cl ions have become possible,[180–183] and the present record for ion stripping has been achieved with a 714-MeV krypton ion beam generated at the Lawrence Berkeley Laboratory's super-heavy-ion linear accelerator (super–Hilac), where Kr ions have been stripped down to two electrons, which results in He-like Kr^{+34}.[184] For such high charge states, the transition energies become very large, the spectra shift into the extreme uv and x-ray regions and atomic lifetimes become extremely short, of the order of a picosecond or less. Therefore, the excited ions leaving the foil decay very rapidly to the ground state, and even for the fast beams of very high-energy accelerators, reaching speeds of order 10^9 cm/s, very high spatial resolution is necessary to perform adequate decay curve measurements. Special spectrometric setups with slits very close to the foil and beam are necessary, and high-precision screws to move the interferometrically controlled foil in steps as small as $0.5\,\mu$m have been applied. Direct decay curve measurements for lifetimes as short as 10^{-12} s have been achieved.[181,182–186] Furthermore, an indirect technique has been developed to make lifetime measurements down to the 10^{-14} s range.[182,183,187]

The beam-foil technique has been applied to a large number of elements and ions. For multiply ionized spectra it is the only experimental technique available to determine atomic lifetimes. Recent measurements include, for example, a number of heavy rare-earth elements (Ce, Pr, Nd, Sm, Tm, Yb, and Lu) in low stages of ionization.[188,189]

The earlier-discussed cascade problem due to the nonselective excitation process has been recognized as the major problem of beam-foil spectroscopy and has received wide attention. Not only have many correction techniques been developed to take this problem approximately into account but several new experimental approaches were developed and tested to obtain cascade-free lifetimes. One method utilizes the phenomenon of quantum beats in a magnetic field,[190] another uses the method of cascade coincidences,[191] and still another combines ion beam and laser excitation (some representative papers are Refs. 192–195). Laser excitation of fast ion beams, originally accomplished by Doppler-tuned beam laser resonance,[192,193] appears to have quite a bit of future potential with the advent of dye lasers and frequency multiplication techniques that should allow an appreciable range of transitions to be excited.[194,195] This technique produces essentially two-stage excitation, the first stage being the usual nonselective beam-foil or beam-gas interaction and the second stage being the selective excitation of a chosen atomic level by tuning a dye-laser to the relevant transition. Harde,[195] for example, produced two-step excitation in Ne I by first populating the metastable $3s$ states by beam-gas interaction and by then populating the $3p$ levels via selective laser excitation, whose decay is subsequently observed. Andrä[193] excited the Ba^+ resonance line by a Doppler-tuned Ar-ion laser and obtained, after a very careful study, considering numerous sources of uncertainties, a lifetime for the upper resonance level within $\pm\frac{1}{4}$% ($\tau = 6.312 \pm 0.016$ ns), which appears to be one of the most precise experimental f values to date, in excellent agreement with other accurate data. Further details are given in the contribution by Andrä (Chapter 20) in this book.

6.3.2. The Delayed-Coincidence Technique and Related Methods

In its basic modern form, atoms contained in some very low-pressure vessel are raised into excited levels by a photon or an electron pulse from a laser or an electron gun, respectively. The radiative decay is then either directly observed[196] or analyzed by studing the temporal distribution of delayed coincidence (see, e.g., Refs. 30, 165, 166). Further details are given by Fowler in his contribution to this work (Chapter 26).

Much of the recent activity and progress on the delayed-coincidence technique has involved the introduction (a) of selective excitation schemes in order to produce cascade-free lifetimes and (b) of new high-spectral-resolution instrumentation to produce unambiguously identified decays from unblended single levels. Furthermore, the use of atomic-beam ovens has extended applications over a wider range of elements. Usually, applications address neutral and singly ionized spectra; extensions to higher stages of ionization are difficult.

Selective excitation of atomic levels was first achieved by Bennett *et al.*[30,165] with electron-beam excitation just above threshold (≈ 0.1 eV). To overcome the very small excitation cross sections in this energy range, a specially designed electron gun was employed. This successful technique has been applied to several noble gases.[167,197] Some of the very accurate results were presented earlier in Table 8.

The advent of the tunable dye laser has led to new activity, especially on complex spectra of heavier elements. This excitation source is indeed very suitable to complex spectra for two reasons: First, because of its small spectral width (≈ 0.01 nm), single levels can be unambiguously excited, which results in selectivity not only with respect to cascade-free excitation, but also with respect to high purity of the atomic state (no blending). Second, excitation energies in these spectra are not large, which places most transitions into a conveniently accessible wavelength range without resorting to such processes as frequency doubling. For example, Figger *et al.*[198-200] performed a series of lifetime measurements on the Fe I spectrum, which have produced some very accurate lifetime data for complex spectra and tie down the absolute scale for Fe I within a few percent.

An even more complex spectrum is that of U I, for which lifetime data have become very important because of the interest in uranium isotope separation by lasers. Measurements of several U I levels have been recently performed by Klose[201] with the conventional delayed-coincidence technique. One of the principal difficulties encountered was that of producing a stable uranium atomic beam. Carlson *et al.*[29] used one tunable dye laser to excite a selected level and then another delayed laser pulse to photoionize from this level. They determined the lifetime by measuring the ion current as a function of the variable time delay between the two laser pulses since the decrease in the ion current is proportional to the population decay in the excited level. Other recent work employing tunable dye lasers (partly with two-step excitation) is that of Lundberg and Svanberg,[202] Gounand et al.,[203] Havey et al.,[168,204,205] and Deech et al.[206] Two other interesting recent techniques are the photon–photon delayed-coincidence method (photon cascades) fully developed by King et al.,[207] which also avoids cascading effects, and Erman's high-frequency deflection (HFD) technique (see, e.g., Ref. 208, in which much of the work with this method is mentioned). In the HFD technique, a specially designed electron gun produces an intense high-voltage ($\approx$10-keV) electron beam which collides perpendicularly with a gas jet. High-frequency sweeping of the intense electron beam optimizes the light output from the interaction region and thus allows the use of high-resolution spectrometers. This makes the technique especially suitable for complex spectra, since it essentially eliminates the problem of line blending; however, it does not avoid the cascading problem. Further references and details on delayed-coincidence work are found in the article by Fowler in this book (Chapter 26).

6.3.3. The Hanle-Effect or Zero-Field Level-Crossing Technique

This well-known technique (discussed in more detail in Chapter 9 of this work by Happer and Gupta) makes use of polarized resonance radiation to excite atoms in the presence of a known variable magnetic field. The excited atoms precess about this field, and the re-emitted radiation has a field dependence that involves the atomic lifetime, which thus may be derived from it.

In early Hanle-effect experiments all those elements were studied which could be readily produced as gaseous vapors in a resonance cell by thermal evaporation. But the number of species which could be studied in this way is quite limited, and the atomic levels are furthermore restricted to the upper states of the principal resonance lines. In view of these restrictions, recent work has been primarily concerned with extending the range of this accurate technique: For example, extensions to other elements have been achieved by using atomic beams[209] or by producing atoms from cathode sputtering.[210] Furthermore, Rambow and Shearer[211,212] produced singly ionized species (Mg II, Ca II, Zn II, Sr II, Cd II, Ba II, and Yb II) by letting a fast-flowing helium afterglow stream into the region where it interacts with the alkaline-earth atoms, which are generated in a small oven. A large number of metastable atoms is generated in the helium flow by suitable microwave excitation, and these metastables are used to ionize the alkaline-earth atoms by collisions (Penning ionization). Andersen *et al.*[213] measured lifetimes for some of the same ions by combining a fast ion beam with the Hanle technique. Bulos *et al.*[214] extended Hanle-effect measurements to additional atomic states by first populating higher levels which then spontaneously decay into the states which are studied ("cascade" Hanle-effect technique).

6.3.4. Special Lifetime Techniques

A method which has been used for a long time and has recently found new applications is that of measuring natural linewidths. As is well-known, these are equal to the sum of the radiative decay constants of upper and lower states. When the lower state is the ground state (or a long-lived metastable state), one can directly determine the inverse lifetime of the upper state. To measure the natural linewidth, which is normally quite small, one has to reduce the other sources of line broadening to insignificance, i.e., Doppler and pressure broadening as well as instrumental broadening. This requires working at very low pressures as well as low temperatures and the application of instrumentation of very high spectral resolution (e.g., a Fabry–Perot interferometer). Recent work has been applied to some Ar II and He II levels.[215,216]

Radiative decay rates have also been recently measured by using certain properties of gas lasers, for example, by utilizing the width of the magnetic field power dip of Zeeman laser systems or the Lamb dip. Results have been reported for He–Ne, He–Xe$^+$, He–Se$^+$, and He–Cd$^+$ laser systems.[217–221]

7. Forbidden Transitions

In the past, most of the work on forbidden transitions, usually calculations, has been concerned with lines of neutral species and low stages of ionization of astrophysical interest. The large majority of these transitions is located in the visible region of the spectrum and are of importance to ground-based astronomy. Some theoretical work is continuing in this area (e.g., Ref. 150), and some interesting and difficult emission intensity measurements closely confirming the theoretical results have been performed on forbidden lines of neutral He[222] and O.[223]

Space astronomy with satellites has recently yielded x-ray and extreme uv spectra of the solar corona, which exhibit many forbidden lines of highly ionized spectra. With this stimulus, transition probabilities of forbidden lines for highly ionized spectra of light elements, especially H- and He-like

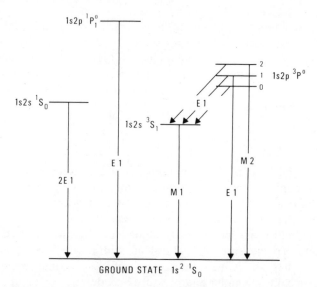

Figure 13. Schematic energy-level diagram and transitions between low-lying levels of He-like ions.

ions, have become of considerable interest. The radiative decay rates for metastable states, which are very small for neutral He and H, grow strongly with the nuclear charge Z. As will be noted below, the scaling with Z occurs with exponents varying from 6 to 10, while decay rates for allowed lines scale with Z^4 for $\Delta n \neq 0$ transitions and with Z for $\Delta n = 0$ transitions. For highly ionized species the rates thus move into a range which is very well-suited for beam-foil spectroscopy observations. (In contrast, times for allowed decays often become extremely short, as was seen earlier[182–187].)

Marrus and co-workers[184,224–228] as well as Cocke and co-workers[180–182,229,230] have studied the very interesting variety of forbidden decays encountered for the singly excited $n = 2$ levels of He-like ions. Figure 13 illustrates the six $n = 2$ levels, of which four decay to the ground state via single-photon emission. These decays were investigated for a number of highly ionized species along the He sequence, mainly noble gas ions, like Ar^{+16} or Kr^{+34}. The principal radiative decays are now separately discussed.

(a) *Decay from the $2\,^1S_0$ State.* Single-photon decay is strictly forbidden by the selection rule $J = 0 \leftrightarrow J = 0$, but decay via two electric dipole (E1) transitions with a continuous distribution of the two-photon energies is allowed and has been calculated by Victor and Dalgarno.[231] In order to detect and isolate this unusual decay mode, which scales according to Z^6, Marrus and Schmieder[225] performed some unique beam-foil measurements on the Berkeley HILAC accelerator by carrying out coincidence measurements with a pair of detectors which counted the simultaneously emitted two photons as a function of distance from the foil. The results are in excellent agreement with the calculations.

(b) *Decay from the $2\,^3S_1$ State.* Single-photon decay from $2\,^3S_1$ to $1\,^1S_0$ can occur as an M1 transition within the framework of relativistic theory.[232,233] This transition was first detected by Gabriel and Jordan[234] in soft x-ray satellite spectra of the solar corona. This prompted a number of transition probability calculations, of which the most extensive ones are by Drake[232] and Johnson and Lin.[233] Several beam-foil measurements[180,226,227,229] are in excellent agreement with these theoretical results and confirm the predicted Z^{10} scaling of the transition probability. An additional possible decay by a two-photon process has been shown to be much slower for the entire He sequence.[235]

(c) *Decay from the $2\,^3P_2$ Level.* This level has a special position among the $n = 2$ levels insofar as both the allowed (E1) transition to $2\,^3S_1$ and the forbidden magnetic quadrupole (M2) transition to the ground state $1\,^1S_0$ contribute to the total decay. The allowed transition is an "in-shell" transition ($\Delta n = 0$) and thus its transition probability grows slowly with Z along the isoelectronic sequence, while the M2 transition increases asymptotically with Z^8.[236,237] For neutral He, $A(\text{E1}) = 0.102 \times 10^8\,\text{s}^{-1}$ and $A(\text{M2}) = 3.27 \times 10^{-1}\,\text{s}^{-1}$, according to Drake's variational calculations.[237] At some large Z, the two A's become comparable, and for higher Z, $A(\text{M2}) > A(\text{E1})$. The crossover occurs at approximately $Z = 18$. Marrus and Schmieder[225] measured the lifetime of the $2\,^3P_2$ level for Ar^{+16}, i.e., just at the crossover point, and arrived at a total rate $A(\text{M2}) + A(\text{E1}) = (5.9 \pm 1.0) \times 10^8\,\text{s}^{-1}$, while the calculated A values, $A(\text{E1}) = 3.55 \times 10^8\,\text{s}^{-1}$ and $A(\text{M2}) = 3.14 \times 10^8\,\text{s}^{-1}$, sum up to $6.7 \times 10^8\,\text{s}^{-1}$, in full agreement within the estimated errors.

References

1. B. M. Glennon and W. L. Wiese, *Bibliography on Atomic Transition Probabilities*, National Bureau of Standards monograph No. 50 (1962).
2. E. W. Foster, *Rep. Progr. Phys.* **27**, 469 (1964).
3. W. Neumann, in *Progress in Plasmas and Gas Electronics*, Eds. R. Rompe and M. Steenbeck, Vol. 1, pp. 1–512, Akademie-Verlag, Berlin (1975).
4. W. L. Wiese, in *Methods of Experimental Physics*, Vol. 7A, Eds. B. Bederson and W. L. Fite, pp. 117–141, Academic Press, New York (1968).
5. M. C. E. Huber, in *Modern Optical Methods in Gas Dynamic Research*, Ed. D. S. Dosanjh, pp. 85–112, Plenum Press, New York (1971).
6. N. P. Penkin, *J. Quant. Spectrosc. Radiat. Transfer* **4**, 41 (1964).
7. R. J. S. Crossley, in *Advances in Atomic and Molecular Physics*, Eds. D. R. Bates and I. Estermann, Vol. 5, pp. 237–296, Academic Press, New York (1969).

8. D. Layzer and R. H. Garstang, in *Annual Reviews of Astronomy and Astrophysics*, Eds. L. Goldberg, D. Layzer, and J. G. Phillips, Annual Reviews, Inc., Palo Alto, California (1968).

9. A. Hibbert, *Rep. Progr. Phys.* **38**, 1217 (1975).

10. W. L. Wiese and A. W. Weiss, *Phys. Rev.* **175**, 50 (1968).

11. W. L. Wiese, in *Beam Foil Spectroscopy*, Ed. S. Bashkin, pp. 147–178, Springer-Verlag, Berlin (1976).

12. A Corney, in *Advances in Electronics and Electron Physics*, Ed. L. Marton, Vol. 29, pp. 115–231, Academic Press, New York (1970).

13. K. Ziock, in *Methods of Experimental Physics*, Vol. 4B, Eds. V. W. Hughes and H. L. Schultz, Academic Press, New York (1967).

14. L. J. Curtis, in *Beam Foil Spectroscopy*, Ed. S. Bashkin, pp. 63–110, Springer-Verlag, Berlin (1976).

15. C. L. Cocke, in *Methods of Experimental Physics*, Vol. 13B, Ed. D. Williams, pp. 213–272, Academic Press, New York (1976).

16. J. R. Fuhr, B. J. Miller, and G. A. Martin, *Bibliography on Atomic Transition Probabilities*, National Bureau of Standards Special Publication No. 505 (1978).

17. W. L. Wiese, M. W. Smith, and B. M. Glennon, *Atomic Transition Probabilities*, Vol. I, H through Ne, NSRDS-NBS 4, US GPO, Washington, D.C., (1966).

18. W. L. Wiese, M. W. Smith, and B. M. Miles, *Atomic Transition Probabilties*, Vol. II, Na through Ca, NSRDS-NBS 22, US GPO, Washington, D.C. (1969).

19. M. W. Smith and W. L. Wiese, *J. Phys. Chem. Ref. Data* **2**, 85 (1973).

20. W. L. Wiese and J. R. Fuhr, *J. Phys. Chem. Ref. Data* **4**, 263 (1975).

21. S. A. Younger, J. R. Fuhr, G. A. Martin, and W. L. Wiese, *J. Phys. Chem. Ref. Data*, **7**, 495 (1978).

22. T. Garz, M. Kock, J. Richter, B. Baschek, H. Holweger, and A. Unsöld, *Nature* **223**, 1254 (1969).

23. J. E. Ross and L. H. Aller, *Science* **191**, 1223 (1976).

24. A. H. Gabriel and C. Jordan, in *Case Studies in Atomic Collision Physics 2*, Eds. E. W. McDaniel and M. R. C. McDowell, pp. 211-291, North-Holland, Amsterdam (1971).

25. R. W. Nicholls, *Ann. Geophys.* **20**, 144 (1964).

26. H. Nubbemeyer, *J. Quant. Spectrosc. Radiat. Transfer* **16**, 395 (1976).

27. R. C. Preston, *J. Quant. Spectrosc. Radiat. Transfer* **18**, 337 (1977).

28. W. L. Wiese and S. M. Younger, in *Beam Foil Spectroscopy*, Eds. I. A. Sellin and D. J. Pegg, Vol. II, p. 951, Plenum Press, New York (1976).

29. L. R. Carlson, J. A. Paisner, E. F. Worden, S. A. Johnson, C. A. May, and R. W. Solarz, *J. Opt. Soc. Am.* **66**, 846 (1976).

30. W. R. Bennett, Jr., P. J. Kindlmann, and G. N. Mercer, *Appl. Opt.*, Suppl. 2 of Chem. Lasers, 34 (1965).

31. A. Einstein, *Verhandl. Deut. Physik. Ges.* [2] **18**, 318 (1916).

32. H. R. Griem, *Plasma Spectroscopy*, McGraw-Hill, New York (1964).

33. W. L. Wiese, in *Methods of Experimental Physics*, Vol. 7B, Eds. B. Bederson and W. L. Fite, pp. 307–353, Academic Press, New York (1968).

34. H. J. Kunze, in *Plasma Diagnostics*, Ed. W. Lochte-Holtgreven, pp. 550–616, North-Holland, Amsterdam (1968).

35. J. M. Bridges and W. R. Ott, *Appl. Opt.* **16**, 367 (1977).

36. W. L. Wiese, in *Plasma Diagnostic Techniques*, Eds. R. H. Huddlestone and S. Leonard, pp. 265–318, Academic Press, New York (1965).

37. P. D. Scholz and T. P. Anderson, *J. Quant. Spectrosc. Radiat. Transfer* **8**, 1411 (1968).

38. D. van Houwelingen and A. A. Kruithof, *J. Quant. Spectrosc. Radiat. Transfer* **11**, 1235 (1971).

39. D. Garz, *Z. Naturforsch. A* **28**, 1459 (1973).
40. R. A. Nodwell, J. Meyer, and T. Jacobson, *J. Quant. Spectrosc. Radiat. Transfer* **10**, 335 (1970).
41. G. Pichler and V. Vujnovic, *Phys. Lett. A* **40**, 397 (1972).
42. P. Ranson and J. Chapelle, *J. Quant. Spectrosc. Radiat. Transfer* **14**, 1 (1974).
43. J. B. Shumaker and C. H. Popenoe, *J. Res. Natl. Bur. Stand. A* **76**, 71 (1972).
44. J. M. Bridges and R. L. Kornblith, *Astrophys. J.* **192**, 793 (1974).
45. M. Martinez-Garcia, W. Whaling, D. L. Mickey, and G. M. Lawrence, *Astrophys. J.* **165**, 213 (1971).
46. P. L. Smith and W. Whaling, *Astrophys. J.* **183**, 313 (1973).
47. C. L. Cocke, A. Stark, and J. C. Evans, *Astrophys. J.* **184**, 653 (1973).
48. W. N. Lennard, W. Whaling, J. M. Scalo, and L. Testerman, *Astrophys. J.* **197**, 517 (1975).
49. W. Whaling, J. M. Scalo, and L. Testerman, *Astrophys. J.* **212**, 581 (1977).
50. W. N. Lennard, W. Whaling, R. M. Sills, and W. A. Zajc, *Nucl. Instrum. Methods* **110**, 385 (1973).
51. M. May, J. Richter and J. Wichelmann, *Astron. Astrophys. Suppl.* **18**, 405 (1974).
52. M. Klemt, *Astron. Astrophys.* **29**, 419 (1973).
53. J. R. Roberts, P. A. Voigt, and A. Czernichowski, *Astrophys. J.* **197**, 791 (1975).
54. S. J. Wolnik and R. O. Berthel, *Astrophys. J.* **179**, 665 (1973).
55. B. Woodgate, *Mon. Not. R. Astron. Soc.* **134**, 287 (1966).
56. G. D. Bell, D. R. Paquette, and W. L. Wiese, *Astrophys. J.* **143**, 559 (1966).
57. A. Goly, J. Moity, and S. Weniger, *Astron. Astrophys.* **38**, 259 (1975).
58. H. Heise, *Astron. Astrophys.* **34**, 275 (1974).
59. R. A. Roig and M. H. Miller, *J. Opt. Soc. Am.* **64**, 1479 (1974).
60. S. J. Wolnik, R. O. Berthel, G. S. Larsen, E. H. Carnevale, and G. W. Wares, *Phys. Fluids* **11**, 1002 (1968).
61. S. J. Wolnik, R. O. Berthel, E. H. Carnevale, and G. W. Wares, *Astrophys. J.* **157**, 983 (1969).
62. J. Musielok and T. Wujec (to be published).
63. J. R. Roberts, T. Andersen, and G. Sorensen, *Astrophys. J.* **181**, 567 and 587 (1973).
64. See, e.g., H. G. Kuhn, *Atomic Spectra*, p. 59 ff., Academic Press, New York (1972).
65. D. E. Blackwell and B. S. Collins, *Mon. Not. R. Astron. Soc.* **157**, 255 (1972).
66. D. E. Blackwell, P. A. Ibbetson, and A. D. Petford, *Mon. Not. R. Astron. Soc.* **171**, 195 (1975).
67. D. E. Blackwell, P. A. Ibbetson, A. D. Petford, and R. B. Willis, *Mon. Not. R. Astron. Soc.* **177**, 219 (1976).
68. T. M. Bieniewski, *Astrophys. J.* **208**, 228 (1976).
69. U. Becker, H. Bucka, and A. Schmidt, *Astron. Astrophys.*, **59**, 145 (1977).
70. J. Marek and J. Richter, *Astron. Astrophys.* **26**, 155 (1973).
71. J. Marek, *Astron. Astrophys.* **44**, 69 (1975).
72. C. L. Cocke, B. Curnutte, and J. H. Brand, *Astron. Astrophys.* **15**, 299 (1971).
73. P. Reinke, *Z. Astrophys.* **66**, 234 (1967).
74. G. D. Bell, L. B. Kalman, and E. F. Tubbs, *Astropnys. J.* **200**, 520 (1975).
75. K. Mie and J. Richter, *Astron. Astrophys.* **25**, 299 (1973).
76. G. D. Bell and E. F. Tubbs, *Astrophys. J.* **159**, 1093 (1970).
77. G. D. Bell and E. F. Tubbs, *Rev. Sci. Instrum.* **41**, 435 (1970).
78. G. D. Bell, M. H. Davis, R. B. King, and P. M. Routly, *Astrophys. J.* **127**, 775 (1958).
79. D. S. Rozhdestvenskii, *Ann. Physik (Leipzig)* **39**, 307 (1912).
80. S. A. Korff and G. Breit, *Rev. Mod. Phys.* **4**, 471 (1932).
81. M. C. E. Huber and W. H. Parkinson, *Astrophys. J.* **172**, 229 (1972).
82. F. P. Banfield and M. C. E. Huber, *Astrophys. J.* **186**, 335 (1973).

83. N. P. Penkin and V. A. Komarovskii, *J. Quant. Spectrosc. Radiat. Transfer* **16**, 217 (1976).
84. M. C. E. Huber and R. J. Sandemann *Proc. R. Soc. London Ser. A*, **357**, 355 (1977)
85. W. H. Parkinson, E. M. Reeves, and F. S. Tomkins, *Proc. R. Soc. London A* **351**, 569 (1976).
86. K. Miyazaki, T. Watanabe, and K. Fukuda, *J. Phys. Soc. Jpn.* **40**, 233 (1976).
87. G. Boldt, *Z. Naturforsch.* **18a**, 1107 (1963).
88. F. Labuhn, *Z. Naturforsch.* **20a**, 998 (1965).
89. A. W. Weiss, *Phys. Rev.* **162**, 71 (1967).
90. C. A. Nicolaides, *Chem. Phys. Lett.* **21**, 242 (1973).
91. J. V. Mallow and P. S. Bagus, *J. Quant. Spectrosc. Radiat. Transfer* **16**, 409 (1976).
92. J. V. Mallow and J. Burns, *J. Quant. Spectrosc. Radiat. Transfer* **12**, 1081 (1972).
93. G. M. Lawrence and B. D. Savage, *Phys. Rev.* **141**, 67 (1966).
94. J. Bromander, *Phys. Scr.* **4**, 61 (1971).
95. W. L. Luken and O. Sinanoglu, *J. Chem. Phys.* **64**, 4680 (1976).
96. S. L. Davis and O. Sinanoglu, *J. Chem. Phys.* **62**, 3664 (1975).
97. D. Stuck and B. Wende, *Phys. Rev. A* **9**, 1 (1974).
98. F. Weinhold, *Phys. Rev. Lett.* **25**, 907 (1970).
99. M. T. Anderson and F. Weinhold, *Phys. Rev. A* **9**, 118 (1974).
100. J. S. Sims, S. A. Hagstron, and J. R. Rumble, *Phys. Rev. A* **13**, 242 (1976).
101. J. S. Sims and R. C. Whitten, *Phys. Rev. A* **8**, 2220 (1973).
102. D. J. G. Irwin, A. E. Livingston, and J. A. Kernahan, *Nucl. Instrum. Methods* **110**, 105 (1973).
103. G. Beauchemin, J. A. Kernahan, E. Knystautas, D. J. G. Irwin, and R. Drouin, *Phys. Lett. A* **40**, 194 (1972).
104. L. Barrett and R. Drouin, *Can. J. Spectrosc.* **18**, 50 (1973).
105. E. J. Knystautas and R. Drouin, *J. Phys. B* **8**, 2001 (1975).
106. E. H. Pinnington, D. J. G. Irwin, A. E. Livingston, and J. A. Kernahan, *Can. J. Phys.* **52**, 1961 (1974).
107. I. Martinson, H. G. Berry, W. S. Bickel, and H. Oona, *J. Opt. Soc. Am.* **61**, 519 (1971).
108. P. D. Dumont, *Physica* **62**, 104 (1972).
109. J. P. Buchet, M. C. Poulizac, and M. Carre, *J. Opt. Soc. Am.* **62**, 623 (1972).
110. J. A. Kernahan, A. E. Livingston, and E. H. Pinnington, *Can. J. Phys.* **52**, 1895 (1974).
111. L. Heroux, *Phys. Rev.* **180**, 1 (1969).
112. M. C. Poulizac and J. P. Buchet, *Phys. Scr.* **4**, 191 (1971).
113. M. C. Buchet-Poulizac and J. P. Buchet, *Phys. Scr.* **8**, 40 (1973).
114. I. Martinson, W. S. Bickel, and A. Ölme, *J. Opt. Soc. Am.* **60**, 1213 (1970).
115. J. A. Kernahan, E. H. Pinnington, A. E. Livingston, and D. J. G. Irwin, *Phys. Scr.* **12**, 319 (1975).
116. S. Hontzeas, I. Martinson, P. Erman, and R. Buchta, *Phys. Scr.* **6**, 55 (1972).
117. I. Martinson, A. Gaupp, and L. J. Curtis, *J. Phys. B* **7**, L 463 (1974).
118. I. Bergström, J. Bromander, R. Buchta, L. Lundin, and I. Martinson, *Phys. Lett.* **28A**, 721 (1969).
119. T. Andersen, K. A. Jessen, and G. Sorensen, *Phys. Rev.* **188**, 76 (1969).
120. A. W. Weiss (private communication).
121. A. Hibbert, *J. Phys. B* **7**, 1417 (1974).
122. Y.-K. Kim and J. P. Desclaux, *Phys. Rev. Lett.* **36**, 139 (1976).
123. L. Armstrong, W. R. Fielder, and D. L. Lin, *Phys. Rev. A* **14**, 1114 (1976).
124. D. L. Lin, W. Fielder, and L. Armstrong Jr., *Phys. Rev. A*, **16**, 589 (1977).
125. K. T. Cheng and W. R. Johnson *Phys. Rev. A*, **16**, 263 (1977).
126. A. W. Weiss, *J. Quant. Spectrosc. Radiat. Transfer* **18**, 481 (1977).
127. W. R. Johnson and C. D. Lin, *Phys. Rev. A* **14**, 565 (1976).

128. C. D. Lin and W. R. Johnson *Phys. Rev. A* **15**, 1046 (1977).
129. C. D. Lin, W. R. Johnson, and A. Dalgarno, *Phys. Rev. A* **15**, 154 (1977).
130. P. Shorer and A. Dalgarno, *Phys. Rev.* **A16**, 1502 (1977).
131. R. D. Cowan and D. C. Griffin, *J. Opt. Soc. Am.* **66**, 1010 (1976).
132. S. M. Younger and A. W. Weiss, *J. Res. Natl. Bur. Stand. A* **79**, 629 (1975).
133. D. R. Bates and A. Damgaard, *Phil. Trans. R. Soc. London A* **242**, 101 (1949).
134. E. Biemont and N. Grevesse, *At. Data Nucl. Data Tables* **12**, 217 (1973).
135. R. L. Kurucz and E. Peytremann, *Smithsonian Astrophys. Obs. Spec. Rep.* 362 (1975), Parts 1–3.
136. P. L. Smith, *Mon. Not. R. Astron. Soc.* **177**, 275 (1976).
137. A. W. Weiss, *Phys. Rev.* **188**, 119 (1969).
138. D. J. Pegg, S. B. Elston, P. M. Griffin, H. C. Hayden, J. P. Forester, R. S. Thoe, R. S. Peterson, and I. A. Sellin, *Phys. Rev. A* **14**, 1036 (1976).
139. D. J. G. Irwin, A. E. Livingston, and J. H. Kernahan, *Can. J. Phys.* **51**, 1948 (1973).
140. J. A. Kernahan, A. Denis, and R. Drouin, *Phys. Scr.* **4**, 49 (1971).
141. W. S. Bickel, *Phys. Rev.* **162**, 7 (1967).
142. L. Heroux, *Phys. Rev.* **153**, 156 (1967).
143. P. D. Dumont, E. Biemont, and N. Grevesse, *J. Quant. Spectrosc. Radiat. Transfer* **14**, 1127 (1974).
144. R. B. Hutchinson, *J. Quant. Spectrosc. Radiat. Transfer* **11**, 81 (1971).
145. D. J. Pegg, L. W. Dotchin, and E. L. Chupp, *Phys. Lett.* **31A**, 501 (1970).
146. A. Hese and H. P. Weise, *Z. Phys.* **215**, 95 (1968).
147. S. A. Chin-Bing and C. E. Head, *Phys. Lett. A* **45**, 203 (1973).
148. Z. Sibincic, *Phys. Rev. A* **5**, 1150 (1972).
149. C. A. Nicolaides and D. R. Beck, *Chem. Phys. Lett.* **35**, 202 (1975).
150. O. Sinanoglu, *Nucl. Instrum. Methods* **110**, 193 (1973).
151. O. Sinanoglu and W. Luken, *J. Chem. Phys.* **64**, 4197 (1976).
152. U. I. Safronova, A. N. Ivanova, and V. N. Kharitonova, *Theor. Exp. Chem. (USSR)*, 209 (1969).
153. E. Träbert, P. H. Heckmann, and H. von Buttlar, *Z. Phys. A*, **281**, 333 (1977).
154. W. Dankwort and E. Trefftz, *Astron. Astrophys.* **47**, 365 (1976).
155. H. Krellmann, E. Siefart, and E. Weihreter, *J. Phys. B* **8**, 2608 (1975).
156. H. A. Schüssler, *Z. Phys.* **182**, 289 (1965).
157. J. Z. Klose, *Astrophys. J.* **198** 229 (1975).
158. S. Rydberg and S. Svanberg, *Phys. Scr.* **5**, 209 (1972).
159. D. Einfeld, J. Ney, and J. Wilken, *Z. Naturforsch. A* **26**, 668 (1971).
160. A. Lurio, R. L. DeZafra, and R. J. Goshen, *Phys. Rev.* **134**, A1198 (1964).
161. H. J. Kluge and H. Sauter, *Z. Phys.* **270**, 295 (1974).
162. A. Lurio and R. Novick, *Phys. Rev.* **134**, A608 (1964).
163. A. Lurio, *Phys. Rev.* **140**, A1505 (1965).
164. G. A. Martin and W. L. Wiese, *Phys. Rev. A* **13**, 699 (1976).
165. W. R. Bennett, Jr., and P. J. Kindlmann, *Phys. Rev.* **149**, 38 (1966).
166. J. Z. Klose, *Phys. Rev.* **141**, 181 (1966).
167. K. E. Donnelly, P. J. Kindlmann, and W. R. Bennett, Jr., *J. Opt. Soc. Am.* **65**, 1359 (1975).
168. M. D. Havey, L. C. Balling, and J. J. Wright, *Phys. Rev. A* **13**, 1269 (1976).
169. H. P. Mühlethaler and H. Nussbaumer, *Astron. Astrophys.* **48**, 109 (1976).
170. R. J. S. Crossley, L. J. Curtis, and Ch. Froese Fischer, *Phys. Lett.* **57A**, 220 (1976).
171. S. M. Younger and W. L. Wiese, *Phys. Rev.* **A17**, 1944 (1978).
172. L. J. Curtis, *Am. J. Phys.* **36**, 1123 (1968).
173. Ch. Froese Fischer, *J. Phys.* **B10**, 1241 (1977).
174. M. Druetta and J. P. Buchet, *J. Opt. Soc. Am.* **66**, 433 (1976).

175. D. J. G. Irwin, J. A. Kernahan, E. H. Pinnington, and A. E. Livingston, *J. Opt. Soc. Am.* **66**, 1396 (1976).
176. E. J. Knystautas and R. Drouin, *J. Quant. Spectrosc. Radiat. Transfer* **17**, 551 (1977).
177. G. Astner, L. J. Curtis, L. Liljeby, S. Mannervik, and I. Martinson, *Z. Phys. A* **279**, 1 (1976).
178. S. Bashkin, in *Beam-Foil Spectroscopy*, Eds. I. A. Sellin and D. J. Pegg, pp. 129–146, Plenum Press, New York (1976).
179. C. Lanczos, *Applied Analysis,* pp. 272–280, Prentice-Hall, Englewood Cliffs, New Jersey (1956).
180. J. A. Bednar, C. L. Cocke, B. Curnutte, and R. Randall, *Phys. Rev. A* **11**, 460 (1975).
181. S. L. Varghese, C. L. Cocke, and B. Curnutte, *Phys. Rev. A* **14**, 1729 (1976).
182. S. L. Varghese, C. L. Cocke, B. Curnutte, and G. Seaman, *J. Phys. B* **9**, L387 (1976).
183. H. Panke, F. Bell, H.-D. Betz, W. Stehling, E. Spindler, and R. Laubert, *Phys. Lett. A* **53**, 457 (1975).
184. H. Gould and R. Marrus, in *Beam-Foil Spectroscopy*, Eds. I. A. Sellin and D. J. Pegg, pp. 305–316, Plenum Press, New York (1976).
185. L. Barrette and R. Drouin, *Phys. Scr.* **10**, 213 (1974).
186. L. Barrette, D. Sc. Thesis, Departmente de Physique, Université Laval (1975).
187. H.-D. Betz, F. Bell, H. Panke, G. Kalkoffen, M. Welz, and D. Evers, *Phys. Rev. Lett.* **33**, 807 (1974).
188. T. Andersen and G. Sorensen, *Solar Phys.* **38**, 343 (1974).
189. T. Andersen, O. Poulsen, P. S. Ramanujam, and A. P. Petkov, *Solar Phys.* **44**, 257 (1975).
190. C. H. Liu and D. A. Church, *Phys. Rev. Lett.* **29**, 1208 (1972).
191. K. D. Masterson and J. O. Stoner, Jr., *Nucl. Instrum. Methods.* **110**, 441 (1973).
192. H. J. Andrä, A. Gaupp, and W. Wittmann, *Phys. Rev. Lett.* **31**, 501 (1973).
193. H. J. Andrä, in *Beam-Foil Spectroscopy*, Eds. I. A. Sellin and D. J. Pegg, pp. 835–851, Plenum Press, New York (1976).
194. H. Harde and G. Guthöhrlein, *Phys. Rev. A* **10**, 1488 (1974).
195. H. Harde, in *Beam-Foil Spectroscopy,* Eds. I. A. Sellin and D. J. Pegg, pp. 859–872, Plenum Press, New York (1976).
196. T. M. Holzberlein, *Rev. Sci. Instrum.* **35**, 1041 (1964).
197. K. E. Donnelly, P. J. Kindlmann, and W. R. Bennett, Jr., *J. Opt. Soc. Am.* **63**, 1438 (1973).
198. H. Figger, K. Siomos, and H. Walther, *Z. Phys.* **270**, 371 (1974).
199. H. Figger, J. Heldt, K. Siomos, and H. Walther, *Astron. Astrophys.* **43**, 389 (1975).
200. K. Siomos, H. Figger, and H. Walther, *Z. Phys. A* **272**, 355 (1975).
201. J. Z. Klose, *Phys. Rev. A* **11**, 1840 (1975).
202. H. Lundberg and S. Svanberg, *Phys. Lett. A* **56**, 31 (1976).
203. F. Gounand, P. R. Fournier, J. Cuvellier, and J. Berlande, *Phys. Lett. A* **59**, 23 (1976).
204. M. D. Havey, L. C. Balling, and J. J. Wright, *J. Opt. Soc. Am.* **67**, 488 (1977).
205. M. D. Harvey, L. C. Balling, and J. J. Wright, *J. Opt. Soc. Am.* **67**, 491 (1977).
206. J. S. Deech, R. Luypaert, and G. W. Series, *J. Phys. B* **8**, 1406 (1975).
207. G. C. King, K. A. Mohamed, F. H. Read, and R. E. Imhof, *J. Phys. B* **9**, 1247 (1976).
208. P. Erman, in *Beam-Foil Spectroscopy*, Eds. I. A. Sellin and D. J. Pegg, pp. 199–215, Plenum Press, New York (1976).
209. H. Liening, *Z. Phys.* **266**, 287 (1974).
210. E. E. Gibbs and P. Hannaford, *J. Phys. B* **9**, L225 (1976).
211. F. H. K. Rambow and L. D. Shearer, *Phys. Rev. A* **14**, 738 (1976).
212. F. H. K. Rambow and L. D. Shearer, *Phys. Rev. A* **14**, 1735 (1976).
213. T. Andersen, O. Poulsen, and P. S. Ramanujam, *J. Quant. Spectrosc. Radiat. Transfer* **16**, 521 (1976).

214. B. R. Bulos, R. Gupta, and W. Happer, *J. Opt. Soc. Am.* **66**, 426 (1976).
215. F. A. Korolev, V. V. Lebedeva, A. I. Odintsov, and V. M. Salimov, *Radio Eng. Electron. Phys. (USSR)* **14**, 1318 (1969).
216. B. van der Sijde, J. W. H. Dielis, and W. P. M. Graef, *J. Quant. Spectrosc. Radiat. Transfer* **16**, 1011 (1976).
217. M. B. Klein and D. Maydan, *Appl. Phys. Lett.* **16**, 509 (1970).
218. A. Dienes and T. P. Sosnowski, *Appl. Phys. Lett.* **16**, 512 (1976).
219. S. Watanabe, K. Kuroda, M. Chihara, and I. Ogura, *Jpn. J. Appl. Phys.* **14**, Suppl. 14-1, 99 (1975).
220. S. Watanabe, M. Chihara, and I. Ogura, *Jpn. J. Appl. Phys.* **13**, 164 (1974).
221. A. Taszner, A. Kowalski, and J. Heldt, *Appl. Phys.* **11**, 203 (1976).
222. J. R. Woodworth and H. W. Moos, *Phys. Rev. A* **12**, 2455 (1975).
223. J. A. Kernahan and P. H.-L. Pang, *Can. J. Phys.* **53**, 455 (1975).
224. R. Marrus, in *Beam Foil Spectroscopy*, Ed. S. Bashkin, pp. 209–236, Springer-Verlag, Berlin (1976).
225. R. Marrus and R. W. Schmieder, *Phys. Rev. A* **5**, 1160 (1972).
226. H. Gould, R. Marrus, and R. W. Schmieder, *Phys. Rev. Lett.* **31**, 504 (1973).
227. H. Gould, R. Marrus, and P. Mohr, *Phys. Rev. Lett.* **33**, 676 (1974).
228. H. Gould and R. Marrus, in: *Beam-Foil Spectroscopy*, Vol. 1, Eds. I. A. Sellin and D. J. Pegg, pp. 305–316, Plenum Press, New York and London (1976).
229. C. L. Cocke, B. Curnutte, and R. Randall, *Phys. Rev. Lett.* **31**, 507 (1973).
230. C. L. Cocke, B. Curnutte, J. R. MacDonald, and R. Randall, *Phys. Rev. A* **9**, 57 (1974).
231. G. A. Victor and A. Dalgarno, *Phys. Rev. Lett.* **25**, 1105 (1967).
232. G. W. F. Drake, *Phys. Rev. A* **3**, 908 (1971).
233. W. R. Johnson and C. Lin, *Phys. Rev. A* **9**, 1486 (1974).
234. A. H. Gabriel and C. Jordan, *Nature* **221**, 947 (1969).
235. G. W. F. Drake, G. A. Victor, and A. Dalgarno, *Phys. Rev.* **180**, 75 (1969).
236. R. H. Garstang, *Publ. Astron. Soc. Pac.* **81**, 488 (1969).
237. G. W. F. Drake, *Astrophys. J.* **158**, 1119 (1969).

26

Lifetime Measurement by Temporal Transients

RICHARD G. FOWLER

1. Introduction

The straightforward method of measuring the average radiative relaxation time of quantized systems is to excite a population of the systems, terminate the excitation abruptly, and observe its decay. Until the mid-1950s it was impossible to realize all (almost any, in fact) of the technology needed to carry out such an experiment except with the rather slow 2^3P state of Hg.[1] Radiation sources which could be manipulated quickly enough were never bright enough to be detected. Detectors which were fast enough did not produce a sufficiently strong signal to record, and no recording techniques approached the needed response times. All of these limitations were removed in a decade, so that now it is possible to make this direct form of attack on an arbitrary problem with a reasonable chance of success.

2. Excitation Methods

Excitation of states only up to a particular instant can be most generally accomplished by electrons, but has also been done by photons and by ions, notably protons. When electrons are used, their degree of organization is a matter of importance. Thermalized electrons such as are present in a gas discharge may provide the intensity and other favorable

RICHARD G. FOWLER • University of Oklahoma, Norman, Oklahoma 73069.

characteristics needed, but at a considerable sacrifice in sharpness of excitation cessation and a broad spread in the energy range of the levels excited. Monoenergetic electrons can be more easily controlled at cutoff, but usually imply thermionic cathodes which are only compatible with certain subject gases. Photon excitation was not a particularly useful technique until the advent of the tunable dye laser because, when the low transmission of Kerr cells is involved, the number of states excited in a single absorption act is very small using ordinary sources and because the transitions are also subject to the uncertainties of radiation trapping, ending as they do upon the ground state. With dye-laser intensities available, double excitation and the exploration of the subordinate states of atoms has become a possibility.

The first direct radiation decay measurements were made with a grid-controlled electron beam and a single-photon counting technique by Heron, McWhirter, and Rhoderick[2] in 1956. They used an electron gun consisting of a small plane oxide-coated cathode, a focusing electrode, an accelerating electrode, a collision-viewing space, and an electron-trapping anode. A triangular accelerating pulse of 30-V amplitude and 20-ns half-width was applied at a repetition rate of 10 kHz. The amplitude of the current in each pulse was ~0.1 mA. Currents from conventional gun configurations are usually very low, resulting in weak excitation and long observation periods. Attempts to increase the current by increasing the applied voltage result in an increase of the complexity of the populating processes for the subject state. Bennett et al.[3] recognized the desirability of suppressing cascade by using threshhold electron energies for excitation. Owing to the smallness of threshhold cross sections, they needed much larger currents, which they obtained by using an elongated triode structure with a cathode 1 cm × 20 cm in size, a vaned grid 1.5 mm away, and an anode 7 mm from that. This configuration developed currents ~100 mA.

In order to compensate for a less sensitive method of radiation detection, Holzberlein[4] devised an inverted form of thermionic diode which proved capable of very large excitation currents. A cylindrical oxide-coated nickel cathode was induction heated to emission temperatures. Inside the cylinder, in close proximity to the cathode, was a cylindrical accelerating tungsten, tantalum, or platinum screen anode. Depending on the balance desired between the magnitude and the monoenergeticity of the current, the screen could range from merely six or eight parallel rods to a fine mesh. The region inside the anode cylinder was a field-free space (Faraday cage) in which the excitation took place. Cathode currents as large as 100 A were observed, but with sufficiently open cathode structures their effective value was greatly enhanced by the recirculation of those electrons which did not make an inelastic collision on their first pass across the excitation space, and so traversed it again and again until they finally transferred their energy to a gas target or the anode. In Holzberlein's original use, single square-pulse switching of the current was employed, leading to a peculiar phenomenon in the form of a charge injection or onset wave which moved quite slowly to the center of the Faraday cage formed by the anode cylinder.[5] This onset wave had no bearing on the sharpness of current cessation and was obviated in later, highly repetitive, cyclic usage of the device,[6] where a low average level of ionization maintained from one cycle to another within the cage permitted unrestricted injection of cathode electrons into the cage.

Because the grid acts as a Faraday cage, cutoff is free of inductive effects in the discharge proper, and when careful impedance matching was made between the invertron and its energy sources, cutoff was measured at less than 0.25 ns. 2D21 xenon thyratrons have been employed for switches when 2.5-ns cutoff times are acceptable, and permit repetition rates of 10 kHz. Mercury-wetted contact relays must be used to achieve briefer cutoff, but they involve a sacrifice of 10^2 in repetition rate. The energy source of choice is a charged coaxial cable the length of which is tailored to the desired on-time for the excitation pulse. Gas densities as low as 0.1 μ Hg have been possible with strong transitions.

The heating of a large cathode cylinder presents a problem, but once accomplished the cathode becomes a furnace in which many elements, usually metals, can be vaporized sufficiently to be examined.[7] These various advantages of the invertron and other thermionically supplied devices are lost, however, with numerous substances which either decompose (molecules) at cathode temperature or suppress the cathode emission. Oxygen, manganese, and iron are notorious poisons. Two devices have been employed which do not depend on thermal emission. The first, an electrodeless discharge,[8] depended for its usefulness as a lifetime method upon the brevity of the natural conduction period between breakdown and polarization of the glass walls (which serve as electrodes in this discharge[9]). The second was the cold cathode version of the invertron devised by Copeland and Fowler.[10] This device imitates the Holzberlein invertron in having a large hollow cathode inside of which the discharge takes place, but the anode configuration must be altered into a central collector rod, so that the activity takes place *between* these electrodes, with the accompaniment of an intrinsic inductancelike behavior that substantially limits the discharge cutoff brevity. Cutoff times ~10 ns are variable with gas pressure, which itself must be quite high (10–100 μ Hg) for operation. Breakdown of the gas upon application of voltage to the anode is slow and limits the repetition rate to about 3 kHz. At low pressures a weak axial magnetic field must be applied to keep the electrons inside the discharge space during breakdown (PIG discharge mode). Nevertheless the device has proved useful in difficult situations,[11] and has not yet been, but might be, exploited with an inductively heated tungsten cathode as the basis for studying otherwise intractable materials such as iron, manganese, and uranium. Electron avalanches[12] and breakdown waves[13] are other gas discharges in which it has been observed that sufficiently fast rise and decay times occur that they could be used as sources for lifetime studies. So far the reverse use only has been made. Decay times have been needed to unfold and understand the structural observations about these phenomena. Wagner did, however, deduce approximate lifetimes of argon states for reflexive use in structural analysis of the avalanche.

3. Observation Methods

Observation of decays in the significant time range 1–100 ns waited on developments in photomultipliers. Post[14] showed that the performance of available nine-stage photomultipliers could be enhanced by proper circuitry and higher operating voltages to reduce transit times and space-charge limitation so that this range of response could be reached. In the decade that followed a rapid development of photomultipliers resulted in a general decrease of electron straggle, increase in cathode efficiency, and increase in gain, to a point at which individual photon counting became practical. Care must still be used in use of gassy and over-age photomultipliers to avoid data falsification.[14a]

Coupled with this development came an improvement of registering devices which began with low-gain real-time oscilloscopes capable of 0.2-ns rise times and continued through a series of artificial time devices in which this limit could be reached with arbitrary signal amplification obtained by accumulation over repeated measurements. Such devices are the sampling oscilloscope, the multichannel analyzer, and the digital computer.

One critical problem with any method of observation is the establishment of a meaningful time instant for the beginning of the decay period. Usually this has been accomplished by developing an electrical signal from the switching pulse which terminates the discharge and using it to trigger the record. Since this signal is uncorrelated with the photons eventually observed during decay, the statistical analysis to be applied varies depending upon whether all photons are recorded or only the first. If the actual excitation process for the state which is subsequently observed can be detected, either because it is populated by a cascade whose photon is detected and used to generate a trigger or by an exciting electron that is recognized by its energy loss and developed into a trigger, then the trigger and the eventual photon are correlated, and simple statistical analysis applies.

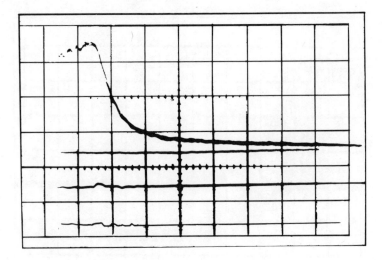

Figure 1. Direct observation decay curve for 4^1D state in $100\,\mu$ Hg of helium. Taken on Tektronix 555 oscilloscope with 100 ns/division.

The brilliance of the invertron permitted direct observation with a real-time oscilloscope for strong transitions.[4] Among methods employing arbitrary triggering this possesses the advantage that all photons emitted after start are seen and recorded, so that the probability of recording a photon is proportional to the simple instantaneous intensity of the decaying discharge. The relatively limited number of states which are bright enough to observe directly quickly led to the abandonment of this method in favor of indirect methods. By the use of sampling oscilloscopes[15] a factor of about 5 in sensitivity can be achieved while maintaining the simple relation of probability to discharge intensity. Effective use requires careful consideration of the sampling parameters. If T is the dwell time at each point sampled, τ is the desired lifetime, and s is the smoothing factor, then the accuracy of the measurement is of order $T/[\tau \ln s/(s-1)]$. The smoothing factor s is the fraction of the difference between each new piece of information and the stored information which is added to the memory. Examples of the better records which were obtained by these two methods are shown in Figures 1 and 2.

Both of these methods have been supplanted by the delayed-coincidence or photon-counting technique of Heron *et al.*[2] as further developed by many others.[3,6,16] An oscilloscope record of a typical data accumulation by this method is given in Figure 3. In practice the data are read out digitally for computational analysis. The data are developed by

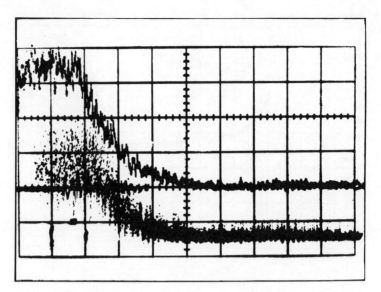

Figure 2. Noisy direct observation decay curve for $N_2^+(0, 0)$ state compared with smooth curve using Tektronix 1S1 sampling unit. 50 ns/division.

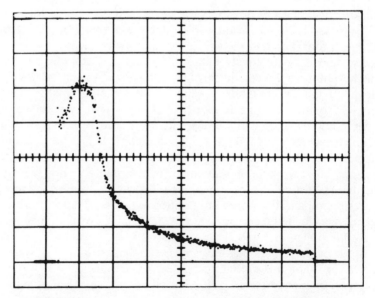

Figure 3. Oscillogram of data points in 256-channel ($= 200$ ns) delayed-coincidence study of L_α line in hydrogen at 60 μ Hg.

generating a start signal (pulse) and using it to initiate the charging of a capacitor. A pulse generated from the first photon seen by the detector is then used to stop the charging, and the voltage on the capacitor is read and recorded in the memory of either a multichannel analyzer or a computer. Repeated cycling of the system then generates a statistical distribution of occurrence frequency for the interval between the two events. When the start signal is correlated with the expected photon in a one-to-one correspondence, this statistical record of intervals is proportional to the probability of observing a photon and yields the true decay curve for the state subject to a correction for the channel width, which needs to be considered only to the extent that it is difficult to construct analyzers with channels of equal width. Neglect of these small inequalities introduces error into both the ordinates and abscissae of the decay curves. If p_n is the normalized probability of a count in the nth channel, then for a single exponential decay, the corrected ordinates are

$$\frac{p_n}{[1 - \exp(-T_n/\tau)]} = \exp\left(-\frac{t_n}{\tau}\right)$$

where τ is the true decay time, T_n is the time width of the nth channel, and t_n is the time lapsed when the accumulation in the nth channel is complete. The expression

$$t_n = \sum_1^{n-1} T_i$$

gives the corrected abscissas. When the channel widths are judged to be adequately equal, then $t_n \sim nT$. The product of channel width and channel efficiency can be measured by triggering the system with the normal trigger mechanism, but by using a steady light as the source of photons to provide random stop pulses. Channel efficiencies can be obtained by using cables to return the start pulse as a delayed stop pulse with the delay being adjusted to center in each channel. Simultaneously an accurate multipoint calibration of the time base can be made by calibrating the cables interferometrically with a frequency-calibrated oscillator and an oscilloscope. Individual channel widths can thus be obtained to normalize the decay probabilities. In practice, most analyzing equipment available today is sufficiently reliable that only the simplest of time-base calibrations (a full-range calibration) need be made.

When the start signal is uncorrelated with the photon emission, the statistical record reflects the joint probability that (1) the detected photon was emitted and (2) that no other photons were emitted and detected before it was detected. Even with a simple single-state decay process this is a far more complex probability function than one obtains in the correlated case. If that fact is ignored, large erroneous shortenings of the decay time can result.[6] If M is the average number of counts generated by the apparatus per cycle of operation, then the probability per unit time (after cutoff) that a count will be registered is

$$p(t) = \frac{M^2}{\tau[1 - \exp(-M)]} e^{-t/\tau} \exp[-M(1 - e^{-t/\tau})]$$

If N is the number of cycles and $C(T)$ is the total number of counts recorded for an arbitrary value of M, then

$$C(T) = \frac{NM\{1 - \exp[-M(1 - \exp(-T/\tau))]\}}{1 - \exp(-M)}$$

where T is the time interval in the cycle during which the system is alert. This relation can be used to evaluate M only if $T \gg \tau$, because otherwise τ is an unknown. Let this special value of $C(T)$ be called C. Then $M = C/N$.

Even knowledge of M is not especially helpful to an effort to evaluate τ by computer fitting to a sum of exponentials, and so one of three courses of action must be followed. First, one may adjust M to a small value, with a considerable sacrifice in sensitivity. Second, one may sacrifice the early portion of the decay curve and rely on the rather rapid decay of the distorting factor. Third, one may perform a renormalization of the data before analysis.

In the first case the lifetime deduced will be in systematic error, being low by an average of $50M\%$. Parenthetically, this is unfortunate because the natural tendency in any experimental situation is to maximize the number of counts one obtains per cycle, especially in low-intensity and low-count-rate situations. This can result in errors that wholly falsify the data, a situation which can easily account for early discrepancies between

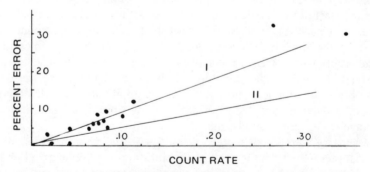

Figure 4. Percent error resulting from excessive count rate. Curve I, maximum error; curve II, probable error with two e foldings of pertinent data. Data by Paske[17] on $N_2^+(0, 0)$ state.

various workers. Paske[17] has treated this source of error experimentally (Figure 4).

In the second approach if the intensity of the transition is very large so that one or more e foldings can be disregarded, then to a first approximation, the initial effect is to simulate a lifetime of $\tau/(1+M)$, while ultimate effect of the distortion term is to add an exponential of half the lifetime and relative amplitude M to the time decay. If, then, enough of the record can be sacrificed to permit these faster decaying exponentials to die away, reasonable accuracy on the true decay can be obtained if cascade is negligible.

In the third (renormalizing) procedure[18] one takes advantage of the fact that in the completed record of N cycles, if $C(t)$ is the integral of the number of counts in all channels from cutoff up to t, then

$$C(t) = \frac{MN(1 - \exp\{-M[1 - \exp(-t/\tau)]\})}{1 - \exp(-M)}$$

and this contains precisely the factor which is falsifying the analysis of the decay. Then in terms of C, defined earlier,

$$p(t)\{1 - C(t)[1 - \exp(-C/N)]/C\} \propto \exp(-t/\tau)$$

so that by dividing the number of counts in each channel by a computable factor, simplicity can theoretically be restored to the decay curve. Cascade, as will be discussed later, does not seriously affect the renormalization. The correction on $p(t)$ remains the same. All that changes is the evaluation of C and the possibility of such an evaluation.

Imhof and Read[19] crossed a beam of the subject gas with a beam of electrons. By velocity analyzing the electron beam they were able to select that electron which had just made an excitation of the desired state and to develop a trigger pulse from it which alerted a timing circuit to await a photon of the expected frequency from the atomic beam. As might be anticipated, intensity limitations of the technique are severe, and even for rather strong transitions exposures of from 4 h to 4 days are needed for statistically significant data accumulations. Because of the correlation between the trigger and the event, however, the data are free of observational bias. Imhof and Read gained a maximum of useful information from their experiment by pulsing the electron beam and fitting an analysis developed by Gale[20] to the full growth and decay of the corresponding excitation pulse.

Habib et al.[21] used a cascade photon emitted prior to the photon from a desired state to generate a trigger pulse. This is the classic technique employed in nuclear physics for lifetime measurement and is free from observational bias also. In atomic physics it suffers from many disadvantages, however. Intensities are very low and as much as a thousand hours of exposure with relatively strong transitions may be needed. This militates against the study of highly ionized states. For normal atoms, where intensities can be large enough, only a few accidental states fulfill the requirement that both transitions must lie in a photometrically usable region of the spectrum. The method thus, at present, is largely useful as a confirmation of the reliability of data achieved by more sensitive methods.

4. Data Interpretation

Presuming that observational difficulties such as those mentioned in the last section have been removed, various natural interventions inherent in the excitation process can complicate both the analysis and the interpretation of the analysis. The factors which play a role in data interpretation are cascading, quenching, transfer of excitation, and resonant trapping. All can be minimized to some degree. Cascade can be controlled either by use of threshold energies or abbreviated excitation times. The other three effects are all pressure-dependent and can be extrapolated away by pressure-dependence studies, providing data at the same time for the rates of the processes. One of the serious dangers to good measurement is that with weak sources and limited statistical information, longer-lived cascades can be lost in the noise at the end of the decay. Ignoring such a longer-lived cascade will always unduly shorten a lifetime measurement. The decay equation system for an ordered hierarchy of interdependent states is easily written down and easily solved formally.[1] As a practical matter, such a general solution is not very useful to guide data interpretation. Some restricted cases are more instructive and by and large adequate. It is important to note that although the problem treated in atomic physics stems from the same set of differential equations as the nuclear decay problem,[22] the two differ drastically in solution because the nuclear problem always begins with a *decay equilibrium population* in all states, while the atomic problem begins with an arbitrarily adjustable distribution of populations.

All the processes we shall include are linear in the populations of excited states, and hence parallel channels of excitation lead to superposable solutions. Decay channels arranged in series fall into two distinctive cases: (1) two upper states in succession feed the observed state, (2) one state is reversibly connected to the observed state.

The equations for the two state cascade case $3 \rightarrow 2 \rightarrow 1$ are

$$\dot{n}_1 = -\alpha_1 n_1 + \alpha_{21} n_2 + P_1$$

$$\dot{n}_2 = -\alpha_2 n_2 + \alpha_{32} n_3 + P_2$$

$$\dot{n}_3 = -\alpha_3 n_3 + P_3$$

It has been assumed that there is no direct coupling of state 3 to state 1. This is normally true, and even if not true only introduces a parallel channel of the single-stage cascade type. (Single-stage cascade will appear as a special case of two-stage cascade.) It has also been assumed that the third state is influenced only by its losses to other states. Such a truncation is fairly reasonable insofar as its effect upon state 1 (the observed state) is concerned. As will be evident subsequently, introduction of a fourth state above the third adds no new behaviors to the system. In the above equations the N_i are the instantaneous state population densities and the P_i are the production rate densities for the respective states from the ground state. The additional populating and depopulating processes are hidden in the decay constants α. Thus

$$\alpha_i = \sum_{j=0}^{i-1} (A_{ij} + \nu_{ij}p + \nu_{ij}^* \eta p)$$

and

$$\alpha_{ji} = A_{ji} + \nu_{ji}p + \nu_{ji}^* \eta p$$

where A_{ij} is the spontaneous radiation rate from the ith state to the jth state, ν_{ij} is the collision frequency for transfer from state i to a state j by neutral atoms at unit gas particle density p, ν_{ij}^* is the average collision frequency for transfer from state i to a state j at unit background electron particle density, ν_{ji} and ν_{ji}^* are corresponding transfer frequencies into state i, and η is the ratio of background electron particle density to gas particle density. It should be noted that in this notation quenching is regarded as transfer to the ground state.

The slow electron fraction η has a pressure dependence of the form $(1 + Lp/\lambda_0)$, where L is a collision space scale factor and λ_0 is the electron free path at unit gas density. Electron quenching is often indistinguishable from gas quenching by pressure dependence alone.

For electron excitation for an interval from $t = 0$ to $t = T$ the production rate densities ideally take the form $P_i = \sigma_i N_0 pi/e$ for monoenergetic electrons and $P_i = \langle \sigma_i v \rangle N_0 p^2 \eta$ for thermalized electrons. Here σ_i is the cross section for excitation at speed v, N_0 is the number of atoms per unit volume at unit density p, and i is the electron current density.

When this requirement is inset into the equations for the decay, the following solution is obtained:

$$N_1 = \left[P_1 + \frac{\alpha_{21}}{\alpha_2 - \alpha_1} \left(P_2 + \frac{\alpha_{32}}{\alpha_3 - \alpha_1} P_3 \right) \right] \frac{1 - e^{-\alpha_1 T}}{\alpha_1} e^{-\alpha_1 t}$$

$$- \frac{\alpha_{21}}{\alpha_2 - \alpha_2} \left(P_2 + \frac{\alpha_{32}}{\alpha_3 - \alpha_2} P_3 \right) \frac{1 - e^{-\alpha_2 T}}{\alpha_2} e^{-\alpha_2 t}$$

$$+ \frac{\alpha_{21}\alpha_{32}}{(\alpha_3 - \alpha_1)(\alpha_3 - \alpha_2)} P_3 \frac{1 - e^{-\alpha_3 T}}{\alpha_3} e^{-\alpha_3 t}$$

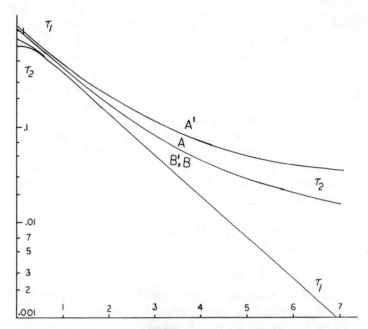

Figure 5. Single-stage cascade. Sums of two exponentials with various amplitudes and lifetimes.

It is seen at once that all exponentially decaying states lying above any given state play a role in the total decay pattern of that state. The problem of experimenters is then to recognize their existence and unravel them, if they hope to isolate the decay constant specific to the observed state. If they are successful in this, they may even obtain unexpected extra data in the form of specific decay constants of experimentally unobservable upper states.

This result is plotted in Figure 5 for a few cases in the two-state case. The curves display the well-known behavior for two states in which if the observed state's specific exponential has a shorter decay time τ_1 than does that of the upper or first cascade state, the coefficients of both are positive and the cascade appears as in the curves labeled A and A' as a "tail" following the decay of the desired specific exponential. When, however, the cascade state's decay τ_2 is faster than that of the observed state, the result is a rounding of the leading edge of the total decay curve, as in the curves labeled B and B'. This has been observed and referred to in the literature as "nonexponential" behavior. It is not, in point of fact. Rather it implies an exponential with a negative coefficient. If such a possibility is allowed for in the curve fitting program used for data analysis, the fast-cascade constant can sometimes be drawn out of the data.

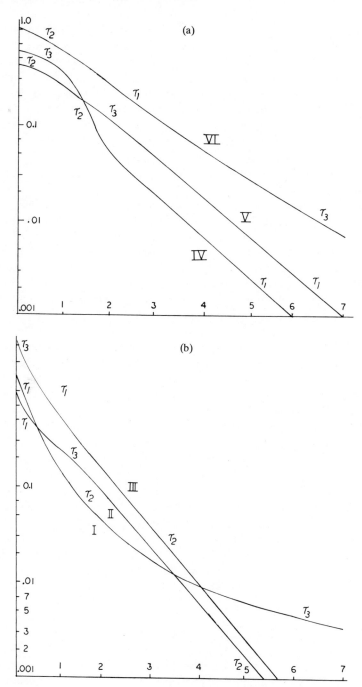

Figure 6. Two-stage cascade. Sums of three exponentials with various lifetimes.

Figure 5 also shows how control of the onset time can be used to adjust the balance of population between the observed and the cascade state, so that we can lay stress on the specific decay constant we desire to measure, whichever it may be. This takes advantage of the sometimes overlooked fact that the degree to which a state becomes populated is governed by its decay constant also. For times of the order of the shorter of the two decay times, the population ratio between the fast state and the slow state is roughly in inverse proportion to the decay times. Curves A and B favor the exponential τ_1. Curves A' and B' have been drawn with both exponentials saturated. Extensive data are needed to separate cascades accurately.

When two successive cascades are present in the observed state's decay a new effect appears. Whereas in the case of single cascade one may correctly assume that the first downward-going (positive coefficient) exponential detectable relates to the specific exponential of the observed state (or said in other words, the fastest *decay* observed) is the true decay, with a third state present this need not be the case. The fastest decay may be that of the second removed upper state. Six identifiable cases have been plotted in Figures 6(a) and 6(b). Only one, curve I, correctly fills the conventional picture of cascade, with three positive coefficient exponentials succeeding one another in an order corresponding to the specific decays of the first, second, and third states. Curve III, on the other hand, simulates curve I in every respect and yet the decay constant of the shortest lifetime exponential is not that of the true state, but of the second removed state. In curves III and IV, the desired specific lifetime of the observed state lies in the middle region of the decay, and is generally lost in the process of analysis. In curves IV and V the desired state could probably be isolated, but might easily be passed over as spurious. Curve II is unique in presenting a bulge in the middle which looks to the experimenter as if there had been a belated pulse of excitation. This behavior as well as that of curve IV can easily be misinterpreted as poor functioning of the cutoff circuitry, a belief that can carry over in lesser degree to curves IV and VI as well. To increase the confusion it will subsequently be shown that this interpretation can be a justifiable one even with proper functioning of the circuitry. All of these cases can be and have been observed.[23]

Before it is assumed that the situation is wholly intractable it should be said that the chief case in nature in which curve III appears is where the coupling from state 3 to state 2 is not by radiation but rather by collisional transfer, with state 3 being a state in the principal series. Thus the process is extremely sensitive to pressure, and this can be used as a decisive parameter. Thompson and Fowler used pressure dependence to analyze such upper-state radiations as that from 7^1D in helium. A spurious fast exponential was found which showed an increase of lifetime with pressure, suggesting resonance trapping rather than quenching. Subsequent measurement of the 9^1P state showed it in fact to have these observed values of lifetimes over the pressure range used.

The second way in which the shortest-lived exponential with positive coefficient need not be the specific decay of the state under observation is when a second state is transfer-coupled with the first. In this case

$$\dot{n}_1 = -\alpha_2 n_1 + \alpha_{21} n_2 + P_1$$

$$\dot{n}_2 = -\alpha_2 n_2 + \alpha_{12} n_1 + P_2$$

when $\alpha_2 > \alpha_1$, both exponentials can have positive coefficients if $\alpha_1 + \alpha_{12} P_1/P_2 > \alpha_2 + \alpha_{21} P_2/P_1$. It is dubious whether or not this condition is met except in rare circumstances, however. The suggestion by Thompson and Fowler that this situation prevailed in the 7^1S state of helium seems unlikely. Since they did in fact perceive a fast first exponential associated with 7^1S, which coincided in time constant with the decay time of the 7^1P state, it seems more probable that this is another example of a three-stage succession of exponential decays. If so, the twice-removed exponentially controlled population must be the population of resonance photons (7^1P-1^1S) which is interacting with the 7^1P state.

Pressure is not the only parameter available for evaluation of cascade. Variation of the electron energy can be used to regulate the production rates P_i in favor of the desired state. This was the technique advocated and employed by Bennett et al. Thompson[24] adapted the invertron to this technique by developing a circuitry in which a two-step excitation pulse could be applied, with an interval at high electron velocity being followed by an interval near threshold, followed by cutoff.

In summary, then, cascade can be either a nuisance or a desirable source of additional information. To identify cascades one can and must analyze the effect of gas density, electron energy, and excitation time on the observed lifetimes.

The partial or complete trapping of resonance radiation lengthens the apparent lifetimes of states which radiate to the ground state of the atom and requires that measurements be made at extremely low gas densities where accuracy suffers from weak signals and poor statistics. Phelps[25] has discussed the effect of trapping on the lifetimes of partially resonant states, and Fowler et al.[26] measured the lifetime of the 3^1P state of helium over a wide range of pressures and found it to have the expected behavior. Considering the complications to accurate interpretation presented by radiation trapping, the Hanle effect is a much to be preferred method of dealing with these energy levels.

Rounding of the leading edge of a decay curve has been shown above to indicate a cascade from a faster-decaying remote state. Normally, however, it is thought of as symptomatic of poor cutoff. There is a sound reason for this confusion, which a more explicit analysis of the production function will explain. Cutoff of the excitation is a circuit problem. The transient currents and voltages in a circuit are governed by the exponential family of functions if the circuit is linear. (Switching elements, such as 2D21 thryratrons, are often not linear, however, and may distort the exponentiality of the transients, at least initially.) The simplest interaction of the circuit transients with the observations would occur if the circuit behaved as an equivalent RLC circuit, and to a first order this is probably a rather good analysis, although it does not preclude the presence of additional decay loops with other time constants. A simple RLC circuit is governed by an expression of the form

$$A e^{-t/\tau_1} + B e^{-t/\tau_2}$$

where τ_1 and τ_2 are the roots of the equations

$$L\tau^2 + R\tau + \tau/C = 0$$

If i_0 is the current at the actual instant of cutoff and $di/dt = 0$ at this instant, then

$$A = \frac{i_0\tau_1}{\tau_1 - \tau_2}, \qquad B = \frac{i_0\tau_2}{\tau_1 - \tau_2}$$

if $R > 2(L/C)^{1/2}$ and the exponentials are real. Otherwise, the current takes the form

$$i \propto e^{-Rt/2L} \sin\left(\frac{1}{LC} - \frac{R^2}{4L^2}\right)^{1/2}(t + t_0)$$

and an oscillation will appear.

The two exponential solutions of the circuit equation behave like two parallel-channel upper states of a superatomic system because production P is proportional to the current density i. Let $P_i = \beta_i i$. If we consider, for example, a simple two-state atom, the differential equations are

$$\frac{dn_1}{dt} = -\alpha_1 n_1 + \alpha_{21} n_2 + \beta_1 i$$

$$\frac{dn_2}{dt} = -\alpha_2 n_2 + \beta_2 i$$

$$L\frac{d^2 i}{dt^2} + R\frac{di}{dt} + \frac{i}{C} = 0$$

To make more explicit the statement that the current behaves like two parallel-channel upper states of the atom, we factor the second-order equation to obtain the two equivalent first-order equations:

$$\frac{di_1}{dt} + \frac{i_1}{\tau_1} = 0, \qquad \frac{di_2}{dt} + \frac{i_2}{\tau_2} = 0$$

where $i = i_1 + i_2$. It is then apparent that either of these current component equations takes the place of the equation for the state n_3 in the set of equations previously discussed in detail for a three-consecutive-state cascade, the only important differences being that a direct coupling exists between the remote state i and state 1 which was not included (but could have been included) in the previous analysis and that the initial conditions on i require both i_1 and i_2 to be present in the final solution.

It is therefore not surprising that the entire range of possible curves seen in Figure 6 can equally well be caused by the energy supply circuitry as by cascade, and the implication is that a full examination of any experiment should include explicit variation of the stray inductance and capacitance parameters of the circuit as well as gas density, on-time, and electron energy. When the circuit is strongly overdamped, with $R \gg 2 (L/C)^{1/2}$, the two time constants are $\tau_1 = L/R$ and $\tau_2 = RC$. Since line terminations are nominally in the order of 50 Ω τ_2 is generally significantly longer than τ_1, and is the channel (of the two) which must be reckoned with chiefly.

Interaction of cascade with count-rate distortion can introduce what seem to be whole families of exponentials into the uncorrelated decay curve. If the probability of photon emission is proportional to $\sum_i Q_i \exp(-\alpha_i t)$, then the normalized probability of a count being recorded by the coincidence system at time t is

$$p(t) = \frac{M^2}{1 - e^{-M}} \frac{\sum_i Q_i e^{-\alpha_i t}}{\sum_i (Q_i/\alpha_i)} \exp\left[-M\left(1 - \frac{\sum_i (Q_i/\alpha_i) e^{-\alpha_i t}}{\sum_i (Q_i/\alpha_i)}\right)\right]$$

The total number of counts in an N-times repeated observing period T is

$$C(T) = \frac{NM}{1 - e^{-M}}\left\{1 - \exp\left[-M\left(1 - \frac{\sum_i (Q_i/\alpha_i) e^{-\alpha_i T}}{\sum_i (Q_i/\alpha_i)}\right)\right]\right\}$$

If C is the number of counts when the interval T used is of sufficient length that

$$\sum_i \left(\frac{Q_i}{\alpha_i}\right) e^{-\alpha_i T} < \sum_i \left(\frac{Q_i}{\alpha_i}\right)$$

then renormalization in exactly the same way as with a single exponential again develops the true decay curve, i.e.,

$$\frac{p(t)}{1 - C(t)(1 - e^{-C/N})/C} \propto \sum_i Q_i e^{-\alpha_i t}$$

Ivanov[26a] has discussed a peculiar systematic error that is possible with linear electron-beam excitation. It arises from the Hanle effect in the earth's magnetic field or any stray field which may pervade the apparatus. He finds that if the observed state is polarized and the magnetic field is perpendicular to the line of the beam, fields of the order of the earth's can engender errors of up to 10% in the observed lifetimes.

5. Data Processing

Analysis of data may be made in two general ways, by graphical reduction or curve fitting. In either case the process may be assigned to a computer with an unlimited increase in the precision of the result and usually a valuable gain in processing time, although the accuracy of the result is not generally increased and sometimes even impaired unless the entire process is critically scrutinized. Graphical reduction involves a process of subtracting successively the exponential of largest time constant as indicated by a linear fit to the data near the end of the time interval of the observation. Since the interval of observation is always finite, there is always an uncertainty about the existence of additional exponentials. The error in any one exponential involved in neglecting any other longer-lived exponential is of order

$$\varepsilon = \frac{a_2}{a_1}\left(1 - \frac{\tau_1}{\tau_2}\right)\left(1 + \frac{a_2\tau_1}{a_1\tau_2}\right)$$

and is more dependent on their relative amplitudes a_2/a_1 than upon their time constants. The assumption that the data at the end of the region indicate a longest-lived exponential correctly without consideration of the shorter-lived components is a dubious one, adequate only when there is a large difference between time constants and amplitudes in favor of the shorter-lived exponentials. Numerous computer programs exist for this procedure of peeling off exponentials successively. An effective one devised by Thompson[27] operates on the following algorithm. Logarithms of the data points are taken. A least-squares fit is made to the last 20 points. The correlation coefficient is calculated. The number of data points is increased by an increment, a new fit is made, and the coefficient is recalculated. This is continued until the coefficient is maximized. A constant, chosen as a given fraction of the last data point is subtracted from all points, new logarithms taken, and the entire procedure repeated to find a new maximum correlation coefficient. Iterations are made with increasing values of the constant until a maximum of the maxima is reached. The value of the corresponding constant and exponential are then subtracted from the data, and the operation is moved to the next apparent exponential and repeated until all present have been recovered. Computerization does not remove the fundamental difficulties of the unpeeling process; the accuracies achieved are not impressive. The program has proved most useful for making estimates with which to begin a nonlinear least-squares fit to the data.

Curve fitting is a more reliable method of analysis, but it cannot supply the missing information beyond the end of the observed data or respond correctly to cascade exponentials with small negative coefficients, especially when they occur in the middle of the decay curve. A direct fitting can be made by comparing an oscillogram of the decay curve with a controllable sum of electronically generated high-accuracy exponentials displayed on the same oscillograph. This may be done either by superimposing traces on a double-beam oscilloscope or by optically superimposing an image of a photograph of the data upon the synthetic curve as displayed by an oscilloscope.

The comparator method does not use the data points with proper weights, and in fact stresses the middle points of the curve, essentially giving an inverse weighting to the early points with their low random errors. Use of a digital computer provides a better control over the weighting of points.

Bennett et al.[3] described their successful experience in finding two exponentials with a least-squares program (called FRANTIC) using Gaussian iteration devised by Rogers[28] of MIT. Another program (called LASL) designed by Norris[29] of the Los Alamos Scientific Laboratory and modified extensively by Thompson to provide for iteration along a path of steepest descent has also had thorough testing and is designed for three

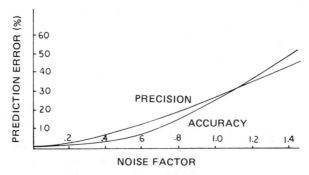

Figure 7. Percent accuracy and precision of predictions of lifetime as a function of random error in the data points. Amplitude ratio 1 : 1, lifetime ratio 2 : 1.

exponentials and a constant (seven parameters). Both programs have been validated by application to synthetic data generated from mathematical exponentials with errors added by a random number procedure to each point so that a desired standard deviation was set in overall. The results showed that when the least-squares programming works well, it works very well. As would be expected, errors are strong functions of the relative values of the lifetimes that must be found and of the relative amplitudes of these exponentials as well as the noiseness of the data. Thompson investigated the magnitude of the error with two exponentials and a constant as a function of these variables for a range of weight functions. Weight is not something which can be established *a priori*, and although its supposed effect is to favor the determination of one exponential over another, the actual experience was that the errors were uncorrelated with the weight for either exponential. The errors in prediction have therefore been averaged over the weight function used in preparing the graphs in Figures 7, 8, and 9. In Figure 7, amplitude ratio and lifetime ratios are held constant and a noise factor defined as the ratio of the standard deviation of the data divided by the square root of the number of counts in the first channel is used to plot (1) the average percent difference between the prediction and the known mathematical value originally chosen (the accuracy) for the lifetime of either exponential, and (2) to plot the percentage standard deviation of the prediction of these lifetimes as calculated from the curve-

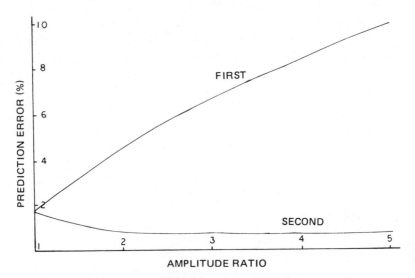

Figure 8. Percent error in lifetime prediction as a function of the amplitude ratio (long-lived/short-lived) of the exponentials. Noise factor 1.35, lifetime ratio 10:1.

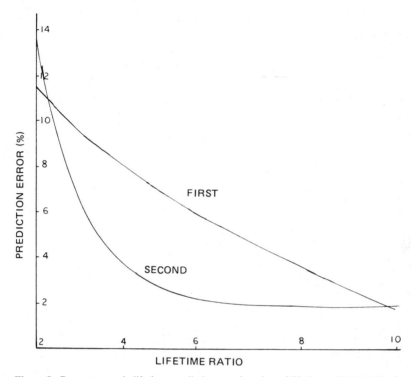

Figure 9. Percent error in lifetime prediction as a function of lifetime ratio. Amplitude ratio 1:1, noise factor 1.35.

fitting process (the precision). In Figure 8, the mean of the accuracy and the precision of the several lifetimes is plotted against the ratio of the amplitudes of the two exponentials at constant lifetime and constant noise. In Figure 9 the mean value of the accuracy and the precision of the several lifetimes is plotted against the ratio of the lifetimes at constant amplitude ratio and constant noise. These results are in excellent agreement with the test reported by Bennett *et al.* Such a family of curves is at best only indicative of the errors to be expected from curve fitting. The time require for a complete nonlinear least-squares analysis of a 256-channel data record is about 7 s on an IBM 360-158.

6. Results

The temporal cutoff of excitation has now been widely tested over almost the entire span of the periodic table. A compendium of the results to date is given in Table 1. So far no single researcher has brought all the

Table 1. Compendium o

Part A

He I	UDC; eI (Ref. 23)	UDC: eI (Ref. 24)	CDC; eB (Ref. 30)	CDC; eB (Ref. 19)	CDC; eB (Ref. 19a)	UDC; eI (Ref. 6)	UDC; eB (Ref. 31)	UDC; pB (Ref. 32)
3^1S	56 ± 2							
4^1S				75 ± 5	73 ± 2	89 ± 3	118 ± 8	
5^1S	160 ± 3							
6^1S	210 ± 4							
7^1S	350 ± 100							
8^1S	450 ± 150							
3^1P							60 ± 20[b]	73 ± 5[b]
4^1P								
5^1P								
7^1P								
3^1D	20 ± 1						15.5 ± 1.5	
4^1D		37 ± 1				41 ± 3	38 ± 1.5	43 ± 5
5^1D	56 ± 10							
6^1D	72 ± 3							
3^3S	57 ± 1							
4^3S				63.5 ± 1.2		62 ± 3	64.5 ± 1	65 ± 3
5^3S	120 ± 20							
2^3P		125 ± 5						
3^3P						122 ± 5		
4^3P	125 ± 10							
5^3P	180 ± 30							
6^3P	360 ± 20							
7^3P	500 ± 150							
3^3D	19.4 ± 0.5			13.4 ± 0.5			14 ± 1	
4^3D	28 ± 10							
5^3D								
6^3D								
7^3D	160 ± 40							
8^3D	200 ± 50							
9^3D	300 ± 100							
10^3D	340 ± 60							
$4F$								
$7F$	400 ± 50							
$9F$	750 ± 100							
$10F$	1100 ± 200							
$11F$	1350 ± 300							

[a] Method code: UDC = uncorrelated delayed coincidence; CDC = correlated delayed coincidence; DO = direct observation; SDO = sampled direct observation; eI = electrons, invertron; eB = electrons, beam; eD = electrons, discharge; pB = protons, beam; $h\nu$ = photon excitation.
[b] Fully trapped.

Lifetimes by Temporal Transients[a]

Helium

UDC; eB (Ref. 33)	UDC; eB (Ref. 34)	UDC; eB (Ref. 3)	DO; eI (Ref. 35)	SDO; eB (Ref. 36)	DO; eD (Ref. 8)	DO; eI (Ref. 37)	SDO; eB (Ref. 15)	UDC; eB (Ref. 2)
60±3		55±3			65.3±0.5			
75±4	83±5	84±10		97±2				
115±5	133±15	140±30		144±3				
	66±7[b]	67±10[b]	70±3 (800 μ)	72±1[b]				
		140±50						
		230±50						
		350±80						
16±1		22±10		16±2		18±5		
30±2	39±5	34±5		47±5		35±4		
46±3	63±9	59±25		79±6				
47±3		44±2						
69±3	65±4	65±4		68±1		59±1	78	68±1
106±5	99±6	100±10		113±4				
115±5		99±6						
	100±3	105±5		115±2	127±1	91±8	106	115±5
			145±2	165±1				153±2
			170±20	235±3				
13±2			17±1	15±2				10±5
23±2			37±10	37±6				
35±3			53±14					
			145±40					
				135				

Table 1 (continued)

Part B. Oxygen, Neon, Sodium

O I	UDC; eD (Ref. 11)	O II	UDC; eD (Ref. 11)	Ne I	UDC; eB (Ref. 16b)	Ref. 3	UDC; eB (Ref. 16a)	Na I	CDC; hν (Ref. 21)
$(6s)^5S^0$	46±4	$(3p)^4S^0$	8.3±1	$2s_2$		96		3^2P^0	16.3±4
$(4p)^3P$	49±4	$(3p)^4P^0(3/2)$	9.6±1	$2s_3$		160			
$(4d)^5D^0$	43±4	$(3p)^4P^0(5/2)$	14.1±2	$2s_4$		98			
$(5d)^5D^0$	47±4	$(3p)^4D^0(1/2)$	13.2±2	$2s_5$		110			
		$(3p)^4D^0(3/2)$	12.8±2	$2p_1$	14.7±0.6	15±1	15±1		
		$(3p)^4D^0(5/2)$	12.3±2	$2p_2$	16.5±0.6	19±2	16±2		
		$(3p)^4D^0(7/2)$	12.1±2	$2p_3$	23±2		32		
		$(3p')^2F^0(7/2)$	10.5±1	$2p_4$	22±1	18.5±1			
				$2p_5$	19±1		40		
				$2p_6$	22±1	29±10			
				$2p_7$	20.3±0.6				
				$2p_8$	24.2±0.8				
				$2p_9$	22.5±0.9				
				$3p_1$	63.5±0.7				
				$3p_{10}$	65±3				
				$4d_1$	480±30				

Table 1 (continued)
Part C. Magnesium

Mg I	UDC; eI (Ref. 7)	UDC; eB (Ref. 37)	Mg II	UDC; eI (Ref. 7)
$(5s)^1S$	163 ± 8		$(3p)^2P$	6.2 ± 0.4
$(6s)^1S$	201 ± 4		$(4s)^2S$	6.2 ± 0.5
$(3d)^1D$	57.0 ± 3.6		$(4f)^2F$	7.1 ± 0.7
$(4d)^1D$	54.9 ± 1.4	59 ± 4		
$(5d)^1D$	44.3 ± 2.4			
$(6d)^1D$	50.3 ± 2.2			
$(7d)^1D$	73.1 ± 2.7			
$(8d)^1D$	85 ± 10			
$(9d)^1D$	99 ± 9			
$(10d)^1D$	66 ± 15			
$(11d)^1D$	37 ± 3			
$(12d)^1D$	32 ± 3			
$(4f)^1F^0$	125^c	50^c		
$(5f)^1F^0$	200^c	90^c		
$(5p)^1P^0$		49^c		
$(4s)^3S$	14.8 ± 0.7			
$(5s)^3S$	26 ± 2			
$(6s)^3S$	52 ± 6			
$(4p)^3P$	80^c			
$(3d)^3D$	11.3 ± 0.8			
$(4d)^3D$	18.4 ± 0.7			
$(4f)^3F$	150^c			

^{c} Cascade inferences.

Table 1 (continued)
Part D. Argon

A I	UDC; eB (Ref. 38)	UDC; eB (Ref. 39)	UDC; eB (Ref. 40)	UDC; eB (Ref. 41)	A II	CDC; $h\nu$ (Ref. 42)	UDC; eB (Ref. 3)
$2p_1$			20 ± 0.5	21 ± 2	$(4p)^2S^0$		8.8 ± 0.3
$2p_2$			28 ± 1.7	25 ± 2	$(4p)^2P^0(1/2)$		8.7 ± 0.3
$2p_3$			25 ± 2.5	26 ± 2	$(4p)^2P^0(3/2)$		9.4 ± 0.5
$2p_4$			28 ± 1.4	31 ± 2	$(4p)^2D^0(5/2)$		9.8 ± 0.2
$2p_6$			22 ± 0.8		$(4p)^2D^0(3/2)$		9.1 ± 0.6
$3p_1$		71 ± 2	106 ± 5		$(4p)^4D^0(3/2)$		7.7 ± 0.2
$3p_3$			174 ± 13		$(4p)^4D^0(5/2)$		7.5 ± 0.5
$3p_5$		97 ± 2	97 ± 4		$(4p)^4D^0(7/2)$	5.06 ± 0.12	
$3p_6$		124 ± 3	200 ± 16		$(4p)^2D^0(5/2)$	6.27 ± 0.06	
$3p_7$		149 ± 3	190 ± 15				
$3p_8$		166 ± 3	148 ± 12				
$3p_9$		141 ± 2					
$3p_{10}$		190 ± 3					
$15d_4$	242 ± 15						

Table 1 (continued)

Part E. Calcium, Iron, and Krypton

Ca I	UDC; eB (Ref. 137)	Fe I	UDC; eB (Ref. 43)	Kr I	UDC; eB (Ref. 44)	Kr II	UDC; eB (Ref. 44)
$(7s)^1S$	116	$z\,^5F_5^0$	61.5 ± 0.4	$6p[1\frac{1}{2}]_2$	198 ± 4	$(5p)^2D^0(5/2)$	7.8 ± 0.5
$(5d)^1D$	45	$z\,^5F_4^0$	67.1 ± 1.1	$6p'[1\frac{1}{2}]_2$	128 ± 4	$(5p)^2D^0(3/2)$	8.5 ± 0.5
$(6d)^1D$	85	$y\,^5F_5^0$	11.5 ± 1.5	$6p'[0\frac{1}{2}]_0$	111 ± 4	$(6s)^4P(3/2)$	6.5 ± 1
$(7d)^1D$	137			$6p[0\frac{1}{2}]_1$	210 ± 4	$(5p)^4D^0(3/2)$	8.4 ± 0.5
$(4f)^1F^0$	33 ± 1.5			$6p[0\frac{1}{2}]_0$	74 ± 1	$(5p)^2P^0(3/2)$	6.5 ± 0.3
				$6p'[1\frac{1}{2}]_2$	127 ± 2	$(5p)^4D^0(7/2)$	8.5 ± 0.3
				$6p[1\frac{1}{2}]_1$	186 ± 2	$(4d)^2D(5/8)$	7.7 ± 1
				$6p[2\frac{1}{2}]_2$	199 ± 4	$(5p)^2F(5/2)$	9.9 ± 0.5
						$(5p)^4P^0(3/2)$	9.8 ± 0.7
						$(5p)^4P^0(5/2)$	8.3 ± 0.1
						$(5p)^2P^0(1/2)$	8.5 ± 0.6
						$(5p)^2S^0(1/2)$	8.6 ± 0.7

Table 1 (continued)

Part F. Cadmium

Cd I	UDC; eI (Ref. 45)	UDC; eB (Ref. 46)	Cd II	UDC; eI (Ref. 45)
$(6s)^1S$		8.7 ± 0.8	$(5s^2)^2D(5/2)$	990 ± 50
$(7s)^1S$	115 ± 10	20.2 ± 0.5	$(5s^2)^2D(3/2)$	300 ± 10
$(8s)^1S$	230 ± 2	44 ± 2	$(7d)^2D(5/2)$	11.7 ± 0.5
$(9s)^1S$	327 ± 8		$(6s)^2S(1/2)$	5.7 ± 0.9
$(5d)^1D$	27.6 ± 1.4	19.8 ± 1.8		
$(6d)^1D$	85.2 ± 1.2	37.5 ± 2		
$(7d)^1D$	84 ± 5	65 ± 5		
$(8d)^1D$	95 ± 7			
$(6s)^3S$	18.5 ± 2	10.6 ± 8		
$(7s)^3S$	29.9 ± 5			
$(5p)^3P^0$	2490 ± 130			
$(5d)^3D$	14.7 ± 2.2	14.6 ± 0.2		
$(6d)^3D$	18.7 ± 2.4			
$(4f)^3F^0$		112 ± 6		

Table 1 (continued)

Part G. Xenon and Barium

Xe I	UDC ; eB (Ref. 31)	Ba I	UDC; eI (Ref. 47)	Ba II	UDC; eI (Ref. 47)
$6p[0\frac{1}{2}]_0$	40 ± 12	$(6p')^1P^0$	29 ± 2	$(7s)^2S(1/2)$	9.4 ± 1.4
$6p[1\frac{1}{2}]_2$	33 ± 20	$(6p')^3D^0$	11 ± 2	$(6p)^2P(3/2)$	11.5 ± 0.6
$6p'[0\frac{1}{2}]_1$	43.5 ± 1.5	$(4f)^3F_2^0$	35 ± 3	$(6p)^2P(1/2)$	13.7 ± 0.6
$6p'[0\frac{1}{2}]_0$	38.5 ± 1.5	$(4f)^3F_4^0$	31 ± 3		
$6p'[1\frac{1}{2}]_2$	39.0 ± 1.0				
$7p[0\frac{1}{2}]_0$	87 ± 5				
$7p[0\frac{1}{2}]_1$	143 ± 4				
$7p[1\frac{1}{2}]_1$	101 ± 6				
$7p[1\frac{1}{2}]_2$	150 ± 10				
$7p[2\frac{1}{2}]_2$	200 ± 12				
$7p[2\frac{1}{2}]_3$	140 ± 10				
$7d[0\frac{1}{2}]_1$	87 ± 12				
$8p[2\frac{1}{2}]_3$	275 ± 23				
$8d[3\frac{1}{2}]_4$	135 ± 5				
$11d[3\frac{1}{2}]_3$	333 ± 27				
$13d[3\frac{1}{2}]_4$	700 ± 200				
$5d'$	150 ± 20				

Table 1 (continued)

Part H. Mercury

Hg I	CDC; eB (Ref. 48)	CDC; eB (Ref. 49)	CDC; $h\nu$ (Ref. 42)	CDC; $h\nu$ (Ref. 50)	UDC; eB (Ref. 4)
7^1S		31 ± 0.5			
8^1S		85 ± 2			
10^1P		51 ± 4			
7^1D		38.5 ± 0.5			
7^3S			8.4 ± 0.4	9.7 ± 0.2	10.4 ± 0.4
6^3P	120 ± 0.7				
6^3D		10.8 ± 0.2			8 ± 6
5^3F					104 ± 5

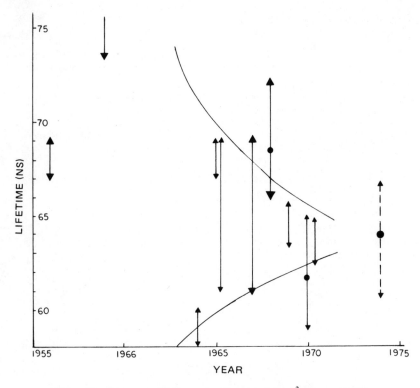

Figure 10. Progress in lifetime measurement of the 4^3S state of helium.

factors that can influence results under control at the same time over a broad investigation. Nevertheless the power of the method is apparent, and the trend is toward greater accuracy. In Figure 10, the history of the observation of the 4^3S line of helium is depicted. One sees a progressive narrowing of the range of prediction and convergence toward the theoretical value, which is exciting but perhaps accidental, since there is some uncertainty in it also. One sees too that at nearly every stage either the precision of the measurements has exceeded their accuracy or the researchers have been overly optimistic about their precision. In view of the large number of systematic errors which can go either way, it seems probable that the former was the case. In the early days of the method, interference filters were commonly used, and their low spectral resolution resulted in numerous accidental overlaps, which are indistinguishable from cascades. This is the probable origin of such major discrepancies as are seen for example in the 7^1S and 8^1S states of cadmium. When the best procedures are employed, there is every reason to suppose that the true lifetime of any atomic state below about $n = 10$ can now be measured to $\frac{1}{2}\%$ accuracy.

In summary, the method of temporal cutoff is the most effective way of achieving the measurement of subordinate state lifetimes in neutral atoms, whereas the Hanle effect serves best for states in the principal series, and the spatial cutoff method (beam-foil) is paramount in the measurement of the ionized states. The Hanle effect avoids the difficulties of resonance radiation trapping, while the beam-foil method reaches levels of excitation well-nigh impossible with electron beams. The Hanle effect is, however, affected as strongly as the lifetime measurement by cascade, and it is relatively difficult to correct for cascade owing to the slowly varying character of Lorentzian profiles. Beam-foil studies have been made on neutral atoms, but there is great uncertainty about where the excitation reaches its final state and what the velocity of the particles is, since they cannot be measured by time of flight as the ions can. There is some indication that these excited neutral atoms are subsequent knock-on excitations of previous beam ions which were temporarily stalled in the target and became neutralized. If so, their velocities will not only be uncertain, but also will be distributed over a considerable range. Each of the three basic methods has, therefore, its domain of greatest worth.

References

1. R. G. Fowler, *Handbüch der Physik*, Ed. S. Flügge, Vol. 22, pp. 209–253, Springer-Verlag, Berlin (1956).
2. S. Heron, R. W. P. McWhirter, and E. H. Rhoderick, *Proc. R. Soc. London* **234**, 565 (1956).
3. W. R. Bennett, Jr., P. J. Kindlemann, and G. N. Mercer, *Appl. Opt. Suppl.* **2**, 34 (1965).
4. T. M. Holzberlein, *Rev. Sci. Instrum.* **35**, 1041 (1964).
5. G. Russell and T. M. Holzberlein, *J. Appl. Phys.* **40**, 3071 (1969).
6. A. W. Johnson and R. G. Fowler, *J. Chem. Phys.* **53**, 65 (1970).
7. A. R. Schaeffer, *Astrophys. J.* **163**, 411 (1971).
8. M. Jeunehomme and A. B. F. Dundan, *J. Chem. Phys.* **41**, 1692 (1964).
9. K. Chandrakar and A. von Engel, *Proc. R. Soc. London* **284A**, 442 (1965).
10. G. E. Copeland and R. G. Fowler, *Rev. Sci. Instrum.* **41**, 1422 (1970).
11. G. E. Copeland, *J. Chem. Phys.* **54**, 3482 (1971).
12. K. Wagner, *Z. Naturforsch* **19A**, 716 (1964).
13. R. N. Blais and R. G. Fowler, *Phys. Fluids*, **16**, 2149 (1973).
14. R. F. Post, *Nucleonics* **10**, 46 (1952).
14a. W. C. Paske, *Rev. Sci. Instrum.* **45**, 1001 (1974).
15. R. G. Bennett and F. W. Dalby, *J. Chem. Phys.* **31**, 434 (1959).
16. J. Klose, *Astrophys. J.* **141**, 814 (1965).
16a. J. Klose, *Phys. Rev.* **141**, 181 (1966).
17. W. Paske, Ph.D. Dissertation, University of Oklahoma (1974).
18. C. A. DeJoseph (private communication).

19. R. E. Imhof and F. H. Read, *Nucl. Instrum. Methods* **90**, 109 (1970).

19a. R. E. Imhof and F. H. Read, *Chem. Phys. Lett.* **3**, 652 (1969).

20. N. H. Gale, *Nucl. Phys.* **38**, 252 (1962).

21. E. E. Habib, B. P. Kibble, and G. Copley, *Appl. Opt.* **7**, 673 (1968).

22. R. P. Evans, *The Atomic Nucleus*, p. 470 ff, McGraw-Hill, New York (1955).

23. R. T. Thompson and R. G. Fowler, *J. Quant. Spectrosc. Radiat. Transfer* **15**, 1017 (1975).

24. R. T. Thompson, *J. Quant. Spectrosc. Radiat. Transfer* **14**, 1179 (1974).

25. A. V. Phelps, *Phys. Rev.* **110**, 1362 (1958).

26. R. G. Fowler, T. M. Holzberlein, and C. H. Jacobson, *Phys. Rev.* **140**, A1050 (1965).

26a. E. Ivanov and M. Chaika, *Opt. Spectrosc.* **29**, (1970).

27. R. T. Thompson, Ph.D. Dissertation, University of Oklahoma (1972).

28. P. C. Rogers, FRANTIC Program for Analysis of Exponential Growth and Decay Curves, MIT Laboratory for Nuclear Science technical report No. 76 (1962).

29. J. Morris (private communication), University of California, Los Alamos Scientific Laboratory, Los Alamos, New Mexico.

30. J. Peresse, A. Pochat, and A. La Nadan, *C.R. Acad. Sci. B* **274**, 791 (1972).

31. L. Allen, D. G. C. Jones, and D. G. Schofield, *J. Opt. Soc. Am.* **59**, 842 (1969).

32. L. L. Nichols and W. E. Wilson, *Appl. Opt.* **7**, 167 (1968).

33. A. L. Osherovich and Ya. F. Verolainen, *Opt. Spectrosc.* **24**, 81 (1968).

34. K. A. Bridgett and T. A. King, *Proc. Phys. Soc. London* **92**, 75 (1967).

35. W. R. Pendleton, Jr. and R. H. Hughes, *Phys. Rev.* **138**, A683 (1965).

36. R. G. Fowler, T. M. Holzberlein, C. H. Jacobson, and S. J. B. Corrigan, *Proc. Phys. Soc. London* **84**, 539 (1964).

37. F. Karstensen and J. Schramm, *Z. Astrophys.* **68**, 214 (1968).

38. J. Campos and B. Zurro, *An. Fis.* **69**, 299 (1973).

39. J. Klose, *J. Opt. Soc.* **58**, 1509 (1968).

40. Ya. F. Verolainen and A. L. Osherovich, *Opt. Spectrosc.* **25**, 258 (1968).

41. J. Klose, *J. Opt. Soc. Am.* **57**, 1242 (1967).

42. C. Camby, A. M. Dumont, M. Dreux, and R. Vitry, *Phys. Lett.* **32A**, 233 (1970).

43. J. Klose, *Astrophys. J.* **164**, 637 (1971).

44. A. Delgado, J. Campos, and C. Sanchez, *Z. Phys.* **257**, 9 (1972).

45. A. R. Schaefer, *J. Quant. Spectrosc. Radiat. Transfer* **11**, 197 (1971).

46. Ya. F. Verolainen and A. L. Osherovich, *Opt. Spectrosc. (USSR)* **20**, 517 (1966).

47. A. R. Schaefer, *J. Quant. Spectrosc. Radiat. Transfer* **11**, 499 (1971).

48. G. C. King and A. Adams, *J. Phys. B* **7**, 1712 (1974).

49. G. C. King, A. Adams, and D. Cvejanovic, *J. Phys. B* **8**, 365 (1975).

50. J. Pardies, *C.R. Acad. Sci. B* **266**, 1586 (1968).

27

Line Shapes

W. Behmenburg

1. Introduction

Shapes of spectral lines throughout the range of the electromagnetic spectrum have, since the days of Michelson, been the subject of numerous studies, both experimental and theoretical, and of both physicists and chemists. The main reason for this interest is that line contour investigations not only yield information about atomic (molecular) structure and fundamental processes (e.g., atomic collisions or light scattering from atoms) but also may serve as analytical tools in many fields of application such as plasma diagnostics or analytical spectroscopy.

There are in general several mechanisms causing the perturbation of a spectral line emitted, absorbed, or scattered by an ensemble of atoms: the Doppler effect due to thermal motion, interaction with the radiation field, and collisions. Furthermore, the shape of optical lines emitted from thick layers is strongly influenced by reabsorption and incoherent scattering inside the layer (radiation diffusion). If there are collective oscillations in a dense gas or solid, the frequency distribution of light incoherently scattered by the modes will be governed by their finite lifetimes.

The broadening mechanism of collisions is very different for various types of interactions. If charged particles take part in the collision process, the line profile will be determined essentially by the interatomic Stark effect. If the particles are neutral, the perturbation of the line is governed by van der Waals forces (except for lines with very high quantum numbers).

In this review we will restrict ourselves to effects of neutral atom collisions and interactions with the radiation field. The Stark effect as well as Doppler broadening have long been used as successful tools in plasma

---

W. Behmenburg • Physikalisches Institut I der Universität, Düsseldorf, 4000 Düsseldorf, BRD.

diagnostics and have recently been reviewed by Griem.[1] The current status of the complicated theory of radiative transfer in optically thick layers, of great interest to astrophysicists, has been presented by Jeffries.[2] The theory and application of light scattering by plasmas has been surveyed recently by Kunze,[3] and the spectroscopy of Brillouin, and Rayleigh-scattering from liquids has been reviewed by Schönes.[4]

In considering spectral profiles we shall further confine ourselves to usual atomic absorption and fluorescence lines; perturbation of anomalous dispersion has not so far been studied in the atomic case, and very little has been done for molecules.[5]

Broadening of magnetic resonances, like level crossings or optical double resonances, the theory of which is similar to that of the optical case, will be treated elsewhere in this book.

2. Experimental Methods

Studying line perturbation due to a single mechanism such as atom interaction or interaction with the radiation field generally requires a choice of experimental conditions such that all other perturbation mechanisms are small or negligible. If they are small and statistically independent, they may be taken into account by a proper deconvolution procedure.

2.1. Influence of Instrumental Effects

In spectroscopic investigations essentially two factors determine the accuracy of measurements: intensity and spectral resolution. Since in general the resolution of the apparatus can be increased only at the expense of the intensity flux accepted by the detector, a compromise has to be found. In the particular case of line-profile scanning this often leads to the experimental situation where the width due to the apparatus is comparable or even larger than the linewidth; in this case the line profile will be severely distorted by the apparatus profile, and a careful deconvolution of the two is necessary.

Generally, the observed profile $P'(\omega)$ is the result of convolution of the true line profile $P(\omega)$ with the apparatus function $A(\omega)$:

$$P'(\omega) = \int P(\omega' - \omega) A(\omega') \, d\omega' \tag{1}$$

or

$$P' = P * A$$

Detailed discussions of the various methods of obtaining $P(\omega)$ from the measured functions $P'(\omega)$ and $A(\omega)$ are given in Refs. 6 and 7.

   (a) *Iteration Method.* In this procedure[8] no special assumption is made regarding the functional form of $P(\omega)$ and $A(\omega)$ and it is therefore generally applicable. $P(\omega) \equiv P$ is obtained from the known functions $P'(\omega) \equiv R_0$ and $A(\omega) \equiv A$ as the limit

$$P = \lim_{p \to \infty} R_p$$

of the set

$$R_p = R_0 + (\delta - A) * R_{p-1} \tag{2}$$

where δ is Dirac's delta function. A detailed discussion of errors involved is given in Ref. 7. The accuracy of the results of the method increases with the signal-to-noise ratio.

   (b) *Method of Fourier Transformation.* This method is based on a well-known theorem which states that the convolution of two functions corresponds to the multiplication of their Fourier transforms. Using this theorem, the convolution of analytical functions, like those of the Lorentz, Gauss, or Voigt type that often occur in line-profile analysis, becomes a simple matter. Its application, however, to the problem of deconvoluting $P'(\omega)$ with respect to $A(\omega)$ requires the knowledge of the functional form of $P(\omega)$. Denoting the respective half-widths by $\omega'_{1/2}$, $\Delta\omega^A_{1/2}$, and $\Delta\omega_{1/2}$, the value of $\Delta\omega_{1/2}$ may then be obtained by its variation such that $\Delta\omega'_{1/2}$ as obtained from (1) agrees with the observed $\Delta\omega'_{1/2}$ value.

2.2. Doppler-Limited Spectroscopy

   Traditional experiments on collisional line broadening are conducted under conditions where Doppler broadening is always present. Consequently, collision effects can be studied only at comparatively high particle densities. This restriction, however, is no longer valid, when the technique of Doppler-free spectroscopy is used (see Section 2.3).

   For the investigation of line shapes, within certain characteristic limits observation in both emission and absorption is suitable.

2.2.1. Observation in Emission and Fluorescence

   This technique may be applied to line excitation both in a plasma (gas discharge, arc, flame) and in a cold gas by selective absorption of radiation. For information on neutral atom collision effects, however, the light sources have to be operated under conditions where Stark effect broadening is negligible and the Doppler width $\Delta\omega_D$ is at most of the order of the collision width $\Delta\omega_c$. If $\Delta\omega_D$ is of the same order as $\Delta\omega_c$, full Voigt profile analysis has to be applied to evaluate $\Delta\omega_c$ from the total half-width.

Scanning of the line profile is usually performed by means of a high-resolution Fabry–Perot interferometer (FPI). For the study of line wings, where less resolution is required in general, a grating instrument may be used as well. Since, however, signal levels in the wings may be as low as 10^{-6}–10^{-8} of the line center, a FPI is preferred because its luminosity is by about 100 times larger than a grating monochromator of the same resolution. For the wings of alkali resonance lines perturbed by noble gases, a triple FPI has been applied with a finesse of 300 and an extremely low transmission outside its bandpass.[9] For quantitative suppression of the very strong center emission, a narrow-band alkali vapor absorption filter has been used.

It is important that only radiation from thin layers is observed since otherwise the line shape would be distorted by radiation trapping. This requirement has to be obeyed especially for resonance transitions and in the line center. Since the optical depth τ is proportional to the number density N of active atoms, this sets a severe limit to N and thus to the line intensity. For wing measurements, however, N may be raised so that for radiation from the wing region τ is still $\ll 1$, whereas in the line center τ may be much larger than unity.

(a) *Emission from Gas Discharges.* For the study of collisional broadening effects at number densities below 10^{18} cm^{-3}, electrodeless discharge or hollow-cathode discharge lamps are suitable. To eliminate Stark effect broadening the discharge has to be driven at very low current densities (typically $\leqslant 5$ mA/cm^2). In addition, cooling by liquid N_2 or He may be necessary for reduction of Doppler broadening. With this technique foreign gas and self-broadening of many noble gas lines[10] as well as the Na diffuse series perturbed by noble gases[38] have been studied in the density range 3×10^{15}–10^{18} cm^{-3} at temperatures below 500 K.

(b) *Emission from Thermal Sources.* Arcs or chemical flames offer themselves for investigation at larger densities and temperatures. A gas-stabilized high-pressure arc, running in an Ar atmosphere of up to 100 atm at $\approx 10,000$ K has been used to investigate neutral atom broadening of Ti lines.[11] An acetylene–oxygen flame at atmospheric pressure and temperatures around 2500 K has been applied to the study of broadening of alkali and alkaline-earth lines perturbed by foreign gases.[12] Because of operation conditions, however, these light sources allow variation of the state parameters only over a limited range. In addition, because of their continuous background, the investigation of line wings is difficult. On the other hand, they easily facilitate excitation of lower levels of a great variety of elements.

(c) *Fluorescence*. Fluorescence cells, operated at ordinary temperatures, are almost ideal light sources for the study of neutral atom collision effects on line shapes: There is no continuous background radiation, no Stark effect broadening, and the residual Doppler width may be greatly reduced by the technique of Doppler-free spectroscopy. Cells have been extensively used for the study of alkali resonance line wings perturbed by noble gases.[9] An extension of these investigations to other transitions suggests itself with the use of continuous-wave (cw) dye lasers.

2.2.2. Observation in Absorption

Observation in absorption is particularly well-suited to studies of the line wings: By means of an absorption tube the optical thickness of the absorbing layer may, in a defined manner, be varied over many orders of magnitude; thus the spectral range of measurable absorption may easily be shifted from the line center to the far wings. An absorption tube may be a simple quartz cell, a heat pipe,[13] or, for measurements at higher temperatures, a King's furnace.[14]

Scanning of the line profile is conventionally achieved by passing continuous background radiation through the absorbing layer and measuring the transmission spectrum by use of a wavelength selector; for measurements of half-widths, shifts, and near wings, high-dispersion grating instruments have been used.[105,106]

Method of Integrated Line Absorption. A different method for line-profile scanning consists of using as background source a line of known profile and measuring its integrated transmission through the absorbing layer while it is frequency-shifted across the absorption profile under study. Denoting by $A(\omega - \omega')$ and $k(\omega - \omega_0)$, respectively, source profile and absorption profile with center frequencies ω' and ω_0, the integrated transmission $T(\omega' - \omega_0)$ is given by

$$T(\omega' - \omega_0) = \frac{\int d\omega\, A(\omega - \omega')\, e^{-k(\omega - \omega_0)l}}{\int A(\omega - \omega_0)\, d\omega} \tag{3}$$

If $A(\omega - \omega')$ is known, $k(\omega - \omega_0)$ may be evaluated from the measured transmission function $T(\omega' - \omega_0)$, using Eq. (3). Conversely, if $k(\omega - \omega_0)$ is known, $A(\omega - \omega')$ may be determined. An advantage of this method, compared to the conventional procedure, is its much larger luminosity at the same resolution since no monochromator is required.

Shifting of the center frequency of either $A(\omega - \omega')$ or $k(\omega - \omega_0)$ may be achieved by application of a magnetic field (Zeeman scanning) or an electric field (Stark scanning). In this case, with the fields available in laboratories, a spectral range of at most a few cm^{-1} may be covered; this means that the source atoms and absorbing atoms have to be identical and that only the central part of a profile (half-width and shift) at $n < 10^{19}$ cm^{-3} may be scanned.

These restrictions no longer exist if a dye laser line is used as background source. In this case, because of its much larger spectral density, the fluorescence spectrum $I_F(\omega)$ (rather than the transmission spectrum) may be scanned. If, in particular, the laser linewidth is much smaller than the half-width of $k(\omega)$ and if the absorbing layer is optically thin, then $I_F(\omega) \sim k(\omega)$.

Zeeman shifting of the background line has been applied to vapor cell measurements of the Hg^{198} transition 6^1S_0–6^3P_1 ($\lambda\,2537$ Å)[7] and 6^1S_0–6^1P_1 ($\lambda\,1849$ Å)[15] as well as Zn and alkaline-earth resonance transitions under flame conditions.[16] Hollow cathode discharge or microwave-driven electrodeless discharge lamps placed between the poles of an electromagnet have been used as background sources. In the case of the resonance transitions the special problem of the apparatus profile $A(\omega - \omega')$ changing with the magnetic field strength due to reabsorption arises. The evaluation of half-width and shift, however, is greatly simplified by using the fact that, at the inflection points ω_i of the unperturbed transmission profile $e^{-k(\omega - \omega_0)l}$, the transmission value $\exp[-k(\omega_i - \omega_0)l]$ equals, to a good approximation, the measured value $T(\omega_i - \omega_0)$, independent of the apparatus profile.[7] This problem does not arise if ω_0 is shifted at a fixed ω', i.e. if the magnetic field is applied to the absorbing layer rather than to the background source.[17]

2.3. Doppler-Free Spectroscopy

By this term is designated a class of techniques, which eliminate or reduce the effect on the resolution limit of Doppler broadening of a line: use of atomic beams, level crossing, frequency-selective nonlinear absorption, and two-photon absorption. Whereas beam- and level-crossing techniques have been used for a long time, saturation and two-photon spectroscopy have been introduced only recently, with the development of narrow-band and frequency-stabilized laser sources. Under optimum conditions (no collisions, source power levels not too high) the maximum resolution obtained by these techniques is given by the natural linewidth ($\approx 10^6$ Hz). Together with the large signal-to-noise values obtained with laser sources, Doppler-free spectroscopy surpasses by far the limitations of conventional methods.

Since beam- and level-crossing experiments are discussed elsewhere in this book, we will restrict ourselves to saturation and two-photon absorption. Both methods have been used for several years for atomic and

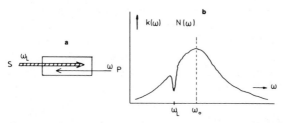

Figure 1. (a) Schematic arrangement of a saturated absorption experiment. (b) Saturation dip in a Doppler-broadened absorption line profile caused by absorption of an intense monochromatic traveling wave (saturating beam).

molecular excited state, fine, and hyperfine structure analysis.[18] However, so far only a few applications to line shape studies have been reported.

(a) *Saturation Spectroscopy.* In principle the experimental arrangement (Figure 1) consists of an intense saturating laser beam S of frequency ω_L and a weak probing laser beam P of variable frequency ω directed in opposite direction through an absorbing vapor in a cell. Light from S will be nonlinearly and selectively absorbed by initial-state atoms of velocity v_Z related to ω_L by

$$v_Z = (\omega_0 - \omega_L)c/\omega_0 \tag{4}$$

This causes selective reduction of the number N of these atoms, leading to a hole around $\omega = \omega_L$ within the Doppler-broadened absorption profile $k(\omega)$. Scanning of the shape of the hole (Benett hole, Lamb dip) may then be achieved by P. The width of the hole is determined solely by the width of the upper and lower atomic levels of the absorbing transition. If S and P are counterpropagating and of the same frequency ω_L, then by scanning of ω_L over the absorption profile, the dip occurs at $\omega_L = \omega_0$.

For the purpose of scanning the hole shape, extreme frequency stabilization of S is necessary. Among the available stabilization techniques, that of saturated absorption of atomic or molecular transitions as stabilizing elements[19] is often used. Frequency variation of the probing laser beam is usually effected by continuously changing the laser cavity length. For frequency control a Fabry–Perot etalon may be used. In order to keep its intensity constant over the profile to be scanned, the laser discharge current may be servo-controlled using a reference beam.[20]

The saturation absorption technique was first applied to measure the collisional half-width of the Ne absorption line $\lambda = 6328$ Å at a pressure of 0.1 atm. This was achieved by placing excited Ne in a plasma tube inside the cavity of a single-frequency He–Ne 6328-Å laser and measuring the output power as a function of laser frequency.[22] Saturation spectroscopy using dye lasers has been proposed by Hänsch et al.[23] With a cw laser source it is possible to measure collisional linewidths of the two Na D_2 hyperfine components in the presence of Ar at 3 Torr as buffer gas.

Since in a saturated absorption experiment only atoms within a certain velocity group interact with the saturating beam, this technique can be used to study the velocity dependence of collisional broadening and shift cross sections. This has been demonstrated already for self-broadening of an infrared transition of NH_3 gas at room temperature.[21] An extension to atomic transitions using dye lasers is possible and is being considered in several laboratories.

(b) *Two-Photon Spectroscopy*. If in a two-photon transition experiment two photons of the same frequency are absorbed each from two counterpropagating light beams, there is no net photon momentum transfer to the absorbing atoms. Under this condition all atoms undergo transitions at the same values of the photon energies, whatever their velocities may be. Consequently, Doppler broadening of the transition is eliminated. This is the basis of two-photon spectroscopy.

Detection of the two-photon transition may be made by collecting photons spontaneously reemitted from the excited level. Cagnac et al.[24] showed that the light power necessary to produce a visible effect is not too high if the half-width of the exciting laser line is smaller than the natural width of the excited level; some experiments can be made with power much less than 1 W. They showed further that in many cases light shifts (or the ac Stark effect) will be negligible.

With currently available dye lasers two-photon spectroscopy opens the possibility of investigating broadening and shifts of highly excited and metastable states. It has recently been applied to collisional broadening and shift studies of the Na transition $3S–4D$ with Ne as perturbing gas.[25] Effects at Ne pressures as low as 1 Torr have been measured with an accuracy of 1 MHz.

3. Atomic Interactions

It is convenient to divide the line shapes into two regions (Figure 2): (1) the impact theory region, corresponding to the Lorentzian-shaped center of the line, and (2) the wing region, where the quasistatic theory is applicable. The semiclassical oscillator model of the atom is sufficient to explain the distinction between these different regions. According to this model the resonance frequency ω_0 of the oscillator is perturbed to $\omega(t)$ during a collision, so that $\hbar[\omega(t)-\omega_0]$ may be considered as the perturbation in the spacing of the atomic energy levels during the collision.[26] The absorption (or emission) at ω' is proportional to the squared modulus

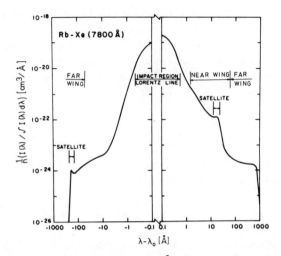

Figure 2. The normalized spectrum of Rb ($5^2P_{3/2}$) in Xe of density n. The data are from Refs. 51, 62, and 77. $T = 315$ K. From Ref. 50.

of the Fourier amplitude of the time-varying oscillator:

$$I(\omega') \sim \left| \int dt \exp\left\{ -i(\omega'-\omega_0)t + i \int^t [\omega(t')-\omega_0] \, dt' \right\} \right|^2 \tag{5}$$

Validity Criteria of Eq. (5). One may show,[27] that Eq. (5) describes correctly the intensity distribution of a collision-broadened atomic line with the following assumptions: (1) Electronic and nuclear motion of the colliding atoms are decoupled (Born–Oppenheimer approximation); (2) the dipole transition probability A_{if} is independent of the internuclear distance R; (3) the motion of the perturber may be described quasiclassically, i.e., by means of WKB wave functions. From assumptions (1) and (2) one obtains the intensity distribution

$$I(\omega') \sim A_{if} \left| \int \psi_i^*(R, E^i) \psi_f(R, E^f) \, dR \right|^2 \tag{6}$$

where the wave functions $\psi_{i,f}(R, E^{i,f})$ describe the translatory motion of the perturber in the field of the active atom in the electronic states i, f with total energies E^i, E^f, respectively. From Eq. (6) one obtains (5) using assumption (3) (see, however, Section 3.3). The above analysis shows that line broadening is essentially caused by partial transfer of electronic energy into kinetic energy of nuclear motion.

Velocity-Changing Collisions. Apart from atomic interactions the Doppler shift due to thermal motion may influence the line shape. In the absence of collisions the Doppler shifts lead to a Gaussian profile with a linewidth of the order of $(\omega_0/c) \Delta v$, where Δv is the velocity bandwidth of the oscillators. Collisions tend to reduce this Doppler width by averaging out the Doppler phase factor $\mathbf{k}vt$ ($\mathbf{k}$ and $\mathbf{v}$ wave vector and velocity vector, respectively) of the individual oscillators. The reduction becomes significant at pressures where the mean free path is less than or comparable to the wavelength $\lambda = k^{-1}$. Collisional narrowing has been predicted by Dicke[28] and recently reviewed by Berman.[29]

3.1. Line Center

The typical duration τ_c of a long-range thermal collision is $\approx 10^{-12}$ s. If one considers a line center region, where

$$\omega' - \omega_0 \ll 1/\tau_c \qquad \text{(impact approximation)} \tag{7}$$

the second exponential in Eq. (5) changes much more rapidly than the first during a collision. The oscillator may then be approximated as being unperturbed except for abrupt phase shifts $\Delta\phi$ from each collision.

In addition we assume that the average time τ between collisions is much larger than τ_c:

$$\tau_c \ll \tau \qquad \text{(binary collision condition)} \tag{8}$$

From Eqs. (7) and (8) Fourier analysis yields, for each velocity class v of the active atoms, a Lorentzian line

$$I(\omega', v) \sim \frac{\gamma(v)}{[\omega' - \omega_0 - \beta(v)]^2 + \gamma^2(v)} \tag{9}$$

of width $2\gamma(v)$, shifted from ω_0 by $\beta(v)$. The final line shape is obtained by averaging over velocities:

$$I(\omega') \sim \int \frac{dv f(v) \gamma(v)}{[\omega' - \omega_0 - \beta(v)]^2 + \gamma^2(v)} \tag{10}$$

where $f(v)\, dv$ is the velocity distribution. Since the shift is v-dependent, the line shape can be asymmetric. It may be shown that the v dependence of γ and β will be important only if $v \gg v_p$ (v_p perturber velocity). In the remaining discussion this effect will be neglected and $I(\omega')$ assumed to be Lorentzian.

It should be remarked that the Lorentzian profile is characteristic of the impact theory, which is valid if assumptions (7) and (8) are fulfilled, independent of the atomic model, the type of collision, and the interatomic potential governing the collision dynamics. The half-width and shift of $I(\omega')$ are proportional to the number density n of the perturbers.

3.1.1. Impact Theory

In collision theory it is customary to distinguish between elastic and inelastic collisions. For our purpose we may define those collisions to be elastic which only affect the shape of a line without reducing its integrated intensity. In this sense not only ordinary phase-shift collisions but also m-mixing collisions in zero magnetic fields would be classified as elastic.

Phase-Shift Theory. In the particular case of pure phase-shift collisions ($\Delta\phi$ real) of an atomic oscillator one obtains for the half-width and shift[30]

$$\gamma + i\beta = n \int dv_r \, v_r f(v_r) \int 2\pi b \, db \{1 - \exp[-i\Delta\phi(b, v_r)]\} \qquad (11)$$

where n, $f(v_r)$, b are, respectively, the perturber density, distribution of relative velocities, and impact parameter. The phase shift $\Delta\phi$ is given by

$$\Delta\phi(b, v_r) = \int_{\tau_c} \{[\omega(t) - \omega_0]\} \, dt$$

$$= \frac{1}{\hbar} \int_{\tau_c} dt [V^i(r(t)) - V^f(r(t))] \qquad (12)$$

with $V^i(r)$ and $V^f(r)$ denoting the adiabatic potentials connected to the initial and final states of the transition. The second integral in (11) may be interpreted as a total cross section for line broadening:

$$\sigma(v_r) = \int 2\pi b \, db \, \sigma(b, v_r) \qquad (13)$$

with the line-broadening probability given by

$$\sigma(b, v_r) = 1 - \exp[-i\Delta\phi(b, v_r)] \qquad (14)$$

An analysis of the oscillatory behavior of $\sigma(b, v_r)$ shows, that the linewidth is mainly determined by "strong" collisions, with $b < b_0$, whereas the line shift is mainly produced by "weak" collisions, with $b > b_0$, where b_0 is defined by[27] $\Delta\phi(b_0) = 1$.

Semiclassical Theory. Phase-shift theory is strictly valid only for a two-level system. Such a system is approximately realized if the separation in energy of the levels is large compared to h times the reciprocal of the collision time, and if the degeneracy of the levels is not raised by the interaction (example: the D_1 component of the Cs resonance line perturbed by thermal collisions with noble gas atoms. In other cases, however, collisions inducing transitions between magnetic substates (disorienting or m-mixing collisions) or to nearby energy levels (J-mixing, configuration-mixing collisions) may, in addition to phase-shift collisions, considerably contribute to the broadening and shift of the line.

These effects may be taken into account by means of time-dependent perturbation theory.[31] For this purpose a time evolution operator $T(t)$ is introduced which is connected to the interaction Hamiltonian $H_c(t)$ by means of the Schrödinger equation

$$i\hbar\dot{T} = \hat{H}_c T \tag{15}$$

Here

$$\hat{H}_c = \exp\left[(i/\hbar)H_0 t\right]H_c \exp\left[-(1/\hbar)H_0 t\right]$$

where H_0 is the Hamiltonian of the unperturbed atoms.

In terms of matrix elements of T defined with respect to the collision frame, one obtains[31,32] for the line-broadening probability

$$\sigma_{J_i} = 1 - \frac{1}{2J_i + 1} \sum_{\text{all } M's} C(J_f 1 J_i; M_f M M_i) C(J_f 1 J_i; M'_f M M'_f)$$
$$\times \langle J_i M_i | T^{i-1} | J_i M'_i \rangle \langle J_f M'_f | T^f | J_f M_f \rangle \tag{16}$$

In Eq. (16), J_f, J_i and M_f, M_i designate the angular momenta and orientation quantum numbers of the respective states, and $C(\cdot\cdot)$ are the Wigner coefficients.

In obtaining (16) the angular average over all orientations of the collision frames relative to the laboratory frame has been made. Determination of $T(t)$ from Eq. (15) amounts to solving a large number of coupled differential equations given by

$$i\hbar\dot{T}^n_{M'M} = \sum_{M''} V^n_{M'M''} T^n_{M''M} \tag{17}$$

where $V^n_{M'M''}$ may be interpreted as a matrix element of some sort of "effective" interaction Hamiltonian according to $V^n_{M'M''} \equiv (H^{n\text{ eff}}_c)_{M'M''}$.[33] This has been done numerically in the case $J_f = 0$, $J_i = 1$ for dipole–dipole interaction, both resonant and nonresonant.[34] In more complicated cases two complementary approximate approaches have been discussed: (1) the "scalar" or "sudden" approximation, valid for weak collisions and (2) the "adiabatic" approximation, valid for strong collisions.[33]

Starting from (17) in matrix form

$$i\hbar\dot{T}^n = V^n T \tag{17'}$$

the *scalar approximation* assumes the solution of (17') to be

$$T = \exp\left(-\frac{i}{\hbar}\int_{-\infty}^{t} dt'\, V(t')\right) = \exp\left[-P(t)\right] \tag{18}$$

with the condition that the matrix elements of $P(+\infty)$ satisfy

$$P^n_{M'M}(+\infty) = \frac{i}{\hbar}\int_{-\infty}^{+\infty} dt\, V^n_{M'M}(t) \tag{19}$$

Ansatz (18) is equivalent to assuming that $V(t)$ commutes with itself at all times.

Within the *adiabatic approximation* a matrix ε is defined describing the interaction energy with respect to the internuclear axis. The elements of T are then given by

$$T^n_{M'M}(+\infty) = D^J_{M'M} \exp\left(-\frac{i}{\hbar} \int_{-\infty}^{+\infty} dt\, \varepsilon^n_M(t)\right) \tag{20}$$

where D^J is the operator describing the rotation of the collision frame into the moving frame (quantization axis parallel to the internuclear axis). The adiabatic approximation is valid for impact parameters $b \ll b_0$, where b_0 is defined by

$$\frac{1}{\tau_c} = \frac{1}{\hbar}[\varepsilon_M(b) - \varepsilon_{M'}(b)] \tag{21}$$

For the calculation of the various matrix elements it has been customary to depict the collision by assuming the perturber to move along a straight-line path at constant velocity:

$$r(t) = (b^2 + v^2 t^2)^{1/2} \tag{22}$$

This assumption will certainly not be valid for strong collisions. The effect of curved trajectories on the results has been investigated.[35]

In calculating total cross sections one should note that the function $\sigma(b)$ has oscillatory behavior in the region of small b, with an average value of unity for both the sudden and adiabatic approximation. In this region, therefore, one approximation may be replaced by the other with almost no modification to the result.[33]

3.1.2. Dependence on Interactions

The line-center parameters (e.g. half-width, magnitude, sign of the shift, and the temperature dependence of these quantities) depend very much on the nature and magnitude of the interaction. Phenomenologically, one may distinguish between *self-broadening* (active and perturbing atom of same kind) and *foreign gas broadening* (active and perturbing atom of different kind). Foreign gas broadening is governed mainly by second-order dipole–dipole interaction $\sim r^{-6}$ and exchange (overlap) interaction, collectively termed nonresonant interaction. On the other hand, the central part of a self-broadened line is determined by first-order dipole–dipole interaction $\sim r^{-3}$, termed resonant interaction, if the resonance level is involved in the transition considered; if it is not involved, nonresonant interaction may also contribute.

Table 1. Broadening and Shift Constants for Various Systems (in Units of 10^{-9} rad s^{-1} cm^3)^a

Line	Perturber	T/K	Ref.	$2\gamma/n$	β/n	$\beta/2\gamma$
Kr 7601 Å	He	80	36	2.60	+0.205	+0.078
$4p^5(^2P_{3/2})5s[\frac{3}{2}]_2$		295	36	4.84	+0.53	+0.110
$-4p^5(^2P_{3/2})5p[\frac{3}{2}]_2$	Ne	80	36	1.29	−0.43	−0.33
		295	36	2.15	−0.40	−0.185
	Ar	80	36	2.22	−0.60	−0.27
		295	36	4.60	−1.38	−0.30
	Kr	80	36	2.46	−0.84	−0.34
		295	36	3.65	−1.23	−0.34
Hg^{198} 2537 Å	He	293	78	1.66	+0.090	+0.055
6^1S_0–6^3P_1	Ne	293	78	1.06	−0.158	−0.151
	Ar	293	79	2.19	−0.49	−0.225
	Kr	293	79	1.47	−0.38	−0.255
	Xe	293	79	2.11	−0.47	−0.225
Cs 4593 Å	He	400	80	16.6	+2.82	+0.171
$6^2S_{1/2}$–$7^2P_{1/2}$	Ar	400	80	12.6	−3.07	−0.245
Cs 4555 Å	He	380	80	13.0	+1.37	+0.106
$6^2S_{1/2}$–$7^2P_{3/2}$	Ar	380	80	10.9	−2.92	−0.267
K 4047 Å	Kr	470	81	12.4	−3.71	−0.29
$4^2S_{1/2}$–$5^2P_{1/2}$						

^{a} The corresponding cross sections are obtained from the relations $\sigma_r = (2\gamma/n)(1/\bar{v}_r)$; $\sigma_i = (\beta/n)(1/\bar{v}_r)$ where $\bar{v}_r$ is the average relative velocity of the colliding atoms.

Nonresonant Interaction. Phase-shift theory predicts with a quasistatic potential ansatz $V(r) = -C_6 r^{-6}$ for half-width 2γ and shift β:

$$\gamma = c_\gamma \bar{v}^{3/5}(C_6/\hbar)^{2/5} n$$
$$\beta = -c_\beta \bar{v}^{3/5}(C_6/\hbar)^{2/5} n \tag{23}$$

($\bar{v}$ is mean relative velocity, c_γ, c_β are constants.) Characteristic for this ansatz are (1) negative sign of β (red shift), (2) constant value for the ratio β/γ, independent of system and temperature, (3) proportionality of γ and β to $T^{3/10}$

Many experimental results on foreign gas broadening, however (Table 1), do not conform to these predictions. In particular, for the blue shift of many lines having He as perturber, the widely differing β/γ values and their change with temperature for a given system cannot be explained. The occurrence of blue shifts strongly indicates the effectiveness of repulsive terms in the potential.

A much better agreement with experiment is obtained by assuming a Lennard–Jones (LJ) 6-12 potential. On the basis of the phase-shift theory one obtains[33]

$$\gamma = (2\pi)^{3/5} h_{\pm}(\alpha) \bar{v}^{3/5} (C_6/\hbar)^{2/5} n$$

$$\beta = (2\pi)^{3/5} v_{\pm}(\alpha) \bar{v}^{3/5} (C_6/\hbar)^{2/5} n \tag{24}$$

where the parameter $\alpha(\bar{v})$ is related to the LJ constants C_6 and C_{12} and the broadening and shift functions $h_{\pm}(\alpha)$ and $v_{\pm}(\alpha)$ are defined and tabulated in Ref. 33. Comparison with experimental data for the Kr emission line[36] $\lambda 7601$ Å measured at 295 and 80 K (Table 2) shows good agreement except for Ar as perturber. The over-all agreement, however, is much better than in case of r^{-6} interaction, where the theoretical ratio γ_{295}/γ_{80} equals that of $\beta_{295}/\beta_{80} = 1.43$, independent of the nature of the perturber.

Further improvement in agreement is obtained by adding to the LJ ansatz a dipole–quadrupole term $\sim r^{-8}$ (Table 3).[37] With values of C_8/C_6 calculated with the use of radial wave functions, the C_6 values, determined from experimental values of half-width and shift, show markedly improved agreement with calculated C_6 values over those found by interpreting the measurements without including the r^{-8} term[37] With a similar ansatz, but replacing the r^{-12} term by a r^{-10} term, it is possible to predict half-width and shift of Na diffuse series lines broadened by noble gases with an accuracy of 10%–20%.[38]

Phase-shift theory, even in connection with realistic (but averaged) potentials is, however, unable to explain the effects of disorienting and energy transfer collisions on optical cross sections. In particular, the different broadening and shift behavior of alkali fine-structure components (Table 4) cannot be understood. The framework of semiclassical collision theory, outlined in Eqs. (15)–(20), has to be applied for the interpretation of these effects.

First attempts along these lines[39] assuming dipole–dipole interaction and neglecting J-mixing collisions yielded good agreement with respect to the shifts of the Na D components perturbed by Ar. The theoretical ratio of the half-widths, however, was much too low. Considerable improvement was obtained by using the realistic adiabatic molecular potentials of Bay-

Table 2. Comparison between Observed and Predicted Temperature Dependence of Shift β and Broadening γ for the Kr $\lambda 7601$ Line Perturbed by Inert Gases[36]

	β_{295}/β_{80}		γ_{295}/γ_{80}	
Perturber	Observed	Predicted	Observed	Predicted
He	2.62	2.16	1.87	1.73
Ne	0.93	0.84	1.67	1.70
Ar	2.29	1.79	2.07	1.64
Kr	1.47	1.43	1.48	1.43

Table 3. Values of C_6 from Measurements of Broadening and Shift Constants, Assuming Lennard-Jones-12-6 Interaction (Column 6) and 12-8-6 Interaction (Column 7), Respectively[a]

Line	Ref.	Transition	Perturber	C_6 (theor) (10^{-59} erg cm^6)	C_6 (exp) (12-6) potential (10^{-59} erg cm^6)	C_6 (exp) (12-8-6) potential (10^{-59} erg cm^6)
Ca 6573	82	$4s\,^1S_0-4p\,^3P_1$	He	0.37	2.0±0.2	1.03±0.2
Ca 6573	82	$4s\,^1S_0-4p\,^3P_1$	Ne	0.67	1.8±0.2	1.03±0.2
Ca 6573	82	$4s\,^1S_0-4p\,^3P_1$	Ar	2.76	10.8±0.8	2.3±0.2
Ca 6573	82	$4s\,^1S_0-4p\,^3P_1$	Kr	4.2	25±2	6.1±0.5
Ca 4227	83	$4s\,^1S_0-4p\,^1P_1$	He	3.1	3±1	1.8±0.6
Ar 6965	84	$4s[1\tfrac{1}{2}]_2-4p'[\tfrac{1}{2}]_1$	Ar	54.1	90±40	28.5±30
K 4047	85	$4s\,^2S_{1/2}-5p\,^2P_{1/2}$	Kr	381	520±150	300±90

[a] From Ref. 37.

Table 4. Broadening and Shift Constants for the Na Transitions $3^2S_{1/2}-3^2P_{1/2}$ and $3^2S_{1/2}-3^2P_{3/2}$ (in units of 10^{-9} rad s^{-1} cm^3)^{a,b}

		$2\gamma/n$		β/n	
		Experiment	Theory	Experiment	Theory
He	D_1	1.90 ± 0.1	1.90	0.038 ± 0.057	0.06
		1.54 ± 0.1	1.74	0.0 ± 0.05	0.094
	D_2	2.19 ± 0.2	1.90	0.07 ± 0.07	0.3
		1.62 ± 0.1	1.78	0.0 ± 0.05	0.043
Ne	D_1	1.14 ± 0.1	1.49	-0.62 ± 0.04	0.20
	D_2	1.44 ± 0.1	1.55	-0.64 ± 0.03	-0.21
Ar	D_1	2.77 ± 0.2	2.48	-1.42 ± 0.03	-1.93
	D_2	2.27 ± 0.2	2.14	-1.51 ± 0.08	-1.56
Xe	D_1	2.63 ± 0.2	3.57	-1.10 ± 0.08	-1.37
	D_2	2.32 ± 0.24	2.93	-1.40 ± 0.08	-1.45

^{a} From Ref. 42.
^{b} The corresponding cross sections are obtained from the relations $\sigma_r = (2\gamma/n)(1/\bar{v}_r)$, $\sigma_i = (\beta/n)(1/\bar{v}_r)$, where $\bar{v}_r$ is the average relative velocity of the colliding atoms.

lis[40] and Pascal and Vanderplanque.[41] In deriving these potentials the interaction of induced dipole of the noble gas atom with all multipoles of the alkali was included in the electrostatic part of the interaction; pseudo-potentials were used for representing the effect of the Pauli exclusion principle on overlapping electron states. A survey on the methods of *ab initio* calculations of interatomic potentials is given by Baylis in Chapter 6, Part 4, of this work. Using these potentials, the broadening and shift of the $P_{1/2}$ and $P_{3/2}$ components of the alkali resonance lines perturbed by noble gases have been calculated.[42,43] Table 4 draws the comparison with experimental data for the Na case as an example; good overall agreement is seen for both the magnitude and the detailed variation of the broadening constants. A very satisfactory feature of these calculations is that the ratio of widths shows a variation with perturbing gas closely similar to that of the experimental measurements. This ratio is >1 for He and Ne and <1 for Ar and Xe.

Resonant Interaction. Many experiments on resonant collision broadening are characterized by a three-level emitter structure

$$
\begin{array}{ll}
b & \text{---}J=0 \\
a & \text{---}J=1 \\
s & \text{---}J=0
\end{array}
$$

where the radiative transition $b \to a$ is observed, whereas radiation from the resonance transition $a \to s$ is completely trapped.

If it is assumed that the radiative decay of the emitter will be uneffected by the resonant radiation transfer process,

$$\lambda_{a,s}/v \ll (\gamma_{a,s})^{-1} \tag{25}$$

[v is the emitter velocity, $(\gamma_{as})^{-1}$ is the emitter lifetime, and $\lambda_{a,s}$ is the wavelength of the resonance line], then the half-width $2\gamma_{ab}$ of the line $b \to a$ becomes[34]

$$2\gamma_{ab} = \gamma_{b,a} + \gamma_{a,s} + 2\Gamma_{ab} \tag{26}$$

where Γ_{ab} is the collisional decay rate of the average electric dipole moment operator associated with the emitter states a and b, and $\gamma_{b,a}$ and $\gamma_{a,s}$ are the emitter radiative decay rates. There is no line shift, a result characteristic of resonant interaction. The collision width $2\Gamma_{ab}$ of the line is given by

$$2\Gamma_{ab} = \frac{\pi}{4} K_{J_a,J_s} N r_0 c f_{a,s} \lambda_{a,s} \tag{27}$$

where N, r_0, c, $f_{a,s}$ are, respectively, perturber density, classical electron radius, velocity of light, and average oscillator strength for the resonance transition; $K_{J_a J_s}$ is a numerical constant depending on the total angular momenta J_a, J_s of the states a and s. Theoretical values of K_{10} obtained in the dipole–dipole approximation by numerically solving Eq. (17) are $1.48 \pm 2\%$,[44] $1.54 \pm 2\%$,[45] $1.54 \pm ?$,[46] and $1.54 \pm 1\%$.[34] For the dipole interaction of two equal classical oscillators one obtains $K = 2/\pi$,[47] a result useful for rough estimations. Γ_{ab} is independent of v and thus of temperature, which is a characteristic of resonant interactions.

Neither a shift nor a temperature effect on the width has been observed in cases where a is the resonance state.[48] On the other hand, in cases where a is metastable both effects are observed to be present, which indicates the effectiveness of nonresonant interaction. This is to be expected, since for metastable states $f_{a,s}$ and thus the contribution of resonant interaction to the linewidth, given by Eq. (27), is small. On the basis of (27) with known values of K_{J_a,J_b}, obtained from measured values of Γ_{ab}, values of $f_{a,s}$ have been determined.[49]

For studying resonance broadening, as an alternative to the transition $b \to a$, the resonance transition $a \to s$ of an isotope may be observed if perturbed by a different isotope of the same element.[46] The emitted radiation is never trapped if the difference in resonance frequencies of both isotopes is much larger than the Doppler width but is much smaller than the inverse collision time. For resonance broadening of transitions other than $J = 0 \leftrightarrow J = 1$, see Ref. 108.

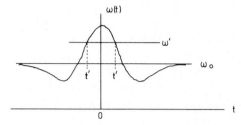

Figure 3. Time dependence of the frequency perturbation during a collision. The zero of the time scale is chosen to coincide with the time when the internuclear separation equals the distance of closest approach.

3.2. Line Wings

The *wing region* is fixed by the condition

$$\omega' - \omega_0 \gg 1/\tau_c \qquad (28)$$

In this case the integrand in (5) oscillates rapidly at all times except when $\omega(t) \approx \omega'$ during a collision (Figure 3) so that the major contribution to the intensity at ω' is very small except when $\omega(t) \approx \omega'$. The fraction of the total line intensity in the ω' to $\omega' + d\omega'$ interval then becomes the ensemble-averaged fraction of time, in which $\omega' \leqslant \omega(t) \leqslant \omega' + d\omega'$. This is the same as the probability that a perturber is between the corresponding internuclear distance R to $R + dR$. One thus obtains

$$I(\omega') \sim n(R) 4\pi R^2(\omega') \, dR(\omega')/d\omega' \qquad (29)$$

where $n(R)$ is the perturber density at R and the oscillator frequency ω' is related to one initial- and one final-state molecular adiabatic potential by

$$\omega' = [V_\Omega^i(R) - V_{\Omega'}^f(R)]/\hbar$$

or

$$\Delta\omega' = \omega' - \omega_0 = \{[V_\Omega^i(R) - V_{\Omega'}^f(R)] - [V_\Omega^i(\infty) - V_{\Omega'}^f(\infty)]\}/\hbar \qquad (30)$$

3.2.1. Quasistatic Theory

Equation (29) represents the intensity distribution in the one-particle approximation of the quasistatic theory (QST). In calculating the density $n_\Omega(R)$ of the adiabatic molecular state Ω, two cases have to be distinguished:

(a) *Nonthermal Distribution of Free Perturber States.* This situation applies in the zero-perturber density limit to an emission spectrum. In this case one has[50]

$$n_\Omega(R) = n_0 \frac{g_\Omega}{g_i} \int_0^\infty dv \, v f(v) \int_0^{b_{max}(v,R)} db \, 4\pi b v_R^{-1}(R, b)$$

$$v_R(R, b) = \{(2/\mu)[E - V_\Omega^i(R) - \mu v^2 b^2/2R^2]\}^{1/2} \qquad (31)$$

where v_R is the radial velocity component, μ is the nuclear reduced mass, $f(v)$ is the normalized velocity distribution, and g_Ω, g_i are the molecular and atomic statistical weights. If $V_\Omega^i(R)$ is repulsive, then b_{max} is the value at which the bracketed expression is zero. For a Maxwellian $f(v)$ one obtains from Eq. (31) $n_\Omega(R) = n_0 \exp\{-[V_\Omega^i(R) - V_\Omega^i(\infty)]/kT\}$. If $V_\Omega^i(R)$ is attractive, all bound states inside the rotational barriers are unpopulated ; $b_{max}(v, R)$ is then the b at which the rotational barrier height equals E.

(b) *Thermal Equilibrium Distribution of Bound and Free States*. This applies to absorption or emission spectra at high perturber densities typically > 1 atm, where the collision rates are much larger than the radiation rates. In this case one obtains for the classical distribution

$$n_\Omega(R) = n_0 \frac{g_\Omega}{g_i} \exp\{-[V_\Omega^i(R) - V_\Omega^i(\infty)]/kT\} \tag{32}$$

Quantum-mechanical evaluation of $n_\Omega(R)$ leads to a distribution which is close to the classical distribution when kT is much greater than the vibrational spacing; the quantum-mechanical $n_\Omega(R)$, however, is more spread out at repulsive barriers, and it may retain some undulations.[51]

3.2.2. Validity Criteria of the Quasistatic Theory

As shown below, the intensity distribution, Eq. (29), of the QST may be derived from the Fourier spectrum, Eq. (5), under certain assumptions. These assumptions, in addition to those made in the derivation of (5) constitute the validity conditions of the QST.

(a) *Stationary-Phase Approximation*. In deriving (29) from (5) it is assumed that only during a small interval Δt around the point t' of stationary phase (Figure 3), defined by $\omega(t') = \omega'$, does the oscillator contribute to the intensity. It may be shown that this approximation is equivalent to the Franck–Condon principle.[52] One may then expand $\omega(t)$ into a series near t'

$$\omega(t) = \omega(t') + \dot{\omega}(t')(t - t') + \ddot{\omega}(t')(t - t')^2 \tag{33}$$

and perform the time integration in (5) from $-\infty$ to $+\infty$. Formula (29) is then obtained under the additional assumption that in (33) the term of second order in $t - t'$ is small compared to the first-order term and thus negligible:[53]

$$\ddot{\omega}(t')/|\dot{\omega}(t')|^{3/2} \ll 1 \tag{34}$$

Condition (34) is not fulfilled for large t', i.e., in the line center and near extrema of $\omega(t)$ corresponding to extrema in $V_{\Omega,\Omega'}^{i,f}(R)$ for $t \neq 0$. In case of a monotonic $V_{\Omega,\Omega'}^{i,f}(R)$ condition (34) implies the existence of a limiting frequency ω_L, so that it is fulfilled in the line wing $|\omega' - \omega_0| \gg |\omega_L - \omega_0|$.

(b) *Adiabatic Approximation.* Since each atomic state n splits generally into several molecular levels Ω, they can be nonadiabatically mixed. The condition for mixing to be negligible is

$$\frac{\hbar}{\tau_c} \ll V_\Omega(R) - V_{\Omega'}(R) \equiv \Delta V_{\Omega,\Omega'}(R) \tag{35}$$

where Ω and Ω' may belong to the same or different n. Condition (35) implies, in case of monotonic $\Delta V_{\Omega,\Omega'}(R)$, the existence of another limiting frequency ω_A, different from ω_L, so that the spectral region $|\omega' - \omega_0| \gg |\omega_A - \omega_0|$ may be considered as formed by adiabatic collisions.

3.2.3. Effects of Pressure and Temperature

Two different wing regions may be distinguished: (1) Near wings (small $|\Delta\omega'|$), corresponding to large R, where $n_\Omega(R) \approx n_0 = $ const. In this region the spectrum is simply proportional to n_0 and independent of temperature. This part of the spectrum is determined exclusively by the difference potential. (2) Far wings (large $|\Delta\omega'|$), corresponding to small R, where $n_\Omega(R) \neq n_0$. Since $n_\Omega(R)$ depends in general on temperature and the

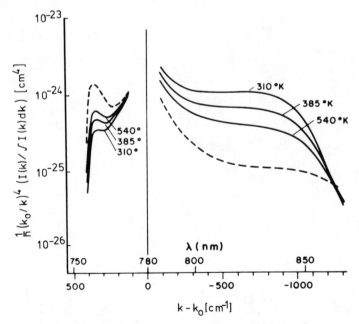

Figure 4. Normalized emission spectrum of the Rb 7800-Å D line perturbed by Kr at 10^{19} cm^{-3}. The gas temperatures in K are indicated. From Ref. 51.

initial-state potential, the same holds also for the spectrum. Furthermore, in the case of an emission spectrum, the intensity will in general no longer be proportional to n_0.

Near wings. In the case of nonresonant interaction one has $V(R) \sim R^{-6}$ and for the spectrum one obtains $I \sim \omega'^{-3/2}$. Such a $\omega'^{-3/2}$ law has been observed in the red wings of many foreign gas-broadened and weak self-broadened lines.[54] In the case of resonant interaction, on the other hand, one has $V(R) \sim R^{-3}$ and $I \sim \omega'^{-2}$. This is characteristic of, and has been observed at, self-broadened resonance lines.[55,106]

Far Wings. Figure 4 gives an example of the temperature effects on the fluorescence spectrum of Rb perturbed by Kr.[51] Opposite temperature dependencies are observed in the red and blue wing. For an explanation of this behavior, in Figure 5 the same spectrum is shown on the left and the adiabatic potentials forming the red wing are shown on the right.[50] Since the $A\,^2\pi_{3/2}$ potential is attractive in the corresponding R region, the perturber density, and thus the fluorescence, increases with temperature. The blue wing, on the other hand, is formed by the $B\,^2\Sigma_{1/2} \to X\,^2\Sigma_{1/2}$ transition, where the initial-state potential $B\,^2\Sigma_{1/2}$ (not shown in the figure), is repulsive, causing an opposite temperature effect on the spec-

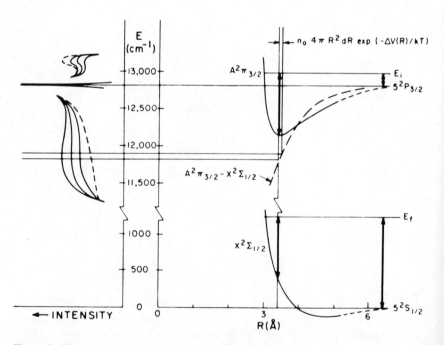

Figure 5. The model used to interpret the far-wing spectra in Figure 4, based on the Franck–Condon principle. From Ref. 50.

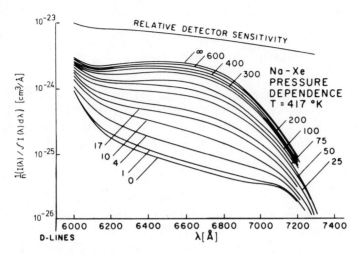

Figure 6. Normalized emission spectrum of the Na *D* lines perturbed by Xe. The Xe pressures are indicated in Torr. From Ref. 56.

trum. The blue satellite is caused by the maximum in the $B\ ^2\Sigma_{1/2} \rightarrow X\ ^2\Sigma_{1/2}$ difference potential (Section 3.3).

Figure 6 shows an example of pressure measurements at a single temperature for the Na–Xe system.[56] The dependence of the spectrum on Xe number density may be explained as follows: In the low-pressure limit the radiation is exclusively due to free pairs Na* + Xe with density $\sim n_{Xe}$; in the high-pressure limit there is an additional contribution from molecular Na*Xe, with equilibrium density $\sim n_{Xe}$. In the intermediate-pressure region, however, the teratomic recombination process Na* + 2Xe → Na*Xe + Xe competes with the free Na* radiative decay, so that the Na*Xe density is $\sim n_{Xe}^2$.

3.2.4. Determination of Potentials

It should be remarked that interaction constants derived from near wings generally refer to averaged potentials for two reasons. First, there is often more than one combination of $V_\Omega^i - V_{\Omega'}^f$ that radiates at the same ω'. Second, the various V_Ω^n can be nonadiabatically mixed if they are not separated by much more than $\hbar/\tau_c$ [Eq. (35)]. These complications in interpreting near-wing spectra are largely avoided in the small-*R* region corresponding to the far wings; the large separation of the various $V_\Omega^n(R)$ here tends spectrally to separate the radiation from the various $V_\Omega^i - V_{\Omega'}^f$ combinations and makes nonadiabatic mixing negligible. With the equilibrium perturber density distribution [Eq. (32)] the quasistatic spectrum

(29) may be rewritten in terms of potentials (using (30)), as

$$I(\omega') \sim n_0 R^2 \left(\frac{dV^{i,f}_{\Omega,\Omega'}(R)}{dR} \right)^{-1} \exp\left[-\frac{V^i_\Omega(R) - V^i_\Omega(\infty)}{kT} \right] \qquad (36)$$

On the basis of (36), measurements of the spectra at constant temperature allow a pointwise mapping of the difference potential $V^{i,f}_{\Omega,\Omega'}(R)$, and measurements of the temperature dependence of the spectra map out $V^i_\Omega(R)$. Then, since

$$V^{i,f}_{\Omega,\Omega'}(R) \equiv V^i_\Omega(R) - V^f_{\Omega'}(R)$$

$V^f_{\Omega'}(R)$ is also determined.[11,57] It should be pointed out that this method does not rely on any assumption regarding the general shape of the potentials. It is comparable and complementary to atomic-beam differential scattering cross-section studies, from which ground-state potentials have been obtained.

From absorption measurements at a single temperature of the Hg 2537 line perturbed by Ar the difference potential for the $X\,^1\Sigma \to A\,^3\Pi_0$ transition of the Hg–Ar system has been inferred,[57] using the known semiempirical $X\,^1\Sigma$ state potential of Heller.[58] On the other hand, from temperature-dependent Cs– and Rb–noble gas fluorescence spectra, both $V^i_\Omega(R)$ and $V^f_{\Omega'}(R)$ have been evaluated.[9,51] In this case, the starting separation $R_0(\omega'_0)$ entering the analysis, was fixed by comparison with atomic scattering data. Figure 7 gives as an example the potentials for the Rb–Xe system resulting from the spectra shown in Figure 4.

3.3. Satellites

In many systems one observes single intensity peaks (Figure 2) and/or an oscillatory structure (Figures 8–11) superimposed to wing continua. Such satellites may be of quite different origin: They may be formed by bound–bound or bound–free transitions, reflecting the discrete energy spacing of molecular states, or they may be connected to free–free transitions if an extremum exists in the difference potential. Finally, they may appear if, in free–free transitions, radial variations in the transition dipole moment are to be expected [collision-induced dipole absorption (emission)].

Free–Free Transitions. Typical features of satellite shapes caused by free–free transitions can be seen in the blue Rb (7800 Å)+Kr satellite (Figure 8): a central maximum, an exponential intensity decrease beyond the maximum, and undulations between the satellite and the unperturbed line.

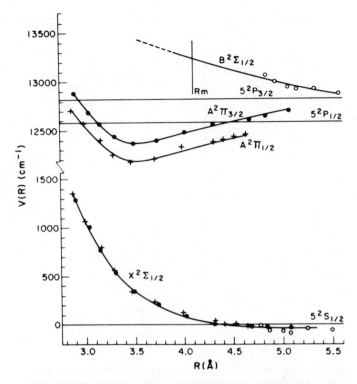

Figure 7. The Rb–Kr adiabatic potentials evaluated from the data in Figure 4. The classical satellite radius is labeled R_m. The open circle points and the $B^2\Sigma$ potential are obtained from the D_2 blue-wing analysis. For the $X^2\Sigma_{1/2}$ potential the dots are obtained from analysis of the D_2 spectrum and the pluses from the D_1 spectrum. From Ref. 51.

All these features may be qualitatively understood by Fourier analysis of the perturbed atomic oscillator. QST predicts a narrow-intensity peak at the extremum photon energy $\hbar\omega'_s = V(R_m)$, since there a large ΔR interval contributes to a narrow-frequency region [$dR/d\omega' \to \infty$ in (29)]. Because of the finite collision time, however, the actual satellite shape is broadened. The exponential decrease beyond the satellite is characteristic of Fourier analysis at a frequency that is never reached by $\omega(t)$.[59] Finally, the undulations between ω'_s and ω_0 are caused by interference between contributions to the radiation amplitude at the two radii where $V(R) = \hbar\omega'$. These undulations are also obtained from Fourier analysis [Eq. (5)] if in the stationary-phase approximation [Eq. (33)] the second-order term in $t-t'$ is retained, causing the appearance of Airy functions, which are analogous to the oscillations near the rainbow angle in the angular distribution of atomic-beam scattering (rainbow oscillations).

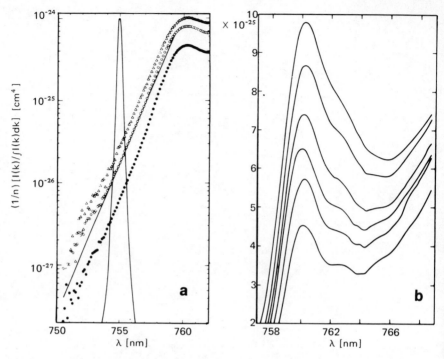

Figure 8. Normalized emission spectrum of the blue satellite of Rb 7800-Å line perturbed by 10^{19} cm^{-3} of Kr: (a) Logarithmic plot at temperatures, from top to bottom, of 740, 530, and 330 K. The spectrometer instrument function, centered at 755 nm is also shown. (b) Linear plot of the peak region at temperatures, from top to bottom, of 706, 603, 488, 423, 377, and 323 K. From Ref. 62.

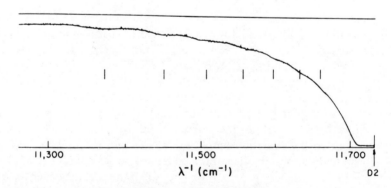

Figure 9. The measured absorption spectrum on the red wing of the D_1 line of Cs–Ar from Ref. 68. This spectrum is due to the $X^2\Sigma_{1/2}$–$A^2\Pi_{3/2}$ transition. The upper line is the transmission signal without Cs in the cell.

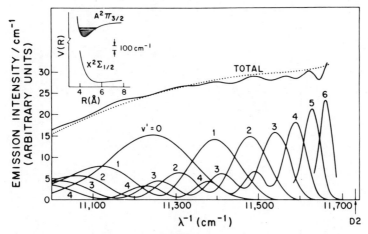

Figure 10. Calculated Cs–Ar bound–free $A^2\Pi_{3/2}$–$X^2\Sigma_{1/2}$ emission spectrum for $T = 490$ K, the potentials shown in the inset, and a constant dipole moment. Contributions to the total spectrum from individual vibrational levels (ν') after summing over bound rotational states are shown for $\nu' = 0$–6. The dots are classical spectrum from Ref. 7, using Eqs. (29) and (32). From Ref. 70.

An approximate analytical expression for satellite shapes, valid for the vicinity of ω'_s, has been derived[61,109] on the basis of semiclassical theory.*[60] If the difference potential near R_m is approximated by a quadratic, one obtains[61]

$$I(\omega', T) \sim \exp\{-[V^i_\Omega(R_m) - V^i_\Omega(\infty)]/kT\}T^{-1/6}T^\pm(U, \infty) \qquad (37)$$

Here $T^\pm(U, \infty)$ is a tabulated universal function describing the satellite shape, and U is the shift $\omega' - \omega'_s$ in reduced units. Opposite T-dependences of the satellite peak intensity I_s are predicted depending on whether $V^i_\Omega(R)$ is attractive or repulsive at the satellite radius R_m. The blue satellites of the Rb D_2 noble gas[62] and that of the Hg 2537 Ar system[63] are examples of this opposite behavior. The predicted approximately exponential change of I_s with T has also been observed.[62,63]

In addition to rainbow oscillations, oscillations in the spectra of free–free transitions may sometimes simply reflect the oscillatory structure of the excited-state wave function of internuclear motion.[64] The mechanism is very similar to that causing oscillations in bound-free spectra (see next section) except that the initial states are now continuous and must be averaged over positive energies as well summed over angular momenta. The bandlike structure observed in the continuous radiation emitted by

* Starting from Eq. (6) and using both the WKB and stationary-phase approximation.

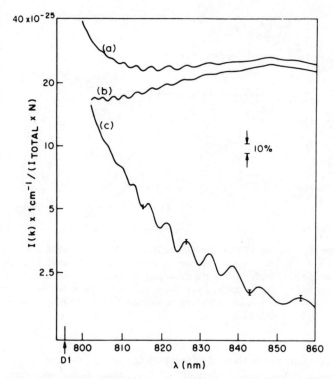

Figure 11. The free–free (c) bound–free (b), and total (equilibrium) emission spectrum (a) measured for Rb–Xe ($A^2\Pi_{1/2}$–$X^2\Sigma_{1/2}$) at 300 K. N is the Xe density in cm^{-3}. The statistical uncertainty in equilibrium and bound–free spectra is about 1%. From Ref. 70.

metastable He ($1s2s\ ^1S_0$) in collision with ground state He is an example[65] of this type of oscillation.

Satellites may also be formed as a consequence of a radial variation in the dipole transition probability A. In this case a factor $A(R)/A(\infty)$ has to be added to the right-hand side of (29). An intensity maximum may then be expected if $A(R)$ increases with decreasing nuclear separation, whereas the Boltzmann factor $\exp[-V_\Omega^i(R)/kT]$ decreases. Considerable variation of A may occur if the transition is forbidden for the separated atoms but allowed in the molecule. In addition, radial variation of A is expected due to changes in the molecular-state coupling scheme. Recently $A_\Omega(R)$ calculations have been reported of Rb $5S$–$5P$ and $5S$–$4D$ transitions perturbed by Xe, using adiabatic perturbed wave functions.[66] Good agreement was found with the experimentally observed intensity of satellites associated with the forbidden $5S$–$4D$ transitions. It was suggested that, in general, satellite bands of forbidden lines[67] may be interpreted as collision-induced absorption (emission).

Bound–Free and Quasibound–Free Transitions. An example of observed oscillatory structure mainly due to bound (rotational–vibrational) states is seen in the spectrum of the $X\ ^2\Sigma_{1/2} \to A\ ^2\Pi_{3/2}$ transition of the Cs–Ar system[68] (Figure 9), where the $X\ ^2\Sigma_{1/2}$ state is repulsive and the $A\ ^2\Pi_{3/2}$ state is attractive.

In the Franck–Condon (FC) factors $|\int \psi_{K'J'}\psi_{v''J''}\,dR|^2$ determining the spectrum the wave functions $\psi_{K'J'}$ associated with the repulsive state may be approximated by δ functions.[69] Thus, the spectral oscillations reflect to some degree the nodal structure of the rotational–vibrational wave functions $\psi_{v''J''}$ associated with the attractive state.

Figure 10 shows the result of FC factor calculations for the CsAr $A^2\Pi_{3/2} \to X\ ^2\Sigma_{1/2}$ transition,[70] where Morse potentials fitted to experimental potentials[11] have been used. By comparison with Figure 9 it is seen that amplitudes and spacings of the undulations are quite similar to those of the observed structure; differences in the positions must be attributed to uncertainties in the potentials and the neglect of contributions from quasibound states (resonances) in the calculated spectrum.[70]

Figure 11, curve (c) shows an example of oscillations exclusively due to quasibound states observed in the emission spectrum of the Rb D_1/Xe system at Xe pressures $\leqslant 10$ Torr. At these small pressures RbXe molecule formation rates, and thus bound-state densities, are negligibly small. The contribution of bound states to oscillations observed in the equilibrium spectrum at pressures > 100 Torr [curve (a)] is seen in curve (b).

3.4. High-Pressure Effects

At number densities above $\approx 10^{19}$ cm^{-3} corresponding to pressures above ≈ 1 atm, the line center becomes asymmetric: the "red" half half-width γ_r differs from the "blue" half half-width γ_b, i.e., the asymmetry ratio $a \equiv \gamma_r/\gamma_b \neq 1$. Furthermore, half-width γ, shift β, and asymmetry ratio a become, in general, nonlinear functions of the perturber density n. In particular, $a(n)-1$ and $\beta(n)$ may pass an extremum and change sign, indicating a change from attractive to repulsive interaction with decreasing average internuclear distance (Figures 12 and 13).

For the explanation of these observations classical phase-shift theory is sufficient. Assuming noninteracting and scalar additivity of perturbations, one obtains for the line profile $I(\omega')$, in terms of the correlation function $\varphi(\tau)$[33]

$$I(\omega') = \int \varphi(\tau)\exp(i\omega'\tau)\,d\tau \tag{38}$$

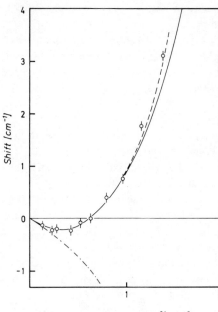

Figure 12. Comparison of theory and experiment in the pressure broadening of the Cs D_1 absorption line perturbed by Ne. ∘-∘-∘-, experiment (Ref. 74); ———, theory assuming Lennard–Jones interaction ($\alpha = 0.32$, $\kappa = 1.18$, $C_6 = 1.59 \times 10^{-58}$ erg cm^6); ---, theory assuming van der Waals interaction ($\alpha = \infty$, $\kappa = 1.7$, $C_6 = 1.59 \times 10^{-58}$ erg cm^6). (From Ref. 33.)

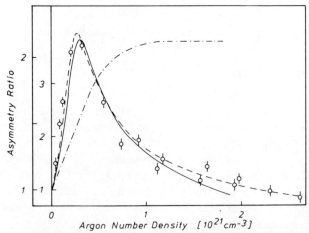

Figure 13. Comparison of theory and experiment in the pressure broadening of the Cs D_1 absorption line perturbed by Ar. ∘-∘-∘-, experiment (Ref. 74); ———, theory assuming Lennard–Jones interaction ($\alpha = 10$, $\kappa = 1.04$, $C_6 = 6.95 \times 10^{-58}$ erg cm^6); ---, theory assuming van der Waals interaction ($\alpha = \infty$, $\kappa = 1.7$, $C_6 = 6.95 \times 10^{-58}$ erg cm^6). From Ref. 33.

with

$$\varphi(\tau) = \exp\left[-n\phi(\tau)\right]$$

$$= \pi^{1/2}\left(\frac{2m}{kT}\right)^{3/2} \int dR\, R^2 \int dv_R \int dv_\phi\, v_\phi$$

$$\times \exp\left\{-\frac{m}{kT}\left[v_R^2 + v_\phi^2 + \frac{2}{m}V_i(R)\right]\right\}\left\{1 - \exp\left[-i\int_0^\tau \Delta\omega(t)\,dt\right]\right\}$$

where τ is the correlation time and R, v_R, v_ϕ are the internuclear distance and the components of the initial velocities in the collision plane. Under the above assumptions Eq. (38) is valid in the whole range of frequency and gas density.

For computation of line shapes on the basis of (38) it is useful to introduce additional simplifying assumptions, such as straight-line perturber trajectories and constant-frequency perturbation of the oscillator during the collision time.[71,72] Computations using these simplifications and assuming Lennard-Jones 6–12 interaction have been performed[73] and compared with measurements of the Cs resonance line perturbed by Ne and Ar.[74] Figures 12 and 13 demonstrate the degree of agreement with respect to $\beta(n)$ and $a(n)$ for the Cs + Ne and Cs + Ar cases, respectively.

It may be shown that the theory for general pressures [Eq. (38)] contains the impact theory as limiting case for large τ and the quasistatic theory as limiting case for small τ.[33] Approximate analytical expressions for $I(\omega')$ valid in the special case of binary collisions ($n \leq 10^{19}$ cm^{-3}) have been derived, and the dependence of the line shape parameters on the interaction potential have been investigated.[75]

3.5. Collisional Line Broadening and Depolarization

Collisional line broadening, like collisional depolarization, is a relaxation phenomenon. It may be interpreted as relaxation of the electric dipole moment $\mu(t)$ of the active atom colliding with perturbers[33]:

$$\frac{d}{d\tau}\{\langle M_i|\mu_z(\tau)|M_f\rangle\}_c = -\Gamma\{\langle M_i|\mu_z(\tau)|M_f\rangle\}_e \tag{39}$$

M_i and M_f are substates of the initial and final levels; $\{\ \}_e$ is the average over the ensemble corresponding to the time interval τ. The relaxation rate Γ is related to half-width γ and shift β of the line by

$$\Gamma = \gamma + i\beta \tag{40}$$

and the corresponding line-broadening probability, calculated on the basis of semiclassical collision theory, is given by Eq. (16). Similarly, collisional depolarization may be described in terms of relaxation of the orientation and alignment[76]:

$$\left(\frac{d\rho_M^J(t)}{dt}\right)_{\text{coll}} = -\Gamma_J\rho_M^J \tag{41}$$

In (41), $\rho_M^1(M = 0, \pm 1)$ represent the three components of orientation, which are proportional to the components of the total angular momentum; $\rho_M^2(M = 0, \pm 1, \pm 2)$ represent the five components of the alignment, which are proportional to average values of quadrupole components, and Γ_1 and Γ_2 denote the corresponding relaxation rates.

According to semiclassical theory the depolarization probability is given by[33]

$$\sigma_J(b) = 1 - \frac{1}{2J_f + 1} \sum_{\text{all } M's} C(J_fJJ_f; M_2MM_1)C(J_fJJ_f\ M_2'M_1')$$

$$\times \langle J_fM_1'|T^f|J_fM_1\rangle\langle J_fM_2|T^{f\dagger}|J_fM_2'\rangle \tag{42}$$

This expression is similar to that for the line-broadening probability [Eq. (16)]. Comparison shows that line broadening is caused by different perturbations of the initial and final levels i, f of the transition. On the other hand, collisional depolarization is possible only if the sublevels of i or f are subject to different perturbations during the collision. Depolarization of $S_{1/2}$ and $P_{1/2}$ state atoms is not possible in first order (See Baylis in Chapter 28 of this work).

Calculations using the framework of the semiclassical theory have been performed on the broadening constants as well as on the depolarization constants for alkaline-earth ion–He collisions.[86] The van der Waals potential and the exchange model potential[107] have been used in these calculations, and the latter turned out to be predominant in the processes considered. The calculated cross sections were found to be in reasonable agreement with available experiments. Similar calculations have been carried out for alkali–rare-gas collisions using the potentials of Pascale and Vandeplanque.[86]

A comprehensive presentation of the topic of collisional depolarization in excited states is given by Baylis in Chapter 28 of this work.

4. Interaction with Strong Radiation Fields

Like collisions (internal fields), coupling of an atom to an external radiation field may also give rise to a broadening and shift of energy levels, thus affecting line shapes. In addition, a splitting of the levels may occur under certain conditions. These effects, collectively called "dynamic Stark effect," were investigated many years ago by means of rf and NMR spectroscopy. The pure optical observation of the phenomena, however, was not possible until recently because of the Doppler width. This difficulty has now been overcome with the development of high-power lasers and the methods of Doppler-free spectroscopy. The theoretical background of the dynamical Stark effect is presented by Stenholm in Chapter 3 of Part A of this book.

4.1. Power Broadening

If a strong radiation field is resonant to an atomic system (i.e., the frequency of the radiation field coincides with that of an atomic transition), then both levels involved are, in addition to their natural widths, broadened due to the finite lifetimes caused by the radiation-induced transitions. This is essentially the cause of "power broadening." This broadening may be observed in a simple fluoresence experiment if the system is excited with broad-band radiation (frequency width large compared to the inverse of the radiative transition probabilities) at sufficiently large intensity (so that the power width becomes comparable to or larger than the Doppler width). The power broadening is proportional to the power of the incident radiation. It is calculated according to the "Rabi solution" (see, e.g., Refs 87 and 88).

4.2. Line Splitting: Linear Stark Effect

If the radiation field is resonant and narrow band (frequency width small compared to the inverse radiative transition probabilities) a splitting of the levels also occurs. This may be observed for one of the levels if it is probed by means of a weakly coupled transition to a third level (Figure 14). The nature of this splitting may be understood if the atom and the electromagnetic field are regarded as a single quantum system.[89] We assume a two-level atom with zero-order energies $E_1^{(0)}$ and $E_2^{(0)}$ ($E_1^{(0)} > E_2^{(0)}$) and a field consisting of N_λ photons in the definite mode λ of frequency ω. If $\hbar\omega = E_1^{(0)} - E_2^{(0)}$, the system is degenerate, since

$$E_{N,1}^{(0)} = E_1^{(0)} + N_\lambda \hbar\omega = E_{N_\lambda+1,2}^{(0)} = E_2^{(0)} + (N_\lambda + 1)\hbar\omega$$

Due to the interaction between the two states the degeneracy is lifted and the system splits into two states, the energies of which are displaced relative to the original energy by $\Delta E^{(1)} = \pm \mathbf{E}_0 \mathbf{P}_{12}/2$, $\mathbf{E}_0$ being the field amplitude and $\mathbf{P}_{12}$ the dipole transition matrix element. The transition to a third level with energy $E_3^{(0)}$ is therefore not a single line with frequency $\omega_1^{(0)} = (E_3^{(0)} - E_1^{(0)})/\hbar$. Instead it splits into two groups: The first group has frequencies $(E_3^{(0)} - E_1^{(0)} - |\Delta E^{(1)}| \pm 2k\hbar\omega)/\hbar$, corresponding to one-photon

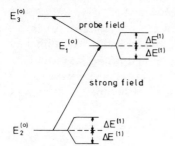

Figure 14. Level splitting of a two-level atom due to interaction with a strong resonant electromagnetic field.

($k = 0$), three-photon ($k = 1$), five-photon ($k = 2$), and so on, resonances; these frequencies are shifted by $\Delta\omega_1 = \frac{1}{2}\mathbf{E}_0\mathbf{P}_{12}/\hbar$, relative to the position of the undisplaced levels. The other group has the frequencies $(E_3^{(0)} - E_1^{(0)} + |\Delta E^{(1)}| \pm 2k\hbar\omega)/\hbar$, shifted by $\Delta\omega_1' = -\frac{1}{2}\mathbf{E}_0\mathbf{P}_{12}/\hbar$. The total splitting is thus $\mathbf{E}_0\mathbf{P}_{12}/\hbar$. The whole situation is illustrated in Figure 14 for the case of one-photon resonance.

In a field of nonresonant radiation the splitting is given by $[(E_1^{(0)} - E_2^{(0)} - \hbar\omega)^2 + \mathbf{E}_0\mathbf{P}_{12}^2]^{1/2}$, which corresponds to the Rabi nutation frequency. In rf spectroscopy this Stark splitting was observed for the first time in the spectrum of OCS by Autler–Townes[90] and is often called the Autler–Townes effect. The splitting is proportional to the amplitude of the applied electromagnetic field and is thus also denoted as the linear Stark effect.

In the optical region Doppler broadening impedes the direct observation of Stark splitting if typical powers of cw lasers are used. With a three-level system, however, where the nonlinear interaction between the two coupled transitions[91] allows a partial cancellation of the Doppler shifts, the splitting has been observed at rather small laser intensities.[92]

By use of an atomic beam for reduction of Doppler width, direct observation of the splitting has been recently achieved in the case of the Na level $3^2P_{3/2}$[93]: An intense laser radiation (laser 1) was locked to the transition 3^2S_{12}, $F = 2$–$3^2P_{3/2}$, $F = 3$, while a weak laser radiation (laser 2) probed the coupled transition $3^2P_{3/2}$, $F = 3$–$5^2S_{1/2}$, $F = 2$. Figure 15 shows the splitting of the probe transition for various intensities I_1 of the strong resonant radiation. The splitting starts at relatively low I_1. Furthermore, the predicted proportionality of the splitting to $I_1^{1/2}$ could be crudely verified. The asymmetry of the experimental doublet must be attributed to the mixing of the $3^2P_{3/2}$, $F = 3$ and $F = 2$ states by the strong field.

The Stark splitting can also be observed in the frequency distribution of fluorescent light induced by monochromatic resonant radiation. In such an experiment Doppler broadening of the scattering atoms has to be reduced by means of an atomic beam, and the fluorescence spectrum has to be studied under high resolution using a Fabry–Perot interferometer. Figure 16 shows as an example the spectrum for the Na transition $3^2P_{3/2}$, $F = 3 \rightarrow$

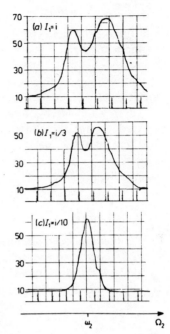

Figure 15. Experimental probe spectra, i.e., population of the Na state $5^2S_{1/2}$, $F = 2$ plotted against frequency Ω_2 of the probe laser for various intensities I_1 ($i \approx$ 0.9 W cm^{-2}) of laser 1 locked to the Na transition $3^2S_{1/2}$, $F = 2$–$3^2P_{3/2}$, $F = 3$. The frequency scale is given by pulses recorded in intervals of 37.5 MHz. From Ref. 94.

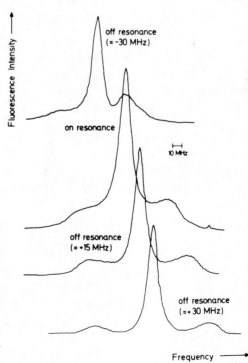

Figure 16. Fluorescence spectra for the transition $3^2P_{3/2}$, $F = 3$–$3^2S_{1/2}$, $F = 2$ of Na measured with a dye laser on resonance and with selected detunings. The laser had a bandwidth of 1 MHz and frequency drift < 1 MHz/min. The maximum excitation power was 3500 W/cm^3. From Ref. 95.

$3^2S_{1/2}$, $F = 2$.[94,95] It exhibits a three-peak structure resulting from the splitting of the upper and lower state. The ratio of the heights of the side maxima to that of the main maximum is about $1:6$, instead of $1:3$, as predicted by most of the theories for the high-power limit.[96-98] The deviation may be due to spatial inhomogeneity of the laser beam or the influence of elastic scattering.

For small laser intensities the spectral width of the fluorescence, if excited by monochromatic and resonant radiation, is expected to be smaller than the natural width.[99] This prediction has been recently supported by observation of the $6^1P_1 \rightarrow 6^1S_0$ transition of ^{138}Ba using an atomic beam.[100]

4.3. Line Shift: Quadratic Stark Effect

If the applied radiation field is nonresonant with respect to an atomic system, then, besides a splitting (in case of narrow-band radiation), a shift of the levels is also obtained. Time-independent perturbation theory gives the following shift of an atomic level induced by an optical field:

$$\mathbf{E} = \mathrm{Re}\, \mathbf{E}_0 \exp i(kz - \omega t)$$

$$\Delta E_n = \frac{1}{4} \sum_{\substack{m \\ n \neq m}} \left(\frac{E_0^2 |P_{mn}|^2}{E_n^{(0)} - E_m^{(0)} - \hbar\omega} + \frac{E_0^2 |P_{mn}|^2}{E_n^{(0)} - E_m^{(0)} + \hbar\omega} \right) \tag{43}$$

where the summation has to be taken over all atomic states $|m>$ with energy $E_m^{(0)}$. These shifts are essentially caused by non-energy-conserving virtual transitions between atomic levels induced by nonresonant light. This is analogous to the calculation of transition probabilities of multiphoton transitions. Level shifts are therefore intrinsically connected to these multiphoton processes. The level shifts are proportional to the square of the amplitude of the electromagnetic wave. It is therefore termed "quadratic Stark effect."

For a two-level atom Eq. (43) reduces to (neglecting the nonresonant term)

$$\Delta E_1 = -\Delta E_2 = \frac{1}{4} \frac{|\mathbf{E}_0 \mathbf{P}_{12}|^2}{E_1^{(0)} - E_2^{(0)} - \hbar\omega} \tag{44}$$

It follows from (44) that the sign of the shifts is determined by the sign of $E_1^{(0)} - E_2^{(0)} - \hbar\omega$. If $\hbar\omega < E_1^{(0)} - E_2^{(0)}$, the levels are shifted so that their distance is increased; if $\hbar\omega > E_1^{(0)} - E_2^{(0)}$, it is decreased.

Equation (43) is valid only in the limits of $|\mathbf{E}_0\mathbf{P}_{12}|^2/(E_n^{(0)} - E_m^{(0)} - \hbar\omega)^2 \ll 1$ and $(E_n^{(0)} - E_m^{(0)} - \hbar\omega)/\Gamma \gg 1$ (Γ is the sum of natural widths of levels n, m). For a two-level atom an expression for the level shift more general than (44) can be obtained by diagonalization of the matrix of the Hamilton operator[89]:

$$\Delta E_1 = -\Delta E_2$$
$$= \tfrac{1}{2}\{(E_1^{(0)} - E_2^{(0)} - \hbar\omega)$$
$$\pm [(E_1^{(0)} - E_2^{(0)} - \hbar\omega)^2 + |\mathbf{E}_0\mathbf{P}_{12}|^2]^{1/2}\} \tag{45}$$

From (45) it follows that for resonant radiation $(E_1^{(0)} - E_2^{(0)} = \hbar\omega)$ the level shifts are zero, and the remaining level splitting is $\mathbf{E}_0\mathbf{P}_{12}$. In the limit of $|\mathbf{E}_0\mathbf{P}_{12}|^2/(E_1^{(0)} - E_2^{(0)} - \hbar\omega)^2 \ll 1$, on the other hand, (45) reduces to (44).

The quadratic Stark effect was first observed as a small change in the nuclear resonance of optically orientated ^{199}Hg.[101] In the first optical experiment on the quadratic Stark effect[102,103] the level shift was observed by means of the change of the D-line absorption of K vapor when the atoms were irradiated with ruby laser light ($\lambda 6933$ Å), which is nearly resonant with the $4^2P_{3/2}$–$6^2S_{1/2}$ transition at 6943 Å.

In the first experiment on the Stark shift using cw tunable lasers[104] the Doppler effect was excluded by the use of double-quantum transitions. The experiment was performed with Na atoms in a cell which were excited by two lasers. The first laser had a wavelength of about 5890 Å, which is close to the $3S$–$4P$ resonance, and the other had one about 5690 Å, close to $4P$–$4D$. The cell was irradiated by the two lasers in opposite directions so that, for the induced double-quantum transition $3S$–$4D$, the Doppler width was determined by the frequency difference $\omega_1 - \omega_2$ of the two lasers (residual Doppler broadening was 62 MHz). The resonance denominators

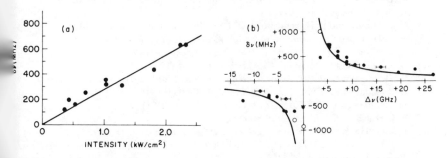

Figure 17. Energy-level shift of the $3S$, $F = 2$ level of Na versus (a) laser intensity with the two 5890-Å laser modes tuned 4.0 and 5.77 GHz from the $3S$, $F = 2$–$3P_{3/2}$ intermediate-state resonance and (b) frequency detuning from the $3S$, $F = 2$–$3P_{3/2}$ intermediate-state resonance. (The two 5890-Å laser modes have mistunings of $\Delta\nu$ and $\Delta\nu - 1.77$ GHz.) Solid circles give shifts for 5890-Å intensity of 2.15 kW/cm^2. The measurements with open circles are taken with reduced intensity and are then normalized to 2.15 kW/cm^2. The solid lines represent theoretically calculated curves using no adjustable parameters. From Ref. 104.

in (43) assured that the shifts of the $3S$ state were mainly induced by the 5890 Å light while the $4D$ state was shifted primarily by the 5890 Å light. By adjusting independently the intensities of the two lasers, either the $3S$ or the $4D$ levels was essentially shifted, and not both. The two-photon transitions were detected by monitoring the $4P$–$3S$ fluorescence which results from the decay of the $4D$ level.

Figure 17 shows the dependence of the shift of the $3S$, $F = 2$ state on light intensity and the frequency mistuning $\Delta\nu$ from the $3P_{3/2}$ intermediate-state resonance. The measurements give good agreement with theory for $|\Delta\nu| > 4$ GHz. For $|\Delta\nu| < 4$ GHz no good agreement with (43) was to be expected due to the possible breakdown of perturbation theory. It is also found that, in this region, the shifts were no longer linearly dependent on the light intensity. Furthermore, a broadening in the double-quantum transitions is observed, which is due to the Doppler effect. This has to be avoided, e.g., by use of a well-collimated atomic beam if the effects of the mistunings from the intermediate-state resonances on the shifts of the $3S$ and $4D$ sublevels are to be studied in a more quantitative manner.

References

1. H. R. Griem, *Spectral Line Broadening by Plasmas*, Academic Press, New York (1974).
2. J. T. Jeffries, *Spectral Line Formation*, Blaisdell, Waltham, Massachusetts (1968).
3. H. J. Kunze, in *Plasma Diagnostics*, W. Lochte Holtgreven, Ed., North-Holland, Amsterdam (1968).
4. F.-J. Schönes, *Habilitationsschrift*, Saarbrücken (1973).
5. H. J. Liebe and T. H. Dillon, *J. Quant. Spectrosc. Radiat. Transfer.* **11**, 1803 (1971).
6. A. Unsöld *Physik der Sternatmosphären*, 2nd. ed., Springer, Berlin (1968).
7. J. Butaux, Thesis, Paris (1972).
8. H. C. Burger and P. H. van Cittert, *Z. Phys.* **79**, 722 (1932); **81**, 428 (1933).
9. R. E. Hedges, D. L. Drummond, and A. Gallagher, *Phys. Rev. A* **6**, 1519 (1972).
10. H. G. Kuhn and J. M. Vaughan, *Proc. R. Soc. London A* **277**, 297 (1963).
11. H. J. Kusch and G. Meinold, *Z. Astrophys.* **66**, 364 (1967).
12. W. Behmemburg and H. Kohn, *J. Quant. Spectrosc. Radiat. Transfer* **4**, 163 (1964).
13. C. R. Vidal and J. Cooper, *J. Appl. Phys.* **40**, 3370 (1969).
14. M. Dahmen and H. J. Kusch, *Z. Astrophys.* **68**, 445 (1968).
15. E. Leboucher, Thesis, Paris (1970).
16. T. J. Hollander and H. P. Broida, *J. Quant. Spectrosc. Radiat. Transfer* **7**, 965 (1967).
    T. J. Hollander, B. J. Jansen, J. J. Plaat and C. Th. J. Alkemade, *J. Quant. Spectrosc. Radiat. Transfer* **10**, 1301 (1970).
17. G. Schulz and W. Stopp, *Z. Phys.* **207**, 470 (1967).
18. P. Jacquinot, in *Atomic Physics*, Vol. 4, Plenum Press, New York, (1975).
19. J. L. Hall, *IEEE J. Quant. Electr.* **QE-4**, 638 (1968).
20. J. Brochard and R. Vetter, *J. Phys.* **B7**, 315 (1974).
21. A. T. Mattick, A. Sanchez, N. A. Kurnit, and A. Javan, *Appl. Phys. Lett.* **23**, 675 (1973).
22. P. H. Lee and M. L. Skolnick, *Appl. Phys. Lett.* **10**, 303 (1967).
23. T. W. Hänsch, I. S. Shahin, and A. L. Schawlow, *Phys. Rev. Lett.* **27**, 707 (1971).
24. B. Cagnac, G. Grynberg, and F. Biraben, *J. Phys. (Paris)* **34**, 45 (1973).

25. F. Biraben, B. Cagnac, and G. Grynberg, *J. Phys. (Paris) Lett.* **36**, 41 (1975).
26. S. Y. Chen and M. Takeo, *Rev. Mod. Phys.* **29**, 20 (1957).
27. W. Weisskopf, *Z. Phys.* **75**, 287 (1932).
28. R. H. Dicke, *Phys. Rev.* **89**, 472 (1953).
29. P. R. Berman, *Appl. Phys.* **6**, 283 (1975).
30. E. Lindholm, Thesis, Uppsalla (1942).
31. P. W. Anderson, *Phys. Rev.* **76**, 647 (1949).
32. F. Schuller and B. Oksengorn, *J. Phys.* **30**, 531 (1969).
33. F. Schuller and W. Behmenburg, *Phys. Rep.* **12**, 273 (1974).
34. P. R. Berman and W. E. Lamb, Jr., *Phys. Rev.* **187**, 221 (1969).
35. F. Schuller, Thesis, Paris (1962).
36. J. M. Vaughan and G. Smith, *Phys. Rev.* **166**, 17 (1968).
37. W. R. Hindmarsh, A. N. Du Plessis, and J. M. Farr, *J. Phys. B* **3**, L5 (1970).
38. J. F. Kielkopf and R. B. Knollenberg, *Phys. Rev. A* **12**, 559 (1975).
39. F. Schuller and B. Oksengorn, *J. Phys. (Paris)* **30**, 919 (1969).
40. W. E. Baylis, *J. Chem. Phys.* **51**, 2665 (1969).
41. J. Pascal and J. Vanderplanque, *J. Chem. Phys.* **60**, 2278 (1974).
42. N. Lwin, D. G. McCartan, and E. L. Lewis, *J. Phys. B* **9**, L161 (1976).
43. E. L. Lewis, L. F. McNamara, and H. H. Michels, *Phys. Rev. A* **3**, 1939 (1971).
44. Y. A. Vdovin, and V. M. Galitskii, *Sov. Phys.-JETP* **25**, 894 (1967).
45. A. P. Kazantsev, *Sov. Phys.-JETP* **24**, 1183 (1967).
46. A. Omont and J. Meunier, *Phys. Rev.* **169**, 92 (1968).
47. V. Weisskopf, *Phys. Z.* **34**, 1 (1933).
48. H. G. Kuhn and E. L. Lewis, *Proc. R. Soc. London A* **299**, 423 (1967).
49. J. M. Vaughan, *Phys. Lett.* **21**, 153 (1966); *Phys. Rev.* **166**, 13 (1968).
50. A. Gallagher, in *Atomic Physics*, Vol. 4, Plenum Press, New York (1975).
51. D. L. Drummond and A. Gallagher, *J. Chem. Phys.* **60**, 3426 (1974).
52. A. Jablonski, *Phys. Rev.* **68**, 78 (1945).
53. I. I. Sobelman, *Fortschr. Phys.* **5**, 175 (1957).
54. H. G. Kuhn, *Proc. R. Soc. London A* **158**, 212 (1937). K. Niemax and G. Pichler, *J. Phys. B* **8**, 2718 (1975).
55. D. Gebhard and W. Behmenburg, *Z. Naturforsch.* **30a**, 445 (1974).
56. G. York, R. Scheps, and A. Gallagher, *J. Chem. Phys.* **63**, 1052 (1975).
57. W. Behmenburg, *Z. Naturforsch.* **27a**, 31 (1972).
58. R. Heller, *J. Chem. Phys.* **9**, 151 (1941).
59. T. Holstein, *Phys. Rev.* **79**, 744 (1950).
60. A. Jablonski, *Phys. Rev.* **68**, 78 (1945).
61. Kenneth M. Sando and J. C. Wormhoudt, *Phys. Rev. A* **7**, 1889 (1973).
62. C. G. Carrington and A. Gallagher, *Phys. Rev. A* **10**, 1464 (1974).
63. C. Petzold, Diplomarbeit, Düsseldorf 1977, unpublished.
64. F. H. Miles, *J. Chem. Phys.* **48**, 482 (1968).
65. Y. Tanaky and K. Yoshino, *J. Chem. Phys.* **39**, 3081 (1963).
66. J. Granier, R. Granier, and F. Schuller, *J. Quant. Spectrosc. Radiat. Transfer* **15**, 619 (1975).
67. R. Granier, *Ann. Phys. (Paris)* **4**, 383 (1969).
68. C. L. Chen and A. V. Phelps, *Phys. Rev. A* **7**, 470 (1973).
69. J. G. Winans and E. C. G. Stuckelberg, *Proc. Natl. Acad. Am.* **14**, 867 (1928).
70. C. G. Carrington, D. Drummond, A Gallagher, and A. V. Phelps, *Chem. Phys. Lett.* **22**, 511 (1973).
71. P. W. Anderson and J. D. Talmann, Proceedings of the Conference on Broadening of Spectral Lines, p. 29 (1956).
72. E. Lindholm, *Ark. Mat. Astr. Fys.* **32A**, 1 (1945).

73. W. Behmemburg, *Z. Astrophys.* **69**, 368 (1968).
74. S. Y. Chen and R. O. Garrett, *Phys. Rev.* **144**, 59 (1966). S. Y. Chen, R. O. Garret, and E. C. Looi, *ibid.* **156**, 48 (1967).
75. E. Czuchaj and J. Fuitak, *Acta Phys. Polon. A* **40**, 165 (1971).
76. A. Omont, *J. Phys.* **26**, 26 (1965).
77. Ch. Ottinger, R. Scheps, G. W. York, and A. Gallagher, *Phys. Rev. A* **11**, 1815 (1975).
78. J. Butaux and R. Lennuier, *C.R. Acad. Sci. (Paris)* **261**, 671 (1965).
79. J. Butaux, F. Schuller, and R. Lennuier, *J. Phys.* **33**, 635 (1972); **35**, 361 (1974).
80. F. Rostas and J. L. Lemaire, *J. Phys. B* **4**, 555 (1971).
81. D. C. McCartan and W. R. Hindmarsh, *J. Phys. B* **2**, 1396 (1969).
82. G. Smith, *Proc. R. Soc. London A* **297**, 288 (1967).
83. W. R. Hindmarsh, *Mon. Not. R. Astron. Soc.* **119**, 11 (1959).
84. W. R. Hindmarsh and K. A. Thomas, *Proc. Phys. Soc.* **77**, 1193 (1961).
85. D. G. McCartan and W. R. Hindmarsh, *J. Phys. B* **2**, 1395 (1969).
86. A. Guisti-Suzor and E. Roueff, *J. Phys. B* **8**, 2708 (1975); *ibid.*, **8**, L429 (1975).
87. L. Allen and J. H. Eberly, *Optical Resonance and Two-Level Atoms*, Wiley, New York (1975).
88. M. Sargent, M. O. Scully, and W. E. Lamb, Jr., *Laser Physics*, Addison-Wesley, Reading, Massachusetts (1975).
89. A. M. Bonch-Bruevich and V. A. Khodovoi, *Sov. Phys. Usp.* **10**, 637 (1968).
90. S. H. Autler and C. H. Townes, *Phys. Rev.* **100**, 703 (1955).
91. P. Hänsch and P. Toschek, *Z. Phys.* **236**, 213 (1970).
92. A. Schabert, R. Keil, and P. E. Toschek, *Appl. Phys.* **6**, 181 (1975).
93. J. L. Picqué and J. Pinard, *J. Phys. B* **9**, L77 (1976).
94. F. Schuda, C. R. Stroud Jr., and M. Herscher, *J. Phys. B* **7**, L198 (1974).
95. R. Schieder, W. Hartig, V. Wilke, and H. Walther, in *Proceedings of the Second International Laser Spectroscopy Conference*, Megève, 1975, Springer, New York (1975).
96. B. R. Mollow, *Phys. Rev.* **188**, 1969 (1969).
97. H. J. Carmichael and D. F. Walls, *J. Phys. B* **8**, L77 (1975).
98. C. Cohen-Tannoudji, in *Proceedings of the Second International Laser Spectroscopy Conference*, Megève, 1975, Springer, New York (1975).
99. W. Heitler, *Quantum Theory of Radiation*, 3rd ed., Oxford University Press, London (1964).
100. R. Schieder, W. Råsmussen, W. Hartig, and H. Walther, *Z. Phys. A* **278**, 205 (1976).
101. C. Cohen-Tannoudji, *Ann. Phys. (Paris)*, **7**, 423, 469 (1962).
102. E. B. Aleksandrov, A. M. Bonch-Bruevich, N. N. Kosten, and V. A. Khodovoi, *JETP Lett.* **3**, 53 (1966).
103. A. M. Bonch-Bruevich, N. N. Kosten, and V. A. Khodovoi, *JETP Lett.* **3**, 279 (1966).
104. P. F. Liao and J. E. Bjorkholm, *Phys. Rev. Lett.* **34**, 1 (1975).
105. D. G. McCartan and W. R. Hindmarsh, *J. Phys. B* **2**, 1396 (1969).
106. K. Niemax and G. Pichler, *J. Phys. B* **7**, 1204 (1974).
107. E. Roueff, *Astron. Astrophys.* **7**, 4 (1970).
108. C. G. Carrington, D. N. Stacey, and J. Cooper, *J. Phys. B* **6**, 417 (1973).
109. J. Szudy, and W. E. Baylis, *J. Quant. Spectrosc. Radiat. Transfer* **15**, 641 (1975).

28

Collisional Depolarization in the Excited State

W. E. BAYLIS

1. Introduction: Purpose and Scope

The field of collisional depolarization in excited states has reached a plateau of maturity: the basic experimental results and their theoretical explanation have been well-established. For example, it is known that in the first excited states of the heavier alkalis, disorientation in the $P_{1/2}$ state is considerably slower than in the $P_{3/2}$ state, and further that the $P_{1/2}$ disorientation rate tends to be smaller when the fine-structure splitting is larger. This is now understood as the result of time-reversal symmetry, which requires the $P_{1/2}$ disorientation to proceed by virtual transitions to another (the $P_{3/2}$) state. In addition, full quantal calculations, performed with realistic potentials, give quantitatively good agreement with measurements obtained, leaving little doubt that depolarization phenomena are fundamentally understood.

Future experimental and theoretical work will still be usefully performed in the field, but will concentrate on filling in details or in extending the work to new areas. The foundation has been firmly laid.

The study of depolarization is useful to a much broader group than those pursuing work within the field, however, essentially because depolarization is the simplest of all inelastic collision processes. Theoretical and experimental techniques largely developed for the study of atomic polarization find wide application elsewhere. In particular density matrix

W. E. BAYLIS • Department of Physics, University of Windsor, Windsor, Ontario, Canada N9B 3P4. On sabbatical leave 1976–1977 at the Max-Planck-Institut für Strömungsforschung, D 3400 Göttingen, West Germany.

methods with expansions in irreducible tensor operators can well be introduced and illustrated in the context of collisional depolarization. Both simple and sophisticated examples of their application are readily available.

In view of the status of collisional depolarization, the emphasis in this chapter is on well-established effects, at least as seen from my admittedly biased standpoint. Simple, illustrative applications of density matrices and irreducible tensor operators have been favored and generally employ excited alkali atoms colliding with noble-gas perturbers as model systems.

The discussion is restricted mainly to effects of atom–atom collisions on optically induced polarization in the excited state. Many aspects of collisional depolarization omitted here can be found elsewhere in the book. In particular, coherence transfer and collisionally induced fine-structure transitions are handled by Elbel (Chapter 29); time-resolved coherence and quantum beats, including beam-foil excitation, by Dodd and Series (Chapter 14); ground-state polarization, by Kluge (Chapter 17); collisional effects in laser spectroscopy, by Stenholm (Chapter 3); and the relation of excited-state polarization to line broadening, by Behmenburg (Chapter 27).

Many experimental and theoretical details, such as descriptions of apparatus or messy derivations have been omitted or greatly abbreviated. Let me apologize to all whose work has thus been slighted. More on experimental methods may be found in the review articles by Krause[1,2] (mainly sensitized fluorescence, energy transfer, and quenching), Alkemade and Zeegers[3] (flame techniques), and Callear and Lambert[4] (collisions of excited atoms with molecules). Many theoretical details can be found in Omont's excellent review[5] about irreducible components of the density matrix and their application to optical pumping and in references therein.

2. Characterizing the Excited State

2.1. Polarization and Multipoles

In order to discuss the effect of collisions on an ensemble of atoms, a means is needed of characterizing the states of atoms in the ensemble. Sometimes it is sufficient to give simply the probability of finding an atom in any one of its basis states. If n distinguishable eigenstates are accessible to the atoms, an n-dimensional vector of probabilities would then satisfactorily describe the ensemble. Although more information must generally be given (see below), the situation in which only relative populations or occupation probabilities are needed serves as a simple example with which important concepts can be introduced.

Consider the nontrivial case of atoms with total angular momentum $j = \frac{3}{2}$, say, an alkali atom in a $P_{3/2}$ level, undergoing collisions with perturbers, say, noble-gas atoms. If we need to keep track of only the occupation probabilities of the substates $|m\rangle$, $m = \pm\frac{1}{2}, \pm\frac{3}{2}$, then the four-dimensional vector

$$\mathbf{N} = \begin{pmatrix} N_{3/2} \\ N_{1/2} \\ N_{-1/2} \\ N_{-3/2} \end{pmatrix} \tag{1}$$

is adequate, where N_m is the density of atoms in the state $|m\rangle$. The time evolution of the ensemble is governed by a source vector $\mathscr{S}$, a collisional relaxation matrix or dyad $\boldsymbol{\gamma}$, and, if the $j = \frac{3}{2}$ level is not the ground state, a natural decay time Γ^{-1}. The components $\mathscr{S}_m$ of $\mathscr{S}$ give the rate at which (alkali) atoms are excited per cm^3 into sublevel $|m\rangle$; the elements $-\gamma_{mm'}$ give the average rate at which an (alkali) atom in the sublevel $|m'\rangle$ makes a collisionally induced transition to $|m\rangle$. The net rate of change of the density vector $\mathbf{N}$ is thus

$$\dot{\mathbf{N}} \equiv \frac{d\mathbf{N}}{dt} = \mathscr{S} - \Gamma\mathbf{N} - \boldsymbol{\gamma} \cdot \mathbf{N} \tag{2}$$

The vectors can of course be expanded as a linear combination of the Cartesian basis vectors

$$\begin{pmatrix} 1 \\ 0 \\ 0 \\ 0 \end{pmatrix}, \quad \begin{pmatrix} 0 \\ 1 \\ 0 \\ 0 \end{pmatrix}, \quad \begin{pmatrix} 0 \\ 0 \\ 1 \\ 0 \end{pmatrix}, \quad \begin{pmatrix} 0 \\ 0 \\ 0 \\ 1 \end{pmatrix}$$

However, a basis with which possible symmetries in the system can be better exploited is the "spherical" basis of vectors

$$\hat{\mathbf{T}}_0 = \frac{1}{2} \qquad \hat{\mathbf{T}}_1 = \frac{1}{2 \cdot 5^{1/2}} \begin{pmatrix} 3 \\ 1 \\ -1 \\ -3 \end{pmatrix},$$

$$\hat{\mathbf{T}}_2 = \frac{1}{2} \begin{pmatrix} 1 \\ -1 \\ -1 \\ 1 \end{pmatrix}, \quad \hat{\mathbf{T}}_3 = \frac{1}{2 \cdot 5^{1/2}} \begin{pmatrix} 1 \\ -3 \\ 3 \\ -1 \end{pmatrix} \tag{3}$$

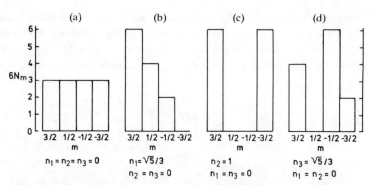

Figure 1. Excited-state multipoles in the population representation. Part (a) shows state densities when only the monopole n_0 ("occupation") component is present. Parts (b)–(d) show densities N_m when, in addition to the monopole component $n_0 = 1$, also a dipole (n_1), quadrupole (n_2), or octupole (n_3) component is present.

It is trivial to verify that the $\hat{\boldsymbol{T}}_i$ are real $(\hat{\boldsymbol{T}}_i = \hat{\boldsymbol{T}}_i^*)$ and orthonormal $(\hat{\boldsymbol{T}}_i^* \cdot \hat{\boldsymbol{T}}_j = \delta_{ij})$. In the expansion of $\mathbf{N}$

$$\mathbf{N} = \sum_{L=0}^{3} n_L \hat{\mathbf{T}}_L \tag{4}$$

the coefficients $n_L \equiv (\hat{\mathbf{T}}_L \cdot \mathbf{N})$ contain all the necessary information about the system. In fact, n_L represents the $2L$th multipole moment of $\mathbf{N}$. See Figure 1. Thus for $L = 0$,

$$n_0 = \tfrac{1}{2}(N_{3/2} + N_{1/2} + N_{-1/2} + N_{-3/2}) \tag{5}$$

is proportional to the total density of $j = \tfrac{3}{2}$ atoms. It is called the "occupation" (or monopole) component of N. For $L = 1$,

$$n_1 = \tfrac{1}{2}5^{-1/2}(3N_{3/2} + N_{1/2} - N_{-1/2} - 3N_{-3/2})$$

$$= 5^{-1/2}\sum_m mN_m = 2n_0\langle J_z\rangle/5^{1/2} \tag{6}$$

is proportional to the magnetic dipole density in the sample and is known as the "orientation" (or dipole) component of $\mathbf{N}$. Similarly

$$n_2 = \tfrac{1}{6}\sum_m [3m^2 - j(j+1)]N_m = \tfrac{1}{3}n_0\langle 3J_z^2 - J^2\rangle \tag{7}$$

is the "alignment" (or quadrupole) component of $\mathbf{N}$ and

$$n_3 = \frac{1}{3\cdot 5^{1/2}}\sum_m m[5m^2 - 1 - 3j(j+1)]N_m = \frac{2n_0}{3\cdot 5^{1/2}}\langle J_z(J_z^2 - 1 - 3J^2)\rangle \tag{8}$$

is the octupole moment of **N**. The source vector $\mathscr{S}$ has an analogous expansion in the basis $\{\hat{\mathbf{T}}_L\}$.

If the energy separation of the sublevels $|m\rangle$ is small ("nearly degenerate") compared to the thermal energy kT of the system, then the rates $\gamma_{mm'}$ and $\gamma_{m'm}$ are equal, and the only net effect of collisions is to make the densities N_m more nearly equal and thus to decrease all moments n_L with $L > 0$. It is possible to invent collision cross sections which can cause one n_L to *increase* as a higher moment relaxes. (For example, if $m = \frac{1}{2} \leftrightarrow m = \frac{3}{2}$ and $m = -\frac{1}{2} \leftrightarrow m = -\frac{3}{2}$ transitions dominate, an octupole moment n_3 will relax to an n_1 moment.) If the collisions are isotropically distributed, however, i.e., if there is no preferred orientation of the collision plane, then the relaxation metrix γ is invariant under rotations (see Section 2), and the multipoles n_L, $L > 0$, simply decrease while n_0 remains constant. Thus the time evolution [Eq. (2)] results in *independent* relaxation of each of the components n_L:

$$\dot{n}_L = \mathscr{S}_L - \Gamma n_L - \gamma_L n_L \tag{9}$$

with

$$\mathscr{S}_L \equiv \hat{\mathbf{T}}_L \cdot \mathscr{S} \quad \text{and} \quad \hat{\mathbf{T}}_L \cdot \gamma \cdot \hat{\mathbf{T}}_{L'} = \delta_{LL'} \gamma_L, \ \gamma_0 = 0$$

For constant $\mathscr{S}_L$, Eq. (9) has a trivial solution

$$n_L(t) = n_L(\infty) + [n_L(0) - n_L(\infty)] \exp\left[-(\Gamma + \gamma_L)t\right] \tag{10}$$

with the steady-state value

$$n_L(\infty) = \mathscr{S}_L / (\Gamma + \gamma_L) \tag{11}$$

2.2. Liouville Space and the Density Matrix ρ

The population representation of the excited ensemble used in the above example is not generally applicable. In order to represent the same physical ensemble in terms of states defined with a different axis of quanization, for example, a density vector such as that in Eq. (1) is not sufficient. To handle atoms with n distinguishable accessible states generally, one uses instead an $n \times n$ density matrix ρ. Of course the matrix may be rewritten as a vector, but as a vector in a new space ("Liouville space") of n^2 dimensions,[7] and the spherical basis vectors in this space have as components matrix elements of irreducible tensor operators $\mathbf{T}_{LM}$ (see Chapter 2, and the recent review by Omont[5]). The orthonormality of the basis vectors in Liouville space is equivalent to the relation

$$\text{Tr}\left[\mathbf{T}_{LM}^{\dagger} \mathbf{T}_{L'M'}\right] \equiv \mathbf{T}_{LM}^{*} \cdot \mathbf{T}_{L'M'} = \delta_{LL'} \delta_{MM'} \tag{12}$$

hey are futhermore so defined that they behave under rotation like spherical harmonics[6] Y_{LM}. As in the example above, the multipoles ρ_{LM} of the system are the coefficients for the expansion of the density in terms of the T_{LM}:

$$\rho = \sum_{LM} \rho_{LM} T_{LM} \tag{13a}$$

$$\rho_{LM} = \text{Tr }(T_{LM}^{\dagger}\rho) \equiv T_{LM}^{*} \cdot \rho = \langle T_{LM}^{\dagger} \rangle \text{ Tr }\rho \tag{13b}$$

where $\langle T_{LM}^{\dagger} \rangle$ is the average of $T_{LM}^{\dagger}$ over the ensemble described by ρ.[8] If the density matrix is normalized, $\text{Tr }(\rho) = 1$. It is often more convenient to take $\text{Tr }\rho = N$, the number of atoms per cm³.

Isotropic collisions in a nearly degenerate multiplet of states causes all ρ_{LM} to relax independently with rates γ_L which do not depend on M. The special case considered in Section 2.1 (only relative populations were used) occurs when all off-diagonal elements of ρ are zero or may be neglected (see Section 2.6 for a discussion of when it is justified to neglect them); then only the $M = 0$ components ρ_{LM} need be considered. The physical meaning and nomenclature for the multipoles ρ_{LM} is in any case similar to that for n_L. Thus a spherically symmetric system is unpolarized and has only an occupation ($L = 0$, $M = 0$) component of the density matrix. If any ρ_{LM} is nonzero for $L > 0$, we say the system is "polarized." If $\rho_{1M} \neq 0$, it is "oriented"; if $\rho_{2M} \neq 0$, it is "aligned." A system which is cylindrically symmetric about the quantization axis has only $M = 0$ multipole components. A system with $M \neq 0$ components is not cylindrically symmetric and is said to possess "coherence" (see Section 2.5). For the case of

Table 1. Multipoles of the Density for Systems with a Single Angular-Momentum Quantum Number j

L	M	$T_{LM}(\rho_{LM} = \langle T_{LM}^{\dagger} \rangle)^a$	Name	$C_L(j)$
0	0	$C_0(j)$	Population (monopole)	$(2j+1)^{-1/2}$
1	0	$C_1(j)J_0$		
1	±1	$C_1(j)J_{\pm 1}$	Orientation (dipole)	$2\left[\dfrac{3(2j-1)!}{(2j+2)!}\right]^{1/2}$
2	0	$\frac{1}{2}C_2(j)(3J_0^2 - J^2)$		
2	±1	$\frac{1}{3}3^{1/2}C_2(j)(J_{\pm 1}J_0 + J_0 J_{\pm 1})$	Alignment (quadrupole)	$4\left[\dfrac{5(2j-2)!}{(2j+3)!}\right]^{1/2}$
2	±2	$((\frac{3}{2}))^{1/2}C_2(j)J_{\pm 1}^2$		
3	0	$\frac{1}{2}C_3(j)J_0(5J_0^2 - 3J^2 + 1)$		
3	±1	$\frac{1}{4}((\frac{3}{2}))^{1/2}C_3(j)[5(J_0^2 J_{\pm 1} + J_{\pm}J_0^2) - 2J^2 J_{\pm 1} - J_{\pm 1}]$	(octupole)	$8\left[\dfrac{7(2j-3)!}{(2j+4)!}\right]^{1/2}$
3	±2	$\frac{1}{2}((\frac{15}{2}))^{1/2}C_3(j)(J_0 J_{\pm 1}^2 + J_{\pm 1}^2 J_0)$		
3	±3	$((\frac{5}{2}))^{1/2}C_3(j)J_{\pm 1}^3$		

$^a J_0 \equiv J_z$, $J_{\pm 1} \equiv \mp (J_x \pm iJ_y)/2^{1/2}$.

a manifold with a total angular momentum j, the tensors $\mathbf{T}_{LM}(jj)$ and their average values ρ_{LM} can be given in terms of the angular momentum operators $\mathbf{J}_0 \equiv \mathbf{J}_z$ and $\mathbf{J}_{\pm 1} \equiv \mp 2^{-1/2}(\mathbf{J}_x \pm i\mathbf{J}_y)$ as in Table 1.

If states of more than one value of j are considered, then somewhat more general irreducible tensors $\mathbf{T}_{LM}(j'j)$, with matrix elements $\langle j'm | T_{LM}(j'j) | jm \rangle$ as defined in Chapter 2 must be used. The presence of nuclear spin I further complicates the situation. The dimensionality of Liouville space increases by a factor of $(2I+1)^2$, and the various j multipoles $\rho_{LM}(jj')$ are coupled through the hyperfine term in the Hamiltonian. See Section 3.6 of the present chapter for details.

2.3. Time Evolution of ρ due to Radiation and Collisions

The time development of ρ is similar to that given in Eq. (2) for $\mathbf{N}$. Formally in the *Schrödinger picture* (see Chapter 2)

$$i\hbar\dot{\rho} = [\mathbf{H}_0, \rho] + [\mathbf{V}, \rho] \tag{14}$$

where $\mathbf{H}_0$ is the unperturbed Hamiltonian and $\mathbf{V}$ is the total perturbation, including collisions and interaction with both static and radiation fields. The stationary-state time dependence is removed from ρ and transferred to $\mathbf{V}$ in the *interaction picture*:

$$\rho_{\text{int}} \equiv e^{i\mathbf{H}_0 t/\hbar} \rho \, e^{-i\mathbf{H}_0 t/\hbar} \tag{15}$$

$$\mathbf{V}(t)_{\text{int}} \equiv e^{i\mathbf{H}_0 t/\hbar} \mathbf{V} \, e^{-i\mathbf{H}_0 t/\hbar} \tag{16}$$

Substitution gives

$$i\hbar\dot{\rho}_{\text{int}} = [\mathbf{V}(t)_{\text{int}}, \rho_{\text{int}}] \tag{17}$$

First consider only the radiative perturbation in $\mathbf{V}(t)$.[9-11] To handle the effects of radiation in the excitation and radiative decay of the excited-state ensemble, one usually applies perturbation theory. First one integrates Eq. (17) and substitutes it back into itself to obtain (we omit the subscript "int" for the sake of notational sanity) in the interaction picture:

$$\hbar^2\dot{\rho} = \int_0^t dt' [\mathbf{V}(t)\rho(t')\mathbf{V}(t') + \mathbf{V}(t')\rho(t')\mathbf{V}(t)]$$

$$- \int_0^t dt' [\mathbf{V}(t)\mathbf{V}(t')\rho(t') + \rho(t')\mathbf{V}(t')\mathbf{V}(t)] \tag{18}$$

A term $-i\hbar[\mathbf{V}(t), \rho(0)]$ has been dropped on the grounds that it oscillates with roughly the frequency of the radiation and thus averages to zero over the time scale of any practical observation. In the spirit of perturbation theory, we replace $\rho(t')$ on the right-hand side of Eq. (18) by $\rho(t)$ and write the relation in the suggestive form for excited-state density matrix elements,

$$\dot{\rho} = \mathscr{S} - \Gamma \cdot \rho(t) + \text{collision terms} + \text{static-field terms} \qquad (19)$$

where the "source term" is (interaction picture)

$$\hbar^2 \mathscr{S} \equiv \int_0^t dt' [V(t)\rho(t)V(t') + V(t')\rho(t)V(t)] \qquad (20)$$

and the "radiative decay" term is (interaction picture)

$$\hbar^2 \Gamma \cdot \rho(t) \equiv \int_0^t dt' [V(t)V(t')\rho(t) + \rho(t)V(t')V(t)] \qquad (21)$$

For the usual case of electric dipole radiation and a lower state with one photon of angular frequency ω and polarization $\hat{\varepsilon}$ per unit volume, the matrix element of $V(t)$ between an upper ("excited") state $|m\rangle$ and a lower ("ground") state $|\mu\rangle$ is (in the interaction picture)[9,11]

$$\langle m|V(t)|\mu\rangle = i\, e^{i(\omega_{m\mu} - \omega)t} (2\pi\hbar\omega_{m\mu})^{1/2} \langle m|\mathbf{d} \cdot \hat{\varepsilon}|\mu\rangle$$
$$= \langle \mu|V(t)|m\rangle^* \equiv e^{i(\omega_{m\mu} - \omega)t} V_{m\mu} \qquad (22)$$

where $\mathbf{d} = \Sigma_i e\, \mathbf{r}$, summed over all electrons of the atom, is the electric dipole operator, and $\hbar\omega_{m\mu}$ is the unperturbed energy separation of the states. The principal contribution to the source term $\mathscr{S}$ [Eq. (20)] is assumed to come from ground-state matrix elements of ρ.

Substitution of Eq. (22) in the expression (21) for Γ gives, after the sum over all final photon states is performed and the ground-state energy separations $\hbar\omega_{\mu\mu'}$ are ignored compared to $\hbar\omega_{n\mu}$ or $\hbar\omega_{m\mu}$,

$$\Gamma_{nm}^{n'm'} = \delta_n^{n'} \delta_m^{m'} (\Gamma_n + \Gamma_m)/2 \qquad (23)$$

Here the states $|m\rangle$, $|n\rangle$ are assumed to be angular momentum eigenstates. Also the notation used is such that

$$(\Gamma \cdot \rho)_{mn} = \sum_{m'n'} \Gamma_{mn}^{m'n'} \rho_{m'n'}$$

and $\Gamma_n \equiv \Gamma_{nn}^{nn}$ is the spontaneous decay rate for level $|n\rangle$:

$$\Gamma_n = \frac{2}{3} \frac{1}{\hbar c^3} \sum_\mu (2j_\mu + 1)\omega_{n\mu}^3 |d_{\mu n}|^2$$
$$= \frac{e^2}{mc^3} \sum_\mu \omega_{n\mu}^2 f_{\mu n} \qquad (24)$$

where the reduced matrix element $d_{\mu n} = \langle \mu \| \mathbf{d} \| n \rangle$ is as defined in Chapter and $f_{\mu n} = \frac{2}{3}(m/\hbar e^2)(2j_\mu + 1)\omega_{n\mu}|d_{\mu n}|^2$ is the emission oscillator strength or number. One sees that Γ is diagonal in Liouville space, i.e., no excited states are mixed by radiative decay.

The source term $\mathscr{S}$ [Eq. (20)] is found similarly. After a time large compared to $\omega_{m\mu}^{-1}$ the interaction picture gives

$$\mathscr{S}_{mn} = 2\pi \sum_{\substack{\text{photon} \\ \text{states}}} \sum_{\mu,\nu} V_{m\mu}\rho_{\mu\nu}V_{\nu n} \, e^{i(\omega_{mn}-\omega_{\mu\nu})t}[\delta_+(\omega_{n\nu}-\omega)+\delta_-(\omega_{m\mu}-\omega)]$$

(25)

where $\delta_\pm$ are related to the Dirac δ function and principal value P by[9]

$$2\pi\delta_\pm(x) = \pi\delta(x) \pm iP(1/x)$$ (26)

and no Doppler broadening has been included. If the exciting light is nearly "white" (i.e., nearly of constant intensity per unit frequency) over the range of resonant frequencies, then the summation over initial photon states in Eq. (25) yields (interaction picture)

$$\mathscr{S}_{mn} = (2\pi)^2 \sum_{\mu\nu\hat{\varepsilon}} u_{\hat{\varepsilon}} \langle m|\mathbf{d}\cdot\hat{\varepsilon}|\mu\rangle \rho_{\mu\nu} \langle \nu|\mathbf{d}\cdot\hat{\varepsilon}^*|n\rangle \, e^{i(\omega_{mn}-\omega_{\mu\nu})t}$$ (27)

where $u_{\hat{\varepsilon}}$ is the energy per unit volume per unit photon energy of exciting radiation of polarization $\hat{\varepsilon}$. {The expression in the Schrödinger picture is the same except that the exponential factor $\exp[i(\omega_{mn}-\omega_{\mu\nu})t]$ is dropped.}

To handle *collisional* perturbations, we solve Eq. (17) for a single collision and then average the effects over all collisions. The formal solution to Eq. (17) is (interaction picture)

$$\boldsymbol{\rho}(t) = \exp\left[-i\hbar^{-1}\int_{t_0}^t dt' \, \mathbf{V}(t')\right]\boldsymbol{\rho}(t_0)\exp\left[i\hbar^{-1}\int_{t_0}^t dt' \, \mathbf{V}(t')\right]$$ (28)

as is readily verified by differentiation. The exponential operators are time-ordered and the operators (or matrices) $\mathbf{V}(t)$, $\mathbf{V}(t')$ generally do not commute when $t \neq t'$ (see Chapter 6). The effect of a *single* collision is to change $\boldsymbol{\rho}$ into $\mathbf{S}\boldsymbol{\rho}\mathbf{S}^\dagger$, where the $\mathbf{S}$ matrix is

$$\mathbf{S} = \exp\left[-i\hbar^{-1}\int_{-\infty}^{\infty} dt' \, \mathbf{V}(t')\right]$$ (29)

and $\mathbf{V}$ is now the perturbation due to a single collision. The change in $\boldsymbol{\rho}$ is thus[12-14]

$$\boldsymbol{\Delta}\cdot\boldsymbol{\rho} \equiv \mathbf{S}\boldsymbol{\rho}\mathbf{S}^\dagger - \boldsymbol{\rho}$$ (30)

The net effect of the collisions on $\boldsymbol{\rho}$ is found by adding the effects of all collisions. We assume here that the collisional *duration* is sufficiently small compared to the time *between* collisions that the collisions may be treated as independent of one another. The $\mathbf{S}$ matrix depends on both the initial angular momentum L of relative motion and the initial velocity v. Equation (30) must be summed over collisions of all L, v, and collision-plane orientations $\Omega \equiv (\hat{\mathbf{b}}, \hat{\mathbf{v}})$, where the unit vector $\hat{\mathbf{b}}$ lies in the collision plane and is normal to $\hat{\mathbf{v}}$. The resultant average rate of change in ρ is

$$-\gamma \cdot \rho = 2\pi n \int_0^\infty b \, db \langle v\Delta \cdot \rho \rangle_{v,\Omega} \qquad (31)$$

where n is the perturber density and $\langle \cdots \rangle_{v,\Omega}$ indicates the average over relative velocities v and over orientations Ω of the collision frame. To put Eq. (31) in impact-parameter form, the impact parameter b has been identified as $b = \hbar(\mu v)^{-1}(L + \frac{1}{2})$ and the sum over L has been approximated by a continuous integration over b:

$$\sum_L (2L+1) \to 2\left(\frac{\mu v}{\hbar}\right)^2 \int_0^\infty b \, db$$

The Liouville matrix γ represents the collisional "relaxation" of ρ. It contains all the information about collisionally induced transitions in the ensemble. In particular, γ embodies the dynamics of collisional depolarization which concerns us in the present chapter. Calculations of γ and its properties are discussed in more detail in Section 3. For now it suffices to mention an important symmetry property: If the collisions are isotropic, i.e., if there is no "preferred" orientation Ω, then γ is rotationally invariant and

$$\gamma \cdot \mathbf{T}_{LM} = \gamma_L \mathbf{T}_{LM} \qquad (32)$$

where γ_L is a single number which gives the relaxation rate of the multipoles ρ_{LM}.[8,12–14]

Combining collisional and radiative effects, we find the time development [Eq. (19)] of the excited state to be given in the interaction picture by

$$\dot{\rho} = \mathcal{S} - \Gamma \cdot \rho - \gamma \cdot \rho - i\hbar^{-1}[\mathbf{V}_0, \rho] \qquad (33)$$

where $\mathcal{S}$ is the source term [Eq. (25) or (27)], Γ represents radiative decay [Eq. (23)], γ describes collision effects* [Eq. (31)], and $\mathbf{V}_0$ is the perturbation due to static external fields.

* By taking the relaxation matrix in Eq. (33) to be the same as γ [Eq. (31)] in the *absence* of external fields and radiative interactions, we have followed most authors [for example, Refs. 5, 11, 12, 97] in ignoring coupling between various terms in the time evolution of ρ. That this is *not* correct may be seen from the above derivation of γ in which it was assumed that the collisional interaction $V(t)$ was the total perturbation. To include the effects of radiative and static-field effects, we should in Eq. (33) use the relaxation matrix with elements

$$\gamma_{mn}^{m'n'} \exp\left[i(\omega_{mn} - \omega_{m'n'})t\right]$$

where $\omega_{mn} \equiv \omega_m - \omega_n$ and $\hbar\omega_m$ is the energy shift in state m due to radiative effects ("ligh

2.4. Effect of External Fields

If the basis states are all eigenstates of the unperturbed Hamiltonian H_0 with eigenenergies ε_m and if they are also eigenstates of the static external field perturbation V_0 with eigenenergies $\hbar\omega_m$:

$$(H_0 + V_0)|m\rangle = (\varepsilon_m + \hbar\omega_m)|m\rangle \tag{34}$$

then $\dot{\rho}$ [Eq. (33)] becomes

$$\dot{\rho} = \mathscr{S} - \mathbf{\Gamma} \cdot \rho - \gamma \cdot \rho - i\mathbf{L}_0 \cdot \rho \tag{35}$$

where $\mathbf{L}_0$ is the Liouville operator whose operation on ρ yields matrix elements[7,15,16]

$$(\mathbf{L}_0 \cdot \rho)_{mn} \equiv [V_0, \rho]_{mn} = \omega_{mn}\rho_{mn} \equiv (\omega_m - \omega_n)\rho_{mn} \tag{36}$$

In a *weak magnetic field* $\mathbf{B}_0$ for example, V_0 can be written[17]

$$V_0 = g_j\mu_B\mathbf{j} \cdot \mathbf{B}_0 \tag{37}$$

where $\mathbf{j}$ is the total angular momentum, g_j the Landé g factor, and μ_B the Bohr magneton. If $\mathbf{B}_0$ lies along the quantization axis, then $\mathbf{j} \cdot \mathbf{B}_0 = j_z B_0$ and

$$\omega_m = m\omega_0, \qquad \omega_0 \equiv g_j\mu_B B_0 \tag{38}$$

For example, the time dependence of the average value $\langle\mathbf{j}\rangle$, which is proportional to the orientation of the system (see Table 1), is given in the interaction picture by {we assume, after the first equality of Eq. (39), that the unperturbed Hamiltonian H_0 is spherically symmetric: $[H_0, \mathbf{j}] = 0$}

$$\frac{d}{dt}\langle\mathbf{j}\rangle = \mathrm{Tr}\,(\mathbf{j}\dot{\rho} + i[H_0, \mathbf{j}]\hbar^{-1}) = \mathbf{j}^* \cdot \dot{\rho}$$

$$= \omega_0 \times \langle\mathbf{j}\rangle + \mathbf{j}^* \cdot \mathscr{S} - \langle\mathbf{j}^* \cdot (\mathbf{\Gamma} + \gamma)\rangle \tag{39}$$

where the term $\omega_0 \times \langle\mathbf{j}\rangle$ gives the precession of $\langle\mathbf{j}\rangle$ about $\mathbf{B}_0$ with frequency

---

shift"[10,18]) and to the external fields. If we ignore light shifts, the proper relaxation matrix for use in Eq. (33) is given by

$$\gamma(t) = e^{-i\mathbf{L}_0 t}\gamma\, e^{i\mathbf{L}_0 t}$$

where $\mathbf{L}_0$ is the Liouville operator defined in the following section [Eq. (36)]. For states all characterized by a single total angular momentum j, axial (or spherical) symmetry of the relaxation process will ensure that γ and $\mathbf{L}_0$ commute, and hence that $\gamma(t) = \gamma$. However, in systems where excited states of various j values must be considered, the time dependence of $\gamma(t)$ can be significant (see also Section 4.3). Indeed elements of $\gamma(t)$ which oscillate rapidly compared to the time scale of the observation have a vanishing net effect and can often be set equal to zero (see Section 4.3).

$\omega_0 = g_j \mu_B B_0$, and the remaining terms, with $\mathbf{j}^* = \bar{\mathbf{j}}$ equal the transpose of $\mathbf{j}$, describe the evolution of $\langle \mathbf{j} \rangle$ due to pumping and relaxation.

If $\mathbf{j}$ is the sum of angular momenta, say $\mathbf{j} = \mathbf{l} + \mathbf{s}$, then Eq. (37) is only an approximation to the more precise relation,

$$V_0 = \mu_B(g_l \mathbf{l} + g_s \mathbf{s}) \cdot \mathbf{B}_0 \tag{40}$$

The approximation is justified by the Wigner–Eckart theorem (Chapter 2) for elements between states $|jm\rangle$ and $|jm'\rangle$ of a single j value, since then

$$\langle jm|(g_l \mathbf{l} + g_s \mathbf{s})|jm'\rangle = \langle jm|\mathbf{jj} \cdot (g_l \mathbf{l} + g_s \mathbf{s})|jm'\rangle [j(j+1)]^{-1}$$

$$= g_j \langle jm|\mathbf{j}|jm'\rangle, \qquad g_l = 1 \qquad g_s \simeq 2 \tag{41}$$

Noting the operator relations $2\mathbf{s} \cdot \mathbf{j} = j^2 + s^2 - l^2$ and $2\mathbf{l} \cdot \mathbf{j} = j^2 + l^2 - s^2$, one thereby finds the Landé formula[17]

$$g_j = \tfrac{1}{2}(g_s + g_l) + \tfrac{1}{2}(g_s - g_l)\frac{s(s+1) - l(l+1)}{j(j+1)} \tag{42}$$

The obvious physical picture at low fields is that of $\mathbf{l}$ and $\mathbf{s}$ precessing about their resultant $\mathbf{j}$ which in turn precesses about the field. However at high fields the precession rate $\boldsymbol{\omega}_0$ about $\mathbf{B}_0$ exceeds that of $\mathbf{s}$ and $\mathbf{l}$ about $\mathbf{j}$. (For a fine-structure term $\lambda \mathbf{l} \cdot \mathbf{s}$ the precession rate of $\mathbf{l}$ and $\mathbf{s}$ about $\mathbf{j}$ is λj and hence is independent of B_0.) The appropriate picture then is of decoupled angular momenta $\mathbf{l}$ and $\mathbf{s}$ precessing independently about $\mathbf{B}_0$. An effect analogous to this Paschen–Back decoupling occurs at lower fields (tens instead of thousands of gauss) between electronic angular momenta $\mathbf{j}$ and nuclear angular momenta $\mathbf{I}$ (Back–Goudsmit decoupling). Decoupling can also occur due to collisions (see Section 3.6).

When ρ is expanded in T_{LM} [Eq. (13)], the coefficients ρ_{LM} contain the time dependence. From Eq. (35),

$$\dot{\rho}_{LM} = \mathscr{S}_{LM} - \Gamma \rho_{LM} - \gamma_L \rho_{LM} - iM\omega_0 \rho_{LM} \tag{43}$$

where we assume isotropic collisions and a weak magnetic field along the quantization axis. The steady-state ($\dot{\rho}_{LM} = 0$) solution

$$\rho_{LM} = \mathscr{S}_{LM}[\Gamma + \gamma_L + iM\omega_0]^{-1} \tag{44}$$

describes the Hanle effect (see Chapter 9) in the presence of collisional depolarization (see also Section 4.3 of this chapter). Steady-state solution are usually appropriate when the source term $\mathscr{S}_{LM}$ is constant or when any variations in $\mathscr{S}_{LM}$ are slow compared to the precession and relaxation times.

2.5. Excited-State Coherence

Consider the time development of ρ_{LM} [Eq. (43)] for a pulsed source $\mathcal{S}_{LM}$. A polarization $\rho_{LM}(0)$, $L>0$, is created at $t=0$ and then left to evolve with $\mathcal{S}_{LM} = \text{const}$, $t>0$. Solution of Eq. (43) then yields [compare Eq. (10)]

$$\rho_{LM}(t) = [\rho_{LM}(0) - \rho_{LM}(\infty)] \exp[-(\Gamma + \gamma_L)t - iM\omega_0 t] + \rho_{LM}(\infty) \quad (45)$$

where $\rho_{LM}(\infty)$ is the steady-state value [Eq. (44)]. In addition to the decay of the (L, M) *multipole moment to* $\rho_{LM}(\infty)$, there will be, in a magnetic field, an *oscillatory* dependence $\exp(-iM\omega_0 t)$ for $M \neq 0$. As mentioned in Section 2.2, when moments ρ_{LM} exist with $M \neq 0$, the density matrix has nonvanishing off-diagonal elements: there is *coherence* in the system.[11,18-21]

Of course at any time the density matrix can be diagonalized, i.e., there is always a set of basis states in which all off-diagonal elements of ρ are zero. The existence of coherence therefore depends on the basis or representation used. Usually, however, the basis states are taken to be stationary eigenstates of $H_0 + V_0$, where H_0 is the unperturbed Hamiltonian and V_0 the perturbation to due static external fields [see Eq. (34)]. The existence of coherence between states $|n\rangle$ and $|m\rangle$ of different energies results in an oscillating time dependence $\exp(-i\omega_{mn}t)$ for elements $\rho_{mn}(t)$.

Any system with coherence is obviously polarized. The $\boldsymbol{\rho}$ of an unpolarized system is structureless and incoherent: it is diagonal (indeed, proportional to the unit matrix) in every basis. A system may be polarized without having coherence. One may then be able to treat it with a fully diagonal population representation (Sections 2.1 and 2.6].

Consider a system with an axially symmetric Hamiltonian $\mathbf{H}_0 + \mathbf{V}_0$:

$$[\mathbf{j}_z, \mathbf{H}_0 + \mathbf{V}_0] = 0 \quad (46)$$

It is *coherent if $\boldsymbol{\rho}$ is not* axially symmetric, i.e., if

$$[\mathbf{j}_z, \boldsymbol{\rho}] \neq 0 \quad (47)$$

Since (see Chapter 2)

$$[\mathbf{j}_z, \mathbf{T}_{LM}] = M\mathbf{T}_{LM} \quad (48)$$

condition (47) is equivalent to the existence of ρ_{LM}, with $M \neq 0$.

The simplest example of coherence is a nonaxial orientation, i.e., $\langle \hat{\mathbf{z}} \times \mathbf{j} \rangle \neq 0$. In Section 2.4 we investigated the time dependence of such an orientation in a magnetic field $\mathbf{B}_0 = B_0 \hat{\mathbf{z}}$ and found that $\langle \mathbf{j} \rangle$ precesses with angular velocity $\omega_0 = g_j \mu_B B_0$ about $\hat{\mathbf{z}}$. From Eq. (45) we can demonstrate a similar behavior for all coherent $(M \neq 0)$ multipole moments. First note

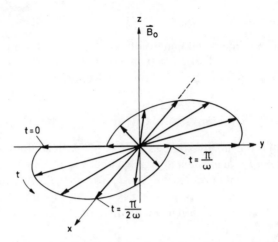

Figure 2. Atomic coherence shown as it precesses about the magnetic field B_0 at the Larmor frequency $\omega = g_J \mu B_0 / \hbar$ and decays with a lifetime $\tau = \pi/\omega$. The double-headed arrows represent the electric quadrupole moment excited at $t = 0$ by light linearly polarized along the y direction: $\hat{\epsilon} = \hat{y}$. Other than ρ_{00} the only nonvanishing density matrix components are $\rho_{2\pm2}(t) = \rho_{2\pm2}(10) = \exp[\pm 2i(\varphi - \omega t) - \Gamma t]$.

that the rotation of $\mathbf{T}_{LM}$ by $\varphi \hat{z}$ gives $\mathbf{T}_{LM}$ times the phase factor $e^{-iM\varphi}$:

$$e^{-ij_z\varphi} \mathbf{T}_{LM} e^{ij_z\varphi} = e^{-iM\varphi} \mathbf{T}_{LM} \tag{49}$$

Consequently, the time dependence of a component $\rho_{LM}(t)\mathbf{T}_{LM}$ of $\boldsymbol{\rho}$ [Eq. (45)] in a constant magnetic field $\mathbf{B}_0$ simply describes its precession about B_0 with angular velocity ω_0 in addition to its decay.* (See Figure 2.) Because of its M-fold symmetry about the z axis, ρ_{LM} repeats itself M times per precessional period.

2.6. When Can Coherence Be Neglected?

If coherence can be ignored, only diagonal elements of the density matrix remain. These diagonal elements represent the *populations* of the various levels, and in fact a significant simplification is achieved by using what we above (Sections 2.1 and 2.2) referred to as the "population representation." The state of the ensemble is then represented by an N-dimensional vector (N is the number of states considered) instead of by an $N \times N$ matrix (or N^2-dimension Liouville space vector). Here we investigate the circumstances under which such a simplification is justified.

Clearly it is justified to ignore coherence if (1) the excitation, (2) the detection, and (3) the intermediate development, as due, say, to collisions

---

* A half a century ago Hanle[22] explained the precession of $M = 2$ alignment in a magnetic field as being a result of the net Lorentz force on an oscillating electron.

and external static fields are all axially symmetric about the quantization direction, since there is then no excited-state coherence. Even if coherence *is* created, it can be ignored if *any two* of steps (1)–(3) are cylindrically symmetric, as will now be shown: If no coherence is excited, then initially ρ is axially symmetric:

$$[J_z, \rho] = 0 \qquad \text{at } t = 0 \tag{50}$$

where J_z is the z component of the total angular momentum of the atom. (For example, in the presence of hyperfine structure, the operator $F_z = J_z + I_z$ is to be used, even if f is not a good quantum number.) If, in addition, the unperturbed Hamiltonian H_0 plus the net collisional perturbation V_{coll} plus the effect V_0 of external fields is axially symmetric

$$[J_z, H_0 + V_{\text{coll}} + V_0] = 0 \tag{51}$$

then the time development of ρ [Eqs. (14) or (17)] ensures that ρ remains axially symmetric up to the time of radiative decay and detection. Even if the detector is capable of measuring coherence, none is present. On the other hand if coherence is excited but not detectable, then an axially symmetric $H_0 + V_{\text{coll}} + V_0$ ensures that the diagonal elements actually detected have evolved independently of the off-diagonal ones. The argument is essentially the time reverse of that for the case that no coherence is excited although it may be detectable. Finally, in the case that no coherence is excited and none detectable, even though off-diagonal elements of ρ may appear in the time development as given by Eq. (17), it will always be possible to eliminate the coherent terms and work only with the diagonal ones.

Coherence can also be neglected (the "secular approximation") between levels that are so widely spaced that corresponding elements of ρ oscillate rapidly compared to the time scale of observation.[21]

2.7. Making and Monitoring the Multipoles

Polarization of a system in the excited state is observed by detecting the intensity of polarized fluorescence. Because of the relation between the polarization vector $\hat{\varepsilon}'$ and the wave vector $\hat{k}'$, namely, $\hat{\varepsilon}' \cdot \hat{k}' = 0$, it can also be measured from the anisotropy of the radiation. The intensity of light of polarization $\hat{\varepsilon}'$ emitted in the decay of atoms to the ground state λ with magnetic sublevels $|\lambda\mu\rangle$ is proportional to [5,11,23–25]

$$\mathbf{D}_\lambda(\hat{\varepsilon}') \cdot \boldsymbol{\rho} \equiv \sum_{m,n,\mu} \langle \lambda\mu | \mathbf{d} \cdot \hat{\varepsilon}'^* | m \rangle \rho_{mn} \langle n | \mathbf{d} \cdot \hat{\varepsilon}' | \lambda\mu \rangle \tag{52}$$

where $\mathbf{D}_\lambda(\hat{\varepsilon}')$ is the "detection operator" (a Liouville vector) for decay to state λ:

$$\mathbf{D}_\lambda(\hat{\boldsymbol{\varepsilon}}') = \sum_\mu \hat{\boldsymbol{\varepsilon}}' \cdot \mathbf{d}|\lambda\mu\rangle\langle\lambda\mu|\mathbf{d}\cdot\boldsymbol{\varepsilon}'^* \tag{53}$$

Just those multipoles of ρ can be observed for which there are nonvanishing components of $\mathbf{D}_\lambda$:

$$\mathbf{D}_\lambda(\hat{\boldsymbol{\varepsilon}}') = \sum_{\substack{LM \\ jk}} D_{\lambda LM}(jk;\hat{\boldsymbol{\varepsilon}}')\mathbf{T}_{LM}(jk) \tag{54}$$

where j and k label the total angular momentum of the excited states m and n;

$$D_{\lambda LM}(jk;\hat{\boldsymbol{\varepsilon}}') = \mathbf{T}^*_{LM}(jk)\cdot\mathbf{D}_\lambda(\hat{\boldsymbol{\varepsilon}}')$$

$$= B_L(jk,\lambda)\Phi_{LM}(\hat{\boldsymbol{\varepsilon}}') \tag{55}$$

and, in accord with the notation of Omont,[5] all the angular dependence is given by

$$\Phi_{LM}(\hat{\boldsymbol{\varepsilon}}) = (2L+1)^{1/2}\sum_{rs}\hat{\boldsymbol{\varepsilon}}\cdot\hat{\mathbf{r}}(\hat{\boldsymbol{\varepsilon}}\cdot\hat{\mathbf{s}})^*(-)^{1-s}\begin{pmatrix}1 & L & 1 \\ r & M & -s\end{pmatrix} \tag{56}$$

($\hat{\mathbf{r}}$ and $\hat{\mathbf{s}}$ span the unit spherical tensors of rank 1, namely, $\pm 1 = \mp(\hat{\mathbf{x}}\pm i\hat{\mathbf{y}})/2^{1/2}$ and $\hat{\mathbf{0}}=\hat{\mathbf{z}}$) and the dynamics are determined by

$$B_L(jk,\lambda) = (-)^{\lambda+L+j+1} d_{j\lambda}d^*_{k_\lambda}\begin{Bmatrix}1 & L & 1 \\ j & \lambda & k\end{Bmatrix} \tag{57}$$

Note that

$$\sum_{\hat{\boldsymbol{\varepsilon}}}\Phi_{LM}(\hat{\boldsymbol{\varepsilon}}) = 3^{1/2}\delta_{L0}\delta_{M0}, \qquad \Phi_{00}(\hat{\boldsymbol{\varepsilon}}) = 1/3^{1/2}$$

$$\Phi_{LM}(\hat{\boldsymbol{\varepsilon}}) = (-)^M\Phi^*_{L,-M}(\hat{\boldsymbol{\varepsilon}}) = (-)^L\Phi_{LM}(\hat{\boldsymbol{\varepsilon}}^*)$$

Tables of $\Phi_{LM}(\hat{\boldsymbol{\varepsilon}})$ are given in terms of Cartesian components of $\hat{\boldsymbol{\varepsilon}}$ by Omont.[5]

From the $3-j$ symbol in Eq. (56) we see that only $L = 0, 1$, and 2 components of ρ are observable if the splitting of the Zeeman sublevels of the ground state λ is not resolved. Similarly, if the ground state is unpolarized* and the Zeeman levels unresolved, then only $L = 0, 1$, and 2 components can be *excited*. Indeed the source term [Eq. (27)] can be

---

* Multipoles with $L > 2$ can be excited from a polarized ground state. See the example given for D2 optical pumping (Section 4.5).

written in the same form as the detection term (in the Schrödinger picture):

$$\mathscr{S} = \sum_{\substack{LM \\ jk}} \mathscr{S}_{LM}(jk)\mathbf{T}_{LM}(jk) \tag{58}$$

with

$$\mathscr{S}_{LM}(jk) = (2\pi)^2 \sum_{\lambda,\hat{\varepsilon}} u_\varepsilon(\lambda)D_{\lambda LM}(jk;\hat{\varepsilon}) \tag{59}$$

From Eq. (56) $\Phi_{LM}(\hat{\varepsilon}) = \Phi_{LM}(\hat{\varepsilon}\,e^{i\beta})$, where β is any real constant. If there exists a β such that $\hat{\varepsilon}^* = \hat{\varepsilon}\,e^{i\beta}$, the polarization is said to be *linear*. It then follows from Eq. (56) that Φ_{1M} vanishes. Furthermore, if $\hat{\varepsilon}$ lies in the xy plane, then $M = \pm 1$ components of Φ_{LM} vanish.

As a simple example, consider again a noncoherent system in the $j = \frac{3}{2}$ state, which decays to a $\lambda = \frac{1}{2}$ ground state. Only the four diagonal elements of $\mathbf{D}(\hat{\varepsilon}')$ are needed:

$$D_m(\hat{\varepsilon}') = \sum_\mu |\langle jm|\hat{\varepsilon}' \cdot \mathbf{d}|\lambda\mu\rangle|^2$$

$$= |d_{j\lambda}|^2 \sum_{\mu r} |\hat{\varepsilon}' \cdot \hat{\mathbf{r}}|^2 \begin{pmatrix} \lambda & 1 & j \\ -\mu & r & m \end{pmatrix}^2 \tag{60}$$

where $m = -j, -j+1, \ldots, j$, μ is summed over the $2\lambda+1$ values $-\lambda, -\lambda+1, \ldots, \lambda$, and r over $-1, 0, +1$. With $j = \frac{3}{2}$ and $\lambda = \frac{1}{2}$, the four-dimensional detection vector

$$\mathbf{D}(\hat{\varepsilon}') = \frac{1}{12}|d_{j\lambda}|^2 \left[|\hat{\varepsilon}' \cdot -\hat{\mathbf{1}}|^2 \begin{pmatrix} 3 \\ 1 \\ 0 \\ 0 \end{pmatrix} + |\hat{\varepsilon}' \cdot \hat{\mathbf{0}}|^2 \begin{pmatrix} 0 \\ 1 \\ 2 \\ 0 \end{pmatrix} + |\hat{\varepsilon}' \cdot \hat{\mathbf{1}}|^2 \begin{pmatrix} 0 \\ 0 \\ 1 \\ 3 \end{pmatrix} \right] \tag{61}$$

is obtained, which gives the relative transition probabilities from the four levels of j as a function of the polarization $\hat{\varepsilon}'$ of the emitted radiation (see Figure 3).

The detected signal, i.e., the intensity of light of polarization $\hat{\varepsilon}'$, is proportional to the scalar product $\mathbf{D}(\hat{\varepsilon}') \cdot \mathbf{N}$ of the detection vector $\mathbf{D}(\hat{\varepsilon}')$,

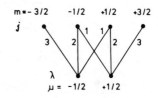

Figure 3. Kastler diagram of relative transition probabilities between a $j = \frac{3}{2}$ excited state and a $\lambda = \frac{1}{2}$ ground state.

with the population density N [compare Eq. (52)]. In particular, the intensity for $\hat{\varepsilon} = \pm 1$ or $\hat{0}$ (observed of course in a direction $\hat{k}$ orthogonal to $\hat{\varepsilon}' : \hat{\varepsilon}' \cdot \hat{k} = 0$) is indicated $I_{\sigma\pm}$ or I_π, respectively:

$$\left. \begin{array}{l} I_{\sigma\pm} = K\mathbf{D}(\pm 1) \cdot \mathbf{N} \\ I_\pi = K\mathbf{D}(\hat{0}) \cdot \mathbf{N} \end{array} \right\} \quad K = \text{const} \tag{62}$$

Expanding N in spherical basis vectors $\mathbf{T}_L$ [see Eqs. (3) and (4)], we find $\mathbf{D}(\hat{\varepsilon}') \cdot \mathbf{N} =$

$$(\tfrac{1}{12}|d_{j\lambda}|^2)[2n_0 + 5^{1/2}n_1(|\hat{\varepsilon} \cdot \hat{1}|^2 - |\hat{\varepsilon} \cdot -1|^2) - n_2(1 - 3|\hat{\varepsilon} \cdot \hat{0}|^2)] \tag{63}$$

Thus, to monitor pure multipoles n_1 and n_2 [note that n_3 does not enter Eq. (63)], we would use*

$$\frac{I_{\sigma^+} - I_{\sigma^-}}{2I_\sigma + I_\pi} = \tfrac{1}{3}(5)^{1/2}\frac{n_1}{n_0} \tag{64}$$

and

$$\frac{I_\sigma - I_\pi}{2I_\sigma + I_\pi} = \frac{n_2}{2n_0} \tag{65}$$

where $I_\sigma \equiv (I_{\sigma^+} + I_{\sigma^-})/2$.

3. Relaxation Matrix γ

3.1. Multipole Relaxation Rates γ_L

The time rate of change of the density matrix ρ due to collisions[26] has been given (Section 2.3) as $-\gamma \cdot \rho$. By directly expanding in irreducible tensors $\mathbf{T}_{LM}$ one finds [see Eq. (13)][8,27]

$$\frac{d}{dt}\bigg|_{\text{coll}} \rho_{LM} = -\mathbf{T}_{LM}^* \cdot \gamma \cdot \rho$$

$$= -\sum_{L'M'} \mathbf{T}_{LM}^* \cdot \gamma \cdot \mathbf{T}_{L'M'} \rho_{L'M'} \tag{66}$$

For relaxation among sublevels of a single j value there are, in the most general or least symmetric case

$$(2j + 1)^2 = \left[\sum_{L=0}^{2j} (2L + 1) \right]^2$$

---

* Except for the value of the numerical coefficient on the right-handside, Eqs. (64) and (65) are valid for any j.

multipole relaxation constants, $\mathbf{T}_{LM}^* \cdot \boldsymbol{\gamma} \cdot \mathbf{T}_{L'M'}$. Usually this number is considerably reduced by symmetry.

A rotation of the coordinates by Ω changes $\mathbf{T}_{LM}^* \cdot \boldsymbol{\gamma} \cdot \mathbf{T}_{L'M'}$ to $\mathbf{T}_{LM}'^* \cdot \boldsymbol{\gamma} \cdot \mathbf{T}_{L'M'}'$, where

$$\mathbf{T}_{LM}' \equiv \mathbf{D}(\Omega)\mathbf{T}_{LM}\mathbf{D}^\dagger(\Omega) = \sum_N \mathbf{T}_{LN} D_{NM}^L(\Omega) \tag{67}$$

where $D_{NM}^L(\Omega)$ are usual rotation matrix elements.[6] Consider, for example, a rotation $\Omega = \varphi \hat{\mathbf{z}}$ about the quantization axis $\hat{\mathbf{z}}$. Since $D_{NM}^L(\varphi \hat{\mathbf{z}}) = \delta_{MN} \exp(-iM\varphi)$, the rotated multipole relaxation constants are

$$\mathbf{T}_{LM}'^* \cdot \boldsymbol{\gamma} \cdot \mathbf{T}_{L'M'}' = \mathbf{T}_{LM}^* \cdot \boldsymbol{\gamma} \cdot \mathbf{T}_{L'M'} \, e^{i(M-M')\varphi} \tag{68}$$

If $\boldsymbol{\gamma}$ is *axially symmetric*, Eq. (68) must be independent of φ and therefore

$$\mathbf{T}_{LM}^* \cdot \boldsymbol{\gamma} \cdot \mathbf{T}_{L'M'} = \delta_{MM'} \mathbf{T}_{LM}^* \cdot \boldsymbol{\gamma} \cdot \mathbf{T}_{L'M} \tag{69}$$

Often collisions are *isotropic*. The $\boldsymbol{\gamma}$ must be independent of coordinate orientation Ω; in particular we can average over Ω:

$$
\begin{aligned}
\mathbf{T}_{LM}^* \cdot \boldsymbol{\gamma} \cdot \mathbf{T}_{L'M'} &= \langle \mathbf{T}_{LM}'^* \cdot \boldsymbol{\gamma} \cdot \mathbf{T}_{L'M'}' \rangle_\Omega \\
&= \sum_{NN'} (8\pi^2)^{-1} \int d\Omega \, D_{NM}^{L*}(\Omega) D_{N'M'}^{L'}(\Omega) \times (\mathbf{T}_{LN}^* \cdot \boldsymbol{\gamma} \cdot \mathbf{T}_{L'N'}) \\
&= (2L+1)^{-1} \delta^{LL'} \delta_{MM'} \sum_N \mathbf{T}_{LN}^* \cdot \boldsymbol{\gamma} \cdot \mathbf{T}_{LN} \\
&\equiv \delta^{LL'} \delta_{MM'} \gamma_L
\end{aligned}
\tag{70}
$$

where $\boldsymbol{\gamma}$ on the right-hand side is the relaxation matrix in any frame and we have used the orthogonality of the rotation matrices.[6] The simplification achieved is considerable. Instead of $(2j+1)^2$ relaxation constants to describe relaxation among the $2j+1$ sublevels $|jm\rangle$, $m = -j, -j+1, \ldots, j$, there are, in the case of isotropic collisions, only $2j+1$ independent constants.

3.2. m_j Mixing Rates

The element $\gamma_{nn'}^{mm'}$ of the relaxation matrix $\boldsymbol{\gamma}$ [Eq. (31)] when multiplied by $\rho_{mm'}$ gives the rate of decrease in $\rho_{nn'}$ which is attributable to $\rho_{mm'}$. The physical interpretation is simpler when only diagonal elements of $\boldsymbol{\rho}$ are involved. Thus $-\gamma_{nn}^{mm}$ gives the collisionally induced transition rate $m \to n$ and is related to the thermally averaged inelastic cross section Q_{nm} (Chapter 6, Section 4.1) for $m \to n$ transitions by

$$n\bar{v} Q_{nm} = -\gamma_{nn}^{mm} \tag{71}$$

where $\bar{v}$ is the mean relative velocity before collision and n is the perturber density. The cross section Q_{mn} for the inverse transition $m \leftarrow n$ is related to Q_{nm} by detailed balance [see Eq. (163), Section 4.1; here m and n label single sublevels]:

$$Q_{mn} = Q_{nm} e^{E_{nm}/kT} \tag{72}$$

where $E_{nm} = E_n - E_m$ is the energy difference between sublevels n and m.

For a system with isotropic collisions, knowledge about the transition rates $-\gamma_{nn}^{mm}$ is equivalent to information about the multipole relaxation rates γ_L. The γ_{nn}^{mm}'s can be fully determined from the γ_L's and vice versa. In this section the relations are derived and illustrated with a simple example.

In Eq. (70), γ_L is given in terms of γ. Writing explicit matrix elements, one has

$$\gamma_L = \sum_{\substack{nn' \\ mm'}} \langle n|T_{LM}|n'\rangle^* \gamma_{nn'}^{mm'} \langle m|T_{LM}|m'\rangle \tag{73}$$

which is valid for any M. The inverse relation is found by expanding γ in $T_{LM}T_{L'M'}^*$ and using Eq. (70):

$$\gamma = \sum_{LM,L'M'} T_{LM}T_{LM}^* \cdot \gamma \cdot T_{L'M'}T_{L'M'}^* = \sum_{LM} T_{LM}\gamma_L T_{LM}^* \tag{74}$$

Taking matrix elements, one obtains

$$\gamma_{nn'}^{mm'} = \sum_{LM} \langle n|T_{LM}|n'\rangle \gamma_L \langle m|T_{LM}|m'\rangle^* \tag{75}$$

Another useful relation between γ and the γ_L is found from Eq. (74) by using the orthonormality [Eq. (12)] of the irreducible tensor operators:

$$\gamma \cdot T_{LM} = \gamma_L T_{LM} \tag{76}$$

Explicitly in terms of matrix elements

$$\sum_{mm'} \gamma_{nn'}^{mm'} \langle m|T_{LM}|m'\rangle = \gamma_L \langle n|T_{LM}|n'\rangle \tag{77}$$

The basis states $|m\rangle$, $|n\rangle$, etc., involved need not be eigenstates of the square of the total angular momentum and its projection on the quantization axis, but *if they are*, with $j^2|n\rangle = \hbar^2 k(k+1)|n\rangle$, $j^2|n'\rangle = \hbar^2 k'(k'+1)|n'\rangle$, $j_z|n\rangle = \hbar n|n\rangle$, etc., *then* (see Chapter 2)

$$\langle n|T_{LM}|n'\rangle = (-)^{k-n}(2L+1)^{1/2}\begin{pmatrix} k & L & k' \\ -n & M & n' \end{pmatrix} \tag{78}$$

permits numerical evaluation of the linear interdependence of γ_L and $\gamma_{nn'}^{mm'}$.

Note that relations between the multipole relaxation rates γ_L and transition rates γ_{nn}^{mm} (i.e., $m' = m$, $n' = n$) involve only tensors T_{LM} with $M = 0$.

Depolarization often occurs within an *isolated* excited-state manifold $\mathscr{E}_j$ with a single total angular momentum quantum number j. The energy separations E_{mn} of the sublevels is then usually small compared to kT so that, according to detailed balance [Eq. (72)], forward and reverse transition rates are equal:

$$\gamma_{nn}^{mm} = \gamma_{mm}^{nn} \tag{79}$$

The $j(2j+1)$ elements γ_{nn}^{mm}, $m < n$ are *mixing rates* in the manifold $\mathscr{E}_j$. If collisions are isotropic, Eqs. (70) and (74) show that not all $j(2j+1)$ (for $j > \frac{1}{2}$) mixing rates are independent. Indeed, if ρ is normalized ($\mathrm{Tr}\,\rho = \mathrm{const}$) or if collisionally induced transitions out of the manifold are ignored, then the monopole relaxation rate $\gamma_0 = 0$ and there are only $2j$ independent mixing rates.

With the many depolarization experiments possible (Section 4), one might imagine it possible to measure all the multipole relaxation rates γ_L and hence, through Eq. (75), all the mixing rates γ_{nn}^{mm}. However, as mentioned in Section 2.7, usual fluorescence experiments on depolarization within an isolated manifold $\mathscr{E}_j$ can determine at most two independent constants, namely, the orientation and alignment relaxation rates γ_1 and γ_2. The mixing rates γ_{nn}^{mm} (for $j > 1$) thus must also remain largely undetermined. To obtain γ_L's for $L > 2$, using electric dipole excitation from unpumped ground states, either manifolds $\mathscr{E}_j$ of different total angular momentum j must be mixed in excitation and by collisions or the excited-state manifold must be spectrally resolved in both source and detector (see Section 4).

As a simple example to display the connection between mixing rates and relaxation rates, consider again the manifold $\mathscr{E}_j$ with $j = \frac{3}{2}$. One can apply the diagonal population representation directly to Eq. (74). The irreducible tensors $\mathbf{T}_{LO}$ are given as the four-dimensional unit spherical vectors $\hat{\mathbf{T}}_L$ (Eq. (3)), and γ becomes the 4×4 matrix of elements γ_{nn}^{mm}. Equation (74) equates γ to a linear combination of dyads $\hat{\mathbf{T}}_L \hat{\mathbf{T}}_L$ and yields immediately

$$\gamma_{3/2,3/2}^{\pm 3/2,\pm 3/2} = (1/4)(\gamma_0 + \gamma_2) \pm (1/20)(9\gamma_1 + \gamma_3) \tag{80a}$$

$$\gamma_{3/2,3/2}^{\pm 1/2,\pm 1/2} = (1/4)(\gamma_0 - \gamma_2) \pm (3/20)(\gamma_1 - \gamma_3) \tag{80b}$$

$$\gamma_{1/2,1/2}^{\pm 1/2,\pm 1/2} = (1/4)(\gamma_0 + \gamma_2) \pm (1/20)(\gamma_1 + 9\gamma_3) \tag{80c}$$

together with the symmetry results

$$\gamma_{nn}^{mm} = \gamma_{mm}^{nn} = \gamma_{-n,-n}^{-m,-m} \tag{81}$$

The usual normalization of $\boldsymbol{\rho}$ is $\mathrm{Tr}\,\boldsymbol{\rho} = 1$. Thus $\rho_{00} = (2j+1)^{-1/2}\,\mathrm{Tr}\,\boldsymbol{\rho}$ is constant and $\gamma_0 = 0$. The remaining three relaxation rates, γ_1, γ_2, and γ_3 determine not only all 12 mixing rates $\gamma_{nn'}^{mm}$, but indeed, through Eq. (75), all 256 elements $\gamma_{nn'}^{mm'}$.

The analysis is easily extended to states of *more than one j value*, e.g., to fine-structure multiplets. There are then generally several independent multipole relaxation rates γ_L for each L, just as there are different tensors $\mathbf{T}_{LM}(kk')$ corresponding to different angular momentum pairs kk' [see Eq. (78)]. With explicit j, k dependence displayed, Eq. (75), for example, becomes

$$\gamma_{kn,k'n'}^{jm,j'm'} = \sum_{LM} \langle kn|T_{LM}(kh')|k'n'\rangle \gamma_L \begin{pmatrix} j & j' \\ k & k' \end{pmatrix} \langle jm|T_{LM}(jj')|j'm'\rangle^* \quad (82)$$

As before, relations of the γ_L to transition rates involve only $M = 0$ tensors and may be found in the diagonal population representation.

In particular, the *fine-structure transition rates* $k = \tfrac{1}{2} \leftarrow j = \tfrac{3}{2}$ can be expressed in terms of the $\gamma_L \binom{j}{k}{}^{j}_{k}$ by expanding $\boldsymbol{\gamma}\binom{j}{k}{}^{j}_{k}$ in the dyads $\hat{\mathbf{T}}_L(kk)\hat{\mathbf{T}}_L(jj)$. Since for $k = \tfrac{1}{2}$, the only nonvanishing spherical vectors $\hat{\mathbf{T}}_L(kk)$ are $2^{-1/2}\binom{1}{1}$ for $L = 0$ and $(2)^{-1/2}\binom{\ 1}{-1}$ for $L = 1$, one finds [see Eq. (3) for $\hat{\mathbf{T}}_L(jj)$, $j = \tfrac{3}{2}$]

$$\boldsymbol{\gamma}\begin{pmatrix} j & j \\ k & k \end{pmatrix} = \sum_L \gamma_L \begin{pmatrix} j & j \\ k & k \end{pmatrix} \hat{\mathbf{T}}_L(kk)\hat{\mathbf{T}}_L^*(jj)$$

$$= \gamma_0 \begin{pmatrix} \tfrac{3}{2} & \tfrac{3}{2} \\ \tfrac{1}{2} & \tfrac{1}{2} \end{pmatrix} \cdot \frac{1}{2(2)^{1/2}} \begin{pmatrix} 1 & 1 & 1 & 1 \\ 1 & 1 & 1 & 1 \end{pmatrix} + \gamma_1 \begin{pmatrix} \tfrac{3}{2} & \tfrac{3}{2} \\ \tfrac{1}{2} & \tfrac{1}{2} \end{pmatrix} \cdot \frac{1}{2(10)^{1/2}} \begin{pmatrix} 3 & 1 & -1 & -3 \\ -3 & -1 & 1 & 3 \end{pmatrix}$$

$$(83)$$

where the 2×4 matrix $-\boldsymbol{\gamma}\binom{j}{k}{}^{j}_{k}$ has as elements the transition rates $-\gamma_{kn,kn}^{jm,jm}$. Thus the transition rate to $|kn\rangle = |\tfrac{1}{2}, \tfrac{1}{2}\rangle$ from $|jm\rangle = |\tfrac{3}{2}, \tfrac{3}{2}\rangle$, for example, is

$$-\gamma_{1/2,1/2;1/2,1/2}^{3/2,3/2;3/2,3/2} = -\frac{1}{2}\left[\frac{1}{(2)^{1/2}} \gamma_0 \begin{pmatrix} \tfrac{3}{2} & \tfrac{3}{2} \\ \tfrac{1}{2} & \tfrac{1}{2} \end{pmatrix} + \frac{3}{(10)^{1/2}} \gamma_1 \begin{pmatrix} \tfrac{3}{2} & \tfrac{3}{2} \\ \tfrac{1}{2} & \tfrac{1}{2} \end{pmatrix}\right] \quad (84)$$

It can be seen that whereas $-\gamma_0\binom{j}{k}{}^{j}_{k}$ gives the net fine-structure transition rate $k \leftarrow j$, $-\gamma_1\binom{j}{k}{}^{j}_{k}$ gives the rate of orientation transfer between the fine-structure levels (see Chapter 29, by Elbel).

3.3. Symmetries and Selection Rules

In the previous section, the spherical symmetry of isotropic collisions was used to relate many transition rate elements $\gamma_{nn'}^{mm'}$ to a few relaxation constants γ_L. Other symmetries of the collision permit further relations to be established among the $\gamma_{nn'}^{mm'}$. Under some conditions *selection rules* result: certain transition rates $-\gamma_{nn}^{mm}$ are found to vanish. In this section such symmetries, some only approximate, are discussed together with their consequences for the relaxation rate matrix γ.

From Eqs. (30) and (31), γ is related to the S matrix by

$$\gamma_{nn'}^{mm'} = -2\pi n \int_0^\infty b \, db \langle v(S_{mn}S_{n'm'}^\dagger - \delta_{mn}\delta_{n'm'})\rangle_{v,\,\Omega} \tag{85}$$

and in particular for the mixing rates $-\gamma_{nn}^{mm}$

$$-\gamma_{nn}^{mm} = 2\pi n \int_0^\infty b \, db \langle v(|S_{mn}|^2 - \delta_{mn})\rangle_{v,\Omega} \tag{86}$$

Now the conservation of particles implies that S is unitary:

$$\sum_n S_{mn}S_{nm'}^\dagger = \delta_{mm'} \tag{87}$$

where the sum extends over all states to which transitions from m or m may be collisionally induced. It follows that for any m and m'

$$\sum_n \gamma_{nn}^{mm'} = \sum_n \gamma_{mm'}^{nn} = 0 \tag{88}$$

With $m = m'$, the result is equivalent to

$$\sum_j (2j+1)^{1/2} \gamma_0 \begin{pmatrix} j & j \\ k & k \end{pmatrix} = 0$$

[see Eq. (82)].

It was shown in Section 4 of Chapter 6 that if, as is practically always possible, a frame is found in which the collisional interaction obeys

$$\mathbf{V}^*(t) = \mathbf{V}(-t) \tag{89}$$

then in this frame $\mathbf{S}$ is symmetric:

$$\mathbf{S} = \tilde{\mathbf{S}} \tag{90}$$

The result is however of limited use since it is frame-dependent.

A generalization can be established by the introduction of the anti-linear *time-reversal* operator, which, with usual phase conventions[17] and a $|jm\rangle$ representation, has the form

$$K = K_0 \, e^{-is_y\pi} \tag{91}$$

where K_0 is the complex-conjugation operator and s_y is the y component of the spin operator. The advantage of using K rather than simply the antilinear K_0 is that rotations $D(\Omega)$ are generally invariant under K but not under K_0:

$$KD(\Omega)K^\dagger = K\,e^{-i\mathbf{J}\,\cdot\,\mathbf{\Omega}/\hbar}K^\dagger = e^{-i\mathbf{J}\,\cdot\,\mathbf{\Omega}/\hbar} = D(\Omega) \tag{92}$$

Equation (92) follows from the effect of K in reversing the angular momentum operator $\mathbf{j}$:

$$K\mathbf{j}K^\dagger = -\mathbf{j} \tag{93}$$

as well as in changing the sign of imaginary parts of complex numbers. The consequent relations between matrix elements of $\mathbf{S}$ and those of $KSK^\dagger$, namely,

$$\langle jm|KSK^\dagger|j'm'\rangle = (-1)^{i-i'+m'-m}\langle j'-m'|S^\dagger|j-m\rangle \tag{94}$$

are independent of the orientation of the collision frame.

The relation takes on significance *if* in *any* frame

$$KV(t)K^\dagger = V(-t) \tag{95}$$

because then (see Section 4 of Chapter 6)

$$KSK^\dagger = S^{-1} = S^\dagger \tag{96}$$

so that from Eq. (94)

$$\langle jm|S|j'm'\rangle = (-1)^{j-j'-m+m'}\langle j'-m'|S|j-m\rangle \tag{97}$$

Thus, time reversal in the sense of Eq. (95) or equivalently in terms of matrix elements [we assume $V(t)$ to be Hermitian],

$$\langle jm|V(t)|j'm'\rangle = (-1)^{j-j'-m+m'}\langle j'-m'|V(-t)|j-m\rangle \tag{98}$$

would imply important relations among the $\mathbf{S}$-matrix elements and hence among the transition rates [Eq. (86)]. In particular, we would find transitions between $|jm\rangle$ and $|j-m\rangle$ to be forbidden for systems with an odd number of electrons:

$$|jm\rangle \not\leftrightarrow |j-m\rangle \qquad j = \tfrac{1}{2}, \tfrac{3}{2}, \tfrac{5}{2}, \ldots \tag{99}$$

The selection rule, Eq. (99), has in the past been asserted with varying degrees of finality.[28-31] It would, if valid, have important consequences for the relative sizes of multipole relaxation rates (see Section 3.4). Measurements,[30-35] however, demonstrate that Eq. (99) is not rigidly obeyed, and hence that there is no frame in which the extended time-reversal relation [Eq. (95)] for classical paths holds.[36-43]

The breakdown of the extended relation [Eq. (95)] may arise because of interactions with external magnetic fields. In the classical path approximation, even a magnetic field created by the relative motion of the perturber is external to the system. Such interactions change sign under time reversal. Thus, for example, an interaction between the electronic and orbital magnetic moments (proportional to $-\mathbf{j}$ and $\mathbf{L}$, respectively)[39]

$$V = A(R)\mathbf{j} \cdot \mathbf{L} \tag{100}$$

although it would be time-reversal-invariant in a fully quantal treatment, actually changes sign under K in a classical path approximation, because the orbital motion and hence $\mathbf{L}$ is external to the states considered and thus fixed, whereas $\mathbf{j}$ changes sign under K [Eq. (93)].

On the other hand, interactions depending only on coordinate positions (e.g., Coulomb and exchange interactions) do obey "instantaneous" time-reversal symmetry:

$$KV(t)K^\dagger = V(t) \tag{101}$$

The equivalent matrix relations

$$\langle jm|V(t)|j'm'\rangle = (-1)^{j-j'm+m'}\langle j'-m'|V(t)|j-m\rangle \tag{102}$$

predict, for example, that atomic interactions $V(t)$ can not directly couple states $|jm\rangle$ and $|j-m\rangle$ for $j = \frac{1}{2}, \frac{3}{2}, \ldots$. The result [Eq. (102)] is simply an extension of Kramer's degeneracy.[17] The additional relation necessary, however, in order to ensure Equation (97) and related selection rules hold, is $V(t) = V(-t)$, which is valid only for spherically symmetric interactions V or for head-on ($b = 0$) collisions.

The selection rules may nevertheless be weakly obeyed. In particular, Eq. (101) guarantees that transitions $|jm\rangle \leftrightarrow |j-m\rangle$, $j = \frac{1}{2}, \frac{3}{2}, \ldots$, can occur only through second- and higher-order interaction terms. All first-order calculations including exponentiated ones[44] such as the Magnus approximation[45-48] (see Chapter 6, Section 4) must give zero amplitude for such transitions. If $j = \frac{1}{2}$, only terms connecting the $|jm\rangle$ state with a $j = \frac{3}{2}$, $\frac{5}{2}, \ldots$ state are able to violate extended time reversal [Eq. (98)] and cause $m = \frac{1}{2} \leftrightarrow -\frac{1}{2}$ transitions. The low-lying $S_{1/2}$ or $P_{1/2}$ states in an alkali, for example, can be collisionally depolarized only to the extent that the collision induces virtual transitions to the $P_{3/2}$ or higher levels. Of course the $j = \frac{1}{2}$ depolarization rates can still be considerably larger than the direct transfer rates for $S_{1/2} \to P_{3/2}$ or $P_{1/2} \to P_{3/2}$, since virtual transitions are "off the energy shell" (i.e., do not require conservation of energy), but they are generally smaller than $P_{3/2}$ disorientation rates.

In addition to time-reversal symmetry, there exists a reflection symmetry through the collision plane. This symmetry is dependent on the choice of the space-fixed "collision axis" and yields no selection rules once the average over collision-frame orientation has been performed. Nevertheless, it may give useful relations among elements of the collision-frame $\mathbf{S}$ matrix.

Let $\hat{\mathbf{z}}$ be normal to the collision plane and Σ_z be the reflection operator which for vector quantities such as a position vector $\mathbf{r} = (x, y, z)$

gives

$$\Sigma_z \mathbf{r} \Sigma_z^\dagger = \mathbf{r} - 2\hat{\mathbf{r}} \cdot \hat{\mathbf{z}}\hat{\mathbf{z}} \tag{103}$$

and for pseudovectors such as angular momentum $\mathbf{J}$ gives

$$\Sigma_z \mathbf{J} \Sigma_z^\dagger = -\mathbf{J} + 2\mathbf{J} \cdot \hat{\mathbf{z}}\hat{\mathbf{z}} \tag{104}$$

With usual phase conventions[17] we find

$$V = \Sigma_z V \Sigma_z^\dagger \quad \text{and} \quad H_0 = \Sigma_z H_0 \Sigma_z^\dagger \tag{105}$$

implies

$$S = \Sigma_z S \Sigma_z^\dagger \tag{106}$$

and hence

$$\langle jm|S|j'm'\rangle = (-1)^{j-j'-m+m'}\langle jm|S|j'm'\rangle \tag{107}$$

In particular, elements for which $(j - m) - (j' - m')$ is odd vanish.

Similarly, if $\hat{\mathbf{y}}$ instead of $\hat{\mathbf{z}}$ is the collision-plane normal, then

$$S = \Sigma_y S \Sigma_y^\dagger \tag{108}$$

and

$$\langle jm|S|j'm'\rangle = (-1)^{j-j'-m+m'}\langle j-m|S|j'-m'\rangle \tag{109}$$

3.4. Calculations of γ

Some model calculations and approximations yield especially simple relations among the multipole relaxation rates γ_L. We first treat these and then compare various numerical treatments. The models we consider are based on classical path treatments of the collision (see Chapter 6, Section 4).

(a) *Collisions with "Extended" Time-Reversal Symmetry.* As discussed in Section 3.3, in first-order approximations such as Born or Magnus approximations, instantaneous time-reversal symmetry [Eqs. (101) and (102)] also implies the extended results [Eqs. (95)–(98)] including, for $j = \frac{1}{2}, \frac{3}{2}, \ldots,$ the selection rule $|jm\rangle \leftrightarrow |j-m\rangle$. In addition these results are *exact* for spherically symmetric interactions V or for head-on collisions. For isolated $j = \frac{1}{2}, \frac{3}{2}$ levels, the γ_L are then related as follows [see, for example, Eq. (73)]

$$j = \tfrac{1}{2}: \quad \gamma_0 = \gamma_1 = 0 \tag{110a}$$

$$j = \tfrac{3}{2}: \quad \gamma_0 = 0, \quad \gamma_1 = \gamma_3 = \tfrac{1}{2}\gamma_2 \tag{110b}$$

(b) *j Randomization.*[28] In hard collisions, one might expect all $m \leftrightarrow m'$ transitions of a given j level to be equally probable. This is the j-

randomization limit, in which all transition rates $-\gamma_{nn}^{mm}$ in an isolated j manifold are equal: $-\gamma_{nn}^{mm} = k$, all $m \neq n$. The γ_L are then

$$\gamma_0 = 0 \tag{111a}$$

$$\gamma_L = (2j+1)k \qquad L > 0 \tag{111b}$$

(c) γ *in Terms of Collision-Frame S Matrix.* Other models to be discussed require explicit calculations of the S matrix. The elements of $\mathbf{S}$ in one frame contain sufficient information to determine $\mathbf{S}$ in any frame and hence, with integration over b and v, to determine the multipole relaxation rates γ_L. We establish here the relations between the collision-frame elements of $\mathbf{S}$ and the γ_L.[112,48-52]

It is convenient to use the Liouville operator $\boldsymbol{\Delta}$ [Eq. (30)], related to γ by Eq. (31):

$$\gamma = -2\pi n \int_0^\infty b \, db \langle v\boldsymbol{\Delta} \rangle_{v,\Omega} \tag{112}$$

Here $\boldsymbol{\Delta}$ may be calculated in the collision frame, and the average over orientations $\Omega = (\hat{\mathbf{b}}, \hat{\mathbf{v}})$ of this frame can be performed by expanding $\boldsymbol{\Delta}$ in $\mathbf{T}_{LM}\mathbf{T}_{L'M'}^*$ [see Eq. (67) and compare Eq. (70)]:

$$\langle \boldsymbol{\Delta} \rangle_\Omega = \sum_{LL'MM'} \mathbf{T}_{LM}\mathbf{T}_{L'M'}^* \langle \mathbf{T}_{LM}^* \cdot \boldsymbol{\Delta} \cdot \mathbf{T}_{L'M'} \rangle_\Omega$$

$$= \sum_L \Delta_L \mathbf{T}_{LM}\mathbf{T}_{LM}^* \tag{113}$$

where

$$\Delta_L \equiv \frac{1}{2L+1} \sum_N \mathbf{T}_{LN}^* \cdot \boldsymbol{\Delta} \cdot \mathbf{T}_{LN} \tag{114}$$

A cross section σ_L for relaxation of ρ_{LM} can then be defined

$$\sigma_L = -2\pi \int_0^\infty b \, db \, \Delta_L \tag{115}$$

with $\gamma_L = n\langle v\sigma_L \rangle$. Equation (114) contains the essential relation between collision-frame and multipole quantities. In terms of S-matrix elements [see Eqs. (30) and (78)] it yields

$$\Delta_L \begin{pmatrix} j & j' \\ k & k' \end{pmatrix} = \sum_{\substack{Mnn' \\ mm'}} (-1)^{j-k+n-m} \begin{pmatrix} k & L & k' \\ -n & M & n' \end{pmatrix} \begin{pmatrix} j & L & j' \\ -m & M & m' \end{pmatrix}$$

$$\times \langle kn|S|jm\rangle \langle k'n'|S|j'm'\rangle^* - \delta_{kj}\delta_{k'j'} \tag{116}$$

A useful sum rule follows:

$$\sum_L (2L+1)\, \Delta_L\begin{pmatrix} j & j' \\ k & k' \end{pmatrix} = \delta_{jk}\, \delta_{j'k'} \sum_{m,m'} [\langle jm|S|jm\rangle\langle j'm'|S|j'm'\rangle^* - 1]$$

(117)

An alternative form of Eq. (116) may be found by expanding the $\mathbf{S}$ matrix in $\mathbf{T}_{LM}$:

$$\mathbf{S} = \sum_{LMkj} \mathbf{S} \cdot \mathbf{T}_{LM}^* \mathbf{T}_{LM}(kj) \equiv \sum_{LMkj} S_{LM}(kj)\mathbf{T}_{LM}(kj)$$

(118)

where

$$S_{LM}(kj) = \sum_{mn} (-1)^{j-m}(2L+1)^{1/2}\begin{pmatrix} k & L & j \\ -n & M & m \end{pmatrix}\langle kn|S|jm\rangle$$

(119)

Substitution of Eq. (118) into (116) gives

$$\Delta_L\begin{pmatrix} j & j' \\ k & k' \end{pmatrix} + \delta_{kj}\, \delta_{k'j'} = \sum_{L'} (-1)^{L'-L-j-k'}\begin{Bmatrix} j & j' & L \\ k & k' & L' \end{Bmatrix} \sum_{M'} S_{L'M'}(kj)S_{L'M'}^*(k'j')$$

(120)

where a sum over four $3-j$ symbols has been contracted into a $6-j$ symbol.[6]

(d) $\mathbf{j} \cdot \mathbf{L}$ *Coupling.* Collisional mixing among the degenerate states $|jm\rangle$, $-j \leq m \leq j$, by $\mathbf{j} \cdot \mathbf{L}$ coupling of the electronic angular momentum $\mathbf{j}$ with the orbital angular momentum L associated with the relative motion of the colliding atoms can be easily treated in the time-dependent classical-path approach. The interaction is

$$\mathbf{V}(t) = A[R(t)]\mathbf{j} \cdot \mathbf{L}$$

(121)

The relative sizes of the perturbation matrix elements is determined by $\mathbf{j} \cdot \mathbf{L}$ whereas the absolute size of all elements varies as $A(R)$. A general state $\boldsymbol{\psi}$ in the degenerate manifold evolves in the interaction picture (Chapter 6, Section 4) according to

$$i\hbar\dot{\boldsymbol{\psi}} = \mathbf{V}(t)\boldsymbol{\psi}$$

(122)

The matrix, say, $\mathbf{B}$, which diagonalizes $\mathbf{V}(t)$

$$\mathbf{B}\mathbf{V}(t)\mathbf{B} = \boldsymbol{\nu}(t), \quad \text{a diagonal matrix}$$

(123)

is independent of t. The time evolution of $\boldsymbol{\phi} \equiv \mathbf{B}\boldsymbol{\psi}$ can consequently be written

$$i\hbar\dot{\boldsymbol{\phi}} = \boldsymbol{\nu}\boldsymbol{\phi}$$

(124)

The physical meaning of the diagonality of $\boldsymbol{\nu}$ in Eq. (124) is that the various components of $\mathbf{\Phi}$ evolve independently: the $2j + 1$ channels are completely decoupled.*

The only effect of the collision is to change the *phase* of each channel of $\mathbf{\phi}$:

$$\mathbf{\phi}(\infty) = e^{-2i\mathbf{\eta}}\mathbf{\phi}(-\infty) \tag{125}$$

where $\mathbf{\eta}$ is a diagonal matrix with elements

$$\eta_{nn} = \int_0^\infty dt \frac{\nu_{nn}(t)}{\hbar} \tag{126}$$

The interaction $\mathbf{V}(t)$ [Eq. (121)] is diagonal if the quantization axis $\hat{\mathbf{z}}$ is taken parallel to the classical vector $\mathbf{L}$:

$$V_{nm}(t) = \delta_{nm}A[R(t)]\hbar m|\mathbf{L}| \tag{127}$$

The $\mathbf{S}$ matrix in this frame is thus

$$\mathbf{S} = e^{-2i\mathbf{\eta}}, \qquad \eta_{nm} = \delta_{nm}m|\mathbf{L}| \int_0^\infty dt\, A[R(t)] \equiv \delta_{nm}m\alpha/2 \tag{128}$$

From Eq. (116) we then find the simple result $(j = j' = k = k')$

$$\Delta_L = \sum_{Mnn'} \begin{pmatrix} j & L & j \\ -n & M & n' \end{pmatrix}^2 e^{-i(n-n')\alpha} - 1$$

$$= (2L+1)^{-1} \sum_{M=-L}^{L} e^{-iM\alpha} - 1 \tag{129}$$

for relaxation of the Lth multipole. Rearranging the terms in Eq. (129), we obtain

$$\Delta_L = -(L + \tfrac{1}{2})^{-1} \sum_{M=1}^{L} (1 - \cos M\alpha) \tag{130}$$

To obtain the cross section σ_L, Δ_L must be integrated over impact parameter b; the final result depends on the detailed dependence of $A[R(t)]$. From Eq. (115) we can write

$$\sigma_L = (L + \tfrac{1}{2})^{-1} \sum_{M=1}^{L} \mathfrak{d}_M \tag{131}$$

where

$$\mathfrak{d}_M = 2\pi \int_0^\infty b\, db (1 - \cos M\alpha) \tag{132}$$

---

* A similar simplification can be made in the fully quantal approach (Chapter 6, Section 4) if and only if the L values are taken to be the same for all coupled channels. The same approximation is often made in the study of rotational excitation of molecules by collisions with atoms.[53]

The integrand $(1 - \cos M\alpha)$ usually oscillates rapidly about an average value of 1 out to some critical impact parameter b_M after which it drops rapidly to 0. Thus, approximately,

$$\sigma_M \simeq \pi b_M^2 \gtrsim \sigma_{M-1} \tag{133}$$

In the hard collision limit, the b_M are all equal and

$$\sigma_1 = \frac{5}{6}\sigma_2 = \frac{7}{9}\sigma_3 = \left(\frac{2}{3} + \frac{1}{3L}\right)\sigma_L \tag{134}$$

More generally, the dependence of α on b [see Eq. (128)] must be calculated:

$$\alpha(b) = \mu v b \int_{-\infty}^{\infty} dt\, A[R(t)] \tag{135}$$

Suppose $A(R) = C_n R^{-n}$. Then, taking straight trajectories

$$\alpha(b) = 2\mu C_n K_n b^{-n+2} \tag{136}$$

where

$$K_n = \int_0^{\pi/2} d\theta\, \cos^{n-2}\theta = \pi^{1/2} \frac{\Gamma[(n+1)/2]}{2\Gamma(n/2)} \tag{137}$$

Putting $M\alpha(b_M) \equiv 1$, we find

$$(L + \tfrac{1}{2})\sigma_L = \tfrac{3}{2}\sigma_1 \sum_{M=1}^{L} M^{2/(n-2)} \tag{138}$$

and

$$\sigma_1 = \tfrac{2}{3}\pi(2\mu C_n K_n)^{2/(n-2)} \tag{139}$$

With $n \to \infty$ the hard collision limit [Eq. (134)] is obtained. With $n = 6$

$$\sigma_1 = \tfrac{1}{3}\pi(2\mu C_6 K_6)^{1/2}$$

$$= (2L+1)\left(3 \sum_{M=1}^{L} M^{1/2}\right)^{-1} \sigma_L \tag{140}$$

$$= 0.6903\sigma_2 = 0.5328\sigma_3$$

By using an adiabatic basis (see Section 4.1.11 of Chapter 6) consisting of $A\Pi_{\pm 1/2}$ states, Gordeyev et al.[36] obtained disorientation cross sections for isolated $P_{1/2}$ states as in Eqs. (128–132), where, in terms of the amplitude $\sin \chi$ of $P_{3/2}$ mixed into the $\Pi_{1/2}$ state,

$$A[R(t)] = 3(\mu R^2)^{-1} \sin^2 \chi \tag{141}$$

See also Ref. 50.

(e) *Rotating Frame.* The perturbation $V(t)$ often is most simply expressed in the rotating frame in which $\hat{z}$ is always directed along the internuclear axis. Let $V^{(rot)}(t)$ be the potential in such a rotating frame. The space-fixed $V(t)$ is related by

$$V(t) = D[\Omega(t)]V^{(rot)}(t)D^+[\Omega(t)] \tag{142}$$

where Ω is the rotation from the space-fixed to the rotating collision frame. Let $U(t, -\infty)$ be the time-evolution matrix relating the initial $\psi(-\infty)$ to $\psi(t)$ in the space-fixed frame:

$$\psi(t) = U(t, -\infty) \cdot \psi(-\infty) \tag{143}$$

From the time-evolution equation for $\psi(t)$ [Eq. (122)], $U(t, -\infty)$ obeys

$$i\hbar\dot{U}(t, -\infty) = D[\Omega(t)]V^{(rot)}D^+[\Omega(t)]U(t, -\infty) \tag{144}$$

Substituting for

$$W \equiv D^+[\Omega(t)]U(t, -\infty)D[\Omega(t)] \tag{145}$$

we find

$$i\hbar\dot{W} = V^{(rot)}W - [J \cdot \dot{\Omega}, W] \tag{146}$$

The last term represents rotational coupling and can usually be treated with perturbation theory. After W is found by integrating Eq. (146), the S matrix is regained by the similarity transformation

$$S \equiv U(\infty, -\infty) = D[\Omega(\infty)]WD^+[\Omega(\infty)] \tag{147}$$

(f) *Adiabatic Representation.* The above analysis is carried one step further by the substitution [see Eq. (119) of Section 4.1 of Chapter 6]

$$V(t) = A(t)\varepsilon(t)A^{-1}(t) \tag{148}$$

where $\varepsilon(t)$ is a diagonal matrix whose elements are the adiabatic energies of the system and A transforms vectors from the adiabatic to the space-fixed frame. We assume here that the manifold of states before and after the collision is degenerate: $\varepsilon(t) \to 0$ as $t \to \pm\infty$. The analysis proceeds as in (e), and one finds

$$S = A(\infty)WA^+(\infty) \tag{149}$$

where now W is the solution of

$$i\hbar\dot{W} = \varepsilon W + i\hbar[W, A^+\dot{A}] \tag{150}$$

In an *adiabatic approximation*, the commutator in Eq. (150) is dropped.

Noting that $A(\infty) = D^{+}[\Omega(\infty)]$, one then obtains

$$\mathbf{S} = \mathbf{D}[\Omega(\infty)]\, e^{-2i\eta}\mathbf{D}^{+}[\Omega(\infty)] \tag{151}$$

where the elements of the diagonal phase-shift matrix are the phase shifts for elastic scattering on the various adiabatic surfaces

$$\eta(\infty) = \tfrac{1}{2}\int_{-\infty}^{\infty} dt\; \varepsilon(t) \tag{152}$$

Essentially the same result can be obtained from Eq. (147) for a single multiplet by ignoring $\dot{\Omega}$ and expanding $\mathbf{V}^{(\mathrm{rot})}$ in the adiabatic energies. Note that if the various phase shifts η_{ii} are equal, then $\mathbf{S}$ [Eq. (151)] is proportional to the identity matrix and no transitions are predicted.

For a $j = \tfrac{3}{2}$ ($l = 1$, $s = \tfrac{1}{2}$) state, there are two adiabatic surfaces. The S matrix and hence the cross sections σ_L can thus be calculated in terms of the difference $\Delta\eta$ in elastic scattering phase shifts for the two.

One finds

$$\sigma_0 = 0$$
$$\sigma_1 = \sigma_3 = \tfrac{1}{2}\sigma_2 = \frac{4\pi}{5}\int_0^{\infty} b\, db[1 - \cos\Delta\eta(b)] \tag{153}$$

The relative sizes of the σ_L are as given by Eq. (110), as indeed one could have predicted from the time-reversal symmetry of Eq. (150) when the commutator is set to zero.

(g) *Spin-Decoupled Approximation.* If the collision time is short compared to the precessional period of $\mathbf{l}$ and $\mathbf{s}$ about their resultant $\mathbf{j}$, i.e., if the energy uncertainty due to the finite duration of a collision is large compared to the fine-structure splitting, then $\mathbf{l}$ and $\mathbf{s}$ can be considered uncoupled during the collision: $\mathbf{s}$ will remain a fixed "bystander," while $\mathbf{l}$ is jerked about by the collisional perturbation.[54] The collision cross sections can then all be expressed as a linear combination of the multipole relaxation cross sections for l; $\sigma_L({}_{ll}^{ll})$, $0 \le L \le 2l$. The relations are given for a $P_{1/2}$, $P_{3/2}$ doublet ($l = 1$, $s = \tfrac{1}{2}$) in Table 2. If quenching from the doublet can be ignored, then the further simplification $\sigma_0({}_{ll}^{ll}) = 0$ can be made.

The spin-decoupled approximation for cross sections $\sigma_L({}_{kk'}^{ji'})$ is certainly valid for Li collisions at and above room temperature[43] but can lead to errors of roughly 20% in cross sections for Na (3^2P) depolarization at 400 K.[43,55,56] When Grawert first determined $P_{1/2}$ and $P_{3/2}$ mixing cross sections in the spin-decoupled approximation,[54] he introduced parameters a_0, a_1, and a_2. The relations of these parameters to the $\sigma_L({}_{ll}^{ll})$ are also given in Table 2.

An "l-randomization" approximation [compare (b) above] can be used to reduce the number of free parameters to 1: all $\sigma_L({}_{ll}^{ll})$, $L > 0$ are then equal, and $\sigma_0({}_{ll}^{ll}) = 0$.

Table 2. Coefficients of the Expansions $\sigma_L({}^{j\,j}_{k\,k}) = \sum_{L'=0}^{2} C_{L'}\sigma_{L'}({}^{l\,l}_{l\,l})$ and $a_x = \sum_{L'=0}^{2} d_{L'}\sigma_{L'}({}^{l\,l}_{l\,l})$ for $l = 1$; j and $k = \frac{1}{2}, \frac{3}{2}$

L	j	k	C_0	C_1	C_2
0	1/2	1/2	2/3	$-2/3$	0
1	1/2	1/2	26/27	$-2/9$	$-20/27$
0	1/2	3/2	$-\sqrt{2}/3$	$\sqrt{2}/3$	0
1	1/2	3/2	$\sqrt{10}/27$	$-\sqrt{10}/9$	$2\sqrt{10}/27$
0	3/2	3/2	1/3	$-1/3$	0
1	3/2	3/2	17/27	$-5/9$	$-2/27$
2	3/2	3/2	7/9	$-2/3$	$-1/9$
3	3/2	3/2	1	0	-1

x			d_0	d_1	d_2
0			1/9	1/3	5/9
1			1/54	1/36	$-5/108$
2			1/18	$-1/12$	1/36

An analogous spin-decoupled approximation can be used to treat the effects of hyperfine structure on the relaxation process (Section 3.6).

(h) *Numerical Calculations Using Classical Paths.* Numerous calculations of depolarization cross sections have been reported in which the time-dependent Schrödinger equation in the interaction picture [Eq. (122)] is numerically integrated with straight classical collision paths.[56–64] In such impact-parameter calculations (see Section 4.1 of Chapter 4), each integration over a path gives the **S** matrix for one impact parameter b. Then an integration over b and perhaps an average over initial relative velocity v must be performed.

Often simple power potentials $\mathbf{V}(t) = -\mathbf{C}(\hat{\mathbf{R}})R^{-m}$ are used with straight paths $R = b + vt$. The coefficient matrix $\mathbf{C}(\hat{\mathbf{R}})$ depends only on the direction of the internuclear axis $\mathbf{R}$. Thus for resonant dipole–dipole interactions, $n = 3$, and for nonresonant van der Waals potentials, $m = 6$. The velocity dependence of the cross section and a crude estimate of its size is given by πb_c^2, where b_c is a generalized Weisskopf radius at which $\mathrm{Tr}\,(\mathbf{C})\int_{-\infty}^{\infty} dt\, R^{-m} \simeq 1$. One thereby finds relaxation rates $\gamma = \langle nv\sigma \rangle \simeq n\pi[K\,\mathrm{Tr}\,(\mathbf{C})]^{2/(m-1)}\langle v^{(m-3)/(m-1)} \rangle$, where K is a numerical constant of order unity. Consequently, resonant R^{-3} interactions give rates which are velocity-independent, whereas van der Waals forces give rates varying as the 3/5 power of v. Perturbation[12,64] and Magnus approximation[48,65,66] calculations as well as numerical studies[52,58,64] verify these results. Such power-law calculations offer an attractive simplification: to treat a new atom pair one needs only scale the coefficient matrix $\mathbf{C}$ accordingly. Unfortunately, for more general interaction potentials scaling is not sufficient; the integration must again be performed "from scratch."

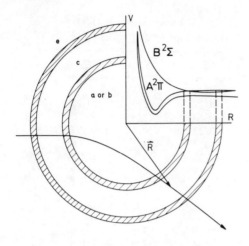

Figure 4. Regions of different coupling traversed during the collision of an alkali atom with a noble-gas perturber. Regions corresponding to Hund's cases a or b, c, and e are shown. The adiabatic potentials correlating to the lowest $P_{1/2}$ and $P_{3/2}$ states of Na are shown for the diatomic Na–Xe system.

Because of the tedium involved, attempts to simplify the calculations have been made, in particular by dividing the collision paths into regions amenable to different approximations, e.g., into regions where the rotation of the internuclear axis is insignificant and those where it is important.[63,67] Nikitin[68–73] in particular has exploited such techniques. He divides the classical paths for alkali–noble-gas collisions, for example, into regions characterized by Hund's coupling cases e, c, a or b, c, and e (see Table 3 of Chapter 6) as first the electronic angular momentum $\mathbf{j}$ is coupled to the orbital angular momentum $\mathbf{L}$ of the colliding pair, then to the internuclear axis $\hat{\mathbf{R}}$, then as $\mathbf{l}$ and $\mathbf{s}$ are coupled separately to $\mathbf{R}$, and finally back through the $\mathbf{j} \cdot \mathbf{R}$ and $\mathbf{j} \cdot \mathbf{L}$ coupling regions. Within each region, the system evolves adiabatically along the appropriate states and transition or matching matrices relate the states and representations from one coupling region to the next. Such calculations provide an easily accessible physical picture of the inelastic scattering process (see Fig. 4).

Results of the calculations depend of course not only on the approximations made to the scattering theory, but also, and more important for their physical significance, on the interatomic potentials assumed. Calculations have been performed for van der Waals forces alone, but it appears that the short-range repulsive part of the potential is also important,[23,36,60,61] and calculations with more realistic potentials[74–76] are certainly more reliable (see also Chapter 6). For resonant collisions, a simple R^{-3} potential is usually adequate, however (see Section 5).

Calculations of depolarization rates are always byproducts of good calculations of impact-parameter line broadening[58,66,77,78] (Chapter 27) or of fine-structure transitions (Chapter 29). The use of *curved* classical paths is better justified for depolarizing collisions—especially if only one adiabatic potential surface is directly involved (e.g., depolarization in a $P_{1/2}$ alkali state)—than for line broadening or fine-structure changing. There appears to be evidence[60,61,79] that such curved paths yield improved results, and in any case the difference between results with curved and with straight paths indicates a lower limit for the accuracy of the straight-path results.

Comparisons of theories and experimental results are given in Section 5. For further discussion the reader is referred to recent reviews.[5,68] The one by Nikitin[68] also contains a discussion of semiclassical calculations of transitions in nonadiabatic regions of classical paths.

(i) *Full Quantum Treatments.* Numerical integration of the time-independent Schrödinger equation and the determination of cross sections by matching multichannel wave functions to given asymptotic forms (see Section 4.3 of Chapter 6) has been performed for only a few alkali–noble-gas cases.[43,55,80,81] Such work gives the most reliable cross sections and indicates that impact-parameter results for depolarization cross sections are probably correct to about 10% in most cases. Detailed comparisons are given in Section 5.

3.5. Magnetic Field Dependence

Magnetic fields can influence the observed depolarization in several ways. The precession of multipoles ρ_{LM} about a field **B** at a rate ω_0 [ω_0 = the Larmor frequency (see Section 2.4)] can of course change the intensity and polarization of the resonance fluorescence, and indeed such changes allow the observation of natural decay and collisional depolarization through measurements of the Hanle effect (see Chapter 9). Coherence excited between levels which cross at higher fields can be used for similar measurements (again, see Chapter 3, and also Section 4.3 of the present chapter.)

In addition, Back–Goudsmit decoupling of the nuclear spin **I** and the electronic angular momentum **j** is often a convenient method of removing the influence of nuclear spin on depolarization (see the following section). At higher fields, similar effects can occur with the decoupling of the electronic orbital and spin angular momenta **l** and **s** in the Paschen–Back effect.

At fields where the Larmor precession period is roughly equal to the lifetime of the collision complex, the magnetic splitting of the levels becomes large enough to make the Massey parameter larger than unity (see Section 4.1g of Chapter 6). The collisions then become nearly adiabatic, and the inelastic cross sections drop rapidly with increasing field.[82-87] The critical field strength is about 10–100 G for complexes lasting as long as the natural lifetime of the excited state; it is roughly 10^6 G for a normal "fly-by" collision; and it lies somewhere in between for orbiting collisions or quasibound resonances.

At fields of several hundred kilogauss, Zeeman splittings become comparable to thermal energies. As a result of the Boltzmann factor in the detailed balance relation (Eq. 72), transitions $m = 0 \rightarrow m = 1$ and $m = 0 \rightarrow m = -1$ are no longer equal; collisions of an initially aligned state (say pure $m = 0$) may therefore induce orientation.

Another effect of applied magnetic fields is to change the absorption and emission spectrum of the sample. The restriction of optical excitation and detection to multipoles ρ_{LM} with $L \leq 2$ (see Section 2.7) is then lifted. Indeed direct measurements of γ_3 have recently been made with a strong field to enable optical resolution of the various Zeeman components[88] (see also Section 4). A shift in the absorption profile relative to that of the exciting light can, by means of the Doppler effect, also change the average velocity of the excited atoms and destroy the isotropic character of the collisions in a fluorescence cell.[89,90]

Of these magnetic field effects, the influence of a small degree of Paschen–Back decoupling on a nearly forbidden transition is, perhaps, the least well-understood. The influence is discussed further in the remainder of this section, with special reference to disorientation in alkali $P_{1/2}$ states.

As discussed in Section 3.3, general symmetry considerations lead to the conclusion that, although the $j = \frac{1}{2}$, $m = \frac{1}{2} \leftrightarrow j = \frac{1}{2}$, $m = -\frac{1}{2}$ transition is not strictly forbidden, it proceeds only in second order through virtual excitation to the $P_{3/2}$ state. This explains the dependence $\sigma_1 \propto \Delta\varepsilon^{-2}$ of the disorientation cross section σ_1 in the $j = \frac{1}{2}$ state on fine-structure splitting $\Delta\varepsilon$ for the metals Tl, Rb, and Cs, and it also explains why for ^2P states with large $\Delta\varepsilon$, σ_1 is always significantly smaller in the $P_{1/2}$ state than in the $P_{3/2}$ one (Section 5).

It became apparent some years ago that $\sigma_1(\frac{1}{2})$ values measured for the $6P_{1/2}$ state of Cs in collisions with noble–gas atoms are found to be significantly larger at fields of about 10 kG[34] than in low fields,[30] even after corrections[91] for nuclear spin (see next section) had been applied to the low-field results. The theory proposed[92] to explain the results suggested that the small admixture of $P_{3/2}$ state (amplitude of about 10^{-3} at 10 kG) into the $P_{1/2}$ state could explain the enhancement of $\sigma_1(\frac{1}{2})$ for $P_{1/2}$. The $\sigma_1(\frac{3}{2})$ cross sections for $P_{3/2}$ range from about 100 to 400 Å^2. For impact parameters $b \leq b_1$ where $\sigma_1(\frac{3}{2}) \simeq \pi b_1^2$, transitions $j = \frac{3}{2}$, $m = \frac{1}{2} \leftrightarrow \frac{3}{2}$, $-\frac{1}{2}$ are saturated. For smaller b, we may think of the transitions as occurring faster, many times during a collision, so that even a small admixture of $P_{3/2}$ in the $P_{1/2}$ state could have significant results. Some doubt has recently been cast[93] on the ability of the theory to account for the measurements, and indeed it has been suggested that the measurements themselves may be in error. However, recent work[94-96] indicates the effect exists not only in Cs, but also in K and Rb.

3.6. Complications of Nuclear Spin

Most atoms have a nuclear spin **I** which can significantly complicate the interpretation of experimental data.[91,97-112] Sometimes the experiment can be performed in a sufficiently strong magnetic field to decouple **I** from the electronic spin, but at other times it is not possible (e.g., when a Hanle measurement is made) or such a magnetic field induces new, perhaps not fully calculable effects (see Section 3.5). It is then desirable to make the straightforward but often tedious corrections for the nuclear spin discussed in this subsection.

An important simplification is made possible by the relative sizes of the collision time and precessional period of **I** and **j** about their resultant **f**. The hyperfine precession periods are typically 10^{-7} to 10^{-9} s. The collisions, in comparison, are practically instantaneous: they last from 10^{-12} to 10^{-13} s. Furthermore, the direct collisional interaction is with **j**; **I** participates only through its coupling with **j**. As a result, **I** moves negligibly during the collision whereas **j** may be jerked about. Thus **I** and **j** are effectively decoupled when the collision occurs.[28] Only afterwards, once **I** and the new **j** again process about their resultant, does **I** indirectly "feel" the collision.

The nuclear spin acts like a flywheel which tends to maintain the angular momentum of the system in the presence of many short perturbations. Depolarization rates are thus usually reduced by the presence of a nonvanishing **I**, from 10% to 20% for $\langle J_z \rangle$ to about a factor of 3 for $\langle J_y \rangle$.

It is straightforward to find hyperfine relaxation rates $\gamma_L \binom{f \ f'}{f_1 f_1'}$ in terms of the electronic ones $\gamma_L \binom{j \ j'}{j_1 j_1'}$. One uses an equation analogous to Eq. (70) but with tensor operators $T_{LM}(ff')$ which can be expanded[5]

$$\mathbf{T}_{LM}(ff') = \sum_{\substack{kk' \\ m,m'}} [(2f+1)(2f'+1)(2k+1)(2k'+1)]^{1/2}$$

$$\times \langle kk'mm'|LM \rangle \begin{Bmatrix} j & I & f \\ j' & I & f' \\ k & k' & L \end{Bmatrix} \mathbf{T}_{km}(jj') \mathbf{T}_{k'm'}(\mathrm{II}) \tag{154}$$

Then from the spherical symmetry of γ and orthogonality of the $\mathbf{T}_{k'm'}(\mathrm{II})$ and of the Clebsch–Gordan coefficients, one finds directly

$$\gamma_L \binom{ff'}{f_1 f_1'} = \mathbf{T}_{LM}^*(f_1 f_1') \cdot \gamma \cdot \mathbf{T}_{LM}(ff')$$

$$= [(2f+1)(2f'+1)(2f_1+1)(2f_1'+1)]^{1/2} \sum_{kk'} (2k+1)(2k'+1)$$

$$\times \begin{Bmatrix} j & I & f \\ j' & I & f' \\ k & k' & L \end{Bmatrix} \begin{Bmatrix} j_1 & I & f_1 \\ j_1' & I & f_1' \\ k & k' & L \end{Bmatrix} \gamma_k \binom{j \ j'}{j_1 \ j_1'} \tag{155}$$

To apply the hyperfine relaxation rates, the time evolution of the density matrices $\rho_{LM}(ff')$ must be derived, as must expressions for the source and detection operators $\mathscr{S}$ and $\mathbf{D}$ in the hyperfine basis. Multipole moments of $\mathscr{S}$ and $\mathbf{D}$ are found by simple applications of Eq. (154), and to find the time evolution of $\rho_{LM}(ff)$ one simply takes components on the $T_{LM}(ff)$ of the general evolution equation (see Section 2.3)

$$\dot{\boldsymbol{\rho}} = -i\mathbf{L}_0 \cdot \boldsymbol{\rho} - \Gamma\boldsymbol{\rho} - e^{i\mathbf{L}_0 t} \cdot \boldsymbol{\gamma} \cdot e^{i\mathbf{L}_0 t} \cdot \boldsymbol{\rho} + \mathscr{S} \tag{156}$$

Details were first worked out for the $j = \frac{1}{2}$ case[91] and then generalized.[97,112] In both cases, coherence between levels of different f values was ignored, as prescribed by the secular approximation when the level separation is large compared to the natural linewidth (see Section 4.3). For the $j = \frac{1}{2}$ case, for example, one obtains*

$$\langle J_M \rangle = \frac{\mathscr{S}_{1M}}{3(2I+1)^2} \frac{\Gamma[1 + \frac{1}{2}(2I+1)^2] + 3\gamma_1 + \frac{3}{2}i\omega M(2I+1)}{(\Gamma+\gamma_1)[\Gamma + 2\gamma_1(2I+1)^{-2}] + \omega^2 M^2 - i\omega M\gamma_1(2I+1)^{-1}} \tag{157}$$

where $\mathscr{S}_{1M}$ is the source term for $\langle J_M \rangle$, γ_1 is the electronic disorientation rate in the $j = \frac{1}{2}$ state, and ω is the hyperfine precession rate in the magnetic field $\mathbf{B}$: $\omega = g_j\mu_B B[\hbar(2I+1)]^{-1}$. The dependence of $\langle J_y \rangle = -\mathrm{Im}\langle J_{-1}\rangle 2^{1/2}$ on B is seen not to have quite the Lorentz shape typical of a Hanle signal. The deviation in shape is not great, however, and the effect of nuclear spin can be approximated by an effective rate $\gamma_{\mathrm{expt}} = \alpha\gamma_1$, which is smaller than γ_1 by a factor $\alpha = \frac{1}{3} + \frac{2}{3}(2I+1)^{-2}$ at low perturber pressures.

The longitudinal component $\langle J_z \rangle$ ($M = 0$) is seen from Eq. (157) to relax with two rates, a fast one $\Gamma + \gamma_1$ and a slow one $\Gamma + 2\gamma_1(2I+1)^{-2}$. Results for $\langle J_z \rangle$ can—and in fact first were—derived more simply from a diagonal representation (see Section 2.6).[104,111]

If states of more than one j level are considered, transfer between states of different j may alter the results, especially when transfer rates are comparable to depolarization rates, e.g., for Li, Na: $P_{1/2} \leftrightarrow P_{3/2}$ (see following chapter). The factor α and similar parameters for the $P_{3/2}$ state

---

* The result [Eq. (157)] is sensitive to the spectral profile of the source. Here a "white" profile was assumed so that $\mathscr{S} = \mathscr{S}_0 \mathbf{1} + \mathscr{S}_1 \cdot \mathbf{J}$ where $\mathscr{S}_0$ and $\mathscr{S}_1$ are constants. Equation (157) is readily derived from the relations in the hyperfine basis

$$\mathbf{F} \cdot \boldsymbol{\gamma} = \gamma_1 \mathbf{J}$$

$$\mathbf{J} \cdot \boldsymbol{\gamma} = \gamma_1[1 + 2(2I+1)^{-2}]\mathbf{J} - 2\gamma_1(2I+1)^{-2}\mathbf{F}$$

$$\mathbf{J} = \mathbf{J} \cdot \mathbf{F}\mathbf{F}F^{-2}$$

The results for $\boldsymbol{\gamma}$ can be proved using z components and a diagonal population representation (FM). Results for general components follow from the isotropic symmetry of collisions.

have been calculated for alkali systems considering population transfer but ignoring coherence transfer[113,114] and data for Na and Rb accordingly have been reanalyzed.[112]

The approximation of ignoring coherence between levels of different f values of course fails near high-field crossings of such levels. If such a crossing can be isolated, however, the secular approximation offers a great simplification: only one element of the matrix $\boldsymbol{\gamma}$ in the hyperfine basis need be considered. As discussed further in Section 4.3, measurements of such crossings[97,115–117] can sometimes provide an indirect measurement of $\gamma_L, L > 2$.

Frequently when nuclear spin I is present, an experiment must be performed in an intermediate magnetic field, where $\mathbf{I}$ and $\mathbf{j}$ are neither fully coupled nor fully uncoupled. A direct extension of the techniques discussed above for handling hyperfine structure at low fields, e.g., the expansion Eq. (155), may be difficult to apply, especially if (as is unfortunately usually true) the coupling changes significantly over the region of field spanned by the measurement. An alternative approach is then to use a field-independent basis of nonstationary states, probably $|jIm_jm_I\rangle$ are most convenient. The Liouville operator $\mathbf{L}$ is then no longer diagonal in the product space, but the matrices which must be inverted to find the steady-state solutions to $\dot{\boldsymbol{\rho}} = 0$ [see Eq. (156)] are nevertheless limited in size by the axial symmetry of $H_0 + V$ and $\boldsymbol{\gamma}$, since only states of like $m_j + m_I$ can couple. Furthermore, the product $\boldsymbol{\gamma} \cdot \boldsymbol{\rho}$ as well as $\mathcal{S}$ and $\mathbf{D}$ are especially simple in the basis $|jIm_jm_I\rangle$.

4. Experimental Methods

4.1. Traditional Measurements in Fluorescence Cells and Flames

The most widely employed method of measuring the effects of collisions on excited states is to excite atoms A in a vapor, usually by irradiating them with resonance radiation but sometimes by electron bombardment,[118] and then to follow their decay by observing the emitted fluorescence.[1,119,120] If N_i is the total number of quanta spontaneously radiated from the initial excited state and N_f that from another excited state, then under the usual experimental condition that $N_f \ll N_i$, the ratio N_f/N_i gives directly the rate γ_{fi} of collisional transfer $i \rightarrow f$ in terms of the spontaneous decay rate Γ_i of the initial state:

$$N_f/N_i = \gamma_{fi}/\Gamma_i \tag{158}$$

It is usually assumed that both emitters and perturbers have thermal isotropic Maxwell–Boltzmann distributions of velocity

$$f(\mathbf{v}) \, d^3v = \pi^{-3/2} u^{-3} \exp\left(-v^2/u^2\right) d^3v \tag{159}$$

where u is the most probable velocity

$$u = (2kT/m)^{1/2} \tag{160}$$

in which case the *relative* velocity distribution has the same form but with m replaced by the reduced mass μ. (For flames, the velocities of perturber and emitter are taken relative to a mean flame velocity.) Results are often presented in terms of a thermal cross section $Q_{fi}(T)$ defined by

$$\gamma_{fi} = n\bar{v} Q_{fi}(T) \tag{161}$$

where $\bar{v} = 2\pi^{-1/2}(2kT/\mu)^{1/2}$ is the mean relative velocity.

Such experiments are typically run under steady-state conditions whereby the excitation and fluorescent rates are constant and the number of quanta from a given atomic transition is proportional to the intensity divided by the frequency of the radiation. The same simple analysis [Eq. (158)] is equally valid for pulsed or modulated excitation, however.

The inverse reaction $f \to i$ can also be observed, and indeed if thermodynamic equilibrium could be reached between the two levels, the total number of transitions $i \to f$ would be just balanced by the inverse reaction. Thus if N_i and N_f are the density of A atoms in state $|i\rangle$ and $|f\rangle$, respectively, then

$$\gamma_{fi} N_i = \gamma_{if} N_f \tag{162}$$

But then also $N_i = (g_i/g_f)N_f \exp\left[-(E_i - E_f)/kT\right]$, where E_i and E_f are the energies of the states and g_i and g_f are the degeneracies. Consequently,

$$\gamma_{if} = \gamma_{fi}(g_i/g_f) \exp\left[-(E_i - E_f)/kT\right] \tag{163}$$

This relation between the forward and reverse transition rates (and hence between thermal cross sections) depends only on the temperature, the energies, and the degeneracies. It is valid whether or not equilibrium is established between the levels. In fact it can be proved by detailed balance, i.e., by using microreversibility of the quantum-mechanical transition operator (see Chapter 6). Its validity *is* contingent on having a thermal relative velocity distribution in both states, however. The relation [Eq. (163)] is frequently used to check results for systematic errors.

If the polarization of the fluorescence is monitored for various incident polarizations, depolarization and coherence transfer rates can be found. The quantity measured is often the "degree" of circular (P_{circ}) or linear (P_{lin}) polarization

$$P_{\text{circ}} = \frac{I_{\sigma^+} - I_{\sigma^-}}{I_{\sigma^+} + I_{\sigma^-}} \tag{164}$$

or

$$P_{\text{lin}} = \frac{I_\| - I_\perp}{I_\| + I_\perp} \tag{165}$$

where I_{σ^+} and I_{σ^-} are intensities of fluorescence which is circularly polarized with helicities ± 1, and $I_\|$ and $I_\perp$ that which is linearly polarized parallel and perpendicular to a given axis. For a $j = \frac{1}{2}$ system, the circular polarization has a simple Stern–Volmer[121] form:

$$P_{\text{circ}} = P_0(1 + \gamma_1/\Gamma)^{-1} \qquad j = \frac{1}{2} \tag{166}$$

Similarly, for $j = 1$, P_{lin} has the same form:

$$P_{\text{lin}} = P_0(1 + \gamma_2/\Gamma)^{-1} \qquad j = 1 \tag{167}$$

In both cases, P_0 is the degree of polarization in the absence of collisions, and γ_1 and γ_2 are the collisional relaxation rates for orientation and alignment, respectively. In other cases, however, the degree of polarization has a more complicated pressure dependence, and some other measure of the polarization may be preferable (see Sections 2.1 and 4.2).[122]

The primary advantage of these traditional vapor cell and flame measurements is the simplicity of the experimental apparatus and analysis. The collisions are usually isotropic and have velocity distributions well characterized by a thermal distribution [Eq. (159)]. Furthermore, the density N of perturbers, which is frequently the factor which limits the accuracy to which the thermal cross section $Q_{fi}(T)$ can be determined, can be measured with relatively small error, typicalty 5% to 10%.

4.2. Radiation Trapping and Other Problems

Even relatively simple experiments often offer ample opportunity for difficulty and misinterpretation. The following is a brief discussion of some of the common problems. Their occurrence is not restricted to cell and flame experiments.

(a) *Density of Atoms A.* Analyses usually consider a *single* atom A in a bath of perturbers B, whereas the experiment itself usually requires a few more, typically $\geq 10^8$ atoms of type A. The actual density of atoms A is not important *providing* it is low enough that radiation at the excitation and detection frequencies can pass unhindered through the sample. More precisely, one requires that the mean free path l_{ν_0} of photons at the center of both excitation and emission lines be large compared to the path length through the sample.

The mean free path l_ν for a spectral line of oscillator strength f at the frequency ν is

$$l_\nu^{-1} = \frac{\pi N e^2}{mc} f b(\nu) \tag{168}$$

where N is the density of lower-state atoms and $b(\nu)$ is the normalized profile. For a Doppler line shape characterized by the most probable velocity $u = (2kT/m)^{1/2}$,

$$b(\nu_0) = \pi^{-1/2}\lambda_0/u, \qquad \lambda = c/\nu_0 \tag{169}$$

The optical depth of a point a distance l into the sample is simply $\tau_\nu = l/l_\nu$.

If the *incident* radiation is partially absorbed, then the spatial and velocity distribution of excited atoms will be different than for ground-state ones due to the intensity and profile dependence of the radiation as a function of position [see (c) below].

The effect of nonnegligible optical depth on the detected radiation causes more serious problems. Additional absorption and reemission of the radiation lengthens the apparent lifetime Γ^{-1} of the excited state, a phenomenon called *radiation trapping* or *imprisonment* and studied both theoretically[123-131] and experimentally.[130-134] Of course, the polarization of the reemitted radiation is not random, but generally favors the polarization of the radiation absorbed. For example, in a $j = 0$ to $j = 1$ transition, twice-scattered radiation has on the average 7/9ths the degree of linear polarization and 5/9ths the degree of circular polarization possessed by the direct (once-scattered) resonance fluorescence. Thus the lifetime of the polarization can also be enhanced by radiation trapping. The phenomenon is known as *coherence narrowing* and has been observed in numerous experiments.[130,131,134-139] An accurate treatment of radiation trapping and coherence narrowing is complex and geometry-dependent. The optical depth, exciting radiation, and density of excited states are all functions of the position, and there is spherical symmetry for at most one point in the sample. Theoretical treatments are usually limited to the simpler case of an infinitely extended, isotropic medium.[129] Then, as with isotropic collisions (Sections 2 and 3), multipoles of the density matrix relax independently, and the effects of trapping can be described by one decay constant for each multipole excited. If asymmetries are taken into account, additional relaxation rates play a role.[128,130]

As the density of atoms A is made higher, resonant collisions cause depolarization and broadening which eventually dominate the trapping and coherence narrowing. Depolarization and line-broadening cross sections have been determined for various resonant collisions (see Section 5 and Chapter 27), and as a rule of thumb, they become important relative to the effects of trapping when the density of atoms A is roughly λ_0^{-3}. One should recognize that the resonant cross sections with A may be orders of magnitude larger than the nonresonant ones with B, and a relatively low density of A atoms may have a significant effect. If, as in the case of Sr and Ba, a metastable state lying a short distance below the excited state is optically pumped during the excitation, it may well be more important than the ground state in inducing depolarization or fine-structure transitions.[135,136]

For studies of nonresonant collisions, an extrapolation to zero density of atoms A can be made. Such an extrapolation is, however, not generally feasible when resonant collisions themselves are the subject of study. One method, then, of circumventing radiation trapping effects is to use two or more different isotopes with nonoverlapping spectral lines. The isotope to be optically excited is kept at low densities while another may have relative high densities to provide large relaxation rates.[64,97] Another method is to measure the excited state by means of transitions to other excited levels, the population of which is much less than of the ground state.[137,139] Note that there is less of a problem with coherence narrowing when a rich nonoverlapping hyperfine spectrum is present.[132]

(b) *Density of Atoms B.* The rates and cross sections cannot be measured more accurately than the relative density values. Pressures of nonreactive gases in a cell may be directly measured by oil manometers or mercury McLeod gauges. (Of course one takes care to pump the mercury free of contaminent gases and to ensure that the gauge is really at the same pressure as the cell.) For vapors of liquids or solids it is tempting to use standard vapor-pressure tables[142] to determine the pressure from the temperature. However, these are often not sufficiently reliable at low pressures[143] or in the presence of other gases, and other measurements, such as of the absorption profile,[144] may be necessary to obtain the desired accuracy.

The role of perturbers in broadening fluorescent profiles should not be overlooked. Three-body collisions, for example, have been observed to reduce resonant depolarization rates through their line-broadening effect.[139-141]

(c) *Light Source.* The traditional light source is typically an rf or microwave discharge, a hollow cathode or a commercial spectral lamp. At low intensities the emitted spectral line has roughly a Voigt profile, corresponding to a mixture of Doppler and Lorentz (or resonant or Stark) broadening. At intermediate intensities, the profile flattens somewhat, and at higher intensities the wings become quite broad and there is typically a marked self-reversal, i.e., a decrease in intensity at line center due to reabsorption by cooler vapor in the outer layers. Only a "white" light source, i.e., one with constant intensity for frequencies near the spectral line, gives an excited-state velocity distribution equal to that in the ground state. A profile peaked at line center excites atoms with an effectively lower temperature than that in the ground state, whereas a self-absorbed line shape excites atoms with an effectively higher temperature. In fact, recent experiments[89,90] with laser light sources have used such effects to select the collision velocity and to eliminate effects of Doppler broadening. The profile of the radiation incident on the sample should be monitored to control the effective temperature of the excited atoms A.

(d) *Background Radiation.* Background radiation poses a problem when its intensity depends roughly linearly on perturber pressure, as for example a background arising from the far pressure-broadening wing of a second spectral line. Since the spectral line to be observed usually has a very different profile than the background, the latter can generally be determined and eliminated by making measurements with two spectral filters of different bandwidths.[145]

(e) *Polarization.* Essentially any excitation (or detection) of a $j > \frac{1}{2}$ state with either unpolarised or linearly polarized electric dipole radiation creates (or monitors) not only the population ($L = 0$) component of the ensemble, but also the alignment ($L = 2$) component, and these components generally behave differently under trapping [see (a) above] and collisional perturbation. Neglect of polarization and depolarization can cause serious error in measurements of, say, rates of energy transfer or of population decay.

One way to avoid problems with polarization in such experiments is to polarize the incident radiation in such a way that the observed fluorescent intensity is independent of the extent of collisional disalignment. Since the radiation pattern for an electric dipole varies as $\sin^2 \theta$, where θ is the angle between the alignment direction (i.e., the polarization direction) and the direction of the detector, one would choose for unpolarized detection $\sin^2 \theta = \frac{2}{3}$, hence $\theta \simeq 55°$, the so-called "magic angle."[130] The presence of stray magnetic fields will of course further complicate the measurement by causing any coherent multipole, whether intended or not, to precess about the field direction.

A problem that may arise when depolarization rates are measured is that the traditional measurements of the polarization P [Eqs. (164) and (165)] generally mix the excited-state orientation and alignment. This actually presents no difficulty unless one ignores the correct multipole dependence and interprets the data according to, say, the Stern–Volmer formula [Eqs. (166) and (167)]. It may be preferable to use other ratios of fluorescent intensities which do isolate the multipoles. (See the example of a $j = \frac{3}{2}$ state above, in Section 2.7.)

(f) *Other Limitations.* In addition to the possible pitfalls discussed above, the traditional cell and flame measurements have several inherent limitations. They measure thermal averages of cross sections, and more often than not, measurements are reported at only one temperature. Yet many cross sections have strong velocity dependences which often need to be known in order to distinguish between competing theories. Measuring over a large range of temperatures[146] is almost as useful as the velocity dependence providing one does not try to observe features such as orbiting resonances or glory oscillations (see Chapter 6) which vary rapidly with velocity. Measurements are frequently restricted at high temperatures by background radiation or, for all measurements, by the stability of the container, although lock-in detection helps the former, and special cell or heat-pipe designs and the use of flames themselves relieve the latter restriction. At low temperatures, measurements may be limited by the vapor pressure of one of the components.

Another limitation is that only total cross sections averaged over all angles with respect to the quantization axis are measured. The Doppler shift further limits measurements by spreading the spectral line so that one may not be able to resolve hyperfine or Zeeman components. A result is that observation of radiation emitted in spontaneous decay (or excitation from an unpolarized ground state) permits the detection (or excitation) of only $L = 0$, 1, and 2 multipoles of ρ. Observation of, for example, octupole relaxation rates can then occur only indirectly through hyperfine mixing of the multipoles. Finally, the elements for which fluorescent cells can be used are of course limited to those which are easily vaporizable. Indeed this is the obvious experimental reason for the concentration on alkali metals and a few other elements, such as mercury, cadmium, lead, and thalium for measurements in cells. Flames have much less limitation in this regard, although there too, many elements have too low a vapor pressure to be reasonable candidates for investigation.

Largely to circumvent such limitations, other methods have been developed. In the rest of this section we consider some of these.

4.3. Broadening of Hanle, Level-Crossing, and Double-Resonance Signals

In the Hanle effect[197] (Chapter 9), the decay and precession of a coherent $L = 1$ or 2 multipole gives, in a plot of fluorescent intensity against magnetic field B, a Lorentz and/or dispersion shape curve, the width of which is proportional to the decay rate. A plot of decay rate vs. perturber density yields directly the thermally averaged cross section for relaxation of the observed multipole (see also Sections 2.4, 2.5, and 3.6).

In addition to the description in terms of precessing coherence, a time-independent picture can also be used: overlapping stationary states which are coherently excited can interfere during decay, since one cannot determine even in principle from which state the observed radiation eminates. As B is increased, the interference disappears as the overlap vanishes, i.e., as the separation of the stationary levels becomes large compared to the natural linewidth. In the time-dependent description, the coherence then precesses many times during the radiative lifetime and has a nearly vanishing average value. In both descriptions, the coherence or interference of two levels is only important when the stationary energies are within a few level widths of each other.

The effective level width is larger in the presence of collisions. It is the sum of natural decay and collisional-relaxation rates for the observed multipole. Thus, two levels which are widely enough separated in the absence of collisions that their coherence (or interference) can be ignored, may contribute significant coherence when collisions are present. In the time-dependent description, the coherent multipole may not be able to precess a full cycle before a collision occurs.

Coherence can be important whenever two upper levels which can be excited from the same lower level come with a few level widths of each other. This can occur not only at zero-field crossing, but also at high-field crossings and at "anticrossings,"[148] achieved with either electric or magnetic fields. It can also occur by "dressing" one level with a rf field so that the total energy (of the atom plus n photons of the rf field) nearly equals that of the other level (see Chapters 3 and 9 for a further discussion of such double-resonance methods). If the coherence is measured as a function of perturber density, the collisional relaxation cross sections can be found.

High-field level crossings can occur between fine-structure levels, but more often occur with hyperfine structure (see Section 3.6). Although the general analysis in terms of relaxation rates is complex, *isolated* high-field crossings permit a significant simplification. A crossing is isolated if all other pairs of interfering levels are separated by many level widths. Then, providing the angular momentum coupling does not change significantly, the level-crossing signal in the region of the crossing can be written as the sum of a nearly constant background and a term due to the coherence between the two stationary states that cross.[52,97,115-117]

Formally, one first recalls that the collision matrix γ is defined in the interaction picture, where under constant collision conditions it is time-independent. In the Schrödinger picture, its time dependence is given by

$$\gamma_{mn}^{m'n'}(t)_{\text{Schr}} = \gamma_{mn}^{m'n'} e^{-i(\omega_{mn}-\omega_{m'n'})t} \tag{170}$$

The evolution of ρ_{mn} is therefore given by

$$\dot{\rho}_{mn}\mathscr{S}_{mn} - [\Gamma_{mn} + i\omega_{mn}]\rho_{mn} - \sum_{m'n'} \gamma_{mn}^{m'n'} e^{-i(\omega_{mn}-\omega_{m'n'})t}\rho_{m'n'} \tag{171}$$

From the axial symmetry of collisions, the difference in azimuthal quantum numbers for states n and m is equal to that for n' and m'. Consequently, if the Landé g factor is the same for all four states m, n, m', and n' (as, for example, when all are Zeeman sublevels with the same total angular momentum), the difference $\omega_{m'n'} - \omega_{mn}$ vanishes.

For an isolated crossing of levels n and m the exponential factor in Eq. (171) averages all these terms to zero unless $m' = m$ and $n' = n$. The steady-state solution of Eq. (171) is thus

$$\rho_{mn} = \mathcal{S}_{mn}[i\omega_{mn} + \Gamma_{mn} + \gamma_{mn}^{mn}]^{-1} \tag{172}$$

so that the width of the isolated level-crossing signal is directly proportional to the average natural decay rate of the levels, $\Gamma_{mn} = (\Gamma_m + \Gamma_n)/2$ (see Section 2.3) plus the coherence relaxation rate γ_{mn}^{mn}. Assuming the nuclear spin $\mathbf{I}$ to be decoupled from $\mathbf{j}$ during collision (see Section 3.6), the measured rate γ_{mn}^{mn} for relaxation of coherence between levels $|m\rangle$ and $|n\rangle$ is given in terms of elements in the $|jm\rangle$ basis by

$$\gamma_{mn}^{mn} = \sum_{\substack{m_4 m_2 m_3 m_4 \\ M_4 M_2 M_3 M_4}} \langle m|j, Im_1 M_1 \rangle$$

$$\times \langle j_2 Im_2 M_2|n\rangle\langle m|j_3 Im_3 M_3\rangle\langle j_4 Im_4 M_4|n\rangle \gamma_{m_3 m_4}^{m_1 m_2}\begin{pmatrix} j_1 & j_2 \\ j_3 & j_4 \end{pmatrix} \tag{173}$$

Relations to the multipole relaxation rates (Sections 3.1 and 3.2) and to S-matrix elements (Section 3.3) are readily found.[52,97] Such isolated crossings have been recently used to measure coherence relaxation rates γ_{mn}^{mn} in Rb.[97,115–117]

4.4. Time-Resolved Measurements; Delayed Coincidence

Time-resolved observations provide a direct measurement of multipole decay rates and hence of the influence of collisions on such rates. In a typical experiment the excitation source is pulsed on for a time shorter than the lifetime to be measured, [for this purpose either a pulsed laser, an electrooptical cell (e.g., Kerr cell or KDP crystal), or, for long lifetimes, a mechanical shutter can be used] and the fluorescence is measured as a function of time after excitation. (See also Chapter 14.)

In the simplest measurement of excited population ($L = 0$) decay, one sees an exponential decrease of intensity with time corresponding directly to the loss of excited-state atoms. Such measurements are useful for determining natural lifetimes,[130,149,150] lifetimes of metastable states against collisional destruction,[151] quenching rates of excited states,[152] and radiation trapping.[132,133]

In the absence of collisions or trapping, all multipoles of ρ relax with the same decay rate as the population ($L = 0$) component. The different decay rates which generally occur due to trapping or collisional relaxation can be sorted out for $L = 0, 1, 2$ by exciting and observing with variously polarized components of the radiation or by performing the experiment in the presence of an external magnetic field which causes $|M| \neq 0$ parts to precess.[130,153]

Observation of level-crossing signals in delayed fluorescence permits a larger average rotation of the coherent moments excited for a given level spacing, thus enabling a better resolution of nearly overlapping crossings.[154-156] By optically pumping the ground state, resolution can also be increased by changing the relative amplitudes of competing level crossings.[157,158] Any analysis of collisional broadening of the level-crossing signals would unfortunately, have to include the effect of collisions on the efficiency of optical pumping.

4.5. D_2 Optical Pumping

A more direct application of optical pumping (see Chapter 9) to the measurement of excited-state relaxation can be made when the efficiency or even the direction of the pumping depends strongly on collisional mixing of excited levels. The scheme has been successfully applied to alkali $P_{3/2}$ states.[111,159-164] (The transition $P_{3/2}$ to the $S_{1/2}$ ground state involves what is historically known as "$D2$" radiation.) Ignoring nuclear spin for the moment, we have the $P_{3/2}$–$S_{1/2}$ relative transition probabilities shown in Figure 3. Consider excitation from the two ground-state levels induced by the σ^+ (i.e., circularly polarized with positive helicity directed parallel to the magnetic field) pumping light. Without collisional mixing in the $P_{3/2}$ state, one obviously pumps atoms into the $\mu_\lambda = +\frac{1}{2}$ level of the ground state where they are held in a cycle between the $\mu_\lambda = \frac{1}{2}$ and $m_j = \frac{3}{2}$ states. On the other hand, with a sufficient number of collisions that the $P_{3/2}$ levels are completely redistributed and achieve nearly equal populations within a fraction of a lifetime after excitation, the pumping works in the reverse direction: both ground-state levels are repopulated with equal probabilities but excitation occurs three times more rapidly from $\mu_\lambda = \frac{1}{2}$ than $\mu_\lambda = -\frac{1}{2}$, so that the $\mu_\lambda = -\frac{1}{2}$ is pumped. At some intermediate perturber pressure the pumping rate, and thus the net ground-state polarization itself, passes through zero. The condition of zero polarization is accurately fixed by comparing the absorption of weak σ^+ and σ^- beams of D_1 light (corresponding to the $P_{1/2} \leftrightarrow S_{1/2}$ transition).

To describe and analyze the experimental results, we can write down the time evolution of ground- and excited-state density matrices as indicated in Section 2 of this chapter and applied above to level crossings (see also Chapter 9). This is a case, however, where only diagonal elements of the density matrix are important and the conceptually simple population-density formulation can be used (see Section 2.6).[165]

Let $\mathbf{N}'$ be the density of the excited-state ($j = \frac{3}{2}$) levels as in Eq. (1), and $\mathbf{N}''$ the density of the ground-state ones. The rate equations for $\mathbf{N}'$ and $\mathbf{N}''$ in the presence of pumping radiation, natural decay (with rate Γ), and collisions (described again by γ) can be written immediately (we ignore the relatively small ground-state relaxation):

$$\dot{N}''_\mu = \Gamma \sum_n u_{\mu n} N'_n - \Gamma c_p u_{\mu,\mu+1} N''_\mu \qquad (174a)$$

$$\dot{N}'_m = -\Gamma N'_m + \Gamma c_p u_{m-1,m} N''_{m-1} - \sum_n \gamma'_{mn} N'_n \qquad (174b)$$

where $u_{\mu n}$ is the branching ratio ($\Sigma_\mu u_{\mu n} = 1$) for the transition to ground-state μ in the decay of excited-state n, and c_p is proportional to the intensity of the incident radiation. Putting $\mu = m - 1$ and adding Eqs. (174) yields the steady-state ($\dot{N}''_\mu = \dot{N}'_m = 0$) relation

$$(\mathbf{A} - \Gamma^{-1}\boldsymbol{\gamma}) \cdot \mathbf{N}' = \mathbf{N}' \qquad (175)$$

where $\mathbf{A}$ is the matrix with elements $A_{mn} \equiv u_{m-1,:n}$. For the $j = \frac{3}{2}$, $I = 0$ case pictured in Figure 1,

$$\mathbf{A} = \frac{1}{3}\begin{pmatrix} 3 & 2 & 1 & 0 \\ 0 & 1 & 2 & 3 \\ 0 & 0 & 0 & 0 \\ 0 & 0 & 0 & 0 \end{pmatrix} \qquad (176)$$

To solve, expand $\mathbf{N}'$ in the unit vectors $\hat{\mathbf{T}}_L$ (see Section 1). We note again that because of collisional isotropy, $\boldsymbol{\gamma} \cdot \hat{\mathbf{T}}_L = \gamma_L \hat{\mathbf{T}}_L$. Direct calculation gives furthermore

$$\mathbf{A} \cdot \hat{\mathbf{T}}_0 = \hat{\mathbf{T}}_0 + 2(5^{-1/2})\hat{\mathbf{T}}_1 - 5^{-1/2}\hat{\mathbf{T}}_3 \qquad (177a)$$

$$\mathbf{A} \cdot \hat{\mathbf{T}}_1 = \tfrac{1}{3}\hat{\mathbf{T}}_1 + \tfrac{1}{3}(5^{1/2})\hat{\mathbf{T}}_2 + \tfrac{2}{3}\hat{\mathbf{T}}_3 \qquad (177b)$$

$$\mathbf{A} \cdot \hat{\mathbf{T}}_L = 0 \qquad \text{for } L > 1 \qquad (177c)$$

Comparison of the tensor components in Eq. (175) then gives

$$2 \cdot 5^{-1/2} n_0 + \tfrac{1}{3}n_1 = (1 + \gamma_1/\Gamma)n_1 \qquad (178a)$$

$$\tfrac{1}{3}5^{1/2} n_1 = (1 + \gamma_2/\Gamma)n_2 \qquad (178b)$$

$$-5^{1/2} n_0 + \tfrac{2}{3}n_1 = (1 + \gamma_3/\Gamma)n_3 \qquad (178c)$$

The condition for no ground-state polarization is $N_{1/2} = N_{-1/2}$, which from Eq. (174) implies

$$n_1 = \frac{3}{2 \cdot 5^{1/2}} n_0 \qquad (179)$$

which combined with Eq. (178) requires

$$\gamma_1 = \tfrac{2}{3}\Gamma \tag{180a}$$

$$(1 + \gamma_2/\Gamma)n_2 = \tfrac{1}{2}n_0 \tag{180b}$$

$$n_3 = 0 \tag{180c}$$

For our hypothetical alkali with no hyperfine structure, then, the D_2 optical pumping passes through zero when γ_1, the dipole relaxation rate due to collisions, is fixed at two-thirds the natural decay rate Γ [Eq. (180a)]. In such a measurement only γ_1 would thus be determined.[165] By measuring the alignment n_2/n_0 at the zero crossing, however, the rate γ_2 can also be found [Eq. (180b)]:

$$\gamma_2 = \Gamma(n_0/2n_2 - 1) \tag{181}$$

It is also apparent [Eq. (180c)] that when the ground-state polarization vanishes, so does the excited-state octopole moment n_3. At pressures *other* than the zero crossing, i.e, where the ground-state polarization is *not* zero, the more general Eqs. (178) can be applied to show that n_3 does not vanish. A similar analysis can be made in the "weak-pumping" limit, i.e., when ground-state relaxation is essentially complete.[159,160]

The presence of nuclear spin can couple the multipole moments and thus complicate the calculations (see Section 3.6). For this reason analyses of $D2$ optical-pumping zeros have generally employed a simple mixing model of the collisional relaxation, such as one in which all off-diagonal elements of $\boldsymbol{\gamma}$ in the jm basis are equal ("j randomization").[111,161]

4.6. Isolating the Zeeman Sublevels; High-Field Measurements

By splitting the Zeeman sublevels of the vapor in a magnetic field of, say, a few thousand gauss, individual sublevels can be excited. If emission from single levels is observed, then all the collisional transition rates (Section 3.2) γ_{mm}^{nn} between sublevels m and n can be measured, and from these, of course all the multipolar relaxation rates γ_L found. Although the rates γ_L for $L > 2$ can be measured indirectly through hyperfine coupling (Section 4.3),[166] resolution of Zeeman sublevels appears to offer the only possibility of direct measurements of these rates.

Excitation of a single level can be achieved by use of a resonance lamp in a different field than that of the fluorescence cell[32,167] or with the aid of a narrow laser line.[88] The field of the cell can be "scanned" until one of the frequencies of the lamp coincides with a transition to the desired sublevel. The transition may sometimes be better isolated by appropriately polarizing the light from the lamp. Coincidences of transitions in different isotopes can also be employed. Detection of fluorescence from a single

Zeeman level can be accomplished by sufficient spectral resolution of the emission say with a Fabry–Perot etalon or with a selective filter consisting of an optically thick sample of the vapor perhaps of a different isotope or in yet a different magnetic field. Measurements at high fields have also demonstrated the effects of energy-level spacing and angular-momentum decoupling on depolarization cross sections (see Section 3.5).

4.7 Atomic-Beam Measurements

Atomic beams permit better control of velocity and scattering-angle parameters than is possible in fluorescent cells or flames. Velocities much lower than thermal can be achieved in merging-beam experiments[168] or for collisions within a single beam.[135,136,169] With sputtering sources or ionic sources perhaps with charge-exchange cells,[170] greatly superthermal velocities can also be achieved. Velocity resolutions of parts in a hundred or better enable observations of fine structures in the velocity dependence, for example, those of glory oscillations or those due to orbiting or quasi-bound resonances.[171–175] The angular dependence of differential cross sections also provides much more information than available in the total, thermally averaged cross sections measured in most cell or flame experiments. (See also the discussion in Section 2.1 of Chapter 6.)

Although beam experiments have been made with metastable atoms since the mid 1960s,[176,177] scattering with atoms in short-lived excited states has only recently been successful.[178–180] A laser-excited Na beam is in one experiment crossed with a noble-gas or Hg beam,[178,179] and in another, scattered by noble-gas atoms in a gas cell[180] (see Section 2.2 of Chapter 6). Scattered atoms are detected and not correlated with fluorescence. Even though there is no final state selection, the results depend not only on elastic scattering in the ground and excited states but also on the inelastic processes of depolarization and fine-structure transitions. Future experiments in which the scattered atoms are detected in coincidence with fluorescent photons will yield more detailed information about depolarization phenomena. Atomic beams can of course be coherently excited by electronic or atomic impact or by collisions in thin foils (see Chapters 2 and 20).

Measurements of resonant collisions in a *single* beam are made by observing scattered fluorescent radiation.[135,136] In a well-collimated thermal beam, the collisions occur with relative velocities nearly parallel to the beam axis and with an average relative velocity only about one-third the thermal velocity of the beam.[169]

Because of the richer detail available from beam studies, the results can no longer be fully described in terms of the multipole relaxation rates γ_L. In general, differential cross sections for transitions $|jm\rangle \rightarrow |j'm'\rangle$ can be used. For collisions within a single thermal beam, however, rates can again be employed but now with axial and inversion symmetrics rather than the spherical symmetry common in fluorescent cells or flames. For example, in Hanle experiments with linear polarization on a single beam, one finds[135] three disalignment rates $\gamma_{L,|M|}$, where $L = 2$ and $|M| = 0, 1,$ *or* 2. With the beam parallel to the magnetic field ($\hat{z}$ direction), the width of the Hanle signal (excitation and observation polarization vectors equal to $\hat{x}$ and $\hat{y}$) is proportional to $\Gamma + \gamma_{22}$; with the beam halfway between $\hat{x}$ and $\hat{y}$, γ_{21} is found; and with the beam along $\hat{x}$ or $\hat{y}$, the width is no longer linear in the density, but proportional to

$$\Gamma + \tfrac{1}{4}(\gamma_{22} + 3\gamma_{20}) + \frac{\tfrac{3}{16}(\gamma_{20} - \gamma_{22})^2}{\Gamma + \tfrac{1}{4}\gamma_{22} + \tfrac{3}{4}\gamma_{20}} \tag{182}$$

The value γ_2 measured for isotropic collisions is related to the $\gamma_{2,|M|}$ by

$$\gamma_2 = \tfrac{1}{5}(2\gamma_{22} + 2\gamma_{21} + \gamma_{20}) \tag{183}$$

4.8. Laser Methods of Investigating Depolarization

Several applications of the intense, narrow-band light available from lasers to depolarization studies have been mentioned above. In addition to its use simply as a convenient replacement of noncoherent light sources, it permits the excitation of atomic beams[178–182] and also the partial velocity selection of atoms in a vapor.[189,190] In the latter, only the velocity component along the beam is selected. If high velocities are selected, the motion is aligned with the laser beam, and the average relative velocity v of collision can be many times the ground-state thermal velocity. If low velocities are chosen, v is roughly perpendicular to the laser beam, and the relative excited-state collision velocity cannot be reduced below two-thirds of the average value for ground-state atoms. In this section a few more applications are briefly mentioned.

Reduction of Doppler broadening permits better resolution of Zeeman or Stark splitting and thus selective excitation and detection of particular Zeeman or Stark sublevels. The Doppler broadening is reduced or eliminated by saturation-absorption spectroscopy,[183] the use of two-photon techniques[184] or the "bunching" of the longitudinal velocity component of an accelerated ion beam.[185]

The intensity available from a laser makes it suitable for detecting excited-state multipoles by absorption or scattering of radiation,[186] possibly an important technique when, perhaps due to branching to a metastable level and inconvenient emission frequencies, the usual fluorescent decay is difficult to detect or does not lend itself to reliable detection of the multipoles. Such detection, as well as the laser excitation, can be pulsed to permit direct measurements of the time evolution of the multipoles.[183] Various nonlinear effects, such as Hanle signals in saturation resonances, can also be used to determine multipole relaxation rates.[187–191]

5. Comparison of Results

Experimental and theoretical results have been referred to throughout the chapter, but it seemed best to collect the numbers together and present them here for easy reference and comparison. Additional references are given in the recent review by Omont.[5]

5.1. Alkali Depolarization in Collisions with Noble Gases

Table 3 is a comparison of some alkali–noble-gas depolarization data. Except where otherwise noted, $P_{1/2}$ values have been corrected so as to remove the effects of nuclear spin (Section 3.6). The list is far from exhaustive; only a roughly representative sample of recent results is given. References to other measurements and calculations are given in the table. The last column presents an abbreviated indication of how the results were obtained. To facilitate a rapid overview, T has been used to indicate a primarily theoretical evaluation, E a primarily experimental one. Where error limits are given, they are as specified by the authors.

As discussed above (Section 3.5), the disorientation cross sections for the $P_{1/2}$ state of the heavier alkalis appear to display a marked dependence on magnetic field beyond that expected for decoupling of j and I. In addition to the description of the experiment in the right-hand column, values determined at high magnetic fields have been marked with an asterisk (*) in the second column.

Various authors have at times used different nomenclature for disorientation cross sections. Thus the "disorientation cross section" in Ref. 32 is actually $\sigma(\frac{1}{2}\frac{1}{2}\leftrightarrow\frac{1}{2}-\frac{1}{2})$, i.e., the cross section for Zeeman mixing in the $P_{1/2}$ state. Other authors [35,96,192] have identified $2\sigma(\frac{1}{2}\frac{1}{2}\leftrightarrow\frac{1}{2}-\frac{1}{2})^{1/2}$ as the $P_{1/2}$ disorientation cross section $\sigma_1\binom{1/2\ 1/2}{1/2\ 1/2}$ [see Eqs. (115), (116)]. This identification would be correct if $\sigma_0\binom{1/2\ 1/2}{1/2\ 1/2}$ were negligibly small. However for Na and K $\sigma_0\binom{1/2\ 1/2}{1/2\ 1/2}$ [$=\frac{1}{2}\sigma_0\binom{3/2\ 3/2}{3/2\ 3/2})=-2^{-1/2}\sigma_0\binom{1/2\ 1/2}{3/2\ 3/2}$ if the fine-structure splitting $\Delta\varepsilon\ll kT$ and if transitions other than those among ^2P sublevels are ignored] is large and can make the actual disorientation cross section

$$\sigma_1\begin{pmatrix}\frac{1}{2} & \frac{1}{2}\\ \frac{1}{2} & \frac{1}{2}\end{pmatrix}=2\sigma(\tfrac{1}{2}\tfrac{1}{2}\leftrightarrow\tfrac{1}{2}-\tfrac{1}{2})+\sigma_0\begin{pmatrix}\frac{1}{2} & \frac{1}{2}\\ \frac{1}{2} & \frac{1}{2}\end{pmatrix}$$

Table 3. Depolarization Cross Sections in Alkali–Noble-Gas Collisions

Atom pair	$^2P_{1/2}$ σ_1/A^2	$^2P_{3/2}$ σ_1/A^2	σ_2/A^2	σ_3/A^2	T/E	Reference(s)	method, and comments[a]
Li(2^2P)+He	86.5	•65.3	99.2	85.0	T	Reid[43] 1975	FQ, pots,[74] $E=0.03$ eV
Na(3^2P)+He	112	98	138	116	T	Wils[55] 1975	FQ, pots,[74,76] $T=400$ K
	99	86	123	104	T	Reid[81] 1973	FQ, pots,[74] $E=0.05$ eV
	114	94	122	86	T	Lewi[65] 1972	IA, MA, pots,[76] $E=0.051$ eV
	133±7	112±6	154±8	129±12	E	Gay[88] 1976	ReZee, $B=6$ kG, $T\approx450$ K
	146±22	128±13	167±17		E	Elbe[105,194,195] 1972, 1973, 1974	Hanle, D2 OP, Depol at $B=270$ G
+Ne	110±9	105±5	134±8	109±12	E	Gay[88] 1976	ReZee, $B=6$ kG, $T\approx450$ K
	137±21	107±11	174±18		E	Elbe[105,194,195] 1972, 1973, 1974	Hanle, D2 OP, Depol at $B=270$ G
+Ar	191±10	198±8	228±13	186±17	E	Gay[88] 1976	ReZee, $B=6$ kG, $T\approx450$ K
	260±39	205±21	308±31		E	Elbe[105,194,195] 1972, 1973, 1974	Hanle, D2 OP, Depol at $B=270$ G
+Kr	244±9	259±10	290±15	254±22	E	Gay[88] 1976	ReZee, $B=6$ kG, $T\approx450$ K
	306±46	243±25	341±34		E	Elbe[105,194,195] 1972, 1973, 1974	Hanle, D2 OP, Depol at $B=270$ G
+Xe	244±12	271±10	308±15	277±24	E	Gay[88] 1976	ReZee, $B=6$ kG, $T\approx450$ K
	356±53	281±28	376±38		E	Elbe[105,194,195] 1972, 1973, 1974)	Hanle, D2 OP, Depol at $B=270$ G, See also Refs 33, 35, 41, 56, 67, 73, 112, 159, 196, 197
K(4^2P)+He	24±4				E	Niew[96] 1975	Depol. Neglects transfer to $P_{3/2}$, $T=393$ K
	92*	86±14	127±37		E	Berd[32,167] 1968, 1971	ZS, $B=3\text{–}8$ kG, $T=368$ K[b]
	35.3–45.3				T	Gord[36] 1969	IA, MA

+Ne	21±3				E	Niew[96] 1975	Depol. Neglects transfer to $P_{3/2}$, $T = 393$ K
	78*	86±9	120±31		E	Berd[32,167] 1968, 1971	ZS, $B = 3$–8 kG, $T = 368$ K[b]
	39.1–51.2				T	Gord[36] 1969	IA, MA
+Ar	37±5				E	Niew[96] 1975	Depol. Neglects transfer to $P_{3/2}$, $T = 393$ K
	104*	164±20	240±34		E	Berd[32,167] 1968, 1971	ZS, $B = 3$–8 kG, $T = 368$ K[b]
	59.5–73.4				T	Gord[36]	IA, MA
+Kr	51±7				E	Niew[96] 1975	Depol. Neglects transfer to $P_{3/2}$, $T = 393$ K
	160*	248±34	301±30		E	Berd[32,167] 1968, 1971	ZS, $B = 3$–8 kG, $T = 368$ K[b]
	71–84				T	Gord[36] 1969	IA, MA
+Xe	69±9				E	Niew[96] 1975	Depol. Neglects transfer to $P_{3/2}$, $T = 393$ K
	214*	251±25	336±61		E	Berd[32,167] 1968, 1971	ZS, $B = 3$–8 kG, $T = 368$ K[b]
	89–103				T	Gord[36] 1969	IA, MA
					T	See also Refs. 71 and 93	
$\text{Rb}(b^2P)$ +He	24				E	Gall[30,91] 1967	Hanle, $T = 295$ K
	23.0	107			E	Bulo[91] 1971	Hanle; Papp[111] 1972 D2 OP with model[52]c
	33.1*	126±6	157±8	118	E	Kamk[94] 1975	Depol, $B = 1.1$ kG, $T = 317$ K
	23.2	111	136		T	Roue[50] 1974	IA, analytic pots
+Ne	16	57	100		E	Gall[30,91] 1967	Hanle, $T = 295$ K
	18.75*	121±6	156±8		E	Kamk[94] 1975	Depol, $B = 1.1$ kG, $T = 317$ K
	19.6				T	Wils[80] 1974	FQ, pots[74]
+Ar	25.9	130	210		E	Gall[30,91] 1967	Hanle, $T = 295$ K
	31.9*	243±12	314±16		E	Kamk[94] 1975	Depol, $B = 1.1$ kG, $T = 317$ K
+Kr	28.3				E	Gall[30,91] 1967	Hanle, $T = 295$ K
	35.95*	343	410		E	Kamk[94] 1975	Depol, $B = 1.1$ kG, $T = 317$ K
+Xe	47.9*	391	492		E	Kamk[94] 1975	Depol, $B = 1.1$ kG, $T = 317$ K
						See also Refs 36, 52, 57, 60, 93, 105, 111, 112, 115–117, 146, 163, 199	

continued overleaf

Table 3 (*Continued*)

Atom pair	$^2P_{1/2}$ σ_1/A^2	$^2P_{3/2}$ σ_1/A^2	σ_2/A^2	σ_3/A^2	References, method, and comments[a]	
$Cs(6^2P)$ + He	4.9 ± 0.7	100 ± 19			E	$P_{1/2}$: Niew[192] 1976	Depol, $T=313$ K, $P_{3/2}$: Fric[161] 1967 D2 OP
	6.0 ± 0.4				E	Guir[95] 1975	Depol, $T=316$ K
	6.1				E	Gall[30,91] 1967	Hanle, $T\simeq295$ K
	$11.8\pm0.5^*$	80 ± 12	86 ± 21		E	$P_{1/2}$: Guir[34] 1972	Depol at $B=9.8$ kG, $P_{3/2}$: Guir[122] 1976 ZS
+Ne	$10-17$		150	134	T	Roue[50] 1974	IA, analytic pots
	2.1 ± 0.3				E	$P_{1/2}$: Niew[192] 1976	Depol, $T=313$ K, $P_{3/2}$: Fric[161] 1967 D2 OP
	2.3				E	Gall[30,91] 1967	Hanle, $T\simeq295$ K
	$4.7\pm0.3^*$	80 ± 12	88 ± 22		E	$P_{1/2}$: Guir[34] 1972	Depol at $B=9.8$ kG, $P_{3/2}$: Guir[122] 1976, ZS
+Ar	5.6 ± 0.8	188 ± 38			E	$P_{1/2}$: Niew[192] 1976	Depol, $T=313$ K, $P_{3/2}$: Fric[161] 1967, D2 OP
	5.0				E	Gall[30,91] 1967	Hanle, $T\simeq295$ K
	$10.7\pm0.6^*$	234 ± 34	288 ± 72		E	$P_{1/2}$: Guir[34] 1972	Depol at $B=9.8$ kG, $P_{3/2}$: Guir[122] 1976 ZS
+Kr	5.8 ± 0.9	289 ± 60			E	$P_{1/2}$: Niew[192] 1976	Depol, $T=313$ K, $P_{3/2}$: Fric[161] 1967, D2 OP
	$37.9\pm3.2^*$				E	Guir[34] 1972	Depol at $B=9.8$ kG, $T=316$ K
+Xe	6.3 ± 0.9	354 ± 72	668 ± 167		E	$P_{1/2}$: Niew[192] 1976	Depol, $T=313$ K, $P_{3/2}$: Fric[161] 1967, D2 OP
	$71.7\pm7.2^*$	397 ± 60			E	$P_{1/2}$: Guir[34] 1972	Depol at $B=9.8$ kG, $P_{3/2}$: Guir[122] 1976, ZS

| Cs(7^2P) + He | 146 | 461 | 629 | 667 | T Roue[50] 1974 See also Refs 36, 52, 92, 93, 95, 200 | IA, analytic pots |

---

[a] Abbreviations

Depol: A direct measurement of the depolarization of resonance fluorescence

E: A predominantly experimental determination

FQ: A fully quantum close-coupling calculation

Hanle: Broadening of a Hanle-effect or level-crossing signal

IA: Impact approximation with classical paths

MA: Matching approximation: idealized coupling is assumed in different regions and solutions are matched at the boundaries

OP: Optical pumping

pots: Indicates the use of computed interatomic potentials

ReZee: Resolved Zeeman levels

T: A predominantly theoretical determination

ZS: Zeeman scanning technique

*: $P_{1/2}$ measurements at kilogauss magnetic fields

[b] The values quoted as "disorientation cross sections" by Berdowski and Krause[32] for the $P_{1/2}$ state of K are in fact cross sections for $j, m = (\frac{1}{2}, \frac{1}{2}) \leftrightarrow (\frac{1}{2}, \frac{1}{2})$ mixing. The values given here are twice their values. They still represent lower limits to $\sigma_1(1/2\,1/2)$ and should be increased by $\sigma_0(1/2\,1/2) = \sigma(j = \frac{1}{2} \rightarrow \frac{1}{2})$ which ranges from 14 (Ne) to 104 Å^2 (Xe) at 368 K.[198]

[c] Paopp and Franz[111] give a cross section for "electronic collisional relaxation" σ which, by means of an impact-parameter theory[52] using van der Waals interactions, is related to σ_L by $\sigma_L = C_L\sigma$, where $C_1 = 0.92$, $C_2 = 1.14$, and $C_3 = 1.02$.

Table 4. Depolarization Cross Sections for Other Nonresonant Collisions

Atom pair	Quantity measured	Value in A^2		Reference, method, and comments[a]
$Mg^+(3^2P)+Ar$	$\sigma_2(P_{3/2})$	130 ± 25	E	Gall[201] 1967 Hanle
$Ca(4^1P)+He$	$\sigma_2(P_1)$	80 ± 15	E	Gibb[202] 1976 Hanle
$+Ne$	$\sigma_2(P_1)$	65 ± 10	E	Gibb[202] 1976 Hanle
$+Ar$	$\sigma_2(P_1)$	170 ± 10	E	Gibb[202] 1976 Hanle
$+Kr$	$\sigma_2(P_1)$	200 ± 20	E	Gibb[202] 1976 Hanle
$+Xe$	$\sigma_2(P_1)$	200 ± 20	E	Gibb[202] 1976 Hanle
				See also Ref. 203
$Ca^+(4^2P)+Ar$	$\sigma_2(P_{3/2})$	140 ± 20	E	Gall[201] 1967 Hanle, $T=413$ K
$Sr^+(5^2P)+He$	$\sigma_2(P_{3/2})$	72 ± 9	E	Webe[164] 1973 D2 OP
$+Ne$	$\sigma_2(P_{3/2})$	80 ± 10	E	Webe[164] 1973 D2 OP
$+Ar$	$\sigma_1(P_{1/2})$	47 ± 25	E	Gall[201] 1967 Hanle
$+Ar$	$\sigma_2(P_{3/2})$	124 ± 15	E	Webe[164] 1973 D2 OP
$+Kr$	$\sigma_2(P_{3/2})$	148 ± 18	E	Webe[164] 1973 D2 OP
$+Xe$	$\sigma_2(P_{3/2})$	196 ± 18	E	Webe[164] 1973 D2 OP
				See also Ref. 61, 136

System	Cross section	Value	E/T	Reference	Method
Ba$^+$(6^2P) + Ar	$\sigma_1(P_{1/2})$	0 ± 25	E	Gall[201] 1967	Hanle
	$\sigma_2(P_{3/2})$	125 ± 25	E	Gall[201] 1967	Hanle
Zn(4^3P)+He	$\sigma_2(^3P_1)$	49 ± 2	E	Crem[204] 1972	Hanle, $T=658$ K
	$\sigma_2(^3P_1)$	20	T	Crem[204] 1972	IA, vdW
+Ne	$\sigma_2(^3P_1)$	45 ± 10	E	Crem[204] 1972	Hanle, $T=658$ K
	$\sigma_2(^3P_1)$	44	T	Crem[204] 1972	IA, vdW
+Ar	$\sigma_2(^3P_1)$	85 ± 6	E	Crem[204] 1972	Hanle, T 658 K
	$\sigma_2(^3P_1)$	86	T	Crem[204] 1972	IA, vdW
+Kr	$\sigma_2(^3P_1)$	101 ± 8	E	Crem[204] 1972	Hanle, $T=633$ K
	$\sigma_2(^3P_1)$	112	T	Crem[204] 1972	IA, vdW
+Xe	$\sigma_2(^3P_1)$	140 ± 11	E	Crem[204] 1972	Hanle, $T=658$ K
	$\sigma_2(^3P_1)$	136	T	Crem[204] 1972	IA, vdW
				See also Refs. 205–207	
Cd(5^3P)+He	$\sigma_1(^3P_1)$	40	E	Alek[208] 1971	D2 OP, $T=300$ K (Cd^{111})
+He	$\sigma_2(^3P_1)$	51 ± 8	E	Barg[209] 1967	Hanle and LMDR, $T=473$ K
		28	T	Barg[209] 1967	IA, theory of Ref. 65
Cd(5^1P)+He	$\sigma_1(^1P_1)$	141 ± 12	E	Pepp[211] 1970	Hanle, $T=358$ K (Cd^{114})
+He	$\sigma_2(^1P_1)$	109 ± 10	E	Pepp[211] 1970	Hanle, $T=358$ K (Cd^{114})
Cd(5^3P)+Ne	$\sigma_1(^3P_1)$	55	E	Alek[208] 1971	D2 OP $T=300$ K (Cd^{111})
+Ne	$\sigma_2(^3P_1)$	52 ± 8	E	Barg[209] 1967	Hanle and LMDR, $T=473$ K
Cd(5^1P)+Ne	$\sigma_1(^1P_1)$	103 ± 9	E	Pepp[211] 1970	Hanle, $T=358$ K (Cd^{114})
+Ne	$\sigma_2(^1P_1)$	88 ± 9	E	Pepp[211] 1970	Hanle, $T=358$ K (Cd^{114})
				See Refs. 208, 209, 211	
Cd(5^3P)+Xe	$\sigma_1(^3P_1)$	163	E	Alek[208] 1971	D2 OP $T=300$ K (Cd^{111})
+Xe	$\sigma_2(^3P_1)$	121 ± 18	E	Barg[209] 1967	Hanle and LMDR, $T=473$ K

continued overleaf

Table 4 (*Continued*)

Atom pair	Quantity measured		Value in A	Reference, method, and comments [a]	
Cd(5^1P) + Xe	$\sigma_1(^1P_1)$	E	547 ± 49	Pepp[211] 1970	Hanle, $T = 358$ K (Cd[114])
+ Xe	$\sigma_2(^1P_1)$	E	412 ± 39	Pepp[211] 1970	Hanle, $T = 358$ K (Cd[114])
				See also Refs 212–215	
Hg(6^3P) + He	$\sigma_2(P_1)$	E	37.8 ± 0.9	Phan[153] 1973	DelCo, $T = 296$ K
	$\sigma_2(P_2)$	E	63 ± 7	Baum[216] 1968	DoRes, $T = 350$ K
+ Ne	$\sigma_2(P_1)$	E	43.5 ± 1.3	Phan[153] 1973	DelCo, $T = 296$ K
	$\sigma_2(P_2)$	E	84 ± 8	Baum[216] 1968	DoRes, $T = 350$ K
+ Ar, Kr				See Refs. 82, 83, 153, 217–219	
+ Xe	$\sigma_2(P_1)$	E	144 ± 6	Phan[153] 1973	DelCo, $T = 296$ K
	$\sigma_2(P_2)$	E	342 ± 34	Baum[216] 1968	DoRes, $T = 350$ K
	$\sigma_2(P_2)$	E	239 ± 31	Casa[218] 1967	
				See also Refs 82–84, 152, 153, 217–236	
Tl(6^2P)[b] + He	$\sigma_1(P_{1/2})$	E	$(6 \pm 2) \times 10^{-3}$	Gibb[31] 1970	OP, $T = 883$ K
+ Ne	$\sigma_1(P_{1/2})$	E	$(1.14 \pm 0.4) \times 10^{-3}$	Gibb[31] 1970	OP, $T = 883$ K
+ Ar	$\sigma_1(P_{1/2})$	E	$(11 \pm 3) \times 10^{-3}$	Gibb[31] 1970	OP, $T = 883$ K
+ Kr	$\sigma_1(P_{1/2})$	E	$(22 \pm 6) \times 10^{-3}$	Gibb[31] 1970	OP, $T = 883$ K
+ Xe	$\sigma(P_{1/2})$	E	$(62 \pm 19) \times 10^{-3}$	Gibb[31] 1970	OP, $T = 883$ K
Tl(7^2S) + He	$\sigma_1(S_{1/2})$	E	0.04	Hsie[139] 1972	Hanle, $T = 623$ K
+ Ar	$\sigma(S_{1/2})$	E	0.1	Hsie[139] 1972	Hanle, $T = 623$ K
				See also Refs 140, 237–239	

Pb(6³P) + He	$\sigma_2(^3P_1^0)$	28 ± 6	E	Gibb[240] 1972	Depol, $T = 850$ K
+ He	$\sigma_2(^3P_1)$	27 ± 5	E	Gibb[240] 1972	Depol, $T = 813$ K
+ He	$\sigma_2(^3P_2)$	36 ± 7	E	Gibb[240] 1972	Depol, $T = 813$ K
+ Ne, Ar, Kr				See Ref. 240	
+ Xe	$\sigma_2(^5P_1^0)$	105 ± 21	E	Gibb[240] 1972	Depol, $T = 850$ K
	$\sigma_2(^3P_1)$	123 ± 25	E	Gibb[240] 1972	Depol, $T = 813$ K
	$\sigma_2(^3P_2)$	159 ± 32	E	Gibb[240] 1972	Depol, $T = 813$ K
He₂(2³P) + He	$\sigma_2(^3P)$	56.3 ± 2.1	E	Land[241] 1968	Hanle, $T = 303$ K
He(3¹D) + He	$\sigma_1(^1D)$	110 ± 50	E	Pina[110] 1974	Depol, $T = 300$ K
	$\sigma_2(^1D)$	173 ± 20	E	Chie[242] 1972	Hanle, $T = 293$ K
He(4¹D) + He	$\sigma_2(^1D)$	268 ± 15	E	Chie[242] 1972	Hanle, $T = 293$ K
He(5¹D) + He	$\sigma_2(^1D)$	433 ± 20	E	Chie[242] 1972	Hanle, $T = 293$ K
				See also Refs. 110, 118, 170, 241, 243, 244	
Ne(3P) + He	$\sigma_2(^3D_2)^c$	73.7 ± 3.5	E	Carr[23] 1971	Hanle, $T = 85$ K
	$\sigma_2(^3D_2)^c$	49.7	T	Carr[14] 1971	IA, vdW, $T = 85$ K
	$\sigma_2(^3D_2)^c$	79.9 ± 2.7	E	Carr[246] 1972	Hanle, $T = 300$ K
	$\sigma_2(3p_4)$	100 ± 10	E	Kotl[247] 1971	Hanle, $T = 300$ K
+ Ne	$\sigma_2(^3P_1)^c$	12.8 ± 0.6	E	Carr[23] 1971	Hanle, $T = 85$ K
	$\sigma_2(^3P_1)^c$	25.4	T	Carr[14] 1971	IA, vdW, $T = 85$ K
	$\sigma_2(^3P_1)^c$	34.7 ± 1.7	E	Carr[246] 1972	Hanle, $T = 870$ K
	$\sigma_2(^3D_2)^c$	85.2 ± 2.8	E	Carr[246] 1972	Hanle, $T = 315$ K
Ne(3p) + Ne	$\sigma_2(3p_4)$	90 ± 20	E	Kotl[247] 1971	Hanle, $T = 300$ K
				See also Refs 23, 246, 248–255	

continued overleaf

Table 4 (*Continued*)

Atom pair	Quantity measured	Value in A		Reference, method, and comments[a]
Ar(3p^5 4_p)+Ar	$\sigma_2(^3D_3)^c$	214±19	E	Land[241] 1968 Hanle, $T=292$ K
	$\sigma_2(^1D_2)^c$	114±22	E	Land[241] 1968 Hanle, $T=292$ K
	$\sigma_2(^1D_2)^c$	110–140	T	Land[241] 1968 IA, vdW[12]
Kr(4p^5 5_p)+Kr	$\sigma_2(^1D_2)^c$	449±32	E	Land[256] 1973 Hanle, $T=292$ K
	$\sigma_2(^3D_3)^c$	479±27	E	Land[256] 1973 Hanle, $T=292$ K
				See also Refs 14, 23, 250, 257–261
I(5p^4 6s)^4 p +He	$\sigma_2(P_{5/2})$	9.8±1.1	E	Will[262] 1974 Hanle, $T=913$ K
+Ne	$\sigma_2(P_{5/2})$	19.5±1.6	E	Will[262] 1974 Hanle, $T=913$ K
+Ar	$\sigma_2(P_{5/2})$	37.6±3.1	E	Will[262] 1974 Hanle, $T=913$ K
+Kr	$\sigma_2(P_{5/2})$	38.7±3.8	E	Will[262] 1974 Hanle, $T=913$ K
+Xe	$\sigma_2(P_{5/2})$	49.2±3.1	E	Will[262] 1974 Hanle, $T=913$ K

[a] Abbreviations (see also list for Table 3):
DelCo: Delayed coincidence
DoRes: Double resonance
LMDR: Light-modulated double resonance[210]
vdW: van der Waals potential
[b] Disorientation in the *ground* state ($6^2P_{1/2}$) of Tl is included here because of its relation to $P_{1/2}$ disorientation in the heavier alkalis.
[c] The Paschen notation for these levels is $2p_2$ and $2p_8$ for levels $3p\ ^3P_1$ and $3p\ ^3D_2$ of Ne, $2p_6$ and $2p_9$ for levels $4p\ ^1D_2$ and $4p\ ^3D_3$ of Ar, and $2p_6$, $2p_8$ for levels $^1D_2$, $^3D_3$ of Kr. (See Ref. 245.)

as much as four times as large as $2\sigma(\tfrac{1}{2}\tfrac{1}{2}\leftrightarrow\tfrac{1}{2}-\tfrac{1}{2})$. Dashevskaya *et al.*[70,73] refer to $\sigma(\tfrac{1}{2}\tfrac{1}{2}\leftrightarrow\tfrac{1}{2}-\tfrac{1}{2})$ as an "orientation depolarization" cross section and to $\sigma_1(\substack{j\,j\\j\,j})$ as an "orientation relaxation" cross section. Similarly, Schneider[33,193] denotes $\sigma(\tfrac{1}{2}\tfrac{1}{2}\leftrightarrow\tfrac{1}{2}-\tfrac{1}{2})$ by $\sigma_{\text{depol}}(\tfrac{1}{2})$ and $\sigma_1(\substack{1/2\,1/2\\1/2\,1/2})$ by $\sigma_{\text{relax}}(\tfrac{1}{2})$. Measurements at low fields with a coupled nuclear spin sometimes refer to σ_{circ} and σ_{lin} for the relaxation of $\langle J_z \rangle$ and $\langle 3J_z^2 - J^2 \rangle$ in the presence of hyperfine coupling.

Most notational variants for $\sigma_L(\substack{j\,j\\j\,j})$ are readily understood, as for example $\sigma^{(L)}(j)$ (Refs. 55 and 112] or $\sigma_L(j)$ Refs. 43 and 81 but more caution must be exercised when the four angular momenta are not all equal. Transfer cross sections $\sigma_L(\substack{j\,j\\k\,k})$ may differ if the irreducible tensors used in their definition are not normalized. Suppose for example the energy separation of levels j and k is small compared to kT. When normalized irreducible tensors are used as here, or as in Chapter 2, and as is usual elsewhere,[5,6,8,11,12,97] then $\sigma_L(\substack{j\,j\\k\,k}) = \sigma_L(\substack{k\,k\\j\,j})$. Otherwise, however, there will be a difference between $k \to j$ and $j \to k$ cross sections, and only the geometric mean will be identical to our $\sigma_L(\substack{j\,j\\k\,k})$ (see the following chapter by Elbel).[194] It should further be noted that coherence relaxation rates for high-field level crossing (Section 4.3) are often listed as γ_{mn} (Refs. 52 and 115–117), which is however *not* the transfer rate γ^{nn}_{mm} but *rather* γ^{mn}_{mn}.

The "model-independent parameters" of Grawert[54] are easily related to depolarization cross sections. Comparing Eq. (80) to Grawert's table of mixing cross sections, one finds immediately for $j = \tfrac{3}{2}$

$$\sigma_0 = b_0 + b_1 + b_2 + b_3$$

$$\sigma_1 = b_0 + \tfrac{11}{15}b_1 + \tfrac{1}{5}b_2 - \tfrac{3}{5}b_3$$

$$\sigma_2 = b_0 + \tfrac{1}{5}b_1 - \tfrac{3}{5}b_2 + \tfrac{1}{5}b_3$$

$$\sigma_3 = b_0 - \tfrac{3}{5} + \tfrac{1}{5}b_2 - \tfrac{1}{35}b_3$$

Similarly for $j = \tfrac{1}{2}$,

$$\sigma_0 = b_0' + b_1'$$

$$\sigma_1 = b_0' - \tfrac{1}{3}b_1'$$

5.2. Other Depolarization in Foreign-Gas Collisions

Depolarization data for some other nonresonant collisions are summarized in Table 4. Again, the listed values are only meant to present a representative sample of data available, and further references are

Table 5. Resonant Depolarization Cross Sections

Atom pair	Quantity measured	Value		Reference, method, and comments[a]
Rb(5^2P)+Rb	$\langle v\sigma_1(P_{1/2})\rangle$	$(1.2\pm0.1)\times10^{-2}$	E	Gall[97] 1974	Hanle, $\Gamma^{-1}=29.4\pm0.7$ ns, $\lambda_0=794.7$ nm
	$\langle v\sigma_1(P_{1/2})\rangle$	0.57×10^{-2}	T	Gall[97] 1974	IA of Ref. 12 with $I\neq0$
	$\langle v\sigma_1(P_{3/2})\rangle$	$(1.7\pm0.1)\times10^{-2}$	E	Gall[97] 1974	Hanle, $\Gamma^{-1}=27.0\pm0.5$ ns, $\lambda_0=780$ nm
	$\langle v\sigma_1(P_{3/2})\rangle$	1.63×10^{-2}	T	Gall[97] 1974	IA of Ref. 12 with $I\neq0$
	$\langle v\sigma_2(P_{3/2})\rangle$	$(2.0\pm0.1)\times10^{-2}$	E	Gall[97] 1974	Hanle, crossing at 57.6 G
	$\langle v\sigma_2(P_{3/2})\rangle$	1.79×10^{-2}	T	Gall[97] 1974	IA of Ref. 12 with $I\neq0$
Zn(4^3P)+Zn	$\langle v\sigma_2(P_1)\rangle$	7×10^{10} cm^3/s	E	Dumo[264] 1962	From Ref. 134, see also Ref. 263
Cd(5^3P)+Cd	$\langle v^{2/5}\sigma_2(P_1)\rangle$	$(1.9\pm0.3)\times10^{-12}$ cm^2 $\times$(cm/s)$^{2/5}$	E	Byro[265] 1964	DoRes Nonresonant $\Gamma^{-1}=2.4$ μs. See also Refs 65, 266
Hg(^3P)+Hg	$\langle v\sigma_1(P_1)\rangle$	$3.8\pm0.4\times10^{-9}$ cm^3/s	E	Omon[64] 1968	Hanle
	$\langle v\sigma_1(P_1)\rangle$	3.96×10^{-9} cm^3/s	T	Omon[69] 1968	IA
	$\langle v\sigma_2(P_1)\rangle$	$4.0\pm0.4\times10^{-9}$ cm^3/s	E	Omon[64] 1968	Hanle, Isotope mixture
	$\langle v\sigma_2(P_1)\rangle$	4.13×10^{-9} cm^3/s	T	Omon[69] 1968	IA
	$\sigma_1(P_2)$	$2.0\pm0.2\times10^{-14}$ cm^2	E	Jacob[166] 1972	DoRes, nonresonant
	$\sigma_2(P_2)$	$2.6\pm0.3\times10^{-14}$ cm^2	E	Jacob[166] 1972	DoRes
	$\sigma_3(P_2)$	$2.8\pm0.3\times10^{-14}$ cm^2	E	Jacob[166] 1972	DoRes from $f=3/2$, $5/2$ levels of Hg^{199}. See also Refs 267–271
Tl(7^2S)+Tl	$\langle v\sigma_1(S_{1/2})\rangle$	$5.2\pm0.5\times10^{-8}$ cm^3/s	E	Hsie[139] 1972	Hanle, $T=620$–1020 K
	$\langle v\sigma_1(S_{1/2})\rangle$	$5.0\pm0.9\times10^{-8}$ cm^3/s	T	Omon[140] 1972	IA
Pb($7^3P_1^0$)+Pb	$\langle v\sigma_1(P_0^0)\rangle$	4.6×10^{-8} cm^3/s	E	Happ[134] 1967	Hanle, $\Gamma^{-1}=5.75\pm0.20$ ns Pb^{208}
	$\langle v\sigma_2(P_1^0)\rangle$	3.8×10^{-8} cm^3/s	E	Happ[134] 1967	Hanle, transition to metastable level. See also Refs. 135, 137, 141

[a] Abbreviations: See lists for Tables 3 and 4.

indicated. Data for collisions between like atoms have generally been listed in Table 5 (see following Section 5.3) even though not all collisions represented are dominated by the resonant dipole–dipole interaction.

5.3. Depolarization in Resonant Collisions

The dominant interaction in resonant collisions is known to be the dipole–dipole potential

$$V(R) = R^{-3}[\mathbf{d}_A \cdot \mathbf{d}_B - 3\mathbf{d}_A \cdot \hat{\mathbf{R}}\mathbf{d}_B \cdot \hat{\mathbf{R}}] \tag{184}$$

where $\mathbf{d}_A$ and $\mathbf{d}_B$ are the electric dipoles of the interacting atoms. Disorientation and depolarization cross sections can therefore be computed within the impact approximation to fair accuracy.[6,12,58,64] Table 5 contains some comparisons with experiment. In some cases, the oscillator strengths are so small that nonresonant processes dominate. Since there is no sharp dividing line, most collisions between excited and ground-state atoms of the same element have been categorized under the heading of this section. Noble-gas self-interactions are, however, generally nonresonant and have thus been entered in Table 4.

Special tricks have usually been employed to eliminate the effects of radiation trapping (Section 4.2). When different isotopes are used, the differential isotope shift must be small compared to the energy uncertainty arising from the finite duration of the collision; this was indeed generally the case for values reported here. Furthermore, with measurements on different isotopes, the actual depolarization rates $\gamma_L(A)$ can be measured, whereas otherwise the effect of coherence transfer cannot be eliminated (see following chapter) and the sum $\gamma_L(A) + \gamma_L(B)$ is measured, where $-\gamma_L(B)$ is the rate (and phase factor) of coherence transfer from the initially excited atom A to its "perturber" B. [Our $\gamma_L(A)$ and $\gamma_L(B)$ are equivalent to $_{ee}g^L(1)$ and $_{ee}g^L(2)$ of Refs. 64, 139, and 140].

Where the dipole–dipole resonant interaction dominates, γ_L is given within a factor of order unity by

$$\gamma_L \sim 10^{-2} N \lambda_0^3 \Gamma$$

where Γ^{-1} is the natural lifetime, λ_0 the wavelength of the resonant transition, and N the atomic density of perturbers. Experiment and theory then uniformly agree to within roughly 10%. Because γ_L is thus velocity-independent, $\gamma_L/N = \langle v\sigma_L \rangle$ is given rather than the cross section σ_L.

References

1. L. Krause, *Adv. Chem. Phys.* **28**, 267 (1975).
2. L. Krause, in *Physics of Electronic and Atomic Collisions*, Eds. T. R. Grovers and F. J. de Heer, North-Holland, Amsterdam (1972), p. 65.

3. C. Th. J. Alkemade and P. J. Zeegers, *Adv. Anal. Chem. Instrum.* **9**, 3 (1971).
4. A. B. Callear and J. D. Lambert, in *Comprehensive Chemical Kinetics*, Vol. 3, Elsevier, Amsterdam (1969).
5. A. Omont, *Irreducible Components of the Density Matrix. Application to Optical Pumping*, in *Prog. Quantum Electronics* **5**, 69 (1977).
6. D. M. Brink and G. R. Satchler, *Angular Momentum*, 2nd ed., Clarendon Press, Oxford (1968).
7. U. Fano, *Phys. Rev.* **131**, 259 (1963).
8. U. Fano, *Rev. Mod. Phys.* **29**, 74 (1957).
9. W. Heitler, *The Quantum Theory of Radiation*, 3rd ed., Oxford University Press, London (1954).
10. W. Happer and B. S. Mathur, *Phys. Rev.* **163**, 12 (1967).
11. W. Happer, *Rev. Mod. Phys.* **44**, 169 (1972).
12. A. Omont, *J. Phys. (Paris)* **26**, 26 (1965).
13. M. I. D'Yakonov and V. I. Perel', *Zh. Eksperim. Teor. Fiz.* **48**, 345 (1965) [English transl: *Sov. Phys.-JETP* **21**, 227 (1965)].
14. C. G. Carrington and A. Corney, *J. Phys. B* **4**, 869 (1971).
15. A. Ben-Reuven, *Phys. Rev.* **141**, 34 (1966).
16. A. Ben-Reuven, *Adv. Chem. Phys.* **33**, 235 (1975).
17. A. Messiah, *Quantum Mechanics*, North-Holland, Amsterdam (1962).
18. C. Cohen-Tannoudji, in *Advances in Quantum Electronics*, Ed. J. R. Singer, Columbia University Press, New York (1961).
19. J.-P. Barrat and C. Cohen-Tannoudji, *J. Phys. (Paris)* **22**, 443 (1961).
20. J.-P. Barrat and C. Cohen-Tannoudji, *J. Phys. (Paris)* **22**, 329 (1961).
21. C. Cohen-Tannoudji, *Ann. Phys. (Paris)* **7**, 423 (1962).
22. W. Hanle, *Z. Phys.* **41**, 164 (1927).
23. C. G. Carrington and A. Corney, *J. Phys. B* **4**, 849 (1971).
24. C. G. Carrington, *J. Phys. B* **4**, 1222 (1971).
25. M. I. D'Yakonov, *Zh. Eksperim. i Teor. Fiz.* **47**, 2213 (1964) [English transl: *Sov. Phys.-JETP* **20**, 1484 (1965)].
26. G. Nienhuis, *J. Phys. B* **9**, 167 (1976).
27. U. Fano and G. Racah, *Irreducible Tensorial Sets*, Academic Press, New York (1967).
28. F. A. Franz and J. R. Franz, *Phys. Rev.* **148**, 82 (1966).
29. F. A. Franz, G. Leutert, and R. T. Shuey, *Helv. Phys. Acta* **40**, 778 (1967).
30. A. Gallagher, *Phys. Rev.* **157**, 68 (1967); addendum **163**, 206 (1967).
31. M. H. Gibbs, G. G. Churchill, T. R. Marshall, J. F. Papp, and F. A. Franz, *Phys. Rev. Lett.* **25**, 263 (1970).
32. W. Berdowski and L. Krause, *Phys. Rev.* **165**, 158 (1968).
33. W. B. Schneider, *Z. Phys.* **248**, 387 (1971).
34. J. Guiry and L. Krause, *Phys. Rev. A* **6**, 273 (1972).
35. B. Niewitecka, T. Skalinski, and L. Krause, *Can. J. Phys.* **52**, 1956 (1974).
36. E. P. Gordeyev, E. E. Nikitin, and M. Ya. Ovchinnikova, *Can. J. Phys.* **47**, 1819 (1969).
37. E. E. Nikitin, *Comments At. Mol. Phys.* **1**, 122 (1969).
38. M. Elbel, *Phys. Lett.* **28A**, 4 (1968).
39. M. Elbel, *Ann. Phys. (Leipzig)* **22**, 289 (1969).
40. M. Elbel, *Can. J. Phys.* **48**, 3047 (1970); erratum **50**, 66 (1972).
41. M. Elbel, *Z. Phys.* **248**, 375 (1971).
42. F. H. Miles, *Phys. Rev. A* **7**, 942 (1973).
43. R. H. G. Reid, *J. Phys. B* **8**, 2255 (1975).
44. R. D. Levine, *Mol. Phys.* **22**, 497 (1971).
45. P. Pechukas and J. C. Light, *J. Chem. Phys.* **44**, 3897 (1966).
46. J. Callaway and E. Bauer, *Phys. Rev. A* **140**, 1072 (1965).

47. W. Magnus, *Commun. Pure Appl. Math.* **7**, 649 (1954).
48. C. A. Wang and W. J. Tomlinson, *Phys. Rev.* **181**, 115 (1969).
49. V. N. Rebane, *Opt. Spektrosk.* **26**, 673 (1969) [English transl: *Opt. Spectrosc.* **26**, 371 (1969)]; *Opt. Spektrosk.* **24**, 296 (1968) [English transl: *Opt. Spectrosc.* **24**, 155 (1968)].
50. E. Roueff and A. Suzor, *J. Phys. (Paris)* **35**, 727 (1974).
51. A. G. Petrashen, V. N. Rebane, and T. K. Rebane, *Opt. Spektrosk.* **35**, 408 (1973) [English transl. *Opt. Spectrosc.* **35**, 240 (1973)].
52. A. I. Okunevich and V. I. Perel', *Zh. Eksperim. i Teor Fiz.* **58**, 666 (1970) [English transl: *Sov. Phys.-JETP* **31**, 356 (1970)].
53. D. Secrest, *J. Chem. Phys.* **62**, 710 (1975).
54. G. Grawert, *Z. Phys.* **225**, 283 (1969).
55. A. Wilson and Y. Shimoni, *J. Phys. B* **8**, 2393, 2415 (1975).
56. F. Masnou-Seeuws and E. Roueff, *Chem. Phys. Lett.* **16**, 593 (1972).
57. L. Kumar and J. Callaway, *Phys. Lett.* **28A**, 385 (1968).
58. P. R. Berman and W. E. Lamb, Jr., *Phys. Rev.* **187**, 221 (1969).
59. G. Meunier, *J. Phys. (Paris)* **37**, 65 (1976).
60. M. B. Hidalgo and S. Geltman, *J. Phys. B* **5**, 265 (1972).
61. A. Giusti-Suzor and E. Roueff, *J. Phys. B* **8**, 2708 (1975).
62. C. G. Carrington, D. N. Stacey, and J. Cooper, *J. Phys. B* **6**, 417 (1973).
63. E. L. Lewis and L. F. McNamara, *Phys. Rev. A* **5**, 2643 (1972).
64. A. Omont and J. Meunier, *Phys. Rev.* **169**, 92 (1968).
65. F. W. Byron and H. M. Foley, *Phys. Rev.* **134**, A625 (1964).
66. M. Baranger, *Phys. Rev.* **111**, 481; 494; **112**, 855 (1958).
67. F. Masnou-Seeuws and R. McCarrol, *J. Phys. B* **7**, 2230 (1974).
68. E. E. Nikitin, *Adv. Chem. Phys.* **28**, 317 (1975).
69. E. E. Nikitin, in *Atomic Physics*, Vol. 5, Ed. G. zu Putlitz, E. W. Weber, and A. Winnacker, Plenum, New York (1975).
70. E. I. Dashevskaya and E. E. Nikitin, *Can. J. Phys.* **54**, 709 (1976).
71. E. P. Gordeev, E. E. Nikitin, and M. Ya. Ovchinnikova, *Opt. Spektrosk.* **30**, 189 (1971) [English transl: *Opt. Spectrosc.* **30**, 101 (1971)].
72. E. E. Nikitin, *Comments At. Mol. Phys.* **3**, 7 (1971).
73. E. I. Dashevskaya, F. Masnou-Seeuws, R. McCarroll, and E. E. Nikitin, *Opt. Spektrosk.* **37**, 209 (1974) [English transl. *Opt. Spectrosc.* **37**, 119 (1974)].
74. W. E. Baylis, *J. Chem. Phys.* **51**, 2665 (1969).
75. J. Pascale and J. Vandeplanque, *J. Chem. Phys.* **60**, 2278 (1974).
76. M. Krauss, P. Maldonado, and A. C. Wahl, *J. Chem. Phys.* **54**, 4944 (1971).
77. F. Schuller and W. Behmenburg, *Phys. Lett. C* **12**, 273 (1974).
78. A. Ben-Reuven, *Phys. Rev. A* **4**, 2115 (1971).
79. E. I. Dashevskaya, *Chem. Phys. Lett.* **11**, 184 (1971).
80. A. D. Wilson and Y. Shimoni, *J. Phys. B* **7**, 1543 (1974); erratum: **8**, 1392 (1975).
81. R. H. G. Reid, *J. Phys. B* **6**, 2018 (1973).
82. J.-C. Gay and A. Omont, *J. Phys. (Paris) Lett.* **37**, L-69 (1976).
83. J.-C. Gay and W. B. Schneider, *J. Phys. (Paris) Lett.* **36**, L-185 (1975).
84. B. Fuchs, W. Hanle, W. Oberheim, and A. Scharmann, *Phys. Lett.* **50A**, 337 (1974).
85. J.-C. Gay, *J. Phys. (Paris)* **37**, 1135 (1976).
86. J.-C. Gay, *J. Phys. (Paris)* **37**, 1155 (1976).
87. J.-C. Gay, *J. Phys. (Paris) Lett.* **36**, L-239 (1975).
88. J.-C. Gay and W. B. Schneider, *Z. Phys.* **278**, 211 (1976).
89. J. Apt and D. E. Pritchard, *Phys. Rev. Lett.* **37**, 91 (1976).
90. M. Elbel, M. Hühnermann, Th. Meier, and W. B. Schneider, *Z. Phys. A* **275**, 339 (1975).

91. B. R. Bulos and W. Happer, *Phys. Rev. A* **4**, 849 (1971).
92. W. E. Baylis, in *Electronic and Atomic Collisions*, Abstracts of papers of the VIIth International Conference on the Physics of Electronic and Atomic Collisions, Eds. L. M. Branscomb *et al.*, North-Holland, Amsterdam (1971), p. 677.
93. E. E. Nikitin and M. Ya. Orchinnikova, *J. Phys. B* **11**, 465 (1978).
94. B. Kamke, *Z. Phys. A* **273**, 23 (1975).
95. J. Guiry and L. Krause, *Phys. Rev. A* **12**, 2407 (1975).
96. B. Niewitecka and L. Krause, *Can. J. Phys.* **53**, 1499 (1975).
97. A. Gallagher and E. L. Lewis, *Phys. Rev. A* **10**, 231 (1974).
98. V. N. Rebane and T. K. Rebane, *Opt. Spektrosk.* **30**, 367 (1970) [English transl: *Opt. Spectrosc.* **30**, 199 [1970)].
99. V. N. Rebane and A. G. Petrashen, *Opt. Spektrosk.* **37**, 826 (1974) [English transl: *Opt. Spectrosc.* **37**, 472 (1974)].
100. V. N. Rebane, *Opt. Spektrosk.* **37**, 216 (1974) [English transl: *Opt. Spectrosc.* **37**, 123 (1974)].
101. V. N. Rebane, *Opt. Spektrosk.* **36**, 1018 (1974) [English transl: *Opt. Spectrosc.* **36**, 598 (1974)].
102. V. N. Rebane and T. K. Rebane, *Opt. Spektrosk.* **33**, 405 (1972) [English transl: *Opt. Spectrosc.* **33**, 219 (1972)].
103. V. N. Rebane and T. K. Rebane, *Opt. Spektrosk.* **34**, 657 (1973) [English transl: *Opt. Spectrosc.* **34**, 378 (1973)].
104. M. A. Bouchiat, *J. Phys.* **24**, 379; 611 (1963).
105. M. Elbel, A. Koch, and W. Schneider, *Z. Phys.* **255**, 14 (1972).
106. F. Grossetête, *C.R. Acad. Sci.* **258**, 3668 (1964).
107. F. Grossetête, *J. Phys. (Paris)* **25**, 383 (1964).
108. H. M. Gibbs, *Phys. Rev. A* **3**, 500 (1972); **139**, A1374 (1965).
109. M. Pinard and J. van der Linde, *Can. J. Phys.* **52**, 1615 (1974).
110. F. A. Franz and C. E. Sooriamoorthi, *Phys. Rev. A* **8**, 2390 (1973).
111. J. F. Papp and F. A. Franz, *Phys. Rev. A* **5**, 1763 (1972).
112. E. L. Lewis and C. S. Wheeler, *J. Phys. B* **10**, 911 (1977).
113. E. L. Lewis, *Opt. Commun.* **7**, 51 (1973).
114. W. E. Baylis, *Phys. Rev. A* **7**, 1190 (1973).
115. M. Lukaszewski and A. Sieradzan, *Phys. Lett.* **43A**, 227 (1973).
116. M. Lukaszewski, *Acta Phys. Pol. A* **49**, 93 (1976).
117. M. Lukaszewski, *Phys. Lett.* **51A**, 481 (1975).
118. W. Schöck, *Z. Naturforsch.* **27a**, 1731 (1972).
119. A. C. G. Mitchell and M. W. Zemansky, *Resonance Radiation and Excited Atoms*, Cambridge University Press, London (1934).
120. G. F. Kirkbright and M. Sargent, *Atomic Absorption and Fluorescence Spectroscopy*, Academic Press, London (1974).
121. O. Stern and M. Volmer, *Phys. Z.* **20**, 183 (1919).
122. J. Guiry and L. Krause, *Phys. Rev. A* **14**, 2034 (1976).
123. E. A. Milne, *J. London Math. Soc.* **1**, 1 (1926).
124. T. Holstein, *Phys. Rev.* **73**, 1212 (1947); **83**, 1159 (1951).
125. C. van Trigt, *Phys. Rev.* **181**, 97 (1969); *Phys. Rev. A* **1**, 1298 (1970); **4**, 1303 (1971).
126. W. J. Sandle and O. M. Williams, *J. Phys. B* **4**, 531 (1971).
127. W. L. Kennedy, *J. Phys. B* **4**, 841 (1971).
128. W. J. Sandle and O. M. Williams, *J. Phys. B* **5**, 987 (1972).
129. M. I. D'Yakonov and V. I. Perel', *Zh. Eksperim. i Teor. Fiz.* **47**, 1483 (1964) [English transl: *Sov. Phys.-JETP* **20**, 997 (1965)].
130. J. S. Deech and W. E. Baylis, *Can. J. Phys.* **49**, 90 (1971).
131. J.-P. Barrat, *J. Phys. Radium* **20**, 541, 633, 657 (1959).

132. B. P. Kibble, G. Copley, and L. Krause, *Phys. Rev.* **153**, 9 (1967).
133. G. Copley and L. Krause, *Can. J. Phys.* **47**, 533 (1969).
134. W. Happer and E. B. Saloman, *Phys. Rev.* **160**, 23 (1967).
135. F. M. Kelly and M. S. Mathur, *Can. J. Phys.* **55**, 83 (1977).
136. F. M. Kelly, T. K. Koh, and M. S. Mathur, *Can. J. Phys.* **52**, 795; 1438 (1974).
137. W. Happer and E. B. Saloman, *Phys. Rev. Lett.* **15**, 441 (1965).
138. A. Omont, *C.R. Acad. Sci.* **260**, 3331 (1965).
139. J. C. Hsieh and J. C. Baird, *Phys. Rev. A* **6**, 141 (1972).
140. A. Omont, J. C. Heish, and J. C. Baird, *Phys. Rev. A* **6**, 152 (1972).
141. J. B. Halpern, A. Baghdadi, and E. B. Saloman, *Phys. Rev. A* **9**, 668 (1974).
142. A. N. Nesmayanov, *Vapor Pressure of the Chemical Elements*, Elsevier, New York (1963).
143. W. M. Fairbanks, Jr., T. W. Hänsch, and A. L. Schawlow, *J. Opt. Soc. Am.* **65**, 199 (1975).
144. A. Gallagher and E. L. Lewis, *J. Opt. Soc. Am.* **63**, 864 (1973).
145. A. Gallagher, *Phys. Rev.* **172**, 88 (1968).
146. H. Doebler and B. Kamke, *Z. Phys.* **280**, 111 (1977).
147. W. Hanle, *Z. Phys.* **30**, 93 (1924).
148. J. Derouard, R. Jost, and M. Lombardi, *J. Phys. (Paris) Lett.* **37**, L-135 (1976).
149. H. Figger, K. Siomos, and H. Walther, *Z. Phys.* **270**, 371 (1974).
150. J. Heldt, H. Figger, K. Siomos, and H. Walther, *Astron. Astrophys.* **39**, 371 (1975).
151. J. Pitre, K. Hammond, and L. Krause, *Phys. Rev. A* **6**, 2101 (1972).
152. J. S. Deech, J. Pitre, and L. Krause, *Can. J. Phys.* **49**, 1976 (1971).
153. R. A. Phaneuf, J. Pitre, K. Hammond, and L. Krause, *Can. J. Phys.* **51**, 724 (1973).
154. G. Copley, B. P. Kibble, and G. W. Series, *J. Phys. B* **1**, 724 (1968).
155. J. S. Deech, P. Hannaford, and G. W. Series, *J. Phys. B* **7**, 1131 (1974).
156. H. Figger and H. Walther, *Z. Phys.* **267**, 1 (1974).
157. W. E. Baylis, *Phys. Lett.* **26A**, 414 (1968).
158. M. Krainska-Miszczak, *Bull. Acad. Pol. Sci.* **15**, 595 (1967).
159. M. Elbel and F. Naumann, *Z. Phys.* **204**, 501 (1967).
160. M. Elbel and W. Schneider, in *Proceedings of the International Conference on Optical Pumping*, Ed. T. Skalinski, p. 193, Warsaw, Poland (1968).
161. J. Fricke, J. Haas, E. Lüscher, and F. A. Franz, *Phys. Rev.* **163**, 45 (1967).
162. J. Haas, J. Fricke, and E. Lüscher, *Z. Phys.* **206**, 1 (1967).
163. R. A. Zhitnikov, P. P. Kuleshov, and A. I. Okunevitch, *Phys. Lett.* **29A**, 239 (1969).
164. E. W. Weber, H. Ackermann, L. S. Laulainen, and G. zu Putlitz, *Z. Phys.* **260**, 341 (1973).
165. F. A. Franz, *Phys. Lett.* **29A**, 326 (1969).
166. E. Jacobson, *Z. Phys.* **251**, 214 (1972).
167. W. Berdowski, T. Shiner, and L. Krause, *Phys. Rev.* **4**, 984 (1971).
168. R. H. Neynaber, *Adv. At. Mol. Phys.* **5**, 57 (1969).
169. W. E. Baylis, *Can. J. Phys.* **55**, 1924 (1977).
170. W. Bachmann and W. Janke, *Z. Naturforsch.* **28A**, 1821 (1973).
171. R. B. Bernstein and T. J. P. O'Brien, *J. Chem. Phys.* **46**, 1208 (1967).
172. R. B. Bernstein and T. J. P. O'Brien, *Disc. Faraday Soc.* **40**, 35 (1965).
173. W. E. Baylis, *Phys. Rev. A* **1**, 990 (1970).
174. J. P. Toennies, W. Welz, and G. Wolf, *J. Chem. Phys.* **61**, 2461 (1974).
175. J. P. Toennies, W. Welz, and G. Wolf, *J. Chem. Phys.* **64**, 5305 (1976).
176. E. W. Rothe, R. H. Neynaber, and S. M. Trujillo, *J. Chem. Phys.* **42**, 3310 (1965).
177. H. Haberland, C. H. Chen, and Y. T. Lee, in *Atomic Physics*, Vol. 3, Eds. S. J. Smith and K. Walters, p. 339, Plenum Press, New York (1973).
178. R. Düren, H. O. Hoppe, and H. Pauly, *Phys. Rev. Lett.* **37**, 743 (1976).

179. R. W. Anderson, T. P. Goddard, C. Parravano, and J. Warner, *J. Chem. Phys.* **64**, 4037 (1976).
180. G. M. Carter, D. E. Pritchard, M. Kaplan, and T. W. Ducas, *Phys. Rev. Lett.* **35**, 1144 (1975).
181. G. M. Carter, D. E. Pritchard, and T. W. Ducas, *Appl. Phys. Lett.* **27**, 498 (1975).
182. H. J. Gerritsen and G. Nienhuis, *Appl. Phys. Lett.* **26**, 347 (1975).
183. T. W. Hänsch, I. S. Shahin, and A. L. Schawlow, *Phys. Rev. Lett.* **27**, 707 (1971).
184. B. Cagnac, G. Grynberg, and F. Biraben, *J. Phys. (Paris)*, **34**, 845 (1973).
185. Th. Meier, H. Hühnermann, and H. Wagner, *Opt. Commun.* **20**, 397 (1977).
186. R. Schieder and H. Walther, *Z. Phys.* **270**, 55 (1974).
187. P. Berman, *Phys. Rev. A* **13**, 2191 (1976).
188. M. Ducloy, *Phys. Rev. A* **8**, 1844 (1973).
189. M. Ducloy, *Phys. Rev. A* **9**, 1319 (1974).
190. M. Dumont, *J. Phys.* **33**, 971 (1972).
191. M. Gorlicki and M. Dumont, *Opt. Commun.* **11**, 166 (1974).
192. B. Niewitecka and L. Krause, *Can. J. Phys.* **54**, 748 (1976).
193. M. Elbel and W. Schneider, *Z. Phys.* **241**, 244 (1971).
194. M. Elbel, B. Kamke, and W. B. Schneider, *Physica* **77**, 137 (1974).
195. M. Elbel and W. B. Schneider, *Physica* **68**, 146 (1973).
196. S. Tudorache, *Rev. Roum. Phys.* **15**, 269 (1970).
197. H. Soboll, *Z. Naturforsch.* **28a**, 793 (1973).
198. G. D. Chapman and L. Krause, *Can. J. Phys.* **44**, 753 (1966).
199. G. H. Copley and L. Krause, *Can. J. Phys.* **47**, 1881 (1969).
200. G. V. Markova, G. Khvostenko, and M. P. Chaika, *Opt. Spektrosk.* **23**, 835 (1967) [English transl: *Opt. Spectrosc.* **23**, 456 (1967)].
201. A. Gallagher, *Phys. Rev.* **157**, 24 (1967).
202. E. E. Gibbs and P. Hannaford, *J. Phys. B* **9**, L225 (1976).
203. W. W. Smith and A. C. Gallagher, *Phys. Rev.* **145**, 26 (1966).
204. G. Cremer and B. Laniepce, *C.R. Acad. Sci. B* **275**, 187 (1972).
205. D. W. Cook, R. S. Timsit, and A. D. May, *Can. J. Phys.* **47**, 747 (1969).
206. J. Hamel-Garcia, G. Cremer, and B. Laniepce, *Opt. Commun.* **16**, 289 (1976).
207. D. Vienne-Casalta, *C.R. Acad. Sci. B* **271**, 206 (1970).
208. E. B. Aleksandrov and A. P. Sokolov, *Opt. Spektrosk.* **31**, 665 (1971) [English transl: *Opt. Spectrosc.* **31**, 353 (1971)].
209. R. L. Barger, *Phys. Rev.* **154**, 94 (1967).
210. A. Corney and G. W. Series, *Proc. Phys. Soc. London* **83**, 213 (1964).
211. R. Pepperl, *Z. Naturforsch.* **25a**, 927 (1970).
212. B. Laniepce and G. Cremer, *J. Phys. (Paris)*, **33**, 853 (1972).
213. B. Laniepce, *J. Phys. (Paris)* **29**, 427 (1968).
214. B. Laniepce, *J. Phys. (Paris)* **31**, 545 (1970).
215. H. Sausserau and M. Barrat, *C.R. Acad. Sci. B* **268**, 475 (1969).
216. M. Baumann and A. Eibofner, *Z. Naturforsch.* **23a**, 1409 (1968).
217. K. Tittel, *Z. Phys.* **187**, 421 (1965).
218. D. Casalta and M. Barrat, *C.R. Acad. Sci. B* **265**, 35 (1967).
219. D. E. Cunningham and L. O. Olsen, *Phys. Rev.* **119**, 691 (1960).
220. M. Baumann and E. Jacobson, *Z. Phys.* **212**, 32 (1968).
221. M. Baumann, E. Jacobson, and W. Koch, *Z. Naturforsch.* **29a**, 661 (1974).
222. J.-P. Barrat, D. Casalta, J. L. Cojan, and J. Hamel, *J. Phys. (Paris)* **27**, 608 (1966).
223. J.-P. Barrat, J. L. Cojan, and F. La Croix-Des Mazes, *C.R. Acad. Sci.* **261**, 1627 (1965).
224. J.-P. Barrat, J. L. Cojan, and Y. LeCluse, *C.R. Acad. Sci. B* **262**, 609 (1966).
225. J. P. Farroux, *C.R. Acad. Sci. B* **262**, 1385 (1966).
226. J. P. Farroux and J. Brossel, *C.R. Acad. Sci. B* **262**, 41 (1966).

227. J. P. Farroux and J. Brossel, *C.R. Acad. Sci. B* **263**, 612 (1966).
228. J. P. Farroux, *C.R. Acad. Sci. B* **264**, 1573 (1967).
229. J. P. Farroux and J. Brossel, *C.R. Acad. Sci. B* **264**, 1452 (1967).
230. J. P. Farroux, *C.R. Acad. Sci.* **265**, 393 (1967).
231. J. P. Farroux and J. Brossel, *C.R. Acad. Sci. B* **265**, 1412 (1967).
232. P. Jean, M. Martin, and D. Lecler, *C.R. Acad. Sci. B* **262**, 609 (1966).
233. Y. Lecluse, *J. Phys. (Paris)* **28**, 785 (1967).
234. C. A. Piketty-Rives, F. Grossetête-Damidau, and J. Brossel, *C.R. Acad. Sci.* **258**, 1189 (1964).
235. B. Lahaye, *J. Phys. (Paris)* **35**, 541 (1974).
236. B. Lahaye, *J. Phys. (Paris)* **35**, 1 (1974).
237. Yu. V. Evdokimov, M. P. Chaika, and V. A. Cherenkovskii, *Opt. Spektrosk.* **27**, 184 (1969) [English transl: *Opt. Spectrosc.* **27**, 97 (1969)].
238. E. Rityn, M. Chaika, and V. Cherenkovskii, *Opt. Spektrosk.* **28**, 636 (1970) [English transl: *Opt. Spectrosc.* **28**, 344 (1970)].
239. V. N. Rebane, T. K. Rebane, and V. A. Cherenkovskii, *Opt. Spektrosk.* **33**, 377 (1972) [English transl: *Opt. Spectrosc.* **33**, 616 (1972)].
240. H. M. Gibbs, *Phys. Rev. A* **5**, 2408 (1972).
241. D. E. Landman, *Phys. Rev.* **173**, 33 (1968).
242. C. W. T. Chien, R. E. Bardsley, and F. W. Dalby, *Can. J. Phys.* **50**, 116 (1972).
243. L. D. Schearer, *Phys. Rev.* **160**, 76 (1967).
244. L. D. Schearer, *Phys. Rev.* **166**, 30 (1968).
245. C. E. Moore, *Atomic Energy Levels*, NSRDS-NBS 35, U.S. Govt. Printing Office, Washington, D.C. (1971).
246. C. G. Carrington, A. Corney, and A. V. Durrant, *J. Phys. B* **5**, 1001 (1972).
247. E. Kotlikov, G. Todorov, and M. Chaika, *Opt. Spektrosk.* **30**, 185 (1971) [English transl: *Opt. Spectrosc.* **30**, 99 (1971)].
248. C. G. Carrington and A. Corney, *Opt. Commun.* **1**, 115 (1969).
249. L. D. Schearer, *Phys. Rev.* **180**, 83 (1969).
250. L. D. Schearer, *Phys. Rev.* **188**, 505 (1969).
251. T. Hänsch, R. Odenwald, and P. Toschek, *Z. Phys.* **209**, 478 (1968).
252. E. Fournier, M. Ducloy, and B. Decomps, *C.R. Acad. Sci. B* **268**, 1495 (1969).
253. M. Dumont and B. Decomps, *C.R. Acad. Sci. B* **269**, 191 (1969).
254. M. Ducloy, E. Giacobino, and B. Decomps, *J. Phys. (Paris)* **31**, 533 (1970).
255. B. Decomps and M. Dumont, *IEEE J. Quant. Electron.* **QE4**, 916 (1968).
256. D. A. Landman and R. Dobrin, *Phys. Rev. A* **8**, 1868 (1973).
257. J. P. Grandin, *J. Phys. (Paris)* **34**, 403 (1973).
258. J. P. Grandin, D. Lecler, and J. Margerie, *C.R. Acad. Sci. B* **272**, 929 (1971).
259. A. B. Gutner, R. A. Zhitnikov, and A. I. Okunevitch, *Zh. Eksperim. i Teor. Fiz. Pisma* **13**, 420 (1971) [English transl: *Sov. Phys.-JETP Lett.* **13**, 298 (1971)].
260. R. A. Zhitnikov and A. I. Okunevich, *Opt. Spektrosk.* **36**, 438 (1974) [English transl: *Opt. Spectrosc.* **36**, 253 (1974)].
261. X. Husson and J. Margerie, *Opt. Commun.* **5**, 139 (1972).
262. L. G. Williams and D. R. Crosley, *Phys. Rev. A* **9**, 622 (1974).
263. M. Barrat and J. Duclos, *C.R. Acad. Sci. B* **263**, 1170 (1966).
264. M. Dumont, Thesis, Ecole Normale Sup. Paris (1962).
265. F. W. Byron, Jr., M. N. McDermott, and R. Novick, *Phys. Rev.* **134**, A615 (1964).
266. M. Barrat, *C.R. Acad. Sci.* **259**, 1063 (1964).
267. D. Perrin-Lagarde, *C.R. Acad. Sci. B* **263**, 1384 (1966).
268. B. Dodsworth, J.-C. Gay, and A. Omont, *J. Phys. (Paris)* **33**, 65 (1972).
269. M. Baumann, *Z. Phys.* **173**, 519 (1963).
270. M. Baumann and E. Jacobson, *Z. Phys.* **212**, 32 (1968).
271. D. Lagarde and R. Lennuier, *C.R. Acad. Sci.* **261**, 919 (1965).

29

Energy and Polarization Transfer

M. Elbel

1. Introduction: Transfer of Energy and Polarization

Sensitized fluorescence has been under continuous study ever since its discovery by Wood and Cario. Accordingly, a considerable number of review articles have appeared at regular intervals. We quote the books of Mitchell and Zemanski,[1] Pringsheim,[2] and Massey and Burhop[3] and the articles by Krause,[4,5] Seiwert,[6] Kraulinya and Kruglevskij,[7] Lijnse,[8] and Nikitin.[34] However, in all these reviews the main attention has been directed toward energy transfer. Since the mid-1960s an increasing interest has arisen in the study of the transfer of quantities other than energy. These are population structures described by density matrices and tensors. The growing importance of these aspects has made us attempt to give a first survey of activities in this field. We will not omit traditional energy transfer altogether, but restrict ourselves to the most recent developments. In this respect we will emphasize where we stand rather than the completeness of the reported work that has appeared in the last 60 years.

2. Transfer of Energy

2.1. Experimental

Collisional transfer of energy occurs during collision processes involving different atoms, one of which is in an excited state. As a result, one of the atoms is in a changed state. The final state of the pair, in general, possesses

---

M. Elbel • Physikalisches Institut der Universität, 3550 Marburg, Renthof 5, West Germany.

different energy. The energy defect ΔE for the final and initial states is defined by

$$A^* + B \;\leftrightarrow\; \left\{ \begin{array}{l} A + B^* \\ A^{**} + B \end{array} \right\} + \Delta E \tag{1}$$

The property of being excited is denoted by an asterisk (two asterisks indicate when another excited state is assumed). The energy defect ΔE is the amount of energy set free in the process and converted to kinetic energy of the relative motion of the collision partners (negative ΔE accordingly means that this amount of energy, taken from the relative motion, is bound during the process).

Equation (1) demands that atom A be brought to an excited state, a process usually achieved by absorption of photons. The photons must have the wavelength of a suitable transition which reaches state A^* from the ground-state or metastable state of atom A. Hence the vapor of atom A must be exposed to light at the wavelength for that transition; the light is supplied by a spectral lamp (or laser) and filtered through a monochromator, a resonance filter, or Lyot filter. A vapor of atom B is mixed with vapor A at varying pressure. The fluorescence of the vapor of A^* and B^* (or A^{**}) is resolved and measured at right angles to the incident light beam. Rate equations are considered for the populations in the excited levels of species A and B:

$$\frac{dN_A}{dt} = C - N_A\left(\frac{1}{\tau_A} + Z_A\right) + N_B Z_{B \to A} \tag{2}$$

$$\frac{dN_B}{dt} = -N_B\left(\frac{1}{\tau_B} + Z_B\right) + N_A Z_{A \to B} \tag{3}$$

Here Z_A, Z_B are rates of collisional depopulation, and $Z_{A \to B}$, $Z_{B \to A}$ are rates of collisional transfer. The stationary solution of the second equation

$$\frac{N_A}{N_B} = \frac{1}{\tau_B Z_{A \to B}} + \frac{Z_B}{Z_{A \to B}} \tag{4}$$

suggests plotting N_A/N_B against the reciprocal density of atoms B since

$$Z_{A \to B} = N_B \bar{v}_r Q_{A \to B} \tag{5}$$

N_A/N_B corresponds to I_A/I_B, the ratio of the photon count rates of the resonant fluorescence and the sensitized fluorescence. A straight line is expected provided $Z_B \propto N_B$. The slope of the curve yields $Q_{A \to B}$.

If a plot of I_B/I_A versus N_B is chosen, then at least the slope of the curve at very low densities N_B yields $v_r Q_{A \to B} \tau_B$ and hence $Q_{A \to B}$. Care must be taken that N_A be low to the same order of magnitude as N_B. If Z_B

depends slightly on N_A, an extrapolation of the resulting $Q_{A \to B}$ towards vanishing densities N_A must be attempted.

A few pitfalls must be avoided. When the axis of the incident light is chosen as an axis of reference, an observer at 90° without a linear polarizer measures the sum of σ light and π light. This quantity must not be taken as a measure of the population number. As will be shown in Section 3.2, $(I_\sigma + I_\pi)_{90°}$, as well as depending on N, depends on the alignment in the excited state. However, $(I_\sigma + \frac{1}{2}I_\pi)_{90°}$ is a true measure of N, independent of the alignment. It is advisable, therefore, to have a polarizer in front of the multiplier, set at 54.7° to the experimental plane, or to measure I_σ and I_π separately. These provisions are unnecessary if the initial state has $j = \frac{1}{2}$ so that it cannot be aligned.

τ_A and τ_B are the radiative lifetimes of one isolated atom A or B. We understand that the probability of collisional transfer multiplies when the photon, after emission, is reabsorbed by another atom, even more so as these emission–reabsorption cycles can be numerous. We have to keep the density of atoms low to the extent that the optical depth of the photon remains large compared with the distance of the fluorescent region, in the cell, from the window. Krause and co-workers[4] used favorable cell designs suitable for work in the pressure region where radiation imprisonment occurs. Studies of radiation imprisonment, both experimental and theoretical, were done by Milne,[9] Holstein,[10] Barrat,[11] Dyakonov and Perel,[12] Omont,[13] Otten and Winnacker,[14] Kibble et al.,[15] Copley and Krause,[16] and Deech and Baylis.[17] It turns out that radiation imprisonment lengthens the decay time of the various tensor polarizations (N among them) at different rates.

In most experiments the densities N correspond to the saturated pressure curves. A collection of tables of the saturated pressures as a function of temperature was provided by Nesmeyanov.[18] For very low pressures many authors did not use pressure tables applicable to higher densities than are present in most experiments, and they determined instead the vapor density from the self-reversal of resonance light transmitted through the vapor (cf., e.g., Coolen et al.[19]).

2.2. Energy Transfer in Collisions of Alkali Atoms with Inert Gas Atoms

The cross sections for transitions between the lowest alkali P-doublet states, Q_{12} and Q_{21}, pertinent to transitions from $^2P_{1/2}$ to $^2P_{3/2}$, and inverse induced in collisions with rare-gas atoms have been measured by Krause and co-workers (Czajkowski et al.,[20] Pitre et al.,[21] Chapman and Krause,[22] Pitre and Krause[23]), and Jordan and Franken.[24]

The general features of their results can be summarized in the following manner.

(i) There is a drastic increase of Q_{21}/Q_{12} toward the heavier alkali atoms. This must be attributed to the widening of the doublet gap ΔE.

(ii) Q_{12}, as well as Q_{21}, increases toward the heavier noble gases with one exception. The cross section for He usually exceeds the one for Ne. For Rb and Xe, cross sections with He are as much as two orders of magnitude larger than with the heavier rare gases.

(iii) Q suffers a drastic reduction by up to six orders of magnitude in going from Na to Cs. This fact, more than the others, emphasizes the importance of ΔE in the collision process.

Point (i) is well explained by the principle of detailed balancing: Every pair of colliding atoms, one of which is an alkali in the $^2P_{3/2}$ state, is eligible to produce a superelastic collision resulting in the alkali being transferred to $^2P_{1/2}$. But to produce the inverse process the threshold energy ΔE, at least, must be available in the center-of-mass system. This reduces the number of eligible pairs by a factor of $\exp\left(-\Delta E/kT\right)$:

$$\frac{Q_{12}}{Q_{21}} = \frac{g_2}{g_1} \exp\left(\frac{-\Delta E}{kT}\right) \tag{6}$$

of $^2P_{1/2}$ and $^2P_{3/2}$.

Here again it must be borne in mind that Q is an averaged cross section related to the proper cross section σ through

$$Q = \langle \sigma v_r \rangle / \langle v_r \rangle \tag{7}$$

where g_1, g_2 are the statistical weights of the levels, viz., 2 and 4 in the case stand for an averaging process based on the Maxwellian distribution

$$f(\mathbf{v}_r)d\mathbf{v}_r = 4\pi^{-1/2}(\mu/2kT)^{3/2} \exp\left(-\mu v_r^2/2kT\right)v_r^2\, dv_r \tag{8}$$

where μ is the reduced mass of the colliding pair.

In earlier investigations reviewed by Krause,[4] the agreement of Q_{12}/Q_{21} with the predicted value was satisfactory for sodium but rather poor for rubidium and even poorer for cesium and the heavier noble gases. Gallagher[25] proved experimentally that line broadening of the resonant fluorescence constitutes a serious source of error. Specifically, after the vapor absorbs a Doppler-shaped D_2 line it reradiates a collisionally broadened D_2 line due to the inert gas collisions. The D_1 interference filters pass, besides sensitized D_1 light, a 60 Å band on the red side of the broadened D_2 line. The intensity received by the detector is proportional to the inert gas pressure due to both influences, line broadening and sensitizing collisions. Since the line broadening primarily affects the red wing of the D lines, the spurious influence is absent when D_1 light is used for excitation and D_2 light is detected. This discovery allowed, in a large part, the removal of the discrepancy of the Q_{12}/Q_2 values.

Points (ii) and (iii) give evidence of the influence of ΔE and the duration τ_c of the collision of the transition probability. Time-dependent perturbation theory, crudely speaking, reduces the problem of finding the transition amplitude to evaluating a time integral of the perturbation matrix multiplied by $\exp(i\Delta Et/\hbar)$. If the collision occurs over many cycles of $\Delta E/\hbar$, as with the Rb and Cs doublets, the oscillatory function will cause a reduction in the transition probability. Only for sodium the influence of ΔE may be negligible, at least for collisions with helium, where the collision is very short-lived.

Callaway and Bauer[26] solved the collision problem in a simple manner. They treated the two doublet levels as nondegenerate and, accordingly, used an average of the individual perturbation matrix elements. Moreover, they restricted their considerations to van der Waals interactions. Even so, the solution requires the assumption that the interaction matrix $Q(t)$ and the time integral

$$\int_{-\infty}^{t} Q(t')\, dt'$$

commute, which is rather a controversial approximation (Omont and Meunier[27]). But the valuable result of this approach is an analytic expression for the transition matrix element

$$S(t=\infty)=\left(\frac{q_1 B}{hvb^{n-1}}\right)M_{n-1}(x) \tag{9}$$

where b is the impact parameter, v the relative velocity, q_1 the constant of the potential which depends on R through R^{-n}, and

$$x=\frac{\Delta E \cdot b}{\hbar v} \tag{10}$$

We note that b/v equals the duration of the collision. For $x \gg 1$, $M_{n-1}(x)$ behaves like

$$M_{n-1}(x)=\frac{2}{\Gamma\left(\dfrac{n-1}{2}\right)}\left(\frac{x}{2}\right)^{n/2-1}e^{-x} \tag{11}$$

Here the exponential dependence on $-x$ becomes apparent. x is widely known as the Massey parameter, whose importance manifested itself at first in the Landau–Zener theory. Here, however, ΔE meant the residual splitting of the avoided level crossing. From Eq. (11) we get a qualitative insight into why the transfer cross sections with He for Rb, Cs are orders of

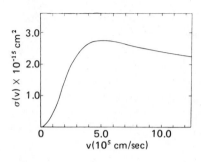

Figure 1. A typical dependence on temperature of the velocity-averaged transfer cross section calculated in the two-level approximation (after Callaway and Bauer[26]).

magnitude larger than with the other noble gases although the potential constant q_1 is usually the smallest. It is a matter of the relative velocity. Also the rapid decrease of Q_{12}, Q_{21} at rising ΔE [point (iii)] is contained in this formula. Callaway and Bauer[26] also succeeded in giving the dependence on the temperature of Q_{12} (Figure 1) analytically. The trend of this curve was qualitatively confirmed by Hidalgo and Geltman[28] and by Masnou–Seeuws[29] (cf. also Kumar and Callaway,[30] Reid and Dalgarno,[31] and Reid[32]). Masnou–Seeuws developed a method of solving Nikitin's[33] coupled equations. These latter equations describe the evolution in time of the interacting sets of states $|\frac{1}{2}\frac{1}{2}\rangle$, $|\frac{3}{2}\frac{1}{2}\rangle$, $|\frac{3}{2}-\frac{3}{2}\rangle$ or $|\frac{1}{2}-\frac{1}{2}\rangle$, $|\frac{3}{2}-\frac{1}{2}\rangle$, $|\frac{3}{2}\frac{3}{2}\rangle$, which are quantized with respect to the collisional plane. Masnou–Seeuws found that the two-state approximation is incomplete in that it does not account for the rotational coupling of states $^2\Pi_{1/2}$ and $^2\Pi_{3/2}$. It deals only with the radial coupling of the states $^2\Pi_{1/2}$ and $^2\Sigma_{1/2}$. Nikitin recently[34] pointed out that both types of interaction happen for fairly large values of internuclear radii. Accordingly, both mechanisms contribute incoherently to the transition probability. He provided analytic expressions for both mechanisms, showing that the dependence on the collisional energy is different.

The rotational and radial couplings become apparent when acting with the Hamiltonian $hi(\partial/\partial t)$ on adiabatic molecular states. Bear in mind that $hi(\partial/\partial t)|_{\text{lab}}$ in the laboratory frame differs from $hi(\partial/\partial t)|_{\text{mol}}$ in the molecular system by two terms:

$$hi\left(\frac{\partial}{\partial R}\,\dot{R} + \frac{\partial}{\partial \Phi}\,\dot{\Phi}\right)$$

The first of these operators couples states with similar angular symmetry; the second one couples states differing in Ω (projection of electronic angular momentum on the internuclear axis) by ± 1.

Dashevskaya et al.[35] and Nikitin[34] have calculated separately the contributions of radial and rotational coupling as a function of collisional energy and compared it with Gallagher's[25] experimental results (Figure

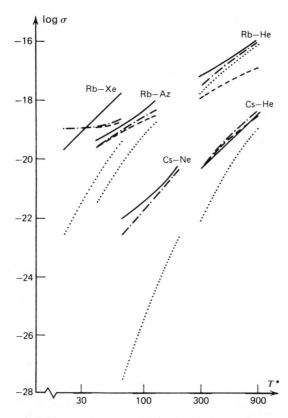

Figure 2. Transfer cross section $\sigma(\frac{1}{2}\rightarrow\frac{3}{2})(\text{Å}^2)$ for different alkali M–noble-gas pairs as a function of the effective temperature $T^* = T(\mu_{M-\text{He}}/\mu_{M-x})$. Solid lines: experimental data; dotted lines: contributions of radial coupling; broken lines: contributions of Coriolis coupling. Sum of both contributions: –·–·–·– (from Nikitin[34]).

2). While the general tendency of the transfer cross section is well reproduced by the theory, it turns out that rotational coupling plays a major part in collisions with helium only. In Rb–He collisions it even dominates the interaction. Siara et al.[36] and Cuvellier et al.[37] have measured the transfer cross sections between the 7^2P levels of Cs in collisions with rare gases, allowing for the temperature dependence as well. It turned out that the latter dependence is less drastic than in the case of Cs 6^2P, a fact which is obvious from the reduced value of ΔE [$\Delta E(6^2P) =$ 554 cm^{-1}, $\Delta E(7^2P) = 181$ cm^{-1}]. Hence it is not surprising that the cross sections between the 8^2P levels of Cs (Pimbert[38]) are even larger (Table

Table 1. Transfer Cross Sections between the ^2P states of Cesium Colliding with the Noble Gases

Transition	ΔE (cm^{-1})	T (K)	Q (Å^2)				
			He	Ne	Ar	Kr	Xe
Cs($6\,^2P_{1/2} \rightarrow 6\,^2P_{3/2}$)[a] Theory[b]	554	311	5.7×10^{-5} 4.6×10^{-5}	1.9×10^{-5} 2.6×10^{-7}	1.6×10^{-5}	8.3×10^{-5}	7.2×10^{-5}
Cs($6\,^2P_{3/2} \rightarrow 6\,^2P_{1/2}$)[a]		450	3.9×10^{-4}	3.1×10^{-4}	5.2×10^{-4}	18.4×10^{-4}	27.4×10^{-4}
Cs($7\,^2P_{1/2} \rightarrow 7\,^2P_{3/2}$)[c]	181		12	1.8×10^{-1}	1.2×10^{-1}	9.1×10^{-2}	6.5×10^{-1}
Cs($7\,^2P_{3/2} \rightarrow 7\,^2P_{1/2}$)[c]			11	1.6×10^{-1}	1.0×10^{-1}	7.7×10^{-2}	5.7×10^{-1}
Cs($8\,^2P_{1/2} \rightarrow 8\,^2P_{3/2}$)[d]	83	420	34	4.4	5.5	4.5	10

[a] Czajkowski et al.[20]
[b] Dashewskaya et al.[35]
[c] Cuvellier et al.[37] (See also: Siara et al.[36].) These provide additional data on transfer to the $6\,^2D_{3/2,5/2}$ levels at much wider energy gap ΔE. They are interpreted in terms of Pascale and Vanderplanque[47] potentials, which show avoided crossings between the respective molecular states for the heavier noble gases. Their influence on the cross sections is made evident. A similar behavior of the cross sections is found for Rb when proceeding from $5p$ to $6p$ (Siara et al.[48]).
[d] Pimbert.[38]

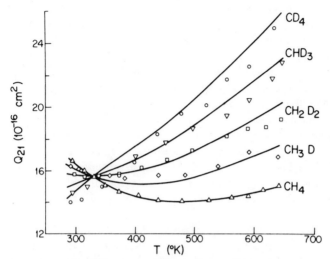

Figure 3. Cross sections for transitions $6\,^2P_{3/2}$–$6\,^2P_{1/2}$ of cesium in collisions with methane. The solid lines are the theory, for which the potential parameters have been adjusted to fit the undeuterated CH_4 data (from Walentynowicz et al.[39]).

1) and that no variation with temperature was observed, in accordance with the almost-resonant character of those transitions.

So far, the time dependence of the Hamiltonian has resulted from the limited duration of the encounter alone. But if molecules are taken as collision partners of the alkalis instead of rare gases, then part of this time dependence may result from motion of the molecules which is other than translational. Concerning the motions, rotations must be mainly considered. Methane represents an example (Baylis et al.[39]). The fifth harmonic of the rotation frequency of CH_4 almost agrees with $\Delta E/h$ of Rb $5p\,^2P$. Methane, although its polarizability is nearly the same as that of Kr, nevertheless has a stationary octopole moment which interacts with the third spatial derivative of the electrostatic potential of the Rb ^2P atom. This interaction depends on the orientation of CH_4 with respect to the internuclear axis and hence oscillates when CH_4 rotates. This rotation can bring to resonance the interaction integral $\int_{-\infty}^{\infty} V(t)\exp(i\Delta Et/h)\,dt$ and explain why the transfer cross section is about 10^5 time larger for CH_4 than for Kr. Moreover, when methane is replaced by deuterated methane CD_4 the moment of inertia is doubled and the frequency is reduced to 50%. Such an effect can change the resonance condition drastically and give rise to sizable altering of the cross section. Figure 3 gives an example. Gradual increase of the deuteration of the methane results in a stepwise decrease of the transfer cross section. Quite different temperature behavior of the cross sections is likewise observed. This isotope effect was discovered earlier for H_2, HD, and D_2 by McGillis and Krause[40] for Cs, but shortly after that verified by Stupavski and Krause[41] for Na and by Hrycyshyn and Krause[42] for Rb. As quenching collisions are not our concern, we just mention that in this case it is the vibrational motion of the molecules which can bring about quasiresonant conditions and, accordingly, enhance their quenching efficiency in comparison with atoms.*

Transfer cross sections and their dependence on the collisional energy can be studied by varying the temperature of the gas. Narrow-banded dye laser excitation provides a novel method proposed by Phillips and Pritchard.[43] Dye laser light whose frequency ν_L is deliberately detuned off the resonant frequency ν_0 of an alkali atom accordingly excites only atoms which move along the laser beam at such a speed v_z that v_z/c equals $(\nu_L - \nu_0)/\nu_0$.

The other velocity components v_x and v_y are not selected and pertain to a Maxwell distribution. The thermal average of the resultant velocity depends on the detuning of the laser having its minimum when $\nu_L = \nu_0$ and increasing when $|\nu_L - \nu_0|$ increases. Transfer rates which have been measured by Apt[44] for Na–Ar and Na–Xe show an increase with increasing average velocity. He obtained good agreement with recent calculations by Pascale et al.[45,47] His results once more confirm that transfer collisions of Na ($3p$), at least with the heavy rare gases, cannot be considered as quasiresonant because then the cross section at rising collision energy would be decreasing instead of increasing.

2.3. Energy Transfer in Collisions between Similar and Dissimilar Alkali Atoms

Energy transfer in collisions between alkali atoms has already been reported by Wood[49] and ever since has fascinated physicists who are in the field of collision processes. This is so in spite of the fact that experiments must be carried out at the lowest vapor pressure ($< 10^{-5}$ Torr) to prevent imprisonment of resonance radiation,† and that the density of the alkali vapors must be inferred from pressure formulas which are meant for the high-pressure range and whose extrapolation to the low-pressure range is rendered doubtful by the undefined absorbing action of the glass walls of the

---

* A recent account on quenching collisions has been given by Krause.[5] More recent work has been published by Hannaford and Lowe.[153]

† Corrections for photon diffusion are possible according to the theory of Holstein[10] and have been applied, e.g., by Seiwert and co-workers[50–52] to measurements where the optical depth was not large compared with the illuminated portion of the cell.

containers. There are two advantages which make the excited alkali–ground-state alkali collisions so attractive. Firstly, a process

$$A^*(^2P_J) + B(^2S_{1/2}) \leftrightarrow A(^2S_{1/2}) + B^*(^2P_J) \tag{12}$$

is mediated by the direct interaction of the transition dipoles d_A and d_B:

$$d_{A,B} = (A^*, B^* \, ^2P_J|er_{A,B}|A, B \, ^2S_{1/2}) \tag{13}$$

where $r_{A,B}$ denotes the radius of the valence electron in the atoms A or B. The potential curves which arise from the different levels of the separated pairs of the alkali atoms accordingly are written in terms of

$$V_{dip} = d_A d_B / R^3 \tag{14}$$

and depend on the quantum numbers of the molecular system. Apparently, the potential curves fall off smoothly as R^{-3}. Therefore, the cross sections for energy transfer are rather large, which facilitates their measurement. Secondly, the transition dipoles d_A, d_B are fairly well known from lifetime measurements or the f sum rule. Moreover, the R^{-3} part of the potential curves has a much longer range than the exchange part (pseudopotentials), which arises from the Pauli exclusion principle and which can be neglected to first order. This even holds true for dissimilar alkali atoms, where $d_A \neq d_B$. For them, there is no degeneracy between $A^* B$ and AB^* levels, so that at large distances the interaction falls off as R^{-6}. However, the gap between the levels $A^* B$ and AB^* for alkalis is rather narrow, so that the dipole–dipole interaction reaches the value of the gap at distances which are still fairly large compared with the dimensions of the atoms. Accordingly, at mean distances there is an extended range in which the potential curves vary as R^{-3}. Therefore, potential curves can be calculated reliably and predictions are possible as to the exact location of crossings in which stepover during the collision process may occur. This makes the case of alkali–alkali collisions a favored playground of theorists.

To start with the resonant excitation transfer $A = B$, $J = J'$, this process can be investigated when transitions to (degenerate) Zeeman sublevels or (almost degenerate) hyperfine sublevels are induced. This has been carried out in the case of mercury by Faroux and Brossel,[53] who studied depolarization. When the degeneracy of the Zeeman sublevels is removed by a magnetic field, the sensitivity of the resonant cross sections to disturbances of the resonance condition can be made evident (Datta,[54] Hanle,[55,154] Gay and Schneider,[56] and Fuchs et al.[57]). If there is no energy gap between the initial and final level, then the order of magnitude of the cross section σ_{res} is determined by the condition (b is the impact parameter)

$$\frac{V(b)\tau}{\hbar} \approx \frac{d_A^2}{b^3} \frac{b}{\hbar v} = 1 \tag{15}$$

Table 2. Transfer Cross Sections in Collisions of Similar Alkali Atoms[a]

Atoms	$\Delta E/(\text{cm}^{-1})$	Experimental			Theoretical	
		Q_1 (Å²)	Q_2 (Å²)	Ref.	Q_2 (Å²)	Ref.[b]
Na* Na	17	532 170	283 110	58 50	101	63, 64
K* K	58	330 370 120	165 250 60	59 60 52	20.4	63, 64
Rb* Rb	238	74 60	46 72	59 61	4.8	63, 64
Cs* Cs	554	6.4 6	31 13	62 51	4.2	63, 64

[a] One is in the ground state, the other is excited to the lowest p doublet. $Q_1 = Q(^2P_{1/2} \rightarrow {}^2P_{3/2})$, $Q_2 = Q(^2P_{3/2} \rightarrow {}^2P_{1/2})$.
[b] See also Orchinnikova.[65]

where

$$\sigma_{\text{res}} = \pi b^2 = \pi d_A^2/\hbar v \qquad (16)$$

If, however, there is a gap ΔE, then the cross section σ_{nonres} for the transition is determined from the condition that the Fourier component of the transition potential $V(t)$ at $\omega = \Delta E/\hbar$ must be significant. This leads to

$$\Delta E \approx V(b) = \frac{d_A^2}{b'^3} \qquad (17)$$

$$\sigma_{\text{nonres}} = \pi \Delta E^{-2/3} d_A^{4/3} = \pi b'^2 \qquad (18)$$

so that

$$\frac{\sigma_{\text{nonres}}}{\sigma_{\text{res}}} = \frac{\hbar v}{\Delta E^{2/3} d_A^{2/3}} = \frac{\hbar v}{b' \, \Delta E} \qquad (19)$$

For sodium, b' is of the order of 50 a.u. and the ratio $\sigma_{\text{nonres}}/\sigma_{\text{res}}$ is approximately 1/40, decreasing towards the heavier alkalis. Thus transfer cross sections must be expected to be two orders of magnitude smaller than depolarizing cross sections. Table 2 gives examples.

Dashevskaya et al.[63] attempted to calculate σ_{nonres} following the lines of Landau–Zener theory. They chose the decreasing cross section Q_2 in Table 2, for example. Figure 4 shows how the levels of separate atoms correlate with the molecular levels (Nikitin[34]) of the combined atoms. Σ and Π orbitals of gerade and ungerade symmetry and triplet and singlet spin states arise. Their energetic position, however, is determined, practically by dipole–dipole interaction and spin–orbit interaction alone. Consider the crossings of the 1_g, 0_g^+ pair and the 1_u, 0_u^+ pair of levels. These levels are not coupled in adiabatic approximation due to their different symmetry. Nonadiabatic coupling, however, is possible due to the term $i\hbar(\partial/\partial\Phi)\dot{\Phi}$; $(\hbar/i)(\partial/\partial\Phi) = J_x$ is apparently the component of the total angular momentum with respect to the axis of the collision plane. It connects the crossing states. Dashevskaya *et al.*[63] suggest that the first of the crossings (cf. correlation diagram) yields small matrix elements for J_x, since the states are still close to the atomic classification in the crossing region. For the latter crossing the Hund's case (a) value $(1u|J_x|0_u) = \hbar/2^{1/2}$ may be almost reached. This value has to be inserted into a Landau–Zener-type

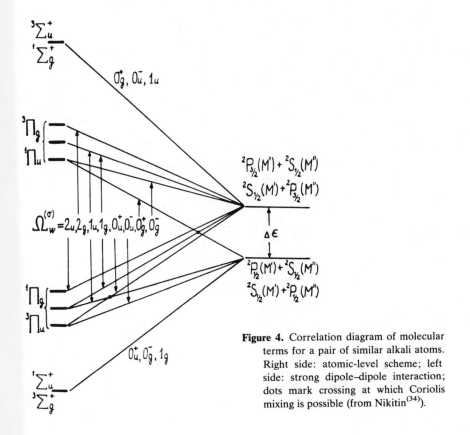

Figure 4. Correlation diagram of molecular terms for a pair of similar alkali atoms. Right side: atomic-level scheme; left side: strong dipole–dipole interaction; dots mark crossing at which Coriolis mixing is possible (from Nikitin[34]).

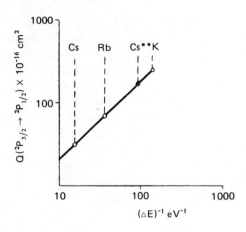

Figure 5. Logarithmic dependence of $Q(^2P_{3/2}-^2P_{1/2})$ on the energy defect ΔE in collisions of similar alkali atoms. Cs** data pertain to the higher $8p$ doublet. Other data are for the lowest doublet. A dependence like $Q \propto \Delta E^{-1}$ would correspond to a slope of 45° (from Kraulinya[7]).

formula for transitions from one crossing level i to the other level f:

$$\sigma_{if} = \frac{16(2)^{1/2}\pi^2}{3(\mu)^{1/2}\hbar} \frac{|\langle i|J_x|f\rangle|^2}{(d/dR)[U_i(R)-U_f(R)]|_{R_s}} \frac{[E-U_i(R_s)]^{3/2}}{E} \quad (20)$$

where $U_i(R)$, $U_f(R)$ are the potential curves, R_s is the crossing radius, μ is the reduced mass, and E is the kinetic energy. The cross section for the transition from one Zeeman multiplet j to the other, j', reads in terms of σ_{if}

$$\sigma(j \to j') = \frac{1}{4(2j+1)} \sigma_{if} \quad (21)$$

where the denominator $4(2j+1)$ results from averaging over the initial sublevels. It accounts for the degeneracy of the initial level $A'\,^2S_{1/2}\,A''\,^2P_j$, $A'\,^2P_j\,A''\,^2S_{1/2}$. Equation (21) answers the question for the dependence on ΔE of $\sigma(j \to j')$. It is

$$\sigma(j \to j') \propto \left\{ \frac{d}{dR}[U_i(R_s)-U_f(R_s)]|_{R_s} \right\}^{-1} \quad (22)$$

We require that

$$U_i(R_s) - U_f(R_s) = 0 \quad (23)$$

$$U_i(\infty) - U_f(\infty) = \Delta E \quad (24)$$

hence we have

$$U_i(R) - U_f(R) = \Delta E - \Delta E\, R_s^3/R^3 \quad (25)$$

and finally

$$\sigma(j \to j') \propto R_s/\Delta E \quad (26)$$

The variation as ΔE^{-1} of $\sigma(j \to j')$ was stated by Krause[4] (Figure 5). The more recent value $\sigma(\frac{3}{2} \to \frac{1}{2})$ for Cs $8p$ (Pimbert et al.[70]) fits well into the curve (Kraulinya and Kruglevskij[7]). Much activity has been devoted to excitation transfer in collisions of dissimilar alkali atoms (Table 3). Dashevskaya et al.[69] gave an analysis of the K–Rb case. Figure 6 shows the levels of the separate atoms and the molecular potential curves which arise from them. In contrast to the case of similar alkali atoms is the coming into play of five avoided crossings (at R_0–R_4) to which another real crossing must be considered (at R_5). The contributions to the transfer cross section of the avoided crossing follow from the Landau–Zener theory, whereas the crossing at R_5 contributes through rotational coupling and gives rise to a formula analogous to Eq. (20). Transitions $K^* \, 4\,^2P_{1/2}$ to $K^* \, 4\,^2P_{3/2}$ are found to arise mainly from the crossing at R_0. The cross section of this process is by far the largest due to the very large value of R_0

Table 3. Cross Sections for Energy Transfer in Collisions of Dissimilar Alkali Atoms

Transition	ΔE (cm^{-1})	Experimental		Theoretical	
		Q (Å^2)	Ref.	Q (Å^2)	Ref.
$K^* \, 4\,^2P_{3/2}$, Rb $\to$ K, Rb* $5\,^2P_{1/2}$	464	3.2	59		
		1.9 ± 0.6	66		
		2.6 ± 0.5	67		
		2.5 ± 0.5	68		
$K^* \, 4\,^2P_{3/2}$, Rb $\to$ K, Rb$^*5\,^2P_{3/2}$	225	27 ± 7	66		
		5.5 ± 1.2	68		
$K^* \, 4\,^2P_{1/2}$, Rb $\to$ K, Rb* $5\,^2P_{1/2}$	409	2.7 ± 0.6	66		
		2.2 ± 0.5	67	4	69
		2.3 ± 0.6	68		
$K^* \, 4\,^2P_{1/2}$, Rb $\to$ K, Rb$^*5\,^2P_{3/2}$	168	40 ± 8	66	10	69
		5.3 ± 0.8	68		
$K^* \, 4\,^2P_{1/2}$, Rb $\to$ $K^* \, 4\,^2P_{3/2}$, Rb	-58	260 ± 65	66	67	69
$K^* \, 4\,^2P_{3/2}$, Rb $\to$ $K^* \, 4\,^2P_{1/2}$, Rb	$+58$	175	66		
Rb* $5\,^2P_{1/2}$, Cs $\to$ Rb, Cs$^*6\,^2P_{3/2}$	847	1.5 ± 0.4	20		
Rb* $5\,^2P_{1/2}$, Cs $\to$ Rb, Cs* $6\,^2P_{1/2}$	1401	0.5 ± 0.1	20		
Rb* $5\,^2P_{3/2}$, Cs $\to$ Rb, Cs* $6\,^2P_{3/2}$	1084	0.9 ± 0.2	20		
Rb* $5\,^2P_{3/2}$, Cs $\to$ Rb, Cs$^*6\,^2P_{1/2}$	1638	0.3 ± 0.1	20		

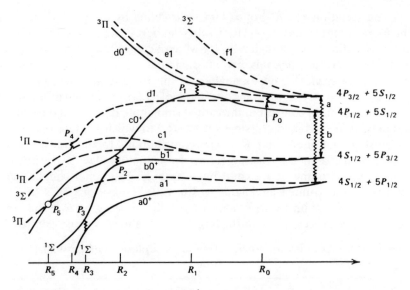

Figure 6. Molecular levels for the (K Rb)* pair R_0 through R_4: radii of avoided crossings in which radial coupling is active; R_5: radius of a crossing where Coriolis coupling comes into play (from Nikitin[34]).

(15–22 a.u.). Transitions from K* $4\,^2P_{1/2}$ to Rb* $5\,^2P_{3/2}$ arise from the crossings at $R_2 = 13$ a.u. and $R_4 = 10$ a.u., which give comparable contributions. The transition from K* $4\,^2P_{1/2}$ to Rb* $5\,^2P_{3/2}$ affords two stepovers between adiabatic potentials. They occur at first at crossing R_2 and then at $R_3 = 12$ a.u. Dashevskaya's results are slightly larger than the experimental values of Stacey and Zare,[68] which lie closest. The ratio of the cross sections of the processes K* $4p\,^2P_{1/2}$ to Rb* $5p\,^2P_{1/2}$ and Rb* $5p\,^2P_{3/2}$ agree, however, with those of Stacey and Zare and show that the theoretical interpretation is basically correct.

2.4. Energy Transfer in Collisions of Other Dissimilar Atoms

Among the large variety of metal pairs that have been studied so far, mercury colliding with other metals bears particular importance. Mercury, cadmium and mercury, and thallium have been under study since the early 1920s, when sensitized fluorescence was discovered for these (Cario[71]). These studies soon revealed that the probability of transfer rapidly decreases as the energy defect $|\Delta E|$ increases (Franck's rule), where ΔE is defined by the process

$$A^* + B \overset{Q}{\leftrightarrow} A + B^* + \Delta E \qquad (27)$$

We consider a process of negative ΔE as inelastic, another of positive ΔE as superelastic. Naturally, the falloff of the cross section Q as a function of ΔE is smoother on the positive side of ΔE than on the negative side, where the energy condition yields a threshold. Many examples are provided by the mercury sodium pair because sodium as the acceptor of excitation has a high-level density at the Hg $6\ ^3P_1$ energy. This example was studied by many authors, among them Beutler and Josephy,[72] Frisch and Kraulinya,[73,74] Rautian and Khaikin,[75] Kraulinya and Yanson,[76] and Czajkowski et al.[77] Notably, the latter authors endeavored to plot their results versus the energy defect and, of course, obtained a smooth curve (Figure 7). It is rather intriguing that other cross sections, such as, for instance, the one for the Hg $6\ ^3P_1 \rightarrow$ Cd $5\ ^3P_1$ process, fit well into this curve (Czajkowski and Krause[78]). The same is true for the cross sections of the Rb $\rightarrow$ Cs or K $\rightarrow$ Rb transfer processes reported in the last section. However, the cross sections reported for the Hg $\rightarrow$ Cd process are strongly at variance, probably due to difficulties in avoiding photon diffusion and accounting for quenching (Table 4). The positive wing of the curve varies roughly as $\exp(-\Delta E)$. Similar curves have been obtained by Friedrich and Seiwert[78] for Cd $\rightarrow$ Cs transfer.

The variation of Q as $\exp(-\Delta E)$ is hard to understand. From a theoretical point of view (Landau,[80] Zener[81]), a strong variation with the

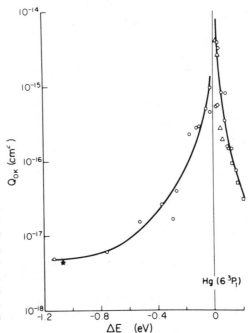

Figure 7. A "resonance curve" showing the dependence of excitation transfer cross sections for collisions between dissimilar atoms on the energy defect ΔE. ★, Hg $\rightarrow$ Cd transfer; ○, Hg $\rightarrow$ Na transfer; □, Rb $\rightarrow$ Cs transfer; △ and ◇, K $\rightarrow$ Rb transfer (from Czajkowski[78]).

Table. 4. Cross Sections for Hg $6\,^3P_1 \to$ Cd $5\,^3P_1$ Excitation Transfer (Q) and Quenching of Cd $5\,^3P_1(q)$ in Collisions with Nitrogen

$Q\,(\text{Å})^2$	$q\,(\text{Å}^2)$	Source
4.6×10^{-2}	1.7	Czajkowski and Krause[78]
9×10^{-1}	—	Morozov and Sosinskij[82]
$1-3 \times 10^{-1}$	—	Kraulinya and Arman[83]
3.4×10^{-1}	—	Chéron[84]
—	6.5×10^{-2}	Lipson and Mitchell[85]

minimum distance V_{12} of the avoided crossings in which the stepover between the states occurs is expected. In particular, the Landau–Zener formula

$$P = 2 \exp\left(-\frac{V_{12}^2}{\hbar[d(W_1 - W_2)/dR]v_r}\right)\left[1 - \exp\left(-\frac{V_{12}^2}{\hbar[d(W_1 - W_2)/dR]v_r}\right)\right]$$

(28)

correlates the transition probability P primarily with the radial velocity. For the latter a connection to the energy E in the centre-of-mass system of the moving particles exists

$$v_r = \left\{\frac{2}{\mu}\left[E + E_{\text{pot}}(R_{\text{crossing}}) - \frac{l(l+1)\hbar^2}{2\mu R_{\text{crossing}}^2}\right]\right\}^{1/2}$$

(29)

which is not unique but also depends on the impact parameter $b = l\bar{\lambda}$. At the crossing point v_r may even vanish when the classical turning point falls in that distance. These trajectory effects have been dealt with by Bates and Williams[86] and Bykhovskij et al.[87] Even after integrating P with respect to the impact parameter b, the result, which still must be averaged over a Boltzmann distribution of the E values, depends in a complex way on $E_{\text{pot}}(R_{\text{crossing}})$, and it is hard to say whether $E_{\text{pot}}(R_{\text{crossing}})$ and the energy defect ΔE are systematically correlated so that by chance a function like

$$Q \propto \exp(-|\Delta E|)$$

(30)

results.

However, at very large values of v_r the Landau–Zener formula approaches the limit

$$P = 2\frac{V_{12}^2}{\hbar(d/dR)[E_{\text{pot},1}(R) - E_{\text{pot},2}(R)]v_r}$$

(31)

and hence varies as v_r^{-1}.

Such a dependence, as already stated, was found by Apt,[44] but also by Bachmann and Janke,[88] who studied the atomic lines emitted from a beam of helium which was fired in ionic form, into a heavy rare-gas target. The helium ions pick up electrons from the rare-gas atoms and, in the same process, get excited to a vast multitude of atomic states. Hence the atomic fluorescence results. In the energy range of the helium ions, 5–30 keV, the rate at which a certain helium state was created turned out to be almost independent of energy

$$v_r Q = \text{const} \tag{32}$$

which agrees with the prediction of Eq. (31) and shows that indeed Landau–Zener theory applies.

On the other hand, at thermal energies Loh et al.,[89] using crossed-beam techniques, studied collisions of $Hg(6\,^3P_{0,2})$ atoms which collided with velocity-selected thallium atoms. The transfer rate to the excited $Tl(7\,^2S_{1/2})$ state was studied as a function of the mean relative velocity $\bar{v}_r$ and a dependence

$$Q[Hg(6\,^3P_{0,2}) \rightarrow Tl(7\,^2S_{1/2})] \propto 1/\bar{v}_r^2 \tag{33}$$

was conjectured. The authors refer to Bates, and Coulson and Zalewski,[90] who have more carefully investigated the high-energy limiting behavior of $Q(v_r)$ and found a $1/v_r^2$ dependence which is also obtained from the Born approximation.

With high-power excitation by pulsed lasers Lidow et al.[91] have given evidence of a quantum-mechanical interference of optical excitation of the host atom and collisional excitation transfer to the guest atom. As a consequence, laser light which is off-resonance with respect to the host level, nevertheless gives rise to strong transfer probabilities when tuned to the excitation energy of the guest levels. Here a new experimental dimension of transfer processes is introduced and appears to be in rapid progress.

3. Transfer of Alignment and Orientation

So far, we have only dealt with collisional transfer of occupation numbers. This is, indeed, the only source of information on the transfer process if there is only one single level which donates atoms and another single level which accepts them. In the vast majority of the studied cases the donating and accepting levels represent multiplets of degenerate Zeeman sublevels and the legitimate question arises whether any of the accepting sublevels, during a collision, is reached from any particular donating sublevel with equal probability or otherwise. In the first case occupational structures in the donating Zeeman multiplet evidently would not be handed down to the accepting multiplet but would be completely

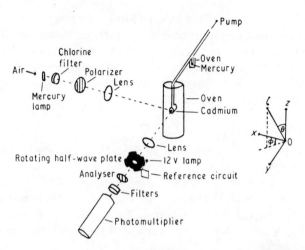

Figure 8. Experimental arrangement by Gough.[92]

wiped out. In the second case, however, transfer of occupational structures would be observed. Some examples are given by Happer and Gupta (Section 3.6 of Chapter 9 of this work) and by Schearer and Parks (Chapter 18 of this work).

The first successful attempt to furnish evidence of transfer of occupational structures was made by Gough[92] with the mercury–cadmium sensitizing collision. An earlier attempt by Mitchell[93] failed probably because the polarizer and analyzer settings were not chosen in the most favorable manner. Early work was also reported by Kraulinya et al.[156,157]

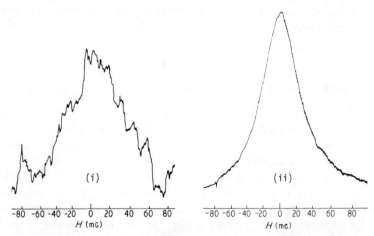

Figure 9a. Early recorder traces obtained by Gough[92] showing (i) alignment in sensitized Cd fluorescence, (ii) alignment in resonant Cd fluorescence.

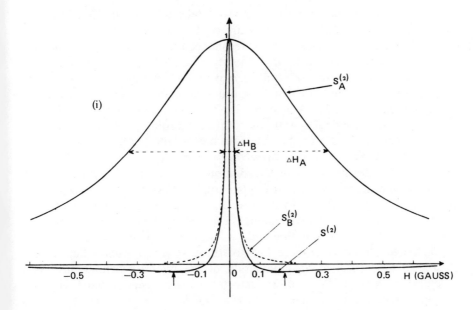

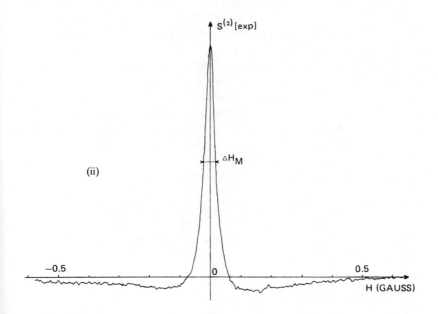

Figure 9b. Signal shape of Hanle curves: $S_A^{(2)}$ alignment in resonant Hg fluorescence, $S_B^{(2)}$ alignment in resonant Cd fluorescence, and $S^{(2)}$ alignment in sensitized Cd fluorescence; (i) calculated, (ii) measured.

Figure 8 demonstrates the experimental setup chosen by Gough.[92] A fused-silica cell containing a mixture of mercury and cadmium vapor is illuminated with mercury 2537-Å resonance radiation incident along the x axis of a laboratory frame. It is linearly polarized along the z axis of the frame. A chlorine filter efficiently removed the mercury 3303-Å line which would otherwise give rise to a background of scattered light. The sensitized fluorescent light 3261 Å due to the cadmium transition $5\,^3P_1$–$5\,^1S_0$ is observed in the y direction. It passes through a rotating linear analyzer and a filter consisting of 3-cm 1 M aqueous nickel sulfate solution, a thin cover glass, and a Wratten 18B ultraviolet filter which absorbs practically all the scattered mercury resonance radiation. The rotating linear analyzer consisted of a rotating half-wave plate followed by a linear polarizer which passed the light parallel to the z axis. Four times per period, the half-wave plate converts light polarized along the x axis into light polarized along the z axis and another four times it passes light polarized along the z axis unchanged. If there is more light of the first type (σ light) than of the second type (π light), a modulation of the intensity behind the linear polarizer is seen. This can be detected by a lock-in detector when an appropriate reference signal is at hand, which in our case is provided by a chopper wheel which interrupts a reference beam at four times the rotating frequency. Here again it is advisable not to have the linear polarizer sheet rotate itself because otherwise the plane of the analyzed light would rotate. This clearly would lead to a spurious signal if the detector was sensitive to polarization as is most often the case.

Figure 9a gives recorder traces obtained when a magnetic field pointing along the y axis was varied. Polarization of the sensitized cadmium fluorescence was found (curve i) which had the same sign as resonant cadmium fluorescence (curve ii) excited when the mercury lamp was replaced by a cadmium lamp. The sensitized and resonant Hanle curves (Figure 9b) had approximately the same width, ensuring that cadmium $5\,^3P_1$ atoms were the common source of both signals. The mercury resonant Hanle curves, by virtue of the 20 times shorter lifetime, were expected to be 20 times broader. Thus mercury atoms could be ruled out in considering possible signal sources. The indirect participation of mercury atoms in the production of the effect could be confirmed, however, by freezing out the mercury vapor from the cell, with the consequence that the signal vanished.

3.1. Gough's Prediction

Gough made a prediction concerning the magnitude of the expected effect. He uses a model for collisional transfer which he takes from an article by Chapman, Krause, and Brockmann[94] and which the latter authors ascribe to Franzen.[95] In short, the model says that during a transfer collision Σm_j is conserved along the collision axis. When it is applied to the Gough experiment, the following situation arises. An incoming photon polarized along the z axis excites a mercury atom to the state $6\,^3P_1$, $m_j = 0$. The state vector is $|m_j = 0\rangle$, the alignment component for $q = 0$ is

$$\langle J_{q=0}^{(2)}\rangle = \frac{\langle m_j = 0|3J_z^2 - J^2|m_j = 0\rangle}{j(2j-1)} = -2 \tag{34}$$

When a collision with a Cd $5\,^1S_0$ atom occurs, the axis of the collision plane with respect to the laboratory frame is, in general oriented at random. We use it as z' axis of a new system, whose x' and y' axes are given by the **k** vector (wave vector before the collision) and a vector perpendicular to **k** and the collision axis. Thus a triple of Eulerian angles $\alpha\beta\gamma$ is defined which is swept out when rotating the primed into the unprimed frame. We shall also indicate state vectors referring to the primed system by a prime. Thus in the primed system the state $|m_j\rangle$ of the excited mercury atom reads like a superposition of primed states with expansion coefficients $D_{m'_j,m_j}^j(-\gamma, -\beta, -\alpha)$, which are representation functions of the rotation in three-dimensional space:

$$|m_j\rangle = \sum_{m'_j} D_{m'_j,m_j}^j(-\gamma, -\beta, -\alpha)|m'_j\rangle' \tag{35}$$

The rule of Franzen claims that after a sensitizing collision with a cadmium atom, the latter will find itself in a state $5\,^3P_1$, which is a superposition of states $|m'_j\rangle$ given by

$$|\text{Cd }5\,^3P_1\rangle = \sum_{m'_j} \exp(i\varphi_{m_j})D_{m'_j,m_j}^j(-\gamma, -\beta, -\alpha)|m'_j\rangle' \tag{36}$$

Here the φ_{m_j} are real phases, which, for many collisions, are randomly distributed over a large angular range. Hence the probabilities of the states $|m'_j\rangle'$ are conserved during the collision, whereas the coherences between these states are completely lost. Hence the density matrix, which, before the collision, is made up from diagonal and nondiagonal elements, after the collision has retained the diagonal elements only:

$$\langle m''_j|\rho|m'_j\rangle^{\text{before c}} = \mathscr{D}_{m'',m_j}^{i*}(-\gamma, -\beta, -\alpha)\mathscr{D}_{m'_j,m_j}(-\gamma, -\beta, -\alpha) \tag{37}$$

$$\langle m''_j|\rho|m'_j\rangle^{\text{after c}} = \langle m''_j|\rho|m'_j\rangle^{\text{before c}}\delta_{m''_j,m'_j} \tag{38}$$

Similarly, for the tensor polarizations in the primed systems, the $q = 0$ components are transferred and the others annihilated. A $\langle \mathscr{T}_0^{(2)}\rangle$ component, which we have prepared in the laboratory frame is represented as a superposition of all the $\langle \mathcal{T}_q^{(2)} \rangle'$ components:

$$\langle \mathcal{T}_0^{(2)} \rangle^{\text{before c}} = \sum_q \mathcal{D}_{q,0}^2(-\gamma, -\beta, -\alpha)\langle \mathcal{T}_q^{(2)} \rangle' \tag{39}$$

From these only $\langle \mathcal{T}_0^{(2)} \rangle'$ survives the collision:

$$\langle \mathcal{T}_0^{(2)} \rangle^{\text{after c}} = \mathcal{D}_{0,0}^2(-\gamma, -\beta, -\alpha)\langle \mathcal{T}_0^{(2)} \rangle'$$

$$= \mathcal{D}_{0,0}^2(-\gamma, -\beta, -\alpha)\left[\sum_q \mathcal{D}_{q,0}^2(\alpha, \beta, \gamma)\langle \mathcal{T}_q^{(2)} \rangle^{\text{before c}}\right] \tag{40}$$

This, when averaged with respect to α, β, γ, is reduced to

$$\langle \mathcal{T}_0^{(2)} \rangle^{\text{after c}} = \tfrac{1}{5}\langle \mathcal{T}_0^{(2)} \rangle^{\text{before c}} \tag{41}$$

For general k, q we similarly obtain

$$\langle \mathcal{T}_q^{(k)} \rangle^{\text{after c}} = \frac{1}{2k+1} \langle \mathcal{T}_q^{(k)} \rangle^{\text{before c}} \tag{42}$$

Only tensor components with equal k and q are coupled. We now define transfer cross sections $\sigma^{(k)}$ by writing down rate equations

$$\frac{d}{dt}[n(\text{Cd } 5\,{}^3P_1)\langle \mathcal{T}_q^{(k)} \rangle_{\text{Cd}}]$$

$$= n(\text{Cd } 5\,{}^3S_0)v_r\sigma^{(k)}[n(\text{Hg } 6\,{}^3P_1)\langle \mathcal{T}_q^{(k)} \rangle_{\text{Hg}}] \tag{43}$$

where n denotes the densities of atoms as indicated and v_r is the mean relative velocity of Hg and Cd atoms. Equations (42) and (43) lead to the conclusion that the transfer cross section $\sigma^{(k)}$ of any tensor component of degree k equals the energy-transfer cross section $\sigma^{(0)}$ divided by $2k+1$.

To render possible an experimental determination of $\sigma^{(k)}$ we at first state that $(I_\sigma - I_\pi)|_{90°}$, the difference of the σ and π intensities detected at 90° to the quantization axis z, provides a measure of the product of the alignment component $q = 0$ in the excited state and the density of atoms n in that state:

$$(I_\sigma - I_\pi)|_{90°} \propto (n\langle \mathcal{T}_0^{(2)} \rangle)_{\text{excited state}} \tag{44}$$

The proof is easy by writing down $I_\sigma - I_\pi$ in terms of atomic densities in the excited Zeeman substates multiplied by the proper spontaneous transition probabilities to the ground state. When considering that the summation over all the substates of the ground state which appears in this expression is complete, the components of the electric dipole operator which constitute the spontaneous transition probabilities may be replaced by the corresponding components of the angular momentum operator. By this substitution the proportionality between $(I_\sigma - I_\pi)|_{90°}$ and $\langle \mathcal{T}_0^{(2)} \rangle$ becomes apparent.

In much the same way another useful proportionality is shown:

$$(I_{\sigma^+} - I_{\sigma^-})_{0°,180°} \propto (n\langle \mathcal{T}_0^{(1)} \rangle)|_{\text{excited state}} \tag{45}$$

The difference between σ^+ and σ^- light emitted parallel or antiparallel to the z axis is proportional to the orientation component $q = 0$ times the overall density of atoms in the excited state. In carrying out the proof, after substitution of the components of the angular momentum operator, commutation rules must be applied. For the proof see Bouchiat,[96] Dyakonov and Perel,[106] and Elbel et al.[98,99]

If an observation along the axis is not feasible, then a small angle ϑ between observational direction and z axis may be allowed. Then, however, right- and left-handed quanta are no longer identical with σ or π quanta. Still an abridged version of Eq. (45) holds if by I_r and I_l the intensities of right- and left-handed light are meant; then

$$(I_r - I_l)|_\vartheta \propto (n\langle \mathcal{T}_0^{(1)} \rangle)|_{\text{excited state}} \cos \vartheta \tag{46}$$

This consideration would not be complete without touching the problem of measuring the entire population n in the excited state. The entire flux of photons into the total solid angle 4π surely provides an appropriate measure but cannot be easily determined. The differential flux $I|_\vartheta = (d\Phi/d\Omega)|_\vartheta$ cannot be used as a measure since an alignment in the excited states regularly leads to an anisotropic angular distribution of photons. There is one exception:

$$(I_\sigma + \tfrac{1}{2}I_\pi)|_{90°} \propto n|_{\text{excited state}} \tag{47}$$

This relation came from the fact that in comparison with the total fluxes, the differential flux of π photons at $90°$ is twice as pronounced as the differential flux of σ photons and that, moreover, π and σ photons at $90°$ can be distinguished by their mutually perpendicular polarization. Relation (47) holds true to the same extent as relation (44). Therefore

$$\frac{I_\sigma - I_\pi}{I_\sigma + \tfrac{1}{2}I_\pi} = \langle \mathcal{T}_0^{(2)} \rangle_{\text{excited state}} \tag{48}$$

The degree of polarization which is in common use among experimentalists, is not identical with the alignment, but rather

$$P = \frac{I_\sigma - I_\pi}{I_\sigma + I_\pi} = \frac{\langle \mathcal{T}_0^{(2)} \rangle}{\tfrac{4}{3} - \tfrac{1}{3}\langle \mathcal{T}_0^{(2)} \rangle} \tag{49}$$

Inserting the transferred alignment according to Eqs. (41) and (42), namely, $-\frac{2}{5}$, one expects

$$P = -\tfrac{3}{11} \qquad (\pi\text{-excitation})$$

If incoherent σ excitation is used, resulting in an alignment of the host of $+1$ instead of -2, one ends up with

$$P = \tfrac{3}{19} \qquad \text{incoherent } \sigma \text{ excitation}$$

These are Gough's predictions. Gough's experimental arrangement applied to the first case. Despite the predicted value, he observed polarization which ranged from -2% to -5%. He attributed the deviation to (i) depolarizing Cd–Hg and Cd–Cd collisions, (ii) photon diffusion, and (iii) the presence of nuclear spins in the uneven Cd and Hg isotopes which are not allowed for in the preceding calculation.

3.2. Alignment and Orientation Transfer in Collisions of the Second Kind

Considerably more light is shed on the true nature of polarization transfer in collisions of the second kind by the work of Chéron.[84,100,101] Chéron improved Gough's work in a twofold way. On the one hand, he not only gave evidence of polarization transfer, but developed methods of quantitative evaluation. In addition, he investigated orientation transfer.

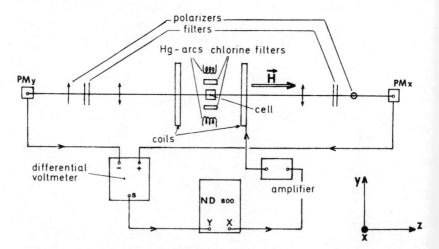

Figure 10. Experimental arrangement by Chéron[84] for the measurement of alignment transfer. For orientation transfer the linear polarizers were to be substituted by one right-handed and one left-handed circular polarizer. In addition circular polarizers of the same sense had to be inserted between the arcs and the cell. The magnetic field had to be turned into an upright position on the drawing plane.

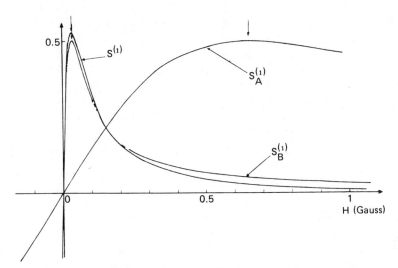

Figure 11. Signal shape of Hanle curves calculated according to the methods of Section 3.5: $S_A^{(1)}$ orientation in *resonant* Hg fluorescence, $S_B^{(1)}$ orientation in *resonant* Cd fluorescence, $S^{(1)}$ orientation in *sensitized* Cd fluorescence (from Chéron[84]).

On the other hand, he interpreted the results in terms of a more realistic collisional model. Figure 10 sketches the experimental arrangement he used. Along the z axis of a laboratory system unpolarized mercury resonance radiation $\lambda = 2357$ Å is incident on a vessel containing mercury and cadmium vapor. Along the y axis cadmium-sensitized radiation $\lambda = 3261$ Å is detected and discriminated by appropriate filtering. Two detectors are used. One of them, placed in the direction 0, y, measures σ quanta; the other one, placed in the direction 0, $-y$, measures π quanta. A differential voltmeter subtracts the dc output of the detectors. The differential signal is then fed into a signal averager. A variable magnetic field pointing in the direction 0, y, serves for depolarizing the transferred alignment. The sweep of magnetic field and signal averager is synchronized. A signal curve obtained in this manner has already been given in Figure 9.

Putting a circular polarizer into the exciting beam and circular analyzers of mutually opposite sense into the detecting beams while turning the magnetic field into an upright position (direction 0, x), one is able to measure orientation transfer. The incident light creates oriented mercury atoms in the state $6\,^3P_1$. The magnetic field makes them precess in the z, y plane while decaying. Collisions of the second kind transfer the orientation to cadmium atoms, in the state $5\,^3P_1$, which continue the precession at equal speed. Like g values of the mercury and cadmium atoms ensure that orientations imparted to various cadmium atoms, at different times, are not dissipated by the action of the magnetic field, but interfere constructively. Hanle curves obtained for the orientation of sensitized and resonant fluorescence of Hg and Cd are shown in Figure 11. Unlike the alignment curves, the orientation curves are not shaped like Lorentzians, but like dispersion curves. For the particulars see Section 3.3. From the transferred polarizations $\langle \mathcal{T}_0^{(2)} \rangle_{Cd}$ and $\langle \mathcal{T}_0^{(1)} \rangle_{Cd}$, the respective transfer cross sections $\sigma^{(2)}$ and $\sigma^{(1)}$ can be derived.

The rate equation (43) can be completed to include terms for creation by collisional transfer and for destruction by radiative and collisional decay:

$$\frac{d}{dt}[n(Cd\,5\,^3P_1)\langle \mathcal{T}_0^{(k)} \rangle_{Cd}] = n(Cd\,5\,^1S_0)v_r\sigma^{(k)}[n(Hg\,6\,^3P_1)\langle \mathcal{T}_0^{(k)} \rangle_{Hg}]$$

$$- (1/\tau_{Cd} + \Gamma_{Cd}^{(k)})[n(Cd\,5\,^3P_1)\langle \mathcal{T}_0^{(k)} \rangle_{Cd}] \qquad (50)$$

τ_{Cd} and $\Gamma_{Cd}^{(k)}$ are the radiative lifetime and the relaxation rate due to nonradiative processes from the $5\,^3P_1$ state of Cd. The solution in steady state

$$n(Cd\,5\,^3P_1)\langle \mathcal{T}_0^{(k)} \rangle_{Cd} = \frac{n(Cd\,5\,^1S_0)v_r\sigma^{(k)}}{1/\tau_{Cd} + \Gamma_{Cd}^{(k)}}\,n(Hg\,6\,^3P_1)\langle \mathcal{T}_0^{(k)} \rangle_{Hg} \qquad (51)$$

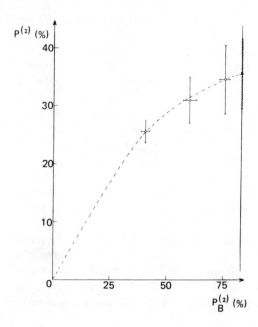

Figure 12. Cd alignment detected in *sensitized* fluorescence versus Cd alignment in resonance fluorescence at varying Cd pressure. The drawn points represent limiting values for the Hg pressure tending to zero. Since for vanishing Cd pressure the resonant alignment was found experimentally to approach 83%, the drawing allows the extrapolation of the limiting value of the transferred alignment for *both* vanishing Hg pressure and Cd pressure. A value of $36 \pm 6\%$ is thus determined (from Chéron[84]).

establishes a proportionality between the polarization of the host (Hg) and the guest (Cd). The constant of proportionality, however, depends itself on the Cd pressure. Figure 12 displays the linear dependence of the alignment of Cd on the alignment of Hg. The parameter is the density of Cd atoms in the ground state. As a measure of this density, the linear polarization of Cd resonant fluorescence produced by direct excitation with a Cd lamp was used. The Cd lamp could be substituted for the Hg lamp. When the Cd density in the cell was lowered, this linear polarization approached a limit of 83%. Thus plotting the slopes of the curves Figure 12 versus the parameter and extrapolating to the limit 83%, one obtains

$$\lim_{n_{Cd} \to 0} \frac{\langle \mathcal{T}_0^{(2)} \rangle_{Cd}}{\langle \mathcal{T}_0^{(2)} \rangle_{Hg}} = \tau_{Cd} \cdot n(Cd\ 5\ {}^1S_0) v_r \sigma^{(2)} \frac{n(Hg\ 6\ {}^3P_1)}{n(Cd\ 5\ {}^3P_1)}$$

$$= \frac{\sigma^{(2)}}{\sigma^{(0)}} \tag{52}$$

because

$$\lim_{n_{Cd} \to 0} \frac{n(Cd\ 5\ {}^3P_1)}{n(Hg\ 6\ {}^3P_1)} = \tau_{Cd} n(Cd\ 5\ {}^1S_0) v_r \sigma^{(0)} \tag{53}$$

In this manner, Chéron determined

$$\frac{\sigma^{(2)}}{\sigma^{(0)}} = 0.47 \pm 0.11, \qquad \frac{\sigma^{(1)}}{\sigma^{(0)}} = 0.42 \pm 0.10 \tag{54}$$

These numbers are to be compared with Gough's predictions 0.20 and 0.33, respectively. The apparent disagreement suggests a revision of Gough's model.

The cross section for excitation transfer was determined by Kraulinya and Arman,[83] Morozov and Sosinskij,[82] and Chéron.[84] The most recent value[78] is

$$\sigma^{(0)} = (0.34 \pm 0.11) \times 10^{-16}\ \text{cm}^2 \tag{55}$$

3.3. Chéron's Analysis of Collisional Transfer

We shall denote density matrices and tensors that pertain to the host atom (Hg) by A and those that pertain to the guest atom (Cd) by B. Again following Chéron,[84] we consider the approach of atoms A and B on their trajectories. In Figure 13 the approach at first takes place on the initial energy level

$$E_i = E(Cd\ 5\ {}^1S_0) + E(Hg\ 6\ {}^3P_1) \tag{56}$$

After the excitation transfer has happened, the atoms fly apart on the final

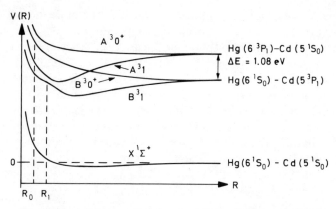

Figure 13. Adiabatic energy levels of the colliding pair Hg–Cd, schematic. The $\lambda = 0$, $\lambda = 1$ crossings lie at distances R_0, R_1, respectively (from Chéron[84]).

energy level

$$E_f = E(\text{Cd } 5\,^3P_1) + E(\text{Hg } 6\,^1S_0) \tag{57}$$

The energy deficit is converted into kinetic energy of the relative motion

$$E_{\text{kin}} = E_i - E_f \tag{58}$$

When at decreasing distance R, the electronic clouds of the atoms begin to merge, the initial as well as the final level tends to split into molecular levels signified by the expectation values of the projection on **R** of the electronic angular momentum. In this case, each level splits into two molecular branches with the molecular quantum number λ equal to 1 and 0, respectively. Hund's case (a) becomes valid. However, the turning motion of the internuclear axis counteracts the establishment of this molecular coupling scheme. If $\dot{\Phi}$ is the angular velocity of the internuclear axis, then the condition which separates the regions of adiabaticity and nonadiabaticity reads

$$\dot{\Phi} \lessgtr \frac{E(R; A \text{ or } B\,^30^+) - E(R; A \text{ or } B\,^31)}{\hbar} \tag{59}$$

If the upper sign is valid, the molecular coupling scheme prevails. If the lower sign is valid, the atomic coupling scheme remains more or less untouched.

When the atoms A and B come closer than a critical distance R_{crit} signified by equality of both sides of relation (59), molecular coupling sets in. The atomic angular momentum begins to precess around the axis and, in doing so, moves around with the turning axis until the critical distance is again passed over. The projection on the axis of the angular momentum during the period between the first and the second passage is conserved. A depolarization of the angular momentum results. Equation (59) provides a rough estimate of the pertinent cross section. Multiplying both sides of Eq. (59) by the duration of the collision τ_c on the left-hand side yields the angle Φ which is swept out during the period of molecular coupling. Assume this angle is sizable, for instance, 1 radian. Assume further a straight trajectory. Then the impact parameter b may be substituted for R on the right-hand side. Equation (59) then provides the critical impact parameter b_{crit}. Impact at smaller impact parameters leads to efficient depolarization. Depolarization will be negligible at $b > b_{crit}$. The depolarizing cross section will then approximately equal πb_{crit}^2. If the impact parameter is small compared with b_{crit} (head-on collision), the incoming atom will penetrate the depolarizing region very deeply. At very small distances the levels with equal λ, $A\lambda$ and $B\lambda$, tend to cross (Figure 13). Similarly the orbital of the valence electron loses the central symmetry with respect to the atom A or B. The valence electron gets increasingly owned by both atoms together. The transformation from the atomic to the molecular orbital evolves adiabatically when the atoms approach each other infinitely slowly. When the approach, however, happens in finite time nonadiabatic processes will become competitive with adiabatic ones. In other words, transitions will occur between the adiabatic levels $A\lambda$ and $B\lambda$ of Figure 13. Clearly, the approach no longer evolves in an adiabatic manner when, during the period at which the adiabatic states $A\lambda$ and $B\lambda$ oscillate relative to each other,

$$\tau \approx \frac{\hbar}{E(R;A\lambda) - E(R;B\lambda)} \tag{60}$$

the energetic distance of both states changes appreciably. More information on the transition amplitudes from $A\lambda$ to $B\lambda$ is yielded by the Landau–Zener–Stückelberg theory, which evidently applies to this case.

We summarize what can be learned from this theory.

(i) Transitions occur between molecular states with equal λ. Consequently, an excited atom A coming in in a Zeeman state m' referred to the internuclear axis before the collision eventually leads to the atom B coming out excited to the analogous Zeeman state m' referred to the internuclear axis after the collision. This is due to the carriage with the rotating axis of the angular momentum.

(ii) Because of the splitting of the molecular states $\lambda = 1$ and $\lambda = 0$, coherences between the states $m' = 1$ and $m' = -1$ are fully transferred. Coherences between states $m' = +1$ or -1 and $m' = 0$ are removed.

(iii) The transition probabilities between $\lambda = 1$ states, on the one hand, and $\lambda = 0$ states, on the other hand, are uncorrelated. Their ratio represents an important experimental parameter. It governs the relative strength of orientation transfer and alignment transfer. *Example*: We assume that transitions between the $\lambda = 0$ states are efficient alone. As a consequence, only the state B, $m' = 0$ is occupied. Alignment but no orientation arises. This restriction propagates to the laboratory frame: transfer of all the alignment components is rendered possible, but no transfer of orientation will take place.

From these statements one directly deduces the structure of the density matrix $\rho(B)$ which describes all the populations and coherences present in state Cd $5p\ ^3P_1$ after the collision due to an arbitrary density matrix $\rho(A)$ pertinent to the state Hg $6\ ^3P_1$ before the collision. Naturally, $\rho(A)$ is referred to a coordinate system x_i, y_i, z_i, with z_i pointing along the internuclear axis *before the collision*, while $\rho(B)$ is referred to a system x_f, y_f, z_f, with z_f pointing along the internuclear axis *after the collision*. Axes x_i, x_f are chosen perpendicular to the scattering plane spanned by z_i, z_f. Hence

$$\rho(B, z_f) = \begin{pmatrix} \Lambda_1\rho_{-1-1}(A, z_i) & 0 & \Lambda_1\rho_{-11}(A, z_i) \\ 0 & \Lambda_0\rho_{00}(A, z_i) & 0 \\ \Lambda_1\rho_{1-1}(A, z_i) & 0 & \Lambda_1\rho_{11}(A, z_i) \end{pmatrix} \quad (61)$$

Here Λ_1, Λ_0 are experimental parameters accounting for the uncorrelated transfer in the $\lambda = 1$ and the $\lambda = 0$ level crossing, respectively. Expanding the density matrices $\rho(A)$, $\rho(B)$ into density tensors (Fano[102]), $\rho_q^k(A)$, $\rho_{q'}^{k'}(B)$, one learns which polarizations are and which are not transferred. The answer is given through the elements of the transfer matrix $M_{qq'}^{kk'}$ defined by

$$\frac{d\rho_{q'}^{k'}(B, z_f)}{dt} = \sum_{kq} M_{qq'}^{kk'} \rho_q^k(A, z_i) \quad (62)$$

It follows from Equations (61) and (62) that the only nonvanishing elements of the matrix M are

$$M_{00}^{00} = \tfrac{1}{3}(2\Lambda_1 + \Lambda_0)$$
$$M_{00}^{20} = M_{00}^{02} = \tfrac{1}{3}(2)^{1/2}(\Lambda_1 - \Lambda_0)$$
$$M_{00}^{11} = \Lambda_1, \qquad M_{22}^{22} = M_{-2-2}^{22} = \Lambda_1 \quad (63)$$
$$M_{00}^{-22} = \tfrac{1}{3}(\Lambda_1 + 2\Lambda_0)$$

We must consider, however, that it is in the laboratory frame that we prepare one density matrix, $\rho(A, z)$, and measure another, $\rho(B, z)$. The laboratory frame x, y, z is in general oriented at random with respect to x_i, y_i, z_i and x_f, y_f, z_f. The transfer coefficients $M_{qq'}^{kk'}$ are therefore not our primary concern. We are, rather, interested in the elements of a transfer matrix which connects density tensors $\rho_q^k(A, z)$ and $\rho_q^k(B, z)$ in the laboratory frame. We define them by

$$\frac{d}{dt}\rho_q^{k'}(B, z) = \sum_{kq} M_{qq'}^{kk'} \rho_q^k(A, z) \tag{64}$$

Taking Eq. (62) and rotating the systems x_f, y_f, z_f through Eulerian angles α_f, β_f, γ_f and x_i, y_i, z_i through Eulerian angles α_i, β_i, γ_i to make them coincident with x, y, z yields the form

$$\frac{d}{dt}\rho_{q''}^{k'}(B, z) = \sum_{kqq'q'''} M_{qq'}^{kk'} \mathscr{D}_{q''q'}^{k'*}(\alpha_f, \beta_f, \gamma_f)\mathscr{D}_{q'''q}^k(\alpha_i, \beta_i, \gamma_i)\rho_{q'''}^k(A, z) \tag{65}$$

Considering that the rotation $\alpha_f\beta_f\gamma_f$ can be carried out in two steps so that a first rotation $(\frac{1}{2}\pi, \beta, -\frac{1}{2}\pi)$ makes x_f, y_f, z_f coincident with x_i, y_i, z_i and then continuing with the rotation $\alpha_i\beta_i\gamma_i$, we obtain

$$\frac{d}{dt}\rho_{q''}^{k'}(B, z) = \sum_{kqq'q''p} M_{qq'}^{kk'} \mathscr{D}_{q''p}^{k'*}(\alpha_i, \beta_i, \gamma_i)\mathscr{D}_{pq'}^{k'*}(\tfrac{1}{2}\pi, \beta, -\tfrac{1}{2}\pi)$$
$$\times \mathscr{D}_{q'''q}^k(\alpha_i, \beta_i, \gamma_i)\rho_{q'''}^k(A, z) \tag{66}$$

On averaging Eq. (66) over a great number of collisions which occur at random sets of α_i, β_i, γ_i, we get

$$\overline{\left(\frac{d}{dt}\rho_{q''}^{k'}(B, z)\right)}^{\alpha_i,\beta_i,\gamma_i} = \rho_{q''}^{k'}(A, z) \sum_{qq'} M_{qq'}^{k'k'} \frac{\mathscr{D}_{qq'}^{k'}(\tfrac{1}{2}\pi, \beta, -\tfrac{1}{2}\pi)}{2k'+1}$$
$$\equiv \rho_{q''}^{k'}(A, z)\mathscr{M}_{q''q''}^{k'k'} \tag{67}$$

Apparently, the $\mathscr{M}$ matrix is diagonal with respect to k' and degenerate in q''. We state the important theorem: In collisions whose scattering planes are oriented at random with respect to the reference frame, density tensor components of the guest atoms are of the same rank q and degree k as those of the host atoms!

We are now able to expand the diagonal matrix element of the $\mathscr{M}$ matrix into matrix elements of the M matrix [Eq. (63)]. The D function makes them dependent on the angle β which the interatomic axis sweeps out during the collison:

$$\mathscr{M}_{q''}^1 = \frac{\Lambda_1}{2\Lambda_0+1}\cos\beta$$

$$\mathscr{M}_{q''}^2 = \frac{1}{10\Lambda_1+5\Lambda_0}(\tfrac{1}{2}\Lambda_1 + \Lambda_0)(3\cos^2\beta - 1)+6\Lambda_1\cos^4\tfrac{1}{2}\beta \tag{68}$$

$\mathscr{M}^1$ and $\mathscr{M}^2$ are represented as functions of β in Figure 14 for $r = \Lambda_1/\Lambda_0$ taking the values 0, $\frac{1}{4}$, 1, 4, and ∞. We state in particular that the orien-

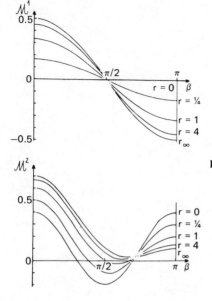

Figure 14. The transfer coefficient for orientation $\mathcal{M}^1$ and for alignment $\mathcal{M}^2$ as a function of the angle β swept out by the internuclear axis during the collision. The reader's attention is directed to the inversion of the transferred orientation for $\beta > \pi/2$ and of the alignment in the vicinity of $\beta = \pi/2$. The relative contribution of the crossings, $r = \Lambda_1/\Lambda_0$, serves as a parameter (from Chéron[84]).

tation transfer occurs at reversal of the sign when the angle β exceeds $\frac{1}{2}\pi$. This follows from the carriage with the turning axis of the angular momentum. In quasiresonant collisions (Omont and Meunier[27] and Chéron and Saintout[103]), where grazing collisions are most efficient in orientation transfer, sign reversal is indeed found. In Hg–Cd collisions the values of $\sigma^{(1)}$ and $\sigma^{(2)}$ [Eq. (54)] show that head-on collisions with a comparably small β yield the bulk of the transfer rates.

To learn about the average value of β, more knowledge about Λ_1 and Λ_0 is demanded. Chéron has made a pioneering experiment to this end. In Hg–Cd transfer collision a considerable amount of kinetic energy is set free. This makes the collision partners fly apart with great force. As a result, the internuclear axis, from the moment of transfer to infinity, will not turn considerably. Moreover, the light emitted from the radiating Cd atom after collision is extremely Doppler-broadened. Filtering this light prior to detection through a Cd vapor of thermal velocity distribution, one is able to receive only the radiation from such Cd atoms which after collision have taken their flight *toward* or *off* the detector. Hence one is able to discriminate between collisions whose axis is parallel to the axis cell detector and those whose axis is perpendicular to it. If the Hg atoms were created in an unpolarized state, aligned Cd atoms could nevertheless arise provided $\Lambda_1 \neq \Lambda_0$. From the sign and size of the Cd alignment observed in the above experiment Chéron inferred $\Lambda_1/\Lambda_0 = 3.25 \pm 0.75$. A glance at Figure 13 gives an understanding of this fact: The $\lambda = 1$ crossing lies at considerably lower energy than the $\lambda = 0$ crossing! The value 3.25 ± 0.75, in turn, together with the results on $\sigma^{(2)}$ and $\sigma^{(1)}$ [Eq. (54)] suggests an average of β in the range $24° \leqslant \beta \leqslant 45°$ by virtue of Eq. (68).

3.4. Transfer between Isotopes of the Same Element

Polarization transfer in collisions of atoms of the same element was predicted as a concomitant of depolarization in those collisions by Omont.[13] A very thorough analysis of this collisional process was given before with the aim of explaining broadening of double-resonance signals on account of self-broadening collisions (Byron and Foley[104]). The process itself reads

$$\mathrm{Hg}((1)6\,^3P_1) + \mathrm{Hg}((2)6\,^1S_0) \rightarrow \mathrm{Hg}((1)6\,^1S_0) + \mathrm{Hg}((2)6\,^3P_1)$$

Virtually no energy deficit exists between the initial and final states, which makes it difficult to distinguish polarization transfer from depolarization. Yet the discrimination of polarization transfer was made possible by application of different Hg isotopes. The isotope shift of, e.g., $400 \times 10^{-3}\,\mathrm{cm}^{-1} \triangleq 50\,\mu\mathrm{eV}$ between the isotopes 198 and 202 hardly represents a restriction on the condition of resonance as far as the collision process is concerned. But it effects easy resolution of Hg 198 and Hg 202 resonance fluorescence the more so as the Doppler width is far below the isotope splitting. Thus Dodsworth and Omont[105] succeeded in verifying orientation transfer in collisions of different Hg isotopes. They observed that orientation is inverted during collision and that approximately one-quarter of the energy transfer collisions are efficient in transferring orientation, too. Alignment transfer could not be verified because their measurements were carried through in retrodiffusing light. This was accomplished, however, by Chéron and Saintout,[103,155] who detected the fluorescent light in a lateral direction. Their experiment, which is displayed in Figure 15, consists of the following elements. A spectral lamp prepared with 99.8% pure 198 mercury is used to illuminate a cell containing a $1:1$ mixture of 198 mercury and 202 mercury. Before entering the cell the light passes through a filter with 202 mercury vapor in a hydrogen buffer of 70 Torr. The absorption profile of the pressure-broadened 202 mercury resonance line efficiently removes those parts of the lamp spectrum which by overlap with the 202 absorption profile could directly excite the 202 mercury component in the cell. In turn, the fluorescent light is filtered through a 198 mercury cell to distinguish the sensitized 202 fluorescence from the resonant 198 fluorescence. With the length being 45 cm and a saturated pressure corresponding to 45°C the absorption coefficient was higher than 99.99%. However, at these conditions also 7% of the sensitized 202 fluorescence was absorbed. Another cell in the fluorescent beam, 3 cm long

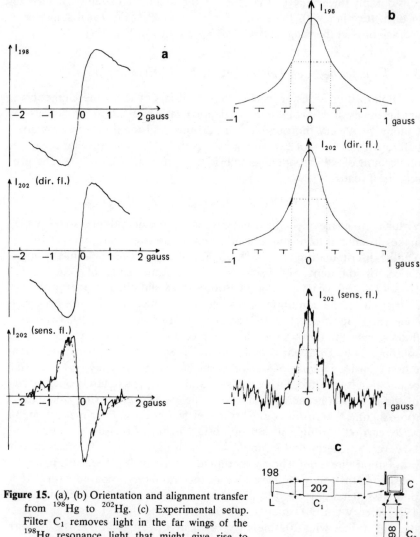

Figure 15. (a), (b) Orientation and alignment transfer from ^{198}Hg to ^{202}Hg. (c) Experimental setup. Filter C_1 removes light in the far wings of the ^{198}Hg resonance light that might give rise to resonant excitation of ^{202}Hg in the cell. In turn, the filter C_2 removes ^{198}Hg resonance light transmitting only the ^{202}Hg sensitized fluorescence. **H** corresponded to the situation exp 1 at excitation-transfer studies and to situation exp 2 at both alignment- and orientation-transfer studies. In the latter case, circular polarizers had to be inserted into the exciting and the detected light beam (from Chéron and Saintout[103]).

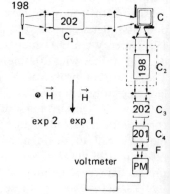

and prepared with isotope 202, allowed the removal of 98% of the 202 sensitized fluorescence when kept at 20°C. However, when a sidearm to this cell was immersed in liquid nitrogen, it became entirely transparent to the sensitized fluorescence. In this way the fraction of sensitized fluorescence in the total fluorescence could be determined. This fraction was typically $1:1000$ when the main cell was kept at $-10°C$. Moreover, a third cell with 201 mercury was used to eliminate almost the entire contributions to the fluorescence of impurities of 199, 201, and 204 mercury. Finally, a Käsemann W78 uv filter was used to preselect the far-ultraviolet spectral range.

When orientation transfer was studied, circular polarizers were inserted in the incident and the fluorescent light beam, and a variable magnetic field was applied perpendicular to the drawing plane. The signal over the magnetic field strength was then shaped like a dispersion curve. When alignment transfer was studied, no polarizers were used neither in the exciting or in the detected light beam. The direction of the magnetic field was chosen parallel to the line cell detector. Using this direction as an axis of reference, the excitation implied a coherence between σ^+ and σ^- transitions. The fluorescent intensity is then decisively determined by this coherence. It drops when the magnetic field strength reaches values comparable to $\hbar/\mu_B\tau$ ($\tau =$ mean life of $6\,^3P_1$). In detail, the signal as a function of H behaves like a Lorentzian.

Hanle signals were measured in the direct fluorescence of 198 and 202 mercury (in the latter case the 198 lamp was substituted for by a 202 lamp) as well as in the sensitized fluorescence. The analysis, which bears strong similarities to the Hg–Cd case already discussed in Section 3.2, leads to a proportion among transfer cross sections

$$\sigma^{(0)}:\sigma^{(1)}:\sigma^{(2)} = 1:(-0.23\pm0.05):(0.082\pm0.016) \qquad (69)$$

whereas Omont and Meunier[27] predicted

$$\sigma^{(0)}:\sigma^{(1)}:\sigma^{(2)} = 1:-0.25:0.05 \qquad (70)$$

There is not enough space to describe fully the very elaborate theoretical methods which Byron and Foley and later Omont used to solve the problem of self-broadening collisions. Also, the problem of depolarization is inseparably connected with the problem of polarization transfer in this type of collision and must be omitted here to receive its proper treatment in Chapter 28 of this volume. We restrict ourselves, therefore, to a simple perturbation treatment to first order, but it is hoped that we will thus facilitate the reader's access to the aforementioned original articles.[13,27,104]

The interaction responsible for excitation transfer from one mercury atom to another is dipole–dipole interaction. Obviously the perturbing

Hamiltonian

$$\mathcal{H}_1(t) = \frac{\mathbf{P}_1 \cdot \mathbf{P}_2}{R^3(t)} - 3\frac{[\mathbf{P}_1 \cdot \mathbf{R}(t)][\mathbf{P}_2 \cdot \mathbf{R}(t)]}{R^5(t)} \tag{71}$$

has the property of connecting an initial state

$$|(1)m\rangle \equiv |(1)\,^3P_{1,m}\,(2)\,^1S_0\rangle \tag{72}$$

with another state where atom (1) is in the ground state and (2) passed into the excited state, which we write, for short,

$$|(2)m'\rangle \equiv |(2)\,^3P_{1,m'}(1)\,^1S_0\rangle \tag{73}$$

The matrix element of $\mathcal{H}_1$ apparently can be reduced to matrix elements of $\mathbf{P}_1$ and $\mathbf{P}_2$ between the ground state $6\,^3S_1$ and excited state $6\,^3P_1$. The product of these matrix elements is proportional to the f value of the electric dipole transition between these states and hence to the reciprocal mean life τ of the state $6\,^3P_1$. The dependence on the Zeeman quantum numbers m, m' of the initial and final state can readily be accounted for by taking the components of $\mathbf{J}$, the total angular momentum operator of the combined atoms, as substitutes for the respective components of $\mathbf{P}_1$, $\mathbf{P}_2$. Omont[13] shows the validity of the substitution

$$\langle(2)m'|\mathcal{H}_1(t)|(1)m\rangle = -\frac{3}{4}\frac{\hbar}{(2\pi/\lambda)^3\tau}\langle(2)m'|\frac{\mathbf{J}^2 - 3\mathbf{J}_R^2}{R^3(t)}|(1)m\rangle \tag{74}$$

where $\mathbf{J}_R$ means the projection on $\mathbf{R}$ of $\mathbf{J}$ and λ the wavelength of the resonance line.

When we denote by $A_{(2)m'}(t)$ the amplitude of the chosen final state, then first-order perturbation theory yields the well-known differential equation

$$\hbar i \dot{A}_{(2)m'}(t) = \langle(2)m'|\mathcal{H}_1(t)|(1)m\rangle \tag{75}$$

whose integration with respect to the time leads to the transition amplitude. It must be assumed that $|A_{(2)m'}|^2$ remains smaller than 1. We find

$$\langle(2)m'|S|(1)m\rangle = A_{(2)m'}(t = \infty)$$

$$= \frac{3}{4}\frac{i}{(2\pi/\lambda)^3\tau}\int_{-\infty}^{\infty}\langle(2)m'|\frac{\mathbf{J}^2 - 3\mathbf{J}_R^2}{R^3(t)}|(1)m\rangle\,dt \tag{76}$$

To calculate the integral it is assumed that the trajectory of the atoms is a straight line and that at $t = 0$ the atoms reach their closest mutual distance, which in straight-path approximation, equals the impact parameter b. The time dependence of $R(t)$ and J_R^2 is then readily obtained in explicit form

$$R^2(t) = b^2 + v^2 t^2 \tag{77}$$

$$J_R^2 = \frac{(bJ_b + vtJ_v)^2}{b^2 + v^2 t^2} \tag{78}$$

J_b and J_v are the projections on the b and v axis of $\mathbf{J}$. With these expressions Eq. (76) is reduced to tabulated finite integrals. One obtains

$$\langle (2)m' | S | (1)m \rangle = -\frac{3}{2} \frac{1}{(2\pi/\lambda)^3 \tau} \frac{1}{vb^2} (J^2 - 2J_b^2 - J_v^2) \tag{79}$$

We choose now a coordinate system (x', y', z') such that x' is parallel to b, y' is parallel to b, and z' is parallel to $(\mathbf{v} \times \mathbf{b})$. The transition matrix between states $|(1)\mu\rangle$ and $|(2)\mu'\rangle$ which are quantized in that system is easily obtained from known matrix elements of the angular momentum. One finds

$$
\begin{array}{c|ccc}
\diagdown \quad 1\mu & & & \\
2\mu' \diagdown & 1 & 0 & -1 \\
\hline
& & & \\
1 & 1/2 & 0 & -1/2 \\
S = \quad 0 & 0 & -1 & 0 \\
-1 & -1/2 & 0 & 1/2 \\
\end{array}
\quad -\frac{3}{2} \frac{1}{(2\pi/\lambda)^3 \tau} \frac{1}{vb^2} \tag{80}
$$

To find the S-matrix elements between states $|(1)m\rangle$ and $|(2)m'\rangle$ in the laboratory frame x, y, z we make use of the fact that S is invariant under rotations of the frame (Dyakonov and Perel,[106] Omont,[97] Grawert[107]). Therefore

$$\langle (2)m' | S | (1)m \rangle = D_{\mu'm'}^{j=1}{}^*(-\gamma, -\beta, -\alpha) \langle (2)\mu' | S | (1)\mu \rangle D_{\mu,m}^{j=1}(-\gamma, -\beta, -\alpha) \tag{81}$$

Here α, β, γ is the angular triple which is swept out when rotating the laboratory frame into the primed position. The products of D functions are expanded into series of D functions according to the contraction theorem (see, for example, Brink and Satchler,[108] Appendix V). In our case these series degenerate into one single element whose angular dependence is given by the function $D_{\mu-\mu',m-m'}^2(\alpha, \beta, \gamma)$. From the S-matrix elements the transition cross sections $m \leftrightarrow m'$ are found according to

$$\sigma(m \leftrightarrow m') = \frac{1}{8\pi^2} \int_0^{2\pi} \int_0^{\pi} \int_0^{2\pi} \sin \beta \, d\beta \, d\alpha \, d\gamma \left\{ \int_0^{\infty} 2\pi b \, db \, |\langle (2)m' | S | (1)m \rangle|^2 \right\} \tag{82}$$

Equation (49) implies averaging with respect to α, β, γ because it is supposed that in gases the scattering planes are oriented at random. We find the matrix of the $\sigma(m \leftrightarrow m')$ to be

$$
\boldsymbol{\sigma} = \quad
\begin{array}{c|ccc}
\diagdown \;\;(1)m & -1 & 0 & 1 \\
(2)m\;\;\diagdown & & & \\
\hline
-1 & 1/15 & 1/5 & 2/5 \\
0 & 1/5 & 4/15 & 1/5 \\
1 & 2/5 & 1/5 & 1/15
\end{array}
\Bigg\} \times \int_{``0"}^{\infty} 2\pi b\, db \left[\frac{9}{4}\frac{1}{(2\pi/\lambda)^6 \tau^2 v^2 b^4}\right] \quad (83)
$$

The quotation marks at the lower limit of the integral indicate that approximation used is not valid for small impact parameters b.

    It is an easy task to derive cross sections for the transfer of alignment and orientation from the cross sections of inter-Zeeman transitions [Eq. (83)]. One has to write down rate equations for the Zeeman state populations $P_{m'}$ of atoms of species (2) and ρ_m of atoms of species (1):

$$
\dot{P}_{m'} = N v_r \sum_m \sigma(m \leftrightarrow m') \rho_m \cdots \quad (84)
$$

N means the density of atoms of species (2) in the ground state. (Relaxation terms which are not our primary concern are omitted here.) Multiplying both sides of Eq. (84) by the matrix element $\langle m'|\mathbf{J}^k_{q=0}|m'\rangle$, where $\mathbf{J}^k_{q=0}$ means the indicated component of the density tensor operator of degree k, and summing the equations of each m' yields an equation of the type

$$
\frac{d}{dt}\langle \mathcal{T}^k_{q=0}(2)\rangle = N v_r \sigma^{(k)}\!\left((1)\leftrightarrow(2)\right)\langle \mathcal{T}^k_{q=0}(1)\rangle \quad (85)
$$

whence it becomes apparent how $\sigma^{(k)}\!\left((1)\leftrightarrow(2)\right)$ is combined from the $\sigma\!\left((1)m \leftrightarrow (2)m'\right)$. In particular,

$$
\sigma^{(2)}\!\left((1)\leftrightarrow(2)\right) = \sigma\!\left((1)-1\leftrightarrow(2)1\right) + \sigma\!\left((1)-1\leftrightarrow(2)-1\right) - 2\sigma\!\left((1)-1\leftrightarrow(2)0\right) \quad (86)
$$

$$
\sigma^{(1)}\!\left((1)\leftrightarrow(2)\right) = \sigma\!\left((1)-1\leftrightarrow(2)-1\right) - \sigma\!\left((1)-1\leftrightarrow(2)1\right) \quad (87)
$$

$$
\sigma^{(\sigma)}\!\left((1)\leftrightarrow(2)\right) = \sigma\!\left((1)-1\leftrightarrow(2)-1\right) + \sigma\!\left((1)-1\leftrightarrow(2)0\right) + \sigma\!\left((1)-1\leftrightarrow(2)1\right) \quad (88)
$$

Insertion of the values of the matrix [Eq. (83)] yields the proportion

$$
\sigma^{(0)}:\sigma^{(1)}:\sigma^{(2)} = 1 : -0.5 : 0.1 \quad (89)
$$

A twofold magnification of $\sigma^{(0)}$ would make coincident this proportion

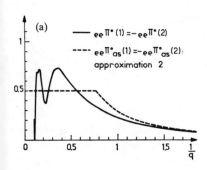

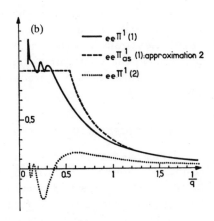

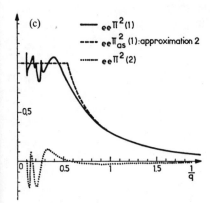

Figure 16. Probabilities for excitation (a), orientation (b), and alignment (c) transfer as a function of a reduced impact parameter $1/q = 4(2\pi/\lambda)^3 vb^2\tau/9$. The host atom is denoted by (1), the guest atom by (2). The quantities $_{ee}\pi^k(2)$ with $k = 0, 1, 2$ for excitation, orientation, and alignment, respectively, are defined in such a way that their negatives correspond to the probability of transfer. Accordingly, at a large impact parameter the transferred orientation assumes negative values, the alignment positive ones [dotted curves in (b), (c); solid curves in (b), (c) are for depolarization and hence not our concern] (from Omont and Meunier[27]).

with the one of Omont and Meunier.[27] Equation (89) agrees, however, with Eq. (34) of Omont's article.[97]

The disagreement lies in the approximate character of the given treatment. The numerical treatment of Omont and Meunier[27] (cf. also Watanabe,[109] Dyakonov and Perel[106]) shows clearly how the transfer probabilites behave at smaller impact parameters where the first-order treatment does not apply. Figure 16 shows this behavior. The transfer probabilities of orientation and alignment get into oscillations about the abscissa, whereas the transfer probability of population remains always positive, oscillating about the statistical mean value 0.5. Therefore, the region of small b values does not significantly contribute to $\sigma^{(1)}$, $\sigma^{(2)}$. This incidentally removes the disagreement between the proportions Eqs. (89) and (70).

We state as a general tendency that in self-broadening collisions the region of small b values (i.e., of the "sudden" collisions) is ineffective in *polarization* transfer although *excitation* transfer occurs with statistical probability. The polarization transfer happens prevailingly in "grazing" collisions. This makes polarization transfer a subordinate effect in these collisions.

3.5. Hanle Signals in Sensitized Fluorescence

Transferred polarizations are subject to magnetic depolarization [Hanle[154]] as are those which are directly created through optical absorption or electronic impact. Magnetic depolarization is the standard method of distinguishing transferred polarizations from spurious effects such as the polarization of reflections at glass walls.

However, Hanle curves in sensitized fluorescence never yield ordinary Lorentzian curves. This is so because more than one precession frequency and lifetime interfere in their production. If there are two levels involved—the host level and the guest level—with precession frequencies ω, Ω and lifetimes τ, T, respectively, line shapes like

$$\mathcal{L}(\omega\Omega, \tau T) = \frac{\Omega + \omega}{(\omega^2 + 1/\tau^2)(\Omega^2 + 1/T^2)} \tag{90}$$

result instead of

$$\mathcal{L}(\omega, \tau) = \frac{\omega}{\omega^2 + 1/\tau^2} \tag{91}$$

which we would obtain in normal Hanle experiments.

The inherent mathematical problem was solved at first by Chéron and Barrat[100] for the special case $\omega = \Omega$, $\tau \neq T$, which applies to Hg 6 $^3P_1$, and for the dual case $\omega \neq \Omega$, $\tau = T$ by Elbel *et al.*,[110] which applies to alkali $^2P_{1/2} \rightarrow {}^2P_{3/2}$ transfer. The problem was then solved in greater generality by Chow-Chiu[111] and Baylis,[112] who also accounted for back transfer from the guest level to the host level which in earlier attempts was neglected. We follow the analysis given by Baylis:

We chose a frame such that the z axis has the direction of the magnetic field H. One of the other axes may be favorably chosen to coincide with the incident light beam or the E vector of the exciting light. The equations of motion of the density matrices P and ρ of the guest and host, respectively, are

$$\dot{\mathbf{P}} = -\mathbf{P}/T - i\Omega[\mathbf{J}_z, \mathbf{P}] + \Gamma\rho \tag{92}$$

$$\dot{\rho} = -\rho/\tau - i\omega[\mathbf{j}_z, \rho] + \gamma\mathbf{P} + \mathbf{S} \tag{93}$$

Here $1/T$ and $1/\tau$ are relaxation rates, involving radiative and collisional relaxation; Γ and γ are transfer rates from host to guest and reverse. $1/T$, $1/\tau$, Γ, and γ themselves are matrices and so is $\mathbf{S}$, the creation rate due to optical excitation of the host. $\mathbf{J}_z$ and $\mathbf{j}_z$ are the z components of the angular momentum operators acting on the guest and the host level. We expand the equations into irreducible tensors defined as

$$\boldsymbol{\rho} = \sum_{l,m} \rho_{l,m} T^{l,m}$$

etc., and similarly for $\mathbf{S}$, to find

$$P_{l,m} = -\frac{P_{l,m}}{T_l} \, im\Omega P_{l,m} + \Gamma_l \rho_{l,m} + S_{l,m} \tag{94}$$

$$\rho_{l,m} = -\frac{\rho_{l,m}}{T_l} - im\omega\rho_{l,m} + \gamma_l P_{l,m} \tag{95}$$

Because of the isotropy of the collisions, m is independent of the pertinent relaxation and transfer rates.

When $M_{l,m}$ are the monitoring efficiencies of the density tensor components $\rho_{l,m}$ (Carrington[113]), we find from the steady-state solutions of Eq. (95) that the detected intensity is

$$\mathscr{L} = \sum_{l,m} M_{l,m}\rho_{l,m}$$

$$= \sum_{l,m} M_{l,m} S_{l,m} \frac{\Gamma_l[(1/T_l)(1/T_l) - \Gamma_l\gamma_l - m^2\omega\Omega]}{|(1/T_l + im\Omega)(1/T_l + im\omega) - \gamma_l\Gamma_l|^2} \tag{96}$$

A typical line shape of this type is displayed in Figure 15. The intensity at rising magnetic field does not steadily decrease but eventually intersects the abscissa and goes through a minimum before vanishing. We emphasize that the preceding considerations are limited to a two-level system. If both levels show hyperfine splitting, formula (96) must be extended to account for interaction between all the hyperfine sublevels of the host and the guest. Even so, the formula may lose validity when approaching the Paschen–Back region of hfs. Then the transfer rates Γ_l may change sign due to decoupling of the nuclear spin. Accordingly, intercepts with the abscissa arise which are not explained by formula (96) but follow from the fading flywheel transfer of the nuclear spin. An instance is discussed by Elbel and Schneider[114] (see Section 3.6).

3.6. Transfer between Hyperfine Sublevels

Until now we have reported on "proper" polarization transfer, which is a consequence of the particular collision dynamics that leads to the transformation of the states. Hence it bears valuable information on the competition of interactions in the collisional process. Another class of transfer phenomena on which we have to report, in the remaining sections of this article, is based on the "flywheel" effect. In some cases other angular momenta like nuclear spins and even spins are coupled so loosely to the angular momentum, which is directly affected by the collisional process, that during the collision they conserve their instantaneous state of motion. A spinning top yields an example of a system which keeps its orientation in space when suddenly the cardanic suspension is moved. After the collision the weak interaction of the flywheel with the disoriented angular momentum sets in again. Consequently, the angular momentum stored in the flywheel is imparted to the disoriented angular momentum. A similar process is polarization transfer in cascading processes where the polarization is partitioned between the photon and the residual atomic state. For more information see Chapter 9 of this work by Happer and Gupta.

After a period which is characteristic of the weak coupling forces, all the interacting angular momenta are oriented as before although to a lesser degree. A condition for the flywheel effect to take place can be formulated such that the period T for the weak coupling of the angular momenta (i.e., their precession time) has to be long compared with the duration (= correlation time) τ_c of the collisional process. This gives rise to a number of phenomena.

(i) The relaxation of the resultant angular momentum is slowed down in comparison with the partial angular momentum which is disoriented directly.

(ii) The sublevels of the resultant angular momentum are mixed up by the collisional process. Energy transfer between them takes place.

(iii) Orientation transfer between the sublevels accompanies the energy transfer.

Only points (ii) and (iii) are our concern. Point (i) is treated, e.g., in the review article on optical pumping by Happer.[115]

Consider an atom in a hyperfine sublevel F of an atomic level J. According to the orientation of its F vector in a chosen coordinate system it is in an initial state

$$|i\rangle = \sum_{m_F} \alpha_{F,m_F} |F, m_F\rangle \tag{97}$$

where the complex amplitudes are the components of a unitary vector. Expansion into functions $|m_I\rangle|m_J\rangle$ of a decoupled basis yields

$$|i\rangle = \sum_{m_F} \alpha_{F,m_F} \sum_{m_I,m_J} \langle m_I m_J |F m_F\rangle |m_I\rangle |m_J\rangle \tag{98}$$

Now we are ready to study the influence of the collision. The latter is supposed to consist of an instantaneous disorientation of $\mathbf{J}$ under conservation of $\mathbf{I}$.

The J randomization model (Bender,[117] Franz and Franz[116]) suggests that any state $|m_J\rangle$ must be substituted for by a superposition of all the states $|m'_J\rangle$ of the Zeeman multiplet with random phase factors:

$$\mathbf{S}|m_J\rangle = (2J+1)^{-1/2} \sum_{m_J} \exp\left(i\varphi_{m_J,m'_J}\right)|m'_J\rangle \qquad (99)$$

where the exponentials are the S-matrix elements. The φ_{m_J,m'_J} are real phases, which ensures that the normalization is conserved (no transitions to other J levels occur). This clarifies the evolution of $|i\rangle$ during scattering:

$$
\begin{aligned}
\mathbf{S}|i\rangle = \sum_{m_F} \alpha_{F,m_F} \sum_{m_I,m_J} \langle m_I m_J | F m_F\rangle \\
\times \sum_{m_J} \exp\left(i\varphi_{m_J,m'_J}\right) \frac{|m_I\rangle|m'_J\rangle}{(2J+1)^{1/2}}
\end{aligned}
\qquad (100)
$$

The amplitude of any desired final state $|f\rangle = |F'm_{F'}\rangle$ in that outgoing state is readily obtained by projection:

$$
\begin{aligned}
\langle f|\mathbf{S}|i\rangle = \sum_{m_F} \alpha_{F,m_F} \sum_{m_I,m_J} \langle m_I m_J | F m_F\rangle \\
\times \langle m_I m'_J | F'm'_F\rangle \frac{\exp\left(i\varphi_{m_J,m'_J}\right)}{(2J+1)^{1/2}}
\end{aligned}
\qquad (101)
$$

The density matrix element $\langle f|\rho|f\rangle$ after collision is obtained from the absolute square of $\langle f|\mathbf{S}|i\rangle$ by averaging, at first, with respect to $\varphi_{m_J, m_{j'}}$, which in many collisions is supposed to be randomly distributed over an angular range large compared with 2π. Secondly, we have to average $\alpha_{Fm_F}\alpha^*_{F'm_F}$ over a great number of atoms undergoing collisions. We assume that in the chosen system neither Zeeman coherences nor hyperfine coherences are present, so

$$
\begin{aligned}
\langle f|\rho|f\rangle = \langle F'm'_F|\rho|F'm'_F\rangle_{\text{after collision}} \\
= \sum_{m_F} |\bar{\alpha}_{Fm_F}|^2 \sum_{m_I,m_J} \langle m_I m_J | F m_F\rangle^2 \\
\times \langle m_I m'_J | F'm'_F\rangle^2 \frac{1}{2J+1}
\end{aligned}
\qquad (102)
$$

Multiplying both sides with $\langle F'm'_F|\mathcal{T}_0^{(k)}|F'm'_F\rangle$, i.e., the matrix element of a standard tensor operator, and tracing with respect to m'_F yields the rate at which the tensor polarization is transferred from level F to level F' (the transfer rate is the same for the component $q = 0$ as for any other component $q \neq 0$ provided the collisions are isotropic in space). Considering that

$$\langle F'm'_F|\mathcal{T}_0^{(k)}|F'm'_F\rangle$$

$$= (-1)^{2k}\langle F'\|\mathcal{T}^{(k)}\|F'\rangle\langle F'm'_F|k0F'm'_F\rangle \tag{103}$$

and starting with the summation with respect to m'_F, we find that we have to sum over three Clebsch–Gordan coefficients. The contraction theorem for Clebsch–Gordan coefficients (Brink and Satchler, Appendix II[108]) yields for that sum

$$\sum_{m'_F} \langle m_I m'_J|F'm'_F\rangle^2 \langle F'm'_F|k0F'm'_F\rangle$$

$$= (-1)^{J-F'-I}(2F'+1)\left(\frac{2F'+1}{2I+1}\right)^{1/2} W(IF'IF', Jk)\langle Im_I|k0Im_I\rangle \tag{104}$$

The second summation runs over m_I. Again we sum over three Clebsch–Gordan coefficients and once more apply the contraction theorem

$$\sum_{m_I} \langle m_I m_J|Fm_F\rangle^2 \langle Im_I|k0Im_I\rangle$$

$$= (-1)^{J-F-I}(2I+1)^{1/2}(2F+1)^{1/2}W(IF, IF, Jk)\langle Fm_F|k0Fm_F\rangle \tag{105}$$

The third summation runs over m_F. It yields the expectation value of $\mathcal{T}_0^{(k)}$ in the initial F state,

$$\sum_{m_F} \overline{|\alpha_{Fm_F}|}^2 \langle Fm_F|k0Fm_F\rangle \propto n_F\langle\mathcal{T}_0^{(k)}\rangle_F \tag{106}$$

This calculation clarifies that the rate at which $\langle\mathcal{T}_0^{(k)}\rangle_{F'}$ is built up in the final state F' is proportional to $\langle\mathcal{T}_0^{(k)}\rangle_F$ in the initial state F.

In particular, for $k = 0$ (population transfer) and $k = 1$ (orientation transfer) we obtain the explicit rate equations:

$$\frac{d}{dt}(n_{F'}) = \frac{2F'+1}{(2J+1)(2I+1)}\frac{n_F}{T} \tag{107}$$

$$\frac{d}{dt}(n_{F'}\langle\mathbf{F}_z\rangle_{F'}) = \frac{2F'+1}{(2J+1)(2I+1)}\frac{I(I+1)+F'(F'+1)-J(J+1)}{2I(I+1)}$$

$$\times \frac{F(F+1)+I(I+1)-J(J+1)}{2F(F+1)}\frac{(n_F\langle\mathbf{F}_z\rangle_F)}{T} \tag{108}$$

Here T is the mean collision time for J-randomizing collisions. The equations allow the following interpretation. Population numbers are statistically distributed over the entire hyperfine multiplet. Then the transfer rate becomes proportional to the statistical weight of the final F' level and inversely proportional to the statistical weight of the entire I, J multiplet. The orientation transfer is understood when considering that Eq. (108) contains two quantum-mechanical cosines, viz., one of the angle between $\mathbf{F}$ and $\mathbf{I}$ and one of the angle between $\mathbf{I}$ and $\mathbf{F}'$. Evidently, the orientation in the I system survives when a collision destroys the one in the J system. Hence the projection of F on I results. After the collison, hyperfine interaction sets in again and states F' are reestablished and receive their orientations from the residual one in the I system. Hence we understand the projection of I on F'.

It is noteworthy that the transfer cross sections to state F' for the quantities n_F and $n_F\langle\mathbf{F}_z\rangle_F$ differ from the ones in the opposite direction. However, reversibility is established when using density tensor operators in Fano's[102] notation:

$$\mathscr{T}_0^{(0)} = (2F+1)^{-1/2}\mathbf{1} \qquad \text{instead of } \mathbf{1} \qquad (109)$$

$$\mathscr{T}_0^{(1)} = \frac{\mathbf{F}_z}{[F(F+1)(2F+1)/3]^{1/2}} \qquad \text{instead of } \mathbf{F}_z \qquad (110)$$

The randomization model does not allow the calculation of absolute depolarizing cross sections because, by assumption, it is not apt to yield the critical impact parameter b_{crit}, whence the reorientation ceases to be random. Omont[97] has shown how to cope with this problem. He calculates the probability of collisional transfer for large impact parameters from realistic interaction potentials. It is then easy to determine b_{crit} from the condition that the probability of transfer $P(b)$ reaches at b_{crit} the value of the randomization model, Eqs. (107) and (108). The transfer cross section then follows from the evaluation of the integral

$$\sigma = \int_0^\infty 2\pi b\, db\, P(b) \qquad (111)$$

by taking the randomization value of P for $0 \leqslant b \leqslant b_{\mathrm{crit}}$ and the approximate value from Omont's model for $b_{\mathrm{crit}} > b$. For potentials which fall off as rapidly as R^{-6}, the contribution of the latter range is negligible. However, for R^{-3} potentials, it may become dominant, so that the randomization model is invalidated altogether.

The flywheel effect was first observed by Faroux and Brossel,[53] who, beginning in 1964, studied the transfer between the hyperfine sublevels of $^{199}\mathrm{Hg}\,6^3P_1$ and $^{201}\mathrm{Hg}\,6^3P_1$. They selectively populated one of the hyperfine sublevels as a result of an incidental resonance with light from another even nuclide which was kept in the lamp. The fluorescent light of one of the odd isotopes was passed through a filter which contained the even nuclide as well. This removed the resonant light and brought only the sensitized light to detection. For the sensitization a helium buffer of 0.1–

Table 5. Relative Relaxation Rates and Relative Orientation and Population
Transfer Rates

| | Omont | | | |
Rate	Approx. I	Approx. II	J randomization	Experiment
$\gamma_{1/2}/^{199}\gamma_2$	0.77	0.92	0.965	0.98 ± 0.02 Faroux
$\gamma_{3/2}/^{199}\gamma_2$	0.69	0.635	0.63	0.75 ± 0.02 Faroux
$\gamma_{1/2,3/2}^{(1)}/^{199}\gamma_2$	-0.24	-0.13	-0.117	-0.09 ± 0.01 Faroux
$\gamma_{1/2,3/2}^{(0)}/^{199}\gamma_2$	0.54	0.48	0.49	0.42 ± 0.08 Barrat

[a] Comparison with the theory (see text). The minus sign of $\gamma_{1/2,3/2}^{(1)}$ indicates the inversion of the orientation during transfer. It is in accordance with our definitions [cf. Eq. (43)], but it is not in the original paper. From Eqs. (107) and (108) together with the definition, Eqs. (109), and (110), the reader can verify readily that $\gamma_{1/2,3/2}^{(1)}/\gamma_{1/2,3/2}^{(0)} = -\frac{1}{9}(5)^{1/2} = -0.25$ close to observation.

1 Torr was used. When orientation transfer was studied, circularly polarized light was used for excitation. The results are given in Table 5.

Magnetic depolarization of the sensitized light allowed the discrimination against spurious background polarization due to reflections at the cell walls. This method, however, was not applicable when population transfer was studied. To get rid of reflections even in this case Faroux and Brossel[53] developed another elegant method. They oriented the Hg atoms in the ground state by optical pumping with a separate beam. Then, by nuclear induction, they made the nuclear spins precess at the Larmor frequency, which entailed modulation of the absorption of circularly polarized resonance light. Any subsequent fluorescence had then to be modulated too, and phase-sensitive detection again allowed elimination of reflections which necessarily were unmodulated.

There is a multitude of phenomena in which the flywheel effect takes part. Dupont-Roc et al.[118] have studied the broadening of Larmor resonances in the hyperfine sublevel of the ^{3}He $(I = \frac{1}{2})\, 2\,^3S_1$ state which results from metastability exchange collisions. In collisions of ^{3}He metastable and ground-state atoms part of the precessing orientation $\langle \mathbf{F} \rangle$ which is in the $F = \frac{3}{2}$ or $\frac{1}{2}$ sublevel is handed down to the other atom which replaces the first one in the metastable state. This part equals the one which is kept in the spin system of the first atom. (The part which is in the nuclear spin system is taken along to the ground state.) Then it is redistributed among the spin and nuclear spin system of the new atom through hyperfine coupling, thus leading to nonvanishing $\langle \mathbf{F} \rangle$ values in the new metastable atom. This effect apparently reduces the broadening effect of metastability exchange collisions, so that the pertinent cross section has to be properly corrected to yield agreement with cross sections obtained from ^{4}He measurements (Colegrove et al.[119]).

Very similar is the explanation which Bulos and Happer[120] give to the reduced pressure broadening of Hanle curves in the alkali $^2P_{1/2}$ states. Here the $\langle\mathbf{F}\rangle$ values of both hyperfine sublevels are coupled due to the transfer of the nuclear orientation between them. The authors not only consider the different speed of precession of $\langle\mathbf{F}\rangle$ on account of the different g_F values, but also consider that a back-transfer of $\langle\mathbf{F}\rangle$ to the initial F level might happen in a subsequent collision. The shape of the Hanle signals, which are, according to the numerous participating interactions, no longer Lorentzian, is successfully explained.

Pavlovic and Laloë[121] discovered that the nuclear spin orientation which by metastability exchange collisions is accumulated in the ground state of optically pumped ^{3}He is transferred by electronic excitation to the levels $3\,^1D$, $4\,^1D$, and $5\,^1D$. There, hyperfine coupling imparts the orientation to the electronic spin. The decoupling of the nuclear spin at moderate magnetic fields curtails this redistribution. Therefore, the measurement of $\langle\mathbf{F}\rangle$ of these levels at varying magnetic fields allows the determination of the strength of hyperfine coupling (i.e., the a values) in these levels.

Leduc and Laloë[122] and Pinard and Leduc[123] have shown that the paramagnetic ground state of ^{3}He ions present in a discharge in which ^{3}He is optically pumped is also oriented due to resonant charge exchange with He atoms. The pertinent cross section was determined. Transfer of nuclear orientation also plays a dominant part in the optical pumping of ions and neutrals of the elements of the first and second group (zu Putlitz and co-workers[124]). Nuclear orientation is also conserved in Penning ionization, studied in collisions of He metastables by Schearer and Holton.[125] More information on the latter process is provided by the article of Schearer and Parks in this book (Chapter 18).

3.7. Transfer between Fine-Structure Sublevels of the Same Element

Three examples have been studied: sodium (Elbel et al.[126]), potassium (Niewitecka and Krause[127]), and thallium (Gelbhaar[128]). The experimental arrangement in these studies was fairly similar. One level of the excited doublet Na $3\,^2P$, K $4\,^2P$, or Tl $6\,^2D$, was excited by irradiating with one resonance line. Orientation was imparted to this level by making the exciting light beam circularly polarized. Buffer gas was admitted to the fluorescent region. Fluorescence from the other level which was populated through collisions was detected at small angles to the backward direction. Right-handed and left-handed quanta were discriminated between in the sensitized fluorescence. From the difference of their rates, the transfer

Table 6. Energy and Orientation Transfer Cross Sections for Na($3p$ ^2P) and K($4p$ ^2P) at Very Low Magnetic Field Strengths[a,b]

Foreign gas	K		Na	
	$\sigma^{(1)}_{1/2 \to 3/2}$ (Å^2)	$\sigma^{(0)}_{1/2 \to 3/2}$ (Å^2)	$\sigma^{(1)}_{1/2 \to 3/2}$ (Å^2)	$\sigma^{(0)}_{1/2 \to 3/2}$ (Å^2)
He	1.7[b]	59.5[c]	22[e]	86[f]
Ne	0.8[b]	14.3[c]	24[e]	67[f]
Ar	<0.5[b]	36.7[c]	26[e]	110[f]
Kr		61.4[c]	20[e]	85[f]
Xe		104[c]	23[e]	90[f]
H$_2$	3.5[b]	76[d]		
CH$_4$	7.0[b]			
CD$_4$	7.7[b]			

[a] Hyperfine interaction causes positive orientation transfer. However, very long hyperfine periods give rise to collisional decoupling of the nuclear spin which could explain the unusually low orientation-transfer cross sections for potassium. Very high foreign gas pressures could even invert the sign of the transferred orientation (Elbel and Schneider[114]).
[b] Niewitecka and Krause.[127]   [c] Chapman and Krause.[60]   [d] McGillis and Krause.[40]
[e] Elbel and Schneider.[98]   [f] Pitre and Krause.[58]

cross section of orientation was inferred. The sum of these rates yielded the energy-transfer cross section. The experiments were carried out at low magnetic fields which were directed parallel to the incident light beam to prevent losses of orientation due to the Hanle effect. Or, when Hanle signals were intended to be measured, the H field was perpendicular to the incident light. When H was swept through zero, the maximum orientation transfer was obtained. Stray fields were carefully compensated for.

It turned out that the transfer cross sections of both energy and orientation had the same sign at low fields (Table 6). An interpretation can be only qualitatively given.

The left-hand side of Figure 17 shows the orientation of the initial $^2P_{1/2}$ level: $F = I + \frac{1}{2}$, $I - \frac{1}{2}$. A sudden collision dephases the molecular σ and π states into which a P state splits when a foreign center of force such as a noble-gas atom is brought into close proximity. If the collision ends after a very short duration τ_c such that

$$\hbar/\tau_c \gg \varepsilon \tag{112}$$

(Massey condition), ε being the gap of the doublet, then the σ–π dephasing results in a randomization of $\mathbf{L}$ alone. The spin $\mathbf{S}$ is not affected, but remains constant throughout the collision. The same holds for the nuclear spin. This situation is displayed at the right-hand side of Figure 17. Note that the encircled parts of the figure refer to the $^2P_{3/2}$ state with spin and orbit parallel! As the sum of both is $\mathbf{J}$, it follows that $\mathbf{J}$ is orientated parallel to z in the $^2P_{3/2}$ part of the outgoing state mixture! It is clear that **J** of the $^2P_{3/2}$ state would be oriented antiparallel to z if there were no nuclear spin. The orientation of **J** which is actually measured in the fluorescent light of the $^2P_{3/2}$ level accordingly would change sign when the nuclear spin is removed from the process. This removal is rendered possible by decoupling in a sufficiently strong magnetic field parallel to the incident light beam (Elbel and Schneider,[98] Schneider[129]). Table 7 gives a survey of the orientation- and energy-transfer cross sections from Na 3 $^2P_{1/2}$ to Na 3 $^2P_{3/2}$. The ratio of both can be predicted from the flywheel effect, Eq. (108), with F, F', I replaced by J, J', s, respectively. L is the angular momentum which is randomized here. The condition that $\hbar/\varepsilon$ is larger than the duration of the collision τ_c is valid for He, but invalid for the heavier noble gases. Accordingly, the transfer cross sections decrease for increasing atomic weight of the noble gases. Moreover, the ratio of both is not constant as one might expect when closer collisions only remain efficient in bringing transfer about. Nikitin and Dashevskaya[130] as well as Wilson and Shimoni[131] even predict that for Rb, where the gap ε is very large, the ratio may be inverted. Grawert[107] has established a general frame to the transfer cross sections. He considered that states with $J = \frac{1}{2}$ and $J = \frac{3}{2}$ are connected by transition operators which behave, under rota-

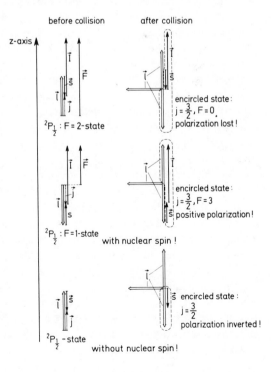

Figure 17. The flywheel transfer of orientation from $^2P_{1/2}$ to $^2P_{3/2}$ when a nuclear spin $I = 3/2$ is present (upper part) or absent (lower part). The coupling of the angular momenta in the respective initial state is given on the left-hand side. On the right-hand side **1** is depolarized (three possible directions drawn) and those parts are encircled in which the mutual orientation of **1** and **s** are parallel, as in the $^2P_{3/2}$ state. The polarization of the $^2P_{3/2}$ part of the final state with respect to the z axis thus becomes apparent. The drawing provides an understanding of the reversal of the transferred orientation which accompanies the decoupling of the nuclear spin.

Table 7. Energy and Orientation Transfer Cross Sections for Na($3p\,{}^2P$) at $H = 900$ Oe ^{a}

Foreign gas	$\sigma^{(0)}_{1/2\to3/2}$ (Å^2)	$\sigma^{(1)}_{1/2\to3/2}$ (Å^2)	Ratio
He	89	-64.0	-0.72
Ne	79	-51.8	-0.65
Ar	116	-80.7	-0.70
Kr	94	-55.5	-0.59
Xe	99	-43.3	-0.44
Theory Eqs. (107)–(108)			$-5/9 = -0.56$

^{a} According to Schneider.[129] Now that the nuclear spin is decoupled strong negative orientation transfer is observed. The l randomization model predicts

$$\frac{\sigma^{(1)}_{1/2\to3/2}}{\sigma^{(0)}_{1/2\to3/2}} = \frac{s(s+1)+j'(j'+1)-l(l+1)}{2s(s+1)} \frac{j(j+1)+s(s+1)-l(l+1)}{2j(j+1)} = -\frac{5}{9}$$

tion, like inner products of first- or second-degree tensors, one acting in the electronic coordinate space and the other acting in the coordinate space of the colliding atoms. Hence two independent dynamic parameters a_1, a_2 arise, which constitute the transfer cross sections. It was attempted (Elbel[132]) to determine these parameters from realistic assumptions about the interaction. It was made evident that a_2 is due to molecular σ, π splitting and consequent dephasing of molecular states whereas a_1 is due to the Coriolis interaction. Therefore, a_2 is already present in an adiabatic approximation which neglects the motion of the interatomic axis, whereas it is this motion alone which determines the Coriolis interaction.

The analysis of transfer collisions was developed by Nikitin[133] and others. Nikitin in his first article stated the importance of the interatomic distance R_1 at which the splitting of the level ${}^2P_{3/2}$ into molecular sublevels ${}^2\Sigma_{1/2}$ and ${}^2\Pi_{3/2}$ reaches the value of the doublet gap ε. When the atoms, in the course of the collision process, come closer than R_1, a sudden (i.e., nonadiabatic) transformation of atomic into molecular states occurs. The latter states are strongly referred to the interatomic axis which sweeps out an angle Φ while the atoms run inside a sphere of radius R_1. When crossing the distance R_1 again on their way out, the atoms change back from molecular to atomic coupling. In the outgoing state substates of ${}^2P_{3/2}$ are present provided the incoming state is a pure ${}^2P_{1/2}$ state. The sum of the squares of their amplitudes provides the transition probability ${}^2P_{1/2} \to {}^2P_{3/2}$ for the particular trajectory in question. It depends on the phase shift which the σ states and π states have suffered due to their stay inside R_1 and on the angle Φ by which the molecular system rotated during its existence. Both the σ, π dephasing and the carriage with the turning axis interfere in the production of L depolarization and, hence, fine-structure transition. Nikitin carefully distinguishes two cases, one in which $\Phi \gg W(^2\pi_{3/2})-W(^2\pi_{1/2})/\hbar$, and another one in which this relation is reversed. The first one applies to the aforementioned model. The second one gives rise to transitions only while the atoms are crossing the critical distance R_1. Then a Landau–Zener-like formula applies. Both approaches yield upper and lower limits for the transition cross section.

Besides these approaches, Nikitin, in terms of time-dependent methods, formulated the evolution of the atomic states $|j, m_j\rangle$. He used the angular momentum of the colliding pair as the z axis of the problem and the relative velocity before the collision as an x axis. The matrix of the Hamiltonian has then elements between those states whose m differs by 0 or ± 2. The six functions accordingly split into two triples which evolve independently in time. A solution of the Schrödinger equation then reads like a superposition of the three interacting states and when entering the Schrödinger equation with this superposition three coupled differential equations for the three time-dependent amplitudes a_1, a_2, a_3 result. Starting with appropriate initial conditions, say, $a_2 (t = -\infty) = a_3 (t = -\infty) = 0$, an integration of the coupled equations up to $t = \infty$ yields finite values $a_2 (t = \infty)$, $a_3 (t = \infty)$, which are to be understood as the transition matrix elements from state 1 to states 2 and 3. The sum $|a_2|^2 + |a_3|^2$ is the probability for a fine-structure transition. The integration was carried out by Masnou–Seeuws.[134] This study was aimed at the dependence on temperature of the fine-structure transition. Later studies (Masnou–Seeuws and Roueff,[135] Masnou–Seeuws and McCarroll[136]) aimed at the detailed inter-Zeeman cross sections whence the orientation transfer cross sections are easy to derive (Elbel et al.[99]).

Parallel to this theoretical development a full quantum treatment of the problem was carried on (Smith,[137] Elbel[138] Reid,[139] Weisheit and Lane,[140] Mies,[141] Wilson and Shimoni,[131,142] Wofsy et al.,[143] Lewis and McNamara,[144] Moskowitz and Thorson,[145] and Preston et al.[146]). Some of these authors take the transfer process as quasielastic, using spin-uncoupled wave functions from the beginning. Others like Reid,[139] on the basis of a rigorous treatment, check some predictions of the spin-uncoupled theory concerning ratios of inter-Zeeman cross sections and find them invalid unless the collision energy exceeds the gap by roughly a factor of 10. Reid also provides angular scattering distributions for transfer collisions which, at present, are being measured in scattering experiments using crossed beams and laser saturation of a resonance transition (Carter et al.[147]) or photon diagnostics of collisionally excited states (Kempter et al.,[148] and Elbel et al.[149]). Moreover, the abundance of inter-Zeeman cross sections [i.e., the $\sigma(jm_j - j'm'_j)$] recently offered by the theory has stimulated attempts to measure them directly. Using an idea of Seiwert and co-workers[150] and an early realization by Berdowski and Krause,[150]

Table 8. Cross Sections for Transitions between Zeeman Sublevels of States $Na(3p\ ^2P_{1/2})$ and $Na(3p\ ^2P_{3/2})$ Induced in Collisions with Helium Atoms[a]

Cross section (Å^2)	Gay and Schneider[151] experiment	Masnou–Seeuws and Roueff[135] theory	Wilson and Shimoni[142] theory	Reid[139] theory
$\sigma(\frac{1}{2}\frac{1}{2}\to\frac{3}{2})$	35.4 ± 2.3	33.5	32.0	27
$\sigma(\frac{1}{2}\frac{1}{2}\to\frac{3}{2}-\frac{1}{2})$	31.5 ± 3.9	27.8	25.7	22
$\sigma(\frac{1}{2}\frac{1}{2}\to\frac{3}{2}\frac{1}{2})$	24.2 ± 3.9	22.2	19.4	17
$\sigma(\frac{1}{2}\frac{1}{2}\to\frac{3}{2}\frac{3}{2})$	15.8 ± 1.3	16.5	13.2	12
$\sigma^{(0)}_{1/2\to3/2}$	107 (Table 7: 89)	100.0	90.3	78
$\sigma^{(1)}_{1/2\to3/2}$	-66 (Table 7: -64)	-56.6	-62.7	-50

[a] These measurements were carried out at 400 K in magnetic fields above 10 kG by Gay and Schneider.[151] Theoretical values are given for comparison. Masnou-Seeuws and Roueff[135] used a semiclassical model, Wilson and Shimoni[142] and Reid[139] used a full quantum-mechanical treatment of the collision process. The last two entries in the table,

$$\sigma^{(1)}_{1/2\to3/2} = 3[\sigma(\tfrac{1}{2}\tfrac{1}{2}\to\tfrac{3}{2}-\tfrac{3}{2})-\sigma(\tfrac{1}{2}\tfrac{1}{2}\to\tfrac{3}{2}\tfrac{3}{2})] + [\sigma(\tfrac{1}{2}\tfrac{1}{2}\to\tfrac{3}{2}-\tfrac{1}{2})-\sigma(\tfrac{1}{2}\tfrac{1}{2}\to\tfrac{3}{2}\tfrac{1}{2})]$$

$$\sigma^{(0)}_{1/2\to3/2} = \sum_{m_j}\sigma(\tfrac{1}{2}\tfrac{1}{2}\to\tfrac{3}{2}\,m_j)$$

are calculated and given for comparison with the values of Table 6.

Gay and Schneider[151] have utilized Zeeman splitting of the sodium D lines in ultrahigh magnetic fields (50 kOe) to completely resolve their Zeeman components by spectroscopic means. By selective excitation of $^2P_{3/2}$, $m = \frac{3}{2}$, using laser light, they were able to measure fluorescent light from all the other Zeeman levels which were populated by transfer collisions with helium atoms. The deduced cross sections not only confirmed the results of low-field measurements but also allowed a direct comparison with the theory. This, as stated by Nikitin,[152] makes the $Na(3\ ^2P)$, He collision, so far, the best-understood collision process involving excited atoms (Table 8).

References

1. A. G. C. Mitchell and M. W. Zemanski, *Resonance Radiation and Excited Atoms*, Cambridge University Press (1971).
2. P. Pringsheim, *Fluorescence and Phosphorescence*, Interscience Publishers, New York (1949).
3. H. S. W. Massey, E. H. S. Burhop, and H. B. Gilbody, *Electronic and Ionic Impact Phenomena*, Vol. III, Clarendon Press, Oxford (1971).
4. L. Krause, *Appl. Opt.* **5**, 1355 (1966).

5. L. Krause, in *The Excited State in Chemical Physics*, John Wiley, New York (1975).

6. R. Seiwert, *Springer Tracts in Modern Physics*, Vol. 47, Springer, New York (1968).

7. E. K. Kraulinya and V. A. Kruglevskij, in *Sensitized Fluorescence of Mixtures of Metal Vapors* (in Russian), Sbornik 2, Riga (1969) and Sbornik 4, Riga (1973).

8. P. L. Lijnse, *Review of Literature on Quenching, Excitation and Mixing Collision Cross-Sections for the First Resonance Doublets of the Alkalies*. Rep. 398, Fysisch Laboratorium, Rijksuniversiteit Utrecht, Netherlands (1972).

9. E. Milne, *J. London Math. Soc.* **1**, 1 (1926).

10. T. Holstein, *Phys. Rev.* **72**, 1212 (1947); **83**, 1159 (1951).

11. J. P. Barrat, *J. Phys. Rad.* **20**, 541, 633, and 657 (1959).

12. M. J. Dyakonov and V. J. Perel, *Sov. Phys.-JETP* **20**, 997 (1965).

13. A. Omont, *J. Phys. Rad.* **26**, 576 (1965).

14. E. W. Otten and A. Winnacker, *Phys. Letts* **23**, 462 (1966).

15. B. P. Kibble, G. Copley, and L. Krause, *Phys. Rev.* **153**, 9 (1967).

16. G. Copley and L. Krause, *Can. J. Phys.* **47**, 533 (1969).

17. J. S. Deech and W. E. Baylis, *Can. J. Phys.* **49**, 90 (1971).

18. A. N. Nesmeyanov, *Vapor Pressure of the Chemical Elements*, Elsevier, Amsterdam (1963).

19. F. C. M. Coolen, L. C. J. Baghuis, H. L. Hagedorn, and J. A. van der Heide, *J. Opt. Soc. Am.* **64**, 482 (1974).

20. M. Czajkowski, D. A. McGillis, and L. Krause, *Can. J. Phys.* **44**, 91 and 741 (1966).

21. B. Pitre, A. G. A. Rea, and L. Krause, *Can. J. Phys.* **44**, 731 (1966).

22. G. D. Chapman and L. Krause, *Can. J. Phys.* **44**, 753 (1966).

23. J. Pitre and L. Krause, *Can. J. Phys.* **45**, 2671 (1967).

24. J. A. Jordan and P. A. Franken, *Phys. Rev.* **142**, 20 (1966).

25. A. Gallagher, *Phys. Rev.* **172**, 88 (1968).

26. J. Callaway and E. Bauer, *Phys. Rev.* **140**, A1072 (1965).

27. A. Omont and J. Meunier, *Phys. Rev.* **169**, 92 (1968).

28. M. B. Hidalgo and S. Geltman, *J. Phys. B* **5**, 265 (1972).

29. F. Masnou–Seeuws, *J. Phys. B* **3**, 1437 (1970).

30. L. Kumar and J. Callaway, *Phys. Lett.* **28A**, 385 (1968).

31. R. H. G. Reid and A. Dalgarno, *Phys. Rev. Lett.* **22**, 1029 (1969); *Chem Phys. Lett.* **6**, 85 (1970).

32. R. H. G. Reid, *J. Phys. B* **6**, 2018 (1973); **8**, 2255 (1975).

33. E. E. Nikitin, *J. Chem. Phys.* **43**, 744 (1965).

34. E. E. Nikitin, in *The Excited State in Chemical Physics*, Vol. V, John Wiley, New York (1975).

35. E. L. Dashevskaya, E. E. Nikitin, and A. J. Reznikov, *J. Chem. Phys.* **53**, 1175 (1970).

36. I. N. Siara, H. S. Kwong, and L. Krause, *Can. J. Phys.* **52**, 945 (1974).

37. J. Cuvellier, P. R. Fournier, F. Gounand, and J. Berlande, *C.R. Acad. Sci. Paris B* **276**, 855 (1973).

38. M. Pimbert, *J. Phys. (Paris)* **33**, 331 (1972).

39. W. E. Baylis, E. Walentynowicz, R. A. Phaneuf, and L. Krause, *Phys. Rev. Lett.* **31**, 741 (1973).

40. D. A. McGillis and L. Krause, *Can. J. Phys.* **46**, 1051 (1968).

41. M. Stupavsky and L. Krause, *Can. J. Phys.* **46**, 2121 (1968); **47**, 1249 (1969).

42. E. S. Hrycyshyn and L. Krause, *Can. J. Phys.* **48**, 2761 (1970).

43. W. D. Phillips and D. Pritchard, *Phys. Rev. Lett.* **33**, 1254 (1974).

44. J. Apt, Doctoral Thesis, Harvard (1976).

45. J. Pascale and P. M. Stone, Proceedings of the Ninth ICPEAC 529 (1975).

46. J. Pascale and R. E. Olsen, *J. Chem. Phys.* **64**, 2528 (1976).

47. J. Pascale and J. Vanderplanque, *J. Chem. Phys.* **60**, 2278 (1974).

48. I. Siara, E. S. Hrycyshyn, and L. Krause, *Can. J. Phys.* **50**, 1826 (1972).
49. R. W. Wood, *Phil. Mag.* **27**, 1018 (1914).
50. R. Seiwert, *Ann. Phys. (Leipzig)* (6) **18**, 54 (1956).
51. H. Bunke and R. Seiwert: *Monatsber. Dtsch. Akad. Wiss. Berlin* **2**, 723 (1960); Tagungsbericht *Optik und Spektroskopie aller Wellenlängen* (Jena 1960) pp. 409 ff., Akademie-Verlag, Berlin (1962).
52. K. Hoffmann and R. Seiwert, *Ann. Phys. (Leipzig)* **7**, 71 (1961).
53. J. P. Faroux and J. Brossel, *C.R. Acad. Sci. Paris* **261**, 3092 (1965); **263B**, 612 (1966); **264B**, 1452 (1967); **262B**, 41 (1966); **262B**, 1385 (1966).
54. G. L. Datta, *Z. Phys.* **37**, 625 (1926).
55. W. Hanle, *Z. Phys.* **41**, 164 (1927).
56. J. C. Gay and W. B. Schneider, *J. Phys. (Paris)* **36**, L185 (1975).
57. B. Fuchs, W. Hanle, W. Oberheim, and A. Scharmann, *Phys. Lett.* **50**, 337 (1974).
58. J. Pitre and L. Krause, *Can. J. Phys.* **46**, 125 (1968).
59. M. A. Thangaraj, Ph.D. Thesis, University of Toronto (1948).
60. G. D. Chapman and L. Krause, *Can. J. Phys.* **44**, 753 (1966).
61. A. G. A. Rea and L. Krause, *Can. J. Phys.* **43**, 1574 (1965).
62. M. Czajkowski and L. Krause, *Can. J. Phys.* **43**, 1259 (1965).
63. E. I. Dashevskaya, A. I. Voronin, and E. E. Nikitin, *Can. J. Phys.* **47**, 1237 (1969).
64. Yu. A. Vdovin, V. M. Galitski, and N. A. Dobrodeev, *Zh. Eksperim. i Teor. Fiz.* **56**, 1344 (1969).
65. M. Ya. Ovchinnikova, *Zh. Teor. Eksperim. Khim.* **1**, 22 (1965) [English transl.: *J. Theor. Experim. Chem.* **1**, 1 (1965)].
66. E. S. Hrycyshyn and L. Krause, *Can. J. Phys.* **47**, 215 (1969).
67. M. H. Ornstein and R. N. Zare, *Phys. Rev.* **181**, 214 (1969).
68. V. Stacey and R. N. Zare, *Phys. Rev. A* **1**, 1125 (1970).
69. E. I. Dashevskaya, E. E. Nikitin, A. I. Voronin, and A. A. Zembekov, *Can. J. Phys.* **48**, 981 (1970).
70. M. Pimbert, J. L. Rocchiccioli, and J. Cuvellier, *C.R. Acad. Sci. Paris B* **270**, 684 (1970).
71. G. Cario, *Z. Phys.* **10**, 185 (1922).
72. H. Beutler and B. Josephy, *Naturwissenschaften* **15**, 540 (1926); *Z. Phys.* **53**, 747 (1929).
73. S. E. Frisch and E. K. Kraulinya, *Dokl. Akad. Nauk SSSR* **101**, 837 (1953).
74. E. K. Kraulinya, *Opt. Spektrosk. Opt. Spectrosc.* **17**, 250 (1964).
75. S. G. Rautian, A. S. Khaikin, *Opt. Spectrosc.* **18**, 406 (1965).
76. E. K. Kraulinya and M. L. Yanson, in *Sensitized Fluorescence of Mixtures of Metal Vapors* (in Russian), Sbornik 3, Riga (1970).
77. M. Czajkowski, G. Skardis, and L. Krause, *Can. J. Phys.* **51**, 334 (1973).
78. M. Czajkowski and L. Krause, *Can. J. Phys.* **52**, 2228 (1974).
79. H. Friedrich and R. Seiwert, *Exp. Tech. Phys.* **5**, 193 (1957); *Ann. Phys. (Leipzig)* **20**, 215 (1957).
80. L. D. Landau, *Phys. Z. Sowjetunion* **1**, 88 (1932).
81. C. Zener, *Proc. R. Soc. London* **A137**, 696 (1932).
82. E. N. Morozov and M. L. Sosinskij, *Opt. Spectrosc.* **25**, 282 (1968).
83. E. K. Kraulinya and M. G. Arman, *Opt. Spectrosc.* **26**, 285 (1969).
84. B. Chéron, Thesis, University of Caen, France (1974).
85. H. C. Lipson and A. C. Mitchell, *Phys. Rev.* **47**, 807 (1935); **48**, 625 (1935).
86. D. R. Bates and D. A. Williams, *Proc. Phys. Soc London* **83**, 425 (1964).
87. V. K. Bykhovskij, E. E. Nikitin, and M. Ya. Ovchinnikova, *Sov. Phys.-JETP* **20**, 500 (1965).
88. W. Bachmann and W. Janke, *Z. Naturforsch.* **27a**, 579 (1972).

89. L. C. H. Loh, C. M. Sholeen, and R. R. Herm, *J. Chem. Phys.* **63**, 1980 (1975).
90. D. R. Bates, *Proc. R. Soc. London A* **257**, 22 (1960); see also C. A. Coulson and K. Zalewski *ibid.* **268**, 437 (1962).
91. D. B. Lidow, R. W. Falcone, J. F. Young, and S. E. Harris, *Phys. Rev. Lett.* **36**, 462 (1976).
92. W. Gough, *Proc. Phys. Soc. London* **90**, 287 (1967).
93. A. C. G. Mitchell, *J. Franklin Inst.* **209**, 747 (1930).
94. G. D. Chapman, L. Krause, and I. H. Brockmann, *Can. J. Phys.* **42**, 535 (1964).
95. W. Franzen, *Phys. Rev.* **115**, 850 (1959).
96. M. A. Bouchiat, *J. Phys. (Paris)* **26**, 415 (1965).
97. A. Omont, *J. Phys.* **26**, 26 (1965).
98. M. Elbel and W. B. Schneider, *Z. Phys.* **241**, 244 (1971).
99. M. Elbel, W. B. Schneider, and B. Kamke, *Physica* **77**, 137 (1974).
100. B. Chéron and J. P. Barrat, *C.R. Acad. Sci. Paris* **266**, 1324 (1968).
101. B. Chéron, *Opt. Commun.* **3**, 437 (1971).
102. U. Fano, *Rev. Mod. Phys.* **29**, 74 (1957).
103. B. Chéron and L. Saintout, *J. Phys. (Paris)* **32**, 731 (1971).
104. F. W. Byron and H. M. Foley, *Phys. Rev.* **134**, 625 (1964).
105. B. Dodsworth and A. Omont, *Phys. Rev. Lett.* **24**, 198 (1970).
106. M. I. Dyakonov and V. I. Perel, *Sov. Phys.-JETP* **21**, 227 (1965).
107. G. Grawert, *Z. Phys.* **225**, 283 (1969).
108. D. M. Brink and G. R. Satchler, *Angular Momentum*, University Press, Oxford (1962).
109. T. Watanabe, *Phys. Rev.* **138**, A1573 (1965); **140**, A135, (1965).
110. M. Elbel, B. Niewitecka, and L. Krause, *Can. J. Phys.* **48**, 2996 (1970).
111. L. Y. Chow-Chiu, *Phys. Rev. A* **5**, 2053 (1972).
112. W. E. Baylis, *Phys. Rev. A* **7**, 1190 (1973).
113. C. G. Carrington, *J. Phys. B* **4**, 1222 (1971).
114. M. Elbel and W. B. Schneider, *Physica* **68**, 146 (1973).
115. W. Happer, *Rev. Mod. Phys.* **44**, 169 (1972).
116. F. A. Franz and J. R. Franz, *Phys. Rev.* **148**, 82 (1966).
117. P. L. Bender, Thesis, Princeton University (1956).
118. J. Dupont-Roc, M. Leduc, and F. Laloë, *Phys. Rev. Lett.* **27**, 467 (1971).
119. F. D. Colegrove, L. D. Schearer, and G. K. Walters, *Phys. Rev.* **132**, 2561 (1963); **135**, 353 (1964).
120. B. R. Bulos and W. Happer, *Phys. Rev.* **A4**, 849 (1971).
121. M. Pavlovic and F. Laloë, *J. Phys. (Paris)*, **31**, 173 (1970).
122. M. Leduc and F. Laloë, *Opt. Commun.* **3**, 56 (1971).
123. M. Pinard and M. Leduc, *J. Phys. (Paris)* **35**, 741 (1974).
124. H. Nienstädt, G. Schmidt, S. Ullrich, H. G. Weber, and G. Z. Putlitz, *Phys. Lett.* **41A**, 249 (1972); more work is reviewed in E. W. Weber, *Phys. Rep.* **32**, 123 (1977).
125. L. D. Schearer and W. C. Holton, *Phys. Rev.* **24**, 1214 (1970); L. D. Schearer and L. A. Riseberg, *Phys. Lett.* **33A**, 325 (1970).
126. M. Elbel, B. Niewitecka, and L. Krause, *Can. J. Phys.* **48**, 2996 (1970).
127. B. Niewitecka and L. Krause, *Can. J. Phys.* **51**, 425 (1972).
128. B. Gelbhaar, *Z. Naturforsch.* **28a**, 257 (1973); *Z. Phys.* **A272**, 53 (1975).
129. W. B. Schneider, *Z. Phys.* **248**, 387 (1971).
130. E. E. Nikitin and E. I. Dashevskaya, *Can. J. Phys.* **54**, 709 (1976).
131. A. D. Wilson and Y. Shimoni, *J. Phys. B* **7**, 1543 (1974).
132. M. Elbel, *Z. Phys.* **248**, 375 (1971).
133. E. E. Nikitin, *J. Chem. Phys.* **43**, 744 (1965).
134. F. Masnou–Seeuws, *J. Phys. B* **3**, 1437 (1970).
135. F. Masnou–Seeuws and E. Roueff, *Chem. Phys. Lett.* **16**, 593 (1972).

136. F. Masnou-Seeuws and R. McCarroll, *J. Phys. B* **7**, 2230 (1974).
137. F. J. Smith, *Planet Space Sci.* **14**, 937 (1966); *Mon. Not. R. Astron. Soc.* **140**, 341 (1968).
138. M. Elbel, *Ann. Phys. (Leipzig)* **22**, 289 (1969); *Can. J. Phys.* **48**, 3047 (1970).
139. R. H. G. Reid, Proceedings of the Seventh ICPEAC 675, Amsterdam (1971); *J. Phys. B* **6**, 2018 (1973); B **8**, 2255 (1975).
140. J. C. Weisheit and N. F. Lane, *Phys. Rev. A* **4**, 171 (1971).
141. F. H. Mies, *Phys. Rev. A* **7**, 942 and 957 (1973).
142. A. D. Wilson and Y. Shimoni, *J. Phys. B* **8**, 2393 (1975).
143. S. Wofsy, R. H. G. Reid, and A. Dalgarno, *Astrophys. J.* **168**, 161 (1971).
144. E. L. Lewis and L. F. McNamara, *Phys. Rev. A* **5**, 2643 (1972).
145. J. W. Moskowitz and W. R. Thorson, *J. Chem. Phys.* **38**, 1848 (1963).
146. R. K. Preston, C. Sloam, and W. H. Miller, *J. Chem. Phys.* **60**, 4961 (1974).
147. G. M. Carter, D. E. Pritchard, M. Kaplan, and T. W. Ducas, *Phys. Rev. Lett.* **35**, 1144 (1975); W. D. Phillips, C. L. Glaser, and D. Kleppner, *Phys. Rev. Lett.* **38**, 1018 (1977).
148. V. Kempter, B. Kübler, and W. Mecklenbrauck, *J. Phys. B* **7**, 2375 (1974). cf. also: H. Alber, V. Kempter, and W. Mecklenbrauck, *J. Phys. B* **8**, 913 (1975).
149. M. Elbel, H. Hühnermann, Th. Meier, and W. B. Schneider, *Z. Phys. A* **275**, 339 (1975).
150. W. Berdowski, R. Seiwert, and L. Krause: *Proceedings of Colloque Ampère XIV*, Amsterdam, North-Holland, Amsterdam (1967); *Phys. Rev.* **165**, 158 (1968).
151. J. C. Gay, and W. B. Schneider, *Z. Phys. A* **278**, 211 (1976).
152. E. E. Nikitin, *At. Phys.* **4**, 529 (1975).
153. P. Hannaford and R. M. Lowe, *J. Phys. B Phys.* **9**, L225 and 2595 (1976).
154. W. Hanle, *Z. Phys.* **30**, 93 (1924).
155. L. Saintout, B. Chéron, and D. Leder, *Opt. Commun.* **8**, 63 (1973).
156. E. K. Kraulinya, A. E. Lezdin, and O. S. Sametis, *Opt. Spectrosc.* **25**, 523 (1968).
157. E. K. Kraulinya, O. S. Sametis, and A. P. Bryukhovetokii, *Opt. Spectrosc.* **29**, 227 (1970).

30

X-Ray Spectroscopy

K.-H. Schartner

1. Introduction

This chapter is concerned with the progress of x-ray spectroscopy of ions with inner-shell vacancies. Although this book deals mainly with the spectroscopy of atoms and ions excited in their outer shells, the editors have allotted space for two limited chapters (5 and 30) for treatment of atoms that are ionized in inner shells, since these systems have become of increasing interest in the last decade. This interest is mainly due to the production of highly ionized systems in atomic collision processes, i.e., in heavy-ion collisions and in high-temperature plasmas.

The progress of x-ray spectroscopy as an analytical tool for studies of the electronic level structure in solids and in connection with problems of chemical analysis is not within the scope of this chapter; the study of emission spectra following the absorption of x-ray radiation or electron impact will also not be discussed. Some excellent reviews and monographs on this subject are drawn to the attention of the reader, the most recent one being that of Azaroff.[1] The topic of x-ray spectroscopy in connection with atomic collisions is treated in recent reviews and books where detailed information about the knowledge and understanding of the process of the collision in connection with changes of the electronic structure of the colliding atomic systems is given (Richard[2]). Below we try to show by examples the application of x-ray spectroscopy in solving these questions. It has been written in accordance with Chapter 5, which provides the reader with the theoretical background for the discussion of the experimental results. It contains two main parts. In the first part the experimental questions connected with the production and observation of highly ionized

K.-H. Schartner • Physikalisches Institut, Justus-Liebig-Universität, D63 Giessen, Germany.

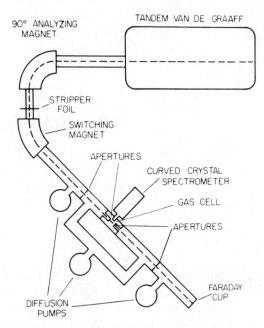

Figure 1. Schematic diagram of an experimental apparatus used for ion impact-induced x-ray emission (Kauffman *et al.*[7]).

atoms are outlined, while the second part gives experimental results and discusses them in connection with Chapter 5. The second part forms the major part of this article. The experimental results are only examples, as are the references.

2. Experimental Techniques

2.1. Sources for Ions with Inner-Shell Vacancies

2.1.1. Accelerators

In experiments using electrons (primary excitation) or photons (secondary excitation) as ionizing particles, the emission spectra are dominated by lines resulting from the decay of a single inner-shell hole (Graeffé *et al.*[3]). The relative amount of multiple ionization increases strongly when one uses protons and heavier highly charged ions. Depending on the values of Z_1/Z_2 and v/u_K, where Z_1 and Z_2 are the nuclear charge of projectile and target and v and u_K are the impact velocity and the velocity of the orbiting electron ionized in the collision, respectively (see Figure 4 of Chapter 5 of Part A), the ion accelerator equipment used in the experiments discussed here covers a wide range of energies corresponding to values for v/u_K between 10^{-3} and 1. There are generators producing voltages up to some hundred keV,[4,5] medium- and high-energy Van de Graaff[6] and tandem accelerators,[7,8] high-energy accelerators like the UNILAC,[9] the SUPER HILAC,[10] and the heavy-ion cyclotron of the JINR.[11]

Figure 1 gives as an example the principal setup of a gas collision experiment.[7] A mass-analyzed beam of singly or multiply charged ions traverses a gas collision chamber and is collected in a Faraday cup. A high-speed pumping system has to be used as the target pressures, which range from 10^{-2} to 10 Torr approximately and are relatively high in comparison to the base pressure of the beam line of about 10^{-6} Torr. Instead of a gas target, target foils or thick targets are also used. From the viewpoint of this book, which treats free atomic systems, the gas targets are of more interest. A curved crystal spectrometer is indicated in Figure 1 as the detection system for the x-ray radiation. Count rates per accumulated projectile charge are registrated as a function of the spectrometer setting, which is usually controlled by a step motor.

2.1.2. Plasma Devices

Multiple ionizing processes take place in high-temperature plasmas where one also observes the emission from highly ionized atomic systems which are produced in successive electron collisions.

X-ray spectrometers have been applied in connection with solar flares,[12] laser-generated plasmas,[13] spark discharges,[14] and high-temperature plasmas applied for CTN.[15] The intensity ratios of different lines can be used to determine the electron temperature and electron density of the observed plasma.[16]

2.2. X-Ray Detectors

2.2.1. Low-Resolution Detectors

Detector systems are used which incorporate an energy-dispersive element, the signal of which delivers the energy information. To this group of detectors belong the proportional counter and the solid-state detector [which is the most commonly used: lithium-drifted silicon detector Si(Li), and lithium-drifted germanium detector Ge(Li)]. The Si(Li) detector is preferentially used for the observation of x-ray radiation between about 0.5 and 30 keV, the Ge(Li) for the higher energies (because of the higher mass of Ge). The resolution of a Si(Li) detector at 6 keV is of the order of 150 eV. Experimental problems connected with the use of a solid-state detector produced by the low level of the signals or the window in front of the crystal are not negligible but are also not too severe and will not be discussed here.

2.2.2. High-Resolution Detectors

To this group of detector systems for x-ray radiation belong the crystal spectrometer, the photoelectron spectrometer, and the Doppler-tuned spectrometer.

Crystal spectrometers make use of Braggs' law for the diffraction of x-rays:

$$n \cdot \lambda = 2d \sin \theta \tag{1}$$

with λ denoting the wavelength of the radiation which is diffracted in the nth order under the angle θ with respect to the surface of the crystal having a lattice constant d, which depends on the selected crystal planes.

Details about plane and curved crystal spectrometers can be found in Refs. 1 and 17 and the references therein. In this short description only two points shall be mentioned: From (1) the resolving power S

$$S = \frac{\lambda}{\Delta \lambda} = \frac{E}{\Delta E} = \frac{\tan \theta}{\Delta \theta} \tag{2}$$

can be deduced; $\Delta \theta$ is determined by the half-width of the total instrumental function and is composed mainly of the half-width of the crystal rocking curve and of instrumental broadening effects such as the angular divergence of Soller slits in plane crystal spectrometers or a mismatching between the curvature of the crystal and the diameter of the focal circle in curved crystal spectrometers.

Approximate numerical values for S to first order range from 500 to 5000 depending upon the quality of the crystals used and the method of mounting. The spectra which are discussed in the following section are in most cases measured with commercial x-ray scanners such as those used in microprobe analyzers. Their resolution to first order is ≤ 1000 at observed energies around 2 keV, but decreases with increasing energy.

The second number of interest is the integrated reflectivity or reflective power P defined by

$$P = \int R(\theta) \, d\theta \tag{3}$$

with $R(\theta)$ being the reflection coefficient of the crystal measured for monochromatic radiation under variation of the angle θ around the Bragg angle (rocking curve). As $R(\theta)$ is a rather narrow distribution, with half-

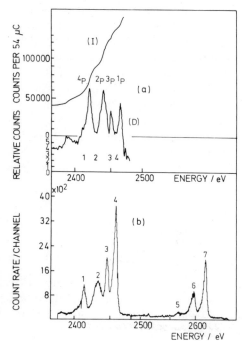

Figure 2. (a) Integrated (*I*) and differentiated (*D*) x-ray spectrum, emitted from S ions of 49.55 MeV traversing a thin foil and registered with a Doppler-tuned spectrometer (Cocke *et al.*[19]). (b) X-ray spectrum emitted from S ions of 92 MeV traversing a carbon foil (150 μg/cm^2) and registered with a plane crystal spectrometer (Bell *et al.*[20]). For classification of lines see Table 1.

width in the order of 1 min and peak values of about 0.8, P is about 10^{-4}, depending also on the mosaic structure, i.e., the microstructure of the crystal. The resolving power and integrated reflectivity have to be matched according to the experimental demands and conditions.

The photoelectron spectrometer (PAX, photoelectrons for analysis of x-rays) is based on the relation

$$E_{kin} = h\nu - E_{ion} \qquad (4)$$

determining the kinetic energy E_{kin} of energy analyzed electrons which result in photoionization processes between the x-ray radiation $h\nu$ to be dispersed and atoms of a converter gas which have the ionization energy E_{ion}.

He, Ne, or Ar are most commonly used as converter gases. Details about PAX can be found in Ref. 18. Photoelectron spectrometers have so far mostly been used in studies of the ionization of inner shells by electrons, although their efficiency and their resolution is comparable to crystal spectrometers and they have the advantage of being able to scan a large energy range (20 eV to about 20 keV), which otherwise needs the application of crystal spectrometers with different crystals and grating spectrometers.

Finally, the *Doppler-tuned* spectrometer is mentioned as applicable to the spectroscopy of beam-foil-excited fast-moving ions.[19] The x-rays emitted by the fast ion are observed at a known angle with respect to the beam through a filter that can be rotated and which has a well-defined absorption edge in the energy range of the observed radiation. By changing the observation angle an integral intensity curve is obtained which can be differentiated to give the peak structure of the spectrum. Figure 2(a) shows the integrated and differentiated spectrum emitted by 49.55-MeV sulfur ions traversing a thin foil (reproduced from Cocke *et al.*[19]). For comparison, a spectrum emitted by 92-MeV sulfur ions traversing a carbon foil of 150-μg/cm^2 thickness, measured with a plane crystal spectrometer, is shown in Figure 2(b).[20]

3. X-Ray Spectra

3.1. Satellite Spectra Emitted in Heavy-Atom Collisions

The spectrum shown in Figure 2(a) can serve as an introduction for the presentation and discussion of the following spectra. It is emitted by sulfur ions having only two or three electrons, one in the K shell and one or two in the L shell, i.e., most of the L-shell electrons have been ionized in collision. This is contrary to the well-known emission spectra obtained by electron impact where the $K_{\alpha 1,2}$ lines dominate. Initial and final states for the lines shown in Figure 2(b) are given in Table 1. At higher impact velocities, like those used to obtain the spectrum shown in Figure 2(b), the

Table 1. Comparison of Experimental and Calculated Transition Energies for the Lines Shown in Figure 2(b) (Panke[20]) (MCF Multiconfiguration Interaction)

					Theory	
Line number	Initial state		Final state	Experiment	*LS* Coupling	*jj* Coupling +MCF
1	$(1s2p^2)\,^4P$	→	$(1s^22p)\,^2P$	2416.6	2419.1	2416.6
	$(1s2s2p)\,^4P$	→	$(1s^22s)\,^2S$			2415.7
2	$(1s2p^2)\,^2D$	→	$(1s^22p)\,^2P$	2434.9	2433.8	2431.3
	$(1s2p^2)\,^2P$	→	$(1s^22p)\,^2P$	2438.9	2437.4	2435.1
	$(1s2s2p)^2P$	→	$(1s^22s)\,^2S$			2436.7
3	$(1s2p)\,^3P$	→	$(1s^2)\,^1S$	2447.8	2447.7	2447.3
4	$(1s2p)\,^1P$	→	$(1s^2)\,^1S$	2460.8	2460.7	2460.2
5	$(2p^3)$	→	$(1s2p^2)$	2573.5	2573.6	
6	$(2p^2)\,^3P$	→	$(1s2p)\,^3P$	2596.5	2.596.7	
7	$(2p)\,^2P$	→	$(1s)\,^2S$	2621.6		2621.6

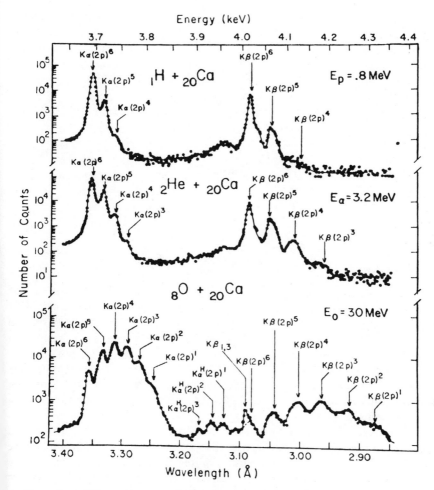

Figure 3. Satellite and hypersatellite lines emitted from solid Ca, bombarded with H, He, and O ions of given energy (McWherter *et al.*[21]).

heliumlike emission dominates and even x-rays from hydrogenlike configurations of S^{15+} are observed to a higher degree. The identification of the lines is achieved via calibration of the energy scale against diagram lines excited by electron or proton impact and comparison with calculated energy values of the respective transitions. The calculations are based on Hartree–Fock methods as discussed in Chapter 5. Theoretical values for the example of sulfur given here are given in Table 1.[20] For a discussion of the theoretical energy calculations, see Chapter 5.

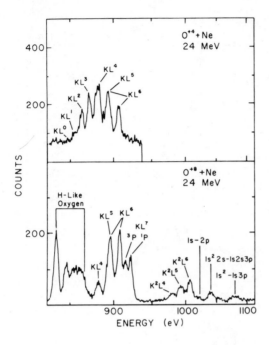

Figure 4. Spectra produced by 24-MeV O^{4+} (top) and O^{8+} (bottom) ions on Ne. The low-energy structure in the bottom spectrum results from the oxygen beam (Kauffman et al.[7]).

   The lines in Figures 2(a) and 2(b) are called *satellite* lines to the *diagram* lines, the K_{α_1} and K_{α_2} lines. *Hypersatellites* are transitions where two K-shell electrons are ionized in the initial state. These are the lines 5, 6, and 7 in Figure 2(b).

   The high degree of ionization seen in Figures 2(a) and 2(b) is not limited to the projectile but is also observed under single-collision conditions in the target. As examples, Figures 3 and 4 show satellite and hypersatellite spectra emitted from solid Ca bombarded with H, He, and O ions (McWherter[21]) and from Ne gas under oxygen bombardment (Kauffman et al.[7]), the oxygen being a bare nucleus or four times ionized. One observes a dramatic increase of the relative contribution of the satellite peaks with increasing mass and charge state of the projectile. Identified also are the hypersatellites and the K_β satellites. The peaks in Figures 3 and 4 are denoted in a different way. Figure 3 denotes a configuration $1s2s^22p^4$ by $K_\alpha(2p)^4$; Figure 4 denotes the same configuration by KL^2. KL^6 of Figure 4 includes $1s2s2p$ and $1s2p^2$ and consequently demands two denotations in Figure 3, namely, $K_\alpha(2s2p)$ and $K_\alpha(2p^2)$, the last one being already used for $1s2s^22p^2$. The information that one would like to obtain from spectra as shown in Figures 2(a), 2(b), and 3 is at least the relative magnitudes of the cross sections σ_{KL^n} for the production of one K and nL vacancies. The intensity I_{KL^n} of one satellite peak KL^n, which leads to the cross section σ_{KL^n}, is proportional to

$$N_{KL^n} = \sum_{\alpha^n} N(\alpha^n) \sum_i g(\alpha^n, J_i)\omega(\alpha^n, J_i) \qquad (5)$$

where $N(\alpha^n)$ denotes the initial population of the configurations α^n which are contained in the unresolved peak KL^n. $g(\alpha^n, J)$ is the relative initial population of the J states. For a statistical population we have

$$g(\alpha^n, J_i) = (2J_i + 1)/\sum_i (2J_i + 1) \qquad (6)$$

where the sum has to be taken over all J_i states within the configuration α^n.

$$\omega(\alpha^n, J_i) = \Gamma_x/\sum_i \Gamma_i \qquad (7)$$

is the fluorescence yield of a state J_i in the configuration α^n. Γ_x denotes x-ray decay rate and $\sum_i \Gamma_i$ the sum of the decay modes of the state J_i which contains the Auger decay rate Γ_A. The elimination of σ_{KL^n} using (5)–(7) contains many assumptions which have so far not been verified experimentally:

(i)   The probability of creating a $2s$ vacancy relative to the one for a $2p$ vacancy has to be assumed.

(ii)   There is no rearrangement in the L shell within the lifetime of the K shell.

(iii)   The assumption of the correct fluorescence yield is aggravated by the fact that $\omega(\alpha^n, J_i)$ is dependent, not only on J_i but, as shown theoretically for the example of LS states $(1s2s^22p^5)^1P_1$, $^3P_1$ of Ne, also strongly on the multiplicity (Bhalla[22]).

(iv)   There is more evidence for a nonstatistical population of the J_i states than for a statistical population,[23] so that (6) probably does not hold.

From the results presented in Figure 4 Kauffmann et al.[7] have derived values for σ_{KL^n}/σ_K, with $\sigma_K = \sum_n \sigma_{KL^n}$ (Figure 5). These relative ionization cross sections are compared with computed values as calculated via application of the binary encounter approximation (BEA) (see Fricke, Chapter 5). To use the concept of the BEA, the ratio of the impact velocity v over the average orbital velocity u, v/u, should be > 1, a condition which is fulfilled for electrons in the L shell. For the calculations used in Figure 5 a knowledge of σ_K is not necessary. Molecular orbital (MO) calculations have so far not been carried out for multiple ionization. For such calculations to hold, the reverse condition, $v/u < 1$, has to be fulfilled. The reproduction of the numbers σ_{KL^n}/σ_K, as shown in Figure 5, is described as reasonable by the authors of Ref. 7.

The sketch of the data analysis as given above is meant to give an impression of the difficulties which make the interpretation of experimental results, as shown in Figures 3 and 4, still rather difficult. This becomes even more evident from the discussion of Figure 6, which shows satellite spectra registered from SiH_4 bombarded with Cl^{7+} and Cl^{12+} in comparison with a satellite spectrum from solid Si ($20\mu g/cm^2$ on $20\ \mu g/cm^2$ carbon backing) (Kauffmann et al.[24]). The energy of the Cl ions was always 45 MeV. The difference between (a) and (b) of Figure 6 is due to the increased charge state of the projectile. The difference between (a), (b), and (c) means the shift of the emission towards configurations having only few L-shell vacancies is at the moment believed to be caused by relaxation processes which fill up vacancies in the L shell prior to the decay of the K hole. The plasmon lifetime for Si, $\tau_p \approx 1.6 \times 10^{-16}$ s (Philipp and Ehren-

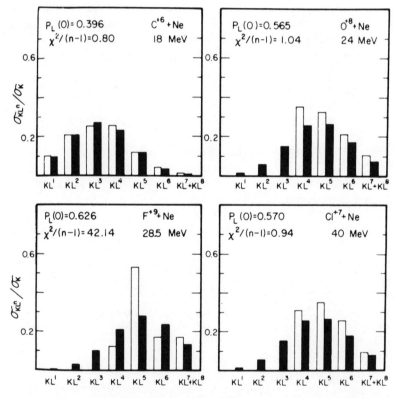

Figure 5. Relative multiple ionization cross sections and their fits to a binomial distribution. The measured cross sections are given by the light bars; the fits are given by the dark bars (Kauffman et al.[7]).

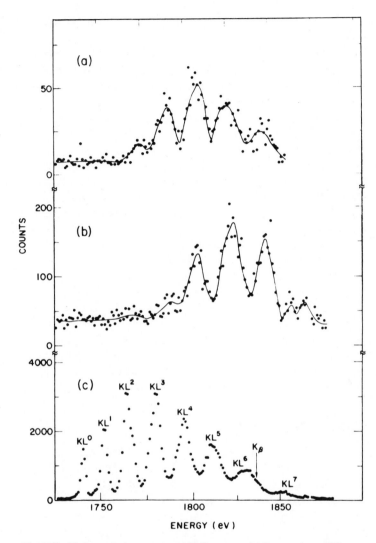

Figure 6. High-resolution spectra of Si K_α x-rays. (a) Spectra from SiH$_4$ gas
produced by Cl^{7+} ions of 45 MeV and (b) by Cl^{12+} ions of 45 MeV; (c)
x-ray spectrum emitted from a thin sample of solid Si bombarded with
Cl ions of 45 MeV (Kauffman *et al.*[24]).

reich[25]), which is taken as a measure for the time scale of the relaxation
process, is almost one order of magnitude smaller than the lifetime of a
K-shell, $\tau_K \approx 1.5 \times 10^{-15}$ s (Walters and Bhalla[26]). In the gaseous phase
the relaxation processes are not possible since the M shell is assumed to be
stripped completely. More work like that presented by Kauffmann *et al.*[24]

is necessary to study the differences in spectra emitted from solid and gaseous material.

This section, concerning collision-induced satellite spectra, will be concluded with a short remark concerning the determination of ionization cross sections. In principle, ionization cross sections σ_i can be deduced from x-ray measurements via the relation

$$\sigma_i = \sigma_x/\omega \qquad (8)$$

where σ_x is the cross section for the production of x-rays. Then the ionization cross section σ_i can be determined provided the fluorescence yield ω is known. As has been shown above, the heavy ions produce a large amount of multiple ionization. The fluorescence yield ω depends upon the different states which are present in one satellite peak KL^n. The deduction of σ_i via (8) demands the assumption of a mean fluorescence yield and consequently leads to a mean value of σ_i. So in light elements, where the Auger effect is the dominating mechanism for filling up the inner-shell vacancies, measurements of the Auger electrons are preferable for the determination of σ_i (see, for example, Woods et al.[27]). Details concerning the determination of ionization cross sections can be found from the contributions of Madison and Merzbacher[28] and of Richard[2] or in the discussion of scaling procedures by Forster et al.[29] In the last paper also the problem of the magnitude of σ_i for the production of K vacancies in U–U collisions is discussed and is mentioned by Fricke in Chapter 5.

3.2. Lifetime Measurements

The production of highly stripped projectile ions with inner-shell vacancies enables one to carry out lifetime measurements. For a discussion of the well-known beam-foil spectroscopy see Andrä (Chapter 20). For rather short lifetimes (of the order of 10^{-14} s) a new method, using the measurement of signal intensity against foil thickness, has been applied by Betz et al.[30] and Varghese et al.[31] As the beginning of a short outline of this method Figure 7 is presented corresponding to Figure 2(b) and showing, for the lines 1–4, the x-ray yield as a function of the foil thickness or the transit time of 92-MeV sulfur ions.[30] One observes a nonproportional dependence of x-ray yield upon target thickness.

The model which explains the behavior shown in Figure 7 is based on the assumption that the observed intensity can be composed from two parts, one emitted while the projectiles are inside the target and one while the excited states decay after passage of the target: $N_i(t)$ denotes the number of ions in the state i, $t = 0$ marking the time of entrance into the foil. There will be a buildup of $N_i(t)$ leading to an equilibrium distribution

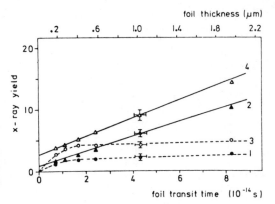

Figure 7. X-ray yield for the He- and Li-like transitions of sulfur shown in Figure 2 and given in Tables 1 and 2 as a function of the target thickness or the foil transit time (Bell et al.,[20] Betz et al.[30]).

N_i^0, so that $N_i(t)$ may be expressed inside the foil as

$$N_i(t) = N_i^0[1 - f_i(t)] \tag{9}$$

where $f_i(t)$ denotes the buildup function. Outside the foil, the natural decay via Auger processes and x-ray emission leads to the exponential decay

$$N_i(t) = N_i(t_0) e^{-t/\tau_i} \qquad t > t_0 \tag{10}$$

with t_0 marking the transit time. The observed intensity $y^i = y^i_{\text{inside}} + y^i_{\text{outside}}$ depends on t_0 and is given by

$$y^i(t_0) = \int_0^{t_0} N_i^0 \Gamma_x^i [1 - f_i(t)] \, dt + \int_0^{\infty} N_i^0 \Gamma_x^i e^{-t/\tau_i} \, dt \tag{11}.$$

with Γ_i and Γ_x^i denoting the transition rates for all decay modes and radiative decay, respectively. Equation (11) leads to

$$y^i(t_0) = N_i^0 \Gamma_x^i(t_0 + \tau_i - \tau) \tag{12}$$

with a buildup time

$$\tau = \int_0^{\infty} f_i(t) \, dt$$

For $\tau < \tau_i$ (because of the considerable chance of electron capture), Eq. (12) can be simplified to

$$y^i(t_0) = N_i^0 \Gamma_x^i(t_0 + \tau_i) \tag{13}$$

For $t_0 \approx \tau_i$, one expects a linear dependence of y^i upon t_0, as found for the lines 2 and 4 of Figure 2(b). For $t_0 \ll \tau_i$, y^i should be independent of t_0, as found for lines 1 and 3 of Figure 2(b). For the first case, τ_i can be determined as the time t_0 for which $y^i(t_0) = 2y^i(t = 0)$. Table 2 gives a

Table 2. Comparison between Experimental and Theoretical Lifetimes $\tau_i/10^{-14}$ s

Line number	Transition	Experimental	Theoretical
1	$(1s2p^2)\,^4P-(1s^22p)\,^2P$	$>30^{(30)}$	$>100^{(33)}$
2	$(1s2p^2)\,^2D-(1s^22p\,^2P$	$0.9\pm0.15,^{(30)}\,1.06\pm0.2^{(31)}$	$0.59^{(31)}$
	$(1s2p^2)\,^2P-(s^22p)\,^2P$		$1.14^{(31)}$
3	$(1s2p)\,^3P-(1s^2)\,^1S$	$>70,^{(30)}\,1.57\times10^{2(32)}$	$1.6\times10^{2(34)}$ for $^3P_1-^1S_0$
4	$(1s2p)\,^1P_1-(1s^2)\,^1S$	$1.7\pm0.3,^{(30)}\,1.25\pm0.3^{(31)}$	$1.49^{(35)}$
7	$(2p)\,^2P-(1s)\,^2S$	$2.3\pm0.4^{(30)}$	$2.44^{(36)}$

comparison of lifetimes determined as described above and theoretical values.$^{(30,31)}$ The lifetime of the hydrogenlike transition, line 7 of Figure 2(b) is known theoretically with good accuracy and can be used for a comparison between experimental and theoretical values. The experimental errors are mostly caused by the uncertainties in the foil thickness.

To obtain more detailed results for the lithiumlike transitions a better spectral resolution is necessary.

3.3. Satellite Spectra from Plasmas

One motivation for the investigation of collisional-produced x-ray spectra as described in Section 3.1 is the comparison of these spectra with spectra emitted from plasmas, such as solar plasmas, laser-induced plasmas, spark plasmas, and fusion plasmas. The temperatures and electron and ion densities in these plasmas are sufficiently high to produce K x-ray radiation from even few-electron systems.

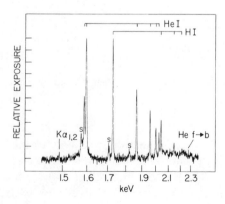

Figure 8. Al K x-ray spectrum from a plasma produced by a 1.8-J, 0.25-ns laser pulse. Rydberg series from Al ions isoelectronic with H and He atoms, and the He-like free-to-bound continuum and satellite lines S constitute the plasma spectrum. The $K\alpha_{1,2}$ lines are from target fluorescence (Nagel *et al.*$^{(13)}$).

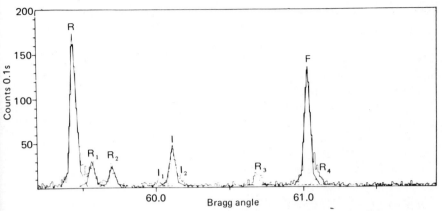

Figure 9. Satellite lines in the solar x-ray spectrum of He-like Mg. R: ($^1P_1-^1S_0$) resonance transition; I: ($^3P_{2,1}-^1S_0$) transitions; F:($^3S_1-^1S_0$) transition; all other lines are Li-like transitions (Parkinson[14]).

This is demonstrated in Figures 8 and 9, which show examples of spectra observed from a laser-induced Al plasma (Nagel *et al.*[13]) and from the solar plasma of Mg (Parkinson[12]). The main interest in these spectra is concentrated on the few-electron systems. Emission from the first excited levels of the heliumlike ions corresponds to the level scheme shown in Figure 10. The $^1P_1$ level decays via an E1 transition to the $^1S_0$ ground state. The respective lines are the resonance line, denoted by R in Figure 9, and the strongest of the HeI lines in Figure 8. The $^3P_1$ level decays via E1 radiation and the $^3P_2$ level decays via M2 radiation to the ground state. The

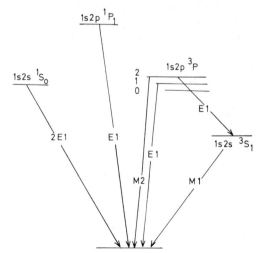

Figure 10. Principal level scheme and decay modes of a two-electron system.

line I in Figure 9 denotes the position of these two unresolved lines; in Figure 8 it is the line at the left side of the above-mentioned HeI line. The $^3S_1$ level decays via relativistic M1 radiation to the ground state. This level can be populated via E1 transitions from the $^3P_{2,1,0}$ levels. The line corresponding to the decay from the $^3S_1$ level is not visible in Figure 8; it is denoted by F in Figure 9.

Different combinations of the described transitions can be used to estimate the density of the plasma which produces the x-ray spectra via the measurement of intensity ratios of the mentioned transitions. The observation of the $^3S_1-^1S_0$ transition is an indication of a low-density plasma. At increasing electron densities secondary processes such as quenching by electron impact and photoexcitation from the $^3S_1$ level will increase the intensity of the transitions from the ^3P levels. At intermediate and high densities the $^1S_0$ level and the ^3P levels become quenched by electron collisions, thereby increasing the intensity of the resonance transition with respect to the intercombination lines from the $^3P_{2,1}$ levels. A strong resonance line therefore indicates a high-density plasma. These qualitative arguments allow the solar plasma producing the spectrum of Figure 9 to be estimated as being of low density ($10^{10}/cm^3$) and the laser-induced plasma of Figure 8 as having higher densities. A detailed discussion of a more quantitative nature, which also includes the lithiumlike satellites (denoted by S in Figure 8) can be found in the work of Gabriel and Jordan.[16]

3.4. Two-Electron–One-Photon Transitions, Radiative Electron Rearrangement, Radiative Auger Effect

So far the two main decay modes for the filling of inner-shell vacancies have been mentioned, i.e., the emission of an x-ray photon or the Auger effect [one Auger electron is most probably emitted, but the emission of two or three electrons has also been observed (Carlson and Krause,[37] Afrosimov et al.[38])]. Weaker decay modes have recently been observed, such as the two-electron–one-photon emission which is observed in the decay of double-K-shell vacancies in competition with the one-electron–one-photon decay and the Auger effect. The observation of the respective x-rays, having an energy slightly larger than twice the one-electron–one-photon energy, was first reported by Wölfli et al.[39] for the decay of double-K-shell vacancies of Fe and Ni. The vacancies were produced by bombarding solid targets with Fe and Ni ions of 40-MeV energy. A theoretical treatment of Wölfli's experiments has been given by Nussbaumer,[40] who showed that high-resolution studies are necessary for a comparison of experimental and theoretical data. Wölfli et al. measured the ratio of the two-electron–one-photon decay to the two-electron–two-photon decay.

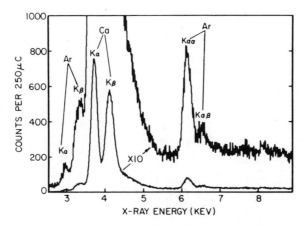

Figure 11. X-ray spectrum obtained with 3.5-MeV Ar$^+$ ions incident on a Ca target. The same spectrum is also shown with the vertical gain increased by a factor of 10 to show the $K_{\alpha\alpha}$ and $K_{\alpha\beta}$ lines more clearly (Knudson *et al.*[42]).

They obtained values of 8×10^{-5} for Ni and 3×10^{-4} for Fe, numbers which are reproduced by the calculations of Nussbaumer.[40] Such calculations are concerned with the order of magnitude but not with the variations from Ni to Fe. The calculation of the magnitude of the cross sections for the production of the double-K-shell vacancy is a different question; this requires the knowledge of the fluorescence yield, which has not yet been determined. The two-electron–one-photon decay mode was also studied by Schuch *et al.*[41] and by Knudson *et al.*[42] for lighter elements using solid targets and by Hoogkamer *et al.*[43] in gas collisions for N, O, and Ne (in the last case rather low energies were used for which the MO model holds). Figure 11 shows the result obtained by Knudson *et al.*[42] for 3.5-MeV Ar ions bombarding Ca. $K_{\alpha\alpha}$ and $K_{\alpha\beta}$ denote the two-electron–one-photon transitions which are discussed here, the two electrons originating from the L and the L and M shell, respectively. Figure 12 gives a comparison between measured and calculated (Hartree–Fock) differences of $K_{\alpha\alpha}$ and $2K_{\alpha}$ energies as a function of the atomic number.[42] Calculations have been done for the electric quadrupole transition $(1s)^{-2} \rightarrow (2p)^{-2}$ and the electric dipole transition $(1s)^{-2} \rightarrow (2s)^{-1}(2p)^{-1}$, a full L shell ($n = 6$) and an L shell with two vacancies ($n = 4$) having been assumed [$(2p)^{-2}$ means here that two $2p$ electrons are absent; this notation is identical to $2p^4$]. Although Figure 12 indicates that the two K-shell vacancies are accompanied by two L-shell vacancies, the interpretation of Figure 12 is hampered by the influence of multiplet splitting and population of the different levels as discussed by Knudson *et al.*[42]

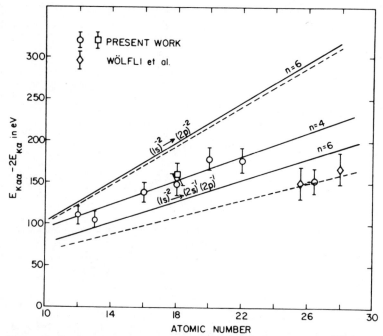

Figure 12. Predicted and measured $K_{\alpha\alpha}$ energy shifts (Knudson *et al.*[45]). The solid lines represent $K_{\alpha\alpha}$ energy shifts predicted by Hartree–Fock calculations. The top solid line assumes an E2 $K_{\alpha\alpha}$ transition, while the lower pair of solid lines assumes an E1 transition with a full L shell ($n = 6$) and with two $2p$ vacancies ($n = 4$) in the initial state. The dashed lines are the shifts for $n = 6$ with multiplet splitting taken into account. The circles represent data obtained with 3.0–3.5 MeV projectiles; the square was obtained with a 7.0-MeV projectile, and the diamonds are the measurements of Wölfli *et al.*[39] at 40 MeV.

Figure 13 presents the result of the gas collision experiment by Hoogkamer *et al.*[43] The x-ray intensity, measured with a Si (Li) detector, is plotted as a function of the energy for the assigned collision processes. In all three cases there is, in addition to the characteristic K transitions, denoted by K_α, a much weaker transition at about twice the K_α energy; this is denoted by $K_{\alpha\alpha}$. Cross sections for the production of the K_α and $K_{\alpha\alpha}$ emission are given in Table 3.

There are more weak-intensity features visible in highly resolved x-ray spectra on the long-wavelength side of the satellite lines. They are considered to be caused by the radiative Auger effect (RAE) and the radiative electron rearrangement (RER). Figures 14(a) and 14(b) show these x-ray lines for the case of the ionization of the K shell of Al by the impact of 1.5-MeV protons (Richard *et al.*[44]) and of 30-MeV oxygen ions (Richard

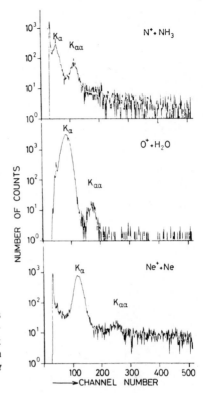

Figure 13. X-ray spectra emitted during collisions of 200-keV $N^+ + NH_3$, $O^+ + H_2O$, and $Ne^+ + Ne$. Next to the K line in each case, the peak attributed to the two-electron–one-photon transition is indicated by $K_{\alpha\alpha}$ (Hoogkamer et al.[43]).

et al.[45]). Figure 14(a) shows the KL^0 and KL^1 satellite peaks and, on their low-energy shoulder, a broad feature, denoted in Figure 14(a) by $KL_{23}L_{23}$, which is caused by the RAE. (a K-shell vacancy is filled from the L_{23} shell and an L_{23} electron is simultaneously ionized to the continuum). The edge of the $KL_{23}L_{23}$ energy is 93 eV below the KL^0 peak. Since the RAE is a fundamental decay mode of an atom, it should also be produced by primary (photon) or secondary (electron-impact) ionization. This is the

Table 3. Experimental Cross Sections for K_α Emission and for $K_{\alpha\alpha}$ Emission in 200-keV Collisions

System	$^\sigma K_\alpha$ (cm^2)	$^\sigma K_{\alpha\alpha}$ (cm^2)
$N^+ + NH_3$	5.8×10^{-21}	1.4×10^{-25}
$O^+ + H_2O$	3.0×10^{-21}	0.8×10^{-25}
$Ne^+ + Ne$	2.2×10^{-21}	5.3×10^{-27}

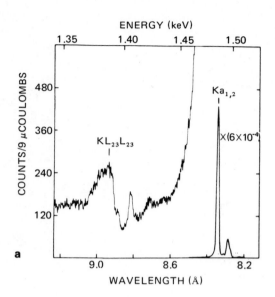

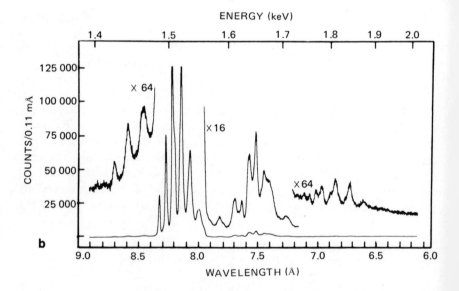

Figure 14. (a) Al K x-ray spectrum following 1.5-MeV H$^+$ bombardment. The calculated energy of the K–$L_{23} L_{23}(^1D_2)$ Auger transition is shown near the RAE edge (Richard *et al.*[44]). (b) Al K x-ray spectrum emitted by Al bombardment with oxygen ions of 30 MeV. The weak low-energy peaks are quoted as being caused by RER processes (Richard *et al.*[45]).

case, as can be seen by a comparison of spectra measured by Siivola et al.[46]

The origin of the second peak in Figure 7(a), at an energy of 1406 eV, and of the three peaks at the long-wavelength side of Figure 14(b), is not fully understood (Jamison et al.[47]). Simultaneous emission of an x-ray quantum and excitation of two $2s$ electrons to the $2p$ shell, called RER, is proposed to be the mechanism responsible for these peaks. The configurations are denoted by $1s2s^22p^n$ for the initial, and $1s^22s^02p^{n+1}$, $n = 1, 2, 3$, for the final state. The RER model is supported by HF calculations and by the experimental observation that the peak sequence seems to mirror the satellite structure of the x-ray spectrum.[47]

3.5. Continuous X-Ray Spectra

Electron-induced x-ray spectra show the characteristic lines and the frequently dominating electron bremsstrahlung continuum. In the course of the investigations of heavy-ion impact-induced x-ray spectra, x-ray continua of remarkable intensity have also been observed and have gained much interest. These continua can be partly discussed in terms of electron bremsstrahlung, which is emitted when almost free or bound target electrons are accelerated via capture processes by the projectile or when fast electrons produced by ionization are decelerated by the target nuclei. These continua are of different origin and yield fundamental knowledge about the collision mechanism; this will be discussed in what follows.

3.5.1. Quasimolecular X-Rays

The ionized atoms have, until now, been regarded as isolated while their vacancies are filled from higher shells. In solid material, and also in molecular gases, there is a considerable probability for the vacancy to be filled while the impinging ion or a recoiling target ion is colliding with a second target atom. The collision times are of the order of 10^{-16} s, which is less than the lifetimes of inner-shell vacancies, which are of the order of 10^{-15} s, but still sufficient for decay modes to be observable, thereby yielding information about the energy structure of the electronic states of the colliding particles via a quasimolecular model (MO model).

The first experimental results for these radiative transitions between transient molecular orbitals were reported by Saris et al.,[48] who observed the emission of an x-ray band around 1 keV when solid Si was bombarded with Ar projectiles of energies between 70 and 600 keV. This experiment initiated a large number of experiments also with heavier particles like I ions of 60 MeV bombarding Au, Th, and U targets (Mokler et al.[49]) for the study of the so-called molecular x-rays.

   As this chapter is intended to guide the reader who is not yet involved with problems of the production of inner-shell vacancies, a more qualitative picture may be given for a short introduction to quasimolecular x-rays. Figure 15 can be deduced from Figure 5 of Chapter 5. The collision time serves as an abscissa representing the incoming system for $t < 0$ and the outgoing system for $t > 0$ instead of the internuclear distance r. Provided that a K-shell vacancy is present in the incoming system, this vacancy can be filled from higher orbitals while the collision system is in the quasimolecular state. In this case a band of x-ray energies will be emitted, as indicated in Figure 15 (MO). On the other hand, if the vacancy survives the collision, a characteristic (Ch) (satellite) line or Auger electrons will be emitted. For the example of Figure 15, K-shell MO x-rays, the energy of the MO band extends to the energy of the characteristic transition of the combined atom and, because of dynamical effects based on the Heisenberg uncertainty principle, even beyond this limit. MO x-ray phenomena are not limited to the decay of K-shell vacancies. The first observations by Saris *et al.*[48] were shown to be L-shell MO x-rays, whereas Mokler *et al.*[49] reported M-shell MO x-rays. The spectral shape of the MO band depends upon the MO diagram and upon the transition rates for transitions between the chosen vacant and filled orbitals. There are calculations which yield very good agreement between the experimental observations and the calculated values, for example, the calculations by Müller[50] for the Ni–Ni K-shell MO x-rays.

   A further discussion of the MO spectra requires the inclusion of those processes which produce, for example, the vacancy in the $1s\sigma$ orbital.

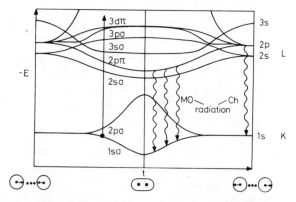

Figure 15. Schematic dependence of molecular orbitals (MO) upon the collision time. The production of a vacancy in the $1s\sigma$ orbital in the incoming collision system, which can decay via emission of MO x-ray or after separation of the system of characteristic (Ch) K_α lines is indicated.

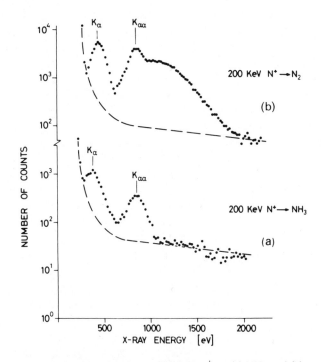

Figure 16. X-ray spectra for 200-keV N^+ on (a) NH_3 and (b) N_2. The $N^+ \to N_2$ spectrum shows, in addition to the K_α and $K_{\alpha\alpha}$ line, a broad continuum extending up to about 2 keV (Hoogkamer *et al.*[51]).

There are the two possibilities: (i) the vacancy which has been produced in the outgoing system survives until a second collision occurs (double-collision process), or (ii) the vacancy has been produced in the incoming system and decays in the same in- or outgoing system (single-collision process).

The examples shown in Figures 16 and 17 are chosen to demonstrate processes (i) and (ii). Figure 16 shows a comparison of the spectra measured by Hoogkamer *et al.*[51] for the emission of N K radiation from NH_3 and N_2 gas bombarded by 200-keV N^+ ions. At this energy the collision occurs under diabatic conditions, so K vacancies are produced via rotational coupling (see Chapter 5) between the $2p\sigma$–$2p\pi$ orbitals in the outgoing system only.

Figure 16(a) therefore shows no remarkable MO radiation; this is contrary to 16(b) obtained for $N^+ + N_2$ collisions. Here a K vacancy is produced in the collision with one of the N atoms of the N_2 molecule and the N^+ projectile as described above; and it then follows the $1s\sigma$ orbital in

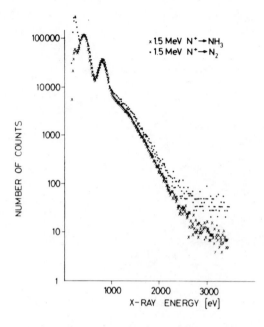

NUMBER OF COUNTS

× 1.5 MeV $N^+ \rightarrow NH_3$
· 1.5 MeV $N^+ \rightarrow N_2$

X-RAY ENERGY [eV]

Figure 17. X-ray spectra for 1.5-MeV N^+ on N_2 and NH_3 (Saris *et al.*[53]).

a second collision between the projectile carrying the K vacancy and the second N atom of the N_2 molecule thereby opening the MO decay channel. A broad continuum of x-rays is observed extending to about 2 keV. 1.8 keV is the energy of the K_α radiation in Si, which is the united atom of the N_2 system. For a comparison of the shape of the measured MO band with respective calculations based on the double-collision mechanism and calculated MO energies, the reader is referred to Saris and Hoogkamer.[52] The result of this comparison strongly supports the double-collision process description. In the review presented by Saris and Hoogkamer[52] heavier systems such as Ni–Ni or Nb–Nb are also discussed. In these solid targets the double-collision process works by successive collisions. The I–Au system has been mentioned in Chapter 5 where the MO spectrum is shown in Figure 6 and is compared with calculations by Fricke *et al.* (see Chapter 5). As an addendum to the discussion of Figure 6 of Chapter 5 it should be noted that the investigation of the anisotropic distributions of the MO radiation mentioned there, which so far are not understood, prove that the MO continua are emitted from a system moving with the center-of-mass velocity of the collision system. This observation strongly supports the MO radiation model.

Figure 17[53] is used for the illustration of the process (ii), the single-collision process. As in Figure 16, the $N^+ + N_2$ and the $N^+ + NH_3$ systems are there compared at higher impact velocities corresponding to 1.5-MeV energy of the N^+ ions. In contrast to Figure 16 there is no severe difference in the shape of the spectra produced in $N^+ + N_2$ and $N^+ + NH_3$ collisions. The double-collision mechanism is proposed so as to add a contribution to the dominating single-collision mechanism; it is thus responsible for the difference between the two spectra of Figure 17.

Increasing the impact velocity even more ($v/u_K \approx 0.3$ in case of Figure 17) leaves the molecular orbitals less defined, and the discussion of continuous spectra in molecular terms has to be replaced by a discussion in atomic terms, thereby introducing processes like radiative electron capture (REC).

3.5.2. Radiative Electron Capture (REC)

When an ion with inner-shell vacancies passes electrons, capture processes are observed. In addition to the known radiationless capture into the projectile ground state, a capture process accompanied by the emission of a photon has also been observed for fast-moving ions. Schnopper et al.[54] observed an x-ray band whose energetic position shifted proportionally to the energy of the fast ions capturing the electron. If it is assumed that the captured electron has been at rest in the laboratory system, then the energy of the emitted photon amounts to

$$E_{h\nu} = E_{\text{ion}} + \tfrac{1}{2}m_e v_i^2 \qquad (14)$$

where E_{ion} is the binding energy of the vacancy which is filled in the projectile and $\mathbf{v}_i$ is the velocity of the projectile in the laboratory system. If the electron has a velocity $\mathbf{v}_e$ in the laboratory system, the energy of the emitted photon becomes

$$E_{h\nu} = E_{\text{ion}} + \tfrac{1}{2}m_e(v_i^2 + v_e^2 + 2\mathbf{v}_i \cdot \mathbf{v}_e) \qquad (15)$$

The REC photon distribution is thus related to the momentum distribution of the target electrons being captured. Figure 18(a) gives an example for a measured REC peak which is obtained when O^{8+} passes through H_2.[55] The term $\tfrac{1}{2}mv_i^2$ of Eq. (14) becomes 2.2 keV and E_{ion} amounts to 0.74 keV if capture occurs into the K shell of oxygen. Figure 18(a) demonstrates the good agreement between the calculated and the measured REC peak value.

It is not necessary for the electron to be captured into a bound state of the projectile; it can also be captured into a continuum state. This process yields a continuum of x-rays extending to the low-energy end point of the REC distribution, which is 2.2 keV in the case of Figure 18(a). Figure 18(b) shows this continuum for protons having a velocity matched to the velocity of the oxygen ions of 65 MeV. The shape and energy position of both continua produced by oxygen or proton impact agree rather well. The Ar K radiation peak results from an impurity and can be used for calibration purposes.

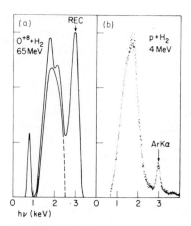

Figure 18. Experimental x-ray spectra obtained when (a) O^{8+} and (b) H^+ ions of the same velocity traverse H_2 gas. The spectral shape, especially at energies below about 1.5 keV, is influenced by window transmission effects. In (a) the REC peak is indicated; the broad curve between 1 keV and 2.5 keV in (a) and (b) is attributed to electron bremsstrahlung, emitted from electrons which are captured into continuum states of the projectile (Schnopper et al.[55]).

At even higher ion velocities and with heavier targets the capture of tightly bound target electrons to projectile continuum states becomes more probable than the REC process. This fact has showed up in investigations of Kienle et al.,[56] who measured x-ray continua extending far beyond the REC energy for Ne ions passing through He and Ne and attributed these continua to the capture process mentioned above, which is called radiative ionization (RI).

4. Concluding Remarks

The contents of the preceding sections have been selected with the intention of giving an introduction to phenomena in the field of x-ray spectroscopy which have shown up in the last decade mostly because of the studies of heavy-ion collision processes. These collision processes led to the formation of atomic configurations of highly stripped atoms which had not been produced in laboratories until then. Since the resolution of the applied spectrometers is at present of the order of 1000, pure spectroscopic investigations had seldom been done. Lifetime measurements of few-electron systems such as He- or Li-like ions where single states can be resolved are discussed above. Investigations of long-living atomic states of the few-electron systems have not been mentioned. These are treated in Chapter 20 of this work by Andrä. Also left unmentioned have been muonic x-rays (see Chapter 31). No details about cross sections, either total or differential, which are derived from the x-ray yield, could be given because of the limited space. Polarization effects, related to a nonstatistical excitation of magnetic substates, have also been left unmentioned, but are gaining more and more interest.

References

1. L. V. Azaroff, *X-Ray Spectroscopy*, McGraw-Hill Book Company, New York (1974).
2. P. Richard, in *Atomic Inner-Shell Processes*, Ed. B. Crasemann, Vol. I, pt. 74, Academic Press, New York (1975).
3. G. Graeffé, J. Siivola, J. Utriainen, M. Linkoaho, and T. Åberg, *Phys. Lett.* **29A**, 464 (1969).
4. F. Saris and D. Onderdelinden, *Physica* **49**, 441 (1970).
5. M. E. Cunningham, R. C. Der, R. J. Fortner, T. M. Kavanagh, J. M. Khan, and C. B. Layne, *Phys. Rev. A* **8**, 2322 (1973).
6. P. G. Burkhalter, A. R. Knudson, D. J. Nagel, and K. L. Dunning, *Phys. Rev. A* **6**, 2093 (1972).
7. R. L. Kauffman, C. W. Woods, K. A. Jamison, and P. Richard, *Phys. Rev. A* **11**, 872 (1975).
8. J. McWherter, D. K. Olsen, H. H. Wolter, and C. F. Moore, *Phys. Rev. A* **10**, 200 (1974).
9. P. Armbruster, in *Invited Papers, Proceedings of the Second International Conference on Inner Shell Ionization Phenomena*, Eds. W. Mehlhorn and R. Brenn, pp. 21, Fakultät für Physik, Freiburg (1976).
10. W. E. Meyerhof, T. K. Taylor, and R. Anholt, *Phys. Rev. A* **12**, 2641 (1975).
11. K. H. Kaun, W. Frank, and P. Manfrass in *Invited Papers, Proceedings of the Second International Conference on Inner Shell Ionization Phenomena*, Eds. W. Mehlhorn and R. Brenn, pp. 68, Fakultät für Physik, Freiburg (1976).
12. J. H. Parkinson, *Nature (London) Phys. Sci.* **236**, 68 (1972).
13. D. J. Nagel, P. G. Burkhalter, C. M. Dozier, J. F. Holzrichter, R. M. Klein, J. M. McMahon, J. A. Stamper, and R. R. Whitlock, *Phys. Rev. Lett.* **33**, 743 (1974).
14. J. L. Schwab and B. S. Fraenkel, *Phys. Lett. A* **40**, 83 (1970).
15. N. Bretz, D. Dimock, A. Greenberger, E. Hinnov, E. Meservey, W. Stodiek, and S. von Godler, Plasma Physics Lab Report No. MATT-1077, Princeton University (1974).
16. A. H. Gabriel and C. Jordon, *Mon. Not. R. Astron. Soc.* **145**, 241 (1969).
17. Y. Cauchois and C. Bonnelle, in *Atomic Inner-Shell Processes*, Ed. B. Crasemann, Vol. II, pp. 84, Academic Press, New York (1975).
18. M. O. Krause, in *Atomic Inner-Shell Processes*, Ed. B. Crasemann, Vol. II, pp. 34, Academic Press, New York (1975).
19. C. L. Cocke, B. Curnutte, and R. Randall, *Phys. Rev. A* **9**, 1823 (1974).
20. F. Bell, H. D. Betz, H. Panke, W. Stehling, and E. Spindler, *J. Phys. B* **9**, 3017 (1976).
21. McWherter, *Z. Phys.* **263**, 283 (1973).
22. Ch. P. Bhalla, *Phys. Lett.* **46A**, 336 (1973).
23. N. Stolterfoht, in *Invited Papers, Proceedings of the Second International Conference on Inner Shell Ionization Phenomena*, Eds. W. Mehlhorn and R. Brenn, pp. 42, Fakultät für Physik, Freiburg (1976).
24. R. L. Kauffman, K. A. Jamison, T. Cray, and P. Richard, *Phys. Rev. Lett.* **36**, 1074 (1976).
25. H. R. Philipp and H. Ehrenreich, *Phys. Rev.* **129**, 1550 (1963).
26. D. L. Walters and Ch. P. Bhalla, *Phys. Rev. A* **3**, 1919 (1973).
27. C. W. Woods, R. L. Kauffman, K. A. Jamison, N. Stolterfoht, and P. Richard, *Phys. Rev. A* **13**, 1358 (1976).
28. D. H. Madison and E. Merzbacher, in *Atomic Inner-Shell Processes*, Ed. B. Crasemann, Vol. I, pp. 2, Academic Press, New York (1975).
29. C. Forster, T. P. Hoogkamer, P. Woorlee, and F. W. Saris, *J. Phys. B.* **9**, 1943 (1976).
30. H.-D. Betz, F. Bell, H. Panke, G. Kalkoffen, M. Welz, and D. Evers, *Phys. Rev. Lett.* **33**, 857 (1974).

31. S. L. Varghese, C. L. Cocke, B. Curnutte, and G. Seaman, *J. Phys. B* **9**, L 387 (1976).
32. C. L. Cocke, in *Beam Foil Spectroscopy*, Eds. I. A. Sellin and D. J. Pegg, pp. 283, Plenum Press, New York (1976).
33. Ch. Bhalla and A. H. Gabriel, in *Beam Foil Spectroscopy*, I. A. Sellin and D. J. Pegg, pp. 121, Plenum Press, New York (1976).
34. G. W. F. Drake and A. Dalgarno, *Astrophysics J.* **157**, 459 (1969).
35. A. Dalgarno and E. M. Parkinson, *Proc. R. Soc. London A* **301**, 253 (1967).
36. H. A. Bethe and E. Salpeter, *Quantum Mechanics of One- and Two-Electron Atoms*, Academic Press, New York (1957).
37. T. A. Carlson and M. O. Krause, *Phys. Rev. Lett.* **14**, 390 (1965).
38. V. V. Afrosimov, Yn. S. Gordeev, A. N. Zinoviev, D. H. Rasulov, and A. P. Shergin, in *Abstracts of Papers of the IXth International Conference on the Physics of Electron and Atomic Collisions*, Ed. J. I. Risley and R. Geballe, pp. 1066, University of Washington Press, Seattle (1975).
39. W. Wölfli, Ch. Stoller, G. Bonani, M. Suter, and M. Stöckli *Phys. Rev. Lett.* **35**, 656 (1975).
40. H. Nussbaumer, *J. Phys. B.* **9**, 1757 (1976).
41. R. Schuch, H. Schmidt–Böcking, R. Schulé, G. Nolte, I. Tserruya, W. Lichtenberg, and K. Stiebing, in *Abstracts of Contributed Papers of the Second International Conference on Inner Shell Ionization Phenomena*, pp. 161, Fakultät für Physik, Freiburg, (1976).
42. A. R. Knudson, K. W. Hill, and D. J. Nagel, *Phys. Rev. Lett.* **37**, 679 (1976).
43. Th. P. Hoogkamer, P. Woorlee, F. W. Saris, and M. Gavrila, *J. Phys. B* **9**, L145 (1976).
44. P. Richard, J. Olten, K. A. Jamison, R. L. Kauffman, C. W. Woods, and J. M. Hall, *Phys. Lett.* **54A**, 169 (1975).
45. P. Richard, C. F. Moore, and D. K. Olsen, *Phys. Lett.* **43A**, 519 (1973).
46. J. Siivola, J. Utriainen, M. Linkoaho, G. Graeffé, and T. Åberg, *Phys. Lett.* **32A**, 438 (1970).
47. K. A. Jamison, J. M. Holl,and P. Richard, *J. Phys. B* **8**, L458 (1975).
48. F. W. Saris, W. F. van der Weg, H. Tawara, and R. Laubert, *Phys. Rev. Lett.* **28**, 717 (1972).
49. P. Mikler, N. Stein, and P. Armbruster, *Phys. Rev. Lett.* **29**, 827 (1972).
50. B. Müller, in *Invited Lectures Review Papers and Progress Reports of the IXth International Conference on the Physics of Electronic and Atomic Collisions*, Eds. J. S. Risley and R. Geballe, pp. 481 University of Washington Press, Seattle (1976).
51. Th. P. Hoogkamer, P. Woerlee, F. W. Saris, and W. E. Meyerhof, *J. Phys. B* **11**, 865 (1978).
52. F. W. Saris and Th. Hoogkamer, in *Atomic Physics 5*, Eds. R. Marrus, M. Prior, and H. Shugart, Plenum Press, New York (1977).
53. F. W. Saris, W. Lennard, I. V. Mitchell, F. Brown, and Th. P. Hoogkamer, to be published.
54. H. W. Schnopper, A. D. Betz, J. P. Delvaille, K. Kalata, A. R. Sohval, K. W. Jones, and H. E. Wegner, *Phys. Rev. Lett.* **29**, 898 (1972).
55. H. W. Schnopper, H. D. Betz, J. P. Delvaille, K. Kalata, A. R. Sohval, K. W. Jones, and H. E. Wegner, *Phys. Lett.* **47A**, 61 (1974).
56. P. Kienle, M. Kleber, P. Povh, R. M. Diamond, F. S. Stephens, E. Grosse, M. R. Maier, and D. Proetef, *Phys. Rev. Lett.* **31**, 1039 (1973).

31
Exotic Atoms

G. BACKENSTOSS

1. Introduction

An "ordinary" atom consists of a positively charged nucleus which interacts electromagnetically with electrons, leading to bound atomic states. Similar systems, in which an electron is replaced by any negatively charged particle other than an electron, are called "exotic" atoms. In 1947, soon after the discovery of the π meson, Wheeler[1] suggested the existence of such atoms. The first species to be observed were muonic atoms at Columbia[2] and π-mesonic (pionic) atoms at Rochester and Pittsburgh[3] in 1952–1953. K-mesonic (kaonic) atoms were first observed in 1965 at Argonne,[4] and Σ hyperonic[5] and antiprotonic atoms[6] were seen and investigated for the first time at CERN in 1970.

Since all the negative particles that replace the electrons have a lifetime of less than $\sim 1~\mu$s and are rather difficult to produce [production is by protons accelerated between 500 MeV (for pions) and $\sim$20 GeV (for antiprotons)], it is not possible to replace more than one electron in a single atom. Furthermore, since the masses of these particles are larger than the electron mass by at least two orders of magnitude, their orbits are correspondingly smaller and hence far inside the remaining electron orbits. Exotic atoms can thus be treated to a very good approximation as hydrogenlike atoms in which the electrons cause only small, mostly negligible, effects and in which there are no differences between neutral and ionized atoms.

The most striking and important features of exotic atoms can be seen from the simple Bohr model of the hydrogen atom, where the energy E_n of

---

G. BACKENSTOSS • University of Basle, Switzerland.

an atomic level with principal quantum number n is

$$E_n = -\frac{\mu c^2}{2}\left(\frac{Z\alpha}{n}\right)^2 = -\mu R_\infty Z^2 \tag{1}$$

with α the fine-structure constant and R_∞ the Rydberg constant; the corresponding Bohr radius r_n is

$$r_n = \frac{\hbar^2}{\mu e^2}\frac{n^2}{Z} \tag{2}$$

where μ is the reduced mass of the orbiting particle. This results in energies for the $n = 0$ state in exotic hydrogen atoms of 2–14 keV. Unlike ordinary atoms, no electronic screening of the nuclear charge occurs; hence the factor Z is fully applicable and yields transition energies in the keV to MeV region. The ground-state Bohr radii are several orders of magnitude smaller than in ordinary atomic orbits and even for a medium-heavy $(Z \approx 50)$ muonic atom comparable to the nuclear radius. It is this latter feature that makes exotic atoms a unique tool in the study of nuclear and elementary particle physics.

In Table 1 the relevant quantities such as the mass and mean life of the particles forming exotic atoms are given, together with the Bohr energy and the Bohr radius for the ground state $(n = 0)$. In the last column the principal quantum number n_0 is given that leads to a Bohr radius r_{n0} equal to the ground-state orbit of the corresponding electronic atom. It can easily be deduced from Eq. (2) that

$$n_0 = (m/m_e)^{1/2} \tag{3}$$

where m is the mass of the particle and m_e the electron mass.

The most simple-minded approach to the formation of exotic atoms pictures the replacement of an electron by the slowed-down particle as occurring at the point where the overlap between the electron wave function with the greatest probability density (1s state) and an atomic state of the particle is largest, i.e., for equal Bohr radii; n_0 is thus the quantum number at which capture takes place. The influence of the $Z-1$ electrons will decrease rapidly as n becomes smaller than n_0.

As for the ordinary hydrogen atom, the unscreened Coulomb potential $\sim 1/r$ is responsible for the energy-level scheme of exotic atoms (as given in Figure 1). However, the orbiting particle approaches the source of the potential much more closely than would be the case in electronic atoms, and hence it is much more sensitive to deviations from the $1/r$ form of the potential. It is hence convenient to distinguish between long- and short-range potentials. We shall call all potentials $\sim r^{-\gamma}(\gamma = 1, 2)$ "long-range," i.e., the potentials produced by the nuclear monopole (Coulomb),

Table 1. Properties of Particles and Their Atoms in the Simple Bohr Model

Particle	Type/Interaction	Spin/Wave equation	Mass (MeV)	Mean life (s)	Bohr energy[a] $E_B = R_\infty(m/m_e)Z^2$ (keV)	Bohr radius[b] $a_B = (m_e/m)a_0/Z$ (fm)	$n_0 = (m/m_e)^{1/2}$
Leptons							
μ^-	Lepton/weak	1/2 Dirac	105.657	2.197×10^{-6}	$2.8\,Z^2$	$256/Z$	14
Hadrons							
π^-	Meson/strong	0 Klein–Gordon	139.567	2.603×10^{-8}	$3.7\,Z^2$	$194/Z$	17
K^-	Meson/strong	0 Klein–Gordon	493.707	1.237×10^{-8}	$13.1\,Z^2$	$54.7/Z$	31
$\bar{p}$	Baryon/strong	1/2 Dirac	938.280	∞	$25.0\,Z^2$	$28.8/Z$	43
Σ^-	Hyperon/strong	1/2 Dirac	1197.35	1.48×10^{-10}	$31.8\,Z^2$	$24.6/Z$	48

[a] $R_\infty = 13.6058$ eV.
[b] $a_B = 5.2917 \times 10^{-9}$ cm.

magnetic dipole, electric quadrupole etc.; this is in contrast to the short-range potentials of the Yukawa or Wood–Saxon type. To the latter type of potential belongs not only the strong interaction potential of hadrons but also the "finite size potentials" (which are the deviations of the potentials of extended multipoles from that of pointlike multipoles). One of the beauties of exotic atoms is that one can study atomic systems by selecting the states that are influenced by the various potentials. Since the long-range potentials are usually of purely electromagnetic origin and hence well-known, the higher atomic states, which are far outside the reach of the short-range potentials, will yield precise information on particle properties. Conversely, the low-lying states will carry information about the finite extension of charge and electromagnetic moments as well as the strong interaction potentials of the nuclei. It should be noted, however, that the strong interaction potentials can only be probed in regions where they contribute moderate perturbations to the Coulomb potential, since otherwise the atomic concept would break down. In Figure 1 the various aspects of exotic

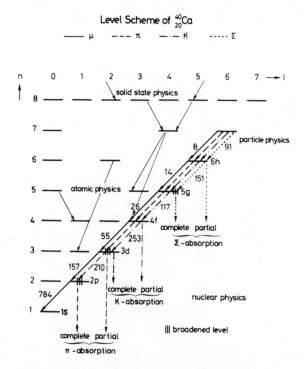

Figure 1. Scheme of the atomic cascade for different orbiting particles and the related fields in physics.

Table 2. Properties of Nuclei, Particles, and Interactions That Can Be Studied with Exotic Atoms

Particle in exotic atoms	μ^-		π^-	K^-	$\bar{p}$	Σ^-
Particle properties (mass m, magnetic moment μ)	—		m	m	m, μ	m, μ
Nuclear properties	Finite size of charge, magnetic moment, and electrical quadrupole moment		Nuclear structure, nuclear surface Neutron distribution $2N$ density Granular structure			
Interaction	Electromagnetic Vacuum polarization Parity conservation		Strong Particle–nucleus and particle–nucleon at zero energy $(\pi, 2N)$ S wave P wave			

atoms are shown schematically. We will discuss these aspects later in more detail.

We may finally ask what kind of information might be extracted from measurements on exotic atoms. Obviously a system consisting of a nucleus and an orbiting particle may yield information about these two substituents as well as about their interaction. Table 2 gives a self-explanatory synopsis of the various aspects and problems in physics which could be expected to profit from suitable studies of exotic atoms.

2. Theoretical Background

2.1. Atomic Cascades

Not many details are known about the formation process of exotic atoms. It is usually tacitly assumed that the capture of the μ^- takes place by a process in which an atomic electron—K-shell electrons are preferred—is replaced by the muon. In this discussion the term "muon" will stand for all particles which form exotic atoms since this capture takes place in a region where only the long-range electromagnetic interaction is significant; all these particles can thus be treated like electrons with heavier masses. From Eq. (3) one thus obtains the values of n_0 given in the last column of Table 1. If one further assumes that the distribution over the $n-1$ substates of different angular momentum quantum number l is given by the statistical weight $2l+1$, one arrives at a simple "initial distribution" which serves as the starting point for a cascade calculation.

The de-excitation of the atom from the initially populated states into the lower states will now be taken into account. In the upper part of the cascade the dominant de-excitation process is the Auger process, which has a transition probability

$$W_A = \frac{2\pi}{\hbar}|\langle \psi_\mu^f \psi_e^f | \frac{1}{r_{e\mu}} | \psi_\mu^i \psi_e^i \rangle|^2 \tag{4}$$

Here ψ_μ^i, ψ_μ^f, ψ_e^i, ψ_e^f are the muon and electron wave functions in the initial and final states, respectively, and $r_{e\mu}$ the distance between muon and electron. In the course of de-excitation to lower levels the electric E1 transitions begin to compete and finally take over completely. The E1 transition probability is given by

$$W_x = \frac{4e^2(\Delta E)^3}{3\hbar^4 c^3}|\langle \psi_\mu^f | r | \psi_\mu^i \rangle|^2 \tag{5}$$

where ΔE is the energy difference between the two levels and r refers to the muon. For both processes, which occur in the central potential of a nucleus, the selection rule $\Delta l = \pm 1$ must be satisfied. It is obvious that the Auger process, which depends on the overlap between the relevant wave functions, favors small values of Δn, whereas the radiative dipole process, which is proportional to $(\Delta E)^3$, favors the largest values of Δn which are allowed by the $\Delta l = \pm 1$ condition. The Auger process therefore tends to keep the l-distribution essentially unchanged, whereas the radiative process favors a population of the orbits with $l = n - 1$. Whereas this cascade calculation is based on the simple initial distribution and describes experimental results surprisingly well, it is, of course, insufficient to deal with more detailed problems. In particular, it cannot deal with problems of crystalline structure or chemical binding. In light atoms, and especially in hydrogen, the contributions of the Stark effect during atomic collisions must be discussed together with the external Auger effect, where the de-excitation energy of the atom is transferred to an electron of another atom during collision.[7] The various chemical effects, such as the Fermi–Teller rule characterizing the distribution of the muon over the various atomic substituents of a crystal or compound, are described in Ref. 8 together with the concept of exotic molecules in which the muon is first captured in molecular orbits and causes the above-mentioned cascade calculations to be radically modified.

2.2. Level Scheme for a Pointlike Nucleus

In order to calculate the transition probabilities, the muonic wave functions ψ_μ must be known. In the upper levels of the cascade, i.e., for all levels except for the lowest ($n \lesssim 5$ for the heaviest muonic atoms and the few last observable levels in hadronic atoms), only the point charge Coulomb potential is important, and the nonrelativistic Schrödinger equation for a point nucleus hydrogen atom can be applied:

$$-(h^2/2\mu)\nabla^2\psi_\mu + V(r)\psi_\mu = E\psi_\mu \tag{6}$$

where μ is the reduced mass, $\mu = m/(1 + m/A)$, with m and A the masses of the muon and nucleus, respectively, and $V(r) = 1/r$ is the Coulomb potential.

The energy eigenvalues are then

$$E_{n,j} = -\frac{\mu c^2}{2}\left(\frac{\alpha Z}{n}\right)^2\left[1 + \left(\frac{\alpha Z}{n}\right)^2\left(\frac{n}{j+1/2} - \frac{3}{4}\right) - \cdots\right] \tag{7}$$

for fermions with $j = l \pm \frac{1}{2}$ and

$$E_{n,l} = -\frac{\mu c^2}{2}\left(\frac{\alpha Z}{n}\right)^2\left[1 + \left(\frac{\alpha Z}{n}\right)^2\left(\frac{n}{l+1/2} - \frac{3}{4}\right) - \cdots\right] \tag{8}$$

for bosons.

For the lower levels, where the short-range potentials become important, the relativistic wave equations, i.e., the Dirac equation for fermions and the Klein–Gordon equation for bosons, must be used.

2.2.1. Fermions

The Dirac equation reduces for a spherically symmetric charge distribution $\rho(r)$ to two coupled first-order radial differential equations:

$$hc\left(\frac{dF}{dr} - \frac{\kappa}{r}F\right) = -[E - mc^2 - V(r)]G$$
$$hc\left(\frac{dG}{dr} + \frac{\kappa}{r}G\right) = +[E + mc^2 - V(r)]F \tag{9}$$

Here G represents the large component and F the small component of the radial wave function;

$$V(r) = V_{FS}(r) = -Ze^2\int_0^\infty \frac{(r^2)\,d^3r'}{|r - r'|} \tag{10}$$

is the Coulomb potential due to an extended charge distribution, and κ is an integer equal to $\pm(j + \frac{1}{2})$ (where the $+$ sign applies for states with $j = l - \frac{1}{2}$ and the $-$ sign for $j = l + \frac{1}{2}$) and describes the fine-structure states.

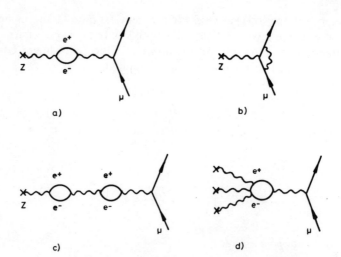

Figure 2. Feynman diagrams. (a) Lowest-order vacuum polarization; (b) self-energy; (c), (d) higher-order vacuum polarization.

2.2.2. Bosons

For bosons the Dirac equation has to be replaced by the Klein–Gordon equation

$$\{h^2\nabla^2 + (1/c^2)[E - V(r)]^2 - \mu^2 c^2\}\psi = 0 \qquad (11)$$

where again $V(r)$ represents the finite size Coulomb potential [Eq. (10)] and μ is the reduced mass. There is, of course, no fine-structure splitting for the spin-zero bosons. According to Eq. (8), which gives the approximate energy eigenvalues of the Klein–Gordon equation, a splitting of the l sublevels proportional to $(\alpha Z/n)^4$ occurs; this splitting manifests itself only by a careful study of parallel transitions, and these usually lack intensity.

2.3. Electromagnetic Corrections

2.3.1. Radiative Corrections

After the eigenvalues have been obtained from the relativistic wave equations for an extended charge, additional effects must be considered. The most prominent effects stem from the radiative corrections where particularly the vacuum polarization [Figure 2(a)], in contrast to the situation in electronic atoms, is very strong. This is essentially due to the μ^{-2} dependence of the self-energy [Figure 2(b)] of the mass of the orbiting particle and the increased nuclear electrical field strength to which the heavier, more closely orbiting particle is exposed. The second-order vacuum

Table 3. Contributions to Radiative Corrections in Muonic Atoms, Nuclear Polarization, and Electron Screening in eV [9c,10]

Atom	Transition	Energy (keV)	Vacuum polarization $(Z\alpha)^n \alpha$				Lamb	Reduced mass	Nuclear polarization	Screening
			$n=1$	$n=3+(\alpha^2 Z\alpha)$	$n=5; n=7$					
^{16}O	$2p-1s$	133.52	763	5	-0		-15	1	8	0
Ca	$2p-1s$	783.85	6049	44	-0		-208	14	187	0
Fe	$2p_{3/2}-1s_{1/2}$	1257.15	10038	69	-0		-445	29	408	0
	$3d_{5/2}-2p_{3/2}$	265.69	1417	8	-0		-6	2	6	-1
Ba	$2p_{3/2}-1s_{1/2}$	3979.80	29029	105	-8		-1706	104	2135	-2
	$3d_{5/2}-2p_{3/2}$	1229.20	9229	9	-3		-156	25	219	-9
	$4f_{7/2}-3d_{5/2}$	433.926	2327.5	-6	-1.2		-8.3	—	12.3	-13.5
Pb	$2p_{3/2}-1s_{1/2}$	5962.77	37303	-51	-54		-2330	110	3791	-9
	$3d_{5/2}-2p_{3/2}$	2500.33	20005	-87	-34		-554	80	1218	-26
	$4f_{7/2}-3d_{5/2}$	937.98	6195	-55	-14		-41	7	60	-53
	$5g_{9/2}-4f_{7/2}$	431.363	2105	-29	-6		-7	6	4	-78

polarization can be described by a potential V_{vp} of the form[9]

$$V_{vp}(r) = -\frac{2}{3} e^2 Z \alpha \lambda_e \int_0^\infty \rho(r') \frac{r'}{r} [Z_1(|r-r'|) - Z_1(r+r')] \, dr' \qquad (12)$$

where

$$Z_1(r) = \int_1^\infty y^{-2} \exp\left(\frac{-2yr}{\lambda_e}\right)\left(1+\frac{1}{2y^2}\right)(1-y^{-2})^{1/2} \, dy$$

and

$$\lambda_e = \hbar/m_e c = 386 \text{ fm}$$

$\rho(r)$ is the charge distribution (normalized to 1) on which V_{vp} depends. V_{vp} can be introduced into the wave equation (9) together with the Coulomb potential (10) and produces an increase of the binding. There are also higher-order contributions to the vacuum polarization [Figures 2(c) and 2(d)] which are being treated in perturbation theory. Table 3 shows examples of the contributions of the different corrections.[10] There is also a small contribution from the relativistic reduced mass effect as well as from the first-order Lamb shift, i.e., the self-energy term, which includes the effect of the anomalous magnetic moment and the vacuum polarization due to $\mu^+\mu^-$ pairs.

2.3.2. Electron Screening

There are further corrections that must be considered because of the screening of the electron cloud, i.e., deviations from the hydrogenlike behavior. It is obvious that this effect decreases with decreasing principal quantum number n. There exists a nearly constant contribution to the screening effect which largely cancels out because the observed transition energies are differences of level energies; it is usually noticeable only for transitions with $\Delta n \gg 1$.

2.3.3. Nuclear Polarization

There exists a shift of the low-lying energy levels which is caused by nuclear polarization which tends to increase the binding energy. Theoretical estimates of the effect exist,[11] but they are less accurate than those for the other corrections. The nuclear polarization shifts are mainly important where the effects of the extended charge distribution are significant; the errors of the nuclear polarization calculations are influencing the accuracy to which charge distributions can be determined.

2.3.4. Finite Size Effect

Particularly for large values of Z and small values of n, where the meson wave function overlaps the charge distribution, the modification of the pure Coulomb potential of a point charge $V_c(r) = Ze^2/r$ [Eq. (10)] becomes important. In this case only a fraction of the charge contributes to the binding, and the binding energy is thus reduced. It is instructive to estimate the effect of a small overlap by a first-order perturbation method. The energy shift due to the finite size effect is then

$$\Delta E_{FS}(n, l) = e \int |\psi_{nl}(r)|^2 [V_{FS}(r) - V_c(r)] \, d^3r \tag{13}$$

It is obvious that the difference of the potentials, which determines the size of the effect, is of short range and of the order of the extension of the charge distribution.

The magnitude of this effect can be quite large and can amount to more than 50% in the $1s$ state of heavy nuclei; this represents an important property of muonic atoms to which we will return later.

2.4. Strong Interaction Effects

The strong interaction between the hadrons and the nucleus is of short-range nature and does not extend significantly beyond the nucleus. The strong interaction is thus also described by a short-range potential.[12] Since the interaction is also absorptive (bosons are not conserved, antiprotons can be annihilated, and Σ hyperons may disappear in strong reactions), a complex optical potential, $V_{St} = \text{Re } V_{St} + i \text{ Im } V_{St}$, is used. Since it is stronger than the short-range part of the electromagnetic potential mentioned above, its resulting effects dominate the finite size effect. If the potential is not too strong (e.g., for pionic atoms) perturbation theory can be used to give

$$E_{St}(n, l) + i\Gamma_{St}(n, l) = \int (\text{Re } V_{St} + i \text{ Im } V_{St}) \phi_{n,l}^2(r) \, d^3r \tag{14}$$

where E_{St} is the energy shift of the state $\langle n, l \rangle$ due to $\text{Re } V_{St}$ and $\Gamma(n, l)$ the broadening of the level due to $\text{Im } V_{St}$.

These effects can be observed experimentally in three ways.[13] In hadronic atoms the atomic cascade (with the exception of the lightest nuclei) stops and does not reach the ground state. As soon as $\Gamma_{St}(n, l) > \Gamma_{E1}(n, l)$, the width due to radiative E1 transitions, the observed x-ray transition has a reduced intensity

$$Y = P(n, l)\Gamma_{E1}(n, l)/[\Gamma_{St}(n, l) + \Gamma_{E1}(n, l)] \tag{15}$$

As a result of the short range of V_{St}, $\Gamma_{St}(n, l)$ increases rapidly with decreasing (n, l) and hence only one or two transitions are observable before nuclear absorption of the hadron causes the complete termination of the atomic cascade process. A measurement of the yield Y thus leads directly to a measurement of Γ_{St} provided the population of the upper level $P(n, l)$ and E1 (n, l) are known. E1 (n, l) can easily be calculated, and $P(n, l)$ can be determined by measuring all the transitions leading to the level (n, l). As a further consequence of the rapid increase of $\Gamma_{St}(n, l)$ by two to three orders of magnitude from (n, l) to $(n-1, l-1)$, the width of an intensity-weakened line for which $\Gamma_{St}(n, l) \approx \Gamma_{E1}(n, l) \approx 0.1 - \sim$ eV causes a width of the lower level $\Gamma_{St}(n-1, l-1)$ in the keV region and hence can be measured directly from the line broadening. Finally, the levels are shifted by $E_{St} = \varepsilon$ with respect to the level energy calculated on the basis of pure electromagnetic interaction, including all the effects mentioned above and taking the hadron to act as a heavy electron. Here only the difference of the energy shifts between upper and lower level can be measured, the shift of the upper level being usually only a fraction of a percent.

It should be mentioned that, with the exception of pions, the perturbation approach breaks down completely for the stronger potentials of the hadrons. In this case the Klein–Gordon equation has to be solved numerically:

$$\{\hbar^2 \nabla^2 + (1/c^2)[E - V(r)]^2 - \mu^2 c^2\}\psi = 2\mu V_{St}\psi \qquad (16)$$

for the complex V_{St}. The direct correspondence between Re V_{St} and ε, and Im V_{St} and Γ_{St} is then also destroyed.

3. Experimental Methods

3.1. Generation of the Atoms

In principle, the experimental technique of atom formation is very simple: If the negative particles are brought to rest in some medium, then they cannot avoid being captured by the atoms in question. In practice this involves the following:

    (i)   production of the particles in question,
    (ii) selection of the desired particles,
    (iii) slowing down of the particles.

Production is usually done by means of proton–nucleus interactions in accelerators. Pions and the subsequent muons (which are produced from pion decay) can be produced by proton accelerators at energies of around 500 MeV, where these particles alone are produced. Although the production by electrons is typically weaker by an order of $\alpha = 1/137$, there are also some electron accelerators producing pions in the more intensive

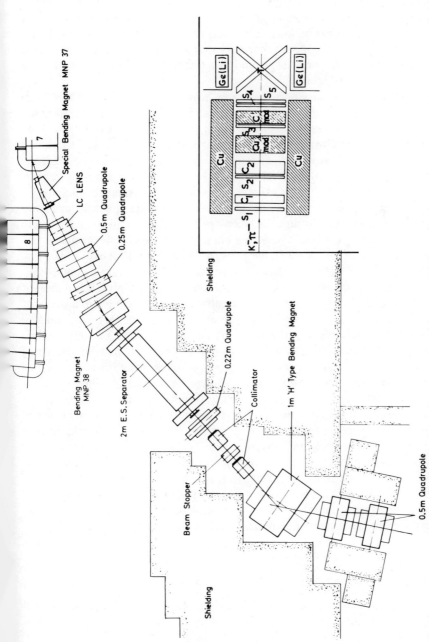

Figure 3. Layout of a typical $K/\bar{p}$ beam at the CERN protonsynchrotron. Insert: counter telescope, $S_1S_2S_3$ scintillation counters, C_1C_2 Čerenkov counters, S_4S_5 scintillation counters dE/dx, Cu and C moderator for slowing down the particles, T target, Ge(Li) detectors.

electron beams. Considerably higher proton energies are required to produce effectively K^- mesons ($\sim$6–8 GeV) and $\bar{p}$ (antiprotons $\sim$15–20 GeV). In these cases pions are also produced at fluxes that are greater by two to three orders of magnitude; it is thus essential that a separation is made (by electrostatic mass separators in a momentum-selected beam) in order to reduce the pion background by a factor of 10–100. A typical beam layout is shown in Figure 3.

Since the secondary particle beams have energies of several hundreds of MeV whereas atomic capture takes place near thermal energies, the particle beam must be slowed down, which is achieved by moderation of the beam via ionization loss in material. The beam quality is, however, worsened because of multiple scattering and energy straggling. Furthermore, hadrons suffer from absorption losses in flight and hence low-energy beams, requiring less moderator thickness, would be desirable. On the other hand, K^- mesons decay rapidly, so that only a few percent survive the transport from the production target through the beam optic elements to the focus (typically about 10–15 m). At higher energies more particles are usually produced and the larger velocity, $\beta = v/c$, implies a time-dilated decay time $T = T_0/(1 - \beta^2)^{1/2}$. A delicate balance is thus required to produce a good beam.

A stop signal is defined by a multiple coincidence of a number of scintillation and Čerenkov counters located in the particle beam in front of the target and an anticoincidence counter behind the target. Čerenkov counters are suitable for the rejection of high-velocity particles (pions being much less slowed down in the moderator than heavier particles) whereas dE/dx counters select the slow particles, which produce bigger pulses. A typical counter telescope is shown in the insert of Figure 3.

The best muon stop rate is presently achieved in the superconducting muon channel at the SIN pion factory and is about 10^5 μ stops per gram of material per second for a proton beam of $\sim$50 μA. Typical kaon and antiproton stop rates amount to $\sim$1 g^{-1} s^{-1}. The lifetime of the Σ^- hyperons is so short that all Σ^- would decay while slowing down and would not form a Σ^- atom. However after K^- capture, about 10% Σ^- are produced, with an energy of only 15 MeV. Many of these Σ^- are captured in the same target, and thus Σ^- spectra with a relative intensity of $\sim$10% as compared to the kaonic spectra are observed simultaneously with them.

3.2. X-Ray Detection

X-ray detection is usually gated with the stop trigger which signals the formation of an atom. With the exception of a few special cases, particularly at very low energies of a few keV where proportional counters have been used, solid-state detectors serve as x-ray spectrometers. At energies

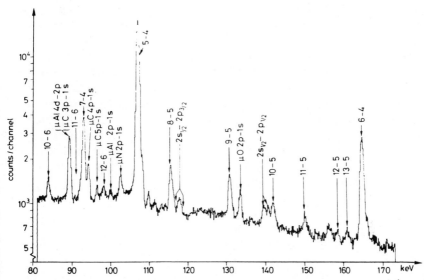

Figure 4. Part of the muonic spectrum of Nb. The lines without special notation originate from Nb [from H. P. Povel, *Nucl. Phys.* A **217**, 573 (1973)].

below about 25 keV Si(Li) detectors are favored, whereas Ge(Li) detectors are useful for energies above about 15 keV. In most cases high resolution is very desirable if not essential; it is obviously needed where natural line-widths are of interest, such as in the hadronic x-ray spectra and also for the determination of fine- and hyperfine-structure patterns in muonic x-ray spectra. High-resolution is also a great advantage where small background lines, which are almost never avoidable, should be discriminated against. At present the attainable resolutions of Si detectors are about 150 eV at 10 keV and those of Ge detectors are 800 eV at 100 keV and 2 keV at 2 MeV. Deterioration of the resolution due to the high in-beam background radiation may, however, be a problem. Typical spectra are shown in Figures 4–7 for muonic, pionic, kaonic, and antiprotonic atoms. In cases where sufficient intensity is available, bent crystal spectrometers have also been used; these reach energy resolutions $\Delta E/E$ close to 10^{-3} between 50 and 100 keV at the expense of achieving only small luminosities. They have been used to measure high-yield pionic x-ray transitions which give values for the π^- mass.

Calibration. Of the greatest importance for quite a number of experiments which are of fundamental interest is a high-precision calibration. The use of γ-ray standards, as used in γ-ray spectroscopy, meets limitations when precisions of about 50 eV or less are aimed at. This is mainly due to the fact that the statistical time distribution of source γ-rays

1400   G. Backenstoss

differs from the distribution of x-rays as produced by particle beams
originating from a pulsed proton beam of an accelerator. Strong precautions must be taken in order to control systematic shifts due to those
effects. Usually it is very advantageous to use built-in calibration lines. For
example, muonic and hadronic x-ray lines are affected only by the electromagnetic interaction for states outside the reach of the finite size or the
strong interaction potentials, and may be calculated to a precision of 1 eV
or better. γ-rays, produced after meson capture by the nucleus, which are
known to a sufficient precision, may also be useful for this purpose. If such
lines do not appear naturally at the desired energies—which is of course a
rare accident—the use of a target containing a suitable element to provide
calibration lines may be of help.

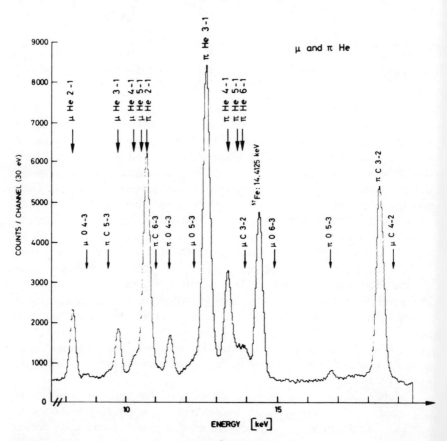

Figure 5. Pionic (muonic) x-ray spectrum of ^{4}He [from G. Backenstoss *et al.*, *Nucl. Phys. A*
232, 519 (1974)].

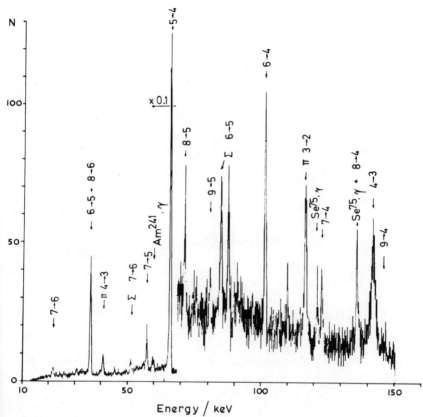

Figure 6. *K*-mesonic (Σ-hyperonic) x-ray spectrum of phosphorus. [from G. Backenstoss *et al.*, *Phys. Lett.* **38B**, 181 (1972)].

4. Results Obtained from Exotic Atom Data

4.1. Atomic Cascades

Atomic cascades have been observed whenever the x-rays of exotic atoms have been measured. In the majority of cases, however, the main interest was devoted to effects caused by the short-range potentials; hence often only the last x-ray transitions of the cascade have been studied in detail. In this chapter we discuss cascades (with the exception of the above-mentioned transitions). The only nontrivial quantity is the intensity of the transitions normalized to the number of stopping particles or relative intensities between different transitions. In cases where more than one type of atom is involved, such as in chemical compounds or in mixtures, the

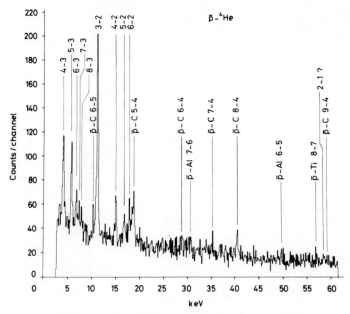

Figure 7. Antiprotonic x-ray spectrum of ^{4}He (Ref. 25).

distribution among the various atoms may be of interest. All these quantities are, however, in one way or another related to the initial capture process and may serve as a tool to reveal details of this process.

4.1.1. Relative Intensities

Relative intensities have been measured for all kinds of exotic atoms. They can be compared with cascade calculations in which one or more free parameters are allowed to vary in order to produce a good fit to the data. The simplest approach is to define the quantum state n_0 according to Eq. (3) and to define the distribution over the l substates as

$$P_n(l) = (2l + 1)^{\alpha l} \tag{17}$$

where α is a free parameter describing the deviation from a purely statistical distribution in which $\alpha = 0$. It should be noted that α has no profound meaning and that a change of α can be compensated, for within certain limits, by a change of the starting level n_0. Also, a suppression of the Auger transition and a corresponding enhancement of the radiative dipole transitions, as could be caused by a depletion of the electronic shells, may appear as a change in α. Most of the x-ray spectra of all the exotic atoms, except the very lightest (in particular H and to a lesser degree He), can be satisfactorily described by such a one-parameter fit with $\alpha \ll 1$. There seem to be some systematic influences on α which are produced by the chemical or physical state of the x-ray source. For example, metals tend to have $\alpha > 0$, which signifies a denser population of the high angular momentum states.[14,15] This could be explained as a suppression of Auger transitions in insulators, which leads to a relative enhancement of radiative E1 transitions and which, by virtue of the strong energy dependence of the transition ($\sim \Delta E^3$), tends to populate the lowest possible n states compatible with $\Delta l = 1$, i.e., $l = l_{max} = n - 1$.

A further systematic variation which occurs is the decrease of lower angular momentum states with respect to higher ones as a function of Z. For example, the $2s$ level is less populated relative to the $2p$ level the larger Z becomes. This effect is also explained as the competition between Auger and radiative transitions. Their relative strength depends upon the transition energy. For larger values of Z the radiative transitions dominate at larger n values, thus increasing the tendency towards the largest possible l states. Hence, e.g., the intensities of the $2s$–$2p$ transitions of interest in muonic atoms are weak ($\sim 1\%$) in heavy atoms.

Since muons originating from the parity-violating π–μ decay are strongly polarized, if the muon beam is momentum-selected, then the degree of polarization is also a characteristic quantity for each atomic state. The depolarization which occurs during the atomic cascade may thus be studied. Until very recently the muon polarization in the stable $1s$ state only was observed by studying the polarization of the μ^- in the $1s$ state via the anisotropy of the decay electrons. Polarizations of up to about 20% have been observed.[16]

The strong increase of the muonic stop rates at the meson factories recently made it possible to investigate also the degree of polarization in different atomic states by means of a measurement of the circular polarization of the x-ray transitions and thus yield additional information about atomic cascades.[16a]

4.1.2. Absolute Intensities

Absolute intensities, normalized to the number of particles stopped in the target (which is the same as the number of atoms formed), have also been measured. In μ, π, and K atoms strong systematic variations of transition intensities which are unattenuated by strong absorption have been observed up to a factor of 4.[17] Since they occur at the same regions of Z (see Figure 8), this effect must be connected with the atomic rather than the nuclear structure or else with interatomic distances in the lattices of the solid elements. A definite and full explanation of these effects has not yet been given.

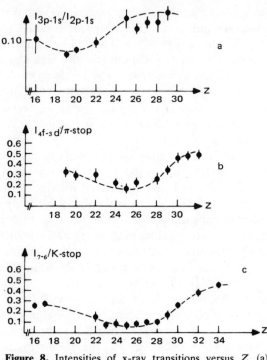

Figure 8. Intensities of x-ray transitions versus Z. (a) $I(3 \to 1)/I(2 \to 1)$ for muons; (b) absolute intensities per stopped pion; (c) absolute intensities per stopped kaon (Ref. 27).

4.1.3. Capture in Chemical Compounds

The relative capture probability $W(Z)$ in binary chemical compounds $Z'_k Z_m$ was first predicted by Fermi and Teller[18] for alloys. According to their derivation

$$\frac{W(Z)}{W(Z')} = \frac{m}{k} \frac{Z}{Z'} \tag{18}$$

are the ratios of the x-ray intensities corresponding to the atomic numbers Z, Z' and the stoichiometric compositions m, k. This so-called Fermi–Teller rule holds only in special cases. We can write more generally

$$\frac{k W(Z)}{m W(Z')} = \left(\frac{Z}{Z'}\right)^n \tag{19}$$

Values of n between -1 and $+1.5$ have been observed where the size of n is connected with the types of chemical binding. $n = 1$ in this case denotes the Fermi–Teller rule and $n = 0$ a capture of the mesons, which is governed only by the ratio of atoms present.

A number of possibilities have been discussed which may help to explain the observed effects. One of the more promising schemes involves the formation of mesomolecules where the mesons (and also muons) nearly replace a valence electron, leaving the molecule in a highly excited molecular state.[8] A rather stable complex $Z_m \mu^- Z_n'$ is thus formed where, however, the meson is not in a central potential. The condition $\Delta l = \pm 1$ is thus not valid any more and may lead to a quite different l distribution in the atomic states when the mesons, in the course of the de-excitation process, switch from a molecular to an atomic orbit. This latter process is therefore responsible for the atomic capture ratio and thus depends on the chemical binding. This may be of particular interest for H bonds. Since the x-ray energies of μH and πH at energies around 2 keV are very hard to observe, one can observe instead in πH the high-energy γ-rays from the reactions

$$\pi^- p \to n + \gamma \qquad (130 \text{ MeV})$$

and

$$\pi^- p \to n + \pi^0$$
$$\downarrow$$
$$2\gamma \qquad (70 \text{ MeV})$$

since in all other nuclei with $A > 2$ the reaction

$$\pi^- + A \to (A-2) + 2N \qquad (\text{no } \gamma)$$

dominates strongly.

Another way of looking at related problems is to study the *transfer* of the muons in collision of muonic atoms with other atoms. The transfer of muons from μH to other gas atoms has particularly been studied. For example, in the μA x-ray cascade the lower l states are much more preferred than the $l = n - 1$ states when the μA is formed by the transfer

$$\mu H + A \to \mu A + H$$

as compared with the direct capture in A gas. This reaction occurs quite readily, and therefore a concentration of 10^{-5} A in H_2 is sufficient to enable a substantial transfer. This method has been used to measure rare isotopes (e.g., ^{36}A) which were mixed in with H_2 at high pressure (1000 atm) in order to obtain a sufficient muon stop rate.[19] Transfer reactions have also been observed in liquid H_2 where the transfer

$$\mu H + D \to \mu D + H$$

occurs easily and is followed by a molecular binding $D\mu^- H$, from which state the nuclear fusion process $d + p \to {}^3\text{He} + \gamma$ occurs, and the muonic x-rays of ^{3}He are observable after an appropriate delay.

4.2. Particle Properties

4.2.1. Masses

At present the most accurate determination of the masses of negative hadrons originate from precise measurements of suitable x-ray transitions. It is obvious from the previous discussions that transitions must be used in which only the well-known long-range Coulomb field plays a role, i.e., in a region where the strong interaction and finite size effects can be ignored as well as the electron screening. The transition energies are adequately described by Eq. (7) or (8) and are directly proportional to the reduced mass μ of the orbiting particle. Radiative corrections and small energy shifts due to the other effects mentioned above may be added if required. The relevant experiments involve making precise measurements of absolute transition energies. Since the relative error of these energies essentially determines the relative mass error, one attempts to measure transitions which satisfy the mentioned requirements and which yield the smallest possible relative errors. The choice of suitable transitions depends on the detection instrument and the availability of calibration lines, as discussed earlier, which practically means energies between 300 and 500 keV for solid-state detectors and between 50 and 100 keV for diffraction spectrometers.

Great care must be taken also in the evaluation of the lines, since the influence of unresolved but weak transitions of the type $(n, l = n - 2) \rightarrow (n - 1, l = n - 3)$ parallel to the main transition $(n, l = n - 1) \rightarrow (n - 1, l = n - 2)$ must be estimated and the absence of any disturbing line of other origin must be guaranteed. For pionic atoms, results of comparable quality are obtained from measurements with Ge(Li) detectors and bent-crystal spectrometers. The latter instruments are restricted to smaller energies (< 100 keV), and as a consequence of their smaller luminosities yield fewer statistics and hence are applicable only for intense meson beams, such as pions or muons. On the other hand, higher resolutions can be achieved and calibration is more straightforward. In Table 4 a list of the best values of the masses obtained by this method is given. All the quoted measurements have been performed with Ge(Li) detectors except for one result which was obtained with a bent-crystal spectrometer.

It should be noted that a precise knowledge of the π^- mass together with a measurement of the momentum of the μ^+ from the $\pi^+ \rightarrow \mu^+ + \nu_\mu$ decay results in the following value for the limit of the mass of the ν_μ neutrino provided CPT (and hence the equality of m_π^+ and m_π^-) is assumed to be correct:

$$m_{\nu_\mu}^2 = \pm 0.22 \pm 0.40 \text{ MeV}/c^2$$

An inequality of the masses of $\bar{p}$ and p of about two standard deviations also appears; this should not, however, be taken too seriously.

Table 4. Particle Masses (m) and Magnetic Moments (μ) Obtained from Atoms

Particle	m (MeV)	μ (Nuclear magnetons)
π^-	139.569 ± 0.006^a	
	139.5657 ± 0.0017^b	
K^-	493.688 ± 0.030^c	
	493.657 ± 0.020^c	
$\bar{p}$	938.179 ± 0.058^c	-2.791 ± 0.021^c
		-2.819 ± 0.046^d
Σ^-	1197.24 ± 0.15^c	$-1.40 \begin{array}{l} +0.41 \\ -0.28 \end{array} \Big\}_c$
		$\text{or } 0.65 \begin{array}{l} +0.28 \\ -0.40 \end{array}$
		-1.48 ± 0.37^e

^{a} CERN, Ge detector, Ref. 25.
^{b} Dubna, crystal spectrometer, V. I. Marushenko *et al.*, *JETP Lett.* **23**, 80 (1976).
^{c} Columbia Ge detector, C. S. Wu *et al.*, *Nucl. Phys. A* **254**, 381, 396, 403 (1975).
^{d} BNL Ge detector, B. L. Roberts *et al.*, *Phys. Rev. Lett.* **33**, 1181 (1974).
^{e} BNL Ge detector, B. L. Roberts *et al.*, *Phys. Rev. Lett.* **32**, 1265 (1974).

4.2.2. Magnetic Moments

Values for the magnetic moments can be deduced from the fine-structure splitting of observed x-ray lines. In the approximation of Eqs. (7) and (8) this energy splitting $\Delta E_{n,l}$ is given by

$$\Delta E_{n,l} = \frac{\mu c^2}{2} \left(\frac{Z\alpha}{n} \right)^4 \frac{n}{l(l+1)} \qquad (20)$$

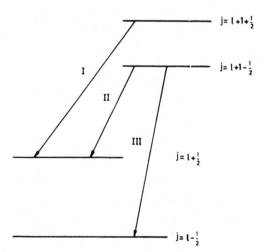

Figure 9. Transitions between two hyperfine doublets.

This relation is valid for a Dirac particle with a magnetic moment $\mu_B = g_0 I_s = eh/2mc$, where g_0 for $I_s = 1/2$ is very nearly equal to 2 and μ_B is the Bohr magneton. Hadrons have an anomalous magnetic moment $\mu_H = (g_0 + g_1)\mu_B$, where g_1 describes the anomalous part of the magnetic moment. The fine-structure splitting of Eq. (20) for hadronic fermions then becomes

$$\Delta E_{n,l} = (g_0 + 2g_1)\frac{\mu c^2}{2}\left(\frac{Z\alpha}{n}\right)^4 \frac{n}{l(l+1)} \tag{21}$$

Since ΔE depends on Z^4, measurements can be performed most easily with heavy atoms. Since, experimentally, transitions between two spin doublets are observed, one has to resolve the three transitions as indicated in Figure 9 for a negative magnetic moment. Assuming the statistical weight of all levels as proportional to $2j + 1$ one can easily calculate the relative intensities (I) of the three components. For $l = 10$ corresponding to the lower doublet of the $n = 11$, $l = 10$ level and the upper doublet $n = 12$, $l = 11$, one finds

$$I_\text{I} : I_\text{II} : I_\text{III} = 252 : 1 : 230$$

That means that for sufficiently large l the line (II) with the lowest energy can be neglected and hence the energy difference $E_\text{III} - E_\text{I}$ yields the difference between the two doublet splittings. It should be noted that the intensity ratios are reversed for a positive magnetic moment, whereas the energy difference depends only on the magnitude of the magnetic moment. $\mu_{\bar{p}}$ has been determined from the $n = 11 \to n = 10$ transitions in Pb and U (see Table 4) and has been found to have the same value as and opposite sign to the proton magnetic moment.

The measurement of the magnetic moment of the Σ^- hyperon is far more difficult because of the very poor statistics and the small energy splitting at large n, which is less than the experimental resolution. One has to assume an intensity pattern for the two main lines and then try to optimize a fit to the data by varying the size of the splitting. It has not yet been possible to determine the sign of the magnetic moment uniquely since both signs lead to fits which are not substantially different in their quality. In Table 4 a set of data is displayed which illustrates the present state of our knowledge.

4.3. Electromagnetic Interactions (Muonic Atoms)

The muonic atom represents an ideal system for testing the electromagnetic interaction since the muon and nucleus can interact only electromagnetically when the weak interaction can be neglected. The basic

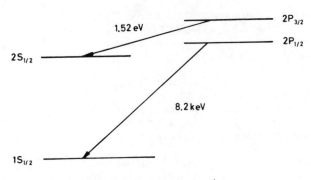

Figure 10. Lamb shift in μ ^{4}He.

interaction is understood to be independent of all details of nuclear structure such as distributions of charges, magnetic moments, etc., and applies
to muonic states which are outside the reach of the short-range finite size
potentials. The two most important examples are the Lamb shift of very
light, and the vacuum polarization of very heavy, muonic atoms. Also of
great interest is the complementary approach in which the electromagnetic
interaction is supposed to be known to a high degree; in this case information on many details of the electromagnetic properties of nuclei can be
obtained.

4.3.1. Basic Electromagnetic Interaction

(a) *Radiative Effects.* Although QED is known to be consistent with
experimental findings to a very high degree, there still remain aspects
where investigations with muonic atoms are rewarding. These are mainly
concerned with the radiative effects in which, as mentioned above, the
vacuum polarization dominates.

The most precise measurement to be performed so far is the
measurement of the Lamb shift in μ ^{4}He.[20] This shift corresponds to the
near-red end of the visible spectrum and uses a novel technique involving a
tunable dye laser.

The metastable $2S_{1/2}$ level in Figure 10 is populated by the muonic
cascade to a few percent, and a tunable dye laser is used to induce the
transition to the $2p_{3/2}$ level. Most of the de-excitation of the μHe atom
leads to the $2p$ level and is observable by the $2p$–$1s$ E1 x-ray transition of
8.2 keV which occurs essentially simultaneously with respect to the μ-stop
signal. A μ-stop signal which is not followed by the 8.2-keV transition is an
indication of the trapping of the μ^- in the $2s$ level. Such a signal is used to
trigger the laser. If the frequency of the laser corresponds to the Lamb
shift, the muon is lifted to the $2p_{3/2}$ level from which it promptly undergoes

Table 5. Contributions to the Energy Difference[a] between the $2P_{3/2}$ and $2S_{1/2}$ Levels in $(\mu^-\,{}^4\mathrm{He})^+$

Vacuum polarization	
First-order in α	+1.6659
Second-order in α	+0.0115
Lamb shift	−0.0143
Finite size of nucleus	$-0.2867 \pm 0.0087 = (-0.1053\langle r^2 \rangle)$
Fine structure	+0.1457
Nuclear polarization	+0.0031
Total	$+1.5251 \pm 0.0087 = (+1.8119 - 0.1053\langle r^2 \rangle)$

[a] Energies in eV, rms radius $= 1.650 \pm 0.025$ fm.

transition to the ground-state level by emission of a 8.2-keV photon. These photons are thus observed in coincidence with the laser trigger as a function of the laser wavelength. A signal has been found for $\lambda = (8117 \pm 5)$ Å corresponding to an energy difference

$$E(2p_{3/2}) - E(2S_{1/2}) = 1.5274 \pm 0.009 \text{ eV}$$

The various theoretical contributions to the calculated shift are given in Table 5.

One notices that the only important uncertainty originates from the uncertainty in the finite size of He nucleus. The difference

$$\Delta E_{\text{expt}} - \Delta E_{\text{theor}} = 0.0023 \pm 0.00087 \text{ eV}$$

is well within the error of the theoretical value and thus provides a testimony to the validity of QED. Provided QED is believed to be correct, a new and better value for the charge rms radius of ^{4}He of $<r^2>^{1/2} = 1.644 \pm 0.005$ fm can be derived.

Similar measurements of the $\mu^- p$ system would be of even more fundamental interest since a most sensitive test of QED information about the proton structure could be obtained. Unfortunately those experiments are even more complicated than those in the case of ^{4}He since the lifetime of the 2s level may be strongly reduced by Stark mixing, and the transition energies of interest near 0.2 eV are located in the ultraviolet region.

A different type of test on QED consists of precise measurements of muonic transition energies in nuclei of high Z. For levels where the finite size correction is small and can easily be scaled from its value at lower levels and where, on the other hand, the electron screening corrections are small, the dominant correction to the level energies is that of the vacuum polarization. Therefore, if the other effects are well understood, the experimentally measured transition energies can be interpreted in terms of the vacuum polarization correction. In particular, the $5g$–$4f$ transitions in $_{82}$Pb and the $4f$–$3d$ transitions in $_{56}$Ba, with energies between 431 and 441 keV, have been measured[21] recently with an accuracy of about 15 eV and agree within these errors with the theoretically predicted energies, which are believed to be accurate to about 8 eV. This agreement has been reached after some doubts[22] which, however, have been the consequence of underestimating experimental accuracy as well as of theoretical errors. Comparison of the results given here with the data of Table 3 therefore shows that the total vacuum polarization term is tested to an accuracy of about 0.22% and the higher terms are tested to about 20%. In spite of the fact that there are more accurate experimental verifications of the validity of QED (as, for example, the very precise g factor of the muon), the present result lives on its own right since there is no other experiment which tests the theory at such high electric field strengths as are present in heavy muonic atoms where field strengths of more than 10^{18} V/cm are accessible.

(b) *Parity in Electromagnetic Interaction.* Electromagnetic interactions have so far been assumed to conserve parity strictly. The existence of neutral currents in weak interactions, as illustrated by neutrino experiments, has stimulated interest in theories which aim at a unified description of weak and electromagnetic interactions. The question now arises of the extent to which neutral currents are parity-violating and of the extent to which parity violation can be observed in processes which are dominated by electromagnetic interactions.

The muonic atom is a system which is very suitable for the study of these questions since it is a bound system in which the muonic wave function overlaps the nucleus much more than in electronic atoms. It has been suggested that the M1(E1) mixing in light or medium muonic atoms should be investigated.[23]

This would occur if, for example, the $2s$ and $2p$ levels were nearly degenerate and a parity-violating potential existed. Since the short-range finite size effect shifts the $2s$ level further upward than the $2p$ level and the vacuum polarization shifts the two levels downward a value of Z can always be found for which near-degeneracy occurs. As a result of a parity mixing the interference term, which manifests itself by a circular polarization or an asymmetry of the emitted photons with respect to the muon spin in a M1 transition of the type $2s$–$1s$, must be observed. Whereas the effects expected here seem at present to be unobservable, the E2(E1) mixing in nearly degenerate $3d$–$2p$ states, where E2 transitions of the type $3d$–$1s$ are being observed, may be slightly more promising. However, the effects (estimated to be of the order of 10^{-6} to 10^{-7}) require 10^{12}–10^{14} events in order to be statistically significant. This would require long running times even at the new meson factories.

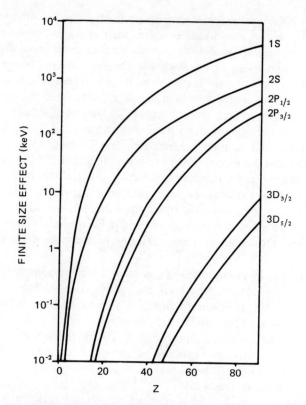

Figure 11. Finite size correction (reduction of binding energy) for various levels vs. Z (Ref. 24).

4.3.2. Electromagnetic Nuclear Properties

(a) *Finite Size.* Since the electromagnetic interaction between the muon and the nucleus is known to high precision, the muon provides an excellent probe for all electromagnetic properties of the nuclei. It is this aspect of muonic atoms which is most widely used in nuclear structure studies and is most readily identified with the field of muonic atoms.[11] This subject has reached a status where it far exceeds the scope of this chapter. Hence we must limit ourselves to a summary of the possibilities for research in this field.

The most straightforward application is the determination of the charge distribution of nuclei. The low-lying levels, i.e., the $1s$, $2s$, $2p$, and $3d$ levels, are predominately influenced by the finite extension of the nuclear charge, which results in a reduction of the binding energy (decreasing with growing n and l) as compared with a point charge. Since the overlap of a $2s$ and $2p$ wave function with the nuclear charge differs substantially, the binding of the $2s$ level is more strongly reduced than that of the $2p$ level, and hence $2s \rightarrow 2p$ transitions may be observed. The population of the $2s$ level, however, decreases with increasing Z from a few percent in light nuclei to less than 1% in heavy nuclei so that this transition is relatively difficult to observe (Figure 4). In Figure 11 the finite size effect is shown for the various levels as a function of the nuclear charge Z.

The interpretation of these level shifts in terms of a nuclear charge distribution is rather involved even for spherical nuclei. As can be seen from the perturbation treatment [Eq. (13)], the finite size shift depends on the wave function of the muon and on the potential which is generated by the charge distribution and hence is equivalent to a measurement of a certain moment of the charge distribution. Each wave function measures a slightly different moment. Therefore, in principle, the measurement of several energy levels leads to the determination of an equivalent number of parameters of an assumed charge distribution. In practice, however, it is only possible to determine two parameters in medium heavy nuclei and three parameters for the heaviest nuclei provided the $2s$ level is accessible. A Fermi-type distribution has usually been used:

$$\rho(r) = \frac{\rho_0[1 + W(r/c)^2]}{\exp[(r-c)/a]+1} \tag{22}$$

with two parameters (half-density c and skin thickness a) or with a third parameter W describing a central depression. This procedure is, however, unsatisfactory since the arbitrariness of the assumed distribution leads to an underestimation of the errors. Various methods have therefore been developed which allow a model-independent evaluation of the charge distribution.[24] Since muonic atoms allow a very precise measurement of $<r^2>$ and only imprecise determinations of higher moments, the best

Table 6. Typical Parameters of Nuclear Charge Distribution Eq. (22) and rms Radius in fm Obtained from e^- Scattering (e) and Muonic Atoms (μ)[24]

Nucleus	Method	Half-thickness c	Skin thickness a	Central parameter W	$\langle r^2 \rangle^{1/2}$
$_6$Ca	e, μ				2.45
$^{40}_{20}$C	e, μ	3.697	0.587	-0.083	3.53
$^{48}_{20}$Ca	e, μ	3.797	0.534	-0.048	3.52
$_{56}$Ba	μ	5.771	0.496	0	4.84
	e	5.83	0.407	0	4.77
$^{208}_{82}$Pb	μ	6.7202	0.5045	-0.061	5.50
	e	6.628	0.544	-0.062	5.4

results are obtained from combined evaluations of muonic atom and electron scattering data. A few examples of charge density parameters are given in Table 6.

For deformed nuclei the situation is even more complicated. Muonic atoms have the advantage that, via the nuclear spin, a direction is given, whereas in electron–nuclear scattering experiments the nuclei must be aligned. With one exception, therefore, no electron scattering data exist. Since the muon magnetic moment is 200 times smaller than that of the electron, hyperfine effects due to electric quadrupole moments exceed those of the magnetic dipole moment. The effect is so large that even second-order effects can be observed and hence splittings in spin-zero nuclei can be observed.[10]

(b) *Isotope Shifts.* In contrast to optical hyperfine-structure measurements, muonic x-ray energies yield absolute nuclear sizes. Here also relative measurements of x-ray energies between isotopes yield higher accuracies than do absolute measurements. In muonic atoms isotope shifts have, therefore, been measured for a considerable number of isotopes. For light and medium heavy atoms the energy increase due to an increased reduced mass in the heavier isotope is comparable to the isotope shift and hence the shift due to the reduced mass is subtracted from the measured isotope shift, and the remaining effect is interpreted in terms of the nuclear size. The isotope shift can be typically measured with an accuracy of a few percent for $\Delta A = 2$, yielding a precision for a model-independent size parameter of a few percent for $\Delta A = 2$ in the mass region 100 and of about 1% for the best measured isotope pair ^{206}Pb/^{208}Pb. If the results of the change in size thus obtained are compared with calculations assuming an $A^{1/3}$ dependence of the radial parameter of the charge distribution, the measured change in nuclear size is, in the majority of cases, considerably smaller (by 50%–80%) than the calculated difference. Exceptions occur in both directions where nuclear shells are filled or where strong nuclear deformations set in.

(c) *Magnetic Moments.* Magnetic hyperfine splitting can be observed in muonic atoms. It occurs as a consequence of the interaction of the magnetic moments of muon and nucleus and is analogous to that in electronic atoms; the hyperfine constant

$$A(I, nlj) = \frac{4}{3} e \mu_0 \mu_I \left\langle \phi_I \middle| \int_0^R f(r) g(r) (r/R)^3 \, dr + \int_R^\infty f(r) \, dr \middle| \phi_I \right\rangle \quad (23)$$

is proportional to the magnetic moments of the muon μ_0 and the nucleus μ_I in the state I. Although $\mu_0/\mu_e = m_e/m_\mu = 1/207$, the overlap between the nuclear wave function ϕ_I, and the Dirac eigenfunctions $f(r)$ and $g(r)$ of the muon are much larger than those for electrons, so that A reaches the order of keV. The value of A depends very sensitively on the finite extension of the distribution of the magnetization. The reduction of the magnetic splitting due to the spatial extension of the magnetic dipole moment, the Bohr–Weisskopf effect, can be therefore determined from muonic atoms. The magnetic hyperfine-structure splitting is given, as for electronic atoms, by

$$\Delta E_m = A(I, nlJ)\frac{F(F+1)-I(I+1)-J(J+1)}{2IJ} \tag{24}$$

with $\mathbf{F} = \mathbf{I} + \mathbf{J}$. Since first-order electric quadrupole splitting occurs only for $J \geq 3/2$, it is preferable to measure the magnetic dipole effects on levels with $J = 1/2$, where F is restricted to $F = I \pm \frac{1}{2}$ and hence

$$\Delta E_m = A\frac{2I+1}{I}$$

Since the transitions $2p_{1/2}$–$1s_{1/2}$ have energies of several MeV for the nuclei of interest, the magnetic hfs can hardly be resolved with present techniques. However, the energy of $2s_{1/2}$–$2p_{1/2}$ transitions occur at ~ 100 keV and hence the hfs is more easily accessible here. Occasionally energy resonances between nuclear and muonic levels also occur. Nuclear excitation is then possible during the muonic de-excitation and the magnetic hyperfine doublet can also be observed on the nuclear γ-ray, which may have a much smaller energy than the $2p$–$1s$ transition energy.

The results deduced for A from these measurements[10] show a reduction up to a factor of 2–3 in most cases as compared to that calculated for a point nuclear magnetic dipole and can be interpreted in terms of the finite size of the distribution of the magnetization; the results can be compared with predictions from various nuclear models.

(d) *Isomer Shift.* It is possible that the nucleus is excited during the process of muonic cascade. This occurs if the energies of nuclear excitations match muonic transition energies and the spin and parity of the muonic states are conserved. For example, the first excited rotational state, 2^+, in a nucleus of ground state 0^+ may mix with the $3d$ muonic state, leading to a nuclear excitation with a probability of up to 50%. The nuclear de-excitation, which is of the order of 10^{-9} s, is four to five orders of magnitude slower than the muonic cascade de-excitation and two orders of magnitude faster than the lifetime of the muon in the $1s$ ground state. The nuclear γ-rays are thus emitted in the presence of the muon in the $1s$ state. This leads to the muonic isomer shift, which can be described in perturbation theory by

$$\Delta E_{is} = e \int V_\mu \, \Delta\rho(r) \, d\tau$$

where the usual roles of nucleus and muon, as given by Eq. (13), are reversed.

V_μ is the potential produced by the 1s muon and $\Delta\rho(r)$ is the difference between the nuclear charge distribution in the excited and the ground states. The isotope shift results in a shift of the energy of nuclear γ-rays, which are usually in the region of ~ 100 keV, by up to a few keV in either direction. This shift must be corrected for the shifts due to the nuclear magnetic moment as discussed above, which is of the same order of magnitude. These shifts can now be interpreted in terms of $\Delta\rho$ and the nuclear charge radius can thus be determined for the several excited states corresponding to the number of γ-transitions available. For example, excited states in ^{153}Eu and ^{181}Tl have been found whose rms radius of an assumed Fermi-type charge distribution is 5–9×10^{-2} times smaller than that of the ground state, corresponding to negative isomer shifts of 3–6 keV, and in ^{209}Bi, where an increase of the rms radius of $\sim 6 \times 10^{-2}$, corresponding to a positive shift of ~ 6 keV, has been observed.[10] The change of the charge distribution in excited nuclei can also be here described in a model-independent way, as mentioned above, and can be compared with Mössbauer measurements.

4.4. Strong Interactions—Nuclear Properties

The study of the basic strong interaction between hadrons and nucleus by means of hadronic atoms turns out to be much more complicated than in the electromagnetic case since here a separation of the interaction from the nuclear structure is intrinsically impossible and hence there are several different ways of interpreting the results.

Some examples of the strong interaction effects in the various hadrons are displayed in Table 7. Obviously the most complete and precise measurements stem from pionic atoms for which the most intense beams are available. Here energy shifts ε and widths Γ for the 1s, 2p, 3d, 4f, and 5g levels have been observed. Width measurements originate from the line shapes of the lower level of a transition as well as from intensity measurements by which the width of the upper level of a transition can be determined. Thus in pionic atoms the 2p level width could be measured from He through to As by these two methods. Energy shifts are pronounced and have been observed for all pionic atoms which have been measured so far, where the 1s level is shifted upward (reduced binding energy) for all nuclei except ^{3}He and all other levels are shifted downward. For all other exotic atoms the situation is different insofar as the shifts are substantially smaller. Furthermore, the decrease of a certain line intensity as a function of Z (a consequence of increasing absorption in the upper level) and the increase of the width of the same line (a consequence of the increasing width of the lower level) is so rapid, that the strong interaction effects can be observed only in certain Z regions, as is illustrated by Figure 12 for kaonic atoms.

Table 7. Examples of Strong Interaction Effects in Hadronic Atoms [a]

Particle	Nucleus	Transition	Energy (keV)	ε_{low} (eV)	Γ_{low} (eV)	Γ_{up} (eV)
π	^{2}H	$2p$–$1s$	2.5928 ± 0.002	4.8 ± 2	—	—
π	^{3}He	$2p$–$1s$	10.692 ± 0.005	$+44 \pm 5$	42 ± 14	
π	^{4}He	$2p$–$1s$	10.698 ± 0.002	-75.7 ± 2.0	45 ± 3	$7.2 \pm 3.3 \times 10^{-4}$
π	Li	$2p$–$1s$	24.038 ± 0.004	-568 ± 4	205 ± 15	
π	^{16}O	$2p$–$1s$	159.95 ± 0.25	-15575 ± 250	7560 ± 500	11 ± 6
π	^{18}O	$2p$–$1s$	155.01 ± 0.25	-20510 ± 250	8670 ± 700	
π	Ne	$2p$–$1s$	238.35 ± 0.50	-33340 ± 500	14500 ± 3000	
π	A	$3d$–$2p$	168.88 ± 0.10	825 ± 100	1170 ± 170	
π	Co	$3d$–$2p$	384.74 ± 0.35	4570 ± 350	7370 ± 700	10.1 ± 2.1
π	Ba	$4f$–$3d$	582.99 ± 0.27	5440 ± 270	4300 ± 900	
π	Pb	$5g$–$4f$	575.56 ± 0.25	1730 ± 250	1100 ± 300	
K	C	$3d$–$2p$	62.73 ± 0.08	-575 ± 80	1730 ± 150	0.98 ± 0.19
K	P	$4f$–$3d$	142.02 ± 0.08	-315 ± 80	1440 ± 120	1.94 ± 0.33
K	S	$4f$–$3d$	161.56 ± 0.06	-460 ± 50	2370 ± 120	3.25 ± 0.41
K	Ni	$5g$–$4f$	231.49 ± 0.07	-180 ± 70	1020 ± 120	6.0 ± 2.3
$\bar{p}$	N	$4f$–$3d$	55.827 ± 0.050	3 ± 50	205 ± 70	0.13 ± 0.03
$\bar{p}$	^{16}O	$4f$–$3d$	73.438 ± 0.036	-124 ± 36	320 ± 150	0.64 ± 0.11
$\bar{p}$	^{18}O	$4f$–$3d$	73.861 ± 0.042	-189 ± 42	550 ± 240	0.80 ± 0.12
$\bar{p}$	S	$5g$–$4f$	140.440 ± 0.040	-60 ± 40	650 ± 100	3.04 ± 0.70
Σ	C	$4f$–$3d$				0.031 ± 0.012
Σ	Ca	$6h$–$5g$				0.40 ± 0.22
Σ	Ba	$9l$–$8k$				2.90 ± 3.5

[a] ε_{low} and Γ_{low} are the shift and width of the lower level, Γ_{up} the width of the upper level.

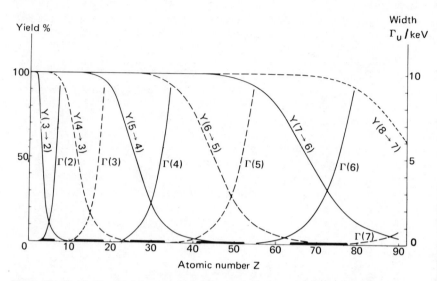

Figure 12. Yields Y and widths Γ of kaonic x-ray transitions vs. Z. The left-hand scale refers to Y, the right-hand scale to Γ.

The same situation holds for $\bar{p}$ and Σ^- atoms. For Σ^- atoms so far only intensity reductions have been observed for a small number of atoms.

4.4.1. Pionic Atoms

The data on strong interactions in pionic atoms have been interpreted rather successfully in terms of an optical potential[13] provided the lightest nuclei with $Z \leq 2$ are omitted. The strong interaction potential V_{St} of Eq. (16) is usually written (according to Ericson and Ericson[12])

$$V(r) = -2\pi \frac{h^2}{2\mu} \left[a_p\rho_p + a_n\rho_n + k_1B_0(\rho_p + \rho_n)^2 + \text{Pauli correlation correction} \right.$$

$$\left. + \text{terms of order } 1/A + \nabla \frac{\alpha(r)}{1 + (4\pi/3)\xi\alpha(r)} \nabla \right] \qquad (25)$$

with

$$\alpha(r) = c_p\rho_p + c_n\rho_n + k_2C_0(\rho_p + \rho_n)^2$$

Here ρ_p and ρ_n are proton and neutron densities; a_p and a_n are π^-p and π^-n s-wave scattering lengths and c_p, c_n are the corresponding p-wave scattering volumes; k_1, k_2 are kinematical factors. B_0 and C_0 are related to the π^-2N scattering amplitudes and, due to the dominating two-nucleon absorption, are complex numbers. The denominator of the last term represents the Lorentz–Lorenz effect, where $0 \leq \xi \leq 1$ is a free parameter describing the degree to which this effect exists. This potential is the result of a description of the π-nucleus interaction in terms of the free π–nucleon interaction at threshold. The first three terms are related to the s-wave interaction and the last term to the p-wave interaction.

One nice feature is that the various measurable quantities depend mainly upon one term. The energy shifts and widths of the $1s$ level thus, respectively, depend on the real and imaginary parts of the first term provided the two following correction terms in Eq. (25) are small. Shifts and widths of p, d, etc., levels depend on real and imaginary parts of the last term. Unfortunately, the real numbers a_p and a_n cancel to a large degree (see below) and hence the repulsive $1s$ level shift originates mainly from the Pauli correlation term and is interpreted as a proof of the granular structure of nuclear matter.

A fit has been made to a large number of experimental data ε_{1s}, Γ_{1s} from $_2$He–$_{12}$Mg; ε_{2p}, Γ_{2p} from $_2$He–$_{33}$As; ε_{3d}, Γ_{3d} from $_{20}$Ca–$_{59}$Pr; ε_{4f}, Γ_{4f} from $_{67}$Ho–$_{92}$U. The following numbers have been obtained:

$$a_n = -0.092 \, m_\pi^{-1} \qquad c_n = 0.385 \, m_\pi^{-3}$$
$$a_p = +0.083 \, m_\pi^{-1} \qquad c_p = 0.037 \, m_\pi^{-3}$$

A better fit could be obtained with $\xi = 1$, i.e., by including the Lorentz–Lorenz effect. It has also recently been concluded[25] that

$$\text{Re } B_0 = -\text{Im } B_0$$

The width Γ_{1s} depends mainly on Im B_0. All the Γ_{1s} values from $_3$Li to $_{11}$Na can be described by this one constant. Im B_0 turns out, however, to be three to four times larger than expected on the basis of the π^- absorption on deuterium.

On the other hand, the π-nucleus scattering length a_π^A of nucleus A can be described[25] as

$$a_\pi^A = a_\pi^{n\alpha} \approx n a_\pi^\alpha$$

where α_π^α is the π–α-particle scattering length, which supports the idea of α-particle clustering of the light nuclei up to $A = 20$.

The dominance of c_n over c_p would, in principle, allow a determination of the neutron distribution by means of ε_{2p}. In practice, however, the measurements have not so far been conclusive (for example ^{40}Ca/^{44}Ca).

4.4.2. Kaonic Atoms

A considerable number of strong interaction data has also been obtained for kaonic atoms. A similar attempt has been made, as for pionic atoms, to describe the effects by an optical potential. Originally the hope existed that the many complications present in pionic atoms would be absent in the kaonic case. It is known that, in contrast to the π–nucleon interaction, the K–nucleon p-wave interaction is weak and hence justifies the omission of the last term in the potential [Eq. (25)]. Furthermore, kaons can be absorbed kinematically by one nucleon through the process

$$K^- + N \to \Lambda(\Sigma) + \pi$$

and hence the two-nucleon absorption, which is exclusively present for pions, can also be neglected here. The potential thus has the simple form

$$V(r) = -\frac{2\pi}{\mu}\left(1 + \frac{m_k}{m_N}\right)(a_p\rho_p + a_n\rho_n) \tag{26}$$

where again μ is the reduced K^-–nucleus mass, a_p and a_n are the K^-p and K^-n free s-wave complex scattering lengths, and m_K and m_N are the kaon and nucleon mass, respectively. a_p and a_n are related to the scattering lengths a_0 and a_1 of the isospin states $I = 0$ and $I = 1$, which are obtained from the K–nucleon scattering amplitude extrapolated to zero energy.

The result of the comparison between the measured ε and Γ with values calculated by using the optical potential, which leaves no free parameter, was rather disappointing. It turned out that only an inversion of the sign of the real part of the potential gave agreement. This means that only an attractive potential could fit the data,[26] whereas strict application of the above concept of using the scattering lengths would result in a repulsive potential.

If one forms an average effective scattering length in order to reduce the number of free parameters, one can define

$$\bar{a} = \frac{1}{N+Z}\left[\left(N + \frac{Z}{2}\right)a_1 + \frac{Z}{2}a_2\right]$$

From a simultaneous fit to the experimental data of the $2p$ levels for $^{10}_5\text{B}$, $^{11}_5\text{B}$, $_6\text{C}$ and the $3d$ levels in $_{15}\text{P}$, $_{16}\text{S}$, $_{17}\text{Cl}$, one finds[26,27] Re $\bar{a} = 0.56$ fm and Im $\bar{a} = 0.69$ fm, whereas the use of the a_1 and a_0 scattering lengths would yield Re $\bar{a} = -0.42$ fm. One can, of course, also try to determine the two complex scattering lengths separately, although this cannot be done unambiguously. One can, however, assume that the problems are mainly located in the K^-p channel, where a resonance $Y_0^*(1405)$ exists just about 27 MeV below the K^-p threshold with a width of 25 MeV, which suggests an attractive interaction (Re $a_p > 0$) near threshold.[28] In Table 8 a comparison is given between the free scattering

Table 8. K^-n and K^-p Scattering Lengths a_n and a_p in fm

Method	Re a_n	Im a_n	Re a_p	Im a_p
Extrapolated from free scattering	−0.07	0.62	−0.90	0.66
Calculated, Ref. 28	+0.13	0.40	+1.00	1.60
Fitted from K^- atom data	+0.13	0.40	+0.99	0.98

length, the values obtained by Bardeen in considering the Y_0^*, and a fit to the data where a_n is fixed to the Bardeen values. It can be seen that the data give an attractive real part for a_p, as does the Bardeen approach; this is not however, without criticism.

It is obvious, that the hope of deducing nuclear structure effects at the surface of the nuclei, as suggested earlier,[29] is fairly modest so long as the basic interaction is not fully understood. An alternative approach may also be of interest. If a nucleus is available which has a known proton and neutron distribution (for example by comparison of an isotope pair), then the data obtained from the strong interaction effects in the corresponding kaonic atoms may yield information about the K^-–nucleus interaction at near-zero energies.

4.4.3. Σ-Hyperonic Atoms

As explained above, the short-lived Σ^- hyperons can form Σ^- atoms only if produced after formation of kaonic atoms and subsequent nuclear capture of the K^-. As a consequence of this, Σ-hyperonic spectra are rather difficult to observe. They appear to be about one order of magnitude weaker than the corresponding kaonic spectra, which occur at the same time, and hence must compete with numerous low-intensity kaonic lines; this may of course disturb the Σ lines of interest. Until now only a very limited number of Σ transitions ($4 \to 3$ in C, $6 \to 5$ in Ca and Ti, $9 \to 8$ in Ba) have been measured[15] and yields of between 0.6 and 0.7 due to nuclear absorption have been obtained. This leads to an absorptive parameter of an optical potential, similarly constructed to that of kaonic atoms [Eq. (26)] of

$$\text{Im } \bar{a} \approx 1.0 \pm 0.35 \text{ fm}$$

A comparison between the intensities of unattenuated transitions of K^- and Σ^- at the same Z is also of interest. One thus obtains a direct measure of the number of Σ^- leaving the nucleus after the K^- capture. It is expected that the number of Σ^- escaping the nucleus depends somewhat on Z. Relative Σ intensities of between 3% and 7% with respect to kaonic intensities have been observed,[15] which enables one to draw conclusions about the Σ^- production and escape probabilities in nuclei.

4.4.4. Antiprotonic Atoms

In antiprotonic atoms all three strong interaction effects have been observed, although the energy shifts are relatively small compared with the absorptive effects and so the measurements are more difficult. Problems also exist in connection with the low stop rate attainable at present with $\bar{p}$ beams. The theoretical situation seems to be similar to that in kaonic atoms in the sense that here an optical potential corresponding to an attractive interaction would also fit the data best,[27] whereas the construction based on the s-wave scattering length of the $\bar{p}p$ and $\bar{p}n$ systems would yield a repulsive potential. Contrary to the kaonic case, no bound state close to threshold has been established yet. If, however, the same explanation is applicable for $\bar{p}$ as for K atoms, then bound states are also suggested in the $\bar{p}p$ system; such states have been anticipated on different grounds in the previous work of Shapiro and collaborators.[30] It might be however, that the entire approach with an optical potential becomes doubtful in the presence of the very strong absorptive interaction.*

In principle, $\bar{p}$ atoms should be very suitable for the study of the neutron distributions of nuclei, particularly at their surface, since by contract to the kaon, where the K^-p interaction dominates the K^-n interaction, the $\bar{p}n$ interaction should be of comparable magnitude. As the mass of the orbiting particle increases, states with larger n, and accordingly higher l, are subjected to the strong interaction. Since the $\psi_{n,l} \sim r^l$, the sensitivity to the peripheral parts of the nuclei increases with increasing l. However, only an expression $\approx (a_p\rho_p + a_n\rho_n)$ is actually determined by the observed effects in the hadronic atoms. Two approaches are attempted at present. The knowledge of the nuclear structure which is needed can be deduced by comparing two isotopes since the differences in nuclear structure are less uncertain than the overall effect. For this case measurements on $\bar{p}{}^{16}O$ and $\bar{p}{}^{18}O$ have been performed,[31] and for the first time clear evidence of an isotope effect has been seen for all three observable quantities (Table 9 and Figure 13). It is possible to deduce some information on the $\bar{p}n$ interaction from an analysis of these data. A more ambitious attempt to gain insight in the $\bar{p}p$ interaction is made by the study of the $\bar{p}p$ system itself. The energy of the $2p-1s$ transition is about 8 keV and

---

* See Note Added in Proof on page 1424.

Table 9. Isotopic Effects in $\bar{p}\ {}^{16}O/{}^{18}O$,
$\delta \equiv [({}^{18}O - {}^{16}O)/{}^{16}O]$

$\delta\varepsilon_{4f} = 0.62 \pm 0.38$
$\delta\Gamma_{4f} = 0.75 \pm 0.31$
$\delta\Gamma_{5g} = 0.34 \pm 0.20$

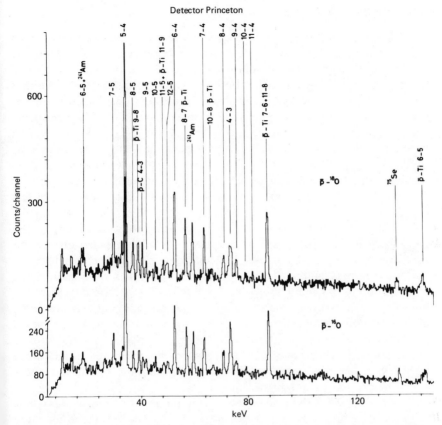

Figure 13. Isotope effect in $\bar{p}\,^{16}O$ and $\bar{p}\,^{18}O$. Lines not specified originate from O. Width, yield, and energy of the $\bar{p}$ $4 \to 3$ transition subject to the isotope effect. A titanium box was used as a container for $D_2^{16}O$ and $D_2^{18}O$ $\bar{p}$ Ti lines serve as calibration (Ref. 31).

hence difficult to measure. The greatest problem, however, originates from the very low intensity which is expected for this line; there might be already considerable absorption from the $2p$ level. Furthermore, Stark mixing is probably rather strong for all the atoms which are formed in liquid hydrogen; they carry no net charge and are hence only weakly repelled by the electrons from the strong nuclear electric fields. The admixture of s states by the Stark effect leads however to considerable nuclear absorption already from high n states for all hadrons and hence to a weak population of the low-lying states. The present state of the art of measuring low-energetic $\bar{p}$ x-rays is demonstrated by the spectrum of $\bar{p}\,^4He$ (Figure 7), where lines with energies as low as 3 keV could be measured sufficiently well to enable the deduction of strong interaction effects.

4.5. Outlook

The progress experienced during the last decade has essentially been achieved as a consequence of the greatly improved energy resolution originating from the use of solid-state detectors. Hyperfine patterns in muonic atoms and strong interaction effects in hadronic atoms have thus been investigated in detail. In the near future a strong impetus may be expected in the muon and pion field from the new meson factories from which pion and muon beams of great intensity and high beam quality have become available, thereby leading to stop densities which are increased by several orders of magnitude. Smaller amounts of material are therefore needed, and the possibility of investigating isotopes and thin targets is facilitated, allowing an improved measurement of the intensities of low energetic lines.

On the other hand, weak transitions such as $2s$–$2p$ transitions can be measured with high precision and the very weak lines in hydrogen isotopes may also be more easily accessible. Coincidence measurements between various members of the x-ray cascade and between x-rays and nuclear γ-rays become possible and open a large field of experiments which lead not only to a better understanding of the cascade but also to more information on strong interaction effects and the absorption mechanism of pions. With higher-intensity beams the measurement of the circular polarization of the muonic x-rays is also possible; hence additional information which is important for many applications, e.g., in μ-capture experiments, may be obtained.

Compared with the possibilities in the muonic and pionic x-ray field the facilities with the heavier particles are rather modest. Here improved beams should also be possible if one is willing to invest in larger-aperture beam elements. Suggestions have recently been made to decelerate antiprotons from the energy at which they are produced (the maximum of the production spectrum is about at 4 GeV for incoming protons of 20 GeV) to energies of a few hundreds of MeV where they could be easily stopped. In this deceleration process no increase of the phase space is expected and hence a beam of extraordinary quality, leading to a $\bar{p}$-stop rate increased by many orders of magnitude as compared to conventional beams, may be obtained. It is clear that such a development would have a similar effect on the antiproton field as the meson factories have had on the pion and muon field.

Note Added in Proof. Recently, possible evidence for strongly bound states of the $\bar{p}p$-system has been found by observing highly energetic ($\sim$100 MeV) γ-transitions.[32] These states are surprisingly narrow ($\Gamma \leqslant$ 10 MeV) if one keeps in mind that the well-known annihilation process for antiprotons $p + \bar{p} \to n\pi$ is a strong process. This has led to speculations about so-called baryonium states made up from a diquark and an antidiquark.

References

1. J. A. Wheeler, *Phys. Rev.* **71**, 370 (1947).
2. V. L. Fitch and J. Rainwater, *Phys. Rev.* **92**, 789 (1953).
3. M. Camac, A. D. McGuire, J. B. Platt, and H. J. Schulte, *Phys. Rev.* **88**, 134 (1952).
4. G. R. Burleson, D. Cohen, R. C. Lamb, D. N. Michael, R. A. Schluter, and T. O. White, *Phys. Rev. Lett.* **15**, 70 (1965).
5. G. Backenstoss, T. Bunaciu, S. Charalambus, J. Egger, H. Koch, A. Bamberger, U. Lynen, H. G. Ritter, and H. Schmitt, *Phys. Lett.* **33B**, 230 (1970).
6. A. Bamberger, U. Lynen, H. Piekarz, J. Piekarz, B. Povh, H. G. Ritter, G. Backenstoss, T. Bunaciu, J. Egger, W. D. Hamilton, and H. Koch, *Phys. Lett.* **33B**, 233 (1970).
7. M. Leon and H. A. Bethe, *Phys. Rev.* **127**, 636 (1962).
8. L. I. Ponomarev, *Ann. Rev. Nucl. Sci.* **23**, 395 (1973).
9. E. H. Wichmann and N. M. Kroll, *Phys. Rev.* **101**, 843 (1956); (b) R. C. Barrett, S. J. Brodsky, G. W. Ericson, and M. H. Goldhaber, *ibid.* **166**, 1589 (1968); (c) J. Blomqvist, *Nucl. Phys. B* **48**, 95 (1972).
10. R. Engfer, H. Schneuwly, J. L. Vuilleumier, H. K. Walter, and A. Zehnder, *At. Nucl. Data Table* **14**, 509 (1974).
11. C. S. Wu and L. Wilets, *Ann. Rev. Nucl. Sci.* **19**, 527 (1969).
12. M. Erickson and T. E. O. Ericson, *Ann. Phys. (N.Y.)* **36**, 323 (1966).
13. G. Backenstoss, *Ann. Rev. Nucl. Sci.* **20**, 467 (1970).
14. G. Backenstoss, J. Egger, H. Koch, H. P. Povel, A. Schwitter, and L. Tauscher, *Nucl. Phys. B* **73**, 189 (1974).
15. G. Backenstoss, T. Bunaciu, J. Egger, H. Koch, A. Schwitter, and L. Tauscher, *Z. Phys. B* **273**, 137 (1975).
16. A. A. Dzhuraev, V. S. Evseev, G. G. Myasischcheva, Yu. V. Obukhov, and V. S. Roganov, *Sov. Phys.-JETP* **35**, 748 (1972).
16a. R. Abela, G. Backenstoss, I. Schwanner, P. Blüm, D. Gotta, L. Simons, and P. Zsoldos, *Phys. Lett.* **71B**, 290 (1977).
17. G. Backenstoss, *Atomic Physics*, Vol. 4, p. 163, Ed. G. zu Putlitz, Plenum Press, New York (1974).
18. E. Fermi and E. Teller, *Phys. Rev.* **72**, 399 (1947).
19. G. Backenstoss, H. Daniel, K. Jentzsch, H. Koch, H. P. Povel, F. Schmeissner, K. Springer, and R. L. Stearns, *Phys. Lett.* **36B**, 422 (1971).
20. A. Bertin, G. Carboni, J. Duclos, U. Gastaldi, G. Gorini, G. Neri, J. Picard, O. Pitzurra, A. Placci, E. Polacco, G. Torelli, A. Vitale, and E. Zavattini, *Phys. Lett.* **55B**, 411 (1975).
21. L. Tauscher, G. Backenstoss, K. Fransson, H. Koch, A. Nilsson, and J. D. Raedt, *Phys. Rev. Lett.* **35**, 410 (1975).
22. R. Rafelski, B. Müller, G. Soff, and W. Greiner, *Ann. Phys. (N.Y.)* **88**, 419 (1974); P. J. S. Watson and M. K. Sundaresan, *Can. J. Phys.* **52**, 2037 (1974); V. W. Hughes, in *Proceedings of the Sixth International Conference on High Energy Physics and Nuclear Structure, Santa Fe, 1975*, p. 515, AIP, New York (1975).
23. J. Bernabeu, T. E. O. Ericson, and C. Jarlskog, *Phys. Lett.* **50B**, 467 (1974); G. Feinberg and M. Y. Chen, *Phys. Rev. D* **10**, 190 (1974).
24. R. C. Barrett, *Rep. Progr. Phys.* **37**, 1 (1974).
25. L. Tauscher, in *Proceedings of the 6th International Conference on High Energy Physics and Nuclear Structure, Santa Fe, 1975*, p. 451, AIP, New York (1975).
26. G. Backenstoss, J. Egger, H. Koch, H. P. Povel, A. Schwitter, and L. Tauscher, *Nucl. Phys. B* **73**, 189 (1974).
27. H. Koch, in *Proceedings of the Fifth International Conference on High Energy Physics and Nuclear Structure, Uppsala, 1973*, p. 255, Almquist and Wiksell, Stockholm (1974).
28. W. A. Bardeen and E. W. Torigoe, *Phys. Lett.* **38B**, 135 (1972). M. Alberg, E. M. Henley, and L. Wilets, *Phys. Rev. Lett.* **30**, 255 (1973).

29. D. H. Wilkinson, *Phil. Mag.* **4**, 215 (1960), in *Proceedings of the International Conference on Nuclear Structure, Tokyo, 1967*, (Suppl. *J. Phys. Soc. Japan* **24**, 469 (1968).
30. L. N. Bogdanova, H. D. Dalkarov, and J. S. Shapiro, *Ann. Phys.* **84**, 261 (1974); C. Dover, in *Proceedings of the Fourth International Symposium on NN̄ Interaction*, Vol. II, p. VIII 37–91, Syracuse University (1975).
31. H. Poth, G. Backenstoss, I. Bergström, P. Blüm, J. Egger, W. Fetscher, R. Guigas, C. J. Herrlander, M. Izycki, H. Koch, A. Nilsson, P. Pavlopoulos, H. P. Povel, I. Sick, L. Simons, A. Schwitter, J. Sztarkier, and L. Tauscher, *Nucl. Phys. A* **294**, 435 (1977).
32. P. Pavlopoulos, G. Backenstoss, P. Blüm, K. Fransson, P. Guigas, N. Hassler, M. Izycki, H. Koch, A. Nilsson, H. Poth, M. Suffert, L. Tauscher, and K. Zioutas, *Phys. Lett.* **72B**, 415 (1978).

Reviews and Summaries

Exotic atoms: E. H. S. Burhop, *High Energy Physics*, Vol. 3, p. 109, Academic Press, New York (1969); Y. N. Kim, *Mesic Atoms and Nuclear Structure*, North-Holland, Amsterdam (1971); Ref. 17.

Muonic atoms: Refs. 11, 10, 22. S. Devons and I. Duerdoth, *Adv. Nucl. Phys.* **2**, 295 (1968).

Pionic atoms: Refs. 13, 25.

Kaonic atoms: R. Seki and C. E. Wiegand, *Ann. Rev. Nucl. Sci.* **25**, 241 (1975); G. Backenstoss and J. Zakrzewski, *Contemp. Phys.* **15**, 197 (1974); Refs. 25, 27.

Σ *Atoms*: G. Backenstoss and J. Zarkrzewski, *Contemp. Phys.* **15**, 197 (1974).

$\bar{p}$ *atoms*: Refs. 25, 27.

32

Positronium Experiments

STEPHAN BERKO, KARL F. CANTER,
AND ALLEN P. MILLS, JR.

1. Introduction

Positronium (Ps), the hydrogenlike bound state between an electron and its antiparticle, the positron, plays a fundamental role in atomic physics. Together with muonium, it serves as an ideal system for testing the predictions of quantum electrodynamics, since the theoretical interpretation is unhindered by the presence of hadrons. In the hydrogen atom, for example, the hyperfine splitting of the ground state is one of the most accurately measured quantities in physics[1] (to one part in 10^{13}). However, the theoretical uncertainties stemming from the finite size and polarizability of the proton have so far precluded a meaningful comparison between theory and experiment to better than a few parts per million.[2] Positronium and muonium, on the other hand, being composed of leptons only, are describable to a very good approximation by lepton and photon fields alone, and are affected by heavy particles only through vacuum polarization terms of negligible magnitude. Compared to the stable atoms, however, Ps presents challenging experimental difficulties attributable to its short annihilation decay times and the impossibility of producing macroscopic quantities for observation.

Since the pioneering work of Deutsch[3] in the early 1950s that started with the discovery of Ps and led to the first measurement of the Ps ground-state splitting, physicists have attempted to measure new quantities, such as the energies of the Ps excited states and their basic annihilation rates. Although great progress has been made in obtaining more and more accurate values of the ground-state fine-structure splitting, the

STEPHAN BERKO and KARL F. CANTER • Brandeis University, Waltham, Massachusetts 02154.   ALLEN P. MILLS, JR. • Bell Laboratories, Murray Hill, New Jersey 07974.

detection of the excited states of Ps has eluded experimentalists for over 20 years. The $n = 2$ states have been detected only recently, owing to a new experimental technique of forming positronium in vacuum. Since there exist a number of excellent reviews covering various stages of Ps physics,[4–7] we shall present mainly a discussion of the recent experimental developments, with particular emphasis on the discoveries which have led to the measurement of one of the fine-structure intervals of the $n = 2$ states of Ps.

Low-energy positron physics is not confined, of course, to the study of positronium in vacuum. During the last two decades the study of the interaction of low-energy positrons with electrons in gases and in condensed matter has been the subject of numerous investigations that led to the development of new fields in atomic physics, in chemistry, and in solid-state physics. In atomic physics,[8] the measurement of low-energy positron–atom collision cross sections of Ps formation yields and of Ps–atom interactions has been of great interest, particularly since the development of slow positron beams. Positron and Ps interactions with complex molecules, particularly in liquids, has become a branch of radiation chemistry.[9] In solid-state physics,[10] the angular distribution between the two photons emitted in the two-quantum annihilation process when thermalized positrons annihilate with electrons leads to the determination of electronic momentum densities and of Fermi surfaces in metals.[11] Recently, the measurement of positron lifetimes and 2γ angular distributions of positrons trapped in various defects has become an important new technique in metallurgy.[12] We refer to these developments of slow positron physics and chemistry because they are relevant to the various ways of producing Ps in the laboratory.

After a brief introduction to the selection rules involved in positron–electron annihilation and a review of the usual Ps detection techniques, we shall discuss the methods of forming Ps for experimental purposes.

2. Annihilation Selection Rules and Positronium Detection Techniques

The usual method of studying positron interactions with matter, including positronium formation, is to detect the annihilation quanta and measure their various properties characteristic of the annihilating electron–positron state. The annihilation rates are given by quantum electrodynamics (QED), and the basic characteristics of the annihilation quanta, such as their number and polarization correlation, are governed by fundamental selection rules due to C and P conservation.[6] One obtains for the number of annihilation quanta N the relation $(-1)^{L+S} = (-1)^N$, where L is the relative orbital angular momentum and S is the total intrinsic spin of the annihilating electron–positron pair. When Ps is formed prior to annihilation, this section rule results in ortho-Ps (spin = 1) S states annihilating into an odd number of photons and para-Ps (spin = 0) S states annihilating into an even number. Single-photon annihilation ($N = 1$) is forbidden, unless there is a nucleus nearby (annihilation with tightly bound electrons) to conserve momentum. Since annihilation takes place only if the positron and electron overlap within a Compton wavelength, the annihilation rate Γ from P states is extremely small; P states of Ps will decay optically to lower-lying S states rather than annihilate. For $n = 2$ Ps, for example, one obtains[13] $\Gamma_{2\gamma}(P) \approx 10^4\,\mathrm{s}^{-1}$ compared to $\Gamma_{2P \to 1S} \approx 10^8\,\mathrm{s}^{-1}$ for the optical rate. The annihilation cross section σ_N into N photons decreases even faster than α^N, where α is the fine-structure constant, due to phase-space considerations: $\sigma_{2\gamma}/\sigma_{3\gamma} \approx 10^3$. Thus for all practical purposes the spin-singlet S states annihilate into two photons (2γ) and the spin-triplet S states into three photons (3γ). The large value of $\sigma_{2\gamma}/\sigma_{3\gamma}$ results in correspondingly large ratios of the annihilation lifetimes (τ) of ortho-Ps ($^3S_1$) and para-Ps ($^1S_0$). For example, in the $n = 1$ state the Ps lifetimes are $\tau_{2\gamma}(^1S_0) \cong 1.25 \times 10^{-10}$ and $\tau_{3\gamma}(^3S_1) \cong 1.4 \times 10^{-7}$ s. The ratio of 2γ to 3γ total annihilation yields Y from these states is $Y_{2\gamma}/Y_{3\gamma} = \frac{1}{3}$, reflecting the statistical factor of the spin states. On the other hand, if positrons annihilate without forming bound states (from "scattering states"), the large value of $\sigma_{2\gamma}/\sigma_{3\gamma}$ produces a large ratio of 2γ to 3γ decays ($Y_{2\gamma}/Y_{3\gamma} \cong 370$ for a spin-averaged system). It is thus easy to detect positronium formation, by measuring the relative yield of 2γ vs. 3γ annihilations. Momentum and energy conservation require the two quanta in 2γ annihilation to be monoenergetic ($E_{1,2} \cong mc^2$) and to be emitted in opposite directions in the center-of-mass system of the annihilating electron–positron pair. On the other hand, the quanta in 3γ annihilation have a characteristic continuous energy distribution.[6]

All measurements in positronium physics are based on the above characteristics of the annihilation quanta: (a) One can measure the 2γ vs. 3γ yield by twofold vs. threefold coincidence experiments between multiple γ-ray detectors. (b) One can obtain the energy distribution of one of the annihilation quanta, and thus distinguish between 2γ vs. 3γ annihilation modes by their spectrum. (c) One can measure the lifetime of the positron. The lifetime measurement is usually performed by using ^{22}Na as a positron emitter: the positron emission leaves the nucleus in an excited state that decays to its ground state by a 1.28-MeV γ-ray with a lifetime of less than 10^{-11} s. The time interval between this nuclear γ-ray and one of the annihilation quanta is then used to determine the lifetime of the positron. Ps is signaled in such experiments by the appearance of the two characteristic lifetimes, that of singlet vs. triplet Ps annihilation. In the presence of atoms or molecules the long-lived triplet Ps atom can collide with electrons of the proper spin and annihilate by the fast 2γ process. The "long-lived" lifetime of triplet Ps is then shortened by this "pick-off" process.[9]

3. Tests of Charge Conjugation Invariance

It is interesting to note that although the triplet $(^3S_1)$ Ps state is strictly forbidden to decay into two photons by total angular momentum conservation, the 3γ decay of the singlet $(^1S_0)$ is only forbidden by charge conjugation (C) invariance. Although C invariance is believed to be obeyed in pure electromagnetic interactions, experiments have been performed to search for possible C-violating terms in the annihilation process. Mills and Berko,[14] for example, searched experimentally for the C-forbidden $^1S_0 \rightarrow 3\gamma$ decay of the Ps ground state. Their experiment looked for the difference in angular distribution between the allowed $^3S_1 \rightarrow 3\gamma$ and the C-forbidden $^1S_0 \rightarrow 3\gamma$ decays. They conclude that $\Gamma_{3\gamma}(1^1S_0)/\Gamma_{2\gamma}(1^1S_0) \leq 2.8 \times 10^{-6}$. Theoretically, C-violating terms would of course be extremely important, but are most unlikely. Some models of such a decay based on a C-nonconserving interaction yield predictions proportional to an unknown coupling constant λ divided by a cutoff mass M raised to a high power. Using $\lambda = 1$ and $M = 2m_e$, the positronium mass, leads to predicted rates $\Gamma(1^1S_0 \rightarrow 3\gamma) \approx 10^{-6}\Gamma(1^1S_0 \rightarrow 2\gamma)$. More believable predictions of $^1S_0 \rightarrow 3\gamma$ rates based on virtual transitions to $K^0 - \bar{K}^0$ states, which do violate time-reversal invariance to a small extent,[15] must yield astronomically small rates because of the light mass of Ps compared to the K^0 mesons. Even smaller rates are expected from weak interaction models because of the enormous mass of the hypothetical W vector boson. Charge-conjugation invariance also forbids the triplet state from decaying into 4γ's, and this annihilation mode has been searched for by Marko and Rich.[16] They find $\Gamma(^3S_1 \rightarrow 4\gamma) < 10^{-5}\Gamma(^3S_1 \rightarrow 3\gamma)$ which is consistent with C conservation. A decay $^3S_1 \rightarrow 4\gamma$ might arise by a C-violating emission of 3γ's followed by the emission of one extra photon at an ordinary C-conserving vertex; or such a decay could be described by a point interaction.[17] As for the $^1S_0 \rightarrow 3\gamma$ decay, the light mass of positronium makes an upper limit on $^3S_1 \rightarrow 4\gamma$ difficult to interpret in terms of known possibilities.

4. Methods of Positronium Formation

Until very recently the only methods available for forming "free" Ps atoms involved fast positrons emitted from β^+ active isotopes (usually

^{22}Na, ^{58}Co, or ^{64}Cu) directed into gases which serve the dual role of slowing down the positrons in a well-defined region of space and providing the electrons necessary for Ps formation. Recent advances in the production of slow positron beams have led to an alternative Ps formation technique that is in some cases (such as excited-state production) more desirable than the gas moderation technique. Nevertheless the conventional technique of Ps formation using fast positrons can be expected to be useful for many ground-state experiments.

4.1. Positronium Formation in Gases

Although fast positrons are emitted into a gas with the usual β^+ decay spectrum ranging over hundreds of keV energy, they are rapidly slowed down into the few-eV region by ionizing collisions with the gas molecules. The annihilation cross section at high velocities is much smaller than the inelastic scattering cross section; thus practically all positrons will slow down to the few-eV region without undergoing annihilation. It is in this energy region where abundant Ps formation can take place by radiationless electron capture of one of the bound electrons of the gas atom or molecule. In order for a positron to form Ps in a gas with single electronic ionization energy E_I, its kinetic energy E_+ must exceed $E_I - E_{Ps}$, where E_{Ps} is the binding energy (6.8 eV for $n = 1$ Ps, 1.7 eV for $n = 2$). $E_I - E_{Ps}$ is the lower limit of the so called "Ore gap,"[18] which refers to the range of positron kinetic energy favorable for stable Ps formation. The upper limit of the Ore gap is less clearly defined. The kinetic energy of the Ps formed by electron capture ($E_+ - E_I + E_{Ps}$) must be less than E_{Ps} in order for the Ps to be stable against dissociative collisions with the gas molecules. This places an upper limit of $E_+ \lesssim E_I$ for stable Ps formation. Ps formation may not be abundant for $E_+ \gtrsim E_{ex}$, the first electronic excitation of the gas molecules, because of the tendency of such excitations by positron impact to dominate over Ps formation. Rough estimates of Ps formation fractions in gases are usually obtained by bracketing the upper limit of the Ore gap between E_{ex} and E_I. For the rare gases one obtains ground-state Ps formation fractions ranging from 10% to 40%, in reasonable accord with the Ore gap picture, with the exception of xenon.[19]

A considerably smaller fraction is predicted for the $n = 2$ Ps formation. In fact, using for the upper limit of the Ore gap values close to E_{ex}, one predicts no Ore gap for $n = 2$ Ps for rare atom gases as well as for most other gases. The lack of Ore gap for $n = 2$ Ps does not completely rule out $n = 2$ Ps formation, but considerably reduces the formation probability, making it very sensitive to the cross section for the competing gas molecule excitation channels. This is perhaps one of the contributing reasons for the null results obtained in various searches for natural $n = 2$ Ps formation in gases.[20–23]

An alternative approach to producing $n = 2$ Ps in a gas is to try to excite the abundantly formed ground-state Ps atoms with external radiation of the proper uv wavelength (2430 Å). This technique will be discussed in Section 5, which deals with $n = 2$ Ps experiments.

4.2. Positronium Formation in Oxide Powders

When fast positrons are injected into certain insulating solids or liquids, they can form Ps after slowing down, in a similar manner as they do in gases. Recently, convincing arguments have been made by Mogensen in support of an alternative Ps formation mechanism in liquids and molecular solids where the positron captures a free electron belonging to the ion trail or "spur" created by the same positron slowing down in the material.[24] The Ps formed in solids, however, is greatly perturbed by the dense medium and differs substantially from the essentially vacuumlike Ps formed in low-density gases. However, if the Ps is formed close to the surface of the solid, there is the possibility that the Ps can escape from the solid by diffusion. Such an effect was first observed by Paulin and Ambrosino using finely divided (≈ 100-Å particle size) metallic oxide powders, most notably MgO.[25] Moreover, their results showed that the Ps that escapes from the oxide resides almost completely in the interparticle voids, as evidenced by an ortho-Ps lifetime close to $\tau_{3\gamma}$, the vacuum lifetime, when the powdered sample is evacuated.

The advantage of forming Ps in powders instead of gases is that the powders offer a greater stopping power in a small region than is available with gases of densities low enough to yield accurate vacuum Ps data. A drawback of the powders is that it is difficult to envision a way of varying the powder density and also maintain reproducible powder surface characteristics, namely, adsorbed contaminants and possible electrostatic charging of the powder particles due to the ionizing effects of the *in situ* fast positron source. It is possible that certain oxides also favor $n = 2$ Ps formation[26]; the detection of the $n = 2$ Ps produced in oxide powders is unlikely, however, because of the $n = 2$ Ps collisions with the powder grains. Curry and Schawlow introduced fast positrons into a thin film of MgO powder and looked for Ps escaping into the free vacuum region away from the film.[27] They observed a high yield ($\approx 2.5 \times 10^{-3}$ Ps emissions per incident fast positron) of $n = 1$ Ps escaping into the vacuum with a kinetic energy of 0.28 ± 0.10 eV. This important result demonstrated that it is energetically favorable for Ps to be outside the powder, as indicated by Paulin's experiments, and provided for the first time a method of obtaining Ps in a vacuum, albeit in the proximity of the fast positron source and its attendant radiation background. Curry and Schawlow searched for $n = 2$ Ps, but were unable to observe Lyman-α emission.[28]

4.3. Slow Positron Beams

Slow positron beam facilities were developed during the last decade in various laboratories in order to study positron–atom collisions. Because of the lack of exchange between e^+ and e^- and the reversal in the sign of the Coulomb interaction, comparison between the measured and calculated scattering cross sections for e^+–atom and e^-–atom collisions often provides highly sensitive tests of various models used in the theory of atomic collisions.[8]

The earliest attempt to obtain slow positrons was made by Madansky and Rasseti[29]; their technique was to moderate fast positrons with various foils of thicknesses comparable to the mean positron range and look for thermal positrons diffusing from the surfaces. Although such an approach was expected to yield approximately one slow positron per 10^3 incident fast positrons, no slow positrons were observed down to one in 10^5, the limit of their detection efficiency. Another experiment, originally carried out by Cherry in 1958 to look at secondary electron emission from metals due to positron impact, yielded the first evidence that a small fraction of the incident positrons might indeed be reemitted at near-thermal energies from surfaces.[30] This unpublished result was not known to most workers in the field for over a decade; in 1969 Madey reproduced the Cherry result with borderline statistics.[31]

The first slow positron beam of sufficient yield to enable any cross-section measurements was produced at Gulf Atomic Research Laboratories in 1969 by Costello, Groce, Herring, and McGowan.[32] The Gulf Atomic group observed an anomalous emission of nearly monoenergetic positrons in the few-electron-volt energy range emerging from the surface of various thin metal foils placed in front of a source of fast (several hundred keV) positrons in an attempt to moderate some of the fast positrons. By allowing helium gas into the slow positron flight tube and measuring the attenuating effect on the slow positrons, they were then able

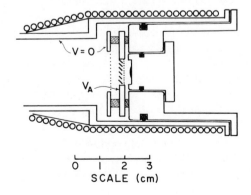

Figure 1. Venetian-blind slow positron converter. The fast positron ^{58}Co source is deposited on the plug to the right of the MgO-coated vanes. The converter is situated in a solenoidal magnetic field and V_A is the acceleration potential for the emerging slow positrons.

to obtain the first directly measured slow-positron–helium scattering cross sections. The Gulf Atomic group suggested that the slow positron emission was due to the "thermalized" positrons in the solid being ejected from the surface by virtue of a negative positron work function.[33] Recent calculations indicated that negative work functions are indeed possible for some metals.[34,35]

A major improvement of $\approx 10^3$ in slow positron yield over the arrangement of Costello *et al.* was achieved by first taking advantage of the increase in fast positron moderation offered by backscattering[36,37] and then by coating the backscattering surface with MgO powder.[38] Figure 1 shows a particular slow positron converter which has the MgO-coated backscattering surfaces arranged in a "venetian-blind" geometry.[38] This converter as presently used at Brandeis emits approximately one slow positron per 3×10^4 fast positron decays from a Co^{58} source placed behind the converter. The choice of MgO as a coating was motivated by the possibility that the very high yield of Ps emitted from a MgO coating, as reported by Curry and Schawlow, may also produce an appreciable yield of slow positrons as well, by virtue of a Ps disassociation mechanism. To date, there have been no systematic direct investigations of the slow positron emission from clean metal or metal oxide surfaces under the necessary ultrahigh vacuum conditions, and further speculation on the detailed physics of the emission mechanism should be withheld until such experiments are carried out.

4.4. Positronium Production with Slow Positrons

Several of the obstacles in forming Ps in powders and gases with fast positrons can be overcome by using a beam of slow positrons. In particular, the use of slow positrons alleviates the need for relatively high densities of the Ps formation medium and also eliminates the high source-related background signals, since the slow positrons can be easily transported well away from the source–converter assembly. In many cases, the gain in these advantages far outweighs the present limitation of the less than 0.003% slow positron production efficiency.

For these reasons experiments were initiated at Brandeis to investigate possible $n = 1$ and $n = 2$ Ps emission from various surfaces as a result of slow positron bombardment. The slow positron beam apparatus used is shown in Figure 2. The first significant result obtained from the Brandeis setup was the discovery that the efficiency with which incident slow positrons (≈ 20 eV) were reemitted as Ps from many materials was surprisingly large.[39] When the target materials were heated to a few hundred degrees Celsius, the Ps production efficiency reached 90%. The first observation of

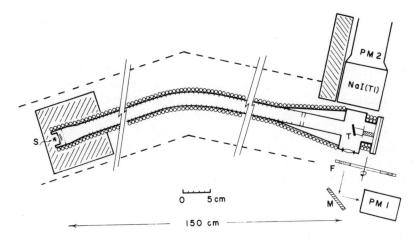

Figure 2. Slow positron beam apparatus. S, ^{58}Co source; T, target; F, filters; M, mirror; PM1, photon detector; PM2, annihilation detector (from Ref. 62).

Ps was performed by detecting the energy spectrum of the annihilation quanta as discussed in Section 2. Figure 3 shows the result of such a measurement. The indicated large increase of (3γ) decay was later verified directly by a 3γ coincidence measurement. The increase of Ps formation at high temperature is attributed to the decrease of water layers on the surface in the rather poor vacuum conditions used in the experiment

Figure 3. Annihilation γ pulse height spectrum for a NaI(Tl) scintillator viewing annihilation of positrons incident on a Ti target. The open circles show the increase of 3γ annihilation over 2γ annihilation as the target temperature is raised (from Ref. 39).

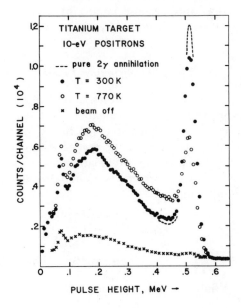

TITANIUM TARGET
10-eV POSITRONS

--- pure 2γ annihilation
• T = 300 K
○ T = 770 K
× beam off

COUNTS/CHANNEL (10^4)

PULSE HEIGHT, MeV →

$(4 \times 10^{-8}$ Torr). Not only does this new technique provide a means of efficiently producing $n = 1$ Ps in a vacuum well away from the fast positron source, but also adds a new dimension to fundamental questions concerning slow positron processes at solid surfaces. Many checks were made to ensure that the observed effects were indeed due to Ps being emitted from the targets into the vacuum, as opposed to a weakly bound Ps-like surface state. Perhaps the most convincing test was to monitor the γ-rays with a pair of detectors viewing a region away from the target. As the target temperature was raised, an increase in detected γ-rays due to ortho-Ps self-annihilations in flight and annihilations as a result of ortho-Ps collisions with the sample chamber walls was observed.

5. Energy-Level Measurements

Although superficially only a "light isotope of hydrogen," positronium differs substantially in several important details from hydrogen: (a) Given the equal masses of its constituents, positronium cannot be handled theoretically to any detailed degree by the Dirac equation, but requires the use of the relativistic two-body equation (Bethe–Salpeter).[40] (b) Because of the large positron vs. proton magnetic moment, the distinction between the easily separable fine vs. hyperfine structure in hydrogen disappears, the spin–orbit vs. the spin–spin interaction being of the same order of magnitude. (c) Due to the particle–antiparticle system, virtual annihilation diagrams become important in the positronium fine-structure splitting, already to order α^2Ry.

5.1. The Fine-Structure Interval of the Positronium Ground State

Since the electron and positron both have spin $\frac{1}{2}$, the lowest state with principle quantum number $n = 1$ is split into a singlet state and a triplet state. The singlet state lies below the triplet state by about α^2Ry owing to the interaction of the magnetic moments of the two particles, the virtual single quantum annihilation of the pair in the triplet state, and various radiative corrections. The current theoretical value of this energy difference is[41]

$$\Delta\nu_{\text{theor}} = \frac{\alpha^4 mc^2}{2h} \left[\frac{7}{6} - \left(\frac{16}{9} + \ln 2 \right) \frac{\alpha}{\pi} + \frac{1}{2} \alpha^2 \ln \alpha^{-1} + O(\alpha^2) \right]$$

$$= 203.4040 \pm 0.0003 \text{ GHz} \tag{1}$$

While the 1.5 ppm uncertainty of this value stems principally from the error in the Sommerfeld fine-structure constant $\alpha = e^2/\hbar c = 1/137.03604 \pm 0.7$ ppm,[42] the neglected terms in the expansion may well represent a much larger uncertainty in $\Delta\nu_{\text{theor}}$ since $\alpha^2 \Delta\nu = 0.0108$ GHz. Recent advances in g-2 measurements for electrons indicate that a value of α accurate to 0.3 ppm is obtainable and should reduce the error in Eq. (1).[43]

All precision measurements of $\Delta\nu$ to date have followed the pioneering experiment of Deutsch and Brown in 1952.[44] These experiments are all based on the annihilation selection rules discussed in Section 2, requiring spin–triplet S states to annihilate into 3γ's, spin–singlet states into 2γ's. Positronium is formed by stopping positrons from a radioactive source in a buffer gas which is contained in a microwave cavity and located in a uniform magnetic induction B. As shown in Figure 4 the triplet ground state is split into two Zeeman sublevels by the magnetic induction, and transitions between the $m = 0$ and the $m = \pm 1$ states may be induced by an rf magnetic field perpendicular to B. The Zeeman effect mixes singlet and triplet states of $m = 0$. Thus at a given field the "triplet" state annihilates by 2γ's proportionally to its singlet admixture. When inducing rf transitions from the $m = \pm 1$ to the $m = 0$ state, further 2γ decays take place. One thus observes a resonant increase in 2γ rays.

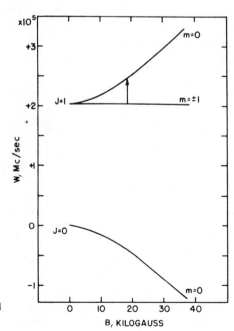

Figure 4. Zeeman splitting of the ground state of positronium.

The Breit–Rabi formula gives the resonant value of the rf frequency f_{01} in terms of B and $\Delta\nu$:

$$f_{01} = \tfrac{1}{2}\Delta\nu[(1+x^2)^{1/2}-1] \tag{2}$$

where $x = 2\mu_B g' B/h\,\Delta\nu$ and g' is the bound-state electron g factor in positronium.[45] If the nuclear magnetic resonance of the protons in a spherical sample of water (proton magnetic moment μ_p') is used to measure B, we may write

$$\Delta\nu = \left(\frac{g'}{g}\right)^2\left(\frac{\mu_e}{\mu_p'}\right)^2\left(\frac{f_p^2}{f_{01}}\right) - f_{01} \tag{3}$$

where f_p is the measured proton resonance frequency in the water, and we

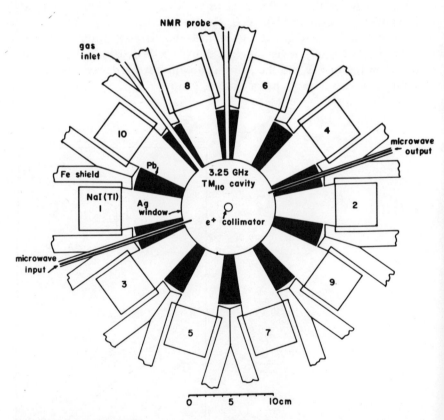

Figure 5. Multiple annihilation detectors used to detect the positronium Zeeman resonance for positronium formed in a buffer gas in a microwave resonant cavity. The magnetic field at the center of the cavity is measured using the removable NMR probe (from Ref. 48).

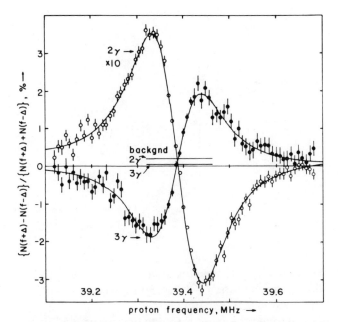

Figure 6. Positronium Zeeman first difference resonance signals obtained using a 500-Torr SF$_6$ buffer gas (from Ref. 48).

use $g'/g = 1 - 11.1 \times 10^{-6}$ and $\mu_e/\mu_p' = 433{,}263.56 \pm .06$.[46] Because of the increased annihilation rate of the mixed $m = 0$ state in a magnetic field, the resonance condition of B and f_{01} causes an increase in the 2γ annihilation yield and a concomitant decrease in the 3γ yield. Since rather large rf fields are required, the experiments use a tuned microwave cavity; and since it is difficult to change the tuning while holding the rf power level constant, a resonance curve is obtained by holding the rf frequency fixed and observing the changes in the annihilation yields as a function of B. The value of f_p at the peak of such a resonance is substituted into Eq. (3) to obtain an experimental value for $\Delta\nu$.

The natural linewidth (full width at half-maximum) of the Zeeman transition is given approximately by[47]

$$\Delta B/B = \Gamma(1^1S_0 \to 2\gamma)/4\pi \, \Delta\nu = 0.312\% \qquad (4)$$

for small B, where $\Gamma(1^1S_0 \to 2\gamma)$ is the annihilation rate of the singlet ground state. To achieve a $\Delta\nu$ precision of 5 ppm the Zeeman line center must be measured to 2.5 ppm or about 10^{-3} of the linewidth. A given signal with an amplitude S on the order of $S = 1\%$ thus dictates that enormous amounts of data have to be collected, the total number of counts required being $N \approx (10^3)^2 S^{-2} \approx 10^{10}$. The most recent experiments to

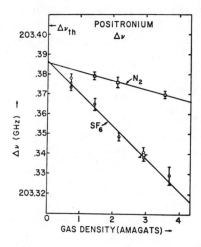

Figure 7. Plot of $\Delta \nu$ vs. gas density. The solid lines are best fits to a linear density dependence (from Ref. 48).

measure $\Delta\nu$ used multiple counter systems (see Figure 5) to obtain the necessary high counting rate.[48,49] Typical Zeeman resonances (essentially the first derivatives) are shown in Figure 6.

Besides obtaining the large amounts of data needed, it is also necessary to make an extrapolation to zero gas density $\Delta\nu(0)$ (see Figure 7) to remove the effects of the pressure shift first observed by Theriot, Beers, and Hughes.[47] Measurements[49] and calculations[50] of the pressure shifts in the rare gases have recently been reported. The experimental $\Delta\nu(0)$ results are shown in Table 1. The discrepancy between these values and the $\Delta\nu_{\text{theor}}$ of Eq. (1) indicates the importance of calculating all the $O(\alpha^2)$ corrections. The present trend towards higher-precision measurements seems to yield a factor of 10 increase in accuracy per 10 years, and it is

Table 1. Experimental Determinations of the Positronium Hyperfine Interval $\Delta\nu$

Authors	Ref.	Date	$\Delta\nu$ (GHz)
Deutsch and Brown	44	1952	203.2(3)
Weinstein *et al.*	51	1954	203.38(4)
Weinstein *et al.*	52	1955	203.35(5)
Hughes *et al.*	53	1957	203.33(4)
Theriot *et al.*	47	1967	203.403(12)
Carlson *et al.*	54	1972	203.396(5)
Mills and Bearman	48	1975	203.3870(16)
Egan *et al.*	49	1975	203.3849(12)
Theory	41		203.404

desirable that further improvements be made. An eventual measurement of $\Delta\nu$ at the 1 ppm level would present a formidable challenge to theorists, and would in principle provide one of the best determinations of the fine-structure constant and a stringent test of QED.

5.2. The Positronium n = 2 State

Soon after the discovery of positronium, Kendall,[55] under the guidance of Deutsch, attempted to form the positronium $n = 2$ states in a gaseous medium by optically exciting the abundantly formed $n = 1$ states. An intense fine line from a tin arc lamp which is within 1 Å of the 2430 Å $1 S \rightarrow 2P$ transition of Ps was used. The technique of observing the excitation of the $2P$ states was based on the different Zeeman mixing of the $n = 2$ states from the $n = 1$ states in the presence of a magnetic field. No statistically significant results were reported by Kendall. More recently, a statistically significant signal [a $(0.19 \pm 0.04)\%$ 2γ coincidence increase] was reported by a group at Yale.[56] A roughly similar setup at Bell Laboratories, however, yielded a null result.[57] Regardless of whether or not the Yale results were indeed due to Ps excitations, it is unlikely that the

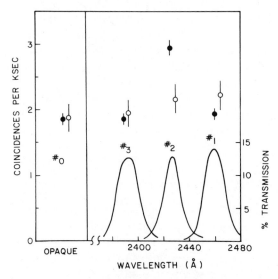

Figure 8. uv photon-annihilation γ coincidence counting rate (100-ns resolving time) using different interference filters in front of the photon detector (see Figure 2). The solid points correspond to an incident positron energy of 25 eV and the open circles correspond to 400-eV incident energy (from Ref. 62).

method would be useful for $n = 2$ Ps spectroscopy because of the large mixing and quenching of the $n = 2$ states in the magnetic field due to Zeeman and motional Stark effects. Several other attempts to observe $n = 2$ Ps were based on looking for $2P$–$1S$ photons from Ps formed in condensed matter,[28,58-61] as well as gases.[20-23]

The first direct observation of the Lyman-α radiation was made at Brandeis using the positronium formation technique by slow positron beams discussed in Section 4.4. Following the discovery of the highly efficient formation of positronium when bombarding surfaces with slow positrons, an intensive search was made to detect Lyman-α radiation expected to be emitted if $n = 2$ states were formed by the same process.[62] An ultraviolet optical setup was used to detect uv single photons by a photomultiplier (see Figure 2). In order to minimize background radiation, the output of the uv photomultiplier was measured in time coincidence with the annihilation γ's observed by a NaI(Tl) detector. A filter wheel changed three filters and an opaque stop to measure the wavelength of the radiation. Figure 8 shows a positive signal obtained in the Lyman-α filter position for 25-eV positrons incident on a Ge target. The fact that the signal vanished when the incident energy was increased to 400 eV served as a check against the Lyman-α signal being due to a spurious scintillation line. The results along with other tests for possible scintillation effects conclusively demonstrated that $n = 2$ states were produced.

The relative abundance of triplet and single $2P$ states was found to be about 3 to 1, as expected, using a time-delayed-coincidence measurement between Lyman-α photon and the subsequent annihilation of the ground state. Based on the reasonable assumption that $2S$ states were also present, an experiment to induce $2S$ to $2P$ transitions was set up. The successful completion of this experiment established conclusively the formation of positronium in the $2S$ and $2P$ states and resulted in the first measurement of a fine-structure interval in an excited state of positronium.[63]

5.3. Fine-Structure Measurement in the $n = 2$ State

The importance of the experimental determination of the $n = 2$ Ps fine structure, which includes Lamb-shift-like terms, is well known.[5,7] In contrast to hydrogen, the positronium $2S$ and $2P$ levels are nondegenerate already in order $\alpha^2 Ry$. The fine structure of these levels has been computed through order $\alpha^3 Ry$ by Fulton and Martin[64] using the Bethe–Salpeter equation.

An energy-level diagram of the $n = 1$ and $n = 2$ states of positronium is shown in Figure 9. Of the 15 possible transitions between the six $n = 2$ energy levels in zero field, the three electric dipole transitions of the long-lived triplet states are the easiest to detect. These three fine-structure

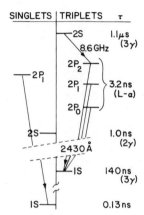

Figure 9. Term diagram of the $n = 1$ and $n = 2$ states of positronium. The annihilation lifetimes of the S states and the Lyman-α decay time of the $2P$ states are also indicated.

intervals are predicted to lie in the common microwave bands ($2\,^3S_1 \rightarrow 2\,^3P_2$, 8625 MHz; $2\,^3S_1 \rightarrow 2\,^3P_1$, 13,010 MHz; $2\,^3S_1 \rightarrow 2\,^3P_0$, 18,496 MHz). From these several accessible fine-structure intervals the $2\,^3S_1$–$2\,^3P_2$ splitting was chosen for study, since it was predicted to lie in the X band and it involves the long-lived $2\,^3S_1$ state. Fulton and Martin obtain theoretically

$$\Delta\nu(2\,^3S_1 - 2\,^3P_2) = \frac{23}{480}\,\alpha^2 \text{Ry}(1 + 3.766\alpha) = 8625.14 \text{ MHz} \qquad (5)$$

and estimate that the neglected terms of order α^3Ry may amount to several MHz. Present uncertainty in α produces only a 0.7 ppm uncertainty in $\Delta\nu$.

The $2\,^3S_1$ state has a 3γ annihilation rate one-eighth of the $1\,^3S_1$ rate to lowest order and has an extremely small rate for optical transitions to the ground state. The $2P$ states, on the other hand, have little probability for annihilation and a rate Γ_p for spontaneous transitions to the $1S$ ground state by Lyman-α emission of

$$\Gamma_p = \Gamma(2P \rightarrow 1S) = (\tfrac{2}{3})^8 \Gamma(1\,^1S_0 \rightarrow 2\gamma) = (3.18 \text{ ns})^{-1} \qquad (6)$$

to lowest order. The $2\,^3S_1$ state is thus metastable compared to the $2P$ states. Applying microwave electric fields at the appropriate frequency should cause transitions $2\,^3S_1 \rightarrow 2\,^3P_2$ followed by Lyman-α emission and finally annihilation of the ground state. The measurement is thus based on the observation of an enhanced Lyman-α emission rate when $2\,^3S_1 \rightarrow 2\,^3P_2$ transitions are induced by an rf electric field at the proper frequency. The experiment was made possible by the production of Ps in vacuum, using a low-energy positron beam, at essentially zero field ($B \approx 50$ G), thus circumventing the problems associated with gas collisions and minimizing Zeeman splitting and motional Stark shifts.[55,65,66]

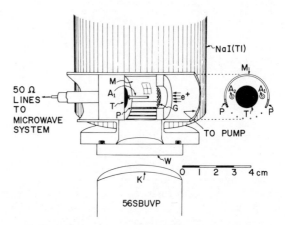

Figure 10. Positron target chamber and microwave cavity for $n = 2$ fine-structure measurement. G, grid; T, copper target; M, aluminized Suprasil quartz mirror; W, Suprasil quartz window; K, CsTe photocathode; P, support posts; A_1, input antenna; A_2, output antenna; NaI(Tl), annihilation detector (from Ref. 63).

A beam of 30-eV positrons is guided into the cylindrical rf cavity shown in Figure 10. One side of the cavity has been replaced by parallel wires to allow the Lyman-α photons to be detected by a uv-sensitive solar blind photomultiplier. The slow positrons collide with the Cu end face of the cavity and form Ps atoms leaving the surface. Approximately 10^{-3}–10^{-4} of the incident positrons form Ps in the $n = 2$ state, leading to the $2P \rightarrow 1S$ 2430-Å photon emission, with the subsequent annihilation of the ground state. The time-delay spectrum between the Lyman-α photons and the subsequent annihilation γ's is recorded, thus registering $2P \rightarrow 1S \rightarrow 2\gamma$ or $2P \rightarrow 1S \rightarrow 3\gamma$ events as a function of time.

Figure 11 shows a lifetime spectrum obtained using the Lyman-α photon as a start pulse and one of the annihilation γ's as the stop signal. The long-lifetime component characteristic of ground-state Ps occurs when a $2\,^3P$ state decays by Lyman-α emission to the $1\,^3S_1$ state which then annihilates via 3γ's. It is important to note that the long component is entirely free of the correlated coincidences that arise in the prompt component due to annihilation γ-induced scintillations detected by the photon counter. For this reason, the folded optics and interference filters could be dispensed with and the detection efficiency was increased nearly 200-fold over the earlier Lyman-α experiments. Any $2\,^3S_1$ states initially present would ordinarily annihilate without the emission of a uv photon and hence are not detected in the delayed coincidence signal. However, if microwaves at the appropriate frequency are used to induce transitions $2\,^3S_1 \rightarrow 2\,^3P_2$, the extra Lyman-$\alpha$ photons from $2\,^3P_1 \rightarrow 1\,^3S_1$ will result in

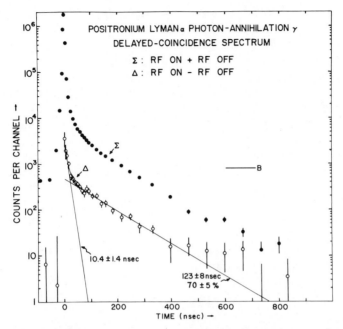

Figure 11. Delayed-coincidence spectra of coincidences between the photon counter start pulse and annihilation detector stop pulse. The constant background due to uncorrelated coincidences is indicated by solid line B and has been subtracted from the raw data (from Ref. 63).

an increase in the intensity of the detected delayed component. This increased signal is shown in Figure 11 as the difference between the "microwaves on" minus the "microwaves off" lifetime spectra. The fact that the long component of ~ 120 ns is somewhat shorter than the vacuum

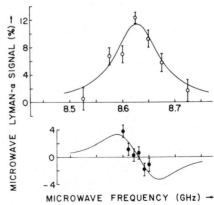

Figure 12. Observed Lyman-α signal S (open circles) and logarithmic first-difference signal S' (solid circles) as a function of microwave frequency (from Ref. 63).

1 $^3S_1$ lifetime (~ 140 ns) is attributed to wall collisions. Indeed, coating the walls of the microwave cavity with MgO smoke increases the long lifetime. In addition, from the observation of the long lifetime of 1 $^3S_1$ Ps in MgO powder[25] one would expect a very small annihilation probability for Ps collisions with MgO powder. The observation of an increase in the long lifetime demonstrates that the $n = 2$ Ps is not stuck to the emitting surface. The appearance of an intermediate lifetime component (~ 10 ns) is not fully explained as yet, but could result from the presence of a high-velocity $n = 2$ Ps component and wall annihilations. This technique of producing Ps in an MgO-coated cavity, but using a more efficient start signal detection method, has recently been used by Gidley *et al.*[67] to obtain a high statistical accuracy determination of 1^3S_1 lifetime and is discussed in Section 6.

Figure 12 shows the "microwave Lyman-α signal" defined as $S(f) = [N_{on}(f) - N_{off}]/N_{off}$, where $N_{on}(N_{off})$ is the counting rate in the time-delayed component of the uv photon annihilation γ time spectrum, with rf on (off). Also plotted is $S'(f)$:

$$S'(f) = [N(f + \Delta) - N(f - \Delta)]/[N(f + \Delta) + N(f - \Delta)]$$

$$= [S(f + \Delta) - S(f - \Delta)]/[2 + S(f + \Delta) + S(f - \Delta)]$$

(7)

obtained in separate runs with $\Delta = 30$ MHz. A Lorentzian line shape $S(f) = \frac{1}{4}A\delta^2[(f - f_0)^2 + \frac{1}{4}\delta^2]^{-1}$ was fitted simultaneously to both sets of data, yielding $A = (11.4 \pm 0.6)\%$, $f_0 = (8628.4 \pm 2.8)$ MHz, and $\delta = (102 \pm 12)$ MHz, with $\chi^2/\nu = 12.1/10$.

The width δ is twice the ≈ 50 MHz natural linewidth ($\Gamma_{2P}/2\pi$) of the 2^3P_2 state. This is attributed to power broadening and, to a smaller extent, to the effects of Ps collisions with the cavity walls. If all substates of the $n = 2$ level are equally populated during production and if there are no geometrical differences in the detection efficiency between the originally produced $2P$ decays and the $2S \rightarrow 2P$ induced Lyman-α decays, one predicts a maximum effect of 33% for $S(f_0)$. Subtracting the background due to accidentals in the time spectrum from $N(f)$ results in an observed effect of $A = (14.1 \pm 1.0)\%$. The change of A with power and the power broadening of the width agrees with an order-of-magnitude computation which includes the natural decay rates, if an additional decay rate of $\approx 6 \times 10^7$ s^{-1}, possibly accounted for by wall collisions, is assumed for the $2S$ state. The details of Ps–wall collisions are too little understood at present for a more detailed model computation. These uncertainties, however, influence mainly the width and magnitude of $S(f)$, and would affect f_0, the resonance frequency, only via a possible asymmetry of the actual line shape, an effect subject to further more precise experimentation.

An rf effect $A = (0.68 \pm 0.09)\%$ at 0.41 mW was also observed in the singles rate of the uv photomultiplier. The small size of the effect is due to the high background of γ induced counts in the uv photomultiplier and is in agreement with an estimate based on the observed delayed coincidence rf effect. It is interesting to note that this constitutes the first observation of Ps which does not rely on the detection of annihilation γ's.

The center frequency $f_0 = 8628.4 \pm 2.8$ MHz obtained from the data of Figure 12 is to be compared with the predicted value given by Eq. (7) only after a correction is made to account for the small Zeeman and motional Stark shifts due to the 54-G rms magnetic induction B in the cavity (produced by the slow β^+ beam solenoid). Estimating that the Ps kinetic energy ranges between 0 and 1 eV implies that the corrected theoretical value of Eq. (7) ranges between 8620 and 8624 MHz compared to the experimental value of 8628.4 ± 2.8 MHz. We note that the quoted ± 2.8-MHz experimental error is strictly due to counting statistics. Given the experimental uncertainties in the constancy in the rf power, as well as the lack of knowledge of the velocity of the $n = 2$ Ps atoms, and thus the uncertainty in the motional Stark shift, we estimate the true uncertainty to be approximately ± 6 MHz. We note, however, that the α^3Ry radiative correction ("Lamb shift") in Eq. (7) corresponds to 231 MHz; thus the experiment already confirms this value to a few percent accuracy and may indicate the need for higher-order corrections.

A resonance measurement using a nonresonant circular waveguide with a traveling wave propagating along the direction of Ps emission was also performed,[63] yielding a resonance of 8652 ± 8 MHz. Given the new experimental conditions ($B \approx 80$ G), this value is consistent with a Zeeman and motional Stark-shifted theoretical value plus a Doppler shift caused by a ≈ 1-eV Ps beam leaving the surface of the target. More detailed knowledge of the Ps beam energy distribution will be required in future high-precision fine-structure measurements.

6. Positronium Annihilation Rates

The predominant mode for $n = 1$ singlet Ps annihilation is via 2γ's. 3γ annihilation is prohibited by the invariance of the electromagnetic interaction under charge conjugation C, as discussed in Section 2. The 2γ rate is calculated to be[68]

$$\Gamma(1^1S_0 \to 2\gamma) = \frac{\alpha^5 mc^2}{2\hbar}\left[1 - \frac{\alpha}{\pi}\left(5 - \frac{\pi^2}{4}\right)\right] = 7.9854 \times 10^9 \, \text{s}^{-1} \qquad (8)$$

This rate has been determined experimentally by Theriot et al.[47] using the method of extrapolating their Zeeman resonance linewidth to zero power.

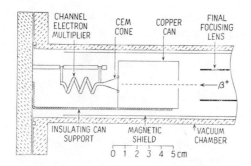

Figure 13. MgO-coated cavity for measuring $1\,^3S_1$ positronium decay rate (Ref. 63). The channeltron (CEM) is used to detect the secondary electron as a result of incident-positron bombardment (from Ref. 67).

They find good agreement with theory: $\Gamma(1\,^1S_0 \rightarrow 2\gamma) = (7.99 \pm 0.11) \times 10^9\,\text{s}^{-1}$. Unfortunately, the timing resolution limitation of conventional delayed-coincidence measurements (between Ps formation and annihilation) prevents a direct and more accurate measurement of this rate. Thus there is presently lacking an experimental verification of the radiative corrections of order α to the $^1S_0$ annihilation rate.

The annihilation rate of the long-lived $1\,^3S_1$ state is more accessible to high-precision measurements by direct timing methods. The first such experiments were performed in gases at various densities by Beers and Hughes[69] and Coleman and Griffith.[70] In both experiments the vacuum annihilation rates were obtained by extrapolation to zero gas density: the results are $\Gamma_{3\gamma} = (7.275 \pm 0.0015) \times 10^6\,\text{s}^{-1}$ and $(7.262 \pm 0.0015) \times 10^6\,\text{s}^{-1}$, respectively. These experimental values are somewhat larger than the lowest-order theoretical value of $7.21 \times 10^6\,\text{s}^{-1}$ computed some time ago by Ore and Powell.[71] The first radiative correction calculations to this value were performed by Stroscio and Holt[72] and Stroscio,[73] who obtain for the total rate

$$\Gamma(1\,^3S_1 \rightarrow 3\gamma) = \frac{2}{9\pi}\frac{\alpha^6 mc^2}{\hbar}(\pi^2 - 9)\left[1 + \frac{\alpha}{\pi}(1.86 \pm 0.45)\right]$$
$$= (7.242 \pm 0.008) \times 10^6\,\text{s}^{-1} \tag{9}$$

In Eq. (9) the term outside the square bracket is the Ore and Powell result.

The apparent agreement between experiment and theory was recently brought into question by the new measurements of the $\Gamma_{3\gamma}$ for Ps formed in uncompressed SiO_2 powders,[74] $\Gamma_{3\gamma} = (7.104 \pm 0.006) \times 10^6\,\text{s}^{-1}$, and Ps formed in vacuum by slow positrons,[67] $\Gamma_{3\gamma} = (7.09 \pm 0.02) \times 10^6\,\text{s}^{-1}$, carried out at the University of Michigan. The agreement between these two different types of experiments and the fact that the values were below the expected theoretical decay rate made it difficult to attribute these results to systematic errors.[75] This situation motivated a reexamination of the theoretical calculation of the first-order radiative correction terms by Caswell, Lepage, and Sapirstein.[76] They obtain a larger correction with opposite sign to the previous computation:

$$\Gamma(1\,^3S_1 \to 3\gamma) = \frac{2}{9\pi}\frac{\alpha^6}{\hbar}mc^2(\pi^2-9)\left[1-\frac{\alpha}{\pi}(10.348\pm 0.070\right]$$

$$= (7.0379 \pm 0.0012) \times 10^6\,\mathrm{s}^{-1} \qquad (10)$$

The $1\,^3S_1$ vacuum annihilation rate measurement performed by the Michigan group[67] used the technique of Ps formation in vacuum by slow positrons developed at Brandeis,[39] combined with a channeltron detector for timing signals. Their setup is illustrated in Figure 13. The slow positrons (400 eV) are focused onto the surface of the entrance cone of the channeltron and form Ps; simultaneously secondary electrons are emitted and are counted by the channeltron. It is this channeltron pulse that is then used as the $t = 0$ timing signal for the lifetime measurement of triplet positronium. They find that approximately 10% of the incident positrons produce Ps which is reemitted into the vacuum chamber. The Ps leaving the channeltron surface was confined to annihilate within a MgO-coated cavity ("copper can," Figure 13). Using spatial collimation in front of the γ-ray detector, the Michigan group confirmed the fact that the Ps atoms leave the surface at which they are produced. They measured directly the maximum Ps kinetic energy of $(0.8\pm0.2)\,\mathrm{eV}$, in qualitative agreement with the estimate for the $n=2$ Ps kinetic energy obtained in the Brandeis Doppler-shift experiment (the $n=1$ and $n=2$ states do not necessarily have to have the same emerging energy). This energy corresponds to an annihilation length of 5 cm, indicating only a few collisions with the MgO-coated enclosure. The fact that in powder experiments the lifetime of Ps annihilating between powder grains ($\approx 10^{15}$ collisions/s) is close to the vacuum lifetime indicates that the few collisions with the MgO-coated container should have a negligible effect on the lifetime. However, since the Ps is not emitted into the cavity at thermal velocities, it is the very lack of collisions that could introduce a small error due to Ps not thermalizing within a reasonable time compared to its lifetime. The possibility of a small admixture of $2\,^3S_1$ Ps being emitted from the channeltron target surface and "contaminating" the measured total decay rate has been recently investigated at Brandeis and found not to be a problem at the 0.1% statistical accuracy level.[77] As of this writing, resolving the $\approx 1\%$ discrepancy between the new theoretical and experimental values must await further experimentation as well as calculation of the next-higher-order corrections $O(\alpha^2 \ln \alpha)$ to the decay rate.

7. Future Developments

New experiments are in progress in various laboratories to improve the intensity of the slow positron beams as well as to understand more thoroughly the Ps formation mechanism at solid surfaces. These developments will no doubt lead to improved measurements of the fundamental properties of Ps and will open up new areas of Ps research in the future. The possibility of producing excited states of Ps by one-photon as well as two-photon excitation of the abundantly formed ground state with high-intensity lasers could lead to a substantial improvement of the fine-structure measurements in the $n = 2$ states. Should the $n = 1$ Ps emission process be better understood and should one be able to produce thermal-energy Ps beams, the two-photon Doppler-free[78] determination of the Rydberg constant, free of hadronic contributions, may become feasible.

Another developing field is the study of Ps–atom and Ps–molecule collisions, an interesting new area of theoretical and experimental atomic physics.

Acknowledgments

We thank the National Science Foundation for their support of the experiments performed at Brandeis University. One of us (S.B.) gratefully acknowledges the hospitality of the Institut Max von Laue–Paul Langevin, Grenoble, France, where parts of this paper were written, and he thanks the J. S. Guggenheim Memorial Foundation for a fellowship during 1976–1977.

Note Added in Proof: A remeasurement of $\Gamma_{3\gamma}$ for Ps formed in a gas yielded $\Gamma_{3\gamma} = (7.056 \pm 0.007) \times 10^6 \text{ sec}^{-1}$.[79]

References

1. R. Vessot, H. Peters, J. Vanier, R. Beehler, J. Barnes, L. Cutler, and L. Bodily, *IEEE Trans. Instrum. Meas.* **IM-15**, 165 (1966).
2. E. deRafael, *Phys. Lett.* **37B**, 201 (1971).
3. M. Deutsch, *Phys. Rev.* **82**, 455, **83** 866 (1951).
4. Discovery of Positronium, *Adventures in Experimental Physics*, Vol. 4, issue δ, pp. 64–127 Ed., B. Maglic, World Science Education, Princeton, NJ (1975).
5. M. A. Stroscio, *Phys. Rep.* (*Phys. Lett. C*) **22**, 215 (1975).
6. S. DeBenedetti and H. C. Corben, *Ann. Rev. Nucl. Sci.* **4**, 191 (1954).
7. V. W. Hughes, *Atomic Physics*, Vol. 3, Plenum Press, New York (1972).
8. H. S. W. Massey, *Phys. Today* **29**, 42 (1976); H. S. W. Massey, E. H. S. Burhop, and H. B. Gilbody, *Electronic and Ionic Impact Phenomena*, Vol. 5, Oxford U. Press (1975).

9. V. I. Goldanskii and V. P. Shantarovich, *Appl. Phys.* **3**, 335 (1974).
10. R. N. West, *Adv. Phys.* **22**, 263 (1973).
11. S. Berko and J. Mader, *Appl. Phys.* **5**, 287 (1975).
12. H. Doyama and R. R. Hasiguti, *Cryst. Lattice Defects* **4**, 139 (1973).
13. A. I. Alekseev, *Sov. Phys.-JETP* **9**, 1020 (1959).
14. A. P. Mills, Jr. and S. Berko, *Phys. Rev. Lett.* **18**, 420 (1967).
15. J. H. Christenson, J. W. Cronin, V. L. Fitch, and R. Turlay, *Phys. Rev. Lett.* **13**, 138 (1964).
16. K. Marko and A. Rich, *Phys. Rev. Lett.* **33**, 980 (1974).
17. H. Mani and A. Rich, *Phys. Rev. D* **4**, 122 (1971).
18. A. Ore, Naturvidenskap Rikke No. 9, U. of Bergen, Årbok (1949).
19. P. G. Coleman, T. C. Griffith, G. R. Heyland, and T. L. Killeen, *J. Phys. B* **8**, L185 (1975).
20. V. W. Hughes, *J. Appl. Phys.* **28**, 16 (1957).
21. W. R. Bennet, Jr., W. Thomas, V. W. Hughes, and C. S. Wu, *Bull. Am. Phys. Soc.* **6**, 49 (1961).
22. B. G. Duff and F. F. Heymann, *Proc. R. Soc. London, A* **272**, 363 (1963).
23. L. W. Fagg, *Nucl. Instrum. Method.* **85**, 53 (1970).
24. O. E. Mogenson, *J. Chem. Phys.* **60**, 998 (1974).
25. R. Paulin and G. Ambrosino, *J. Phys. (Paris)* **29**, 263 (1968); W. Brandt and R. Paulin, *Phys. Rev. Lett.* **21**, 193 (1968).
26. V. J. Goldanskii, B. M. Levin, and A. D. Mokrushin, *JETP Lett.* **11**, 23 (1970).
27. S. M. Curry and A. L. Schawlow, *Phys. Lett.* **37A**, 5 (1971).
28. S. M. Curry, Ph.D. Thesis, Stanford University (1972).
29. L. Madansky and F. Rasetti, *Phys. Rev.* **79**, 397 (1950).
30. W. H. Cherry, Ph.D. Thesis, Princeton University (1958).
31. J. M. J. Madey, *Phys. Rev. Lett.* **22**, 784 (1969).
32. D. G. Costello, D. E. Groce, D. F. Herring, and J. Wm. McGowan, *Can. J. Phys.* **50**, 23 (1972).
33. D. G. Costello, D. E. Groce, D. F. Herring, and J. Wm. McGowan, *Phys. Rev. B* **5**, 1433 (1971).
34. B. Y. Tong, *Phys. Rev. B* **5**, 1436 (1971).
35. R. M. Nieminen and C. H. Hodges, *Solid State Commun.* **18**, 1115 (1976).
36. P. G. Coleman, T. C. Griffith, and G. R. Heyland, *Proc. R. Soc. London A* **331**, 561 (1973).
37. S. Pendyala, P. W. Zitzewitz, J. Wm. McGowan, and P. H. R. Orth, *Phys. Lett.* **43A** 298 (1973).
38. K. F. Canter, P. G. Coleman, T. C. Griffith, and G. R. Heyland, *J. Phys. B* **5**, L167 (1972).
39. K. F. Canter, A. P. Mills, Jr., and S. Berko, *Phys. Rev. Lett.* **33**, 7 (1974).
40. E. E. Salpeter and H. A. Bethe, *Phys. Rev.* **84**, 1232 (1951).
41. T. Fulton, D. A. Owen, and W. W. Repko, *Phys. Rev. A* **4**, 1802 (1971); D. A. Owen, *Phys. Rev. Lett.* **30**, 887 (1973).
42. E. R. Cohen and B. N. Taylor, *J. Phys. Chem. Ref. Data* **2**, 663 (1973).
43. R. S. Van Dyck, Jr., P. B. Schwinberg, H. G. Dehmelt, *Phys. Rev. Lett.* **38**, 310 (1977).
44. M. Deutsch and S. C. Brown, *Phys. Rev.* **85**, 1047 (1952).
45. H. Grotch and R. A. Hegstrom, *Phys. Rev. A* **4**, 59 (1971).
46. B. N. Taylor, W. H. Parker, and D. N. Langenberg, *Rev. Mod. Phys.* **41**, 375 (1969).
47. E. D. Theriot, Jr., R. H. Beers, and V. W. Hughes, *Phys. Rev. Lett.* **18**, 767 (1967).
48. A. P. Mills, Jr., and G. H. Bearman, *Phys. Rev. Lett.* **34**, 246 (1975).
49. P. O. Egan, W. E. Frieze, V. W. Hughes, and M. H. Yam, *Phys. Lett.* **54A**, 412 (1975).
50. G. H. Bearman and A. P. Mills, Jr., *J. Chem. Phys.* **65**, 1841 (1976).

51. R. Weinstein, M. Deutsch, and S. Brown, *Phys. Rev.* **94**, 758 (1954).

52. R. Weinstein, M. Deutsch, and S. Brown, *Phys. Rev.* **98**, 223 (1955).

53. V. W. Hughes, S. Marder, and C. S. Wu, *Phys. Rev.* **106**, 934 (1957).

54. E. R. Carlson, V. W. Hughes, M. L. Lewis, and I. Lindgren, *Phys. Rev. Lett.* **29**, 1059 (1972).

55. H. W. Kendall, Ph.D. Thesis, Massachusetts Institute of Technology (1954).

56. S. L. Varghese, E. S. Ensberg, V. W. Hughes, and I. Lindgren, *Phys. Lett.* **49A**, 415–417 (1974).

57. S. L. McCall, *Bull. Am. Phys. Soc.* **18**, 1512 (1973).

58. R. L. Brock and J. R. Streib, *Phys. Rev.* **109**, 399 (1958).

59. M. Leventhal, *Proc. Natl Acad. Sci.* **66**, 6 (1970).

60. J. F. Kielkopf and P. J. Ouseph, *Bull. Am. Phys. Soc.* **19**, 592 (1974).

61. A. J. Dahm and T. G. Eck, *Phys. Lett.* **49A**, 267 (1974).

62. K. F. Canter, A. P. Mills, Jr., and S. Berko, *Phys. Rev. Lett.* **34**, 177 (1975).

63. A. P. Mills, Jr., S. Berko, and K. F. Canter, *Phys. Rev. Lett.* **34**, 1541 (1975).

64. T. Fulton and P. C. Martin, *Phys. Rev.* **95**, 811 (1954).

65. S. M. Curry, *Phys. Rev. A* **7**, 447 (1973).

66. M. L. Lewis and V. W. Hughes, *Phys. Rev. A* **8**, 625 (1973).

67. D. W. Gidley, P. W. Zitzewitz, K. A. Marko, and A. Rich, *Phys. Rev. Lett.* **37**, 729 (1976).

68. I. Harris and L. M. Brown, *Phys. Rev.* **105**, 1656 (1957).

69. R. H. Beers and V. W. Hughes, *Bull. Am. Phys. Soc.* **13**, 633 (1968); V. W. Hughes, *Physics 1973, Plenarvortren Physikertag, 37th,* pp. 123–135, Physik Verlag, Weinheim, Germany (1973).

70. P. G. Coleman and T. C. Griffith, *J. Phys. B* **6**, 2155 (1973).

71. A. Ore and J. L. Powell, *Phys. Rev.* **79**, 1696 (1949).

72. M. A. Stroscio and J. M. Holt, *Phys. Rev. A* **10**, 749 (1974).

73. M. A. Stroscio, *Phys. Rev. A* **12**, 338 (1975).

74. D. W. Gidley, K. A. Marko, and A. Rich, *Phys. Rev. Lett.* **36**, 395 (1976).

75. G. W. Ford, L. M. Sander, and T. A. Witten, *Phys. Rev. Lett.* **36**, 1269 (1976).

76. W. E. Caswell, G. P. Lepage, and J. Sapirstein, *Phys. Rev. Lett.* **38**, 488 (1977).

77. K. F. Canter, B. O. Clark, and I. J. Rosenberg, *Phys. Lett.* **65A**, 301 (1978).

78. T. W. Hänsch, *Phys. Today* **30**, 34 (1977).

79. D. W. Gidley, A. Rich, P. W. Zitzewitz, and D. A. L. Paul, *Phys. Rev. Lett.* **40**, 737 (1978).

33

Applications of Atomic Physics to Astrophysical Plasmas

CAROLE JORDAN

1. Introduction

The purpose of this chapter is to show, through examples of recent work, the continuing interaction between atomic physics and astrophysics. It will not be an extensive review, since each type of application of atomic data discussed would itself deserve a separate chapter, but rather it will concentrate on some work within the past few years and on areas where further atomic data will be required.

The immediate need on obtaining astrophysical spectra in a newly observed wavelength range is to identify the absorption and emission lines present. In the visible region of the spectrum this stage, for most sources, occurred many years ago and recent work has been concerned either with the euv (extreme ultraviolet) and x-ray regions or, at the other end of the spectrum, with radio wavelengths. Even below 2000 Å the progress over the past 10 years has been so rapid that the strongest lines in solar and stellar spectra have been identified. The examples discussed below, in Section 2, will mainly relate to solar emission lines which have become observable as specialized techniques have made it possible to study spectra of distinctive parts of the solar atmosphere, such as active regions, sunspots, and flares.

Interpretation of the intensities of absorption and emission lines observed in astrophysical sources requires a wide variety of atomic data. Oscillator strengths (and damping constants) are needed to understand the formation of absorption lines in locations as diverse as stellar atmospheres, the interplanetary medium, and quasars. One example, the application of

CAROLE JORDAN • Department of Theoretical Physics, University of Oxford.

absorption oscillator strengths to quasar spectra is discussed in Section 3. Spontaneous transition probabilities can be important in interpreting the intensities of forbidden emission lines and are also required in order to determine branching ratios from excited levels. For many emission lines, however, the most important atomic parameter is the cross section for collisional excitation by electrons. Some of the most useful applications of excitation cross sections are to atomic systems which have term structures which allow the electron temperature and density to be determined from the relative intensities of selected line pairs. Recent work on the (Be I)-like ions which makes use of both transition probabilities and excitation cross sections is discussed in Section 3.

2. Energy Levels and Line Identifications

2.1. Introduction

The process of identifying emission lines in the solar euv and x-ray spectrum has progressed through a combination of theoretical atomic physics, the production of similar spectra in laboratory experiments, and an understanding of the physical conditions likely to be present in the solar atmosphere. Even the early work of Grotrian[1] and Edlén[2,3] on the identification of magnetic dipole transitions in highly ionized iron, observed in the visible spectrum of the solar corona during total eclipses, is an example of this fruitful combination of techniques. The method of classifying lines from high stages of ionization by building up data from different elements along an isoelectronic sequence has now been widely used. A variety of laboratory sources is available for producing the different states of ionization, for example, θ-pinch devices, a range of spark sources, and in recent years, laser-produced plasmas. The latter have been used with particular success, allowing very high stages of ionization to be produced and providing differentiation of ionization stages through the spatial distribution of the emitting ions. The accuracy with which the wavelengths and relative intensities of transitions in high ions can be calculated has also improved and this combination of experimental and theoretical techniques continues to be applied.

For *ab initio* calculations of energy levels the most widely used techniques have been a modified version of the Hartree–Fock method, i.e., the Hartree program developed by Cowan,[4,5] and the multiconfiguration Hartree–Fock program developed by Froese-Fischer.[6] The methods of Edlén[7] for extrapolating level intervals in terms of the Slater energy parameters have been extensively applied in conjunction with observed energy-level intervals.

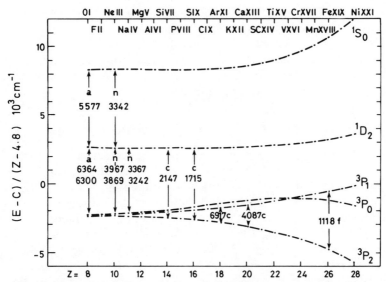

Figure 1. The level structure in the $2s^2 2p^4$ configurations, taken from Edlén.[16] The energies, scaled by $Z - 4.8$, are relative to the center of gravity. The letters indicate the source from which the lines were identified: a, auroral; n, nebular; c, coronal; f, flare.

A review of results obtained up to about 1973 through the combination of experimental and theoretical methods has been given by Fawcett,[8] and references to earlier work can be found there. Also Edlén[9] has recently revised his earlier compilation[10] which gives references to term analyses of laboratory data for elements between hydrogen and nickel carried out since the publication of Moore's *Atomic Energy Levels*.[11]

Other surveys of the current state of the identification of solar lines have been given by Behring *et al.*[12,13] and Doschek,[14] and Feldman[15] has discussed recent progress in the spectroscopy of laser-produced plasmas at high temperatures.

Figure 1, taken from a recent review by Edlén,[16] illustrates the variety of astrophysical sources from which forbidden lines have been identified. It shows the levels of the $2s^2 2p^4$ configuration, with energies, scaled by $Z - 4.8$, relative to the center of gravity. The lines of O I were identified in auroral spectra, the lines of Ne III and Na IV in nebulae, those of Si VII and S IX in the corona, and that of Fe XIX in flare spectra.

In the absence of laboratory or theoretical data the key to identifying new lines in the solar spectrum has often been provided by existing knowledge of conditions in the solar atmosphere. This approach was useful in the identification of the $1s^2 \, {}^1S_0 - 1s2s \, {}^3S_1$ transition in the heliumlike ions[17] observed for the first time in solar x-ray spectra and in the classification of the magnetic dipole transitions observed in the euv coronal spectrum.[18] In both these cases new solar spectra have stimulated further theoretical or experimental atomic physics.

Perhaps the most interesting of more recent work has centered around lines observed during solar flares. Because of the continuing importance of this area, it is discussed further below.

2.2. Spectra of Solar Active Regions and Flares

In the quiet solar atmosphere the electron temperature T_e has values between $\sim 10^4$ and 2×10^6 K, while the electron density N_e varies between $\sim 10^{11}$ cm^{-3} and 10^8 cm^{-3}. Under these conditions the lightest elements (up to carbon) are in all stages of ionization up to the hydrogenlike ion, and heavier elements such as silicon and iron exist in stages of ionization up to about 11 times ionized. The strongest emission lines are electric dipole transitions with $\Delta n = 0$, or ± 1, whose upper levels are excited by electron collisions from the ground term. The spectra of such ions are by now well established.

The term "active region" is used to signify an area of enhanced magnetic flux. Above the photosphere, active regions are composed of sets of magnetically confined flux tubes in which both the density and temperature can be enhanced. They are also the site of solar flares, which give rise to a sudden brightening of most of the electromagnetic spectrum and are caused by the release of additional mechanical or magnetic energy, during which the electron density may reach $\sim 10^{13}$ cm^{-3} and the electron temperature $\sim 2 \times 10^7$ K. Thus the active regions and flares are characterized by the presence of material at far higher temperatures than that in the quiet sun, and elements up to about iron become fully stripped. Recent improvements in the techniques of making solar euv and x-ray observations with good spatial resolution, in particular, the instruments on the Apollo telescope mount (ATM) on the USA's Skylab manned satellite, have made it possible to study active regions and flares through forbidden lines in spectral regions where previously only the quiet sun had been observed. These new observations have given further stimulus to laboratory studies of highly ionized systems, mainly using laser produced plasmas, and considerable progress has been made recently. The interest in the spectra of high ions extends beyond solar physics since similar transitions are observed in high-temperature plasma devices, and emission at the wavelength of the Fe XXV resonance line (~ 6.7 keV) has already been reported from the Perseus clusters of galaxies.[19]

Forbidden Lines in Flare Spectra. One of the most interesting developments in the last few years has been the observation and classification of further forbidden lines from transitions in the $2s^2 2p^n$

ground configurations. (In iron these correspond to Fe XVIII to Fe XXII.) These are magnetic dipole transitions of the same type as the well-known visible-region coronal lines. They are important in the solar spectrum for several reasons: they establish the position of particular energy levels; their intensities, relative to other transitions, may allow the electron density to be determined; and several lie at wavelengths above the Lyman limit at 912 Å. It is this last feature that is particularly valuable. During solar flares the long wavelength of the lines allows the simultaneous study of material at $\sim 10^7$ K and at transition-region temperatures, $\sim 10^5$ K. Also, the line profiles can be measured to determine the ion motions. The lines which lie above 912 Å offer the possibility of observing very hot material in other astrophysical sources, using satellites such as the international ultraviolet Explorer (IUE). A large part of the euv and soft x-ray spectrum of these sources will be unobservable because of the absorption below 912 Å by interstellar hydrogen. Such objects include flare stars, Seyfert, Markarian, and Zwicky galaxies, quasars, and x-ray sources. Feldman[15] has already searched quasar spectra for some of these transitions, and finds statistically significant coincidences, but further observations are needed to confirm the proposals.

The forbidden lines are not directly observable in laboratory sources, but the separation of the energy levels has been found in many cases from the wavelength differences between transitions in the far-ultraviolet and x-ray regions, for example, transitions of the type $2p^n-2p^{n-1}\,3l$ and $2s^2\,2p^n-2s2p^{n+1}$. If only a few suitable transitions can be observed, then this method results in quite large uncertainties in the wavelengths of the forbidden lines, and it is also difficult to connect terms of different multiplicity. The transitions of the type $2p^n-2p^{n-1}\,3l$ and $2s^2\,2p^n-2s2p^{n+1}$ which in iron lie in the wavelength regions 10–17 Å and 90–120 Å, respectively, are themselves of great interest to solar flare studies, but space does not permit a review of recent progress in the identifications of these transitions. *Ab initio* calculations of the energy levels have been used in conjunction with the laboratory observations, and also extrapolations of energy levels within the $2s^2\,2p^n$ configurations have been made using the work of Edlén.[20]

Table 1 gives the wavelengths of the highly ionized forbidden iron lines definitely identified so far in the solar spectrum.

Fe XVIII has the ground configuration $2s^2\,2p^5$, which gives rise to a ^2P term. The $^2P_{1/2}-^2P_{3/2}$ interval can be found by isoelectronic extrapolation using the methods of Edlén.[20] However, the separation has now been derived from the laboratory observations of the wave number difference between several pairs of transitions of the type $2p^5-2p^4 3d$ which lie around 15 Å, e.g., the $2p^5\,^2P_{3/2}-2p^4(^1D)3d\,^2D_{3/2}$ and $2p^5\,^2P_{1/2}-2p^4(^1D)3d\,^2D_{3/2}$ at 14.150 and 14.361 Å, respectively,[21,22] or from $2s^2$

Table 1. Forbidden Lines of Highly Ionized Iron
Identified in Solar Flare Spectra

Ion	Transition	λ (observed) (Å)
Fe XVIII	$2s^2 2p^5 \ ^2P_{3/2}-^2P_{1/2}$	974.8
Fe XIX	$2s^2 2p^4 \ ^3P_2-^3P_1$	1118.1
Fe XXI	$2s^2 2p^2 \ ^3P_0-^3P_1$	1354.1
Fe XXII	$2s^2 2p \ ^2P_{1/2}-^2P_{3/2}$	845.1

$2p^5 \ ^2P_{1/2,3/2}-2s2p^6 \ ^2S_{1/2}$[23] at around 100 Å. The predicted Fe XVIII line was observed in flare spectra obtained with the NRL telescopes on Skylab (Doschek et al.[24]).

Fe XIX $(2s^2 2p^4)$ has an inverted ^3P ground term as well as ^1D and ^1S terms. So far only the $^3P_2-^3P_1$ transition at 1118.1 Å has been definitely identified.[24] Since the intensities of the lines depend on the collisional excitation rates from the lower levels and on branching ratios for spontaneous decay, any transition whose upper level can be reached by collisions from $2s^2 2p^4 \ ^3P_2$ or by cascades from levels in the excited $2s2p^5$ and $2s^2 2p^3 3l$ configurations could be observable. The populations of the $2s^2 2p^4 \ ^3P_0$ and $^3P_1$ levels will be too small for excitations from these levels to $2s^2 2p^4 \ ^1D_2$ and $^1S_0$ to be significant. The $^3P_2-^3P_1$ separation was predicted by direct extrapolation using work by Edlén[25] and from separations in laboratory spectra of transitions between $2s^2 2p^4 \ ^3P$ and $2s2p^5 \ ^3P$.[21] The $^3P_0-^3P_1$ and $^1S_0-^1D_2$ separations were also found in a similar manner.[26] The singlet and triplet terms have recently been connected through laboratory observations of intercombination lines made by Kononov et al.[27] The wavelengths of such intersystem lines have been calculated by Safronova.[28,29] The data of Feldman et al.[22] and Edlén[25] allowed the author to extrapolate intervals in the $2s^2 2p^4$ configuration. In particular the $^3P_1-^1S_0$, $^3P_2-^1D_2$, and $^3P_1-^1D_2$ transitions were predicted to lie at ~424, 589, and 1261 Å, respectively. The recent work of Kononov et al.[27] leads to wavelengths of 423.8, 591.4, and 1256 Å for these transitions. Further observations of solar flare lines have recently been reported by Sandlin et al.[30] in spectra obtained with the NRL's normal incidence spectrograph on Skylab. These spectra contain several lines around 424 and 591 Å. Sandlin et al. discuss possible identifications for the Fe XIX $^3P_1-^1S_0$ transition and tentatively suggest the line at 420.98 Å, based on their extrapolated wavelength of 421.8 Å. If the identification[31] of the line at 423.98 with Ar XV $2s^2 \ ^1S_0-2s2p \ ^3P_1$ is correct, then the nearest other unidentified line is at 424.24 Å. (An error of ±0.02 Å in the x-ray transitions around 90 Å leads to an uncertainty of ±0.8 Å at 424 Å.) If the $^3P_1-^1S_0$ transition is observable, then the $^3P_2-^1D_2$ transition should be at least as intense, and should be near 591 Å. A flare line is reported by Sandlin *et al.* at 592.2 Å. This may not be identical with a flare-enhanced line at 592.8 Å which has been observed by the Harvard group both in OSO–6 and ATM spectra,[32] since there is evidence that a further, lower ionization line may be present also in the vicinity of this wavelength. However, the 592.2-Å line can be tentatively identified with the $^3P_2-^1D_2$ transition in Fe XIX.

The ground configuration of Fe XX ($2s^2 2p^3$) gives rise to ^4S, ^2D, and ^2P terms. Although laboratory spectra and theoretical calculations have enabled doublet–doublet and quartet–quartet transitions between the $2s^2 2p^k$ and $2s2p^{k+1}$ configurations to be classified,[33–35] no intersystem lines have been identified, and consequently only isoelectronic extrapolations are available for the wavelengths of the $^4S-^2D$, ^2P forbidden lines. Doschek *et al.*,[26] Kononov *et al.*,[35] and Fawcett and Cowan[34] have given doublet separations in the $2s^2 2p^3$ configuration. Forbidden lines between ^2P and ^2D may be observable if the excitation rates are sufficiently large and the branching ratios for spontaneous radiative decay favor these transitions rather than the $^4S-^2P$ transitions. Predicted wavelengths are given in Table 2. Sandlin *et al.* discuss the further possible Fe XX identifications.[30] Their extrapolation for the $^4S_{3/2}-^2P_{3/2}$ wavelength agrees well with the author's (323 Å), and a line at 323.53 Å is, they consider, consistent with a stage of ionization such as Fe XX. There is then no line consistent with the $^4S_{3/2}-^2P_{1/2}$ transition predicted at 406 Å. However, one pair of lines of similar ionization class fits the expected $^2P_{1/2}-^2P_{3/2}$ separation. These lines are observed at 326.71 and 411.55 Å. At present a more definite identification is not possible. Sandlin *et al.* also point out the coincidence in wavelength between a strong line at 541.08 Å and the predicted wavelength of the $^2D_{3/2}-^2P_{3/2}$ transition, but do not consider the identification convincing. The extrapolations for the $^4S_{3/2}-^2D_{3/2}$, $^2D_{5/2}$ lines by the author use improved wavelengths for these transitions in Si VIII and S X, as given by Feldman and Doschek[36] and Sandlin *et al.*[37]

Fe XXI ($2s^2 2p^2$) has $^3P_0$ as its ground level, and the $2p^2 {}^3P_0-{}^3P_1$ transition has been identified in solar flare spectra at 1354.1 Å.[24] The $^3P_1-^3P_2$ transition can be predicted from laser plasma laboratory data[33–35] but has not yet been observed. The triplet and singlet terms have not been connected by observations, but the intervals have been extrapolated.[34,38]

Extrapolated intervals for other transitions are given in Table 2. Sandlin *et al.* point out there are many candidates for the $^3P_1-^1S_0$ transition predicted to lie at ~340 Å, and that a line at 338.04 Å has a variation consistent with this stage of ionization. According to the transition probabilities calculated by Nussbaumer,[39] the $^3P_2-^1D_2$ and $^3P_1-^1D_2$ transitions should have an intensity ratio of 1:2, and since the effective excita-

Table 2. Other Possible Identifications for Forbidden Lines in Flares and Extrapolated Wavelengths

Ion	Transition	λ (predicted) (Å)	Ref.	λ (observed) (Å)	Ref.
Fe XIX	$2s^2 2p^4$ $^3P_1 - ^3P_0$	7084	21		
		6977	27		
		7092	26, 33		
	$^3P_1 - ^1S_0$	424		424.2	30
		421.8	30	(420.98)	30
		423.8	27		
	$^3P_2 - ^1D_2$	583		592.2	
		591.4	27	or 592.8	
	$^3P_1 - ^1D_2$	1261			
		1256	27		
Fe XX	$2s^2 3p^3$ $^4S_{3/2} - ^2D_{5/2}$	645			
	$^4S_{3/2} - ^2D_{3/2}$	902			
	$^4S_{3/2} - ^2P_{3/2}$	323		323.53	30
		323.5	30	or 326.71	
				411.55	
	$^4S_{3/2} - ^2P_{1/2}$	406			
	$^2D_{5/2} - ^2P_{3/2}$	679	26		
		677	35		
	$^2D_{5/2} - ^2P_{1/2}$	1188	26		
		1182	35		
	$^2D_{3/2} - ^2P_{3/2}$	541	26, 35	(541.08)	
	$^2D_{3/2} - ^2P_{1/2}$	822	26		
		821	35		
	$^2P_{1/2} - ^2P_{3/2}$	1585	26		
		1588	35		
	$^2D_{3/2} - ^2D_{5/2}$	2674	26		
		2688	35		
Fe XXI	$2s^2 2p^2$ $^3P_1 - ^3P_2$	2304	24, 33		
		2310	38		
	$^3P_1 - ^1S_0$	337–338	34	338.04	30
		339.4	38	or 340.12	30
	$^3P_2 - ^1D_2$	772–846	34		
		796.6	38		
	$^3P_1 - ^1D_2$	578–575	34	576.1	
		592.2	38	or (592.2)	

tion rate to $^1D_2$ is likely to be higher than that to $^1S_0$, if the $^3P_1-^1S_0$ transition is observed, then the lines from $^1D_2$ to $^3P_1$ and $^3P_2$ should be even stronger. On the basis of wavelength[38] and visibility in the flare spectra of Sandlin *et al.*, the transition $^3P_1-^1D_2$ must be a candidate for identification with one of the lines observed at 576.1 Å and 592.2 Å. If 592.2 Å is identified with Fe XIX, then the line at 576.1 Å is the most likely identification for the Fe XXI line and has an appearance consistent with a stage of ionization between Fe XIX and XXII.

The Fe XXII transition $2s^22p\ ^2P_{1/2}-^2P_{3/2}$ at 845.1 Å was the first of the high-ion forbidden lines to be seen in solar flare spectra, and was observed in OSO-6 data by Noyes.[40] More recent laboratory observations have confirmed the identification.[33-35] Intersystem lines in Fe XXII have recently been proposed by Sandlin *et al.* as identification for lines around 250 Å.

The lowest-energy forbidden line in Fe XXIII occurs between the levels of the ^3P term in the $2s2p$ excited configuration. The strength of the line $2s2p\ ^3P_1-^3P_2$ will depend on the collisional excitation rate from $2s^2\ ^1S_0$ and the relative size of the magnetic dipole transition probability for $^3P_2 \rightarrow ^3P_1$ and the magnetic quadrupole transition probability for $^3P_2 \rightarrow {}^1S_0$. From Mühlethaler and Nussbaumer[41] the branching ratio is such that only one in 1.3×10^3 decays occurs to $^1S_0$. The $^3P_2-^3P_1$ transition will therefore be about a factor of 7 less intense than the $^1S_0-^3P_1$ transition. The intersystem line $2s^2\ ^1S_0-2s2p\ ^3P_1$ is observed.[31]

As can be seen from the above discussion, there is an immediate need for collision cross sections and transition probabilities for forbidden and permitted transitions in the ions Fe XVIII–Fe XXIII, so that the observed intensities can be used for understanding conditions in the solar flares.

2.3. Other Forbidden Lines

Edlén[16] has recently reviewed the forbidden transitions observed in astrophysical sources, such as the sun and planetary nebulae. The most recent summary of coronal lines observed above 3000 Å is contained in a paper by Smitt[42] and is reproduced as Table 3. Many of the identifications listed have been known for some years, and only recent changes or additions will be discussed below. The intervals between levels of terms in the $2s^22p^2$, $2s^22p^3$, and $2s^22p^4$ configurations can be found in the work by Edlén.[25] Svensson[43] has predicted wavelengths of transitions within the $3s^23p^2$, $3s^23p^3$, and $3s^23p^4$ configurations. The methods used have been discussed in the introduction to this section.

The most recent identifications given in Table 3 are of transitions between metastable levels in excited configurations of the type $3p^{n-1}3d$, mainly in iron. Edlén[20] proposed that many of the weaker, then uni-

Table 3. Summary of Observed Coronal Lines Above 3000 Å[a]

3000	Fe IX	3601.1	Ni XVI	5533.4	Ar X
3020.1	Fe X	3642.7	Fe IX	5539.1	Fe X
3072.0	Fe XII	3685.5	Mn XIII	5693.6	Ca XV
3124.0	Fe IX?	3800.8	Fe IX	6374.6	Fe X
3167.0	Ni XII	3986.8	Fe XI	6536.3	Mn XIII
3178	Cr XI	3996.8	Cr XI	6701.7	Ni XV
3302.8	Ni XI	4087.1	Ca XIII	6917	Ar XI
3327.5	Ca XII	4231.2	Ni XII	7058.6	Fe XV
3338.5		4311.8	Fe X	7611.0	S XII
3355.1	Fe IX	4351	Co XV?	7891.8	Fe XI
3388.5	Fe XIII	4359.4	Fe IX	8024.1	Ni XV
3454.2	Fe X	4412	Ar XIV	8153.8	Cr XII
3471.6	Fe IX	4566.2	Fe XI?	9911	S VIII
3488.5		4585.3	Fe IX	10746.8	Fe XIII
3502.5	Ca XIII	5115.8	Ni XIII	10797.9	Fe XIII
3533.6	Fe X	5303.4	Fe XIV	14305	Si X
3577.1	Fe X	5444	Ca XV	30275	Mg VIII

[a] Taken from Smitt[42]; the values are for air and are given in angstroms.

dentified, lines were due to such transitions. Wagner[44] and Wagner and House[45] proposed some specific identifications for transitions between levels in $3p^5 3d$ configurations in Fe IX and Ni XI, but several of these have since been revised in the light of more recent work. The observation of intersystem transitions between the $3p^6$ and $3p^5 3d$ configurations in the solar limb spectrum[46] and the forbidden transition $3p^5 3d$ $^3P_2$–$^3D_2$ at 2042.2 Å enabled the $^3D_2$, $^3D_1$, $^3P_2$, and $^3P_1$ levels to be established [in the paper by Firth et al.[46] the energy (cm^{-1}) for $^1D_2$ 457, 276 is misprinted as 457, 726], and it was pointed out that the identification of the euv line at 230.48 Å with $3p^6$ $^1S_0$–$3p^5 3D$ $^3F_2$ excluded the identification of $3p^5 3d$ $^3D_2$–$^3F_2$ with the line observed at 3533.6 Å, proposed by Wagner and House. The $^1F_3$ energy listed by Firth et al. is now known to be incorrect since it was based on the identification of $3p^5 3d$ $^1F_3$–$^3F_2$ with the line at 3167.0 Å as proposed by Wagner and House. Svensson et al.[47] independently determined the Fe IX $3p^5 3d$ energy levels from laboratory observations of $3s^2 3p^5 3d$–$3s 3p^6 3d$ transitions in the region 300–400 Å and suggested the revised identifications for $^1F_3$–$^3F_2$ (3124.0 Å) and $^3D_2$–$^3F_2$ (3471.6 Å) given in Table 3. They confirmed the identification of 4359.4 Å with $^1D_2$–$^3F_2$, and 3642.7 Å with $^1D_2$–$^3F_3$ made by Wagner and House and also proposed the other Fe IX transitions shown in Table 3. Figure 2 shows the term scheme for $3p^5 3d$, with energies from Svensson et al.,[47] with the exception of $^1P_1$, which lies rather higher.

Identification of transitions within the $3p^4 3d$ configuration in Fe X have recently been proposed by Mason and Nussbaumer[48] and by

Smitt.[42] The former use calculated relative intensities to establish which lines should be the strongest and then compared these with the observed intensities discussed by Magnant-Crifo.[49,50] Smitt's identifications are based on further experimental studies of the Cl I isoelectronic sequence. The quartet level intervals, established from transitions between the $3s^2 3p^4 3d$ and $3s 3p^5 3d$ configurations, allow direct confirmation of the line at 3454.2 Å as $3p^4 3d$ $^4D_{7/2}-{}^4F_{9/2}$. Lines from the doublet levels rely on extrapolated laboratory data and energy levels calculated by Cowan (unpublished), but transitions from a common upper level must fit the observed quartet intervals. The two analyses both result in the line at 3533.6 Å being due to $^4F_{7/2}-{}^2G_{7/2}$ and that at 4311.8 Å being due to $^4F_{9/2}-({}^3P)^2F_{7/2}$. In addition, Smitt proposes that further observed lines are due to the following transitions: 3577.1 Å, $^4F_{7/2}-{}^2G_{9/2}$; 3020.1 Å, $^4F_{9/2}-{}^2G_{9/2}$; and 5539 Å, $^4F_{7/2}-({}^3P)^2F_{7/2}$. Mason and Nussbaumer also suggest that the lines at 4566.2 and 5539.1 Å are due to the transitions $3p^3 3d$ $^3F_4-{}^3G_5$ and $3p^3 3d$ $^3F_4-{}^3G_4$ in Fe XI. This latter proposal is excluded by the Fe X identification above. Further laboratory observations in combination with calculated energy levels are needed to classify the lines remaining unidentified, which are almost certainly of a similar origin in Fe XI or higher ions.

Turning now to the region below 3000 Å, the coronal spectrum in the wavelength range 2200–3000 Å was recorded for the first time with the NRL's normal incidence spectrograph on Skylab. Feldman and Doschek[36] and Sandlin et al.[37] have reported new lines between 2400 and 2650 Å arising from transitions in Fe IX, Fe XI, Fe XII, and Fe XIII. Two remain unidentified. The Fe XII lines at 2405.7 Å (air) and 2565.9 Å (air) could be predicted from previously observed euv and visible region

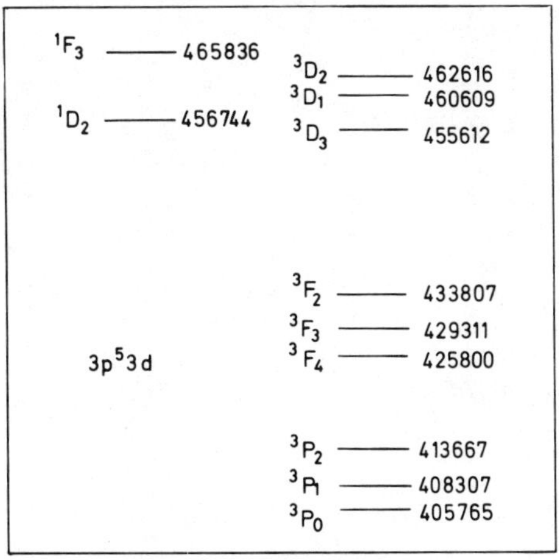

Figure 2. Energy levels in the $3p^5 3d$ configuration in Fe IX. The energies in cm^{-1} are from Svensson et al.[47] The $3p^5 3d$ $^1P_1$ level is not shown since it lies considerably higher.

transitions.[18,43] All the transitions in the ground configuration of Fe XI and Fe XIII expected to be strong enough to be observed have now been recorded.

These authors[36,37] also report many weak new coronal lines between 1189 and 1918 Å and have resolved lines which were blended in the 1970 eclipse data.[51] Lines recently identified are discussed below and are listed in Table 4. The original papers should be consulted for the full list of lines.

In ions with a $2s^2 2p^2$ ground configuration, an improved wavelength is given for Mg VII $^3P_1-{}^1S_0$, at 1189.82 Å, and the line earlier recorded[51] at 1190.2 ± 1.0 Å is resolved into this Mg VII line and a Mg VI line at 1190.07 Å, due to the $2s^2 2p^3\ {}^4S_{3/2}-{}^2P_{3/2}$ transition. The identification

Table 4. Further Forbidden Lines in the Region 1189–2649 Å[a]

λ (solar)[37] (Å)	λ (calculated) (Å)	Ion	Transition	Reference to other previous relevant work
1189.82	1189.7	Mg VII	$2p^2\ {}^3P_1-{}^1S_0$	18, 25
1190.07	1190.1	Mg VI	$2p^3\ {}^4S_{3/2}-{}^2P_{3/2}$	25
1191.62	1191.6	Mg VI	$2p^3\ {}^4S_{3/2}-{}^2P_{1/2}$	25
1196.24	1196.9	S X	$2p^3\ {}^4S_{3/2}-{}^2D_{5/2}$	18, 25
1212.96	1213.6	S X	$2p^3\ {}^4S_{3/2}-{}^2D_{3/2}$	18, 25
1216.43	1217.2	Fe XIII	$3p^2\ {}^3P_1-{}^1S_0$	43
1277.23	1277.0	Ni XIII	$3p^4\ {}^3P_1-{}^1S_0$	43
1324.44	1324.4	Mg V	$2p^4\ {}^3P_1-{}^1S_0$	25
1359.57	1359.5	Mn XI	$3p^3\ {}^4S_{3/2}-{}^2P_{3/2}$	43
1375.95	1376.6	Ca XV?	$2p^2\ {}^3P_2-{}^1D_2$?	25
1392.12	1390.7	Ar XI?	$2p^4\ {}^3P_2-{}^1D_2$?	25
1409.45				18
1428.75				18
1440.50	1440.6	Si VIII	$2p^3\ {}^4S_{3/2}-{}^2D_{5/2}$	25
1463.49		Fe X	$3p^3 3d\ {}^4F_{9/2}-{}^2F_{7/2}$	42, 48
1489.04	1488.9	Cr X	$3p^3\ {}^4S_{3/2}-{}^2P_{3/2}$	43
1564.30	1564.1	Cr X	$3p^3\ {}^4S_{3/2}-{}^2P_{1/2}$	43
1582.56		Fe X	$3p^3 3d\ {}^4F_{7/2}-{}^2F_{7/2}$	42
1603.21		Fe X	$3p^3 3d\ {}^4D_{7/2}-{}^2G_{7/2}$	42, 48
1611.70		Fe X	$3p^4\ {}^4D_{7/2}-{}^2G_{9/2}$	42
1805.94	1805.9	Mg VI	$2p^3\ {}^4S_{3/2}-{}^2D_{3/2}$	25
1841.57	1841.3	Fe IX	$3p^5 3d\ {}^3P_1-{}^3D_2$	47
1917.21	1916.8	Fe IX	$3p^5 3d\ {}^3P_2-{}^1F_3$	47
1918.25		Fe X	$3p^4 3d\ {}^4D_{7/2}-{}^2F_{7/2}$	18, 42
2405.68		Fe XII	$3p^3\ {}^4S_{3/2}-{}^2D_{3/2}$	18, 43
2497.5		Fe IX	$3p^5 3d\ {}^3F_4-{}^1F_3$	47
2565.93		Fe XII	$3p^3\ {}^2D_{3/2}-{}^2P_{3/2}$	18, 43
2578.77		Fe XIII	$3p^2\ {}^3P_1-{}^1D_2$	43
2648.71		Fe XI	$3p^4\ {}^3P_2-{}^1D_2$	43

[a] From Feldman and Doschek[36] and Sandlin et al.[37]

proposed$^{(36,37)}$ for the line observed at 1582.6 Å, Ar XIII $^3P_2-^1D_2$, is unlikely to be correct or the line is blended, since in the 1970 eclipse data the line behaves as if it is formed at lower temperature than expected for Ar XIII. Smitt has proposed this line is due to Fe X. It is difficult to confirm or exclude the identification of Ca XV ($^3P_2-^1D_2$) proposed for the line at 1375.95 Å.

In the isoelectronic sequence with $2s^22p^3$ ground configuration the transitions $^4S_{3/2}-^2P_{3/2,1/2}$ are observed in Mg VI at 1190.1 and 1191.6 Å. The $^4S_{3/2}-^2D_{3/2}$ transition has also been seen in Mg VI at 1805.97 Å, in Si VIII at 1445.75 Å, and in S X at 1213.0 Å. The tentative Al VII proposal$^{(18)}$ for the line at 1603 Å now seems less likely than Fe X (see below). In the same sequence the $^4S_{3/2}-^2D_{5/2}$ transition is considerably weaker but has now been observed in Si VIII (1440.5 Å) and in S X (1196.2 Å). Edlén$^{(20)}$ had previously proposed that the lines reported at 1197 and 1213 Å in the 1970 eclipse data could be due to S X. The improved wavelengths for Si VIII and S X aid the extrapolation of the doublet–quartet interval in higher ions such as Fe XX.

In the O I isoelectronic sequence ($2s^22p^4$), Mg V ($^3P_1-^1S_0$) is observed at 1324.45 Å. A further weak line at 1392.1 Å is proposed as Ar XI ($^3P_2-^1D_2$), in rather poor agreement with the calculated wavelength$^{(25)}$ of 1390.67 Å.

The Fe XIII transition $^3P_1-^1S_0$ in the $3s^23p^2$ configuration was originally proposed for the line observed at 1213.0 Å.$^{(18,51)}$ However, with the new observation of a line at 1216.46 Å, closer to the predicted wavelength$^{(43)}$ of 1217.2 Å for the Fe XIII line, and an improved wavelength of the S X line at ~1197 Å, this earlier proposal must be rejected.

Further lines from the $3s^23p^2$ configuration, isoelectronic with the known Fe XII lines, have also been reported. These are the Cr X $^4S_{3/2}-$ $^2P_{1/2}$, $^2P_{3/2}$ lines at 1564.3 and 1489.0 Å, respectively, and the Mn XI $^4S_{3/2}-^2P_{3/2}$ line at 1359.6 Å. The observed wavelengths agree well with those predicted by Svensson.$^{(43)}$

The Ni XIII line isoelectronic with Fe XI ($3s^23p^4\ ^3P_2-^1S_0$) is observed at 1277.23 Å, in agreement with the predicted wavelength.$^{(43)}$

The remaining lines in Table 4 are unlikely to come from forbidden lines in the ground configuration. As discussed above, the excited configurations $3p^{n-1}3d$ in Fe IX, Fe X, and Fe XI give rise to observed transitions in the visible part of the spectrum. Two further lines of Fe IX can now be identified below 2000 Å from the work of Svensson et al.$^{(47)}$ These are $3p^53d\ ^3P_2-^1F_3$ at 1917.21 Å and $3p^53d\ ^3P_1-^3D_2$ at 1841.57 Å.

In Fe X the line at 1918.25 Å can be confirmed as $3p^43d\ ^4D_{7/2}-^2F_{7/2}$. Mason and Nussbaumer have suggested that lines at 1428.75 and 1603.21 Å are due to $3p^43d\ ^4F_{9/2}-(^1D)^2F_{7/2}$ and $^4D_{7/2}-^2G_{7/2}$. Smitt,

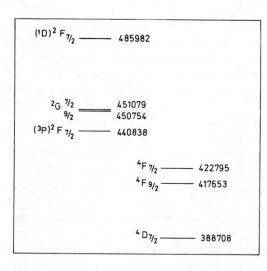

Figure 3. Energies of some levels in the $3p^4 3d$ configuration in Fe X, from which forbidden lines arise. The energies, in cm^{-1}, are from Smitt.[42]

however, does not agree with the former identification and proposes the alternative line at 1463.49 Å. The 1970 eclipse data[18] show that 1463.49 Å and 1603.21 Å behave in a similar manner although it was thought that their temperature class was below that of Fe X. It is difficult to compare these weak lines with the image of the line at 1918 Å, and there is a possibility that the comparison is complicated by different density dependences of the lines. The line at 1428.75 Å does *not* behave like 1603.21 Å, having a distinctly higher temperature. The transition $^4F_{7/2}$–$(^1D)^2F_{7/2}$ has a common upper level with 1603.21 Å, and Smitt identifies

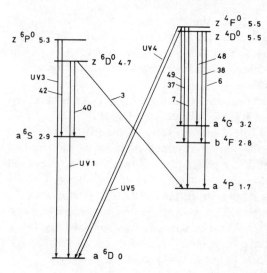

Figure 4. Partial energy-level diagram for Fe II showing some transitions of interest in peculiar galaxies. The energies are given in eV.

this with the line at 1582.56 Å. This line is very weak in the 1970 eclipse data and it cannot be positively said that it behaves like 1603.21 Å. Smitt identifies a line observed by Doschek and Feldman and Sandlin *et al.* at 1611.7 Å with $^4D_{7/2}-^2G_{9/2}$. Figure 3 shows the known metastable levels of the Fe X $3p^43d$ configuration, with energies from Smitt.[42] Levels giving rise to classified permitted transitions lie rather higher.

3. Atomic Data for Plasma Diagnostics

3.1. Oscillator Strengths

Knowledge of oscillator strengths is fundamental to an understanding of the absorption spectra of stellar atmospheres and the interstellar medium. These applications are well known. The example chosen here to illustrate the importance of oscillator strengths is that of the emission lines of Fe II in the spectra of Seyfert galaxies and quasars (Seyfert galaxies and other peculiar galaxies are in many aspects intermediate between normal galaxies and quasars).

During the past few years improved observations of Seyfert galaxies and quasar spectra have become available and there has been some interest in understanding the relative intensities of both permitted and forbidden lines of Fe II in these spectra. In particular Boksenberg *et al.* have published new spectra of the Seyfert galaxy NGC 4151[52] and of the quasar 3C 273,[53] Shectman and MacAlpine[54] of the Seyfert NGC 7469, and Osterbrock[55] of the Seyfert Markarian 376.

Boksenberg *et al.*[53] discuss observations which show that Fe II emission is found preferentially in high-luminosity objects and draw attention to what they regard as the anomalous behavior of multiplet 42, which has lines at 4924, 5018, and 5169 Å in 3C 273. Figure 4 shows a diagram of levels and multiplets which will be discussed below. Multiplet 42 is anomalously strong with respect to multiplets 37, 38 and 48, 49, which are observed in many objects. Boksenberg *et al.* also refer to unpublished observations by Allen which show that hot-emission-line stars with infrared excesses also have intensities in multiplet 42 which can differ in ratio from those of multiplets 37, 38, 48, and 49 by more than a factor of 2. The above authors have not proposed an explanation for the behavior of multiplet 42, but there is a reason such a variation would be expected to occur, which has not to date been mentioned.

As can be seen from Figure 4 the visible region multiplets have upper levels in common with the ultraviolet resonance multiplets, which lie between 2200 and 2700 Å. It was suggested some years ago by Wampler and Oke[56] that the visible-region multiplets are produced by the resonance fluorescence mechanism, i.e., excitations in the uv multiplets produce photons in the visible-region multiplets. Bahcall and Kozlovsky[57] and Adams[58] have also discussed this process. Also Osterbrock[55] points out that a reasonable extrapolation of the observed continuous spectrum of Markarian 376 has more than enough near-ultraviolet photons in the region 2300–2800 Å to produce all the observed Fe II emission by the process.

Since the uv continuum is thought to follow a ν^{-n} ($n \sim 0.4$–1.2) law (ν is the frequency), one would not expect large variations in the *relative* photo-excitation rates from source to source. Nor would *relative* collisional excitation rates vary, in view of the similarity of upper excitation potential of the levels. Also collisional de-excitation in the permitted lines shown does not appear to compete with spontaneous radiative decay at densities of up to $\sim 2 \times 10^8$ cm^{-3} quoted for the sources.

It does appear possible, however, that the variation of multiplet 42 is due to the variation in optical depth of the uv resonance lines from source to source. As is well known, the optical depth of line center can be expressed as

$$\tau_0 = 1.2 \times 10^{-14} \lambda f_{12} M^{1/2} \int_L \frac{N_1}{N(\text{ion})} \frac{N(\text{ion})}{N(E)} N(\text{H}) \, T_{\text{ion}}^{-1/2} \, dl$$

where λ is the wavelength in Å, f_{12} is the absorption oscillator strength, M is the mass; $N(E)/N(\text{H})$ is the element abundance, and N_1, $N(\text{ion})$, and $N(\text{H})$ are the population number densities of the lower level, the ion, and

Table 5. Fe II Transitions

Transition	Multiplet	Wavelength (Å)	gf Multiplet Experimental	gf Multiplet Calculated[62]	Reference for gf experiment
$a\ ^6D$–$z\ ^6P^0$	uv3	2327–2381	4.0	6.1	63, 64
$a\ ^6S$–$z\ ^6P^0$	42	4924–5169	8.3×10^{-2}	9.9×10^{-8}	63
$a\ ^6D$–$z\ ^6D^0$	uv1	2586–2631	14.6	9.3	64, 65
$a\ ^4D$–$z\ ^6D^0$	1	3235–3314	1.8×10^{-2}	1.1×10^{-2}	66
$a\ ^4P$–$z\ ^4D^0$	6	3167–3228	0.23	0.21	66
$b\ ^4F$–$z\ ^4D^0$	38	4508–4648	3.8×10^{-2}	3.4×10^{-2}	63, 67, 68
$a\ ^4G$–$z\ ^4D^0$	48	5267–5435	—	2.5×10^{-3}	—
$a\ ^4P$–$z\ ^4F^0$	7	3142–3196	0.11	0.01	66
$b\ ^4F$–$z\ ^4F^0$	37	4473–4667	2.1×10^{-2}	1.2×10^{-2}	67, 68
$a\ ^4G$–$z\ ^4F^0$	49	5198–5477	0.10	2.4×10^{-2}	63, 68
$a\ ^4P$–$z\ ^6P^0$	5	3316–3425	—	2.1×10^{-3}	—
$b\ ^4F$–$z\ ^6P^0$	36	4993–5036	—	7.6×10^{-4}	—
$a\ ^4G$–$z\ ^6P^0$	47	5691–5932	—	2.2×10^{-5}	—
$a\ ^6D$–$z\ ^4D^0$	uv5	2249–2269	—	0.15	—
$a\ ^6D$–$z\ ^4F^0$	uv4	2237–2280	—	0.12	—
$a\ ^4P$–$z\ ^6D^0$	3	3914–3982	—	1.0×10^{-2}	—

hydrogen, respectively; L is the distance in the line of sight. Thus $\tau_0 \propto f_{12}N_1$, and with reasonable approximations the *relative* optical depths of the lines can be calculated. The oscillator strengths available are given in Table 5, and are discussed further below. A Boltzmann population at $T_e = 2 \times 10^4$ K was used for the relative populations of the lower levels. The calculated relative optical depths cover a range of $\sim 5 \times 10^3$. It is simple to consider the relative intensities of lines from a common upper level. These are given by

$$\frac{E_1}{E_2} = \frac{b_1}{b_2} \frac{\lambda_2}{\lambda_1} \frac{q_1}{q_2}$$

where b is the branching ratio for spontaneous radiative decay against all other decay process and q is the probability that a created photon escapes the atmosphere, without reabsorption.

It can be seen that following processes occur as the optical depths increase, starting from $\tau(\mathrm{uv}1) = 1.0$, and all other lines optically thin. Multiplet uv1 transfers photons to multiplets 1 and 3; uv3 becomes optically thick and transfers photons to 42; when $\tau(\mathrm{uv}1) \sim 100$, multiplets uv4 and uv6 start transferring photons to 6, 7, 37, 38, 48, and 49; multiplets 6 and 7 also start transferring photons. Eventually these other lines become optically thick as well. Thus the relative intensities of multiplets 42 and 37, 38, 48, 49 reflect the absolute value of $\tau(\mathrm{uv}1)$, and although the scale of the variation is difficult to establish without more complex calculations, it is found that the required ratio varies, while the relative intensity of 37 + 38 to 48 + 49 varies little. [Adams[58] has used other arguments to show that $\tau(\mathrm{uv}3)$ could be found from the intensity of multiplet 42 if the exciting uv flux were known].

The importance of knowing the f values is twofold: first to determine the values of τ_0; second to determine the branching ratios from common upper levels. This area of application has become even more important now that the international ultraviolet Explorer satellite is operating. The uv multiplets could be observable simultaneously with the visible-region lines. An independently determined value of τ, in combination with absolute intensities, would be a valuable parameter in models of the sources. Since collisional excitation may be important in some sources, there is also a need for collision cross sections, but this appears to be a long-term problem. Viotti[59,60] has already made some calculations of populations using a very simple approximation for the cross sections.

We return to the oscillator strengths shown in Table 5. At present a limited number of reliable experimental values exists; the hook method, wall-stabilized arcs, shock-tube spectroscopy, and beam-foil time-of-flight method have all been used. Some references are given in Table 5. Smith[61] has also recently discussed available data for Fe II lines. Semiempirical calculations of oscillator strengths, using scaled Thomas–Fermi–Dirac wave functions and intermediate coupling have been made by Kurucz and Peytremann.[62] Their values are also given in Table 5. It can be seen that for the strong lines there is usually, but not always, reasonable agreement between experimental and calculated values. However, for the application discussed above, good values for weak lines are also needed, but it is for these transitions that there is an absence of experimental data and for which the calculated values would be expected to be the least accurate.

3.2. Collision Cross Sections

Ion–electron collision cross sections are required in the interpretation of emission-line intensities: (a) in analyzing *absolute* line intensities to derive information concerning the atmospheric layers producing the line and (b) in analyzing *relative* line intensities to determine parameters such as N_e and T_e.

A rather lower accuracy for the cross sections can be accepted in the former application than in the latter. Considering the simplest possible case of a singlet resonance line excited by electron collisions from the ground state only, the energy emitted in a line can be expressed in the usual way as

$$E = \frac{hc}{\lambda} \int_V N_2 A_{21} \, dV \text{ erg s}^{-1} \tag{1}$$

where N_2 is the number density in the excited level, A_{21} is the spontaneous radiative decay rate, and V is the emitting volume. In statistical equilibrium

$$N_2 A_{21} = N_1 C_{12} N_e \tag{2}$$

Thus

$$E = \frac{hc}{\lambda} \int_V N_1 C_{12} N_e \, dV \tag{3}$$

and it can be seen that the line intensity depends on C_{12}, the collisional excitation rate, rather than on A_{21}. It is usual to assume a Maxwellian distribution for the electron energy distribution function $f(E)$, and then C_{12} is related to the cross section Q_{12} by

$$C_{12} = \int_{E_0}^{\infty} v Q_{12} f(E) \, dE \tag{4}$$

where v and E are the electron velocity and energy and E_0 is the threshold for excitation.

The lower-level population N_1 is usually expressed as

$$N_1 = \frac{N_1}{N(\text{ion})} \frac{N(\text{ion})}{N(E)} \frac{N(E)}{N(H)} 0.8 N_e \tag{5}$$

where in the simplest case $N_1/N(\text{ion}) = 1.0$, $N(\text{ion})/N(E)$ is the ionization equilibrium population for the ion, and $N(E)/N(H)$ is the abundance of the element relative to hydrogen. Because uncertainties of at least $\sim 25\%$ exist in the measured absolute intensities, in $N(E)/N(H)$ and in $N(\text{ion})/N(E)$ at a given T_e there is little demand at present for values of C_{12} of accuracy greater than 25% when the values are needed only for interpreting absolute intensities, and currently used methods such as those involving the distorted-wave approximation are adequate for most ions.

The situation is rather different, however, when cross sections are needed to interpret *relative* intensities. When two lines from the same ion are considered, most of the uncertainties mentioned above are removed or reduced. Further, the sensitivity of a particular line pair to T_e or N_e may not be great and even 25% uncertainties in the excitation rates can lead to order-of-magnitude uncertainties in N_e.

The lines in the (Be I)-like ions have so far received most attention as possible indicators of N_e, and they will be discussed further below. Because limits on N_e can be found from the analyses of solar absolute line intensities, the solar spectrum provides a testing ground for the atomic data. This has led to a lively, but, one hopes, fruitful, debate between those engaged in calculating cross sections and those applying the results to solar observations.

The recent review of Seaton[69] discusses the methods available for calculating excitation cross sections and also long-standing applications such as to visible-region planetary nebulae spectra. The field of diagnostic methods for determining N_e is a rapidly expanding one, both for astrophysical applications and also for understanding conditions in laboratory

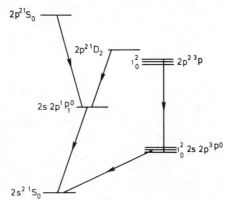

Figure 5. Energy-level diagram for (Be I)-like ions. Transitions observed in either C III or O V are indicated.

Table 6. Wavelengths (Å) of C III and O V Transitions

Transition	C III	O V
$2s^2\,^1S_0-2s2p\,^1P_0$	977.03	629.73
$2s^2\,^1S_0-2s2p\,^3P_1^0$	1908.73	1218.36
$2s2p\,^3P_1^0-2p^2\,^3P_2$	1174.93	758.68
0 1	1175.26	759.44
2 2	1175.71	760.44
1 1	1175.59	760.23
1 0	1175.99	761.13
2 1	1176.37	762.00
$2s2p\,^1P_1^0-2p^2\,^1D_2$	2296.89	1371.29
$2s2p\,^1P_1^0-2p^2\,^1S_0$	1247.38	774.52

fusion plasmas. Feldman and Doschek[70] have discussed some possible applications in this latter area.

The energy levels in the $2s^2$, $2s2p$, and $2p^2$ configurations of the (Be I)-like ions are shown, not to scale, in Figure 5. The wavelengths of the lines observed in C III and O V are given in Table 6.

At the beginning of this section it was shown that in the simplest case the intensity of a line depends only on the collisional excitation rate. However, if a collisional process is comparable in size to the spontaneous decay rate, then a further density term enters the intensity expression. For example, consider a level m populated by collisions from the ground level 1 but depopulated by spontaneous radiative decay A_{m1} and by collision to several levels n. Then

$$E_2 \propto N_m A_{m1}$$

as before but now

$$N_m(A_{m1} + C_{mn}N_e) = N_1 C_{1m}N_e \tag{6}$$

Thus for two lines, one with upper-level population described by Eq. (2) and the other described by Eq. (6) we have

$$\frac{E_1}{E_2} \propto \frac{C_{12}}{C_{1m}}\left(1 + \frac{C_{mn}N_e}{A_{m1}}\right) \tag{7}$$

which for $A_{m1} \sim C_{mn}N_e$ becomes sensitive to N_e. For very large N_e the ratio is usually impractically large to measure reliably. It can be seen that contrary to many statements to the effect, the intensity of a line such as that given by Eq. (6), arising from a level depopulated by collisions does not *decrease* as N_e increases, but first increases with N_e and then eventually remains constant.

In the (Be I)-like ions the density dependence arises from the low value of A for the intersystem transition $2s^2\,^1S_0-2s2p\,^3P_1$. The competing collisional process in the low to moderately charged ions is excitation to $2s2p\ ^1P_1$, as well as collisional de-excitation to the ground level $2s^2\ ^1S_0$. The detailed equations of statistical equilibrium and relative importance of the different rates have been discussed in the review by Gabriel and Jordan[71] and will not be repeated here. However, the large amount of data required will be appreciated when it is said that not only do the electron collision rates between practically all the terms shown in Figure 5 have to be known, but also electron and proton collision rates between the fine-structure terms of $2s2p\ ^3P$ enter the equations. Particular rates of interest and recent changes to earlier data are discussed below.

Early calculations of the density dependence of the (Be I)-like ion lines[72,73] resulted in different values of N_e from a given observed solar line ratio due to the use of different atomic data. As improved calculations using the distorted-wave method became available[74] it was found that there were distinct discrepancies between the calculated and observed line ratios.[75] For C III the density derived from the solar observations corresponded to a much lower value, $\sim 10^8\ cm^{-3}$, than usually associated with the low-transition region. In O V the observed value for the $2s^2\ ^1S_0-2s2p\ ^1P_1/2s2p\ ^3P-2p^2\ ^3P$ intensity ratio was significantly below the calculated value even for $N_e \lesssim 10^6\ cm^{-3}$, showing that not all the problems could be attributed to lack of knowledge about the solar atmosphere. Further calculations of cross sections using the close-coupling method[76] suggested that the atomic data should be reliable to $\sim \pm 25\%$. Jordan[75] investigated the effect of changing the $2s^2\ ^1S_0-2s2p\ ^1P_1$, $2s^2\ ^1S_0-2s2p\ ^3P_1$, and $2s2p\ ^3P-2p^2\ ^3P$ excitation rates by these amounts, and also the effect of reducing the loss from $2s2p\ ^3P$ by collisions to $2s2p\ ^1P$ and $2p^2\ ^1D_2, ^1S_0$ and the effect of increasing the effective rate of mixing the populations of the $2s2p\ ^3P$ levels. It was found that for C III the ratio in spectra of the quiet sun could be accounted for by changes in the atomic data of $\sim 25\%$ provided the rate for the exchange of population between the $2s2p\ ^3P$ levels was also increased. This latter conclusion is supported by the observation by Nicolas (private communication) that the relative intensities of the lines in the 1175-Å multiplet do not change from point to point on the disc. The rather high values observed for active regions were less satisfactorily accounted for, but later observations[77] suggest that these could be due to rapid time changes in line intensities, and the possibility of transient effects in active regions cannot be excluded. Similarly for O V, the ratio in quiet sun spectra could be accounted for provided the $2s2p\ ^3P$ mixing was increased by up to a factor of 4. At the same time Loulergue and Nussbaumer[78] suggested that the problem in C III could be resolved by postulating a much extended transition region of considerably lower density ($\sim 10^8\ cm^{-3}$) than is usually accepted.

Since then Mühlethaler and Nussbaumer[41] have included the effects of further configurations in wave function calculations but find only small

Table 7. Intensities of C III and O V Lines

Ion	Transition	λ(Å)	Log intensity (erg cm^{-2} s^{-1} st^{-1})^a for various h (arc sec)							
			−12	−4	−2	0	+2	+4	+6	+8
C III	$2s2p$ $^3P_1^0$–$2p^2$ $^3P_2$	1174.93	1.83	1.78	1.75	1.81	2.25	2.16	2.09	1.64
	0 1	1175.26	1.76	1.76	1.66	1.72	2.09	2.03	1.98	1.53
	2, 1 2, 1	1175.71, 1175.59	2.18	2.12	2.12	2.12	2.15	2.51	2.45	2.10
	1 0	1175.99	1.81	1.72	1.65	1.74	2.21	2.06	1.93	1.55
	2 1	1176.37	1.83	1.79	1.75	1.80	2.25	2.13	2.06	1.59
	$2s2p$ $^1P_1^0$–$2p^2$ $^1S_0$	1247.38	—	—	—	—	1.35	1.49	1.27	0.74
	$2s^2$ $^1S_0$–$2s2p$ $^3P_1^0$	1908.73	—	3.06	3.08	3.36	3.64	3.51	3.36	3.02
O V	$2s^2$ $^1S_0$–$2s2p$ $^3P_1^0$	1218.36	2.36	2.34	2.39	2.30	2.56	2.66	2.56	2.33
	$2s2p$ $^1P_0^0$–$2p^2$ $^1D_2$	1371.29	1.72	1.69	1.65	1.75	2.10	2.04	2.01	1.63

^{a} Intensity data are from Moe and Nicolas.[83]

changes (~8% for $2s^2$ ^1S–$2s2p$ ^1P and ~6% for $2s2p$ ^3P–$2p^2$ ^3P) in the resultant cross-section values. However, the *direction* of the changes is to reduce the discrepancies. The ionization equilibrium for C III and other carbon ions has been recalculated by Nussbaumer and Storey,[79] including the effects of photoionization. Although at $T_e \sim 5 \times 10^4$ K these results differ insignificantly from earlier values,[80] at low temperatures, $\leqslant 3 \times 10^4$ K, the C III total ion population is enhanced. Thus Loulergue and Nussbaumer[81] now propose that the C III observed ratios can be explained with $N_e \sim 10^{10}$ cm^{-3} but with $T_e = 3 \times 10^4$ K, and again a rather thicker atmosphere layer than in models derived from analyses of other emission lines.

The observational data for C III have been improved by the spectra and spectroheliograms obtained with the Harvard instrument on the Skylab satellite[77] and the further data discussed below. The analysis of the Harvard data led to a reaffirmation that the calculated intensities could not be reconciled with the observations. A temperature of $T_e = 7.0 \times 10^4$ K, corresponding to the peak of the function

$$g(T) = T_e^{-1/2} [N(\text{ion})/N(E)] \, e^{-W/kT_e}$$

was used by Jordan[75] and Dupree *et al.*[76] to avoid the introduction of information on the variation of $\int N_e^2 \, dh$ with T_e. However, a temperature of 5.6×10^4 corresponding to the peak of the function $\int N_e^2 g(T) \, dh$ was used by Gabriel and Jordan.[71] These authors also calculated the temperature dependence of the line ratios, and as can be seen also from the calculation by Loulergue and Nussbaumer (shown in Figure 6), this slightly lower temperature does not cause a significant change in the line ratio discussed above. A drop to ~2.5×10^4 K is needed to reconcile solar observations, calculations, and the usual "pressure" of $N_e T_e = 6 \times 10^{14}$ cm^{-3} K found to be appropriate in the transition region.

The recently published observations of the intensities of lines in the $2s2p\ ^3P-2p^2\ ^3P$ multiplet and of the $2s2p\ ^1P-2p^2\ ^1S$ and $2s^2\ ^1S_0-2s2p\ ^3P_1$ lines by Doschek et al.,[82] Feldman et al.,[83] and Moe and Nicolas[84] allow a further discussion of the C III intensity ratios. An important factor concerning these observations is that they are all made at the solar limb. It was shown by Dupree et al.[77] that the 977-Å line certainly becomes optically thick at the limb. Simple estimates[85] also show that the 1175 multiplet will also become optically thick at the limb. Thus one cannot, as was done by Loulergue and Nussbaumer,[81] use intensities relative to the 1175-Å multiplet to derive N_e. Only the $2s2p\ ^1P-2p^2\ ^1S$ and $2s^2\ ^1S_0-2s2p\ ^3P_1$ lines at 1247 and 1909 Å can be used in this way at the limb since they will certainly be optically thin.

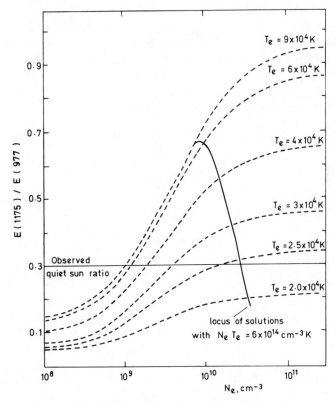

Figure 6. Calculated variation of the C III line ratio $E\ (1175\ \text{Å})/(E\ (977\ \text{Å}))$ as a function of N_e and T_e (from Loulergue and Nussbaumer[81]). The observed ratio and locus of solutions with $N_e T_e = 6 \times 10^{14}$ cm^{-3} K are also shown.

The observed relative intensities of the lines within the 1175-Å multiplet can be compared with values expected in the limiting cases of low and high optical depth. Only the weakest members of the multiplet are sensitive to N_e, and then only if the $2s2p\ ^3P$ levels are far from being fully mixed. Table 7 gives the intensities of the C III lines as a function of height across the limb observed by Moe and Nicolas.[84] These are also plotted in Figure 7, relative to the sum of the $^3P_2^0-^3P_2$, $^3P_1^0-^3P_1$ blend. The limiting values at high and low optical depths are shown. It can be seen that at the limb where the lines are most intense, the relative intensities approximate more closely to the optically thick value than the optically thin values, although the $^3P_0^0-^3P_1$ component is anomalous at +2 arc sec. Moreover, the ratios tend to the optically thin values at $h \gtrsim 10$ arc sec.

Considering now the $2s2p\ ^1P-2p^2\ ^1S/2s^2\ ^1S_0-2s2p\ ^3P_1$ ratio, the observed values are given in Table 7 and the author's calculations at $T_e = 5.6 \times 10^4$ K, and as a function of N_e, are shown in Figure 8. (These calculations give similar results to those which can be derived from Loulergue and Nussbaumer.) The cross sections on which the $2s^2\ ^1S_0-2s2p\ ^3P_1/2s2p\ ^3P-2p^2\ ^3P$ intensity ratio depends have little effect on the 1247/1909 ratio, and the two curves shown represent the range of values

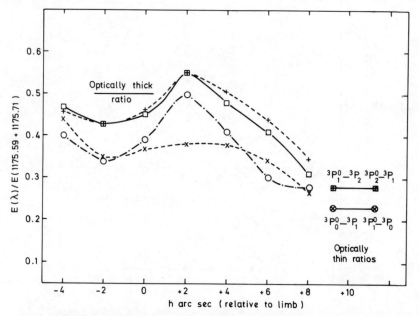

Figure 7. The variation of relative intensities within the $2s2p\ ^3P^0-2p^2\ ^3P$ multiplet in C III, as a function of height above the limb. The heights of the horizontal rules give the ratio for the optically thick case and the ratios expected under optically thin conditions.

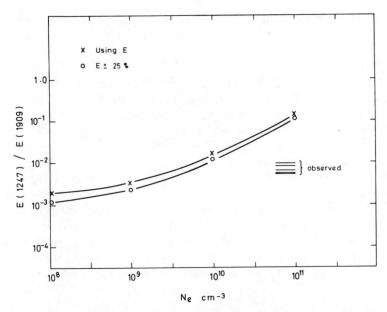

Figure 8. The variation of the intensity ratio of $2s2p\ {}^1P^0-2p^2\ {}^1S$ (1247 Å) and $2s^2\ {}^1S_0-2s2p\ {}^3P_1$ (1909 Å) in C III, as a function of N_e and of the atomic data used. E refers to Eissner's[74] data, as used by Jordan.[75] The observed ratios are from Moe and Nicolas[84] and Doschek et al.[82]

discussed by Jordan.[75] It can be seen that for $N_e \sim 10^{10}-10^{11}$ cm^{-3} this line ratio provides a good method for determining N_e. Moreover, the values derived from the observed ratios agree well with the value of $N_e \sim 10^{10}$ cm^{-3} used in many transition-region models at $T_e \sim 5.6 \times 10^4$ K. The suppression of the 1175-Å multiplet by the preferential escape of photons in the radial direction (or direction of lowest optical depth) appears to be about a factor of 10.

Some further developments have occurred also in the calculation of atomic data for O V. Mühlethaler and Nussbaumer[41] recalculated the $2s^2\ {}^1S_0-2s2p\ {}^3P_1$ transition probability, and Malinovsky[86] has made further distorted-wave calculations for the excitation cross sections, including nine configurations. It is worth comparing these new calculations with those used by Jordan.[75] The $2s^2\ {}^1S-2s2p\ {}^1P$ cross section is not significantly different. The $2s^2\ {}^1S_0-2s2p\ {}^3P_1$ excitation rate must take account of resonances below the $2s2p\ {}^1P$ and higher levels. Malinovsky's calculations give a collision strength Ω near threshold that is about 10% larger than the values found by Osterbrock,[87] which were used by Jordan. However, Malinovsky finds that at $T_e = 2.5 \times 10^5$ K, when integrated over a Maxwellian distribution, the effective rate is increased by a factor of about 2, whereas Jordan finds a factor of more like 30%. Another significant difference, not discussed by Malinovsky, is in the $2s2p\,^3P-$ $2p^2\,^3P$ rate. Malinovsky's nine-configuration value is $\sim 15\%$ *lower* than the three-configuration value calculated by Eissner.[74,88] This lower rate reduces the $2s2p\,^3P-2p^2\,^3P$ intensity and offsets the effect of increasing the $2s^2\,^1S-2s2p\,^3P$ rate. The change is in a direction to improve the agreement between the observed and calculated values of the $2s^2\,^1S-$ $2s2p\,^1P/2s2p\,^3P-2p^2\,^3P$ intensity ratio. One of the points stressed by Malinovsky is that resonances must also be included in calculations of the collision rates between the fine-structure levels of the $2s2p\,^3P$ term. She finds a factor of between 2 and 4 increase in these electron collision rates. However, the inclusion of the resonances does not cause such a large increase in the *total* mixing rate for the $2s2p\,^3P$ levels. When the proton excitation rates and population changes due to excitation and decay in the $2s2p\,^3P-2p^2\,^3P$ multiplet are taken into account, if it is remembered that this latter rate has been reduced by $\sim 15\%$, it is found that the total rate increases by only $\sim 30\%$. The final results of Malinovsky's calculations lead

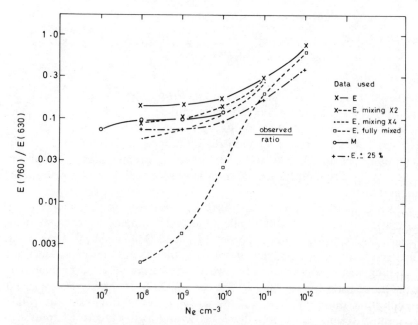

Figure 9. The variation of the intensity ratio of $2s2p\,^3P-2p^2\,^3P$ (760 Å) and $2s^2$ $^1S_0-2s2p\,^1P_0$ (630 Å) in O V, as a function of N_e and atomic data used. E refers to Eissner's[88] data, as used by Jordan.[75] The effects of increasing the rate of mixing between the $2s2p\,^3P$ levels is illustrated. The results from changing Eissner's rates by $\pm 25\%$ and from using Malinovsky's[86] data are also shown.

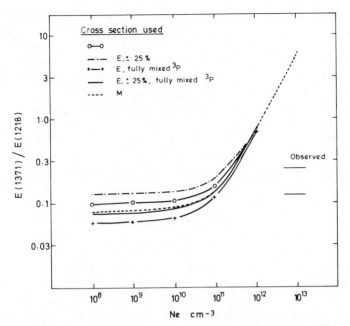

Figure 10. The relative intensity of the $2s2p$ $^1P_1^0$–$2p^2$ $^1D_2$ (1371 Å) and $2s^2$ $^1S_0$–$2s2p$ $^3P_1$ (1218 Å) transitions in O V, as a function of density and cross-section data used. The results using the data of Eissner[88] and of variations from these as in Jordan[75] are shown. Ratios calculated by Feldman et al.,[89] using atomic data from Malinovsky,[86] are indicated. The observed ratios are from Doschek et al.[82] and from Moe and Nicolas.[84]

to a $2s^2$ ^1S–$2s2p$ $^1P/2s2p$ ^3P–$2p^2$ ^3P ratio which has values between the two cases considered by Jordan,[75] using Eissner's cross sections and Eissner's values changed arbitrarily by $\sim 25\%$. These calculations are shown in Figure 9. The observed quiet sun ratio (0.06) is still significantly below the calculated value for $N_e \geqslant 10^7$ cm^{-3}. However, as discussed before, an increase in the effective rate of collisions between $2s2p$ ^3P levels would act in a direction to remove the discrepancy. It is not clear what process could cause this high effective rate. The extreme case where the relative populations of the $2s2p$ ^3P levels have values proportional to their statistical weight is also shown in Figure 9. Further analyses of the Harvard ATM data, comparable with those carried out for C III will be of value in establishing the magnitude of variations in this ratio both in the quiet sun and between quiet and active regions.

A further pair of transitions, the $2s2p$ ^1P–$2p^2$ ^1D line at 1371 Å and the $2s^2$ $^1S_0$–$2s2p$ $^3P_1$ line at 1218 Å, can also be studied from observations made with the NRL's spectrograph on Skylab. Figure 10 shows the ratio

$I(1371)/(1218)$ calculated from Malinovsky's data by Feldman et al.[89] and also the ratios calculated by the author. The effect of collisions between the $2s2p\,^3P$ levels is not so drastic for this line pair. The ratio is not sensitive to density for $N_e \lesssim 10^{10}\,\mathrm{cm}^{-3}$ and depends therefore simply on the relative excitation rates used. It can be seen that the calculated ratio is a little lower than the observed quiet sun values.

The $I(1371)/I(1218)$ ratio may be useful in determining N_e in active regions and flares, but transient effects may dominate in the latter. Feldman et al.[89] have given intensity data for the June 15, 1973 flare, and that paper should be seen for a discussion of the line intensity and width variations as a function of time. However, it should be noted that the maximum value of the ratio that was recordable in the stationary part of the flare (not necessarily an upper limit because the lines were not always strong enough to see) was 0.23, corresponding to $N_e \sim 1.6 \times 10^{11}\,\mathrm{cm}^{-3}$, if the statistical equilibrium calculations are applicable. Ratios as low as ~ 0.055 were recorded later in the flare. This low value would, as discussed by Feldman et al., appear to be difficult to reconcile with the quiet sun values. It can be seen that even with a change to a Boltzmann distribution among the $2s2p\,^3P$ levels a low ratio cannot be obtained with $N_e \gtrsim 10^{10}\,\mathrm{cm}^{-3}$. Yet an alternative method of deducing N_e from other absolute line intensities suggests that N_e remains $\sim 10^{11}\,\mathrm{cm}^{-3}$ while the ratio drops. A change to recombination as the dominant mechanism populating the $2p^2\,^1D$ and $2s2p\,^3P$ levels could in principle increase the triplet population relative to the singlet population, but it is not clear that this could happen without at the same time increasing the $2s2p\,^3P$–$2p^2\,^1D$ collisional excitation rate. An alternative explanation could be that the lines are formed by collisional excitation in a rapidly recombining plasma, and then T_e for optimum line formation could be distinctly lower than the usual value of $\sim 2 \times 10^5\,\mathrm{K}$. At $N_e \sim 10^{11}\,\mathrm{cm}^{-3}$, with the relative populations of the $2s2p\,^3P$ levels given by their statistical weights, a temperature of $T_e \sim 5 \times 10^4\,\mathrm{K}$ would be required to give the observed ratio. A detailed theoretical treatment of line ratios in uv under transient conditions is required before the flare spectra can be analyzed further.

In sunspot spectra, since these enhance transition region lines,[90] the ratio of the $2s^2\,^1S_0$–$2s2p\,^3P_1$ line at 1218 Å, to the $2s2p\,^3P$–$2p^2\,^3P$ multiplet should provide the best way of establishing the relative populations of the $2s2p\,^3P$ levels, since unlike $2p^2\,^1D_1$, the $2p^2\,^3P$ levels are not significantly populated from the ground state $2s^2\,^1S_0$.

Finally, there is some hope that the discrepancies between the observations and calculated intensities may be resolved. Berrington et al.[91] have recently recalculated the $2s^2\,^1S$–$2s2p\,^3P$ and $2s^2\,^1S$–$2s2p\,^1P$ cross sections using the R-matrix method and compared their results with those using the distorted-wave and close-coupling methods. They find that the use of accurate target eigenstates and what they consider to be a more accurate treatment of the collision problem increases the $^1S-^1P$ cross section in C III by about 20% and reduces the $^1S-^3P$ cross section by the same amount. Thus there is now some support for the earlier conclusion[75] that changes of 25% are required to give results consistent with all types of solar observations.

The analyses of (Be I)-like ion line intensities illustrate well the benefits of constant interaction between atomic physics and astrophysics. Without reliable atomic data, progress in understanding astrophysical spectra is limited, but at the same time, through the requirements for consistent solutions for plasma conditions the astrophysical spectra will no doubt continue to provide a good testing ground for atomic theory.

References

1. W. Grotrian, *Z. Astrophys.* **3**, 220 (1931).
2. B. Edlén, *Z. Astrophys.* **22**, 30 (1942).
3. B. Edlén, *Mon. Not. R. Astron. Soc.* **105**, 323 (1945).
4. R. D. Cowan, *Phys. Rev.* **163**, 54 (1967).
5. R. D. Cowan, *J. Opt. Soc. Am.* **58**, 808 (1968).
6. C. Froese-Fischer, *Bull. Astron. Inst. Neth.* **19**, 86 (1967).
7. B. Edlén, in *Handbuch der Physik*, Ed. S. Flügge, Vol. 27, p. 80, Springer-Verlag, Berlin (1964).
8. B. C. Fawcett, in *Advances in Atomic and Molecular Physics*, Vol. 10, p. 229, Academic Press, New York (1974).
9. B. Edlén, in *Proceedings of the Fourth Beam-Foil Conference, Gatlinburg, Beam-Foil Spectroscopy*, Vol. 1, Eds. I. A. Sellin and D. J. Pegg, p. 1, Plenum Publishing Corp., New York (1976).
10. B. Edlén, *Phys. Scr.* **7**, 93 (1972).
11. C. E. Moore, N.B.S. Circular 467, Vols. I and II, U.S. GPO, Washington, D.C. (1949).
12. W. E. Behring, L. Cohen, and U. Feldman, *Astrophys. J.* **175**, 493 (1972).
13. W. E. Behring, L. Cohen, U. Feldman, and G. A. Doschek, *Astrophys. J.* **203**, 521 (1976).
14. G. A. Doschek, *Solar, X, and euv Radiation*, Ed. S. R. Kane, International Astronomical Union, Symposium, No. 68 (1975).
15. U. Feldman, *Astrophys. Space Sci.* **41**, 155 (1976).
16. B. Edlén, *Mem. Soc. R. Sci. Liège*, Ser. 6, **9**, 235 (1976).
17. A. H. Gabriel and C. Jordan, *Nature* **221**, 947 (1969).
18. C. Jordan, *Sol. Phys.* **21**, 381 (1971).
19. R. J. Mitchell, J. L. Culhane, P. J. N. Davison, and J. C. Ives, *Mon. Not. R. Astron. Soc.* **176**, 29P (1976).
20. B. Edlén, *Sol. Phys.* **9**, 439 (1969).
21. U. Feldman, G. A. Doschek, D. J. Nagel, W. E. Behring, and L. Cohen, *Astrophys. J.* **183**, L43 (1973).
22. U. Feldman, G. A. Doschek, R. D. Cowan, and L. Cohen, *J. Opt. Soc. Am.* **63**, 1445 (1973).
23. V. A. Boiko, Yu. P. Voinov, V. A. Gribkov, and G. V. Sklizkov, *Opt. Spectrosc.* **29**, 545 (1970).

24. G. A. Doschek, U. Feldman, K. P. Dere, G. D. Sandlin, M. E. Van Hoosier, G. E. Brueckner, J. D. Purcell, and R. Tousey, *Astrophys. J.* **196**, L83 (1975).

25. B. Edlén, *Sol. Phys.* **24**, 356 (1972).

26. G. A. Doschek, U. Feldman, R. D. Cowan, and L. Cohen, *Astrophys. J.* **188**, 417 (1974).

27. E. Ya. Kononov, A. N. Ryabtsev, U. I. Safronova, and S. S. Churilev, *J. Phys. B* **9**, L477 (1976).

28. U. I. Safronova, *J. Quant. Spectrosc. Radiat. Transfer* **15**, 223 (1975).

29. U. I. Safronova, *J. Quant. Spectrosc. Radiat. Transfer* **15**, 231 (1975).

30. G. D. Sandlin, G. E. Brueckner, V. E. Scherrer, and R. Tousey, *Astrophys. J.* **205**, L47 (1976).

31. K. G. Widing, *Astrophys. J.* **197**, L33 (1975).

32. R. Noyes (private communication, 1974).

33. U. Feldman, G. A. Doschek, R. D. Cowan, and L. Cohen, *Astrophys. J.* **196**, 613 (1975).

34. B. C. Fawcett and R. D. Cowan, *Mon. Not. R. Astron. Soc.* **171**, 1 (1975).

35. E. Ya. Kononov, K. N. Koshelev, L. I. Podobedeva, S. V. Chekalin, and S. S. Churilov, *J. Phys. B* **9**, 565 (1976).

36. U. Feldman and G. A. Doschek (unpublished) (1976).

37. G. D. Sandlin, G. E. Brueckner, and R. Tousey, *Astrophys. J.* **214**, 898 (1977).

38. E. Ya. Kononov, K. N. Koshelev, L. I. Podobedeva, and U. I. Safronova, Publications of the Institute of Spectroscopy, Akademgorodok, Preprint 1 (1975).

39. H. Nussbaumer, *Astrophys. J.* **166**, 411 (1971).

40. R. W. Noyes, in *High Energy Phenomena on the Sun* Eds. R. Ramaty and R. G. Stone, NASA SP-342 (1973).

41. H. P. Mühlethaler and H. Nussbaumer, *Astron. Astrophys.* **48**, 109 (1976).

42. R. Smitt, *Sol. Phys.* **51**, 113 (1977).

43. L. A. Svensson, *Sol. Phys.* **18**, 232 (1971).

44. W. J. Wagner, Ph.D. Thesis, University of Colorado (1969).

45. W. J. Wagner and L. L. House, *Astrophys. J.* **155**, 677 (1969).

46. J. G. Firth, F. F. Freeman, A. H. Gabriel, B. B. Jones, C. Jordan, C. R. Negus, D. B. Shenton, and R. F. Turner, *Mon. Not. R. Astron. Soc.* **166**, 543 (1974).

47. L. A. Svensson, J. O. Ekberg, and B. Edlén, *Sol. Phys.* **34**, 173 (1974).

48. H. E. Mason and H. Nussbaumer, *Astron. Astrophys.* **54**, 547 (1977).

49. F. Magnant-Crifo, *Sol. Phys.* **31**, 91 (1973).

50. F. Magnant-Crifo, *Sol. Phys.* **41**, 109 (1975).

51. A. H. Gabriel, W. R. S. Garton, L. Goldberg, T. J. L. Jones, C. Jordan, F. J. Morgan, R. W. Nicholls, W. H. Parkinson, H. J. B. Paxton, E. M. Reeves, D. B. Shenton, R. J. Speer, and R. Wilson, *Astrophys. J.* **169**, 595 (1971).

52. A. Boksenberg, K. Shortridge, D. A. Allen, R. A. E. Fosbury, M. V. Penston, and A. Savage, *Mon. Not. R. Astron. Soc.* **173**, 381 (1975).

53. A. Boksenberg, K. Shortridge, R. A. E. Fosbury, M. V. Penston, and A. Savage, *Mon. Not. R. Astron. Soc.* **172**, 289 (1975).

54. S. A. Shectman and G. M. MacAlpine, *Astrophys. J.* **199**, L85 (1975).

55. D. E. Osterbrock, *Astrophys. J.* **203**, 329 (1976).

56. E. J. Wampler and J. B. Oke, *Astrophys. J.* **148**, 695 (1967).

57. J. N. Bahcall and B. Kozlovsky, *Astrophys. J.* **155**, 1077 (1969).

58. T. F. Adams, *Astrophys. J.* **196** 675 (1975).

59. R. Viotti, *Astrophys. J.* **204**, 293 (1976).

60. R. Viotti, Mon. *Not. R. Astron. Soc.* **177**, 617 (1976).

61. P. L. Smith, *Mon. Not. R. Astron. Soc.* **177**, 275 (1976).

62. R. L. Kurucz and E. Peytremann, *Smith. Astrophys. Obs. Spec. Rep.* **362** (1975).

63. S. J. Wolnik, R. U. Berthel, and G. W. Wares, *Astrophys. J.* **166**, L31 (1971).

64. G. E. Assousa and W. H. Smith, *Astrophys. J.* **176**, 529 (1972).

65. M. C. E. Hüber, *Astrophys. J.* **190**, 237 (1974).
66. G. L. Grasdalen, M. Huber, and W. H. Parkinson, *Astrophys. J.* **156**, 1153 (1969).
67. B. Baschek, T. Garz, H. Holweger, and J. Richter, *Astron. Astrophys.* **4**, 229 (1970).
68. O. Roder, *Z. Astrophys.* **55**, 38 (1962).
69. M. J. Seaton, in *Advances in Atomic and Molecular Physics*, Vol. XI, Eds. D. R. Bates and B. Bederson, p. 83, Académic Press, New York (1975).
70. U. Feldman and G. A. Doschek, *J. Opt. Soc. Am.* **67**, 726 (1977).
71. A. H. Gabriel and C. Jordan, in *Case Studies in Atomic Collision Physics*, Vol. 2, Eds. E. McDaniel and M. C. McDowell, p. 210, North-Holland, Amsterdam (1972).
72. R. H. Munro, A. K. Dupree, and G. L. Withbroe, *Sol. Phys.* **19**, 347 (1971).
73. C. Jordan, in *Highlights in Astronomy*, Ed. C. de Jager, p. 519, D. Reidel, Dordrecht (1971).
74. W. Eissner, in *Physics of Electronic and Atomic Collisions* VII ICPEAC, p. 460, North-Holland, Amsterdam (1972).
75. C. Jordan, *Astron. Astrophys.* **34**, 69 (1974).
76. D. R. Flower and J. M. Launay, *Astron. Astrophys.* **29**, 321 (1973).
77. A. K. Dupree, P. V. Foukal, and C. Jordan, *Astrophys. J.* **209**, 621 (1976).
78. M. Loulergue and H. Nussbaumer, *Astron. Astrophys.* **34**, 225 (1974).
79. H. Nussbaumer and P. J. Storey, *Astron. Astrophys.* **44**, 321 (1975).
80. C. Jordan, *Mon. Not. R. Astron. Soc.* **142**, 501 (1969).
81. M. Loulergue and H. Nussbaumer, *Astron. Astrophys.* **51**, 163 (1976).
82. G. A. Doschek, U. Feldman, M. E. Van Hoosier, and J.-D. F. Bartoe, *Astrophys. J.* Suppl. **31**, 417 (1976).
83. U. Feldman, G. A. Doschek, M. E. Van Hoosier, and J. D. Purcell, *Astrophys. J.* Suppl. **31**, 445 (1976).
84 O. K. Moe and K. R. Nicolas, *Astrophys. J.* **211**, 579 (1977).
85. W. M. Burton, C. Jordan, A. Ridgeley, and R. Wilson, *Astron. Astrophys.* **27**, 101 (1973).
86. M. Malinovsky, *Astron. Astrophys.* **43**, 101 (1975).
87. D. E. Osterbrock, *Astrophys. J.* **160**, 25 (1970).
88. H. E. Saraph, *Comp. Phys. Commun.* **3**, 256 (1972).
89. U. Feldman, G. A. Doschek, and F. D. Rosenberg, *Astrophys. J.* **215**, 652 (1977).
90. A. K. Dupree, M. C. E. Huber, R. W. Noyes, W. H. Parkinson, E. M. Reeves, and G. L. Withbroe, *Astrophys. J.* **182**, 321 (1973).
91. K. A. Berrington, P. G. Burke, P. L. Dufton, and A. E. Kingston, *J. Phys. B* **10**, 1465 (1977).

34

Wavelength Standards

Kenneth M. Baird

1. Introduction

The measurement of wavelengths has long been a fundamental part of spectroscopy. While it is true that the spectroscopist is interested primarily in energy-level differences, these are observed as emitted radiation frequencies which are related through the velocity of light to wavelengths and traditionally have been measured as wavelengths. With the development of modern spectroscopy, wavelengths are measured over a range extending from picometers to centimeters, requiring a wide variety of techniques, instruments, and detectors, of which no single combination can be used to cover a very large fraction of the spectrum. Because of this, spectroscopy requires, in addition to a primary standard, an extensive system of secondary standards in order to provide a common absolute scale over the whole spectrum. Such a system has resulted from the contributions of a great many individual workers, as well as laboratories such as the U.S. National Bureau of Standards and has benefited from coordination by international bodies such as the International Astronomical Union (IAU), the Inter-Union Commission on Spectroscopy of ICSU, and the International Committee of Weights and Measures (CIPM). These organizations report the continual updating and extension of the system of wavelengths standards.[1-3]

Wavelengths were once routinely given for standard or normal air, but vacuum wavelengths or wave numbers are recently becoming more widely used, at least for the more important standards; all values in this chapter refer to vacuum. Conversion from air to vacuum or vice versa can be made

KENNETH M. BAIRD • National Research Council of Canada, Division of Physics, Ottawa, Ontario, Canada.

by the use of dispersion formulas which are accurate to about one part in 10^8 from 200 to 1700 nm.[4–5]

Early spectroscopy was concerned with the part of the spectrum near the visible, the region then covered by the best detectors: the eye and the photographic plate. Thus, not surprisingly, the primary standard and principal secondary standards traditionally have been in this region; other standards have tended to be less accurately known depending on their separation from the visible standards, and certain regions, such as the vacuum ultraviolet, pose special difficulties because of problems with optical materials. Recently, however, new techniques involving stabilized lasers and electronic frequency comparison methods are changing the pattern, especially in the infrared, as will be discussed later.

In many instances the requirements for wavelength standards in atomic spectroscopy may differ from those used in molecular spectroscopy. However, similar principles apply to both cases and, of course, the absolute scale is the same, so no distinction will be made in this chapter.

Spectroscopic wavelength standards have for many years also provided the most precisely reproducible means for defining a length, and, since the end of the 19th century, have been contenders for the role of primary standard, the international meter. However from 1907 till 1960 the meter remained defined in terms of the separation of lines engraved on a metal bar, while spectroscopists made use of their own unit, the angstrom. The latter was defined in 1907 by the International Union for Cooperation in Solar Research in terms of the Cd line at 644 nm in such a way as to be 10^{-10} m, within the accuracy of several measurements made in the 19th century and confirmed by later redeterminations.[6] In 1960 the meter was redefined in terms of the wavelength of a ^{86}Kr line at 606 nm,[7] and shortly thereafter the angstrom was defined as exactly 10^{-10} m.[8] Brief accounts of the development of the modern system of wavelength standards have been given by Edlén[9] and Terrien.[10]

Wavelengths of spectral lines are usually measured by reference to nearby lines whose wavelengths are known (secondary standards) and the use of some interpolation procedure that takes into account the dispersion characteristics of the spectrometer or other instrument used. The accuracy of the measurement depends on the measurement process, the accuracy of the original measurement of the standard, and the accuracy to which the source of the standard line reproduces the radiation originally measured. The essential considerations concerning sources and methods of measurement for day to day spectroscopy and for the special experiments that are used to measure the secondary standards themselves are similar and will be briefly reviewed before we discuss specific standards.

2. Sources

As the conditions under which a spectral line is excited can have a marked effect on the emitted frequency, the source used to produce wavelength standards is of utmost importance. Variations in the measured wavelength will result from effects which have an unsymmetrical influence on the line profile: examples are pressure broadening, Stark shift, self-reversal, hyperfine structure, the presence of faint unresolved lines, and asymmetrical Doppler effects.[11] The magnitude of the perturbation can vary by several orders, as the following examples illustrate: Some Fe standards produced by an arc in air are shifted to the red by as much as 0.001 nm ($10^{-6}\lambda$) relative to those produced in a cooled hollow cathode source[12]; the Cd red line, chosen in 1907 as the spectroscopists' primary standard, being fortuitously free from significant perturbations, proved to be reproducible to better than 0.0001 nm ($10^{-7}\lambda$); the ^{86}Kr line at 606 nm, used as the present primary standard, has compensating perturbations of about $2 \times 10^{-8}\lambda$, resulting in a demonstrated reproducibility of about 2×10^{-6} nm ($3 \times 10^{-9}\lambda$); commercially stabilized He–Ne lasers, now frequently used in interferometer control for Fourier spectroscopy, have shifts of up to $10^{-7}\lambda$ due to gas cleanup, servo-control defects, etc., but can be used for more accurate work if monitored by reference to the ^{86}Kr standard or to the more precise I_2 or CH_4 lines discussed below.[13]

In addition to accurate reproducibility, other important considerations for a source of wavelength standards are: linewidth, intensity, distribution of lines, and convenience of operation. A detailed description of sources is beyond the scope of this article; but, having noted the importance of source characteristics in establishing standards, it is not inappropriate to mention some of the more widely used sources; descriptions are given in standard texts.[14]

For the optical region, the cooled hollow cathode and electrodeless discharge tube are most used for many-line spectra such as Fe and Th. Hot cathode sources, such as described below for the ^{86}Kr standard, are very precisely controllable and are used to produce some interferometric standards. Electric arcs, sparks, and pulsed high-frequency discharges are used to obtain spectra at higher stages of ionization, and vacuum and sliding sparks, having confined discharges in vacuum, are used to produce sharp lines in the vacuum ultraviolet.[15] In the far infrared, the wavelength standards are commonly absorption lines observed by the use of a blackbody radiator such as the globar or the Nernst glower (electrically heated, bonded ceramics).[16] There is a growing list of precisely known laser radiations in the visible and infrared, as discussed later in some detail.

Lists of wavelength standards usually specify the sources used, and the user is well advised to duplicate that source as nearly as possible if accuracy is of critical importance.

3. Measurement Methods

While there have recently been striking developments in wavelength comparison techniques, such as the infrared frequency measurements described later in this section, most standards in use today have been measured by the use of well-known instruments and techniques described in standard texts.[17]

The largest lists of spectral lines, and in general the least accurate, are produced by the use of measurements made with angular dispersion instruments, mainly grating spectrometers, and such lines are suitable for spectroscopy making use of these instruments. Tables of Fe and Th spectra and the well-known MIT tables[18] have largely been made in this way. Their accuracies in the optical region range from 10^{-6} to $10^{-7} \lambda$, being limited by the available resolution, unaccounted dispersion effects, and the fact that, with such large numbers of lines, a proper study of asymmetry and perturbations (which may be very large with some lines) is not possible. A major source of error in the measurement and use of these lines concerns the problem of ensuring uniform flooding of the instrument apertures, especially in vacuum ultraviolet spectrometry when the source cannot be imaged on the slit; where possible, a common source producing both the line measured and the standard ought to be used, or where separate sources are necessary, one source ought to be imaged in the other. Optical grating spectrometry is used from the centimeter wave region to the nanometer region of the spectrum, and tables of standards covering this range are available, although practical difficulties in obtaining high accuracy become considerable at the ends of the range because of the limitations of suitable optical materials and detectors; in the far infrared, the small size of practical apertures, compared to the wavelength, imposes a low proportional limit $(10^{-4} \lambda)$ on the resolution. The technique of overlapping orders can be used within certain limitations, to extend the use of standards established in the longer-wavelength region to the far ultraviolet.

More accurate wavelengths are given in the smaller lists of so-called interferometric standards, measured mostly by the use of the Fabry–Perot and Michelson interferometer. These instruments are capable of very high resolution but, their use being a more complex operation, are less practical than gratings for large numbers of measurements. Variations of the Michelson have found considerable use recently in the technique of Fourier spectroscopy whereby a spectrum is obtained by automatic Fourier analysis of the output of a continuously scanning interferometer, whose path difference is related to an accurate linear scale provided by interference fringes produced by the use of an accurate standard such as a controlled laser.[19] These instruments have the advantage of the multiplex principle and have been applied to the measurement of standards; they are especially advantageous in the infrared where detectors are not shot-noise-limited.

Recently there have appeared descriptions of specialized interferometers, "Lambda meters," or "σ meters," which make simultaneous counts of fringes produced by the known and the unknown radiations during a relatively fast translation of an interferometer mirror, giving a rapid and accurate value of the unknown wavelength.[20,21]

Interferometric standards are usually lines of simple spectra from sources which produce sharp lines and introduce perturbations small enough or well enough known to justify precise measurements ($\pm 10^{-7}$ to $10^{-8} \lambda$ in the visible). These standards are useful for checking relatively widely spaced points on angular dispersion instruments and for checking interferometers; in principle only one standard is necessary for the latter, but if a broad spectral range is to be covered, several widely spaced standards are required to ensure that effects having dispersion (such as diffraction and phase change), are correctly taken into account. Proper aperture flooding by both the standard and unknown lines is important with interferometers as with gratings, and, in the case of lasers, an additional hazard must be guarded against: the phenomenon of "talk back," by which radiation reflected back into the laser can "pull" its frequency.[13]

As with angular dispersion spectrometry, accurate wavelength measurements by interferometry are most easily made in and near the visible part of the spectrum where detection, optical materials, and surface finishing techniques are highly developed. Measurement becomes difficult in the far ultraviolet and far infrared, where such is not the case. Interferometry with centimeter waves has been used in connection with velocity of light measurements, but even wavelengths as short as 10 μm cause problems because of diffraction corrections that become large and difficult to take into account to an accuracy of $\pm 10^{-8} \lambda$. Interferometry has also been attempted in the LiF region below 200 nm, but practical difficulties are considerable and few accurate standards in this region have been established in this way.[15]

On the other hand, standards for the x-ray and γ-ray region have been measured to an accuracy of $\pm 10^{-6}$ by the use of interferometry.[22] This was accomplished by the use of x-ray sources, in themselves not accurately reproducible, to determine crystal-lattice spacings by simultaneous observation of x-ray interference and visible-light fringes. The γ-rays were then measured in terms of the lattice spacings.

In the techniques of wavelength comparison discussed above, an optical device is used to provide a common distance to which the wavelengths are related. These suffer the defect that the commonality is imperfectly realized because the effective optical distance is wavelength dependent as a result of effects such as dispersion of phase change on reflection, diffraction, etc. The problem can be avoided by the use of the two techniques described next.

A well-established method of checking wavelength standards in one region of the spectrum against those in another is by the use of the Ritz combination principle, according to which the frequency from some given transition will be exactly the sum of frequencies of the series of separate transitions that begins and ends on the same levels as the single transition. Once the energy-level system of an atom or molecule has been established, this principle can be used to extend a system of wavelength standards into regions of the spectrum where accurate direct measurement is difficult or impossible and, by statistical methods, to improve sets of standards from a given source. Many of the more accurate standards in published tables result from this process, and no reason to doubt the validity of the principle has been found within the uncertainty of the energy levels themselves.

Another method of establishing wavelength standards, not involving an optical distance, is by the use of electronic frequency comparison techniques. Use of the latter, in a recent rapid development, has been extended from the microwave almost to the visible region of the spectrum. Coherent radiation from stabilized lasers and very fast, broadband, diode detectors have made possible "chains" connecting frequencies as high as 150 THz (2 μm) to the Cs frequency standard at 10 GHz (3 cm)[23]; while no continuous operating frequency or phase-locked system has yet been reported, active work to this end is in progress, and such systems are likely soon to be operational. Developments reported at recent conferences[24,25] suggest that the use of dye lasers, color center lasers, optical parametric oscillators, spin–flip Raman lasers, etc., will make possible continuously tunable coherent sources for interpolating between accurate frequency standards throughout the spectrum as far as the ultraviolet standards whose frequencies will be accurately known from comparison measurements using heterodyning and harmonic generation in high-speed diodes, Josephson junctions, nonlinear crystals, atomic oscillators, etc.[26] The wavelengths of thermal sources would then be determined by reference to the frequency standards, whose wavelengths (if wavelengths continue to be used) result from dividing their measured frequency into an accepted value for the velocity of light, c (299,792,458 m/s).

The frequencies of a number of standards in the far and near infrared have already been well established to an accuracy better than $\pm 10^{-8} \lambda$, as discussed in the next section. The accuracy of frequency comparison techniques can in general be much greater than direct wavelength comparison because of avoidance of the dispersion effects already mentioned, and the limit tends more often to be in the uncertainty of the lines themselves than in the process of comparison. At the time of writing, accurate frequency comparison by the use of diodes has not been successful to greater than 150 THz (2 μm). However nonlinear mixing in crystals, plasmas, or field emission devices appear to be likely methods for extension to the higher frequencies.

4. The Primary Standard

At present the primary standard of wavelength (and of length) is the wavelength in vacuum of the unperturbed emission resulting from the $2p_{10}$–$5d_5$ transition in ^{86}Kr ($\lambda^{-1} = 1,650,763.73$ m^{-1}), which is the definition of the international meter; the angstrom unit is 10^{-10} m.[7,8] The krypton line was adopted in 1960 after an intensive series of studies of its perturbations and characteristics, largely made by the national standards laboratories, in a program coordinated through the International Committee of Weights and Measures. The latter recommended, as the preferred source, a lamp shown schematically in Figure 1, with the following specifications: ^{86}Kr of isotopic purity not less than 99% in sufficient quantity to ensure presence of solid Kr when the lamp is at 64 K; capillary of internal diameter of 2–4 mm and wall thickness approximately 1 mm, carrying a dc current density of 0.3 ± 0.1 A cm^{-2}; the lower part of the lamp

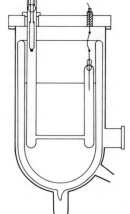

Figure 1. Schematic diagram of officially recommended source for the ^{86}Kr wavelength standard. The hot cathode lamp is immersed in N_2 maintained at its triple point, and the radiation is observed emitted from the anode end of the capillary.

to be immersed in a bath maintained to within ± 1 K of the triple point of N_2 (63.15 K); the radiation to be viewed as emitted from the anode end of the capillary. Under these conditions there exist a number of perturbations amounting individually to about two parts in 10^8, due to effects such as pressure shift and drift of atoms along the capillary caused by electron impact. However, they tend to compensate each other, and the primary standard realized in practice has proven to be reproducible to better than three parts in 10^9, as shown by independent comparisons with precisely stabilized lasers.[10] This reproducibility was a consequence of the use of emission from a specified lamp rather than through the realization of the unperturbed radiation originally envisaged in defining the meter. Further studies of the ^{86}Kr standard that were in progress have now been terminated because of the developments described next.

Recently a number of laser lines stabilized on molecular transitions in the visible and infrared, by the use of techniques for Doppler-free spectroscopy,[27] have produced wavelength standards that are orders of magnitude narrower and more precise than the ^{86}Kr standard line. They have been measured in terms of the ^{86}Kr standard with an accuracy limited by the uncertainty of the latter and are the most precise measurements that have been made in terms of the meter. Furthermore, the extension of the use of electronic frequency comparison techniques into the infrared, discussed above, has yielded values for the frequencies of some of these lines in terms of the Cs standard to an accuracy of $\pm 10^{-9}$ and better. This work had, in 1973, already yielded sufficient accuracy and confidence in the knowledge of the velocity of light to lead the CIPM and the IAU to recommend the value for c of 299,792,458 m/s, with the further recommendation that any new definition of either the meter or of the second should, if possible, be made so as not to change this value.[28,29] Thus, the groundwork has been laid for a definition of the meter, and of the second, in terms of the same atomic or molecular transition and a value of c adopted by convention (and almost certainly to be equal to the agreed value of c given above). At present the Cs standard (at 3.3 cm, $f = 9.2$ GHz), with a width of $10^{-8} f$ and a well-documented absolute reproducibility of better than 1 in 10^{13}, is the most precise known frequency source and will most likely provide the first dual, length–frequency, primary standard.[10] There are already systems, such as the stabilized lasers discussed next, that provide lines that are orders of magnitude narrower than the Cs radiation and potentially more precise. However a great deal of work remains to be done to realize this potential and to determine which is the best of the many possible choices.[30] In the meantime, the situation regarding the primary standard is in a state of flux, as summarized at the end of the next section.

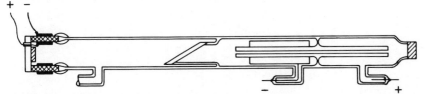

Figure 2. Schematic diagram of a simple form of CH_4-controlled He–Ne laser emitting radiation at 3.39 μm. The cavity length is servo-controlled by the piezoelectric tube supporting the left reflector so as to maintain the emission produced in the capillary on the right at the center of the saturated absorption feature produced by CH_4 in the left portion of the source.

5. Laser Standards

The laser oscillator is not inherently stable except in so far as its output remains within the gain band width. In gas lasers, this corresponds roughly to the Doppler half-height width of the spectral line involved ($\sim 10^{-6} \lambda$). However a laser does serve admirably as an extremely narrow-band oscillator that can be tuned or servo-controlled to some accurate reference frequency.[31] At present the most reproducible references are certain molecular absorption lines in gases at low pressure, observed by the use of Doppler-free techniques such as saturated absorption, two-photon absorption, or the use of beams. A number of systems based on this technique are now operating or under development and will be described next.

The first such system, and the most thoroughly studied, employs a He–Ne laser, emitting at 3.39 μm, servo-stabilized on the F component of the P-7 line of CH_4, by the use of saturated absorption.[32] In its simple form, as in the example shown in Figure 2, it consists of a dc discharge gain tube about 30 cm long and an absorption tube, of about the same length, containing CH_4 at a pressure of less than 1 Pa. The saturated absorption feature has a width of about $10^{-9} \lambda$, limited by the time of transit of CH_4 molecules across the laser beam, and it is reproducible to about $10^{-11} \lambda$, being limited by errors due to wavefront curvature, unresolved hyperfine structure, servo errors, etc. Very large CH_4 stabilized laser structures (the absorption tube dimensions were over 10 m long by 30 cm in diameter) have been used in the USA and USSR to produce linewidths of about 1 kHz ($< 10^{-10}$) and a stability as high as $\pm 10^{-15}$, making possible observation of the resolved hyperfine structure and the shifts due to second-order Doppler effect and photon recoil.[33,34] The magnitudes of these effects were found to be such as to limit the absolute reproducibility with the simple form to about 10^{-12}.

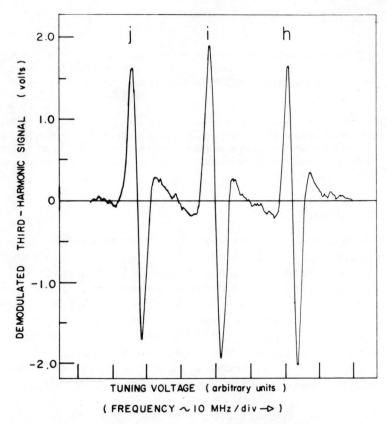

Figure 3. Scan of three of the hyperfine components of the $R(127)$ line of the 11–5 band of $^{127}I_2$ as produced by an I_2-controlled He–Ne laser.

The CH_4 wavelength, as used in the simple stabilized laser shown, has been measured at several laboratories, leading to a CIPM recommended value of $\lambda = 339,223,140 \times 10^{-15}$ m.[35] Its frequency has also been measured at both NBS and NPL, yielding the value of 88,376,181,622 kHz.[36,37] These results provided the basis for the value of the velocity of light given above, and the radiation is considered an equivalent alternative to the ^{86}Kr primary standard.[10]

A second stabilized He–Ne laser system, making use of the $R(127)$ line of $^{127}I_2$ at 0.633 nm as the reference, was developed at about the same time as the CH_2 system.[38] It has also been the object of considerable study, principally at national standards laboratories, resulting in a CIPM recommended wavelength value for a hyperfine component designated "i" (see Figure 3) of $\lambda = 632,991,399 \times 10^{-14}$ m, and a calculated frequency of 473,612,214 MHz.[35]

The I_2-stabilized laser system typically has an overall length of about 30 cm, including the absorption tube of about 10-cm length. The linewidth of about $5 \times 10^{-9} \lambda$ is mainly limited by the natural lifetime of the upper state and, because of the low thermal population of the lower state, the saturated absorption features (shown in the third derivative form in Figure 3) are relatively weak, making the servo control more difficult than is the case with CH_4. On the other hand, the hyperfine structure is clearly resolved and, because of the large mass of the I_2 molecule, the second-order Doppler and photon-recoil effects are less serious than for CH_4. To date, reproducibility approaching 10^{-11} has been demonstrated, and 10^{-12} may be obtainable, taking into consideration pressure, electric field effects, etc. The isotope $^{129}I_2$ has also been used because of an advantage in giving a somewhat stronger signal, but its hyperfine structure has not been analyzed, and there has been evidence of interfering unresolved lines.

The copious spectrum of I_2 provides a large number of lines potentially suitable for stabilizing lasers; those in the green part of the spectrum have longer lifetimes than those at 633 nm by factors of 10 or more and so may make possible stabilized systems of reproducibility comparable to the Cs frequency standard. Studies made by the use of an I_2 molecular beam in conjunction with an argon laser emitting at 515 nm suggested that a reproducibility of the order of 10^{-14} is possible.[39]

A third laser system indicating the potential of sufficiently high reproducibility to merit consideration as a primary standard employs CO_2 laser emission in the 9- and 10-μm bands, controlled by reference either to the same transitions in CO_2 itself,[40] or to nearly coincident transitions in some heavy molecule such as SF_6, SiF_4, or OsO_4.[41,42] The latter can provide extremely sharp reference lines of low limiting perturbations, but their rather complex spectra may cause practical difficulties and such systems will require a great deal of further experimentation and analysis to demonstrate their merits.

The use of the CO_2 transitions themselves as reference frequencies suffers the disadvantage of a weak signal because of the low thermal population of the lower level of the transition. Observation of the line center is usually made by means of saturated fluorescence rather than by saturated absorption because the latter is so weak compared to the laser signal, which results in a poor signal-to-noise ratio. On the other hand, there are several advantages: the reference line is nearly centered on the gain curve for all the transitions; the natural lifetime width is narrow; the spectra are well understood and there ought to be no hyperfine structure; and the transverse Doppler and recoil effects are below 10^{-12}. If solutions to the problem of servo control with a weak signal can be found, the very convenient CO_2 laser emission may well provide the best dual length–frequency standard, lying as it does in a convenient position between the microwave region, so important for communications and time standards, and the optical region, important in spectroscopy and laboratory metrology. Both the wavelengths and frequencies of several CO_2 lines have been accurately measured, results that have provided support in establishing the recommended value of "c".[36,37]

Saturated absorption (or fluorescence) systems, as described above, have so far been the most used of the Doppler-free techniques. Saturated polarization, two-photon absorption, and beam techniques show similar promise for the realization of very precise standards. In addition to the work with I_2 beams already mentioned, systems using dye lasers with Ca and Ba beams have recently been reported.[30] A notable example of two-photon spectroscopy is the recent work of Hänsch with atomic hydrogen including a new determination of the Rydberg constant $(109,737.3143 \text{ cm}^{-1})$.[30,43] The use of an ion trap in Doppler-free spectroscopy is also under study for application to the realization of a primary standard.[30,44]

The effect of the above developments on the situation regarding the primary standard can be summed up as follows: the present ^{86}Kr primary standard, an incoherent (thermal) source, is reproducible to about 3×10^{-9} and at present is adequate for most purposes; some stabilized lasers are more precisely reproducible and have been accurately enough measured in terms of the ^{86}Kr standard to be accepted as standards of equal significance. The basis already exists for using an agreed value of the velocity of light $(299,792,458 \text{ m/s})$ and the Cs frequency standard to define a new standard of length and wavelength, although a formal redefinition of the meter awaits confirming measurements and the schedule of the General Conferences of Weights and Measures. In the meantime the *de facto* definition in terms of c can be used with considerable confidence in its permanence, and any new wavelength comparisons involving lines whose frequencies can be determined ought to be interpreted in a manner consistent with the above value of c, that is, if the frequency is known to 10^{-9} or better, then the wavelength derived from the frequency ought to be given priority.

6. Secondary Standards

Secondary standards include the relatively large lists that serve as reference wavelengths for "day-to-day" spectroscopy over the whole spectrum and, in addition, a more limited number of better-understood, more precisely measured lines at widely separated parts of the spectrum, useful for tying together the larger list into a common absolute scale. Any published wavelength tables may, in a sense, serve as standards, but strictly speaking, a wavelength standard implies a certain degree of confirmed

Table 1. Principal Secondary Wavelength Standards

Source	Transition	Vacuum wavelength (nm)
^{86}Kr	$2p_9\text{-}5d_4'$	645.80720
	$2p_8\text{-}5d_4$	642.28006
	$1s_3\text{-}3p_{10}$	565.11286
	$1s_4\text{-}3p_8$	450.36162
^{198}Hg	$6^1p_1\text{-}6^1D_2$	579.22683
	$6^1p_1\text{-}6^3D_2$	577.11983
	$6^3P_2\text{-}7^3S_1$	546.22705
	$6^3P_1\text{-}7^3S_1$	435.95624
^{114}Cd	$5^1P_1\text{-}6^1D_2$	644.02480
	$5^3P_2\text{-}6^3S_1$	508.72379
	$5^3P_1\text{-}6^3S_1$	480.12521
	$5^3P_0\text{-}6^3S_1$	467.94581
CH_4	F component of $P(7)$ of ν_3 band	3392.23140
$^{127}I_2$	i component of $R(127)$ of 11-5 band	632.991399
CO_2	$R(14)$ of $00°1\text{-}(10°0\text{-}02°0)$ band	9305.38564

accuracy. Many such lines have been formally adopted by international bodies such as the IAU, following the tradition of requiring three independent concordant measurements. However, as pointed out by Kaufman and Edlén,[12] there is much less need for the latter criterion now than formerly, and it is not required for the large lists of reference wavelengths since the absolute scale is established pretty well throughout the spectrum, and statistical methods making use of the Ritz combination principle provide a suitable independent check of experimental results.

The best-established lines produced by thermal sources are the sets of lines of ^{86}Kr, ^{198}Hg, and ^{114}Cd recommended by the CIPM in 1973 and given in Table 1.[2] They were originally estimated to be accurate to 2×10^{-8}, an estimate that is likely overconservative, as proved to be the case with the original estimate of the reproducibility of the primary standard.

Stabilized lasers provide many lines having reproducibilities comparable to, or better than, the primary standard. In addition to the CH_4, I_2, and CO_2 lines mentioned previously (see Table 1) there exists a large number of lines in the region from 2–12 μm that have been accurately measured by frequency comparison with the CO_2 lines. The most accurate are the 9- and 10-μm bands of CO_2, whose values relative to directly measured CO_2 lines are given by the rotational constants determined by Petersen *et al.*[45] Somewhat less accurate values are known for wavelengths in the same region (9–11 μm) produced by lasers using CO_2 of exotic isotopic composition,[46] N_2O,[47] and the hot bands of CO_2.[48] All these refer to the saturated absorption or saturated fluorescence feature and are accurate to better than 100 kHz ($\sim 3 \times 10^{-9}$). Many other lines in the infrared have recently been measured by frequency methods to an accuracy higher than existing standards in the region but, being referred to the Doppler profile, are limited to an accuracy not better than $\pm 10^{-7}$. Examples are the CO bands at 5–6 μm,[49] CO_2 sequence bands,[50] and a number of useful submillimeter laser spectrometer frequencies.[51]

Direct absolute frequency measurement has so far not been reported for higher than 150 THz (2 μm)[23] where high-speed diodes appear to cut off. However, one can expect an eventual extension into the ultraviolet of standards based on frequency comparison by the use of other methods such as the mixing and doubling of frequencies in nonlinear crystals and gases.[26,52]

In addition to the above precise secondary standards, a number of accurate standards have been established by interferometry and the Ritz principle for Ne, Ar, etc.,[12] covering the range into the ultraviolet to about 150 nm. The wavelengths of x-ray lines have been measured, but the radiations themselves are too broad and ill defined to be properly labeled standards. However, some γ-ray lines are much sharper and more precisely reproducible and have recently been measured accurate to $\sim 10^{-6}$.[22]

The results of a large number of measurements of the more traditional type of wavelength standards are included in some recently published lists that contain a large number of lines providing an updated, broad coverage of the spectrum. Crosswhite[53] has published a list of over 4000 wavelengths between 400 and 900 nm for Fe I, Fe II, Ne I, and Ne II produced in a hollow cathode discharge and has given data on intensity and some Ritz standards. Giacchetti[54] and others have used observations made by Connes' method of Fourier transform spectroscopy to produce a list of 3100 classified lines of Th I and Th II in the region between 0.9 and 3μ accurate to about 0.002 cm^{-1}. Kaufman and Edlén[12] have published a compilation of reference wavelengths from atomic spectra in the range 1.5 to 2500 nm, with data on intensities. Humphreys[55] has given descriptions to the first spectra of Ne, Ar, and Xe 136 in the 1.2–4.0 μm region.

The Commission on Molecular Structure and Spectroscopy has been instrumental in publishing infrared standards approved by the Inter-Union Commission of Spectroscopy of ICSU. The latest data are in a book edited by Cole, which lists lines covering the range from 4300 to 1 cm^{-1} for the calibration of spectrometers.[3]

New work is at present under way on improved sources and standards for the far ultraviolet at NBS and NPL.[1] Such work is generally, as it appears in print, summed up by the Committee on Wavelength Standards of the IAU and published in its transactions.

In concluding this description of wavelength standards it is appropriate to note the publication of relevant information, such as the considerable amount of spectroscopic data given in the *Handbook* of the American Institute of Physics,[56] the tables of energy levels and term diagrams published by the US National Bureau of Standards,[57] and the lists of laser lines such as that recently edited by Beck.[58]

References

1. International Astronomical Union (IAU), *Trans.* **16A**, 31 (1976).
2. *Comité Consultatif pour la Définition du Mètre, 3rd Session*, pp. 9–20, Gauthier Villars, Paris (1962).
3. *Tables of Wave Numbers for the Calibration of Infra Spectrometers* 2nd ed., Ed. A. R. H. Cole, Pergamon, London (1976).
4. B. Edlén, *Metrologia* **2**, 71 (1966).
5. E. R. Peck and K. Reeder, *J. Opt. Soc. Am.* **62**, 958 (1972).
6. K. H. Hart and K. M. Baird, *Can. J. Phys.* **39**, 781 (1961).
7. *CIPM Procès Verbaux*, 2nd Ser, **28**, 70 (1960).
8. *IAU Trans.* **11B**, 89 (1961).
9. B. Edlén, in *Polarisation, Matière et Rayonnement*, pp. 219–227, Presses Universitaires de France, Paris (1969).
10. J. Terrien, *Rep. Prog. Phys.* **39**, 1067 (1976).
11. R. G. Breene, Jr., *The Shift and Shape of Spectral Lines*, Pergamon Press, London (1961).
12. V. Kaufman and B. Edlén, *J. Phys. Chem. Ref. Data* **3**, 825 (1974).
13. W. R. C. Rowley, *Comité Consultatif pour la Définition du Mètre, 4th Session*, pp. M108–113, Offilib, Paris (1970).
14. *Methods of Experimental Physics*, Vol. 13A, *Spectroscopy*, Ed. D. Williams, pp. 205–227 and 259–274, Academic Press, New York (1976).
15. B. Edlén, *Rep. Prog. Phys.* **26**, 181 (1963).
16. *The Encyclopedia of Spectroscopy*, Ed. G. L. Clark, p. 453, Reinhold, New York (1960).
17. *Methods of Experimental Physics*, Vol. 13A, *Spectroscopy*, Ed. D. Williams, Academic Press, New York (1976).
18. *MIT Wavelength Tables*, John Wiley, New York (1960).
19. P. Connes and G. Michel, *Appl. Opt.* **14**, 2067 (1975).
20. J. L. Hall and S. A. Lee, *Appl. Phys. Lett.* **29**, 367 (1976).
21. F. V. Kowalski, R. T. Hawkins, and A. L. Schawlow, *J. Opt. Soc. Am.* **66**, 956 (1976).
22. R. D. Deslattes, in *Proceedings of Course No. LXVIII "Metrology and Fundamental Constants," Summer School of Physics—Enrico Fermi, Varenna, Italy, July 1976*.
23. D. A. Jennings, F. R. Petersen, and K. M. Evenson, *Appl. Phys. Lett.* **26**, 510 (1975).
24. *Laser Spectroscopy, Proceedings of the Second International Conference, Megeve, France, June 1975*, Eds S. Haroche, J. C. Pebay-Peyroula, T. W. Hansch, and S. E. Harris, Springer-Verlag, Berlin (1975).
25. *Tunable Lasers and Applications, Proceedings of the Loen Conference, Norway, 1976*, Eds. A. Mooradian, T. Jaeger, and P. Stokseth, Springer-Verlag, Berlin (1976).

26. D. J. E. Knight and P. T. Woods, *J. Phys. E* **9**, 898 (1976).
27. *High Resolution Laser Spectroscopy*, Ed. K. Shimoda, *Topics in Applied Physics*, Vol. 13, Springer-Verlag, Berlin (1976); see also appropriate chapters in the present book.
28. *Quinzieme Conférence Général des Poids et Mesures, Paris, 1975, Comptes Rendus*, Bureau Internal des Poids et Mesures (publishers), Paris.
29. *IAU Trans.* **15B**, 55 (1973).
30. *Proceedings of the Second Frequency Standards and Metrology Symposium, July 1976, Copper Mountain, Colorado*, U.S. Natl. Bur. Sts.
31. K. M. Baird and G. R. Hanes, *Rep. Prog. Phys.* **37**, 927 (1974).
32. J. L. Hall, in *Atomic Physics*, Vol. 3, Eds S. J. Smith and G. K. Walters, pp. 615–646, Plenum Press, New York (1973).
33. J. L. Hall, C. J. Bordé, and K. Uehara, *Phys. Rev. Lett.* **37**, 1339 (1976).
34. V. P. Chebotaev, in *Proceedings of the International Enrico Fermi School of Physics, Metrology and Physical Constants, Varenna, Italy, 1976*, North-Holland, Amsterdam (in press).
35. *Comité Consultatif pour la Definition du Mètre*, 5th Session, Offilib, Paris (1973).
36. K. M. Evenson, J. S. Wells, F. R. Petersen, B. L. Danielson, and G. W. Day, *Appl. Phys. Lett.* **22**, 192 (1973).
37. T. G. Blaney, G. J. Edwards, B. W. Joliffe, D. J. E. Knight, and P. T. Woods, *J. Phys. D.* **9**, 1323 (1976).
38. G. R. Hanes, K. M. Baird, and D. DeRemigis, *Appl. Opt.* **12**, 1600 (1973).
39. L. Hackel, R. Hackel, and S. Ezekiel, in Ref. 30.
40. C. Freed and A. Javan, *Appl. Phys. Lett.* **16**, 53 (1970).
41. M. Ouhayoun and C. J. Bordé, *Metrologia* **13**, 149 (1977).
42. Yu. S. Dimin, U. M. Tatarenkov, and P. S. Shumiatskii, *Sov. J. Quantum Electron.* **3**, 2612 (1975); or O. N. Kompanets, A. R. Kukudzhanov, U. S. Letokov, E. L. Michailov, in Ref. 30.
43. T. W. Hansch, in *Atomic Physics 4, Proceedings of the Fourth International Conference on Atomic Physics*, p. 93, Plenum Press, New York (1975).
44. H. A. Schuessler, *Metrologia* **7**, 103 (1971).
45. F. R. Petersen, D. G. McDonald, J. D. Cupp, and B. L. Danielson, in *Laser Spectroscopy*, Eds. R. G. Brewer and A. Mooradian, pp. 555–569, Plenum Press, New York (1974).
46. C. Freed, R. G. O'Donnell, and A. H. M. Ross, *IEEE Trans.* **IM-25**, 431 (1976).
47. B. G. Whitford, K. J. Siemsen, H. D. Riccius, and G. R. Hanes, *Opt. Commun.* **14**, 70 (1975).
48. B. G. Whitford, K. J. Siemsen, and J. Reid, *Opt. Commun.* **22**, 261 (1977).
49. R. S. Eng, H. Kildal, J. C. Mikkelsen, and D. L. Spears, *Appl. Phys. Lett.* **24**, 231 (1974).
50. B. G. Whitford and K. J. Siemsen, *Opt. Commun.* **22**, 11 (1977).
51. H. E. Radford, F. R. Petersen, D. A. Jennings, and J. A. Mucha, *IEEE J. Quantum Electron.* **QE-13**, 92 (1977).
52. V. Chebotayev, V. Klementyev, Yu. Kolpakov, and Yu. Matyugin, in Ref. 30.
53. H. M. Crosswhite, *J. Res. Natl Bur. Sts* **79A**, 17 (1975).
54. A. Giacchetti, J. Blaise, C. H. Carliss, and R. Zalubas, *J. Res. Natl Bur. Stand.* **78A**, 247 (1974).
55. C. H. Humphreys, *J. Phys. Chem. Ref. Data* **2**, 519 (1973).
56. *American Institute of Physics Handbook*, 3rd ed., Ed. D. E. Gray, McGraw-Hill, New York (1972).
57. National Standards Reference Data Series NBS3, U.S. GPO Washington, DC (1976).
58. R. Beck, W. Englisch and K. Gurs, *Table of Laser Lines in Gases and Vapors*, Springer-Verlag, Berlin (1976).

Index